# Theory of Aerospace Propulsion

# Theory of Aerospace Propulsion

**Pasquale M. Sforza**
University of Florida

AMSTERDAM • BOSTON • HEIDELBERG • LONDON
NEW YORK • OXFORD • PARIS • SAN DIEGO
SAN FRANCISCO • SINGAPORE • SYDNEY • TOKYO
Butterworth-Heinemann is an imprint of Elsevier

Butterworth-Heinemann is an imprint of Elsevier
225 Wyman Street, Waltham, MA 02451, USA
The Boulevard, Langford Lane, Kidlington, Oxford, OX5 1GB, UK



**Notices**
Knowledge and best practice in this field are constantly changing. As new research and experience broaden our understanding, changes in research methods, professional practices, or medical treatment may become necessary.

Practitioners and researchers must always rely on their own experience and knowledge in evaluating and using any information, methods, compounds, or experiments described herein. In using such information or methods they should be mindful of their own safety and the safety of others, including parties for whom they have a professional responsibility.

To the fullest extent of the law, neither the Publisher nor the authors, contributors, or editors, assume any liability for any injury and/or damage to persons or property as a matter of products liability, negligence or otherwise, or from any use or operation of any methods, products, instructions, or ideas contained in the material herein.

**Library of Congress Cataloging-in-Publication Data**
Sforza, P. M.
Theory of aerospace propulsion / Pasquale M. Sforza.
p. cm.
Includes bibliographical references and index.
ISBN 978-1-85617-912-6 (alk. paper)
1. Jet propulsion. I. Title.
TL709.S38 2012
629.1′1–dc23

2011027614

**British Library Cataloguing-in-Publication Data**
A catalogue record for this book is available from the British Library.

For information on all Butterworth–Heinemann publications,
visit our website: www.elsevierdirect.com

Printed in the United States of America

11 12 13 14 15 16 17   10 9 8 7 6 5 4 3 2 1

# Contents

# Preface

This textbook derives from notes that were developed and assembled over many years of teaching propulsion and high-speed airplane design courses at the Polytechnic Institute of Brooklyn, as well as courses in propulsion and aerospace vehicle design at the University of Florida. The material presented is intended as an aid in studying the theory and concepts of propulsion and derives from a diversity of sources. The text strives to provide a fundamental approach suitable for courses at the senior undergraduate and first-year master's level. The problems are intended to promote an appreciation for applications of the theory to problems of practical interest.

There is little attempt to provide a specifically historical or motivational background of aerospace propulsion, as it is expected that readers will come to this level of study with a well-developed sense of interest in and appetite for a quantitative involvement in the field. The first chapter immediately introduces the student to the fundamentals of jet propulsion from the standpoint of the ideal quasi-one-dimensional conservation equations. The second chapter delves deeper into the details of these equations in order to arrive at a consistent set of equations that are broadly applicable to various propulsion systems. The third chapter carries out an extensive set of analyses of the operation of a variety of air-breathing engines under conditions of ideal operation to facilitate keeping interest focused on the underlying concepts. The foundation for such an approach is that if an ideal system works poorly, a practical version of it will not work better. The cases studied offer the equivalent of a set of sample problems to aid the reader in understanding the ideal workings of the various engines. The effect of the efficiencies of a real system is discussed at the end of the chapter in a manner that should facilitate repeating the calculation of all the cases with reasonable concern for losses common to practical operation. Chapters 4, 5, and 6 concentrate on the three engine components basic to jet propulsion principles: combustors, nozzles, and inlets. Fundamentals are discussed in some detail and application to actual hardware is shown. Then Chapters 7 and 8 are devoted to the fundamentals and details of the turbomachinery required for the operation of air-breathing jet engines for flight up to and including supersonic speeds, while integration of the various components into a working unit is the subject of Chapter 9. The last chapter concerned with aircraft flight operations is Chapter 10, which covers the operation of propellers and the application of the gas turbine engine to them. Chapters 11 and 12 are concerned with liquid and solid rocket motors, respectively, while Chapter 13 analyzes the nuclear rocket and its possible role in space propulsion. The area of space propulsion is covered in Chapter 14 with attention given to electric propulsion techniques that are of importance in satellite operations and deep space exploration. Finally, Chapter 15 gives a brief overview of the case for high-speed flight and how propulsion plays a major role in making such operations practical. Seven appendices follow dealing with important auxiliary information, including supersonic flows and shock waves; thermodynamic properties of hydrocarbon–air combustion products; the Earth's atmosphere; orbital operations and staging of rockets; principles of safety, reliability, and risk assessment related to propulsion; flight performance of aircraft; and thermodynamic properties of selected chemical species.

I acknowledge the inspiration provided by Professor Antonio Ferri who taught propulsion courses I took as a graduate student at the Polytechnic Institute of Brooklyn many years ago. Appreciation is due to my close colleagues Professor Marian Visich, Jr. of the State University of New York at Stony Brook and Professor Herbert Fox of the New York Institute of Technology for their long-term cooperation in, criticism of, and support for this book. I also thank the many students who, over the years, provided suggestions and corrections to the class notes that evolved into this book. Final thanks are due to my wife, Anne, for patience and encouragement in this project whose duration and required effort were wildly underestimated.

**Pasquale M. Sforza**
Highland Beach, Florida,
sforzapm@ufl.edu,
August 23, 2011

CHAPTER

# Idealized Flow Machines

1

## CHAPTER OUTLINE



## 1.1 CONSERVATION EQUATIONS

A flow machine is one that ingests a stream of fluid, processes it internally in some fashion, and then ejects the processed fluid back into the ambient surroundings. An idealization of such a generalized flow machine is depicted schematically in Figure 1.1

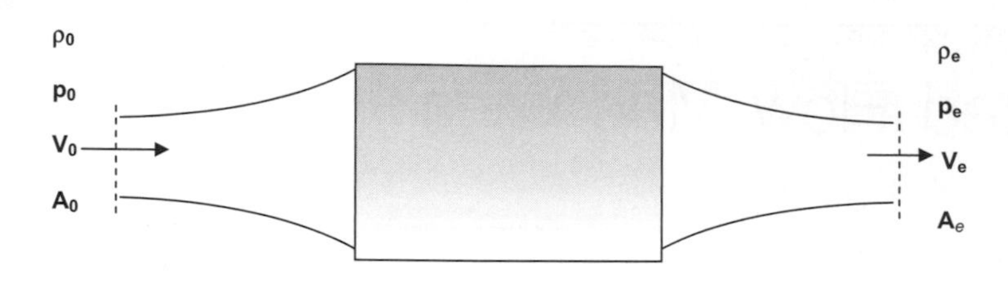


FIGURE 1.1

Schematic diagram of idealized flow machine and associated streamtube control volume.

In order to develop the basic features of operation of the idealized flow machine without introducing unnecessary algebraic complexity, we make the following assumptions:

- Flow through the streamtube entering and leaving the machine is steady and quasi-one-dimensional.
- The entrance and exit stations shown are chosen sufficiently far from the flow machine entrance and exit such that pressures at those stations are in equilibrium with their surroundings, that is, $p_e = p_0$.
- There is no heat transfer across the boundaries of the streamtube or the flow machine into the ambient surroundings.
- Frictional forces on the entering and leaving streamtube surfaces are negligible.
- Mass injected into the fluid stream within the flow machine, if any, is negligible compared to mass flow entering the flow machine.

With these restrictions in mind we may assess the consequences of applying the basic conservation principles to the streamtube control volume. Some implications of the assumptions used are important to understand.

The assumption of steady flow implies that $V_0$ is constant, that is, the idealized flow machine may be considered to be flying at speed $V_0$ through a stationary atmosphere with the ambient environmental values of pressure, density, and temperature denoted in Figure 1.1 by $p_0$, $\rho_0$, and $T_0$, respectively. Alternatively, we may consider our coordinate system to be fixed on the flow machine such that the atmosphere constitutes a free stream flow approaching at speed $V_0$ with static conditions of pressure, density, and temperature denoted in Figure 1.1 by $p_0$, $\rho_0$, and $T_0$, respectively. This (Galilean) transformation of coordinates is possible because the motion is steady.

Another implication arising from the assumption that the flow machine is moving through the atmosphere at constant speed is that there must be no unbalanced force on the machine. Because there will be resistance to motion due to drag $D$, there must be another force applied that can maintain the constant motion, which is net thrust $F_n$. The rate at which work must be done to maintain the motion is $DV_0$, and because $D = F_n$, this required power may also be written as $FV_0$.

### 1.1.1 Conservation of mass

The net change in the mass flow passing through the flow machine is zero, which may be written as

$$-\rho_0 A_0 V_0 + \rho_e A_e V_e = 0. \tag{1.1}$$

This is equivalent to stating that the mass flow $\dot{m} = \rho A V =$ constant throughout the system.

### 1.1.2 Conservation of momentum

The net change in momentum of the fluid passing through the streamtube is equal to the force on the fluid, or

$$(-\rho_0 A_0 V_0)V_0 + (\rho_e A_e V_e)V_e = F.$$

Because thc mass flow is constant, this equation can be abbreviated to the following:

$$\dot{m}(V_e - V_0) = F. \tag{1.2}$$

The force acting on the fluid is denoted by $F$, and, for equilibrium, the force exerted on the control volume by the fluid is $-F$. In general, forces on the streamtube are negligible compared to those on the flow machine proper and are neglected. One important case where this is not necessarily true is that of the so-called additive drag of inlets in supersonic flight, where the force on the entering streamtube surface may not be negligible.

### 1.1.3 Conservation of energy

The net change in the total enthalpy of the flowing fluid is equal to the sum of the rate at which heat and work are added to the fluid, or

$$\dot{m}\left[(h_e - h_0) + \frac{1}{2}\left(V_e^2 - V_0^2\right)\right] = \dot{m}Q + P. \tag{1.3}$$

The quantities $h$, $Q$, and $P$ denote enthalpy, heat addition per unit mass, and power added, respectively. We may consider some extreme cases to illustrate several basic kinds of flow machines.

## 1.2 FLOW MACHINES WITH NO HEAT ADDITION: THE PROPELLER

Here we assume that $Q = 0$ so that

$$P = \dot{m}\left[(h_e - h_0) + \frac{1}{2}\left(V_e^2 - V_0^2\right)\right]. \tag{1.4}$$

However, if no heat is added to the flowing fluid it is reasonable to expect that the enthalpy of the fluid is essentially unchanged in passing through the machine so that $h_e \approx h_0$, which results in

$$P \approx \frac{1}{2}\dot{m}\left(V_e^2 - V_0^2\right) = \frac{1}{2}\dot{m}(V_e - V_0)(V_e + V_0) = FV_{avg}. \tag{1.5}$$

Thus the power supplied to the fluid is approximately equal to the product of the force on the fluid and the average of the velocities entering and leaving the machine.

### 1.2.1 Zero heat addition with $V_e > V_0$

Here $F > 0$ and therefore $P > 0$, so that work is done on the fluid. This is the case of the propeller, the fan, and the compressor, where the device does work on the fluid and produces a force on the fluid in the same sense as the entering velocity. Note that this means that the force of the fluid on the machine is in the opposite sense, that is, a *thrust* is developed.

### 1.2.2 Zero heat addition with $V_e < V_0$

Here $F < 0$ and $P < 0$, so that work is done by the fluid. This is the case of the turbine, where work is extracted from the fluid and the fluid experiences a retarding force, that is, the force on the fluid is in the opposite sense to that of the incoming velocity. The force on the machine is therefore in the same sense as the entering velocity and is therefore a *drag* force.

### 1.2.3 Zero heat addition with $P$ = constant > 0

In this variation, we see that thrust force drops off with flight speed:

$$F = \frac{P}{V_{avg}} = \frac{1}{V_0} \frac{P}{\frac{1}{2}\left(1 + \frac{V_e}{V_0}\right)}.$$

In general, the velocity ratio $V_e/V_0$ is not much greater than unity, thus $V_{avg} \sim V_0$. This is the case of a propeller propulsion system where increases in flight speed are limited by the power available. This effect is illustrated in Figure 1.2.

### 1.2.4 Propulsive efficiency

The total power expended is not necessarily converted completely into thrust power $FV_0$, the rate at which force applied to the fluid does work. Remember that flight speed $V_0$ is constant and therefore the drag on the vehicle is equal to the thrust produced, $D = F$. Then the rate at which work must be done to keep the vehicle at constant speed $V_0$ is $DV_0 = FV_0$. However, it has been shown that the power

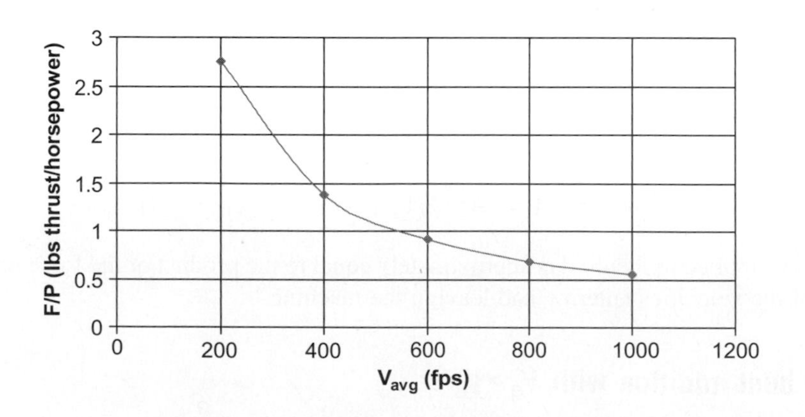


**FIGURE 1.2**

Specific thrust produced by a propeller as a function of average speed.

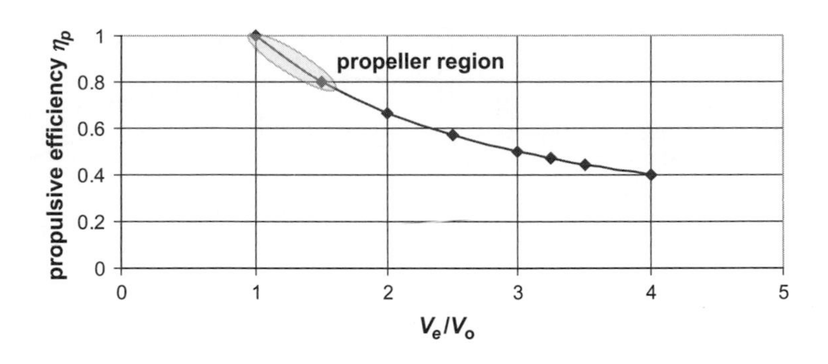


**FIGURE 1.3**

Efficiency of a propeller as a function of the ratio of exit speed to flight speed.

expended is $P = FV_{avg}$ so that propulsive efficiency $\eta_p$ may be defined as the ratio of useful thrust power to total power delivered to the airstream:

$$\eta_p = \frac{FV_0}{P} = \frac{FV_0}{FV_{avg}} = \frac{V_0}{\frac{1}{2}(V_0 + V_e)}. \tag{1.6}$$

This equation shows that at a given flight speed the efficiency drops off with increasing exhaust velocity $V_e$, as shown in Figure 1.3.

## 1.3 FLOW MACHINES WITH $P = 0$ AND $Q =$ CONSTANT: THE TURBOJET, RAMJET, AND SCRAMJET

Here we assume that no net power is exchanged with the fluid so that

$$(h_e - h_0) + \frac{1}{2}(V_e^2 - V_0^2) = Q. \tag{1.7}$$

But we may write the kinetic energy term as

$$V_e^2 - V_0^2 = (V_e + V_0)(V_e - V_0) = (V_e + V_0)\frac{F}{\dot{m}}. \tag{1.8}$$

Substituting this back into the first equation and solving for the thrust yields

$$F = \frac{\dot{m}}{V_{avg}}[Q - (h_e - h_0)]. \tag{1.9}$$

### 1.3.1 Heat addition, $Q > 0$

If sufficient heat is added to the fluid, Equation (1.9) shows that $F > 0$ and thrust is produced on the flow machine. This is the basis of operation of the simple jet engine. The general internal configuration

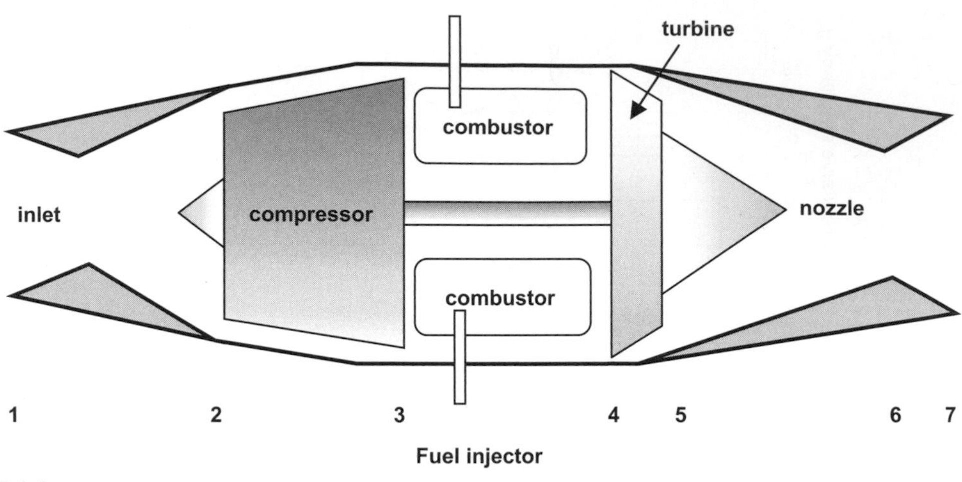


**FIGURE 1.4**

Schematic diagram of a typical turbojet engine showing required turbomachinery components and a common station numbering scheme. The combustor burns the injected fuel supplying heat to flow passing through the turbojet.

of the practical jet engine is dependent on the flight speed. For flight in the range of $0 < M_0 < 3$, the jet engine requires a compressor to increase the pressure of the incoming air before fuel is added and burned, particularly in the low end of the speed range. The compressor requires a shaft power source to drive it. Both these functions are best supplied by turbomachinery, or rotating machinery. The air compressor (turbo-compressor) is powered by a gas turbine, with both being attached to a common driveshaft. Such an arrangement is called a turbojet engine and is illustrated schematically in Figure 1.4.

For supersonic flight in the Mach number range of $2 < M_0 < 5$, the ram pressure produced by the inlet in slowing down the incoming air to subsonic speeds obviates the need for the compressor and therefore its driving turbine. As a consequence, a practical jet engine for this flight regime is called a ramjet and its configuration is very simple, as shown in Figure 1.5, where the turbomachinery no longer needed is indicated by dashed lines. This simplicity comes at a price however, because the ramjet cannot operate effectively at lower speeds. In particular, it generates no thrust at zero flight speed so it cannot provide thrust for take-off. The ramjet must be accelerated to supersonic speeds by some other propulsive means before it can produce enough thrust to sustain flight of the vehicle it powers. It is therefore often used to power missiles launched from an aircraft or a rocket booster.

For hypersonic flight speeds, $M_0 > 5$, the temperature increase accompanying the ram compression to subsonic speeds is so high that little or no additional heat can be added in the combustor by burning fuel. The only alternative is to use the inlet to slow the flow down from the flight speed to some lower supersonic Mach number, thereby not increasing the temperature too much. However, then fuel must be added to a supersonic stream, mixed, and combusted. Achieving this supersonic combustion is very difficult because the high speed in the combustor gives very little time for mixing and combustion to

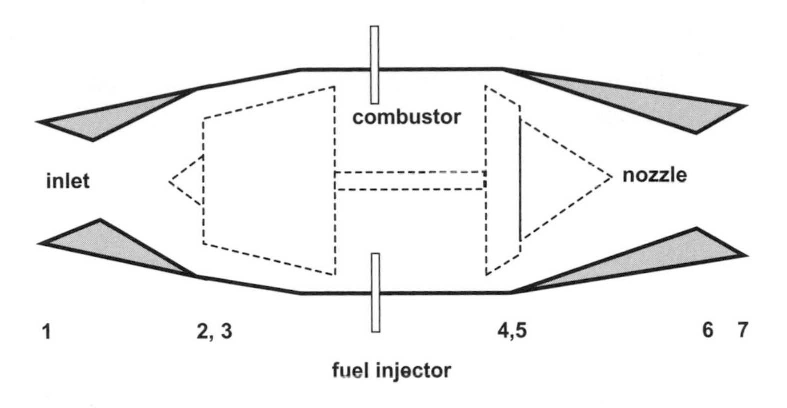


FIGURE 1.5

Schematic diagram of a typical ramjet engine. Dashed lines show the turbomachinery components no longer needed. The combustor burns the injected fuel supplying heat to the flow passing through the ramjet.

take place. Such a supersonic combustion ramjet is popularly known as a scramjet. The general configuration is like that of the ramjet shown in Figure 1.5.

### 1.3.2 Constant heat addition, $Q$ = constant > 0

In this case, we see that the thrust

$$F = 2\rho_0 A_0 \frac{V_0}{V_0 + V_e}[Q - (h_e - h_0)].$$

Therefore the thrust is essentially constant with flight speed $V_0$, as shown in Figure 1.6. This is the basic advantage of the jet engine—its thrust is independent of the flight speed and therefore it is not speed limited, as is the propeller.

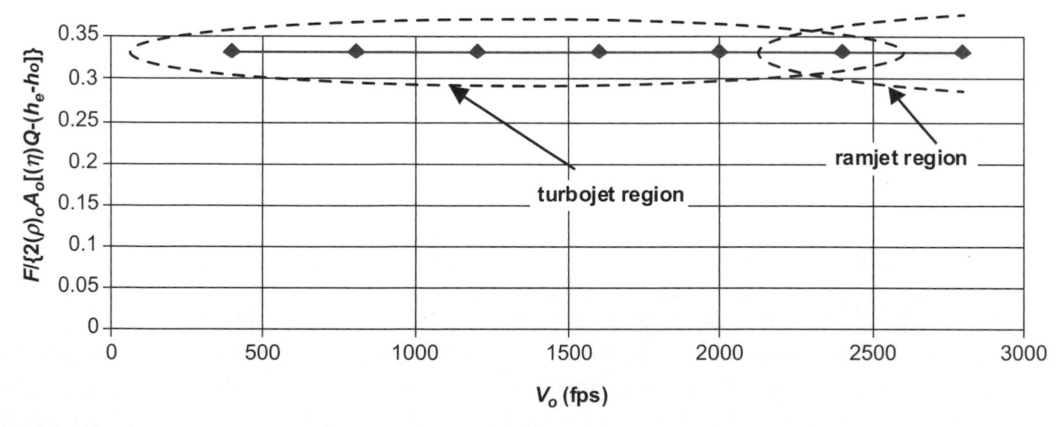


FIGURE 1.6

Thrust of a turbojet engine as a function of flight speed.

### 1.3.3 Overall efficiency

The overall efficiency in the case of thrust produced by means of heat addition may be defined as the ratio of the required thrust power to the rate of heat addition

$$\eta_o = \frac{FV_0}{\dot{m}Q} = \frac{V_0}{V_{avg}}\left[1 - \frac{h_e - h_0}{Q}\right]. \tag{1.10}$$

However, as already shown in Equation (1.6), the propulsive efficiency is

$$\eta_p = \frac{V_0}{V_{avg}}.$$

We may then define the thermal efficiency as

$$\eta_{th} = 1 - \frac{h_e - h_0}{Q}. \tag{1.11}$$

Then the overall efficiency is the product of propulsive and thermal efficiencies

$$\eta_o = \eta_p \eta_{th}. \tag{1.12}$$

Thermal efficiency accounts for the fact that not all the heat added is converted to useable heat power, as some is rejected as increased internal energy in the exhaust gases. Thus, thermal efficiency illustrates the extent to which the flow machine puts the heat added to good use in increasing the kinetic energy of the exhaust stream, that is, increasing $V_e$, whereas propulsive efficiency illustrates the extent to which that increased kinetic energy provides thrust power to maintain flight at the speed $V_0$. In a jet engine, these two efficiencies generally drive in different directions, with the higher exhaust velocities sustainable at high thermal efficiency leading to lower propulsive efficiencies at a given flight speed.

### 1.3.4 Fuel efficiency

The heat addition to the flow is directly proportional to the rate of fuel consumption $\dot{m}_f$ and to the specific energy content of the fuel $HV$ as expressed by

$$\dot{m}Q \sim \dot{m}_f HV.$$

The constant of proportionality is the burner efficiency $\eta_b$ so that the rate of heat addition through combustion becomes

$$\dot{m}Q = \eta_b \dot{m}_f HV. \tag{1.13}$$

This equation is based on the assumption that the rate of heat addition arises from the energy released in the chemical conversion of the fuel added as characterized by $HV$, which is called the heating value of the fuel or the net specific energy of the fuel. More accurately, for combustor applications, we mean the lower heating value or net-specific energy of the fuel. This is because the water in the combustion products of the fuel is expected to be in a gaseous state so that the latent heat of condensation of the water is not included in the energy content of the fuel. The burner efficiency $\eta_b$ represents the ratio of the heat release actually transferred to the flowing combustion gases to the total heat release possible.

**Table 1.1** Lower Heating Values for Typical Liquid Fuels

| Fuel | *HV* (Btu/lb) | *HV* (MJ/kg) | *HV'* (Btu/gal) | *HV'* (kJ/liter) |
|---|---|---|---|---|
| Jet A | 18,660 | 43.4 | 124.5 | 34.7 |
| Methane | 21,500 | 50.0 | 76.06 | 21.20 |
| Ethanol | 11,710 | 27.23 | 77.57 | 21.62 |
| Hydrogen | 51,690 | 120.24 | 30.21 | 8.42 |

Lower heating values, in both the English and the SI system of units, are given for some typical fuels in the liquid phase in Table 1.1 (from Goodger, 1986), along with the heating value per unit volume *HV'*. Note that hydrogen has about three times the heat release per unit mass of Jet A while only about one-quarter the heat release per unit volume. Because the volume of an aircraft influences its drag the quantity *HV'* is an important parameter to consider, as well as *HV* itself, and hydrogen is not generally a good alternative to a hydrocarbon fuel.

It is worthwhile to list some useful conversion factors between the two systems as follows:

$$
\begin{aligned}
1\,\text{Btu/lb} &= 0.002326\,\text{MJ/kg} \\
1\,\text{Btu/gal} &= 0.278717\,\text{kJ/liter} \\
1\,\text{Btu/ft}^3 &= 37.2589\,\text{kJ/m}^3
\end{aligned}
$$

A common measure of fuel efficiency is specific fuel consumption $c_j$, which traditionally was defined as the ratio of the fuel weight flow rate to the thrust produced. This is typically reported as pounds of fuel consumed per hour per pound of thrust, or simply $\text{hr}^{-1}$. In recent years, use of the SI system of units has altered this definition so that specific fuel consumption may also be defined as the ratio of the mass flow rate of fuel consumed to the thrust produced, which has the units kg/hr-N. It must be emphasized that care should be taken to account properly for the units, particularly since both systems of units are in routine use in the aerospace industry. The remainder of this chapter uses the weight-based heating value and the English system of units.

Recall that the set of assumptions made at the outset included the requirement that the amount of fluid, in this case fuel, added to the general stream entering the flow machine is negligible, that is, $\dot{m}_f \ll \dot{m}_0 = \dot{m}$. It will be shown subsequently, when dealing with combustion chambers, that the fuel weight flow rate is indeed much smaller than the air weight flow rate, so no loss in generality is incurred here by not including the added fuel. Using Equations (1.10) and (1.13) and considering the weight-based heating value in English units, the specific fuel consumption is

$$c_j = \frac{\dot{w}_f}{F} = \frac{V_{avg}}{\eta_{th}\eta_b HV}. \tag{1.14}$$

For hydrocarbon fuel ($HV = 18{,}660$ Btu/lb) and velocity in ft/s, the specific fuel consumption in pounds of fuel per hour per pound of thrust becomes

$$c_j = 2.480 \times 10^{-4} \frac{V_{avg}}{\eta_b \eta_{th}}.$$

Burner efficiency is generally quite high, as will be seen in Chapter 4, and for present purposes it may be taken as $\eta_b = 0.95$. However, thermal efficiency bears a bit more consideration. Expanding Equation (1.11) leads to the following representation of thermal efficiency:

$$\eta_{th} \approx 1 - c_{p,e}T_{t,e}\frac{\left(\dfrac{T_e}{T_{t,e}} - \dfrac{c_{p,0}T_0}{c_{p,e}T_{t,e}}\right)}{\left(\dfrac{\dot{w}_f}{\dot{w}_0}\right)\eta_b HV}.$$

Under typical flight conditions, the fuel to air ratio $\dot{m}_f/\dot{m}_0 \sim 0.02$, the specific heat ratio $c_{p,0}/c_{p,e} \sim 0.87$, the ratio of atmospheric temperature to exhaust stagnation temperature $T_0/T_{t,e} \sim 0.25$, $\eta_b \sim 0.95$, and $HV = 18{,}660$ Btu/lb so that

$$\eta_{th} \approx 1 - c_{p,e}T_{t,e}\frac{\left(\dfrac{T_e}{T_{t,e}} - 0.22\right)}{359}.$$

The ratio of static to stagnation temperature in the exhaust, where we may take $\gamma \sim 1.33$, is given by

$$\frac{T_e}{T_{t,e}} = \frac{1}{1 + \dfrac{\gamma_e - 1}{2}M_e^2} = \frac{1}{1 + \dfrac{M_e^2}{6}}.$$

The exhaust Mach number varies from $M_e \sim 1$ for subsonic jet aircraft to perhaps 2 for supersonic jet aircraft so that the range of the temperature ratio is $0.86 > T_e/T_{t,e} > 0.6$. Then

$$\eta_{th} \approx 1 - Kc_{p,e}T_{t,e}.$$

Here the value for $K$ lies in the range $0.0018 > K > 0.0011$, while the exhaust stagnation temperature lies in the range $1250\text{R} < T_{t,e} < 1750\text{R}$, and therefore the thermal efficiency is in the range $45\% < \eta_{th}$

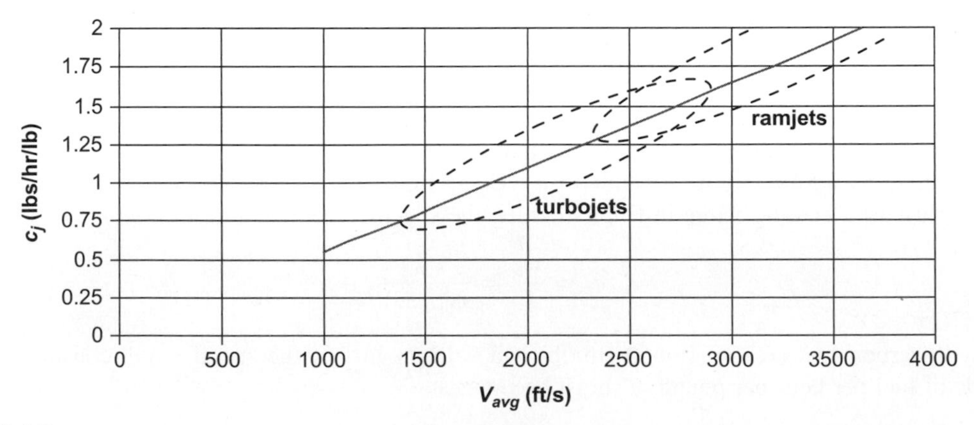


**FIGURE 1.7**

Specific fuel consumption as a function of average velocity for turbojets.

$< 49\%$. Then, taking an intermediate value, say $\eta_{th} = 47\%$, the specific fuel consumption (hr$^{-1}$) becomes

$$c_j = 5.69 \times 10^{-4} V_{avg}.$$

The specific fuel consumption varies linearly with average velocity, as shown in Figure 1.7. The notional regime for ramjets is also shown in Figure 1.7. The only difference between turbojets and ramjets is that the latter needs no compressor-turbine machinery, with the compression process being entirely due to ram pressure.

As shown in Equation (1.14), the specific fuel consumption of a jet engine increases with reductions in propulsive efficiency, that is, with increases in $V_{avg}$ and therefore with the exhaust velocity $V_e$. High heating value fuels will reduce $c_j$, while reduced thermal and burner efficiencies will increase it.

## 1.4 FLOW MACHINES WITH $P = 0$, $Q =$ CONSTANT, AND $A_0 = 0$: THE ROCKET

If no air is taken on board the flow machine, that is, $A_0 = 0$, there is no momentum penalty realized, and the momentum equation is simply

$$F = \dot{m} V_e. \tag{1.15}$$

A schematic diagram of the flow field in this special case is shown in Figure 1.8. Application of the energy equation leads to the following result:

$$F = \frac{2\dot{m}}{V_e}[Q - (h_e - h_i)] = 2\rho_e A_e [Q - (h_e - h_i)]. \tag{1.16}$$

Here all the gas ejected from the flow machine was onboard; the quantity $h_i$ represents the enthalpy of that internal propellant supply. This is the case of the simple rocket, and, like the pure jet, the thrust produced is independent of flight speed. The propulsive efficiency is now given by

$$\eta_p = \frac{FV_0}{FV_e} = 2\frac{V_0}{V_e}. \tag{1.17}$$

The propulsive efficiency of a rocket is even smaller than that of a jet, which, in turn, is smaller than that of a propeller. This lower propulsive efficiency means that the overall efficiency of the rocket is

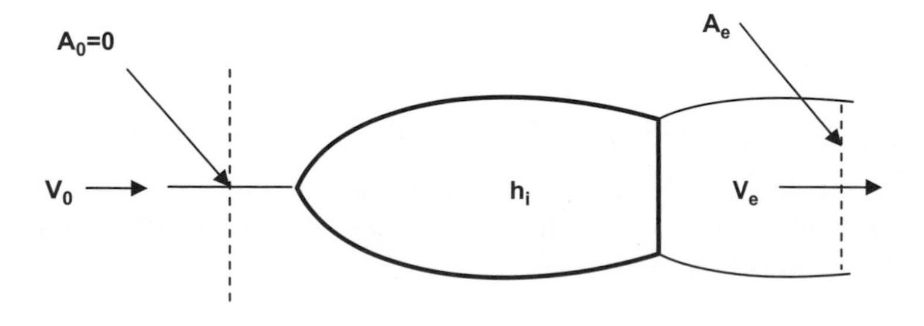


**FIGURE 1.8**

Schematic diagram of idealized flow field of a rocket.

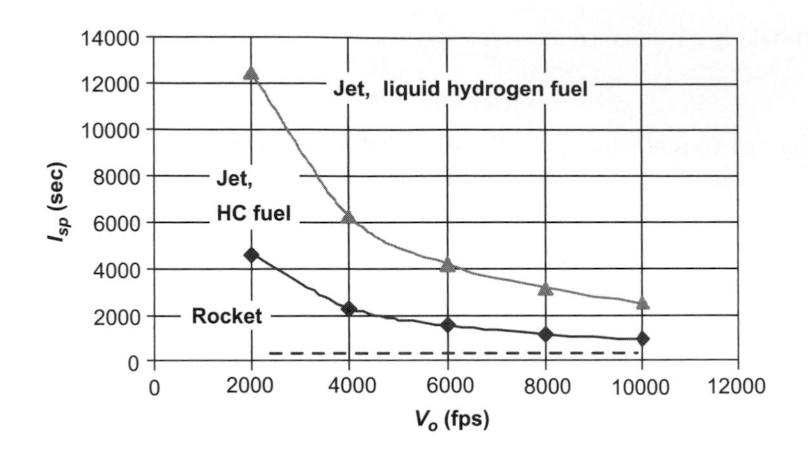


**FIGURE 1.9**

Specific impulse variation for turbojet and rocket engines as a function of average speed. Note that on this scale the rocket $I_{sp}$ is small and independent of $V_0$.

also smaller than the jet and propeller. The metric for fuel efficiency used for rockets is the specific impulse, which is defined as the ratio of the thrust produced to the weight flow rate of propellant, or

$$I_{sp} = \frac{F}{\dot{m}g} = \frac{\dot{m}V_e}{\dot{m}g} = \frac{V_e}{g}. \tag{1.18}$$

The specific impulse is typically measured in pounds of thrust per pound or propellant consumed per second, or seconds. This is essentially the inverse of the specific fuel consumption, but it is based on the weight flow rate of onboard propellants. The specific impulse for jets is based solely on the onboard fuel flow rate and is given by

$$I_{sp,jet} = \frac{F}{\dot{w}_f} = \frac{3600}{c_j}. \tag{1.19}$$

Therefore, the specific impulse of air-breathing jets is greater than that of rockets. A comparison of specific impulse for jets and rockets is shown in Figure 1.9. Note that, according to Equation (1.18), the specific impulse for rockets doesn't depend on flight speed and that therefore we may consider that at some flight speed the specific impulse of the rocket will be equal to or superior to that of a jet, as suggested by Figure 1.9.

## 1.5 THE SPECIAL CASE OF COMBINED HEAT AND POWER: THE TURBOFAN

If we consider that the flow through the idealized flow machine is divided into a central hot stream, denoted by the subscript $h$, and an outer cold stream, denoted by the subscript $c$, the conservation of mass equation may be written as follows:

$$\dot{m}_0 = \dot{m}_h + \dot{m}_c = \dot{m}_h\left(1 + \frac{\dot{m}_c}{\dot{m}_h}\right) = \dot{m}_h(1 + \beta). \tag{1.20}$$

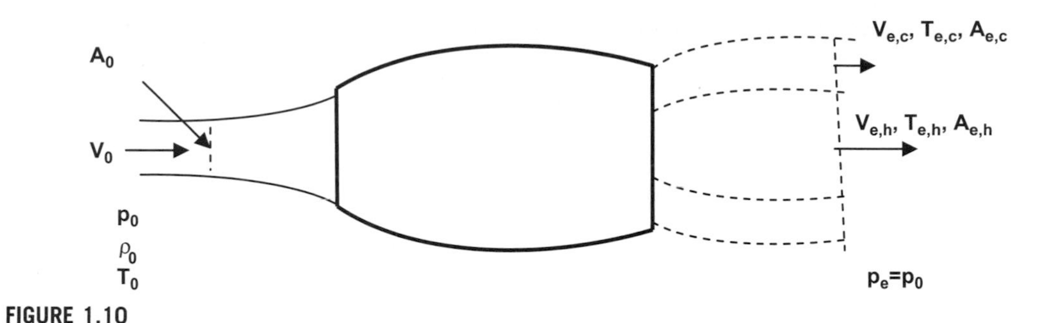


**FIGURE 1.10**

Schematic diagram illustrating an idealized one-dimensional flow machine with a hot gas exit flow surrounding a cold gas core flow.

The quantity $\beta$ measures the ratio of the cold mass flow to the hot mass flow and is called the bypass ratio. This idealized flow field is illustrated schematically in Figure 1.10.

Because we are assuming that flow outside the machine is inviscid, the central hot flow does not mix with the cold outer flow, and the force on the fluid is given by

$$F = F_h + F_c = \frac{\dot{m}}{1+\beta}(V_{e,h} - V_0) + \frac{\dot{m}\beta}{1+\beta}(V_{e,c} - V_0). \tag{1.21}$$

The cold outer flow is assumed to have had power added, but not any heat. Using results obtained previously for the case of flow with power added, but no heat added, we find that

$$F_c = \frac{P}{V_{c,avg}}. \tag{1.22}$$

Similarly, assuming that the hot core flow has had heat added, but has no net power added to it, we find that

$$F_h = \frac{\dot{m}_h}{V_{h,avg}}\left[Q - (h_{e,h} - h_0)\right] = \frac{\dot{m}}{(1+\beta)}\frac{\eta_{th}Q}{V_{h,avg}}. \tag{1.23}$$

The total force on the fluid may then be written as

$$F = \frac{P}{V_{c,avg}} + \frac{\rho_0 A_0 \eta_{th} Q}{1+\beta}\frac{V_0}{V_{h,avg}}. \tag{1.24}$$

This equation may be expanded to show the dependence on flight speed to yield

$$F = \frac{2P}{V_0 + V_{e,c}} + \frac{2\rho_0 A_0 \eta_{th} Q}{1+\beta}\frac{V_0}{V_0 + V_{e,h}}. \tag{1.25}$$

For a high bypass ratio, contribution of the jet to the thrust of the flow machine is reduced, particularly at low flight speeds, compared to the fan contribution. At high speeds, however, the fan contribution drops off substantially, while the jet thrust is approximately constant. Thus high bypass ratios (now as high as 10 on large modern turbofan engines) provide the high thrust needed for acceleration and climb during take-off while also maintaining required thrust levels for a high-speed cruise. Let us consider this case in additional detail by first rewriting Equation (1.24) in the following form:

$$F = \dot{m}_h \left( \frac{\eta_{th} Q}{V_{avg,h}} + \frac{\beta W_c}{V_{avg,c}} \right). \tag{1.26}$$

Here we assume that power for the cold outer stream $\dot{m}_c W_c$ is provided by a turbine in the hot central stream, where $W_c$ denotes the work done on the fluid per unit mass of fluid processed. That turbine, which operates only in the hot stream, also supplies power to a compressor in the hot stream, but the net power in the hot stream remains zero. This total power balance may be written as

$$\dot{m}_h (\beta W_c + W_h) = \dot{m}_h W. \tag{1.27}$$

Total heat added to the flow by burning fuel is given by Equation (1.13) and subtracting from that the work added to the cold outer stream yields the heat available for transformation to thrust

$$Q = \frac{\dot{m}_f g \eta_b HV}{\dot{m}_h} - \beta W_c. \tag{1.28}$$

Solving for $\beta W_c$ in Equation (1.28) and substituting for it in Equation (1.26) gives the following form for the total thrust:

$$F = \dot{m}_h Q \left( \frac{\eta_{th}}{V_{avg,h}} - \frac{1}{V_{avg,c}} \right) + \frac{\dot{m}_f g \eta_b HV}{V_{avg,c}} \tag{1.29}$$

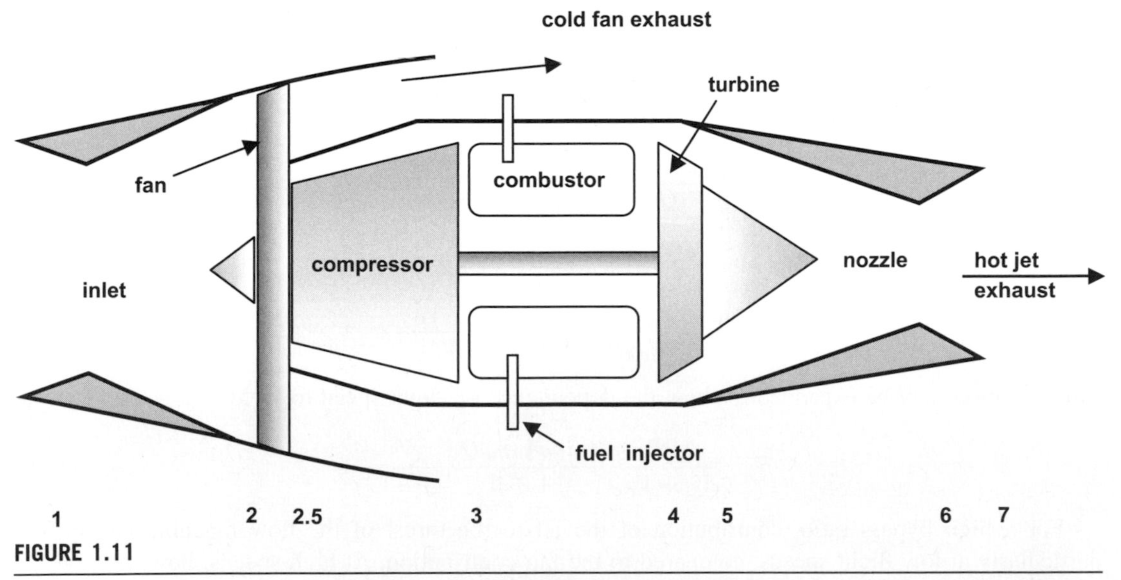


**FIGURE 1.11**

Schematic diagram of a typical turbofan engine showing required turbomachinery components and usual station numbering scheme. The combustor burns the injected fuel supplying heat to the central flow passing out the nozzle, and the fan accelerates the cold outer flow of the turbofan.

Although this particular form is not appreciated easily as it stands, it does provide an easy means to check the limiting cases, which then may make it easier to interpret. A schematic diagram of the internal configuration of a basic turbofan engine is shown in Figure 1.11.

### 1.5.1 Very small bypass ratio, $\beta$ << 1, the turbojet

In this case, there is essentially no cold outer flow and $W_c \sim 0$ so that, from Equation (1.28),

$$Q \approx \frac{\dot{m}_f g \eta_b HV}{\dot{m}_h}. \tag{1.30}$$

Using Equation (1.30) in Equation (1.29) gives the thrust as

$$F \approx \frac{\dot{m}_h Q \eta_{th}}{V_{avg,h}}. \tag{1.31}$$

This is the equation for thrust of a pure turbojet, as suggested by Equation (1.9). The specific fuel consumption is then

$$c_j = \frac{\dot{m}_f g}{F} = \frac{V_{avg,h}}{\eta_b \eta_{th} HV}. \tag{1.32}$$

Note that this is the specific fuel consumption for a turbojet as given in Equation (1.14). In the limit as the bypass ratio approaches zero we recover results for a turbojet.

### 1.5.2 Very large bypass ratio, $\beta$ >> 1, the turboprop

In this case, the cold outer mass flow $m_c$ is much larger than the hot central mass flow $\dot{m}_h$, but $\dot{m}_h$ itself is not especially small because it provides the power for driving the cold flow. However, after driving the outer flow, the heat available in the hot central flow for the jet is much smaller than the work required for the outer flow, $Q \ll W_c$, and Equation (1.28) becomes

$$\beta W_c \approx \frac{\dot{m}_f g \eta_b HV}{\dot{m}_h}. \tag{1.33}$$

Thus Equation (1.29) is approximated by

$$F \approx \frac{\dot{m}_c W_c}{V_{avg,c}}. \tag{1.34}$$

This is the case of the propeller as given by Equation (1.5), and in particular it corresponds to the turboprop engine where the gas turbine drives the propeller and little jet thrust is produced. The specific fuel consumption in this case is

$$c_j = \frac{V_{avg,c}}{\eta_b HV}. \tag{1.35}$$

By comparing Equations (1.32) and (1.35), we see that this case of a very high bypass ratio has the fuel efficiency advantage, as $V_{avg,c} < V_{avg,h}/\eta_{th}$.

### 1.5.3 Finite $\beta$, the turbofan

We may now rewrite Equation (1.29) in terms of specific thrust, that is, thrust per unit total mass flow, obtaining

$$\frac{F}{\dot{m}} = \frac{1}{\beta+1}\left(\frac{f}{a}\right)\frac{(g\eta_{th}\eta_b HV)}{V_{avg,h}} + \frac{\beta}{\beta+1}\left(\frac{1}{V_{avg,c}} - \frac{\eta_{th}}{V_{avg,h}}\right). \tag{1.36}$$

Here $f/a$ is the fuel–air ratio and the product $g\eta_{th}\eta_b HV$ may be considered, for purposes of this discussion, as essentially fixed by the generic engine design. Thus for $\beta \ll 1$, only the first term on the right-hand side of Equation (1.36) contributes to the specific thrust and the engine is a turbojet, whereas for $\beta \gg 1$, only the second term on the right-hand side of the equation contributes to the specific thrust and the engine is a turboprop. As $\beta$ grows from zero, we see an increasing contribution from the propeller-like second term and the engine is a turbofan of increasing bypass ratio.

Specific fuel consumption may be obtained from Equation (1.36) as follows:

$$c_j = \frac{\dot{m}_f g}{\frac{F}{\dot{m}}(\dot{m}_h + \dot{m}_c)} = \frac{g}{\beta+1}\frac{\left(\frac{f}{a}\right)}{\left(\frac{F}{\dot{m}}\right)} = \left[\frac{(\eta_{th}\eta_b HV)}{V_{avg,h}} + \frac{\beta}{g\left(\frac{f}{a}\right)}\left(\frac{1}{V_{avg,c}} - \frac{\eta_{th}}{V_{avg,h}}\right)\right]^{-1}. \tag{1.37}$$

It is evident from Equation (1.37) that as the bypass ratio increases, the specific fuel consumption will decrease, which is the advantage of high bypass turbofan engines in a high fuel cost environment. Note that $\beta = \dot{m}_c/\dot{m}_h$ cannot become arbitrarily large because the central hot mass flow $\dot{m}_h$ provides the power needed to drive the cold flow fan.

## 1.6 FORCE FIELD FOR AIR-BREATHING ENGINES

Consider the air-breathing engine shown schematically in Figure 1.12. Air is taken onboard at approximately flight speed $V_0$, fuel is added and burned, and the products of combustion are exhausted at the exit velocity, $V_7$. Further downstream, pressure $p_e$ achieves equilibrium with the ambient pressure at the flight altitude $p_0$, that is, $p_e = p_0$. Because experimental measurements are generally taken at the engine exit plane, station 7, we confine our attention to stations up to and including that point.

We may apply the conservation of momentum principle to a control volume that extends from station 0 to station 7, as shown in Figure 1.13. Here the control volume is the streamtube, which enters the airframe structure that houses the engine. For simplicity, we will consider a nacelle, or engine pod, bounded by stations 0 and 7. Force $F_1$ is the force exerted on the fluid by the external flow, and force $F_2$ is the force exerted on the fluid by the walls of the nacelle. The momentum theorem states that, in the absence of body forces, the resultant force on the boundary of a control volume fixed in inertial space is equal to the momentum flux passing through that boundary.

Therefore, assuming that there are no body forces and that the flow is quasi-one-dimensional, we have

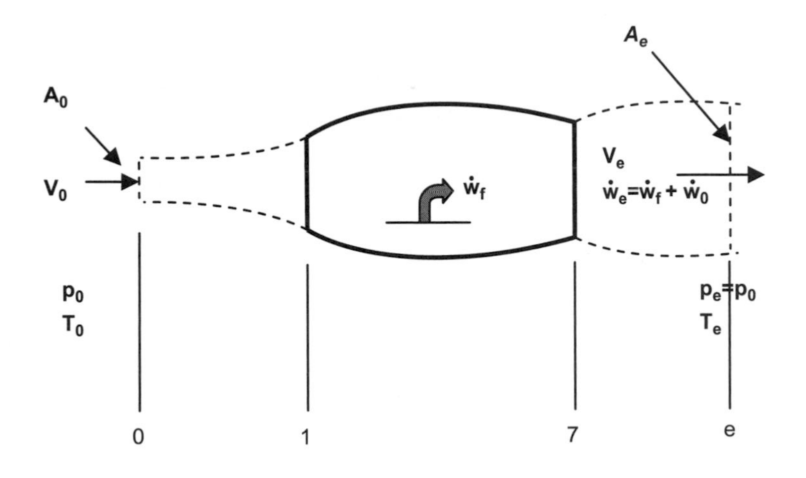


**FIGURE 1.12**

Schematic diagram of the control volume for an air-breathing engine.

$$F_1 + F_2 + p_0 A_0 - p_7 A_7 = \frac{\dot{w}_7 V_7}{g} - \frac{\dot{w}_0 V_0}{g}. \tag{1.38}$$

Now we may consider the forces acting on the nacelle enclosing the engine by examining the free-body diagram in Figure 1.14. Here $F$ is the force on the nacelle exerted by the aircraft through the pylon, $F_2$ is the force exerted by the fluid on the walls of the nacelle, and $F_4$ is the force exerted on the nacelle by the external stream. Equilibrium of these forces requires that

$$F + F_4 - F_2 = 0. \tag{1.39}$$

Then, using results for $F_2$ from Equation (1.38) in Equation (1.39), we obtain

$$F = \frac{\dot{w}_0 + \dot{w}_f}{g} V_7 - \frac{\dot{w}_0}{g} V_0 - p_0 A_0 + p_7 A_7 - F_1 - F_4. \tag{1.40}$$

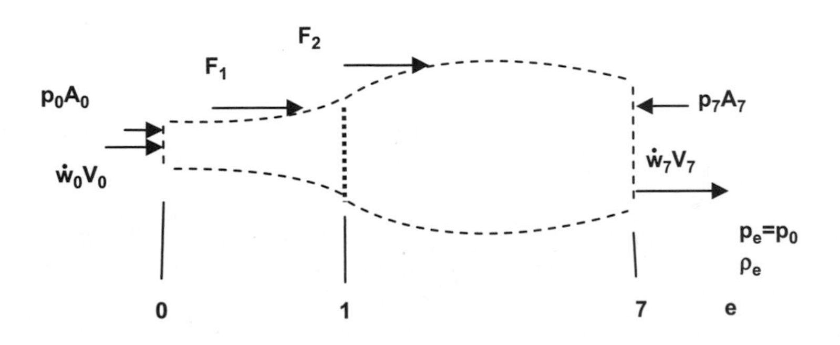


**FIGURE 1.13**

Control volume for one-dimensional analysis of an air-breathing engine.

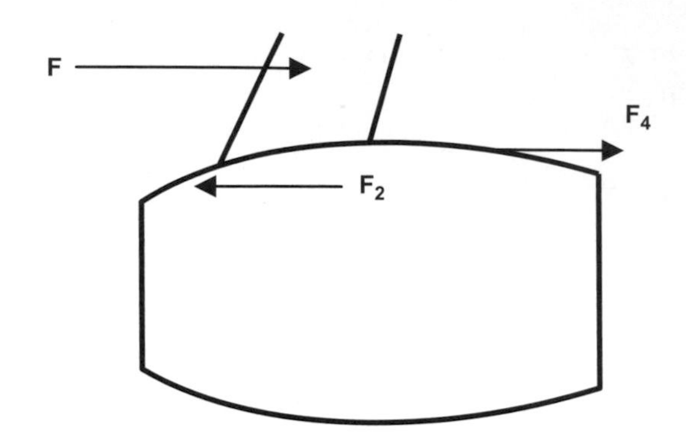


**FIGURE 1.14**

Schematic diagram of the structure containing a flow machine.

Adding and subtracting $p_0A_7$ to Equation (1.40) and rearranging it yields

$$F = \left[\frac{\dot{w}_0 + \dot{w}_f}{g} V_7 - \frac{\dot{w}_0}{g} V_0 + A_7(p_7 - p_0)\right] - [p_0(A_0 - A_7) + F_1 + F_4]. \tag{1.41}$$

Now Equation (1.41) is in the form $F =$ [net thrust] – [nacelle, or airframe, drag], where the first term in square brackets is the engine contribution to resultant force $F$ and the second term in square brackets is the airframe contribution to resultant force $F$. The net thrust provided by the engine is defined as

$$F_n = \left[\frac{\dot{w}_0 + \dot{w}_f}{g} V_7 + A_7(p_7 - p_0)\right] - \frac{\dot{w}_0}{g} V_0. \tag{1.42}$$

The net thrust, written in this form, is defined as $F_n =$ gross thrust – ram drag. Consider now the drag of the airframe housing the engine, which is expressed as

$$D = p_0(A_0 - A_7) + F_1 + F_4. \tag{1.43}$$

Here $F_1$ is the force in the streamwise direction due to pressure of the external flow on the capture streamtube extending from station 0 to station 1 and may be expressed as

$$F_1 = \int_{A_0}^{A_1} p dA. \tag{1.44}$$

Quantity $A$ is the cross-sectional area of the stream tube. Force $F_4$ is the total drag force acting on the nacelle due to external flow. It is composed of both pressure and friction forces and may be represented by

$$F_4 = \int_{A_1}^{A_7} p dA + D_f. \tag{1.45}$$

When flight velocity $V_0 = 0$ and the power is off, the pressure drag should be zero, the pressure and viscous drag must be zero, and therefore force $F_1 = 0$. This means we must define these forces somewhat more carefully and consistently so that they are accounted for accurately under static test conditions. Therefore, introduce

$$F_1' = \int_{A_0}^{A_1} (p - p_0) dA = D_{p,1} \tag{1.46}$$

$$F_4' = \int_{A_1}^{A_7} (p - p_0) dA + D_f = D_{p,4} + D_f. \tag{1.47}$$

Using Equations (1.46) and (1.47) in Equation (1.43), the nacelle drag may be rewritten as

$$D = \int_{A_0}^{A_1} (p - p_0) dA + \int_{A_1}^{A_7} (p - p_0) dA + D_f = D_{p,1} + D_{p,4} + D_f. \tag{1.48}$$

The term $D_{p,1}$ is called the additive drag due to the inlet; it will be seen to be an important factor in the design of inlets for supersonic flight. The nacelle drag has interesting ramifications, particularly for wind tunnel testing and engine-airframe integration. Consider differences between the flow field under power-on and power-off conditions, as illustrated in Figures 1.15 and 1.16. Disruption of the flow field in the power-off case, where the nacelle acts like a bluff body, leads to the result that the nacelle drag is greater than in the power-on case.

For purposes of comparing various jet engines, we may use $V_e$ to determine the thrust as done previously,

$$F_n = \frac{\dot{w}_0}{g}(V_e - V_0). \tag{1.49}$$

The net thrust given by Equation (1.42) is equal to the net thrust given by Equation (1.49), which permits us to calculate $V_e$, which is often called the effective exhaust velocity, as follows:

$$V_e = V_7 + \frac{gA_7}{\dot{w}_0}(p_7 - p_0). \tag{1.50}$$

In obtaining this result, we have assumed that $\dot{w}_f \ll \dot{w}_0$, and because the ratio of the two is typically about 2%, as shown in a subsequent chapter, the approximation is quite a good one. It should be noted

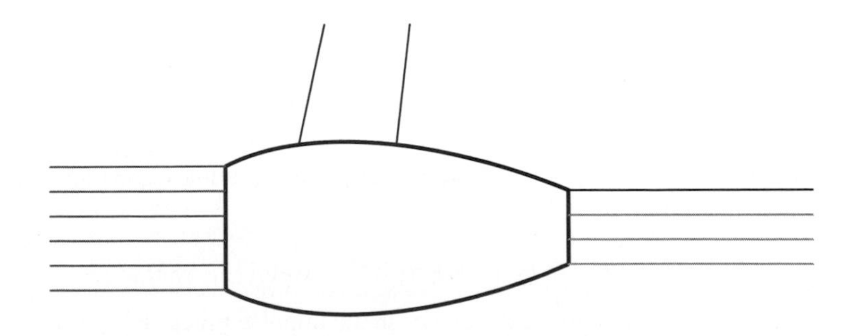

**FIGURE 1.15**

Schematic diagram of flow during powered operation. The flow proceeds smoothly over the nacelle, and the exit station pressure may be adjusted to match the ambient pressure.

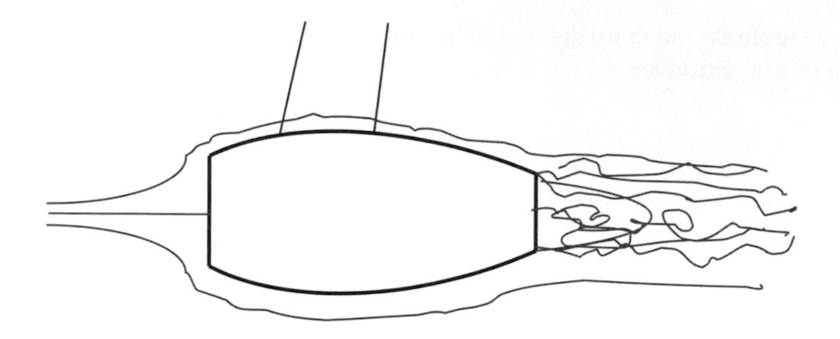

**FIGURE 1.16**

Schematic diagram of flow with power off. The flow at the front face is a stagnation point with high pressure. The flow may separate from the cowl lip and cause irregular turbulent flow over the nacelle, while the exit station pressure is that of a recirculation region and generally lower than the ambient pressure.

that for subsonic exit velocities, that is, $V_7 < a_7$, the pressure is always equilibrated, that is, $p_7 = p_0$, hence the exhaust velocity increases until $V_7 = a_7$. For supersonic flows, $p_7$ is not necessarily equal to $p_0$; this situation will be considered subsequently.

We now turn to considering some of the equations relating the flow properties for jet engines. The gross thrust of Equation (1.42) is given by

$$F_g = \frac{\dot{w}_0 + \dot{w}_f}{g} V_7 + A_7(p_7 - p_0). \tag{1.51}$$

Using the equation for sound speed as $a^2 = \gamma p/\rho$, the momentum flux in Equation (1.51) may be written as

$$\frac{\dot{w}_0 + \dot{w}_f}{g} V_7 = \rho_7 A_7 V_7^2 = \gamma_7 p_7 A_7 M_7^2. \tag{1.52}$$

Substituting this result into Equation (1.51) leads to the following equation for gross thrust:

$$F_g = A_7\left[p_7(1 + \gamma_7 M_7^2) - p_0\right]. \tag{1.53}$$

Similarly, we may write the ram drag as

$$F_r = \frac{\dot{w}_0}{g} V_0 = \gamma_0 p_0 A_0 M_0^2. \tag{1.54}$$

The net thrust would then be given by the difference between Equation (1.53) and Equation (1.52), which may be written as follows:

$$F_n = F_g - F_r = p_7 A_7(1 + \gamma_7 M_7^2) - p_0 A_0(1 + \gamma_0 M_0^2).$$

This shows that the net thrust is equal to the difference in the impulse function between free stream and exit conditions. The impulse function is discussed in more detail in Section 2.10.

If the flow in the exhaust nozzle is assumed to be adiabatic with constant specific heats, then the stagnation temperature is constant through it. If it is further assumed that the flow is also reversible, that is, isentropic, then the stagnation pressure is also constant through it. The quasi-one-dimensional energy equation under these assumptions may be written as

$$c_{p,7}T_7 + \frac{1}{2}V_7^2 = c_{p,7}T_{t,7}. \tag{1.55}$$

This equation shows that the sum of the internal or thermal energy is a constant so that converting the thermal energy into kinetic energy permits the nozzle to accelerate the flow and produce the high exhaust velocity necessary for the production of high levels of thrust. Solving Equation (1.55) for the exit plane velocity yields

$$V_7 = \sqrt{2c_pT_{t,7}\left(1 - \frac{T_7}{T_{t,7}}\right)}. \tag{1.56}$$

The isentropic relation between pressure and temperature is

$$\frac{T_7}{T_{t,7}} = \left(\frac{p_7}{p_{t,7}}\right)^{\frac{\gamma_7-1}{\gamma_7}}.$$

In general then, the exit velocity of the nozzle is given by

$$V_7 = \sqrt{2c_{p,7}T_{t,7}\left[1 - \left(\frac{p_7}{p_{t,7}}\right)^{\frac{\gamma_7-1}{\gamma_7}}\right]}. \tag{1.57}$$

We see here that the nozzle acts as a mechanical accelerator for the flow, converting the thermal energy of the flow to kinetic energy. Note that the ability to convert all the stagnation enthalpy into kinetic energy requires expansion to zero static temperature, which implies zero static pressure, according to Equation (1.57). The density, which is proportional to the ratio of pressure to temperature, of course must still remain finite. We may examine the static to stagnation pressure ratio in Equation (1.57) to get an appreciation for the manner in which the engine components limit performance by writing it as follows:

$$\frac{p_7}{p_{t,7}} = \frac{p_7}{p_0}\frac{p_0}{p_{t,0}}\frac{p_{t,0}}{p_{t,7}}. \tag{1.58}$$

Because using a chain rule to relate the properties across several parts of the engine is often helpful, we expand Equation (1.58) into the following form:

$$\frac{p_7}{p_{t,7}} = \frac{p_7}{p_0}\frac{p_0}{p_{t,0}}\frac{1}{\frac{p_{t,2}}{p_{t,0}}\frac{p_{t,3}}{p_{t,2}}\frac{p_{t,4}}{p_{t,3}}\frac{p_{t,5}}{p_{t,4}}}\frac{p_{t,5}}{p_{t,7}}. \tag{1.59}$$

The first term on the right-hand side of Equation (1.59) is the ratio of the nozzle exit pressure to the free stream pressure at the flight altitude and indicates the extent to which the nozzle is adapted to the pressure in the surrounding ambient. We will subsequently prove that the maximum thrust occurs for a matched nozzle, that is, one where $p_7 = p_0$. The second term expresses the ratio of static to stagnation pressure at the given flight conditions and is a function of the flight Mach number according to the isentropic flow relation:

$$\frac{p_0}{p_{t,0}} = \left(1 + \frac{\gamma_0 - 1}{2} M_0^2\right)^{-\frac{\gamma_0}{\gamma_0 - 1}}.$$

The third term is the inverse of the so-called inlet pressure recovery and is a function of the flight Mach number. The inlet pressure recovery is close to unity until the Mach number $M_0 > 1$ and then decreases as higher Mach numbers are reached. This behavior is covered in detail in Chapter 6. The fourth term is the inverse of the pressure ratio across the compressor, which depends on the amount of work done on the air passing through it; this parameter is discussed in Chapter 7. The fifth term is the inverse of the pressure recovery of the combustor. This parameter is usually close to unity for a well-designed combustor and will be treated in Chapter 4. The sixth term is the pressure drop across the turbine, which represents the work extracted from the combustion products, which is used to drive the compressor. This parameter is also discussed further in Chapter 7. The final term is the stagnation pressure ratio across the nozzle and expresses the efficiency of the nozzle. A well-designed nozzle will have a stagnation pressure ratio close to unity; this parameter is covered in Chapter 5. Many factors influence stagnation pressure losses in an air-breathing engine, and it will be shown that the greatest losses occur in the inlets of engines in supersonic flight. Therefore, good inlet design is crucial for the effective operation of engines for supersonic flight.

## 1.7 CONDITIONS FOR MAXIMUM THRUST

Let us consider the conditions that must be met for the isentropic nozzle of the engine in order that maximum thrust is produced for a given value of total energy in the flow entering the nozzle, that is, a given value of stagnation temperature, $T_{t,7}$, which is constant inside the nozzle. At the same time, we are assuming a given value of the stagnation pressure, which, in our ideal case of isentropic flow, means that stagnation pressure $p_{t,7}$ is also constant throughout the nozzle. We are therefore investigating what conditions must be met by the nozzle so as to make best use of the acceleration capabilities of the nozzle. First we note that the gross thrust given in Equation (1.51) for given engine geometry and flight conditions depends only on the exit pressure of the nozzle. We may search for an extremum in the gross thrust by taking the partial derivative of the gross thrust with respect to the pressure and setting it to zero, leading to the following equation:

$$\frac{\partial F_g}{\partial p_7} = \frac{\dot{w}_7}{g}\frac{\partial V_7}{\partial p_7} + A_7\frac{\partial p_7}{\partial p_7} + (p_7 - p_0)\frac{\partial A_7}{\partial p_7} = 0. \tag{1.60}$$

Here we have used the exact form of the equation in that the weight flow shown is that at station 7. A subsequent chapter will show that the quasi-one-dimensional momentum equation requires that

$$\dot{m}dV = -Adp. \tag{1.61}$$

Incorporating this result into Equation (1.60) yields

$$\frac{\partial F_g}{\partial p_7} = (p_7 - p_0)\left(\frac{\partial A}{\partial p}\right)_7 = 0. \tag{1.62}$$

Gross thrust is an extremum when the exit pressure matches the ambient atmospheric pressure, $p_7 = p_0$, or when $(\partial A/\partial p)_7 = 0$. To determine whether or not this extremum is a maximum, we evaluate the second derivative of the gross thrust with respect to pressure evaluated at the extreme point. Thus we find

$$\left(\frac{\partial^2 F_g}{\partial p^2}\right)_7 = (p_7 - p_0)\left(\frac{\partial^2 A}{\partial p^2}\right)_7 + \left(\frac{\partial A}{\partial p}\right)_7. \tag{1.63}$$

As shown in a subsequent chapter dealing solely with the equations of motion, the variation of pressure with area for a quasi-one-dimensional flow has the behavior shown in Figure 1.17.

It is clear that conditions for an extremum as given by Equation (1.62) are satisfied either at the matched exit case where $p_7 = p_0$ or at the sonic point where $p_7 = p^*$, the pressure for $M = 1$. For the case $p_7 = p_0$, Equation (1.63) reduces to

$$\left(\frac{\partial^2 F_g}{\partial p^2}\right)_{p_7 = p_0} = \left(\frac{\partial A}{\partial p}\right)_7. \tag{1.64}$$

From the shape of the curve in Figure 1.13, it is clear that if the exit flow is supersonic, the right-hand side of Equation (1.64) is negative and the thrust is a maximum. If the exit flow for this case is sonic, the right-hand side of Equation (1.64) is zero, and the extremum is a saddle point. It is tempting to say further that if the exit flow is subsonic, then the right-hand side of Equation (1.64) is positive and the thrust is a minimum. However, that conclusion doesn't hold up under close scrutiny. Recall that we began this section with a consideration of the gross thrust as given by Equation (1.51), which includes the pressure imbalance term. Because a subsonic flow exiting as a free jet from the exit of the nozzle

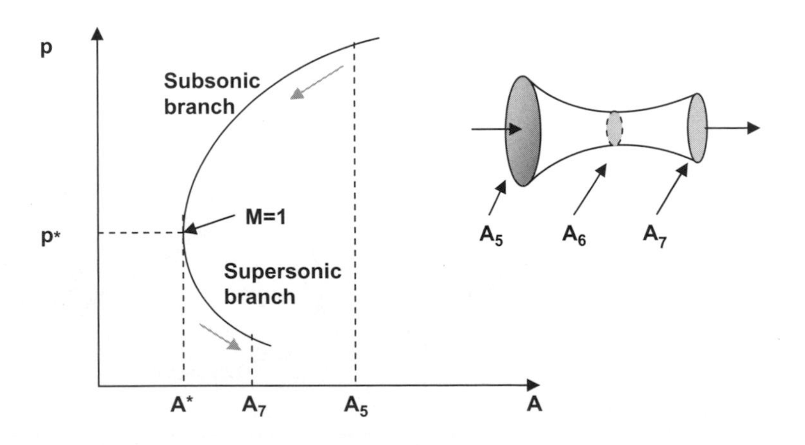


**FIGURE 1.17**

General variation of pressure with cross-sectional area for a quasi-one-dimensional flow.

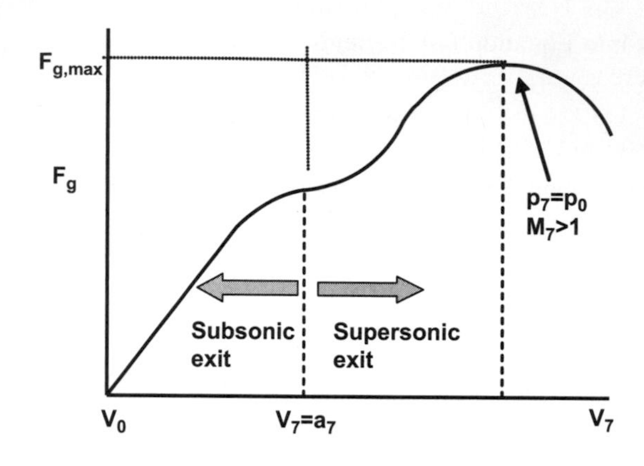


**FIGURE 1.18**

Variation of gross thrust with nozzle exit velocity.

cannot support a pressure difference, the pressure is always matched and the gross thrust for a subsonic jet is just

$$F_g = \frac{\dot{w}_7}{g} V_7. \tag{1.65}$$

As a result, the first derivative of $F_g$ becomes $\left(\frac{\partial F_g}{\partial p}\right)_7 = A_7$ and there is no extremum. The gross thrust may then be illustrated as in Figure 1.18.

We may consider the result that gross thrust is a maximum for matched operation in a somewhat more physical fashion using the illustration in Figure 1.19. The solid black line represents the nozzle

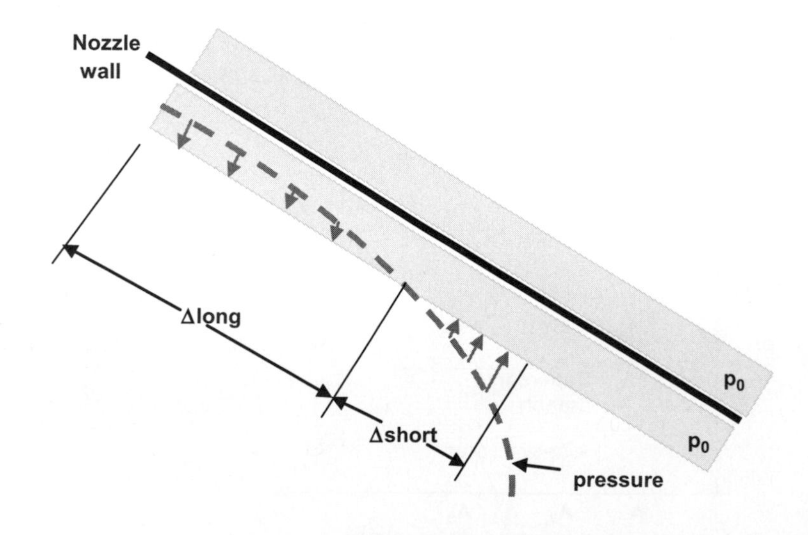


**FIGURE 1.19**

Illustration of effect of making a nozzle longer or shorter than required for matched operation, that is, for the case where $p_7 = p_0$.

wall, and the shaded areas above and below it represent the ambient pressure level. If there were no flow, the ambient pressure force on one side of the nozzle wall would cancel that on the other side. With supersonic flow in the nozzle, pressure distribution is shown as the dashed line, and the point where this line crosses into the shaded region indicates the length of nozzle required for matched operation. If the nozzle is lengthened by an amount shown as Δlong, the pressure on the lengthened portion of the nozzle will be below $p_0$ and the thrust will drop from the matched value. Similarly, if the nozzle is shortened by an amount Δshort, the portion of the nozzle represented by this length, and the positive thrust associated with it, will be lost and again the thrust will drop from the matched value. Therefore, maximum thrust must occur when the nozzle is matched, that is, when $p_7 = p_0$.

The general behavior of the net thrust is shown in Figure 1.20 with a notional illustration of the types of nozzle geometries typical of different exit conditions. Subsonic jet aircraft generally operate with a choked exit station, whereas supersonic operation requires opening the exit area to produce a supersonic flow and perhaps to attempt to match the exit pressure to the ambient atmospheric pressure.

# 1.8 EXAMPLE: JET AND ROCKET ENGINE PERFORMANCE

## 1.8.1 Jet engine performance

The Pratt & Whitney J58 jet engine for the SR-71 has a fuel flow rate of 8000 gal/hr of JP-7 with a fuel to air ratio (f/a) of 0.034. The take-off gross thrust $F_g$ = 34,000 lbs at a speed $V_{t\text{-}o}$ = 200 kts. For a nozzle exit Mach number M7 = 1 and a turbine exit temperature TET = 1580F = Tt,7 (Figure 1.21),

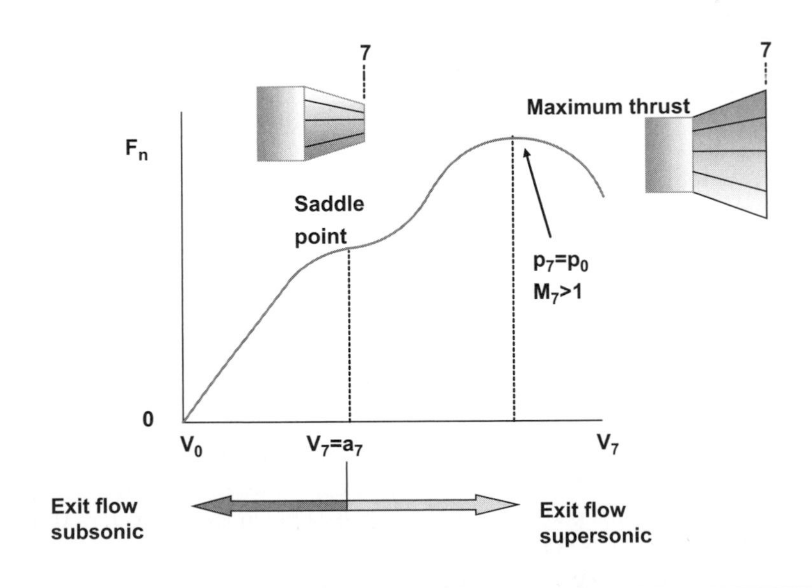


FIGURE 1.20

Net thrust as a function of exit velocity and illustration of nozzle geometries appropriate to the indicated conditions.

**FIGURE 1.21**

The Lockheed SR-71 in flight and its Pratt & Whitney J58 firing on a test stand.

find (a) effective exhaust velocity, (b) net thrust, (c) specific fuel consumption, (d) free stream capture area, (e) useful power, (f) nozzle exit temperature, (g) nozzle exit velocity, (h) nozzle exit area, and (i) nozzle exit pressure:

**(a)** $V_{eff} = gF_g/\dot{w}_f(1 + a/f)$= 2400 fps.
**(b)** Net thrust $F_n = F_g - \dot{m}_0 V_0 = 29{,}400$ lbs.
**(c)** $c_j = \dot{w}_f / F_n = 15{*}3600/29400 = 1.84$ lb.fuel/lb.thrust/hr.
**(d)** $A_0 = \dot{m}_0/\rho_0 V_0 = 17.4$ ft$^2$ ($D_1 = 4.7$ ft).
**(e)** $P_{useful} = F_n V_0 = 14750$ hp (11 MW).
**(f)** $T_7 = T_{t,7}[1 - (\gamma_7 - 1)M_7^2/2]^{-1} = 2040R(0.86) = 1750R$ (*use* $\gamma_7 = 1.33$).
**(g)** $V_7 = a_7 = (\gamma_7 RT_7)^{1/2} = 2000$ fps ($R = 1716$ ft$^2$/s$^2 - R$).
**(h)** Because gross thrust $F_g = A_7[p_7(1 + \gamma_7 M_7^2) - p_0]$ and $\dot{m}_7 = (\rho VA)_7$, then $A_7 = [\dot{m}_7(1 + \gamma)(RT/\gamma)^{1/2}]_7/p_0 - F_g/p_0$ and $A_7 = 7.36$ ft$^2$ ($D_7 = 3.1$ ft).
**(i)** $p_7 = (F_g/A_7 + p_0)/(1 + \gamma_7 M_7^2) = 1.37 p_0 = 2900$lb/sq.ft.

### 1.8.2 Rocket engine performance

The German V-2 rocket burned about 8000 kg of propellant, liquid oxygen (LOX) and alcohol, with an oxidizer-to-fuel ratio (O/F) = 1.3, in 65 seconds with a nozzle exit velocity $V_e \sim 2000$ m/s while the nozzle was operating in the matched mode (Figure 1.22).

Find (a) thrust produced, (b) power produced on the launch pad and at a flight speed of 2 km/s, (c) specific impulse, and (d) overall efficiency:

**(a)** Thrust for a rocket is given by $F = \dot{m}V_e$ so that $F$ = (8000 kg/65 s)(2000 m/s) = 246 kN (or 55.3 klbs).
**(b)** The power $P$ produced on the pad and at $V_0 = 2$ km/s is given by $P = V_0 F = 0$ (on the pad) and 2250 kN/s (2.25 MW or 660,000 hp) at 2 km/s.
**(c)** $I_{sp} = F/\dot{m}g$ = 246 kN/(8000 kg/65 s)(9.81 m/s$^2$) = 203 s.
**(d)** The overall efficiency $\eta_o = FV_0/\dot{m}\Delta Q = FV_0/w_f\eta_b HV$ or $\eta_o = FV_0/\{(\dot{m}g/[1 + O/F])\eta_b HV\}$. Then $\eta_o$ = (492 MN-m/s)/{123 kg/s(1)(27.2 MJ/kg)/(1 + 1.3)} = 33.8%.

FIGURE 1.22

The V-2 rocket during lift-off.

## 1.9 NOMENCLATURE

| | |
|---|---|
| $A$ | cross-sectional area |
| $a$ | sound speed |
| $c_j$ | specific fuel consumption |
| $c_p$ | specific heat at constant pressure |
| $D$ | drag |
| $D_{p,1}$ | inlet additive drag |
| $D_{p,4}$ | nacelle drag |
| $F$ | thrust |
| $F_n$ | net thrust |
| $g$ | acceleration of gravity |

| | |
|---|---|
| $HV$ | heating value of the fuel |
| $h$ | enthalpy |
| $I_{sp}$ | specific impulse |
| $M$ | Mach number |
| $\dot{m}$ | mass flow |
| $h$ | enthalpy |
| $p$ | pressure |
| $P$ | power delivered to the fluid |
| $Q$ | heat added to the fluid |
| $R$ | gas constant |
| $T$ | temperature |
| $V$ | velocity |
| $W$ | work done on the fluid |
| $\dot{w}$ | weight flow $\dot{m}g$ |
| $\beta$ | bypass ratio |
| $\gamma$ | ratio of specific heats |
| $\rho$ | density |
| $\eta$ | efficiency |

### 1.9.1 Subscripts

| | |
|---|---|
| $avg$ | average |
| $b$ | burner |
| $c$ | cold section |
| $e$ | conditions downstream where $p = p_0$ |
| $f$ | fuel |
| $h$ | hot section |
| $th$ | thermal |
| $max$ | maximum condition |
| $o$ | overall |
| $p$ | propusive |
| $t$ | stagnation conditions |
| 0 | conditions in the free stream |
| 1 | conditions at the inlet entrance |
| 7 | conditions at the nozzle exit |

## 1.10 EXERCISES

**1.1** McDonnell Aircraft's F2H Banshee fighter shown in Fig. E1.1 saw wide service in the Korean War. It first flew in 1947 and some served into the 1960s. This fighter, which figured prominently in James Michener's novel *The Bridges of Toko-Ri*, was an improvement on the company's FH-1 Phantom. The aircraft used two Westinghouse J34-WE-30 turbojets delivering 3150 lbs of thrust each, giving a maximum speed of 587 mph and a range of almost 1300 miles. The F2H could climb at about 7380 feet per minute and reach a service ceiling of 48,500 feet. If the airplane is operating at an altitude of 20,000 feet with a choked nozzle exit ($M_7 = 1$, $\gamma_7 = 1.33$) and the nozzle exit area $A_7 = 180$ in$^2$ and the exit pressure is $p_7 = 9$ psia, find the following: (a) gross thrust produced, (b) percentage of gross thrust caused by the pressure imbalance at the nozzle exit, (c) exit weight flow rate if $T_7 = 1000$F, (d) ram drag at a flight velocity of 550 fps if $\dot{w}_f/\dot{w}_0 = 0.02$, (e) net thrust, (f) thrust power produced, (g) exit velocity $V_7$, (h) effective exhaust velocity $V_{eff}$, (i) propulsive efficiency $\eta_p$, (j) $\eta_{th}$ and $\eta_o$ for $HV = 18{,}900$ Btu/lb, and (k) specific fuel consumption.

**FIGURE E1.1**

McDonnell Aircraft's F2H Banshee fighter.

**1.2** A General Electric J79-GE-15 turbojet engine is one of two propelling a McDonnell F4C airplane shown in Fig. E1.2 (wing area $S = 530$ ft$^2$, inlet area for each engine $A_1 = 6.5$ ft$^2$) cruising at a constant Mach number $M_0 = .82$ at an altitude of 35,000 ft (the 1976 U.S. Standard Atmosphere model may be used). The drag coefficient of the aircraft under these conditions is 0.045. Determine (a) net thrust of the engine, (b) gross thrust of the engine, and (c) weight (or mass) flow through the engine. Then, if the exhaust pressure $p_7 = p_0$ (d) estimate the exhaust velocity $V_7$ and, assuming that the turbine exit stagnation temperature, which is the nozzle entry stagnation temperature, $T_{t,5} = T_{t,7} = 1350$ F, (e) estimate the exit static temperature $T_7$, and (f) the exit Mach number $M_7$.

**FIGURE E1.2**

The McDonnell F4C Phantom.

**1.3** The Redstone missile, shown in Fig. E1.3 launching the Freedom 7 spacecraft on May 5, 1961, was the launch vehicle for America's first man in (suborbital) space, Alan Shepard. This rocket can boost a 2850-lb (1300-kg) payload to an altitude of 115 miles (185 km). It is a one-stage rocket vehicle using a single A-6 engine with a burn time of approximately 155 s.

**FIGURE E1.3**

The Redstone missile, shown here launching the Freedom 7 spacecraft on May 5, 1961.

The propellant, which is composed of alcohol and LOX, is consumed at the rate of about 300 lbs/s. Experiments have shown that the exhaust velocity of the combustion products $V_7 = 8400$ fps. Assuming that the exit pressure of the nozzle is matched ($p_7 = p_0$), determine (a) thrust and thrust power under static conditions ($V_0 = 0$), (b) thrust and thrust power at a flight speed $V_0 = 5700$ mph, (c) specific impulse $I_{sp}$, and (d) overall efficiency $\eta_0$ at the given flight speed if the heating value of the propellants $HV = 12{,}800$ Btu/lb. The heating value $HV$ is typically quoted for the fuel alone, and the quoted value of 12,800 Btu/lb is the appropriate value for alcohol. The optimal oxidizer to fuel ratio is $r = 1.47$ for the alcohol–LOX propellant combination.

**1.4** Consider the effect of specific fuel consumption on the cruise range of an aircraft whose total weight at the start of cruise is $W_1 = W_{spp} + W_{fc}$, where $W_{spp}$ is the combined weight of structure, powerplant, and payload and $W_{fc}$ is the weight of fuel available for the cruise segment of the mission. Consider cruise in the stratosphere at constant velocity $V_0 = 500$ mph and lift to drag ratio $L/D = 15$. Assume that at the end of cruise all the fuel is used so that the final weight $W_2 = W_{spp}$. What is the effect of specific fuel consumption on the range $R$ when the ratio $W_{spp}/W_1$ is varied between 0 and 1? What has a more important effect on the cruise range, specific fuel consumption (consider values between 0.5 and 1.5 lbs fuel per hour per pound of thrust) or $W_{spp}/W_1$? Substantiate your conclusions with graphs and/or sensitivity analyses.

**1.5** The 46-ft-long BOMARC interceptor missile shown in Fig. E1.4 was powered by an Aerojet General LR59-AG-13 liquid fuel booster rocket of 35,875 lb (159.6 kN) thrust and two

**FIGURE E1.4**

Boeing's BOMARC long-range interceptor missile.

Marquardt RJ43-MA-3 sustainer ramjets of 11,500 lb (51.2 kN) thrust each (CIM-10A). The maximum speed was 1975 mph (3178 km/hr), and the ceiling was 65,000 ft (19.8 km), with a range of 260 mi (418 km). The nominal inlet diameter of the engine was 24 in. (0.61 m), its length 145 in. (3.68 m), and its weight 485 lb (220 kg). For flight at $M_0 = 1.5$ and an altitude of 50,000 ft (15.2 km) with an effective exhaust velocity of the engine $V_{eff} = 2500$ fps (762 m/s) and assuming the flow conditions at the inlet are equivalent to those of flight, calculate the following: (a) air flow rate $\dot{w}_0$, (b) fuel flow rate $\dot{w}_f$ for a fuel to air ratio of 0.03, (c) exhaust gas flow rate $\dot{w}_7$, (d) gross thrust, (e) ram drag, (f) net thrust, (g) thrust power, (h) propulsive efficiency $\eta_p$, (i) thermal efficiency $\eta_{th}$, (j) specific fuel consumption, and (k) overall efficiency $\eta_o$.

**1.6** The Space Shuttle Transportation System (SSTS) shown in Fig. E1.5 incorporates three space shuttle main engines (SSME, shown in Fig. E1.6) using liquid hydrogen ($LH_2$) as the fuel and LOX as the oxidizer in a ratio of 1:6. The heating value of $LH_2$ is 51,600 Btu/lb (120 MJ/kg). The thrust produced by each engine in near-vacuum conditions of the upper atmosphere is 512,000 lbs (2.28 MN) at a specific impulse of 454 s. The typical burn time for the engines is 8.5 min. Determine the following: (a) effective exhaust velocity of the engine, (b) weight flow rate of fuel and of oxidizer for the engine, (c) total weight of the fuel and oxidizer required for the cluster of three engines, (d) volume for the fuel and oxidizer tanks, (e) length of the fuel and oxidizer tanks, assuming they are cylinders equal to the (46.9 m) long space shuttle external tank, and (f) overall efficiency $\eta_o$ at a flight speed of 6000 fps (1.83 km/s).

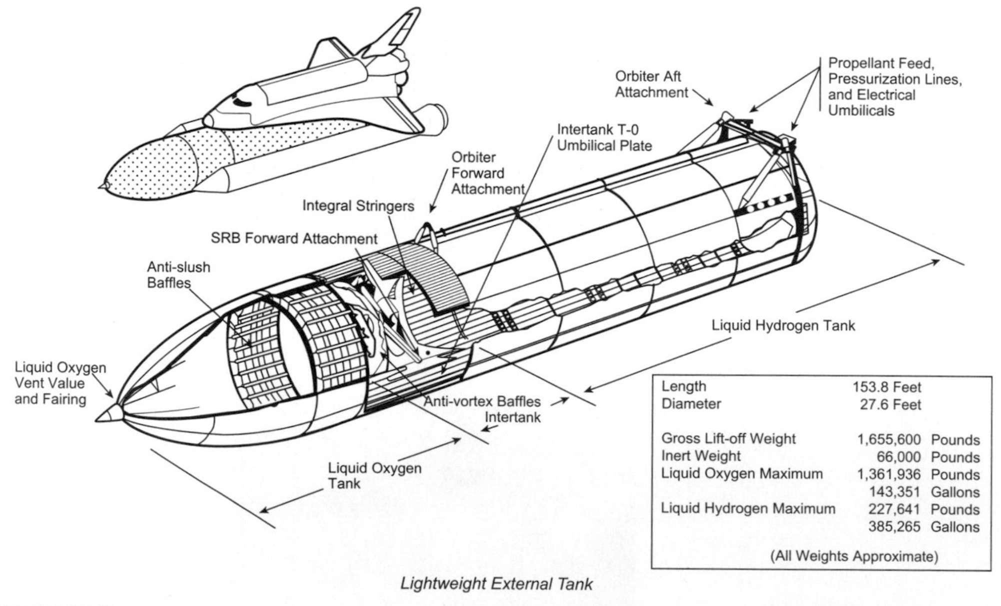


**FIGURE E1.5**

The Space Shuttle Transportation System (SSTS) incorporates the Space Shuttle Orbiter, the large external tank, and two integrated solid rocket boosters.

**FIGURE E1.6**

The Space Shuttle Orbiter incorporates three space shuttle main engines (SSME) using liquid hydrogen ($LH_2$) as the fuel and liquid oxygen (LOX) as the oxidizer.

**1.7** Five F-1 engines were used in the first stage of the Saturn launch vehicle, shown in Fig. E1.7, that lifted astronauts on a flight to the moon in the Apollo program. The F-1, shown in Fig. E1.8, still the most powerful rocket engine ever built, uses RP-1 (kerosene) as fuel and LOX as the oxidizer in a ratio of 1:2.7. The heating value of RP-1 is 18,900 Btu/lb (44 MJ/kg). The thrust developed is $1.52 \times 10^6$ lbs (6.76 MN) at a specific impulse of 265 s with a typical burn time of about 2.5 min. Determine the following: (a) effective exhaust velocity of each engine, (b) weight flow rate of fuel and oxidizer for the engine, (c) total weight of fuel and oxidizer required for the cluster of five engines, (c) weight of fuel and oxidizer required, (d) volume for the fuel and oxidizer tanks, (e) length of the fuel and oxidizer tanks, assuming they are cylinders equal in diameter to that of the first stage, 33 ft (10 m), and (f) the overall efficiency at a flight speed of 6000 fps (1.83 km/s).

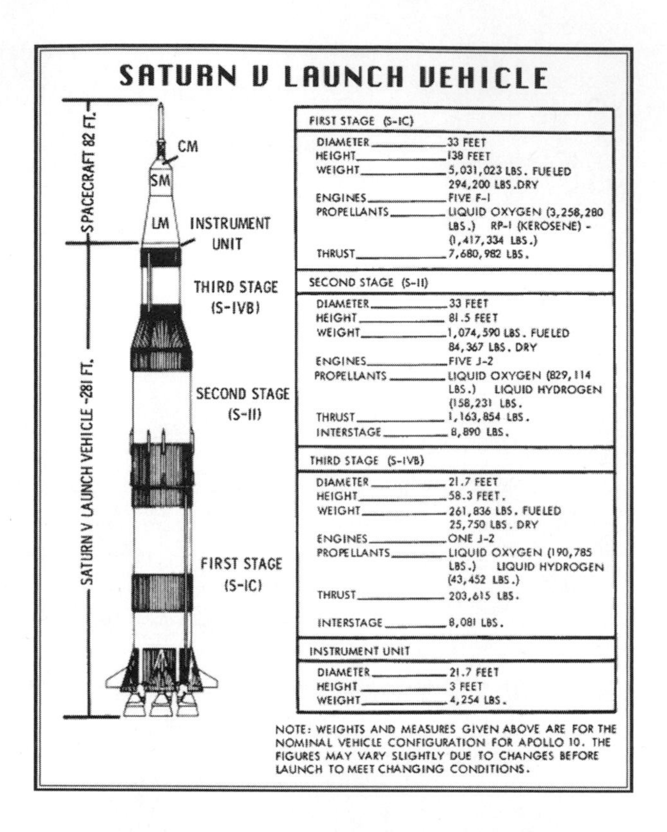


**FIGURE E1.7**

Characteristics of the Saturn V launch vehicle that lifted astronauts on a flight to the moon in the Apollo program.

**FIGURE E1.8**

The Rocketdyne F-1 engine uses RP-1 (kerosene) as fuel and LOX as the oxidizer.

## Reference

Goodger, E. M. (1986). *Alternative Fuel Technology Series, Vol. 2—Comparative Properties of Conventional and Alternative Fuels*. Cranfield, UK: Cranfield Press.

CHAPTER

# Quasi-One-Dimensional Flow Equations

2

## CHAPTER OUTLINE



## 2.1 INTRODUCTION

Equations that describe flow through a duct within the constraints of a steady one-dimensional approximation are developed in this section. The equations will be applied to simpler flow situations in other propulsion subsystems, such as combustors, nozzles, inlets, and turbomachinery cascades. The equations are particularly simple to deal with, yet they are commonly acknowledged to be reasonably accurate, even for fairly complex problems. This attribute makes the equation set to be developed particularly useful for preliminary design purposes. The analysis accounts for the following effects:

- Changes in combustor cross-sectional area
- Variations in gas molecular weight and specific heat
- Exchange of heat with surroundings

**Theory of Aerospace Propulsion.**

- Drag caused by internal bodies or solid particles in the flow
- Losses due to friction on the combustor walls

These equations are developed in such a manner that close correspondence with the combustor problems at hand is maintained without significant loss in generality. The format due to Shapiro (1953) is followed, where the basic equations are treated sequentially and then a final set of equations are developed from them.

## 2.2 EQUATION OF STATE

The basic assumption is that fluid flowing through the system behaves like a perfect gas and therefore follows the perfect gas equation:

$$p = \frac{\rho R_u T}{W}. \tag{2.1}$$

The perfect gas law will be useful here in logarithmic differential form:

$$\frac{dp}{p} = \frac{d\rho}{\rho} + \frac{dT}{T} - \frac{dW}{W}. \tag{2.2}$$

In flows with chemical reactions, as in combustors, afterburners, and rocket nozzles, the molecular weight may change substantially and the effects may need to be taken into account.

## 2.3 SPEED OF SOUND

The speed of sound is the speed at which an infinitesimal pressure disturbance is propagated in a compressible gas. Because the pressure change is slight, the accompanying density change is similarly slight and the process is so rapid little heat can be transferred. Thus, in the limit the sound wave may be considered a reversible and adiabatic process so there is no change in entropy. The entropy may be related to other state variables by the following thermodynamic relation:

$$ds = c_p \frac{dT}{T} - \frac{dp}{\rho T},$$

where $s$ denotes the entropy and $c_p$ the specific heat at constant pressure. Using Equation (2.2), the entropy change becomes

$$\frac{ds}{c_v} = \frac{dp}{p} - \gamma \frac{d\rho}{\rho} + \gamma \frac{dW}{W}, \tag{2.3}$$

where $c_v$ is the specific heat at constant volume and quantity $\gamma = c_p/c_v$ is the ratio of the specific heats of the gas. If the gas is a pure perfect gas or if it is a mixture of pure perfect gases not undergoing chemical reaction, then the molecular weight of the gas is constant. Under these conditions, when $ds = 0$, the process is isentropic and is described by the well-known relation

$$p = C\rho^{\gamma},$$

where quantity $C$ is the constant of integration of Equation (2.3) with $ds = dW = 0$. In such a gas, the sound speed is defined by

$$a^2 = \left(\frac{\partial p}{\partial \rho}\right)_s = \left[\frac{\partial}{\partial \rho}(C\rho^\gamma)\right] = \frac{\gamma p}{\rho} = \frac{\gamma R_u T}{W}. \quad (2.4)$$

The isentropic process exponent $\gamma$ is the ratio of specific heats and is a function of temperature. In flows with nonequilibrium chemical reactions or with condensed phases present, for example, fuel droplets, the isentropic exponent may be different from $\gamma$, the ratio of specific heats. We will not deal with such specialized problems; for a more detailed discussion, see Barrere and colleagues (1960). The logarithmic differential form of the speed of sound is obtained by differentiating Equation (2.4) and is given by

$$\frac{da}{a} = \frac{1}{2}\left(\frac{d\gamma}{\gamma} + \frac{dT}{T} - \frac{dW}{W}\right). \quad (2.5)$$

## 2.4 MACH NUMBER

The Mach number is the ratio of the local fluid speed to the local sound speed:

$$M = \frac{V}{a}.$$

It will be convenient to deal with the square of the Mach number, which may be written in the form

$$\frac{dM^2}{M^2} = \frac{dV^2}{V^2} + \frac{dW}{W} - \frac{d\gamma}{\gamma} - \frac{dT}{T}. \quad (2.6)$$

## 2.5 CONSERVATION OF MASS

The integral form of the mass conservation relation for steady flow is

$$\int_A \rho \vec{V} \cdot \hat{n} dA = 0, \quad (2.7)$$

where $\vec{V}$ is the velocity vector; $A$ is the surface area of the control volume, which is fixed in space; and $\hat{n}$ is the unit outward normal vector to the control surface. Considering the fluid to be a continuum, we may use Gauss' divergence theorem to write the equation in differential form. In cylindrical coordinates, the conservation of mass equation may be written in differential form as follows:

$$\frac{\partial}{\partial r}(\rho r V_r) + \frac{\partial}{\partial \theta}(\rho V_\theta) + \frac{\partial}{\partial z}(\rho r V_z) = 0, \quad (2.8)$$

where the quantities $r$, $\theta$, and $z$ are the radial, azimuthal, and axial coordinates, respectively, and the velocity components in those directions are denoted by the corresponding coordinates, while $\rho$ denotes fluid density. Multiplying by $drd\theta$ and integrating from 0 to $r_w(\theta,z)$ and 0 to $2\pi$, respectively, eliminates the middle term in Equation (2.8), leaving

$$\int_0^{2\pi} (\rho r V_r)_w d\theta + \int_0^{2\pi}\int_0^{r_w} \frac{\partial}{\partial z}(\rho r V_z) dr d\theta = 0. \quad (2.9)$$

Integrating the second term in Equation (2.9) over $r$ using Leibniz' rule yields

$$\int_0^{2\pi}\left[(\rho r V_r)_w + \left\{\frac{\partial}{\partial z}\int_0^{2\pi}(\rho r V_z)dr - (\rho r V_z)_w \frac{dr_w}{dz}\right\}\right]d\theta = 0. \tag{2.10}$$

The boundary condition at the wall for frictionless flow is (see Figure 2.1)

$$\frac{dr_w}{dz} = \frac{V_{r,w}}{V_{z,w}}. \tag{2.11}$$

Then the integrated mass conservation equation reduces to

$$\int_0^{2\pi}\left[\frac{\partial}{\partial z}\int_0^{r_w}(\rho r V_z)dr + \rho_w r_w V_{r,w}\left(\frac{\rho_w r_w V_{r,w}}{\rho_w r_w V_{z,w}} - \frac{dr_w}{dz}\right)\right]d\theta = 0. \tag{2.12}$$

The second term in the square brackets is zero by Equation (2.11), and the remaining term is the rate of change of mass flow through the streamtube so that

$$\frac{\partial}{\partial z}\int_0^{2\pi}\int_0^{r_w}(\rho r V_z)dr d\theta = \frac{d\dot{m}}{dz} = 0. \tag{2.13}$$

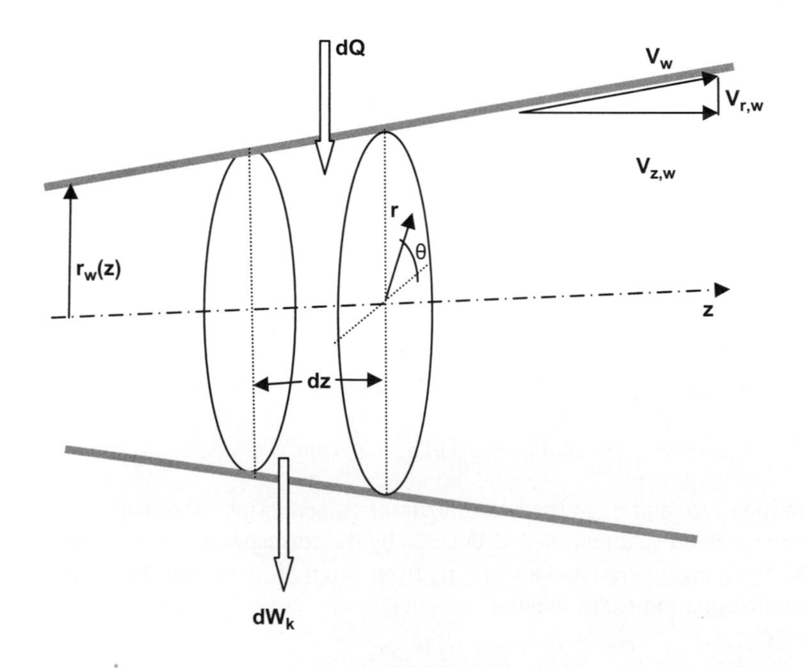


**FIGURE 2.1**

Schematic diagram of a quasi-one-dimensional flow field.

Thus the mass flow is constant, that is,

$$\dot{m} = \int_0^{2\pi} \int_0^{r_w} (\rho r V_z) dr d\theta = const. \tag{2.14}$$

In the quasi-one-dimensional approximation, we assume that the variation of the streamtube cross-sectional area is

$$\frac{\partial r_w(\theta, z)}{\partial z} \ll 1. \tag{2.15}$$

This suggests that it is reasonable to further assume that the flow properties vary little across the streamtube and that we may approximate the mass flow rate as follows:

$$\dot{m}(z) = \rho(z)V(z)A(z) = const. \tag{2.16}$$

Using the perfect gas law and the Mach number definition in Equation (2.16), we may also write the mass flow as

$$\dot{m} = pAM\sqrt{\frac{\gamma}{RT}}.$$

Furthermore, if we introduce the stagnation pressure and temperature relations discussed in Chapter 1, we may put the mass flow in the form

$$\dot{m} = \frac{p_t A}{\sqrt{T_t}}\sqrt{\frac{\gamma}{R}} M \left(1 + \frac{\gamma - 1}{2} M^2\right)^{-\frac{\gamma+1}{2(\gamma-1)}}. \tag{2.17}$$

The quasi-one-dimensional mass conservation [Equation (2.16)] may also be cast in the logarithmic differential form

$$\frac{d\dot{m}}{\dot{m}} = \frac{d\rho}{\rho} + \frac{dV}{V} + \frac{dA}{A} = 0. \tag{2.18}$$

For flows of interest in jet propulsion, the requirement of slowly varying cross-sectional area is closely met. There are some ducts, however, that contain relatively large changes in the cross-sectional area or abrupt turns in the flow path. In such configurations the quasi-one-dimensional approximation would fail locally for some distance downstream of these flow path disturbances. In conventional jet propulsion engines, the flow path is smooth, reasonably straight, and characterized by small changes in flow area, making the application of Equation (2.18) quite appropriate.

There are also situations in which mass is added to the flow in the form of injected liquid or gaseous fuel or coolant, as well as cases where evaporation or condensation occurs. In most propulsion systems, this additional mass is very small in comparison to the main gas flow and has a minor influence on the momentum and energy of the flow. As a consequence, such mass addition is neglected here and Shapiro (1953) may be consulted for further information.

## 2.6 CONSERVATION OF ENERGY

Assuming that the quasi-one-dimensional approximation serves to describe all pertinent variables as being constant across the duct shown in Figure 2.1, we consider the conservation of mass to be expressed as follows:

$$\dot{m}(dQ - dW_k) = \dot{m}\left(dh + d\frac{V^2}{2}\right) = \dot{m}dh_t. \tag{2.19}$$

This equation is applied across the ends of a control volume removed from regions in which mixing occurs and in which $dQ$ represents the incremental amount of heat per unit mass added to the stream from external sources by way of conduction, convection, or radiation. The term $dW_k$ represents the incremental work per unit mass extracted from the stream, and $dh_t$ is the increment in the total enthalpy per unit mass of fluid flowing through the control volume.

### 2.6.1 Thermodynamics of perfect gas mixtures

Consider a perfect gas of species $i$ occupying a volume, $V$, at some temperature, $T$. The perfect gas law states that

$$p_i = \rho_i R_i T. \tag{2.20}$$

Now if $N$ other species, also perfect gases, are introduced into the volume, Dalton's law says that each gas occupies the volume as if the other species were not present, and each species obeys the perfect gas law of Equation (2.20). We assume that there is thermodynamic equilibrium among the gases so that there is just one temperature for the mixture and it is the same for all species present. The mixture pressure is the sum of the partial pressures of each of the species so that

$$p = \sum_{i=1}^{N} p_i = \rho RT.$$

Noting that $R_i = R_u/W_i$ and $R = R_u/W$, we may write the ratio of the partial pressure, $p_i$, of species $i$ to the mixture pressure, $p$, as

$$\frac{p_i}{p} = \frac{\rho_i}{\rho}\frac{W}{W_i} = Y_i\frac{W}{W_i}.$$

Quantity $Y_i$ is called the mass fraction of species $i$. Here $\rho_i/W_i$ is the number of moles $n_i$ of species $i$ per unit volume and $\rho/W$ is the total number of moles in the mixture per unit volume. Thus the ratio of the partial pressure of species $i$ to the pressure of the mixture is equal to the ratio of the number of moles $n_i$ of species $i$ per unit volume to the total number of moles in the mixture per unit volume, or

$$\frac{p_i}{p} = \frac{n_i}{n} = X_i.$$

Here quantity $X_i$ is called the mole fraction of species $i$. Note that

$$X_i = Y_i\frac{W}{W_i}.$$

Summing over all species yields

$$\sum_{i=1}^{N} X_i = 1 = \sum_{i=1}^{N} Y_i \frac{W}{W_i}.$$

This shows that the mixture molecular weight is given by

$$W = \frac{1}{\sum_{i=1}^{N} \frac{Y_i}{W_i}}.$$

### 2.6.2 Fuel–air mixture

The enthalpy per unit mass of the flowing mixture of fuel and oxidizer must be calculated from enthalpy $h_i$ and mass fraction $Y_i$ of the individual species among the $N$ species present in the flowing stream and is given by

$$\begin{aligned} h &= \sum_{i=1}^{N} Y_i h_i \\ Y_i &= \frac{\rho_i}{\rho} \end{aligned} . \tag{2.21}$$

The enthalpy of any individual species may be expressed as

$$h_i = \Delta h^0_{f,i} + \int_{T_r}^{T} c_{p,i} dT.$$

The standard heat of formation $\Delta h^0_{f,i}$ is defined as the energy per unit mass necessary to form the species from the constituent atoms in its standard state at a pressure of 1 atmosphere and a reference temperature $T_r = 298.16\text{K}$. The differential of the enthalpy is

$$dh = \sum_{i=1}^{N} Y_i dh_i + \sum_{i=1}^{N} h_i dY_i. \tag{2.22}$$

The species enthalpy definition yields

$$\begin{aligned} dh_i &= c_{p,i} dT \\ c_p &= \sum_{i=1}^{N} Y_i c_{p,i} \end{aligned} . \tag{2.23}$$

Therefore, the equation for the differential of the mixture enthalpy becomes

$$dh = c_p dT + \sum_{i=1}^{N} h_i dY_i. \tag{2.24}$$

Incorporating these results into the energy relation yields

$$\frac{dQ - dW_k - dH}{c_p T} = \frac{dT}{T} + \frac{dV^2}{2c_p T}. \tag{2.25}$$

Here the heat addition per unit mass due to chemical reaction is

$$dH = \sum_{i=1}^{N} h_i dY_i. \tag{2.26}$$

Then, using the following

$$\begin{aligned} c_p - c_v &= \frac{R_u}{W}, \\ \gamma &= \frac{c_p}{c_v} \end{aligned} \tag{2.27}$$

the energy equation, Equation (2.25), becomes

$$\frac{dQ - dW_k - dH}{c_p T} = \frac{dT}{T} + \left(\frac{\gamma - 1}{2}\right) M^2 \frac{dV^2}{V^2}. \tag{2.28}$$

We can write the heat addition term $dH$ in Equation (2.26) in terms of molar thermodynamic properties as follows:

$$dH = \frac{1}{nW} \sum_{i=1}^{N} H_i dn_i, \tag{2.29}$$

where $H_i$ is the enthalpy per mole of species $i$, $n_i$ is the number of moles of species $i$ per unit volume, $n$ is the number of moles of mixture per unit volume, and $W$ is the molecular weight of the mixture. If we consider a chemical reaction given by

$$\sum_{i=1}^{N} n_i' A_i \rightarrow \sum_{i=1}^{N} n_i'' A_i, \tag{2.30}$$

quantity $A_i$ denotes a general chemical species, whereas $n_i'$ and $n_i''$ denote the number of moles of species $i$ in the reactants and products, per unit volume, respectively. Penner (1957) shows that the heat of reaction at a given temperature and pressure is

$$\Delta H(T) = \sum_{i=1}^{N} n_i'' \left[ \Delta H_{f,i}^0 + \int_{T_r}^{T} C_{p,i} dT \right] - \sum_{i=1}^{N} n_i' \left[ \Delta H_{f,i}^0 + \int_{T_r}^{T} C_{p,i} dT \right]. \tag{2.31}$$

The equation for $\Delta H$ given in Equation (2.31) may also be written on a per unit mass basis in terms of a finite chemical composition change as follows:

$$\Delta H(T) = \frac{1}{nW} \sum_{i=1}^{N} (n_i'' - n_i') H_i. \tag{2.32}$$

We see that this is the finite difference equivalent of the differential heat release due to the reaction given previously in Equation (2.29). Note that this equivalence is only approximately correct and is applicable because the product of the total number of moles and the mixture molecular weight of the mixture of chemical products does not change appreciably from that for the reactant mixture during a reaction.

Thus we may set $\Delta H$ in Equation (2.32) equal to $dH$ in Equation (2.26) as the heat of reaction per unit mass for the fuel and oxidizer reaction under consideration. Because we are not considering chemical kinetics in detail, it would be helpful to use this heat of reaction formula to determine the heat released by the complete combustion of a given fuel, which is generally assumed to be complete by the time the exit of the combustor is reached. This is called the heating value of the fuel, which was introduced in Section 1.3.4. In particular, if the heat of vaporization of water is not included, it is called the lower heating value. As mentioned previously, temperatures of interest in propulsion applications are high enough so that when the term heating value is used it will be understood to mean the lower heating value. The heat release may be considered to be complete at a given axial station in the flow or, if desired, it may be considered to have some prescribed axial variation, which would be modeled empirically.

## 2.7 EXAMPLE: HEATING VALUES FOR DIFFERENT FUEL–OXIDIZER COMBINATIONS

We may apply Equation (2.32) to various reactions to determine the heating value the fuel provides. The simplest case is that of the hydrogen–oxygen reaction typical of rockets for which the reaction is

$$H_2 + \frac{1}{2}O_2 \rightarrow H_2O.$$

It is important to note that Equation (2.31), the expanded form of Equation (2.32), yields a heat of reaction that depends on the temperature specified. However, as pointed out by Penner (1957), Equation (2.31) is essentially independent of the pressure. It is convenient to use the reference temperature $T_r = 298.16\text{K}$ as the temperature at which to evaluate Equation (2.31). This makes the integrals in Equation (2.31) equal to zero and requires only the standard heats of formation to calculate the heat release. Obviously, any excess oxygen would be present on both sides of the chemical reaction, and because its heat of formation is zero, it makes no contribution to the heat of reaction of the fuel. Furthermore, because the standard heats of formation of the oxygen and hydrogen reactants are both zero, using the thermodynamic tables in Appendix G or, for example, in Turns (1996), we find that

$$\Delta H(T_r) = (1mol)\Delta H^0_{f,H_2O} = (1mol)(-241.645MJ/mol).$$

Then if we wish to base this heat release on the mass of the fuel, in this case hydrogen, we may define the heating value ($HV$) of the fuel as follows:

$$HV = \frac{-\Delta H}{n_{H_2}W_{H_2}} = \frac{241.845MJ}{(1mol)(2kg/mol)} = 120.9MJ/kg_{H_2}.$$

Note that the heating value of the fuel, or as it is also known, the heat of combustion, is equal to, but opposite in sign, of the heat of reaction in Equation (2.31). The heat release term in the energy equation [Equation (2.28)], is $-dH$ per unit mass of mixture and may be written in terms of the heating value as follows:

$$-dH = \frac{\dot{m}_f HV}{\dot{m}}.$$

We may also consider paraffin hydrocarbons typical of jet fuel with the chemical formula $C_nN_{2n+2}$. As discussed in Chapter 4, their reaction with air has the following chemical equation:

$$C_nH_{2n+2} + \frac{3n+1}{2}[O_2 + 3.76N_2] \rightarrow (n+1)H_2O + nCO_2 + \frac{3n+1}{2}(3.76N_2).$$

We may consider the case of octane with $n = 8$ and find that

$$\Delta H = 9\Delta H^0_{f,H_2O} + 8\Delta H^0_{f,CO_2} - \Delta H^0_{f,C_8H_{18}}.$$

Once again, using the tables in Appendix G, and for octane, those in Turns (1996), we find

$$\Delta H = 9(-241.845) + 8(-393.546) - (-208.447) = -5,116MJ/mol.$$

Then the heating value of octane is

$$HV = \frac{-\Delta H}{(1mol)(114.23)} = 44.79MJ/kg_{C_8H_{18}}.$$

## 2.8 CONSERVATION OF SPECIES

In the case of reacting flows, one must know the axial evolution of the various species in order to calculate the energy release, which was given previously in Equation (2.26) as

$$dH = \sum_{i=1}^{N} h_i dY_i. \tag{2.33}$$

Conservation of species requires that the net efflux of species $i$ through the bounding surface of the control volume be equal to the net rate of production of species $i$ within the control volume. This conservation law may be expressed in the following way:

$$-\rho_i VA + (\rho_i + d\rho_i)(V + dV)(A + dA) = \dot{m}_i A dz, \tag{2.34}$$

where subscript $i$ denotes the $i$th species and $\dot{m}_i$ is the mass rate of production of the $i$th species. Expanding Equation (2.34) and ignoring second-order infinitesimals yields

$$dY_i = \dot{m}_i A \frac{dz}{m} = \frac{\dot{m}_i}{\rho}\frac{dz}{V}. \tag{2.35}$$

In order to obtain a more complete appreciation of the species conservation equation as written earlier, introduce a scale length $L$ and a timescale $\tau_c$ such that

$$\begin{aligned} \bar{z} &= \frac{z}{L} = O(1) \\ \overline{\left(\frac{\dot{m}_i}{\rho}\right)} &= \tau_c\left(\frac{\dot{m}_i}{\rho}\right) = O(1). \end{aligned} \tag{2.36}$$

Then our species conservation equation, Equation (2.35), may be rewritten as

$$dY_i = \frac{1}{\tau_c}\frac{L}{V}\left(\frac{\dot{m}_i}{\rho}\right)d\bar{z} = \frac{\tau_f}{\tau_c}\left(\frac{\dot{m}_i}{\rho}\right)d\bar{z}. \tag{2.37}$$

The Damkohler parameter $\tau_f/\tau_c$ is the ratio of the characteristic residence time of a fluid particle in the combustor to the characteristic chemical reaction time and thus determines the nature of the chemical reaction:

- $\tau_f/\tau_c \ll 1$: Here $dY_i \sim 0$ because the residence time is very brief compared to the reaction time and no appreciable reaction occurs; thus $dH = 0$. This case is said to be *chemically frozen*. The flow calculation would be carried out as for a single component gas with thermodynamic characteristics of the fixed composition mixture. This is a case where the molecular weight remains unchanged as described in Section 2.3.
- $\tau_f/\tau_c \gg 1$: Here $dY_i$ is undefined because the reaction time is negligible compared to the residence time, and the reaction proceeds instantaneously to the equilibrium composition pertinent to the local temperature and pressure; thus $dH$ is determined using the chemical equilibrium conditions. This case is called *chemical equilibrium* flow. The flow calculation would be carried out for the evolving mixture with the composition and thermodynamic characteristics appropriate to chemical equilibrium conditions. That is, the heat of reaction is

$$\Delta H(T) = \frac{1}{nW}\sum_{i=1}^{N}(n_i'' - n_i')H_i. \tag{2.38}$$

  Now, of course, the concentrations are those of the chemical equilibrium composition of the mixture. In the case of a flow in chemical equilibrium it can be shown that the entropy is again constant; see Turns (1996).
- $\tau_f/\tau_c \sim 1$: Here the chemical and flow evolutions are comparable, and the chemistry of the flow is said to be *rate controlled*. The mass rate of production term $\dot{m}_i$ must be specified in terms of the other variables of the problem for each of the $N$ species present and must be used in the full species conservation equation, Equation (2.35). The chemical heat release is then calculated from Equation (2.29) and used in the full flow equations. Finite-rate chemical calculations are quite involved and beyond the scope of this book. For an introduction to chemical kinetics, see Turns (1996).

## 2.9 CONSERVATION OF MOMENTUM

The integral form of the momentum equation for the control volume is

$$\int_A (\rho\vec{V}\cdot\vec{n})\vec{V}dA = \int_v \rho\vec{F}_b + \int_A \frac{d\vec{F}_s}{dA}dA. \tag{2.39}$$

For the current quasi-one-dimensional application, the momentum conservation equation reduces to

$$\dot{m}dV = -Adp - \tau_w dA_w - dF_z. \tag{2.40}$$

The shear stress may be expressed in a nondimensional form as the skin friction coefficient as defined by the following equation:

$$c_f = \frac{\tau_w}{\frac{1}{2}\rho V^2}. \tag{2.41}$$

The dynamic pressure is defined as

$$q = \frac{1}{2}\rho V^2 = \frac{1}{2}\gamma p M^2.$$

The wetted area may be related to the cross-sectional area by means of hydraulic diameter $D$, where $D/4$ is the ratio of cross-sectional area to the wetted perimeter bounding that area, or

$$\frac{dA_w}{A} = 4\frac{dz}{D}. \tag{2.42}$$

Substituting these expressions into one-dimensional momentum equations yields

$$\frac{dp}{p} + \frac{1}{2}\gamma M^2 \frac{dV^2}{V^2} + \frac{1}{2}\gamma M^2 \left[ 4c_f \frac{dz}{D} + \frac{dF_z}{\frac{1}{2}\gamma p A M^2} \right] = 0. \tag{2.43}$$

## 2.10 IMPULSE FUNCTION

The impulse function is defined as

$$I = A(p + \rho V^2) = pA(1 + \gamma M^2). \tag{2.44}$$

The change in this function is the force on the fluid and is often useful in evaluating forces on components in flow systems. The differential form may be written as follows:

$$\frac{dI}{I} = \frac{dp}{p} + \frac{dA}{A} + \frac{d(\gamma M^2)}{1 + \gamma M^2}. \tag{2.45}$$

This may be shown to reduce to the following:

$$\frac{dI}{I} = \frac{1}{1 + \gamma M^2} \frac{dA}{A}. \tag{2.46}$$

## 2.11 STAGNATION PRESSURE

The stagnation pressure for a compressible flow is defined as

$$p_t = p\left(1 + \frac{\gamma - 1}{2}M^2\right)^{\frac{\gamma}{\gamma - 1}}. \tag{2.47}$$

Then in logarithmic differential form this becomes

$$\frac{dp_t}{p_t} = \frac{dp}{p} + \frac{\gamma M^2}{2\left(1 + \frac{\gamma - 1}{2} M^2\right)} \frac{dM^2}{M^2}. \tag{2.48}$$

## 2.12 EQUATIONS OF MOTION IN STANDARD FORM

We may choose the following as independent variables:

$$\frac{dA}{A}, \frac{dQ - dW_k - dH}{c_p T}, \frac{4c_f dz}{D} + \frac{dF_z}{\frac{1}{2}\gamma p A M^2}, \frac{dW}{W}, \frac{d\gamma}{\gamma}. \tag{2.49}$$

Dependent variables are selected as follows:

$$\frac{dM^2}{M^2}, \frac{dV}{V}, \frac{da}{a}, \frac{dp}{p}, \frac{dT}{T}, \frac{d\rho}{\rho}. \tag{2.50}$$

Manipulation of the equations reduces them to a standard form, which is best represented by a table of influence coefficients as shown in Table 2.1.

The equation for any dependent variable in the first column is formed by the sum of the products of the influence coefficients and the corresponding independent variable in the top row of the table. For example, for the pressure we obtain

$$\begin{aligned} \frac{dp}{p} = {} & \left[\frac{\gamma M^2}{1 - M^2}\right] \frac{dA}{A} + \left[-\frac{\gamma M^2}{1 - M^2}\right] \frac{dQ - dW_k - dH}{c_p T} \\ & + \left[-\gamma M^2 \frac{1 + (\gamma - 1)M^2}{2(1 - M^2)}\right] \left(4c_f \frac{dz}{D} + \frac{dF_z}{\frac{1}{2}\gamma p M^2 A}\right) + \left[\frac{\gamma M^2}{1 - M^2}\right] \frac{dW}{W}. \end{aligned} \tag{2.51}$$

In the simplified case of constant composition and constant isentropic exponent, the matrix of influence coefficients in Table 2.1 is somewhat simplified.

## 2.13 EXAMPLE: FLOW IN A DUCT WITH FRICTION

Air flow through an insulated circular tube with a length to diameter ratio $L/D = 10$ is shown in Figure 2.2. As shown in Figure 2.2, the stagnation pressure and temperature in the convergent entrance of the duct are 0.7 MPa and 60°C, respectively. Frictional effects in the convergent section may be neglected. The maximum mass flow rate of air that can pass through the 5-cm-diameter circular pipe occurs when the exit is choked, that is, when $M_2 = 1$. Determine the Mach number at tube entrance $M_1$ for the case of choked flow using the given stagnation conditions and assuming a constant friction coefficient $c_f = 0.005$. What is the effect of friction on the stagnation pressure in the tube? What is the final stagnation pressure?

**Table 2.1** Table of Influence Coefficients for Quasi-One-Dimensional Flow, Including Variable Molecular Weight and Ratio of Specific Heats

| | $\frac{dA}{A}$ | $\frac{dQ - dW_k - dH}{c_pT}$ | $\left[\frac{4c_f dz}{D} + \frac{dF_z}{\frac{1}{2}\gamma pM^2A}\right]$ | $\frac{dW}{W}$ | $\frac{d\gamma}{\gamma}$ |
|---|---|---|---|---|---|
| $\frac{dM^2}{M^2}$ | $-2\left[\frac{1+\frac{1}{2}(\gamma-1)M^2}{1-\gamma M^2}\right]$ | $\frac{(1+\gamma M^2)}{1-M^2}$ | $\frac{\gamma M^2\left[1+\frac{1}{2}(\gamma-1)M^2\right]}{1-M^2}$ | $\frac{-(1+\gamma M^2)}{1-\gamma M^2}$ | $-1$ |
| $\frac{dV}{V}$ | $\frac{-1}{1-M^2}$ | $\frac{1}{1-M^2}$ | $\frac{\gamma M^2}{2(1-M^2)}$ | $\frac{-1}{1-M^2}$ | 0 |
| $\frac{da}{a}$ | $\frac{(\gamma-1)M^2}{2(1-M^2)}$ | $\frac{(1-\gamma M^2)}{2(1-M^2)}$ | $\frac{-(\gamma-1)\gamma M^4}{4(1-M^2)}$ | $\frac{-(1-\gamma M^2)}{2(1-M^2)}$ | $\frac{1}{2}$ |
| $\frac{dT}{T}$ | $\frac{(\gamma-1)M^2}{1-M^2}$ | $\frac{(1-\gamma M^2)}{1-M^2}$ | $\frac{-(\gamma-1)\gamma M^4}{2(1-M^2)}$ | $\frac{(\gamma-1)M^2}{1-M^2}$ | 0 |
| $\frac{d\rho}{\rho}$ | $\frac{M^2}{1-M^2}$ | $\frac{-1}{1-M^2}$ | $\frac{-\gamma M^2}{2(1-M^2)}$ | $\frac{1}{1-M^2}$ | 0 |
| $\frac{dp}{p}$ | $\frac{\gamma M^2}{1-M^2}$ | $\frac{-\gamma M^2}{1-M^2}$ | $\frac{-\gamma M^2\left[1+(\gamma-1)M^2\right]}{2(1-M^2)}$ | $\frac{\gamma M^2}{1-M^2}$ | 0 |
| $\frac{dI}{I}$ | $\frac{1}{1+\gamma M^2}$ | 0 | 0 | 0 | 0 |

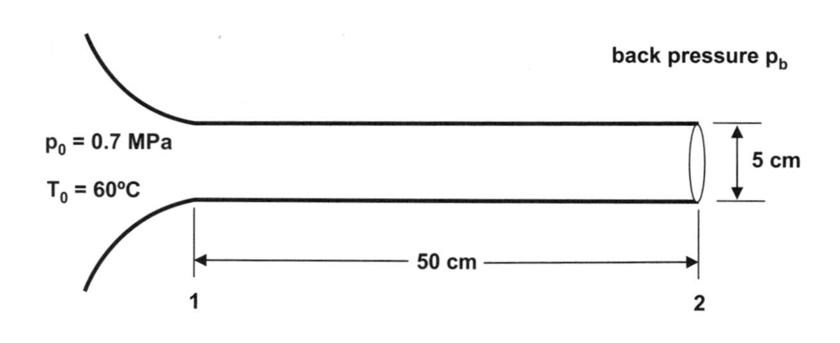


**FIGURE 2.2**

Schematic diagram of a circular tube with frictional flow.

Because the tube is insulated, the flow is adiabatic. The area is constant, no work is done, and no drag forces are acting within the tube; therefore, only friction has an effect on the flow. Selecting the Mach number development as given by the influence coefficient Table 2.1 we may write

$$\frac{dM^2}{M^2} = \frac{\gamma M^2\left[1+\frac{\gamma-1}{2}M^2\right]}{1-M^2}4c_f\frac{dz}{D}.$$

This equation may be integrated between stations 1 and 2 to yield

$$\frac{\gamma+1}{2}\ln\left(\frac{1+\frac{\gamma-1}{2}M_2^2}{M_2^2}\right)-\frac{1}{M_2^2}-\frac{\gamma+1}{2}\ln\left(\frac{1+\frac{\gamma-1}{2}M_1^2}{M_1^2}\right)+\frac{1}{M_1^2}=4\gamma c_f\frac{L}{D}.$$

For maximum mass flow, the exit Mach number $M_2 = 1$. For the relatively low stagnation temperature of the air it is reasonable to use $\gamma = 1.4$ in this equation, which may be solved for the entrance Mach number, which is found to be $M_1 = 0.7$. Then the mass flow may be determined using Equation (2.18):

$$\begin{aligned}
\dot{m} &= \frac{p_t A}{\sqrt{T_t}}\sqrt{\frac{\gamma}{R}}M\left(1+\frac{\gamma-1}{2}M^2\right)^{-\frac{\gamma+1}{2(\gamma-1)}} \\
\dot{m} &= \frac{(0.7\times 10^6 N/m^2)[\pi(.05/2)^2 m^2]}{\sqrt{333K}}\sqrt{\frac{1.4}{287m^2/s^2-K}}M(1+0.2M^2)^{-3}. \\
\dot{m} &= 5.261M(1+.2M^2)^{-3}
\end{aligned} \tag{2.52}$$

Because the stagnation pressure may be assumed constant in the convergent section leading up to station 1, we can apply the conditions there where $M_1 = 0.7$ and determine that $\dot{m} = 2.78$ kg/s. The stagnation temperature is given in terms of $p$ and $M$ in Equation (2.48) and, using the influence coefficients in Table 2.1, yields

$$\begin{aligned}
\frac{dp_t}{p_t} &= \frac{\gamma M^2[1+(\gamma-1)M^2]}{2(1-M^2)}\frac{4c_f}{D}dz+\frac{\gamma^2 M^4}{2(1-M^2)}\frac{4c_f}{D}dz \\
\frac{dp_t}{p_t} &= \left(-\frac{\gamma M^2}{2}\right)\frac{4c_f}{D}dz
\end{aligned} \tag{2.53}$$

The stagnation pressure clearly is not constant in this flow and decreases along the tube as a result of friction, which is an irreversible process. Because the equation for $p_t$ is not a perfect integral, it cannot be readily integrated analytically, although a numerical integration involving the coupled $M$ and $p_t$ equations can be carried out. However, we can determine the final stagnation pressure, $p_{t,2}$, by noting that the mass flow is constant through the tube and that the stagnation temperature is constant throughout because the flow is adiabatic. Then Equation (2.52) may be written as

$$p_{t,1}M_1\left(1+0.2M_1^2\right)^{-3} = p_{t,2}M_2\left(1+0.2M_2^2\right)^{-3}.$$

Using the information developed thus far, the final stagnation pressure is found to be $p_{t,2} = 0.64$ MPa. The change in stagnation pressure from station 2 to station 1 is $-8.62\%$. Without the mass flow information, we might estimate the stagnation pressure drop by simply assuming a constant average Mach number in the duct $M_{avg} = 0.85$ and use that in Equation (2.53). The result is a change in stagnation pressure of $-7.88\%$, which is fairly accurate.

## 2.14 NOMENCLATURE

| | |
|---|---|
| $A$ | cross-sectional area |
| $a$ | sound speed |
| $C$ | constant |
| $c_f$ | skin friction coefficient |
| $c_j$ | specific fuel consumption |
| $c_p$ | specific heat at constant pressure |
| $c_v$ | specific heat at constant volume |
| $dH$ | heat release per unit mass due to chemical reaction |
| $F$ | thrust |
| $F_n$ | net thrust |
| $g$ | acceleration of gravity |
| $H$ | enthalpy per mole |
| $h$ | enthalpy per mass |
| $I$ | impulse function |
| $M$ | Mach number |
| $\dot{m}$ | mass flow |
| $n$ | molar concentration |
| $p$ | pressure |
| $Q$ | heat added to the fluid |
| $q$ | dynamic pressure |
| $R$ | gas constant |
| $R_u$ | universal gas constant |
| $r$ | radial coordinate |
| $s$ | entropy |
| $T$ | temperature |
| $V$ | velocity |
| $W$ | molecular weight |
| $W_k$ | work done on the fluid |
| $w$ | weight flow $mg$ |
| $X_i$ | mole fraction of species $i$, $n_i/n$ |
| $Y_i$ | mass fraction of species $i$, $\rho_i/\rho$ |
| $z$ | axial coordinate |
| $\gamma$ | ratio of specific heats |
| $\Delta h_f^0$ | heat of formation per unit mass |
| $\rho$ | density of a gas mixture |
| $\theta$ | azimuthal coordinate |

| | |
|---|---|
| $\eta$ | efficiency |
| $\tau_c$ | characteristic time to complete chemical reaction |
| $\tau_f$ | characteristic time for flow through duct |
| $\tau_w$ | shear stress on the wall |

### 2.14.1 Subscripts

| | |
|---|---|
| $b$ | burner |
| $e$ | conditions downstream where $p = p_0$ |
| $f$ | fuel |
| $i$ | species $i$ |
| $r$ | radial direction |
| $t$ | stagnation conditions |
| $u$ | universal |
| $w$ | wall condition |
| $z$ | axial direction |
| 0 | free stream conditions |
| $q$ | azimuthal direction |

### 2.14.2 Superscripts

| | |
|---|---|
| $\rightarrow$ | vector quantity |
| $\hat{}$ | unit vector |
| $()'$ | reactant species |
| $()''$ | product species |

## 2.15 EXERCISES

**2.1** Show that the change in the impulse function $I = A\ (p + \rho V^2)$ between two stations in an idealized flow machine yields the force acting on the fluid in that interval (it is suggested to put the impulse function in terms of Mach number). Use this result to determine the net thrust acting between stations 1 and 7.

**2.2** Consider frictionless flow in the constant area duct shown in Figure E2.1, in which work may be added between stations 2 and 3. (a) Using Table 2.1, develop equations for pressure, temperature, density, velocity, and work (per unit mass) done on the fluid as a function of Mach number in the duct and (b) for air entering station 1 at $p_1 = p_2 = 101$ kPa, $T_1 = T_2 = 288$K, and $M_1 = M_2 = 0.4$, find velocity, stagnation pressure, and stagnation temperature at station 2. (c) If 169.3 kJ/kg of work is done on the flow, find Mach number, pressure, temperature, density, velocity, stagnation pressure, and stagnation temperature at station 3.

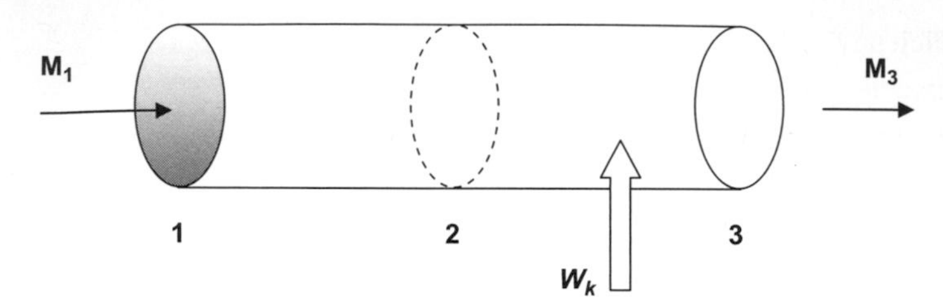


FIGURE E2.1

Flow in a constant area duct with work input.

**2.3** Assume that a simple converging duct is attached to station 3 of Exercise 2.2 and that the exit of that duct is called station 7, with the area ratio $A_{3a}/A_3 = 0.85$ (see Figure E2.2). Using the results of Exercise 2.2 for the conditions of station 3, find Mach number, pressure, temperature, density, velocity, stagnation pressure, and stagnation temperature at station 7. The flow in this duct may be considered frictionless, but because no transfer of heat or work occurs, the flow is isentropic in the converging duct. Determine the force acting on the entire structure if the radius $r_1 = 25$ cm.

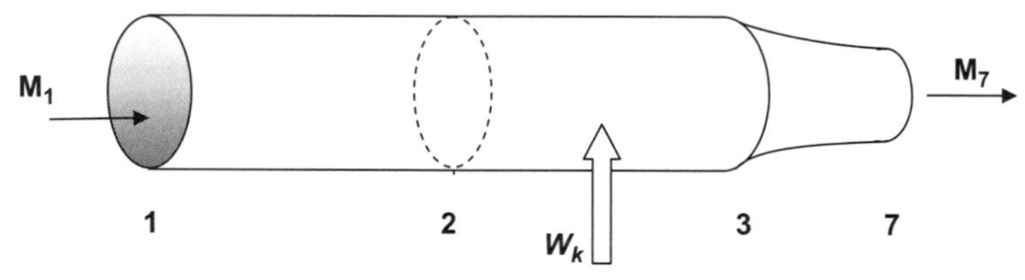


FIGURE E2.2

Constant area duct coupled to converging duct.

**2.4** A constant area duct has heat added to the flow between stations 3 and 4, while work is removed from the flow between downstream stations 4 and 5. Develop an equation for the pressure ratio $p_5/p_3$ in terms of Mach numbers $M_3$ and $M_4$ and $k$. It may be assumed that $k = \gamma =$ constant and that the molecular weight of the gas remains constant.

**2.5** An airplane flying at $M_0 = 0.8$ under standard sea level conditions has instrumentation embedded at the stagnation point on its nose. Assuming that the flow brought to rest at this point has undergone an isentropic process, calculate the temperature and pressure there. Determine the difference between these values and the corresponding static values in the undisturbed atmosphere ahead of the airplane.

**2.6** A diffuser in a jet airplane intake system is instrumented to give readings of stagnation temperature and stagnation pressure. If these readings are 333K and 138 kPa, respectively, and the static pressure is 101.3 kPa, calculate Mach number, static temperature, and air velocity at the instrumented station.

## References

Barrere, M., Jaumotte, A., De Veubeke, B. F., & Vandenkerckhove, J. (1960). *Rocket Propulsion*. New York: Elsevier.

Penner, S. S. (1957). *Chemistry Problems in Jet Propulsion.* New York: Pergamon.

Shapiro, A. H. (1953). *The Mechanics and Thermodynamics of Compressible Fluid Flow, Vol. 1*. New York: Ronald Press.

Turns, S. R. (1996). *An Introduction to Combustion.* New York: McGraw-Hill.

CHAPTER

# Idealized Cycle Analysis of Jet Propulsion Engines

3

## CHAPTER OUTLINE



Theory of Aerospace Propulsion.

## 3.1 INTRODUCTION

In this chapter, several different air-breathing jet engine types are analyzed in some detail. However, they are considered to operate ideally to an extent deemed reasonable in order to provide an appropriate quantitative illustration of their performance potential. Because modern gas turbine engines represent a rather mature technology, efficiencies of the various engine system components are quite high in practice and therefore an ideal analysis can provide a relatively simple introduction to system operation as a whole. Furthermore, the detailed analyses presented here can be generalized readily to incorporate the more accurate performance capabilities of the various engine components dealt with in later chapters to treat practical propulsion problems. Nine different situations are examined and include the following: the turbojet in take-off, high subsonic cruise, and afterburning supersonic cruise, the ramjet in high supersonic cruise, the turbofan engine in take-off, high subsonic cruise, and afterburning supersonic cruise. Although some may feel that there is excessive repetition of the details of the analysis, it is equally possible to consider them to represent a number of sample problems that can be worked out by the student to develop a firm understanding of the principles involved. It should be understood that a substantial effort on the part of the student, both within the boundaries of the text and in outside reading and discussion, is required to arrive at a true appreciation of the subject.

## 3.2 GENERAL JET ENGINE CYCLE

Chapter 1 described the operation of idealized flow machines based on the steady-state quasi-one-dimensional conservation equations of fluid mechanics. In particular, it showed that jet propulsion engines employ the addition of heat to the fluid being processed, with or without net power addition, to raise its energy so that it may be accelerated through a nozzle to produce thrust. We may now apply thermodynamic principles to describe more fully how such engines perform. Consider the schematic diagram of an air-breathing jet propulsion engine as shown in Figure 3.1.

Air is brought on board at the lip of the inlet, denoted as station 1, and is compressed in some fashion within the engine. Fuel is then added at a mass flow rate $\dot{m}_f$ and burned at constant pressure to heat the resulting gas. This hot gas is then exhausted through a nozzle, exiting at high speed at station 7. Chapter 1 showed that the gross thrust is given by

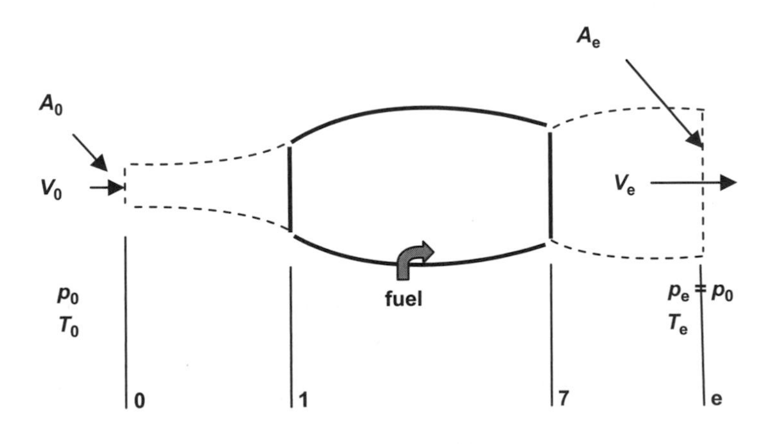


**FIGURE 3.1**

Schematic diagram of an air-breathing jet engine.

$$F_g = \dot{m}_7 V_7 + A_7(p_7 - p_0). \tag{3.1}$$

This may also be written in terms of exit Mach number as

$$\frac{F_g}{A_7} = p_7(1 + \gamma_7 M_7^2) - p_0. \tag{3.2}$$

This form normalizes the gross thrust by the nozzle exit area so that the result is independent of engine size and completely dependent on nozzle performance. Net thrust is the difference between gross thrust and ram drag $F_r$, which is given by

$$F_r = \dot{m}_0 V_0 = \gamma_0 p_0 A_0 M_0^2. \tag{3.3}$$

The net thrust normalized to the nozzle exit area is then

$$\frac{F_n}{A_7} = p_7(1 + \gamma_7 M_7^2) - p_0\left(1 + \gamma_0 M_0^2 \frac{A_0}{A_7}\right). \tag{3.4}$$

This form shows that the net thrust depends on both the nozzle and the inlet. The first term is set purely by the nozzle, while the second term involves the inlet. Inlet area $A_0$ determines the size of the streamtube, and therefore the mass flow, entering the engine.

The thermodynamic cycle of the ideal jet engine is basically that of the Brayton cycle and involves isentropic compression, heat addition at constant pressure, and isentropic expansion, as shown schematically in the *h,s* diagram in Figure 3.2.

A notional turbojet engine showing the major components and a typical, although not exclusive, station numbering scheme is shown in Figure 3.3. The inlet, which occupies segment 1-2, conditions, compresses, and directs air to the compressor, segment 2-3, where additional compression is carried out. Note that at high flight Mach numbers the amount of ram compression delivered by the inlet may

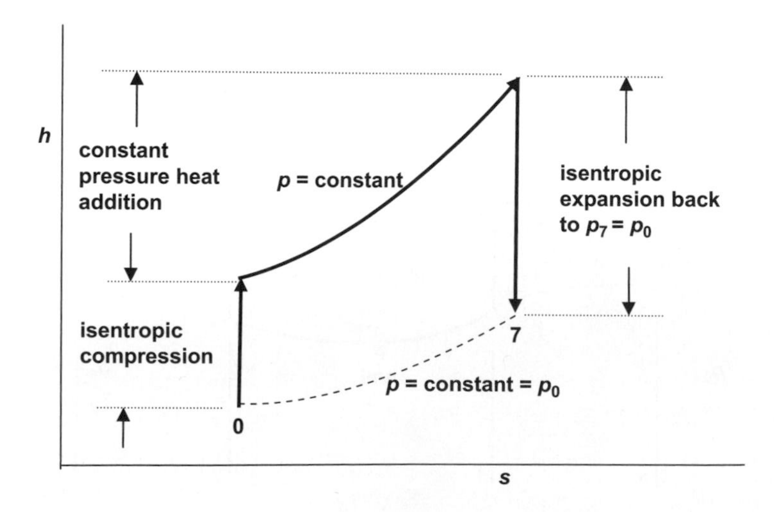


**FIGURE 3.2**

Schematic plot of the Brayton-like cycle of an ideal jet engine.

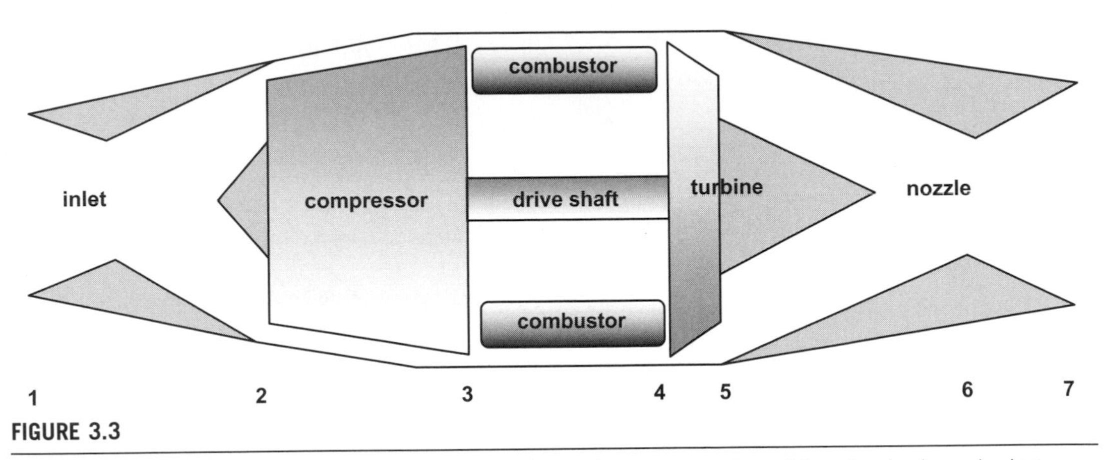


**FIGURE 3.3**

Schematic diagram of a typical turbojet engine showing major components and the standard numbering scheme for engine stations.

be sufficient for operation of the cycle and the compressor–turbine combination may be dispensed with altogether. The high-pressure air then is fed to the combustor, segment 3-4, where fuel is introduced, mixed, and burned to raise the temperature of the airstream. The hot air, with a relatively small amount of combustion products, is then passed through the turbine, segment 4-5, which extracts power to drive the compressor, which is attached to the same shaft. The turbine exhaust is still quite hot when it enters the nozzle, segment 5-6-7, where it accelerates to high speed and exits at station 7 at a pressure near or equal to the ambient pressure $p_0$. Using this generic turbojet engine configuration, a more descriptive $h,s$ diagram can be constructed and would appear, for a turbojet in subsonic flight, $M_0 < 1$, as shown in Figure 3.4.

It should be noted here that in subsonic flight, $M_0 < 1$, the pressure in the captured streamtube at the inlet lip is generally lower than the free stream pressure $p_0$ because the streamtube area usually decreases as it enters the inlet at station 1. In supersonic flight, the pressure at station 1 is always greater than the free stream pressure because of a shock wave in and around the inlet lip, that is, $p_1 > p_0$ when $M_0 > 1$. These details are described fully in Chapter 6 which is devoted to inlets. Compression processes in the inlet, compressor, and turbine are shown to involve entropy increases, signifying losses in those components, whereas the expansion process in the nozzle is shown as being closer to isentropic because real nozzles are operable in a very close to isentropic manner. A further detail is that expansion through the nozzle is shown to end at station 7 with $p_7 = p_0$, that is, the nozzle is perfectly matched.

## 3.3 IDEAL JET ENGINE CYCLE ANALYSIS

In order to appreciate the contribution of the engine components to overall performance, let us consider an idealized jet engine cycle as shown in Figure 3.5 in which all the compression and expansion processes are considered to be isentropic. Furthermore, for simplicity, and without loss of generality, we will consider the thermodynamic properties $c_p$ and $\gamma$ to be constant. Recall that the fuel

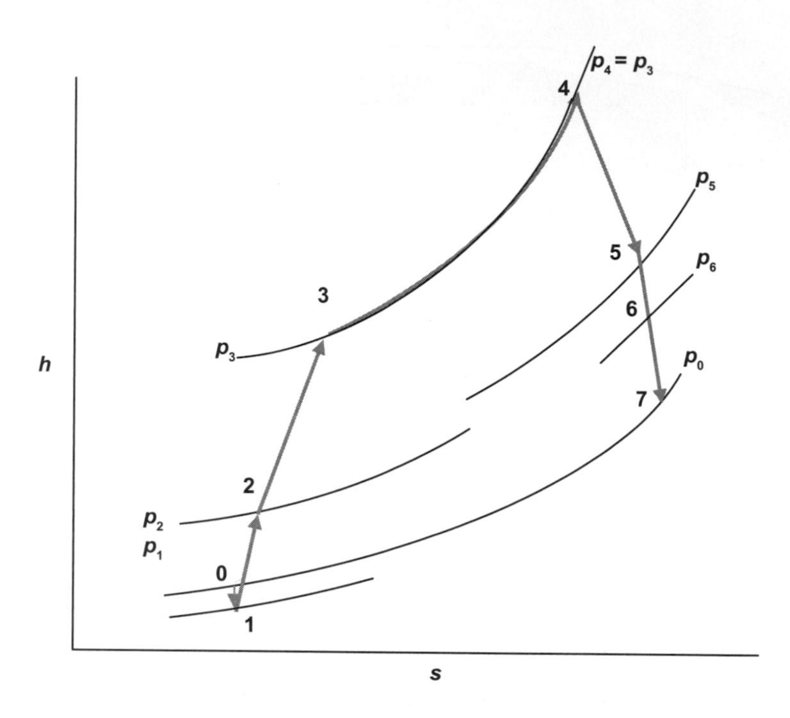


FIGURE 3.4

Generic turbojet cycle diagram for subsonic flight, $M_0 < 1$.

weight flow added for combustion is small enough so as to often be considered negligible with respect to the airflow through the engine, that is, $\dot{m}_f/\dot{m}_0 \ll 1$. The gas processed by the engine is thus considered to be pure air with $c_p = 1.00$ kJ/kg-K, $\gamma = 1.40$, and molecular weight $W = 28.86$ kg/mol. We will consider several important flight conditions for jet engines in order to demonstrate how performance depends on the processes in the cycle.

## 3.4 IDEAL TURBOJET IN MAXIMUM POWER TAKE-OFF

Civil and military aircraft designs are strongly influenced by take-off considerations because this represents the maximum weight condition. The lifting characteristics of the aircraft and take-off distance requirements generally fix its take-off thrust-to-weight ratio. This result influences the choice of engine for the aircraft, and therefore take-off thrust capability is a crucial metric for jet engines. The pacing factor for a jet engine operation is the turbine inlet stagnation temperature $T_{t4}$ because of thermal limitations on turbine materials. Current engines operate in the range 1200K $< T_{t4} <$ 1700K, although materials and cooling techniques permitting higher temperatures are being sought constantly. The basic take-off state is standard day sea-level conditions, that is, altitude $z = 0$, pressure $p_0 = 101.3$ kPa, temperature $T_0 = 288$K, and speed of sound $a_0 = 340.3$ m/s. The typical take-off flight condition is taken here as $M_0 = 0$ ($V_0 = 0$), also called the static thrust condition. The Mach number at aircraft rotation and subsequent lift-off is low enough ($M_0 < 0.25$, that is, $V_0 < 85$ m/s) that the difference

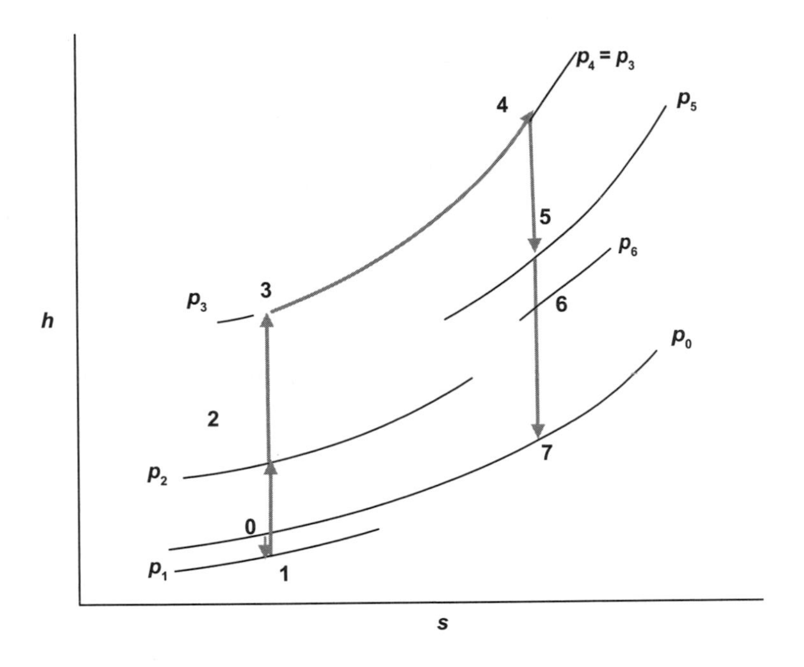


**FIGURE 3.5**

Idealized turbojet cycle for $M_0 < 1$ in which all the compression and expansion processes are considered to be isentropic.

between static and stagnation conditions is relatively small. We will proceed through the cycle sequentially using various values of the turbine inlet stagnation temperature $T_{t4}$ as an independent parameter.

### 3.4.1 Inlet flow, stations 0-2

The inlet is considered to operate isentropically so that the stagnation values of pressure and temperature delivered to the compressor face are equal to those in the free stream, that is, $p_{t2} = p_{t0}$ and $T_{t2} = T_{t0}$. The isentropic flow relations are as follows:

$$\frac{p_t}{p} = \left(1 + \frac{\gamma - 1}{2} M^2\right)^{\frac{\gamma}{\gamma - 1}} \tag{3.5}$$

$$\frac{T_t}{T} = 1 + \frac{\gamma - 1}{2} M^2. \tag{3.6}$$

For the free stream station, with $\gamma = 1.4$, these become

$$\frac{p_{t0}}{p_0} = \left(1 + \frac{M_0^2}{5}\right)^{3.5} \tag{3.7}$$

$$\frac{T_{t0}}{T_0} = 1 + \frac{M_0^2}{5}. \tag{3.8}$$

Using the given take-off conditions, the stagnation conditions at station 2 are readily calculated to be $p_{t2} = p_{t0} = 101.3$ kPa and $T_{t2} = T_{t0} = 288$K.

### 3.4.2 Compressor flow, stations 2-3

The compressor provides an increase in stagnation pressure from the inlet value determined by the compressor pressure ratio $p_{t3}/p_{t2}$. Current engines operate in the range $5 < p_{t3}/p_{t2} < 30$, and we will use the following values for comparison purposes: $p_{t3}/p_{t2} = 5$, 10, 20, and 30. The compressor is assumed to be ideal so that it provides isentropic compression and therefore permits calculation of the temperature ratio from the basic isentropic relation

$$\frac{T_{t3}}{T_{t2}} = \left(\frac{p_{t3}}{p_{t2}}\right)^{\frac{\gamma-1}{\gamma}}. \tag{3.9}$$

Then, using the conditions determined thus far, we find the compressor exit stagnation temperature, in degrees K, to be given by

$$T_{t3} = 288\left(\frac{p_{t3}}{p_{t2}}\right)^{0.286}. \tag{3.10}$$

The stagnation pressure and temperature at the compressor exit for the various compressor pressure ratios are given in Table 3.1.

### 3.4.3 Combustor flow, stations 3-4

Chapter 1 described heat addition to the air flow through the engine in terms of the fuel flow added in Equation (1.13). We may rewrite that equation here for the ideal case of perfect burner efficiency and constant specific heat as follows:

$$\left(1 + \frac{\dot{m}_f}{\dot{m}_0}\right) c_p T_{t4} - c_p T_{t3} = \frac{\dot{m}_f}{\dot{m}_0} HV. \tag{3.11}$$

Here the quantity $\dot{m}_f/\dot{m}_0 = f/a$ is the ratio of fuel flow into the combustor to the air flow through the combustor. We account for it here but note that since, in general, $\dot{m}_f/\dot{m}_a \ll 1$ it is often neglected with

**Table 3.1** Stagnation Pressure and Temperature at Compressor Exit at Take-Off for Various Compressor Pressure Ratios

| $p_{t3}/p_{t2}$ | $p_{t3}$ (MPa) | $T_{t3}$ (K) |
|---|---|---|
| 5 | 0.5065 | 456.7 |
| 10 | 1.013 | 556.8 |
| 20 | 2.026 | 678.9 |
| 30 | 3.039 | 762.3 |

**Table 3.2** Fuel–Air Ratio for Various Compressor Pressure Ratios and Turbine Inlet Stagnation Temperatures at Take-Off

| $p_{t3}/p_{t2}$ | $T_{t3}$ (K) | 1200K | 1300K | 1400K | 1500K | 1600K | 1700K |
|---|---|---|---|---|---|---|---|
| 5 | 457 | 0.01761 | 0.02003 | 0.02246 | 0.02490 | 0.02735 | 0.02981 |
| 10 | 557 | 0.01524 | 0.01765 | 0.02008 | 0.02251 | 0.02496 | 0.02741 |
| 20 | 679 | 0.01235 | 0.01475 | 0.01717 | 0.01960 | 0.02204 | 0.02449 |
| 30 | 762 | 0.01037 | 0.01277 | 0.01518 | 0.01760 | 0.02004 | 0.02248 |

respect to unity in many analyses. The quantity $HV$ is the heating value of the fuel; for Jet-A fuel, Table 1.1 gives $HV = 43.4$ MJ/kg. From Equation (3.11) we may determine the fuel–air ratio that will yield the selected limiting value of $T_{t4}$ from the suggested range of $1200\text{K} < T_{t4} < 1700\text{K}$ as follows:

$$\frac{\dot{m}_f}{\dot{m}_0} = \frac{T_{t4} - T_{t3}}{43.4 \times 10^3 - T_4}. \tag{3.12}$$

Results are shown in Table 3.2, where it is seen that the higher the compressor ratio, the lower the fuel–air ratio required. Obviously this is because the isentropic compression process in the high-pressure ratio compressors has already raised the stagnation temperature a good deal and less heat has to be added to get to the limiting turbine inlet stagnation temperature. This factor becomes of increasing importance in supersonic flight where ram compression raises pressures and temperatures so much that the compressor and its driving turbine become superfluous. The jet engine will then be a ramjet rather than a turbojet, as will be demonstrated subsequently.

### 3.4.4 Turbine flow, stations 4-5

The turbine is part of the engine solely because it is needed to drive the compressor and as such must produce just enough power to do so. Thus there is a matching condition to be met: the turbine power must equal the compressor power. With the ideal condition imposed here we need only have that

$$\dot{m}_0 c_p (T_{t3} - T_{t2}) = (\dot{m}_0 + \dot{m}_f) c_p (T_{t4} - T_{t5}). \tag{3.13}$$

This may be written as follows:

$$\left(1 + \frac{f}{a}\right) T_{t4} \left[1 - \frac{T_{t5}}{T_{t4}}\right] = T_{t2} \left[1 - \frac{T_{t3}}{T_{t2}}\right]. \tag{3.14}$$

Because we are considering only isentropic processes, we may put Equation (3.14) into a form involving the pressure ratios of the compressor and the turbine, resulting in

$$\frac{p_{t5}}{p_{t4}} = \left[1 - \frac{T_{t2}}{T_{t4}} \frac{1}{\left(1 + \frac{f}{a}\right)} \left\{\left(\frac{p_{t3}}{p_{t2}}\right)^{\frac{\gamma-1}{\gamma}} - 1\right\}\right]^{\frac{\gamma}{\gamma-1}}. \tag{3.15}$$

**Table 3.3** Turbine Exit Stagnation Pressures (kPa) for Various Compressor Pressure Ratios and Turbine Inlet Stagnation Temperatures at Take-Off

| $p_{t3}/p_{t2}$ | 1200 | 1300 | 1400 | 1500 | 1600 | 1700 |
|---|---|---|---|---|---|---|
| 5 | 301.3 | 314.8 | 326.8 | 337.4 | 346.9 | 355.5 |
| 10 | 423.7 | 457.8 | 488.6 | 516.4 | 541.7 | 564.7 |
| 20 | 521.0 | 592.7 | 659.5 | 721.6 | 779.2 | 832.8 |
| 30 | 535.5 | 636.9 | 734.1 | 826.6 | 914.2 | 999.7 |

**Table 3.4** Turbine Exit Stagnation Temperatures (K) for Various Compressor Pressure Ratios and Turbine Inlet Stagnation Temperatures at Take-Off

| $p_{t3}/p_{t2}$ | 1200 | 1300 | 1400 | 1500 | 1600 | 1700 |
|---|---|---|---|---|---|---|
| 5 | 1034 | 1135 | 1235 | 1336 | 1436 | 1536 |
| 10 | 935.2 | 1036 | 1137 | 1237 | 1338 | 1438 |
| 20 | 813.8 | 914.7 | 1016 | 1117 | 1217 | 1318 |
| 30 | 730.4 | 831.5 | 932.6 | 1034 | 1135 | 1237 |

For constant pressure combustion $p_3 = p_4$ and at the low speeds required for good combustion we may reasonably set $p_{t4} = p_{t3}$; $p_{t3}$ may be found in Table 3.1. Using these values and Equation (3.15), we find the results for $p_{t5}$ as shown in Table 3.3 for the various limiting turbine inlet stagnation temperatures selected.

The turbine exit stagnation temperature may be found from the isentropic relation

$$T_{t5} = T_{t4}\left(\frac{p_{t4}}{p_{t5}}\right)^{-\frac{\gamma-1}{\gamma}}. \tag{3.16}$$

Results for $T_{t5}$ for the different compressor pressure ratios and limiting turbine inlet stagnation temperatures are shown in Table 3.4.

### 3.4.5 Nozzle flow, stations 5-7

The nozzle considered for take-off will be of the simple converging duct type so that stations 6 and 7 are coincident. The maximum thrust for such a nozzle will be developed when the exit station is choked, that is, $M_7 = 1$. The nozzle is considered to operate ideally so that the stagnation pressure is constant throughout the nozzle, $p_{t5} = p_{t6} = p_{t7}$. Thus the pressure at the exit station $p_7$ may be found using the isentropic flow relation of Equation (3.5) with the condition $M_7 = 1$ and values for the turbine exit stagnation pressure from Table 3.3. Results for $p_7$ are shown in Table 3.5. Note that the ratio of exit pressure to free stream pressure $p_7/p_0 > 1$, indicating that the jet exhaust is always underexpanded under take-off conditions. A variable area diverging section between stations 6 and 7 would permit the

**Table 3.5** Nozzle Exit Pressures (kPa) for Various Compressor Pressure Ratios and Turbine Inlet Stagnation Temperatures at Take-Off

| $p_{t3}/p_{t2}$ | 1200 | 1300 | 1400 | 1500 | 1600 | 1700 |
|---|---|---|---|---|---|---|
| 5 | 159.2 | 166.3 | 172.6 | 178.3 | 183.3 | 187.8 |
| 10 | 223.8 | 241.8 | 258.1 | 272.8 | 286.2 | 298.3 |
| 20 | 275.3 | 313.1 | 348.4 | 381.2 | 411.7 | 440.0 |
| 30 | 282.9 | 336.5 | 387.8 | 436.7 | 483.0 | 528.2 |

exhaust nozzle to get closer to matched conditions, and therefore greater thrust, but the exit flow would be supersonic, and noise considerations would assume greater importance.

In the same fashion, we may calculate nozzle exit temperature $T_7$ using the isentropic flow relation of Equation (3.6) with the condition $M_7 = 1$ and the values for the turbine exit stagnation temperature from Table 3.4. Results for $T_7$ are shown in Table 3.6. The exit velocity can then be found from the relation $V = aM$, where the sound speed

$$a_7 = \sqrt{\gamma R T_7} = \left(20.04 m/s - K^{\frac{1}{2}}\right)\sqrt{T_7}. \tag{3.17}$$

Here $\gamma R = 1.4(287\ m^2/s^2\text{-K})$, and results for $V_7$ are shown in Table 3.7.

**Table 3.6** Nozzle Exit Temperatures (K) for Various Compressor Pressure Ratios and Turbine Inlet Stagnation Temperatures at Take-Off

| $p_{t3}/p_{t2}$ | 1200 | 1300 | 1400 | 1500 | 1600 | 1700 |
|---|---|---|---|---|---|---|
| 5 | 861.9 | 945.6 | 1029.2 | 1112.9 | 1196.6 | 1280.2 |
| 10 | 779.3 | 863.2 | 947.0 | 1030.9 | 1114.8 | 1198.6 |
| 20 | 678.1 | 762.2 | 846.3 | 930.4 | 1014.5 | 1098.6 |
| 30 | 608.6 | 692.9 | 777.1 | 861.4 | 945.7 | 1030.8 |

**Table 3.7** Nozzle Exit Velocities (m/s) for Various Compressor Pressure Ratios and Turbine Inlet Stagnation Temperatures at Take-Off

| $p_{t3}/p_{t2}$ | 1200 | 1300 | 1400 | 1500 | 1600 | 1700 |
|---|---|---|---|---|---|---|
| 5 | 588.3 | 616.2 | 642.9 | 668.5 | 693.2 | 717.0 |
| 10 | 559.4 | 588.8 | 616.7 | 643.4 | 669.1 | 693.8 |
| 20 | 521.9 | 553.3 | 583.0 | 611.3 | 638.3 | 664.2 |
| 30 | 494.4 | 527.5 | 558.7 | 588.2 | 616.3 | 643.4 |

### 3.4.6 Turbojet thrust and fuel efficiency in take-off

Using Equations (3.1) and (3.3), the net thrust is given by

$$F_n = \dot{m}_0\left(1 + \frac{f}{a}\right)V_7 - \dot{m}_0 V_0 + A_7(p_7 - p_0). \tag{3.18}$$

The mass conservation equation requires that

$$\dot{m}_0\left(1 + \frac{f}{a}\right) = \dot{m}_7 = \rho_7 V_7 A_7. \tag{3.19}$$

Using this relation, we see that

$$A_7 = \frac{\dot{m}_0}{\rho_7 V_7}\left(1 + \frac{f}{a}\right).$$

Then the specific thrust may be written as

$$\frac{F_n}{\dot{m}_0} = \left(1 + \frac{f}{a}\right)V_7\left[1 + \frac{1}{\gamma M_7^2}\left(1 - \frac{p_0}{p_7}\right)\right] - V_0. \tag{3.20}$$

Using all the information gathered thus far, the specific thrust in take-off mode for the ideal turbojet engine may be calculated using Equation (3.20), and results are illustrated in Figure 3.6. There is an optimum compressor pressure ratio of between 10 and 15 for the specific thrust, which is related to the fact that the nozzle is choked and thus exit pressures are not matched, which affects the maximum thrust achievable.

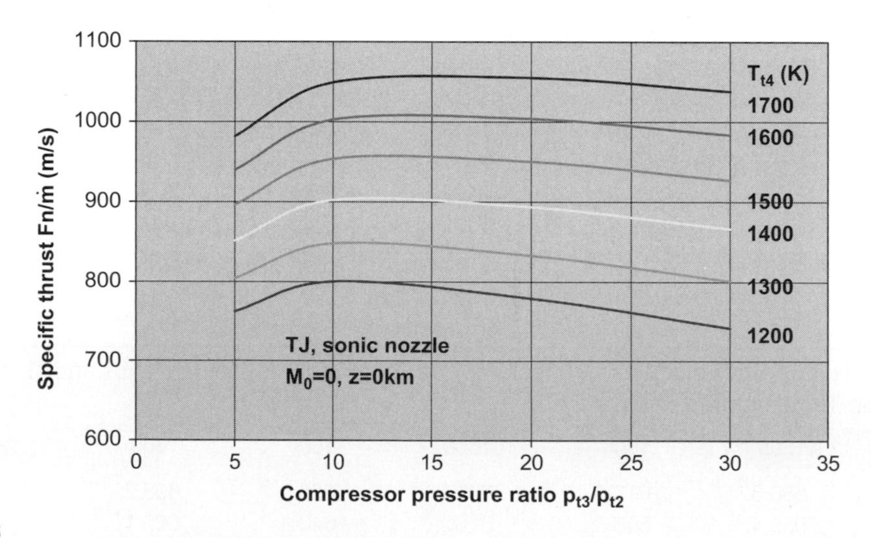


**FIGURE 3.6**

Specific thrust as a function of compressor pressure ratio and turbine inlet stagnation temperature for an ideal turbojet in take-off.

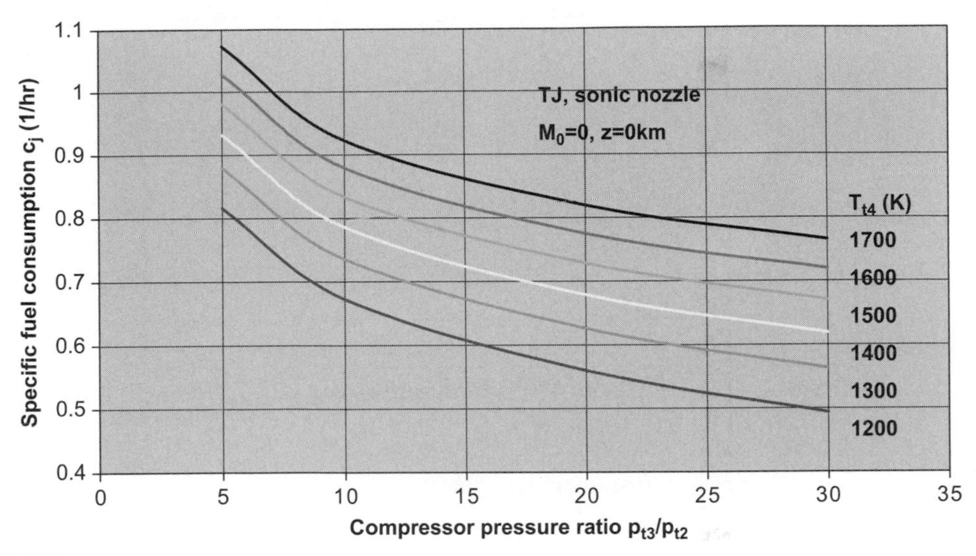


**FIGURE 3.7**

Specific fuel consumption as a function of compressor pressure ratio and turbine inlet stagnation temperature for an ideal turbojet in take-off.

The specific fuel consumption is given by

$$c_j = \frac{\dot{m}_f g}{F_n} = g\left(\frac{f}{a}\right)\left(\frac{F_n}{\dot{m}_0}\right)^{-1}. \tag{3.21}$$

The specific fuel consumption is shown in Figure 3.7.

Results shown in Figures 3.6 and 3.7 demonstrate the value and cost of high turbine inlet stagnation temperature: higher specific thrust but at higher specific fuel consumption. High compressor pressure ratios, $p_{t3}/p_{t2} > 10$, are also beneficial, but more for fuel efficiency than for enhancing thrust production.

### 3.4.7 Real turbojet engine in take-off

For purposes of comparison of our ideal engine with a real engine, we may consider the J57 (Figure 3.8), a popular early military turbojet engine manufactured by Pratt & Whitney, which powered the North American F-100 Super Sabre, the first of the USAF fighter aircraft to exceed the speed of sound in level flight (Figure 3.8). The commercial version of this engine was known as the JT3C and powered early Boeing 707 and Douglas DC-8 commercial airliners. This engine had a thrust capability in the range of 45 to 80 kN. The JT3C-7 had a thrust of 53.4 kN, a pressure ratio of 12.5, a mass flow of 80.7 kg/s, and a turbine inlet temperature of 1140K (*Flight*, March 18, 1960, p. 385).

Locating this pressure ratio and turbine inlet temperature in Figure 3.6, we may infer that $F_n/\dot{m} =$ 730 m/s. Using the actual mass flow for this engine suggests that the engine static thrust is 60 kN, an overprediction of about 12%. From Figure 3.7 we see that the corresponding specific fuel consumption

FIGURE 3.8

Pratt & Whitney J57 turbojet (from Pratt & Whitney classic engines).

$c_j = 0.63\ \text{hr}^{-1}$ at take-off, or static thrust, conditions, that is, at $M_0 = 0$. Because the actual engine at full power under static conditions has $c_j = 0.785\ \text{hr}^{-1}$, the ideal analysis underpredicts the specific fuel consumption by about 20%. This shows that we will have to ultimately take into account the inefficiencies that occur in a real engine in order to make reasonably accurate predictions of performance. However, this is still a reasonably accurate prediction given the fact that this is a cycle analysis for an ideal engine.

## 3.5 IDEAL TURBOJET IN HIGH SUBSONIC CRUISE IN THE STRATOSPHERE

Early commercial jetliners, such as the Boeing B-707 and the Douglas DC-8, would cruise in the lower stratosphere, $z = 10$ km, at a high subsonic Mach number around $M_0 = 0.80$. This ability to fly above the weather at high speed ushered in a wave of rapid growth of the commercial airplane business. Although substantial effort was placed on keeping aerodynamic drag down at these almost sonic speeds, there was also a requirement for reliable high-thrust engines and the turbojet filled that need. As mentioned in the previous section, current engines operate in the range $1200\text{K} < T_{t4} < 1700\text{K}$, although materials and cooling techniques permitting higher temperatures are being sought constantly. The basic cruise state considered here is standard-day stratospheric conditions, that is, altitude $z = 10$ km, pressure $p_0 = 26.5$ kPa, temperature $T_0 = 223\text{K}$, and speed of sound $a_0 = 300$ m/s. The typical cruise flight condition is taken here as $M_0 = 0.80$ ($V_0 = 240$ m/s), although current aircraft cruise in the range $0.75 < M_0 < 0.85$. We will proceed through the cycle sequentially using various values of the turbine inlet stagnation temperature as in the case for take-off.

### 3.5.1 Inlet flow, stations 0-2

Because the flight Mach number is subsonic, the inlet of the ideal turbojet is still considered to operate isentropically so that the stagnation values of pressure and temperature delivered to the compressor face, that is, at station 2, are equal to those in the free stream: $p_{t2} = p_{t0}$ and $T_{t2} = T_{t0}$. Using Equations (3.7) and (3.8) for the given cruise conditions, the stagnation conditions at station 2 are calculated readily to be $p_{t2} = p_{t0} = 40.4$ kPa and $T_{t2} = T_{t0} = 252\text{K}$.

**Table 3.8** Stagnation Pressure and Temperature at Compressor Exit in $M = 0.8$ Cruise at 10 km Altitude

| $p_{t3}/p_{t2}$ | $p_{t3}$ (kPa) | $T_{t3}$ (K) |
|---|---|---|
| 5 | 201.9 | 399.0 |
| 10 | 403.8 | 486.5 |
| 20 | 807.6 | 593.1 |
| 30 | 1211 | 666.1 |

### 3.5.2 Compressor flow, stations 2-3

The compressor provides an increase in stagnation pressure from the inlet value that is determined by the compressor pressure ratio $p_{t3}/p_{t2}$. We will continue to use typical values for comparison purposes: $p_{t3}/p_{t2} = 5$, 10, 20, and 30. The compressor is assumed to be ideal so that it provides isentropic compression and therefore permits calculation of the temperature ratio from the basic isentropic relation. Using the conditions determined thus far, we find the compressor exit stagnation temperature from Equation (3.9) to be given by

$$T_{t3} = 252\left(\frac{p_{t3}}{p_{t2}}\right)^{0.286}. \tag{3.22}$$

The stagnation pressure and temperature at the compressor exit for the various compressor pressure ratios are given in Table 3.8. Upon comparison of Tables 3.1 and 3.8, it is clear that the stagnation pressure being delivered to the combustor is much lower in the cruise state than at take-off because the ambient atmospheric pressure is only one-quarter that at sea level.

### 3.5.3 Combustor flow, stations 3-4

Using Equation (3.12) again we may determine the fuel–air ratio that will yield the selected limiting value of $T_{t4}$ from the suggested range of $1200\text{K} < T_{t4} < 1700\text{K}$. Results are shown in Table 3.9, where it is seen once more that the higher the compressor ratio, the lower the fuel–air ratio required. Note that the stagnation temperature $T_{t3}$ in cruise is somewhat lower than in take-off. As a result, the fuel–air ratio in cruise is a bit greater because a little more heat can be added to get to the limiting turbine inlet

**Table 3.9** Fuel–Air Ratio for Various Compressor Pressure Ratios and Turbine Inlet Stagnation Temperatures in $M = 0.8$ Cruise at 10 km Altitude

| $p_{t3}/p_{t2}$ | 1200K | 1300K | 1400K | 1500K | 1600K | 1700K |
|---|---|---|---|---|---|---|
| 5 | 0.01898 | 0.02140 | 0.02383 | 0.02628 | 0.02873 | 0.03120 |
| 10 | 0.01691 | 0.01932 | 0.02175 | 0.02419 | 0.02664 | 0.02910 |
| 20 | 0.01438 | 0.01679 | 0.01921 | 0.02164 | 0.02409 | 0.02654 |
| 30 | 0.01265 | 0.01506 | 0.01747 | 0.01990 | 0.02234 | 0.02479 |

stagnation temperature. As mentioned previously, this factor becomes of increasing importance in supersonic flight where the ram compression raises pressures and temperatures so much that the compressor and its driving turbine become superfluous and the jet engine becomes a ramjet.

### 3.5.4 Turbine flow, stations 4-5

For constant pressure combustion, we again consider $p_{t4} = p_{t3}$, and $p_{t3}$ may be found in Table 3.8. Using these values and Equation (3.15), we find the results for $p_{t5}$ as shown in Table 3.10 for the various limiting turbine inlet stagnation temperatures selected.

Note that the turbine exit stagnation pressures in cruise are about half the values in take-off. This is due to the reduced free stream pressure at altitude, about one-quarter that at sea level, which is made up only partially through ram pressure increases. The turbine exit stagnation temperature may be found from the isentropic relation of Equation (3.16). Results for $T_{t5}$ for the different compressor pressure ratios and limiting turbine inlet stagnation temperatures are shown in Table 3.11.

Comparing Table 3.11 with Table 3.4 shows the turbine exhaust stagnation temperature in cruise to be somewhat higher than that at take-off. This means that the nozzle expansion can reach higher exhaust velocity, improving thrust.

### 3.5.5 Nozzle flow, stations 5-7

The nozzle considered for take-off is of the simple converging duct type so that stations 6 and 7 are coincident. The maximum thrust for such a nozzle will be developed when the exit station is choked, that is, $M_7 = 1$. The nozzle is considered to operate ideally so that the stagnation pressure is constant

**Table 3.10** Turbine Exit Stagnation Pressures $p_{t5}$ (kPa) for Various Compressor Pressure Ratios and Turbine Inlet Stagnation Temperatures in $M = 0.8$ Cruise at 10 km Altitude

| $p_{t3}/p_{t2}$ | 1200K | 1300K | 1400K | 1500K | 1600K | 1700K |
|---|---|---|---|---|---|---|
| 5 | 127.7 | 132.6 | 136.9 | 140.7 | 144.0 | 147.1 |
| 10 | 188.5 | 201.2 | 212.5 | 222.6 | 231.8 | 240.1 |
| 20 | 250.3 | 278.1 | 303.7 | 327.1 | 348.7 | 368.6 |
| 30 | 275.2 | 316.3 | 354.8 | 390.9 | 424.5 | 455.8 |

**Table 3.11** Turbine Exit Stagnation Temperatures $T_{t5}$ (K) for Various Compressor Pressure Ratios and Turbine Inlet Stagnation Temperatures in $M = 0.8$ Cruise at 10 km Altitude

| $p_{t3}/p_{t2}$ | 1200K | 1300K | 1400K | 1500K | 1600K | 1700K |
|---|---|---|---|---|---|---|
| 5 | 1053 | 1153 | 1253 | 1353 | 1453 | 1553 |
| 10 | 965.1 | 1065 | 1165 | 1265 | 1365 | 1465 |
| 20 | 858.4 | 958.4 | 1058 | 1158 | 1258 | 1358 |
| 30 | 785.4 | 885.4 | 985.4 | 1085 | 1185 | 1285 |

**Table 3.12** Sonic Nozzle Exit Pressures (kPa) for Various Compressor Pressure Ratios and Turbine Inlet Stagnation Temperatures in $M = 0.8$ Cruise at 10 km Altitude

| $p_{t3}/p_{t2}$ | 1200K | 1300K | 1400K | 1500K | 1600K | 1700K |
|---|---|---|---|---|---|---|
| 5 | 67.47 | 70.04 | 72.30 | 74.31 | 76.09 | 77.69 |
| 10 | 99.60 | 106.3 | 112.2 | 117.6 | 122.5 | 126.8 |
| 20 | 132.2 | 146.9 | 160.4 | 172.8 | 184.2 | 194.7 |
| 30 | 145.4 | 167.1 | 187.5 | 206.5 | 224.2 | 240.8 |

throughout the nozzle, $p_{t5} = p_{t6} = p_{t7}$. Thus the pressure at the exit station $p_7$ may be found using the isentropic flow relation of Equation (3.5) with the condition $M_7 = 1$ and the values for the turbine exit stagnation pressure from Table 3.11. Results for $p_7$ for the sonic exit condition $M_7 = 1$ are shown in Table 3.12.

Note that since at the cruise altitude $p_0 = 40.4$ kPa the ratio of exit pressure to free stream pressure $p_7/p_0 > 1$, indicating that the jet exhaust is always underexpanded under cruise conditions. A variable area diverging section between stations 6 and 7 would permit the exhaust nozzle to expand the flow further to supersonic exit Mach numbers and thereby get closer to matched conditions, and therefore greater thrust. This approach is used by military aircraft to reach higher flight speeds, but such variable geometry devices add greatly to engine cost and are not used on subsonic airliners.

As in the take-off case, we may calculate the nozzle exit temperature $T_7$ using the isentropic flow relation of Equation (3.6) with the condition $M_7 = 1$ and the values for the turbine exit stagnation temperature from Table 3.11. Results for $T_7$ are shown in Table 3.13. Exit velocity can then be found from the relation $V = aM$, where the sound speed is calculated from Equation (3.17). Here, $\gamma R = 401.8$ $m^2/s^2$-K, and results for $V_7$ are shown in Table 3.14.

If the nozzle were of variable area so that $A_7$ could be increased over $A_6$ and thereby accelerate the nozzle flow to lower pressures such that a matched condition, $p_7 = p_0$, could be achieved, the nozzle exit Mach number, temperature, and velocity would change. Noting that for an isentropic nozzle $p_{t7} = p_{t5}$ and for a matched exit $p_7 = p_0$, and keeping $\gamma = 1.4$ for this ideal study, we may find $M_7$ shown in Table 3.15 using the isentropic relation

$$\frac{p_{t7}}{p_7} = \frac{p_{t5}}{p_0} = \left(1 + \frac{M_7^2}{5}\right)^{3.5}.$$

**Table 3.13** Sonic Nozzle Exit Temperatures (K) for Various Compressor Pressure Ratios and Turbine Inlet Stagnation Temperatures in $M = 0.8$ Cruise at 10 km Altitude

| $p_{t3}/p_{t2}$ | 1200K | 1300K | 1400K | 1500K | 1600K | 1700K |
|---|---|---|---|---|---|---|
| 5 | 877.2 | 960.6 | 1044 | 1127 | 1211 | 1294 |
| 10 | 804.3 | 887.6 | 970.9 | 1054 | 1138 | 1221 |
| 20 | 715.3 | 798.6 | 882.0 | 965.3 | 1049 | 1132 |
| 30 | 654.5 | 737.8 | 821.2 | 904.5 | 987.8 | 1071 |

**Table 3.14** Sonic Nozzle Exit Velocities (m/s) for Various Compressor Pressure Ratios and Turbine Inlet Stagnation Temperatures in $M = 0.8$ Cruise at 10 km Altitude

| $p_{t3}/p_{t2}$ | 1200K | 1300K | 1400K | 1500K | 1600K | 1700K |
|---|---|---|---|---|---|---|
| 5 | 592.4 | 619.9 | 646.2 | 671.5 | 695.9 | 719.4 |
| 10 | 567.2 | 595.9 | 623.2 | 649.4 | 674.6 | 698.8 |
| 20 | 534.9 | 565.2 | 594.0 | 621.4 | 647.7 | 672.9 |
| 30 | 511.7 | 543.3 | 573.1 | 601.5 | 628.6 | 654.6 |

**Table 3.15** Matched Nozzle Exit Mach Numbers for Various Compressor Pressure Ratios and Turbine Inlet Stagnation Temperatures in $M = 0.8$ Cruise at 10 km Altitude

| $p_{t3}/p_{t2}$ | 1200K | 1300K | 1400K | 1500K | 1600K | 1700K |
|---|---|---|---|---|---|---|
| 5 | 1.685 | 1.710 | 1.731 | 1.749 | 1.765 | 1.779 |
| 10 | 1.940 | 1.982 | 2.017 | 2.047 | 2.073 | 2.096 |
| 20 | 2.122 | 2.190 | 2.246 | 2.294 | 2.335 | 2.370 |
| 30 | 2.183 | 2.272 | 2.346 | 2.408 | 2.461 | 2.506 |

**Table 3.16** Matched Nozzle Exit Velocities (m/s) for Various Compressor Pressure Ratios and Turbine Inlet Stagnation Temperatures in $M = 0.8$ Cruise at 10 km Altitude

| P.R. | 1200K | 1300K | 1400K | 1500K | 1600K | 1700K |
|---|---|---|---|---|---|---|
| 5 | 875.8 | 925.0 | 971.7 | 1016 | 1059 | 1100 |
| 10 | 913.1 | 970.9 | 1025 | 1077 | 1127 | 1174 |
| 20 | 904.6 | 971.4 | 1034 | 1093 | 1149 | 1202 |
| 30 | 878.0 | 951.1 | 1019 | 1083 | 1143 | 1200 |

Note that exit Mach numbers lie in the range $1.69 < M_7 < 2.51$, which require area ratios in the range $1.32 < A_7/A_6 < 2.65$ in order to produce the matched exit flow. The exit velocity $V_7$ in the matched case may be found by first using the isentropic relation Equation (3.6) to find $T_7$ from $M_7$ and $T_{t7} = T_{t5}$ and then using the Mach number definition $M_7 = V_7/a_7$. Table 3.16 gives results for $V_7$ in the case of the matched exhaust nozzle.

### 3.5.6 Turbojet thrust and fuel efficiency in cruise

Using all the information gathered thus far, the specific thrust in the cruise mode for the ideal turbojet engine may be calculated using Equation (3.20). Results for the fixed geometry sonic nozzle are

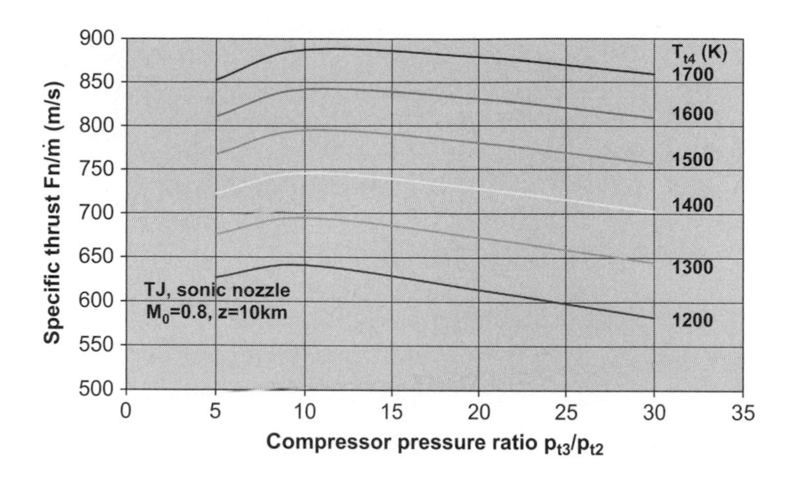


**FIGURE 3.9**

Turbojet specific thrust as a function of compressor pressure ratio and turbine inlet stagnation temperature with a fixed geometry sonic nozzle in $M = 0.8$ cruise at 10 km altitude.

illustrated in Figure 3.9. The ratio of specific thrust for the variable area matched nozzle to that for the fixed nozzle is illustrated in Figure 3.10.

The specific fuel consumption in high subsonic cruise for the sonic nozzle is shown in Figure 3.11. The ratio of specific fuel consumption for the variable area matched nozzle to that for the fixed nozzle is illustrated in Figure 3.12.

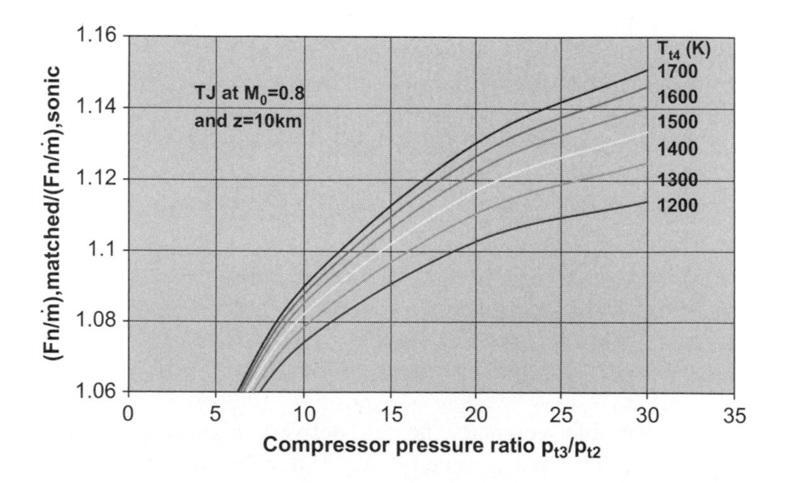


**FIGURE 3.10**

Ratio of specific thrust as a function of compressor pressure ratio and turbine inlet stagnation temperature for a matched nozzle compared to a sonic nozzle in $M = 0.8$ cruise at 10 km altitude.

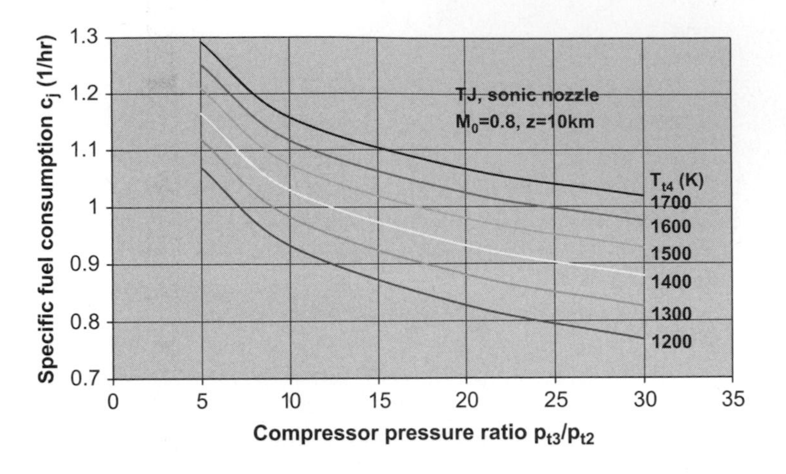


**FIGURE 3.11**

Specific fuel consumption as a function of compressor pressure ratio and turbine inlet stagnation temperature for ideal turbojet with a fixed geometry sonic nozzle in $M = 0.8$ cruise at 10 km altitude.

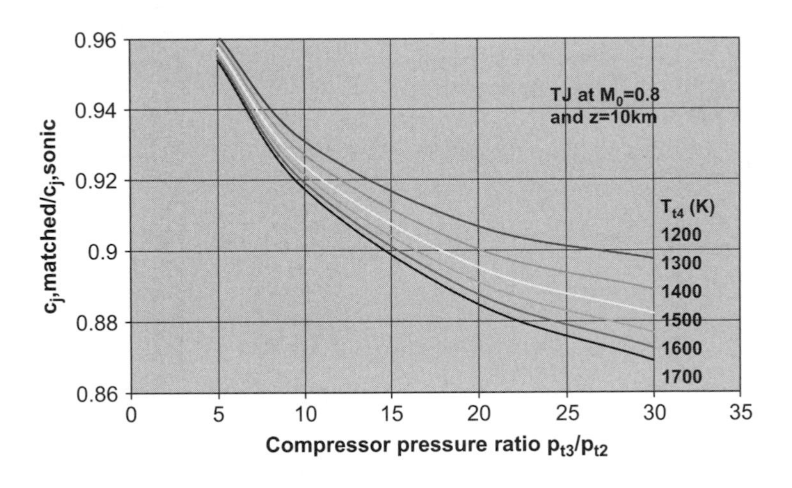


**FIGURE 3.12**

Specific fuel consumption ratio as a function of compressor pressure ratio and turbine inlet stagnation temperature for ideal turbojet with a variable geometry-matched nozzle compared to a fixed geometry sonic nozzle in $M = 0.8$ cruise at 10 km altitude.

It is clear that the variable geometry nozzle delivers significantly better specific thrust performance compared to the fixed geometry sonic nozzle for the same flight conditions. Commercial jet transports use the fixed geometry nozzle for cost and reliability reasons, whereas military aircraft often use variable geometry nozzles in order to achieve maximum performance. For supersonic flight where aircraft drag increases substantially, it is necessary to use variable geometry nozzles. The Concorde commercial supersonic transport needed variable geometry nozzles in order to maintain its Mach 2 cruise speed. However, it bears repeating that the cost, weight, and complexity of variable geometry nozzles make them impractical for use in subsonic commercial applications.

### 3.5.7 Real turbojet engine in subsonic cruise

In steady, level, cruising flight the equilibrium conditions for an aircraft require that the lift equal the weight $L = W$ and the thrust equal the drag $F_n = D$. Therefore, we may write

$$\frac{L}{D} = \frac{W}{F_n}. \tag{3.23}$$

The ratio of lift to drag for commercial airliners in high-altitude $M_0 = 0.8$ cruise is generally in the range $16 < L/D < 19$ so that the ratio of net thrust to aircraft weight is in the range of 0.052 to 0.062, considerably less than in take-off where acceleration requirements force high-thrust levels. For the JT3C-powered Boeing 707 discussed in the take-off case, the cruise thrust is about 15.8 kN with a specific fuel consumption of $c_j = 0.909\ \text{hr}^{-1}$ (*Flight*, March 18, 1960, p. 385).

From Figure 3.11 we see that the specific fuel consumption $c_j \cong 0.87\ \text{hr}^{-1}$ for the sonic nozzle case, whereas for the matched nozzle case, as shown in Figure 3.12, it is slightly lower, $c_j = 0.82\ \text{hr}^{-1}$. Here the prediction of specific fuel consumption for the sonic nozzle is only about 4% less than that observed in practice. This result suggests that the ideal analysis provides about the right ratio of $f/a$ compared to $F_n/\dot{m}$, but each of these two terms may not be predicted as accurately. In general, ideal analysis is found to overpredict the specific thrust, an understandable discrepancy considering that the efficiencies of the various components of the engine have not been taken into account and they affect the actual thrust levels achievable.

## 3.6 IDEAL TURBOJET IN SUPERSONIC CRUISE IN THE STRATOSPHERE

The lure of high-speed commercial air travel led to development of the Concorde supersonic airplane, which entered service in 1976. Military aircraft had been able to fly supersonically since the 1950s, but the Concorde was to cruise supersonically over long ranges. The Concorde would cruise at Mach 2 at higher altitudes than subsonic airliners, around $z = 15$ km. The ability of the Concorde to fly across the Atlantic in about 3 hr ushered in a wave of interest in supersonic commercial airplanes. Although substantial effort was placed on keeping aerodynamic drag down at these supersonic speeds, there was also a requirement for reliable high-thrust engines, and turbojets outfitted with afterburners filled that need. As mentioned in previous sections, current engines operate in the range of turbine inlet temperatures $1200\text{K} < T_{t4} < 1700\text{K}$, although materials and cooling techniques permitting higher temperatures are being sought constantly. The basic cruise state is standard-day stratospheric conditions, that is, altitude $z = 15$ km, pressure $p_0 = 12.1$ kPa, temperature $T_0 = 217\text{K}$, and speed of sound $a_0 = 295$ m/s. The typical cruise flight condition is taken here as $M_0 = 2$ ($V_0 = 590$ m/s). We will again proceed through the cycle sequentially using various values of the turbine inlet stagnation temperature as in the case for take-off and subsonic cruise and then require the afterburner to raise the stagnation temperature in the jet exhaust back up to the selected turbine inlet temperatures. This provides an assessment of afterburning, as well as taking into account the material limitations represented by the selected turbine inlet temperature.

### 3.6.1 Inlet flow, stations 0-2

Although the flight Mach number is supersonic, the inlet of the ideal turbojet is still considered to operate isentropically so that the stagnation values of pressure and temperature delivered to the compressor face are equal to those in the free stream, that is, $p_{t2} = p_{t0}$ and $T_{t2} = T_{t0}$. Using Equations (3.7) and (3.8) for the given supersonic cruise conditions, the stagnation conditions at station 2 are calculated readily to be $p_{t2} = p_{t0} = 94.7$ kPa and $T_{t2} = T_{t0}= 391$K. As shown in Chapter 6, a well-designed inlet can deliver a pressure recovery of more than 90% at $M_0 = 2$.

### 3.6.2 Compressor flow, stations 2-3

The compressor provides an increase in stagnation pressure from the inlet value determined by the compressor pressure ratio $p_{t3}/p_{t2}$. We will continue to use the typical values for comparison purposes: $p_{t3}/p_{t2} = 5$, 10, 20, and 30. The compressor is assumed to be ideal so that it provides isentropic compression and therefore permits calculation of the temperature ratio from the basic isentropic relation. Using the conditions determined thus far, we find the compressor exit stagnation temperature from Equation (3.9) to be given by

$$T_{t3} = 391\left(\frac{p_{t3}}{p_{t2}}\right)^{0.286}. \tag{3.24}$$

The stagnation pressure and temperature at the compressor exit for the various compressor pressure ratios are given in Table 3.17. Upon comparison of Tables 3.17 and 3.8, it is clear that the stagnation pressure being delivered to the combustor is much higher in supersonic cruise than in subsonic cruise even though the ambient atmospheric pressure at $z = 15$ km is only one-half that at $z = 10$ km. This is due to the higher ram pressure at supersonic speed.

### 3.6.3 Combustor flow, stations 3-4

Using Equation (3.12) again, we may determine the fuel–air ratio that will yield the selected limiting value of $T_{t4}$ from the suggested range of 1200K $< T_{t4} <$ 1700K. Results are shown in Table 3.18, where it is seen once more that the higher the compressor ratio, the lower the fuel–air ratio required. Note that

**Table 3.17** Stagnation Pressure and Temperature at the Compressor Exit in $M = 2$ Cruise at 15 km Altitude

| $p_{t3}/p_{t2}$ | $p_{t3}$ (kPa) | $T_{t3}$ (K) |
|---|---|---|
| 5 | 474 | 620 |
| 10 | 947 | 755 |
| 20 | 1894 | 921 |
| 30 | 2841 | 1034 |

**Table 3.18** Fuel–Air Ratio for Various Compressor Pressure Ratios and Turbine Inlet Stagnation Temperatures in $M_0 = 2$ Cruise at 15 km Altitude

| $p_{t3}/p_{t2}$ | 1200K | 1300K | 1400K | 1500K | 1600K | 1700K |
|---|---|---|---|---|---|---|
| 5 | 0.01375 | 0.01616 | 0.01858 | 0.02101 | 0.02345 | 0.02591 |
| 10 | 0.01054 | 0.01294 | 0.01535 | 0.01777 | 0.02020 | 0.02265 |
| 20 | 0.006610 | 0.009001 | 0.01140 | 0.01382 | 0.01624 | 0.01868 |
| 30 | 0.003927 | 0.006311 | 0.008707 | 0.01112 | 0.01353 | 0.01596 |

the stagnation temperature $T_{t3}$ in supersonic cruise is higher than in subsonic cruise. As a result, the fuel–air ratio in cruise is lower in supersonic cruise because less heat can be added to get to the limiting turbine inlet stagnation temperature.

### 3.6.4 Turbine flow, stations 4-5

For constant pressure combustion, we again assume $p_{t4} = p_{t3}$, and $p_{t3}$ may be found in Table 3.17. Using these values and Equation (3.15), we find the results for $p_{t5}$ as shown in Table 3.19 for the various limiting turbine inlet stagnation temperatures selected.

Note that the turbine exit stagnation pressures in supersonic cruise are considerably larger than the values in subsonic cruise. Even though the free stream pressure at 15 km is half that at 10 km altitude, supersonic ram pressure increases more than make up for that reduction.

The turbine exit stagnation temperature may be found from the isentropic relation of Equation (3.16). Results for $T_{t5}$ for the different compressor pressure ratios and limiting turbine inlet stagnation temperatures are shown in Table 3.20.

### 3.6.5 Afterburner flow, stations 5-5b

Because the exit flow from the turbine, station 5, is still quite rich in oxygen, fuel may be injected into the duct leading to the nozzle throat, station 6, and burned, again at constant pressure. Thus the afterburner section, as illustrated in Figure 3.13, acts like a large constant pressure combustor, and the relatively low-temperature gas coming from the turbine exit may be boosted to higher levels, $T_{t5b}$.

**Table 3.19** Turbine Exit Stagnation Pressures $p_{t5}$ (kPa) for Various Compressor Pressure Ratios and Turbine Inlet Stagnation Temperatures in $M_0 = 2$ Cruise at 15 km Altitude

| $p_{t3}/p_{t2}$ | 1200K | 1300K | 1400K | 1500K | 1600K | 1700K |
|---|---|---|---|---|---|---|
| 5 | 226.0 | 240.7 | 253.7 | 265.5 | 276.1 | 285.7 |
| 10 | 266.8 | 299.5 | 329.7 | 357.5 | 383.3 | 407.0 |
| 20 | 246.3 | 302.9 | 358.3 | 411.8 | 463.2 | 512.2 |
| 30 | 193.2 | 260.3 | 329.9 | 400.0 | 469.7 | 538.0 |

**Table 3.20** Turbine Exit Stagnation Temperatures $T_{t5}$ (K) for Various Compressor Pressure Ratios and Turbine Inlet Stagnation Temperatures in $M_0 = 2$ Cruise at 15 km Altitude

| $p_{t3}/p_{t2}$ | 1200K | 1300K | 1400K | 1500K | 1600K | 1700K |
|---|---|---|---|---|---|---|
| 5 | 971.2 | 1071 | 1171 | 1271 | 1371 | 1471 |
| 10 | 835.3 | 935.3 | 1035.3 | 1135 | 1235 | 1335 |
| 20 | 669.6 | 769.6 | 869.6 | 969.6 | 1070 | 1170 |
| 30 | 556.3 | 656.3 | 756.3 | 856.3 | 956.2 | 1056 |

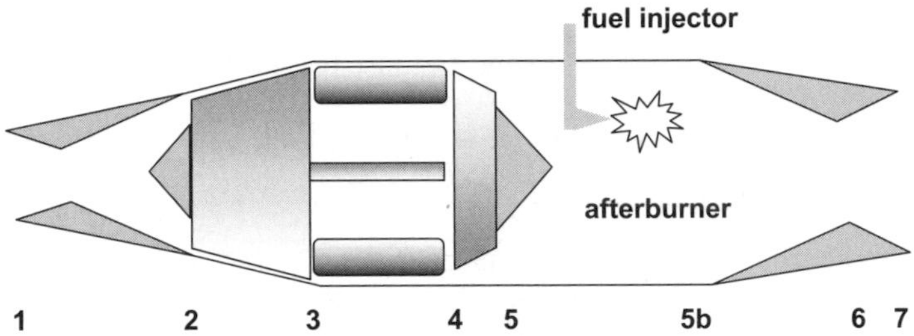

**FIGURE 3.13**

Schematic diagram of an afterburning turbojet engine.

The increased stagnation temperature permits the generation of higher exit velocity $V_7$ and thereby increases the thrust capability of the engine.

We may assume that combustion in the afterburner brings the final temperature $T_{t5b}$ back up to the limiting turbine inlet temperature so that the afterburner exit temperature $T_{t5b} = T_{t4}$. This is the best that can be achieved, as we have assumed the turbine inlet temperature is set by the thermal capability of engine materials. This operation of the afterburner is described schematically on the cycle diagram shown in Figure 3.14.

However, we must take note of the fact that some of the oxygen present in the air entering the engine has been consumed in the combustor and is no longer available for combustion in the afterburner. To account for this, we introduce an effective heating value that depends on the combustor fuel to air ratio and is decreased from the usual value of 43,400 kJ/kg by an amount proportional to the fuel to air ratio in the combustor. A simple approach is to assume that the actual stagnation temperature rise achievable is linearly proportional to some function of the fuel to air ratio used in the core engine $\kappa\,(f/a)$ as given by

$$K = \frac{T_{t6} - T_{t5}}{T_{t6,i} - T_{t5}} = 1 - \kappa\left(\frac{f}{a}\right).$$

Here the actual stagnation temperature rise is assumed to be directly proportional to the ideal stagnation temperature rise (that for nondepleted oxygen in the air flow) with the constant of proportionality depending on the fuel to air ratio in the core engine combustor. Thus one may assume that

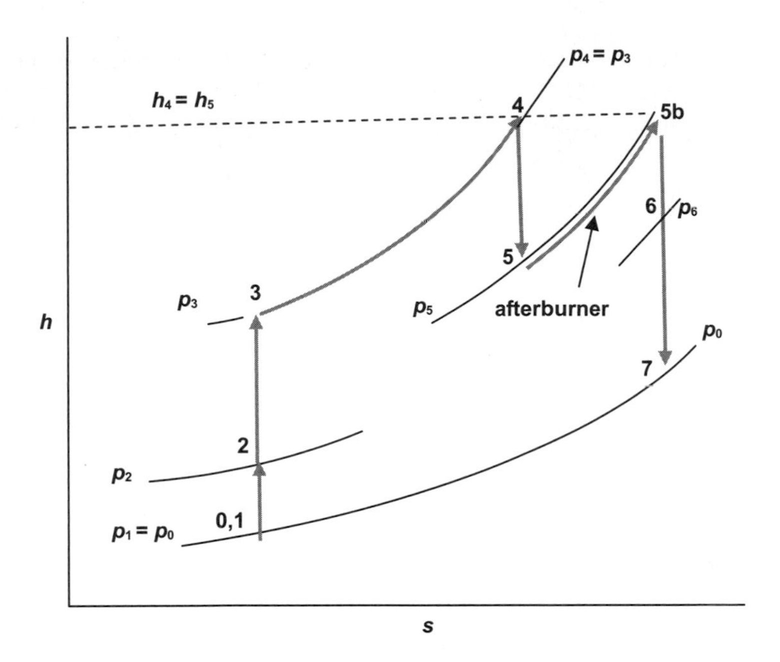


FIGURE 3.14

Afterburner operation shown on an ideal supersonic turbojet cycle diagram.

when the fuel to air ratio in the core engine is stoichiometric, denoting complete consumption of the oxygen in the flow, the absence of any excess oxygen will permit no combustion and therefore no stagnation temperature rise so that $K = 0$. Likewise, if the fuel to air ratio in the core engine combustor were zero, then there would be the full complement of oxygen available and the ideal stagnation temperature rise could be achieved. For this set of assumptions, $\kappa = \phi$, which is called the equivalence ratio or

$$\kappa = \phi = \frac{\left(\frac{f}{a}\right)}{\left(\frac{f}{a}\right)_{stoich}}.$$

The stagnation temperature rise ratio $K$ may also be written as follows:

$$K = \frac{HV_{eff}}{HV} = 1 - \frac{1}{\left(\frac{f}{a}\right)_{stoich}}\left(\frac{f}{a}\right) = 1 - \phi.$$

For a typical hydrocarbon-fueled jet engine, the stoichiometric fuel to air ratio may be taken as $(f/a)_{stoich} = 0.0667$, as described in detail in Chapter 4, which leads to the result that

$$HV_{eff} = \left[1 - 15\left(\frac{f}{a}\right)\right]HV. \quad (3.25)$$

**Table 3.21** Afterburner Fuel–Air Ratio for Various Compressor Pressure Ratios and Turbine Inlet Stagnation Temperatures in $M = 2$ Cruise at 15 km Altitude

| $p_{t3}/p_{t2}$ | 1200K | 1300K | 1400K | 1500K | 1600K | 1700K |
|---|---|---|---|---|---|---|
| 5 | 0.00698 | 0.00736 | 0.00779 | 0.00794 | 0.00882 | 0.00945 |
| 10 | 0.01043 | 0.01097 | 0.01157 | 0.01181 | 0.01299 | 0.01384 |
| 20 | 0.01409 | 0.01477 | 0.01551 | 0.01590 | 0.01726 | 0.01829 |
| 30 | 0.01630 | 0.01705 | 0.01787 | 0.01836 | 0.01978 | 0.02089 |

The energy balance in the afterburner follows that for the combustor as given by Equation (3.11) and leads to the following equation for the fuel to air ratio required to achieve the desired afterburner stagnation temperature $T_{t5AB} = T_{t6}$:

$$\left(\frac{f}{a}\right)_{AB} = \left(\frac{\dot{m}_f}{\dot{m}_0}\right)_{AB} = \frac{\left[1+\left(\frac{f}{a}\right)\right](T_{t6} - T_{t5})}{43.4 \times 10^3\left[1 - 15\left(\frac{f}{a}\right)\right] - T_{t6}}. \tag{3.26}$$

Results for the fuel to (inlet) air ratio are given in Table 3.21.

### 3.6.6 Nozzle flow, stations 5b-7

The nozzle considered for supersonic cruise must be of the variable-area converging–diverging duct type so that it may be considered to operate in the matched pressure mode $p_7 = p_0$, thereby assuring the maximum thrust achievable. Because the stagnation pressure is constant throughout the nozzle, $p_{t5AB} = p_{t6} = p_{t7}$. Keeping $\gamma = 1.4$ for this ideal study, we may find $M_7$ using the isentropic relation of Equation (3.11); results are shown in Table 3.22.

Note that the exit Mach numbers lie in the range $2.56 < M_7 < 3.13$, which require area ratios in the range $2.8 < A_7/A_6 < 4.8$ in order to produce the matched exit flow. The exit velocity $V_7$ in the matched case may be found by first using the isentropic relation Equation (3.6) to find $T_7$ from $M_7$ and $T_{t7} = T_{t5AB}$ and then using the Mach number definition $M_7 = V_7/a_7$.

**Table 3.22** Matched Nozzle Exit Mach Numbers for Various Compressor Pressure Ratios and Turbine Inlet Stagnation Temperatures in $M = 2$ Cruise at 15 km Altitude

| $p_{t3}/p_{t2}$ | 1200K | 1300K | 1400K | 1500K | 1600K | 1700K |
|---|---|---|---|---|---|---|
| 5 | 2.559 | 2.600 | 2.634 | 2.663 | 2.689 | 2.711 |
| 10 | 2.667 | 2.742 | 2.805 | 2.858 | 2.904 | 2.944 |
| 20 | 2.615 | 2.749 | 2.859 | 2.952 | 3.030 | 3.098 |
| 30 | 2.458 | 2.651 | 2.805 | 2.932 | 3.039 | 3.131 |

### 3.6.7 Turbojet thrust and fuel efficiency in supersonic cruise

Using all the information gathered thus far, the specific thrust in the cruise mode for the ideal afterburning turbojet engine may be calculated using Equation (3.20) modified to account for the fuel burned in the afterburner:

$$\frac{F_n}{\dot{m}_0} = \left[1 + \left(\frac{f}{a}\right) + \left(\frac{f}{a}\right)_{AB}\right] V_7 \left[1 + \frac{1}{\gamma M_7^2}\left(1 - \frac{p_0}{p_7}\right)\right] - V_0. \tag{3.27}$$

Results are illustrated in Figure 3.15, and a comparison of the improvement possible with afterburning is shown in Figure 3.16. It is clear from Figure 3.16 that afterburning is crucial to good performance at supersonic Mach numbers. At reasonably high compressor pressure ratios and turbine inlet temperatures, the increase in specific thrust can be considerable, on the order of 50% to 100%. Note that the specific thrust performance for the afterburning turbojet at $M_0 = 2$ is about the same as that for the subsonic turbojet at $M_0 = 0.8$. The specific fuel consumption for the afterburning turbojet is given by

$$c_j = 3600g \frac{\left(\frac{f}{a}\right) + \left(\frac{f}{a}\right)_{AB}}{\frac{F_n}{\dot{m}_0}}.$$

Results for $M = 2$ and 15 km altitude are shown in Figure 3.17, and a comparison to that for no afterburning is shown in Figure 3.18. The price for supersonic performance is fuel efficiency, and Figure 3.17 shows that the specific fuel consumption almost doubles as compared to the standard turbojet in subsonic cruise (Figure 3.11). This factor has delayed the introduction of supersonic commercial jets and hastened the retirement of the Concorde.

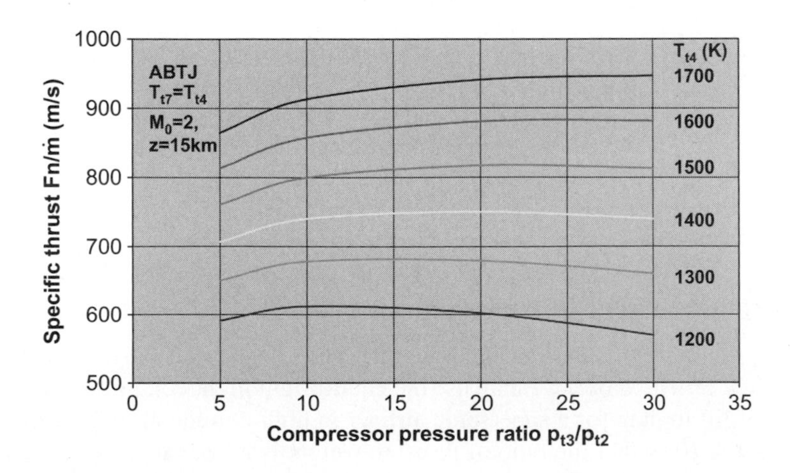


**FIGURE 3.15**

Specific fuel consumption as a function of compressor pressure ratio and turbine inlet stagnation temperature for an ideal turbojet with a variable geometry-matched nozzle in $M = 2$ cruise at 15 km altitude.

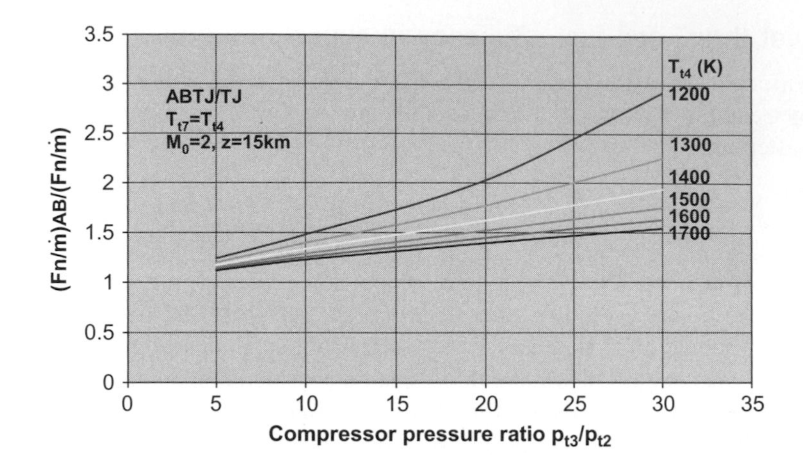


**FIGURE 3.16**

Ratio of specific thrust with and without afterburning as a function of compressor pressure ratio and turbine inlet temperature.

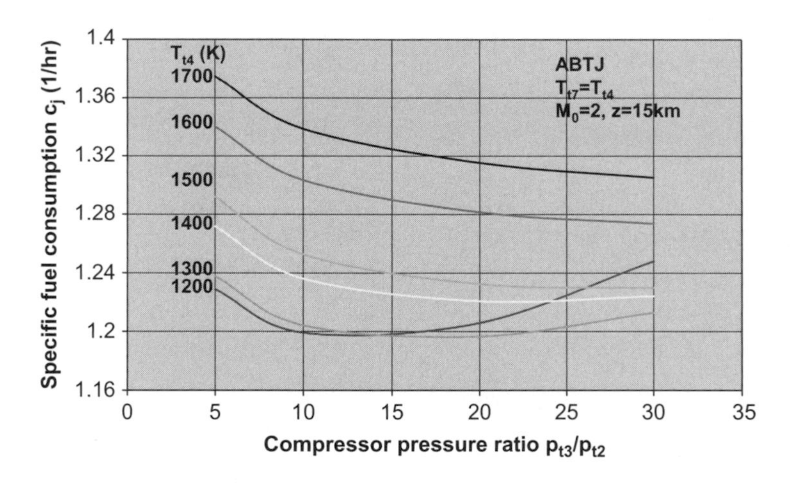


**FIGURE 3.17**

Specific fuel consumption for an afterburning turbojet at $M = 2$ and 15 km altitude.

## 3.6.8 Real turbojet engine in supersonic cruise

In cruising flight, equilibrium conditions require that the lift equal the weight $L = W$ and the thrust equal the drag $F_n = D$. Therefore, we may use the equilibrium flight condition again, that is, $L/D = W/F_n$. The ratio of lift to drag for a supersonic airliner in high-altitude $M = 2$ cruise is generally in the range $8 < L/D < 10$ so that the ratio of thrust to weight is in the range of $0.1 < F_n/W < 0.125$, considerably less than in take-off where acceleration requirements force high thrust levels. For the Rolls-Royce Olympus 593-powered Concorde SST, the weight at the beginning of cruise at 16.6 km altitude is around 1.6 MN so a total thrust level of about 180 kN is required, so each engine needs

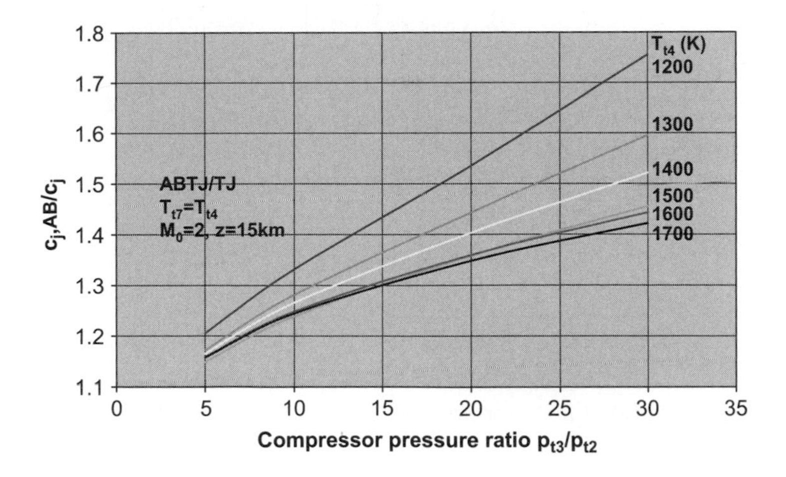


**FIGURE 3.18**

Ratio of specific fuel consumption with and without afterburning as a function of compressor pressure ratio and turbine inlet temperature.

$F_n = 45$ kN. In $M_0 = 2$ cruise, the Olympus 593, with afterburning, is said to produce 44.6 kN with a specific fuel consumption of 1.19 hr$^{-1}$ at an overall pressure ratio of 11.3 (Hill and Peterson, 1992). From Figure 3.17, we see that the specific fuel consumption $c_j = 1.22$ hr$^{-1}$ for the case similar to that of the Olympus 593, assuming the maximum turbine inlet temperature to be 1350K. This appears to be a reasonably accurate prediction given the fact that this is a cycle analysis for an ideal engine. The accuracy of this specific fuel consumption prediction suggests that the ideal analysis provides about the right ratio between $f/a$ and $F_n/\dot{m}$, yet the two terms individually may not be predicted as accurately. We shall see that the ideal analysis typically overpredicts the specific thrust, an understandable discrepancy considering that the efficiencies of the various components of the engine have not been taken into account and they affect the actual thrust levels achievable.

## 3.7 IDEAL RAMJET IN HIGH SUPERSONIC CRUISE IN THE STRATOSPHERE

Military aircraft have been able to fly supersonically since the 1950s, and substantial effort was placed on keeping aerodynamic drag down at supersonic speeds, as well as on developing reliable turbojets outfitted with afterburners to fill that need. However, for very high speeds corresponding to $M_0 = 4$ in the stratosphere at around 24 km altitude, where $T_0 = 226$K and $p_0 = 1.2$ kPa, the stagnation temperature at the end of the inlet $T_{t2} = T_{t0}$ would be given by Equation (3.8) as $T_{t2} = 950$K, whereas the stagnation pressure, for ideal isentropic compression, would be given by Equation (3.7) as $p_{t2} = 182$ kPa. These conditions are suitable in themselves for adding heat through combustion without the need for compressors and turbines. This is clearly the realm of the ramjet, a schematic diagram of which is shown in Figure 3.19.

Several basic cruise states are assumed for the current study under standard-day conditions at a specified altitude $z$ and Mach number $M_0$. These conditions are chosen so as to keep the dynamic pressure $q_0$ constant at about 34 kPa, a value that implies a constant ratio of wing loading to cruise lift

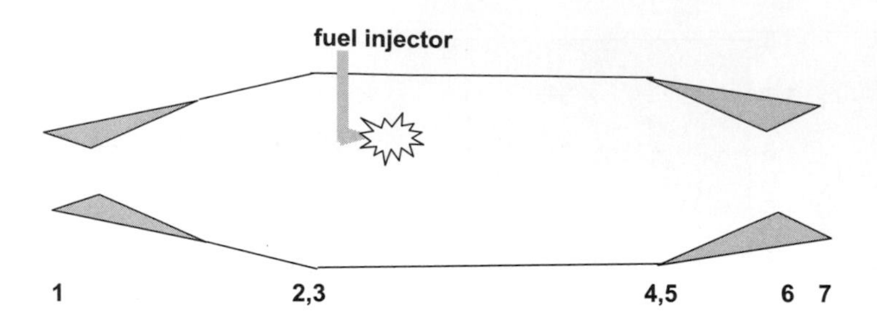


FIGURE 3.19

Schematic diagram of an ideal ramjet for high supersonic cruise.

**Table 3.23** Free Stream Conditions for Various Flight Mach Numbers and Altitudes for an Ideal Ramjet

| $M_0$ | z (km) | $p_0$ (kpa) | $T_0$ (K) | $q_0$ (kPa) | $p_{t0}$ (kPa) | $T_{t0}$ (K) |
|---|---|---|---|---|---|---|
| 2 | 15 | 12.1 | 217 | 33.9 | 95 | 391 |
| 3 | 20 | 5.52 | 217 | 34.8 | 203 | 608 |
| 4 | 24 | 2.97 | 221 | 33.3 | 451 | 928 |

coefficient, $(W/S)/c_L$, at a reasonable level for high-altitude airplane flight. The flight conditions considered are described in Table 3.23.

We will again proceed through the ramjet cycle sequentially, but now using various values of the combustor exit stagnation temperature as a parameter because there is no turbine. This approach provides an assessment of the ramjet engine, as well as taking into account the material limitations represented by the selected burner exit temperature. The ideal ramjet cycle looks the same as the ideal turbojet cycle portrayed in Figure 3.5.

### 3.7.1 Inlet flow, stations 0-2,3

The flight Mach numbers considered are high enough to make it impractical to consider the inlet of the ideal ramjet as operating isentropically, even for an ideal analysis. The stagnation values of pressure and temperature in the free stream, that is, $p_{t0}$ and $T_{t0}$, may be found using Equations (3.7) and (3.8) for the given supersonic cruise conditions; they are shown in Table 3.23. As is shown in Chapter 6, inlet designs for $M \geq 3$ become quite complicated if pressure recoveries of around 90% are desired. For comparison purposes, we may use the military specification for standard pressure recovery (MIL-E-5008B, 1959) in the range $1 \leq M_0 \leq 5$, which is given by

$$\frac{p_{t2}}{p_{t0}} = 1 - 0.075(M_0 - 1)^{1.35}.$$

### 3.7.2 Combustor flow, stations 2,3-4,5

Using Equation (3.12) again we may determine the fuel–air ratio that will yield the selected limiting value of $T_{t4}$ from the suggested range of $1200\text{K} < T_{t4} < 1700\text{K}$. Results are shown in Table 3.24 where

**Table 3.24** Fuel–Air Ratio for Various Flight Mach Numbers $M_0$ and Maximum Stagnation Temperatures $T_{t4}$ (K) for an Ideal Ramjet

| $M_0$ | 1200K | 1300K | 1400K | 1500K | 1600K | 1700K |
|---|---|---|---|---|---|---|
| 2 | 0.01918 | 0.02160 | 0.02403 | 0.02648 | 0.02893 | 0.03140 |
| 3 | 0.01404 | 0.01645 | 0.01887 | 0.02130 | 0.02374 | 0.02620 |
| 4 | 0.00644 | 0.008831 | 0.01123 | 0.01365 | 0.01607 | 0.01851 |

it is seen once more that the higher the flight Mach number, the lower the fuel–air ratio required. Note that the stagnation temperature $T_{t3}$ in supersonic cruise is higher than in subsonic cruise. As a result, the fuel–air ratio is lower in supersonic cruise because less heat can be added to get to the selected limiting stagnation temperature $T_{t5} = T_{t4}$.

### 3.7.3 Nozzle flow, stations 4,5-7

The nozzle considered for supersonic cruise must be of the variable-area converging–diverging duct type so that it may be considered to operate in the matched pressure mode $p_7 = p_0$, thereby assuring the maximum thrust achievable. Because the stagnation pressure is constant throughout the nozzle, $p_{t5b} = p_{t6} = p_{t7}$. Keeping $\gamma = 1.4$ for this ideal study, we may find $M_7$ by using the isentropic relation of Equation (3.11); results are shown in Table 3.25.

Note that the exit Mach number $M_7$ almost equals the flight Mach number. This is a consequence of the adiabatic compression in the inlet that keeps $p_{t2} < p_{t0}$. Then, because combustion takes place at constant pressure, the stagnation pressure in the nozzle is also the same as $p_{t2}$. Finally, because we are considering a matched nozzle $p_7 = p_0$, the ratio $p_7/p_{t7} < p_0/p_{t0}$ and thus $M_7 < M_0$. Of course, the exit temperature $T_7$ is greater than the free stream temperature $T_0$, as shown in Table 3.26, so the nozzle exit velocity is greater than the flight velocity, $V_7 > V_0$, as shown in Table 3.27.

### 3.7.4 Ramjet thrust and fuel efficiency in high supersonic cruise

Using all the information gathered thus far, the specific thrust and specific fuel consumption in the cruise mode for the ideal ramjet engine may be calculated using Equations (3.20) and (3.21); results are illustrated in Figures 3.19 and 3.20.

**Table 3.25** Matched Nozzle Exit Mach Numbers $M_7$ for Various Flight Mach Numbers $M_0$ and Maximum Stagnation Temperatures $T_{t4}$ (K) for an Ideal Ramjet

| $M_0$ | 1200K | 1300K | 1400K | 1500K | 1600K | 1700K |
|---|---|---|---|---|---|---|
| 2 | 1.951 | 1.951 | 1.951 | 1.951 | 1.951 | 1.951 |
| 3 | 2.861 | 2.862 | 2.862 | 2.862 | 2.862 | 2.862 |
| 4 | 3.708 | 3.708 | 3.708 | 3.708 | 3.708 | 3.708 |

**Table 3.26** Matched Nozzle Exit Temperatures $T_7$ (K) for Various Flight Mach Numbers $M_0$ and Maximum Stagnation Temperatures $T_{t4}$ (K) for an Ideal Ramjet

| $M_0$ | 1200K | 1300K | 1400K | 1500K | 1600K | 1700K |
|---|---|---|---|---|---|---|
| 2 | 681.3 | 738.0 | 794.8 | 851.6 | 908.4 | 965.2 |
| 3 | 454.9 | 492.8 | 530.7 | 568.6 | 606.6 | 644.5 |
| 4 | 320.0 | 346.7 | 373.3 | 400.0 | 426.7 | 453.3 |

**Table 3.27** Matched Nozzle Exit Velocity $V_7$ (m/s) for Various Flight Mach Numbers $M_0$ and Maximum Stagnation Temperatures $T_{t4}$ (K) for an Ideal Ramjet

| $M_0$ | 1200K | 1300K | 1400K | 1500K | 1600K | 1700K |
|---|---|---|---|---|---|---|
| 2 | 1021 | 1063 | 1103 | 1141 | 1179 | 1215 |
| 3 | 1223 | 1273 | 1321 | 1368 | 1413 | 1456 |
| 4 | 1330 | 1384 | 1436 | 1487 | 1535 | 1583 |

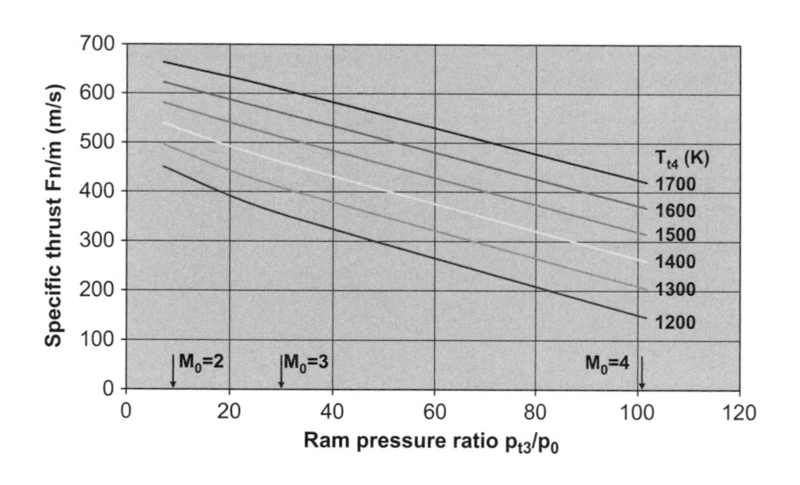


**FIGURE 3.20**

Specific thrust as a function of flight Mach number and maximum stagnation temperature for an ideal ramjet with a variable geometry-matched nozzle.

Results are presented in Figures 3.20 to 3.22 in terms of the ram pressure ratio $p_{t3}/p_0$, which provides an idea of the pressure rise associated with the high-speed ram effect. The flight Mach numbers considered are also located in the figure. Note that the net specific thrust for the ideal ramjet falls off with Mach number and the levels are not very high compared to the afterburning turbojet at lower $M_0$ and altitude. This drop-off in net specific thrust seen in Figure 3.20 results in an increase in specific fuel consumption from a minimum around $M_0 = 3.25$, as can be seen in Figure 3.21.

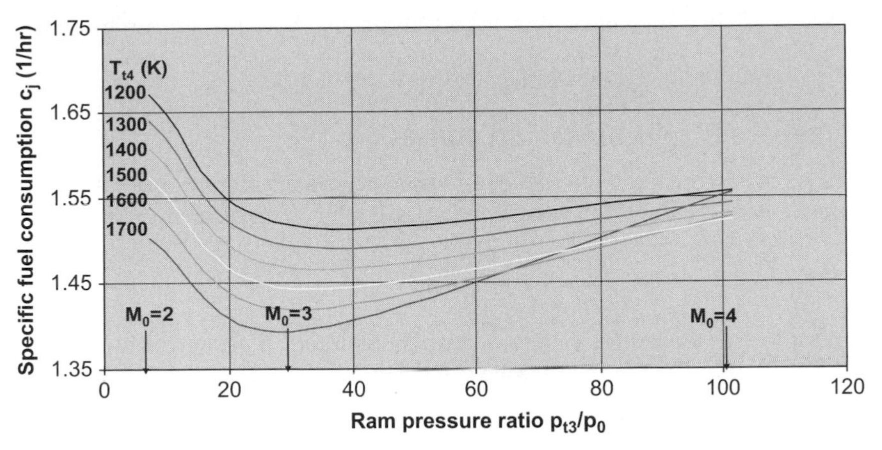


**FIGURE 3.21**

Specific fuel consumption as a function of flight Mach number and maximum stagnation temperature for an ideal ramjet with a variable geometry-matched nozzle.

Interestingly, the quoted supersonic cruise speed for the SR-71 is $M_0 = 3.2$ at an altitude of around 26 km altitude, close to the values considered here. It is often noted that in cruise the SR-71's J-58 engines are operating mainly as ramjets.

It is often useful to consider fuel efficiency in terms of specific impulse $I_{sp} = 3600/c_j$, which has the units of seconds. The variation of specific impulse for the ideal ramjets considered is shown as a function of flight Mach number in Figure 3.22. The curve labeled practical ramjets is based on results

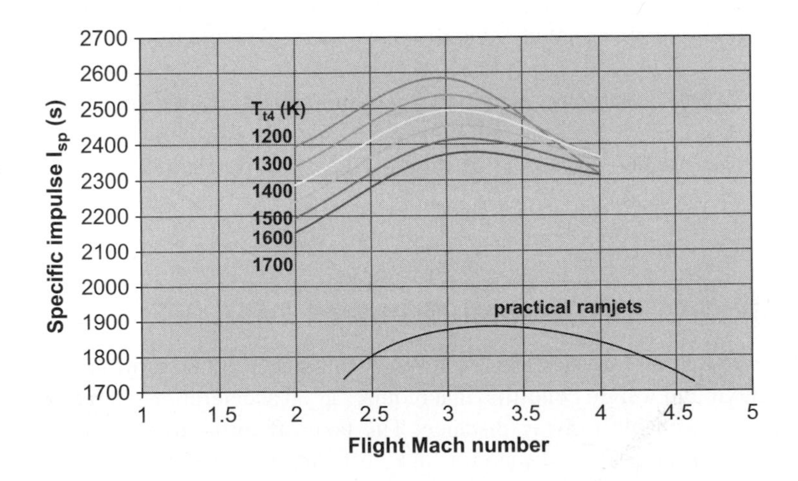


**FIGURE 3.22**

Variation of specific impulse $I_{sp}$ as a function of $M_0$ for the ideal ramjets considered. Practical ramjet data from Segal (2009).

presented in Segal (2009). It is clear that the ideal results show the proper trend, but are quite optimistic.

### 3.7.5 Real ramjet in high supersonic cruise

As shown previously, in cruising flight, the equilibrium conditions require that lift equals the weight $L = W$ and thrust equals the drag $F_n = D$. Therefore, we may again write

$$\frac{L}{D} = \frac{W}{F_n}.$$

The ratio of lift to drag for a supersonic vehicle in high-altitude high supersonic cruise is discussed in Appendix F. For the Pratt & Whitney J58-powered SR-71, the weight at the beginning of $M_0 = 3.0$ cruise at 20 km altitude is around 400 kN so for $L/D \cong 5$ a total thrust level of about 80 kN is required so that each of the two engines needs $F_n = 40$ kN.

The exhaust system of the SR-71 incorporates Hastelloy-X, which can withstand temperatures up to around 1500K. Thus for the $M_0 = 3.0$ case with 1500K assumed for the burner exit temperature, we find $F_n/\dot{m}_0 = 500$ m/s, approximately. Then the mass flow required by the engine is about $\dot{m} = 80$ kg/s. The area of free streamtube captured by engine $A_0$ may be found from the conservation of mass equation

$$A_0 = \frac{\dot{m}_0}{\rho_0 V_0}.$$

For $M_0 = 3$ cruise at 20 km altitude, the captured streamtube mass flux is $\rho_0 V_0 = 78.7$ kg/m$^2$-s and the corresponding captured streamtube area $A_0 \cong 1.02$ m$^2$. At supersonic speed, the maximum free streamtube capture area $A_0$ is never larger than the inlet entry area $A_1$. As shown in Chapter 6, at the cruise speed of $M_0 = 3$, the maximum area ratio $A_0/A_1 = 1$. Thus the required area $A_1 = 1.02$ m$^2$ and the inlet diameter is then about $D_1 = 1.14$ m. The inlet diameter may be scaled from available drawings of the SR-71, resulting in an estimated value of about 1.2 m, which agrees quite well with the ideal result presented.

From Figure 3.21 we see that the specific fuel consumption $c_j = 1.45$ hr$^{-1}$ for the case similar to that of the J58, which, at high Mach number, operates mainly as a ramjet utilizing the air bypassed around the compressor. Available information suggests that the $c_j$ for the J58 lies in the range 0.8 dry to 1.9 under afterburner operation.

## 3.8 IDEAL TURBOFAN IN MAXIMUM POWER TAKE-OFF

As mentioned previously, aircraft designs are strongly influenced by take-off considerations because this represents a maximum weight condition that requires good acceleration characteristics to reach the take-off speed in a reasonable runway distance. The take-off thrust-to-weight ratio influences the choice of engine for an aircraft, and therefore take-off thrust capability is a crucial metric for jet engines. As discussed in Chapter 1, the thrust and resulting acceleration characteristics of a propeller-driven aircraft are superior to those of a turbojet-powered aircraft. A compromise between the two propulsion mechanisms was achieved by introduction of the turbofan engine, which divides its thrust

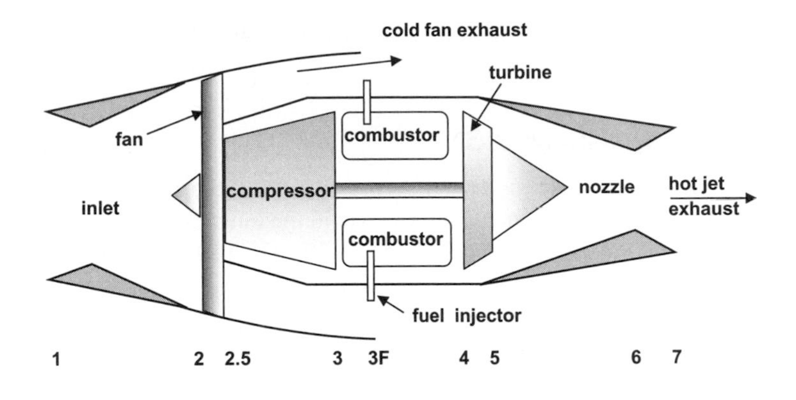


**FIGURE 3.23**

Schematic diagram of a turbofan engine.

production between propeller-like and turbojet-like actions, as illustrated in Figure 3.23. The pacing factor for turbofan engine operation is still the turbine inlet stagnation temperature $T_{t4}$ because of thermal limitations on turbine materials. Current engines operate in the range $1200\text{K} < T_{t4} < 1700\text{K}$, although materials and cooling techniques permitting higher temperatures are being sought constantly. The basic take-off state is standard day sea-level conditions, that is, altitude $z = 0$, pressure $p_0 = 101.3$ kPa, temperature $T_0 = 288\text{K}$, and speed of sound $a_0 = 340.3$ m/s. The typical take-off flight condition is taken here as $M_0 = 0$, that is, the static thrust condition. These are the same conditions discussed in Section 3.3 where the ideal turbojet engine was considered. We will proceed through the cycle sequentially using various values of the turbine inlet stagnation temperature.

### 3.8.1 Inlet flow, stations 0-2

The inlet is considered to operate isentropically so that stagnation values of pressure and temperature delivered to the compressor and fan faces are equal to those in the free stream, that is, $p_{t2} = p_{t0}$ and $T_{t2} = T_{t0}$. The isentropic flow relations were given in Equations (3.5) and (3.6) as follows:

$$\frac{p_t}{p} = \left(1 + \frac{\gamma - 1}{2} M^2\right)^{\frac{\gamma}{\gamma - 1}}$$

$$\frac{T_t}{T} = 1 + \frac{\gamma - 1}{2} M^2.$$

For the free stream station, with $\gamma = 1.4$, as shown in Equations (3.7) and (3.8), these become

$$\frac{p_{t0}}{p_0} = \left(1 + \frac{M_0^2}{5}\right)^{3.5}$$

$$\frac{T_{t0}}{T_0} = 1 + \frac{M^2}{5}.$$

Using the given take-off conditions, the stagnation conditions at station 2 are calculated readily to be $p_{t2} = p_{t0} = 101.3$ kPa and $T_{t2} = T_{t0} = 288$K.

### 3.8.2 Compressor flow, stations 2-3

The compressor provides an increase in stagnation pressure from the inlet value determined by the compressor pressure ratio $p_{t3}/p_{t2}$. Current engines operate in the range $5 < p_{t3}/p_{t2} < 30$, and we use the following values for comparison purposes: $p_{t3}/p_{t2} = 5$, 10, 20, and 30. The compressor is assumed to be ideal so that it provides isentropic compression and therefore permits calculation of the temperature ratio from the basic isentropic relation of Equation (3.9)

$$\frac{T_{t3}}{T_{t2}} = \left(\frac{p_{t3}}{p_{t2}}\right)^{\frac{\gamma-1}{\gamma}}.$$

Then, using the conditions determined thus far, we find the compressor exit stagnation temperature, in degrees K, to be given by Equation (3.10) as

$$T_{t3} = 288\left(\frac{p_{t3}}{p_{t2}}\right)^{0.286}.$$

The stagnation pressure and temperature at the compressor exit for the various compressor pressure ratios are given in Table 3.28.

It should be noted that the results in Table 3.28 are identical to those in Table 3.1 for the compressor of the pure turbojet, as we are considering exactly the same compressor pressure ratio in both cases.

### 3.8.3 Fan flow, stations 2-3F

The fan provides an increase in stagnation pressure from the inlet value as given by the fan pressure ratio $p_{t3F}/p_{t2}$. Current engines operate in the range $1.5 < p_{t3F}/p_{t2} < 3$, and in the subsequent calculations we will use only the lower value of 1.5, which is typical of commercial engines, for comparison purposes. The fan is assumed to be ideal so that it provides isentropic compression and therefore permits calculation of the stagnation temperature ratio from the basic isentropic relation

**Table 3.28** Stagnation Pressure and Temperature at Compressor Exit at Take-Off for Various Compressor Pressure Ratios

| $p_{t3}/p_{t2}$ | $p_{t3}$ (MPa) | $T_{t3}$ (K) |
|---|---|---|
| 5 | 0.5065 | 457 |
| 10 | 1.013 | 680 |
| 20 | 2.026 | 679 |
| 30 | 3.039 | 762 |

$$\frac{T_{t3F}}{T_{t2}} = \left(\frac{p_{t3F}}{p_{t2}}\right)^{\frac{\gamma-1}{\gamma}}.$$

Then, using the conditions determined thus far, we find the compressor exit stagnation temperature, in degrees K, to be given by

$$T_{t3F} = 288\left(\frac{p_{t3F}}{p_{t2}}\right)^{0.286}.$$

The stagnation pressure and temperature at the fan exit for the various compressor pressure ratios are given in Table 3.28. The fan duct is generally a simple converging duct so that the fan exit Mach number $M_{3F} \leq 1$ with the upper limit indicative of a choked fan exit. In the case where the fan exit is subcritical, $M_{3F} < 1$, the exit pressure is equal to that of the local ambient pressure, which we will consider to be the undisturbed free stream pressure $p_0$. However, if the fan exit is critical, $M_{3F} = 1$, then the exit pressure may be higher than the ambient pressure and is determined, using the isentropic relation of Equation (3.5), to be

$$\frac{p_{3F}}{p_{t3F}} = \left(1 + \frac{\gamma - 1}{2}M_{3F}^2\right)^{-\frac{\gamma}{\gamma-1}}.$$

For choked flow at the fan exit $M_{3F} = 1$ and for $\gamma = 1.4$, we have $p_{3F} = 0.5283p_{t3F}$. Note that in Table 3.29 the lowest fan pressure ratio produces a subcritical exit Mach number, while the two higher fan pressure ratios would produce a supercritical exit Mach number and therefore the exit is instead considered to be choked, which results in an attendant increase in fan exit plane static pressure $p_{3F}$. The static temperature may be obtained from the isentropic relation of Equation (3.6) and is given by

$$\frac{T_{3F}}{T_{t3F}} = \left(1 + \frac{\gamma - 1}{2}M_{3F}^2\right)^{-1}.$$

The fan exit velocity may be found from the fan exit Mach number using

$$V_{3F} = M_{3F}\sqrt{\gamma R T_{3F}}.$$

The values for static pressure and temperature, as well as velocity, at the fan exit are also listed in Table 3.29.

**Table 3.29** Stagnation Pressure and Temperature, Mach Number, Velocity, and Static Pressure and Temperature at Fan Exit at Take-Off for Various Fan Pressure Ratios

| $p_{t3F}/p_{t2}$ | $p_{t3F}$ (kPa) | $T_{t3F}$ (K) | $M_{3F}$ | $p_{3F}$ (kPa) | $T_{3F}$ (K) | $V_{3F}$ (m/s) |
|---|---|---|---|---|---|---|
| 1.5 | 152 | 323 | 0.784 | 101 | 288 | 267 |
| 2 | 203 | 351 | 1.000 | 107 | 293 | 343 |
| 2.5 | 253 | 374 | 1.000 | 134 | 312 | 354 |

### 3.8.4 Combustor flow, stations 3-4

Flow passing through the combustor is the same for the turbojet and turbofan cases because we are considering the same compressor pressure ratios and the same turbine entrance total temperatures. Therefore, results for the fuel air ratio as a function of compressor pressure ratio and turbine inlet stagnation temperature, as given in Table 3.2, are the same for both turbojet and turbofan cases. The difference between the cases will be determined by the fan pressure ratio and the fan bypass ratio, as shown in the next section. For constant pressure combustion at low subsonic flow in the combustor, we again assume that $p_{t4} = p_{t3}$, where $p_{t3}$ may be found in Table 3.28.

### 3.8.5 Turbine flow, stations 4-5

The turbine is part of the turbofan engine solely because it is needed to drive the compressor and the fan and as such must produce just enough power to do so. Thus there is a matching condition to be met: the turbine power produced must equal the sum of the required compressor and fan power, which may be written as follows:

$$\dot{m}_F c_p (T_{t3F} - T_{t2}) + \dot{m}_C c_p (T_{t3} - T_{t2}) = \dot{m}_c c_p (T_{t4} - T_{t5}).$$

Note that the turbine must produce the required power using just the core mass flow. In terms of the bypass ratio $\beta = \dot{m}_F / \dot{m}_C$, which was introduced in Chapter 1, this relation becomes

$$\beta c_p (T_{t3F} - T_{t2}) + c_p (T_{t3} - T_{t2}) = c_p (T_{t4} - T_{t5}).$$

With the ideal process imposed here and the conditions that $c_p$ is constant and $\gamma = 1.4$, we may put the temperatures in terms of the pressures in order to arrive at

$$\beta\left[\left(\frac{p_{t3F}}{p_{t2}}\right)^{0.286} - 1\right] + \left[\left(\frac{p_{t3}}{p_{t2}}\right)^{0.286} - 1\right] = \frac{T_{t4}}{T_{t2}}\left[1 - \left(\frac{p_{t4}}{p_{t5}}\right)^{-0.286}\right].$$

Because the bypass ratio, fan and compressor ratio, and turbine inlet temperature are all independent parameters, we may solve for the required turbine pressure ratio in terms of these parameters as follows:

$$\frac{p_{t4}}{p_{t5}} = \left[1 - \frac{T_{t2}}{T_{t4}}\left\{\beta\left(\frac{p_{t3F}}{p_{t2}}\right)^{0.286} + \left(\frac{p_{t3}}{p_{t2}}\right)^{0.286} - (\beta + 1)\right\}\right]^{-3.5}. \qquad (3.28)$$

Therefore, for every selected value of the turbine inlet temperature, the required turbine pressure ratio depends on three independent parameters so that a large number of solutions are possible. Recall that for constant pressure combustion we have $p_{t4} = p_{t3}$, and $p_{t3}$ may be found in Table 3.28. Using these values and Equation (3.28), we find the results for $p_{t5}$ as shown in Table 3.30 for the various limiting turbine inlet stagnation temperatures selected for various bypass and compressor pressure ratios.

The turbine exit stagnation temperature may then be found from the isentropic relation

$$T_{t5} = T_{t4}\left(\frac{p_{t4}}{p_{t5}}\right)^{-\frac{\gamma-1}{\gamma}}.$$

Results for $T_{t5}$ for the different compressor pressure ratios and limiting turbine inlet stagnation temperatures are shown in Table 3.31.

**Table 3.30** Turbine Exit Stagnation Pressures (kPa) for Various Bypass Ratios, Compressor Pressure Ratios, and Turbine Inlet Stagnation Temperatures at Take-Off

| $\beta = 3$ | $p_{t3}/p_{t2}$ | $p_{t4}$ (kPa) | 1200K | 1300K | 1400K | 1500K | 1600K | 1700K |
|---|---|---|---|---|---|---|---|---|
| | 5 | 506.5 | 204.0 | 220.8 | 235.9 | 249.6 | 262.1 | 273.4 |
| | 10 | 1013 | 273.4 | 308.2 | 335.2 | 370.5 | 398.2 | 423.8 |
| | 20 | 2026 | 312.3 | 375.9 | 426.8 | 495.9 | 551.7 | 604.6 |
| | 30 | 3039 | 301.2 | 384.1 | 452.9 | 549.0 | 628.6 | 705.3 |
| $\beta = 6$ | $p_{t3}/p_{t2}$ | $p_{t4}$ (kPa) | 1200K | 1300K | 1400K | 1500K | 1600K | 1700K |
| | 5 | 506.5 | 133.1 | 147.8 | 164.0 | 179.0 | 193.0 | 205.9 |
| | 10 | 1013 | 168.8 | 196.0 | 227.0 | 256.5 | 284.5 | 311.0 |
| | 20 | 2026 | 176.1 | 220.3 | 273.0 | 325.5 | 377.0 | 427.1 |
| | 30 | 3039 | 156.0 | 209.1 | 275.6 | 344.3 | 413.8 | 483.0 |
| $\beta = 9$ | $p_{t3}/p_{t2}$ | $p_{t4}$ (kPa) | 1200K | 1300K | 1400K | 1500K | 1600K | 1700K |
| | 5 | 506.5 | 81.9 | 95.4 | 110.8 | 125.5 | 139.5 | 152.7 |
| | 10 | 1013 | 96.5 | 119.3 | 146.4 | 173.2 | 199.3 | 224.7 |
| | 20 | 2026 | 88.7 | 121.7 | 163.5 | 207.3 | 252.0 | 296.8 |
| | 30 | 3039 | 69.3 | 105.4 | 154.4 | 208.4 | 265.8 | 325.0 |

**Table 3.31** Turbine Exit Stagnation Temperatures (K) for Various Bypass Ratios, Compressor Pressure Ratios, and Turbine Inlet Stagnation Temperatures at Take-Off

| $\beta = 3$ | $p_{t3}/p_{t2}$ | 1200K | 1300K | 1400K | 1500K | 1600K | 1700K |
|---|---|---|---|---|---|---|---|
| | 5 | 925 | 1025 | 1125 | 1225 | 1325 | 1425 |
| | 10 | 825 | 925 | 1020 | 1125 | 1225 | 1325 |
| | 20 | 703 | 803 | 897 | 1003 | 1103 | 1203 |
| | 30 | 620 | 720 | 812 | 920 | 1019 | 1119 |
| $\beta = 6$ | $p_{t3}/p_{t2}$ | 1200K | 1300K | 1400K | 1500K | 1600K | 1700K |
| | 5 | 819 | 914 | 1014 | 1114 | 1211 | 1314 |
| | 10 | 719 | 813 | 913 | 1013 | 1108 | 1213 |
| | 20 | 597 | 689 | 789 | 889 | 984 | 1089 |
| | 30 | 513 | 605 | 705 | 805 | 899 | 1005 |
| $\beta = 9$ | $p_{t3}/p_{t2}$ | 1200K | 1300K | 1400K | 1500K | 1600K | 1700K |
| | 5 | 713 | 806 | 906 | 1006 | 1106 | 1206 |
| | 10 | 612 | 705 | 805 | 905 | 1005 | 1105 |
| | 20 | 490 | 582 | 682 | 782 | 882 | 981 |
| | 30 | 407 | 497 | 597 | 697 | 797 | 897 |

**Table 3.32** Choked Nozzle Exit Pressure (kPa) for Various Bypass Ratios, Compressor Pressure Ratios, and Turbine Inlet Stagnation Temperatures at Take-Off

| $\beta = 3$ | $p_{t3}/p_{t2}$ | 1200K | 1300K | 1400K | 1500K | 1600K | 1700K |
|---|---|---|---|---|---|---|---|
| | 5 | 107.8 | 116.6 | 124.6 | 131.9 | 138.4 | 144.4 |
| | 10 | 144.4 | 162.8 | 177.1 | 195.7 | 210.4 | 223.9 |
| | 20 | 165.0 | 198.6 | 225.5 | 262.0 | 291.5 | 319.4 |
| | 30 | 159.1 | 202.9 | 239.3 | 290.1 | 332.1 | 372.6 |
| $\beta = 6$ | $p_{t3}/p_{t2}$ | 1200K | 1300K | 1400K | 1500K | 1600K | 1700K |
| | 5 | 70.3 | 78.1 | 86.7 | 94.6 | 101.9 | 108.8 |
| | 10 | 89.2 | 103.6 | 119.9 | 135.5 | 150.3 | 164.3 |
| | 20 | 93.0 | 116.4 | 144.2 | 171.9 | 199.2 | 225.6 |
| | 30 | 82.4 | 110.5 | 145.6 | 181.9 | 218.6 | 255.2 |
| $\beta = 9$ | $p_{t3}/p_{t2}$ | 1200K | 1300K | 1400K | 1500K | 1600K | 1700K |
| | 5 | 43.3 | 50.4 | 58.5 | 66.3 | 73.7 | 80.7 |
| | 10 | 51.0 | 63.0 | 77.3 | 91.5 | 105.3 | 118.7 |
| | 20 | 46.9 | 64.3 | 86.4 | 109.5 | 133.1 | 156.8 |
| | 30 | 36.6 | 55.7 | 81.6 | 110.1 | 140.4 | 171.7 |

### 3.8.6 Nozzle flow, stations 5-7

The nozzle for take-off will be of the simple converging duct type so that stations 6 and 7 are coincident. The maximum thrust for such a nozzle will be developed when the exit station is choked, that is, $M_7 = 1$. The nozzle is considered to operate ideally so that the stagnation pressure is constant throughout the nozzle, $p_{t5} = p_{t6} = p_{t7}$. Thus the pressure at exit station $p_7$ may be found using the isentropic flow relation of Equation (3.5) with the condition $M_7 = 1$ and values for the turbine exit stagnation pressure from Table 3.30. Results for $p_7$ are shown in Table 3.32. Note that the ratio of exit pressure to free stream pressure $p_7/p_0 > 1$, indicating that the jet exhaust is always underexpanded under take-off conditions. A variable area diverging section between stations 6 and 7 would permit the exhaust nozzle to get closer to matched conditions, and therefore greater thrust, but the exit flow would be supersonic, and noise considerations would assume greater importance.

In the same fashion, we may calculate nozzle exit temperature $T_7$ using the isentropic flow relation of Equation (3.6) with the condition $M_7 = 1$ and values for the turbine exit stagnation temperature from Table 3.31. Results for $T_7$ are shown in Table 3.33. The exit velocity can then be found from the relation $V = aM$, where the sound speed is given by Equation (3.17):

$$a_7 = \sqrt{\gamma R T_7} = 20\sqrt{T_7}.$$

Here $\gamma R = 1.4(287 \text{ m}^2/\text{s}^2\text{-K})$, and results for $V_7$ are shown in Table 3.34.

### 3.8.7 Turbofan thrust and fuel efficiency in take-off

The thrust of the turbofan is equal to the sum of the thrust of the fan and the nozzle, and the net thrust is given by

$$F_n = \dot{m}_7 V_7 - \dot{m}_C V_0 + A_7(p_7 - p_0) + \dot{m}_F V_{3F} - \dot{m}_F V_0 + A_{3F}(p_{3F} - p_0). \tag{3.29}$$

**Table 3.33** Choked Nozzle Exit Temperature (K) for Various Bypass Ratios, Compressor Pressure Ratios, and Turbine Inlet Stagnation Temperatures at Take-Off

| $\beta = 3$ | $p_{t3}/p_{t2}$ | 1200K | 1300K | 1400K | 1500K | 1600K | 1700K |
|---|---|---|---|---|---|---|---|
| | 5 | 771 | 854 | 938 | 1021 | 1104 | 1188 |
| | 10 | 688 | 771 | 850 | 938 | 1021 | 1104 |
| | 20 | 586 | 669 | 747 | 836 | 919 | 1002 |
| | 30 | 516 | 600 | 677 | 766 | 850 | 933 |
| $\beta = 6$ | $p_{t3}/p_{t2}$ | 1200K | 1300K | 1400K | 1500K | 1600K | 1700K |
| | 5 | 682 | 762 | 845 | 928 | 1009 | 1095 |
| | 10 | 599 | 677 | 761 | 844 | 924 | 1011 |
| | 20 | 497 | 574 | 658 | 741 | 820 | 908 |
| | 30 | 428 | 504 | 587 | 671 | 749 | 837 |
| $\beta = 9$ | $p_{t3}/p_{t2}$ | 1200K | 1300K | 1400K | 1500K | 1600K | 1700K |
| | 5 | 594 | 672 | 755 | 839 | 922 | 1005 |
| | 10 | 510 | 588 | 671 | 754 | 838 | 921 |
| | 20 | 409 | 485 | 568 | 651 | 735 | 818 |
| | 30 | 339 | 414 | 498 | 581 | 664 | 748 |

**Table 3.34** Choked Nozzle Exit Velocity (m/s) for Various Bypass Ratios, Compressor Pressure Ratios, and Turbine Inlet Stagnation Temperatures at Take-Off

| $\beta = 3$ | $p_{t3}/p_{t2}$ | 1200K | 1300K | 1400K | 1500K | 1600K | 1700K |
|---|---|---|---|---|---|---|---|
| | 5 | 557 | 586 | 614 | 640 | 666 | 691 |
| | 10 | 526 | 557 | 585 | 614 | 640 | 666 |
| | 20 | 485 | 519 | 548 | 580 | 608 | 635 |
| | 30 | 455 | 491 | 522 | 555 | 584 | 612 |
| $\beta = 6$ | $p_{t3}/p_{t2}$ | 1200K | 1300K | 1400K | 1500K | 1600K | 1700K |
| | 5 | 524 | 553 | 583 | 611 | 637 | 663 |
| | 10 | 491 | 522 | 553 | 582 | 609 | 637 |
| | 20 | 447 | 480 | 514 | 546 | 574 | 604 |
| | 30 | 415 | 450 | 486 | 519 | 549 | 580 |
| $\beta = 9$ | $p_{t3}/p_{t2}$ | 1200K | 1300K | 1400K | 1500K | 1600K | 1700K |
| | 5 | 488 | 520 | 551 | 581 | 609 | 636 |
| | 10 | 453 | 486 | 519 | 551 | 580 | 608 |
| | 20 | 405 | 441 | 478 | 512 | 543 | 573 |
| | 30 | 369 | 408 | 447 | 483 | 517 | 548 |

The specific net thrust may then be written in terms of the bypass ratio as follows:

$$\frac{F_n}{\dot{m}_0} = \left(1+\frac{f}{a}\right)\frac{V_7}{1+\beta}\left[1+\frac{1}{\gamma M_7^2}\left(1-\frac{p_0}{p_7}\right)\right] + \frac{\beta V_{3F}}{1+\beta}\left[1+\frac{1}{\gamma M_{3F}^2}\left(1-\frac{p_0}{p_{3F}}\right)\right] - V_0. \tag{3.30}$$

The specific net thrust is illustrated for three bypass ratios, $\beta = 3$, 6, and 9, in Figures 3.24 through 3.26. These results are given for one fan pressure ratio, FPR $= p_{t3F}/p_{t2} = 1.5$. Comparing these results to those for the pure turbojet in Figure 3.6 we see that the specific thrust of the turbofan is increasingly less as the bypass ratio increases. Because the turbojet may be considered a turbofan with $\beta = 0$, the increasing bypass ratio makes the engine operate more and more like a propeller, which, as recalled, produces thrust by accelerating a large mass of air a small amount. To produce a given amount of thrust with a turbofan will require a larger mass flow than a turbojet; therefore, the diameter of the turbofan will be increasingly larger than that of a turbojet as the bypass ratio increases.

The specific fuel consumption is given by

$$c_j = \frac{\dot{m}_f g}{F_n} = \frac{\dot{m}_f}{\dot{m}_0}\frac{g}{(F_n/\dot{m}_0)} = \frac{1}{1+\beta}\frac{(f/a)}{(F_n/\dot{m}_0)}. \tag{3.31}$$

Note that this is equivalent to the turbojet result in Equation (3.21) divided by $1+\beta$ so that when the bypass ratio is zero the turbojet result is recovered. The specific fuel consumption is illustrated for three bypass ratios, $\beta = 3$, 6, and 9, in Figures 3.27 through 3.29.

Comparing these results to those for the pure turbojet in Figure 3.7, we see that the specific fuel consumption of the turbofan is increasingly less as the bypass ratio increases. The turbofan operates more and more like a propeller, which, as recalled from Chapter 1, is more efficient than a pure jet engine. High bypass engines currently in use typically have a take-off, or static thrust, specific fuel

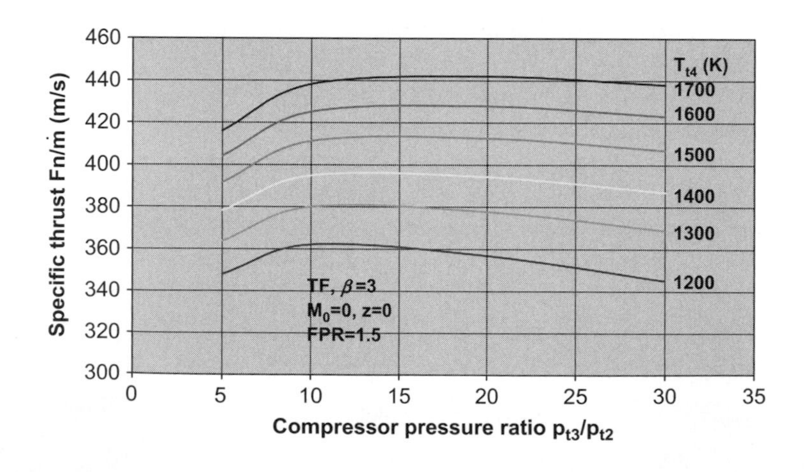


**FIGURE 3.24**

Specific thrust of a turbofan in take-off as a function of compressor pressure ratio for various turbine inlet temperatures, fan pressure ratio FPR $= 1.5$, and bypass ratio $\beta = 3$.

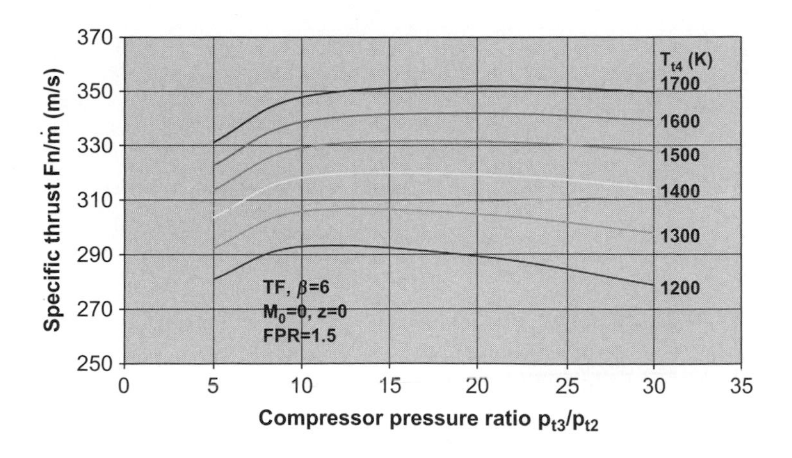


**FIGURE 3.25**

Specific thrust of a turbofan in take-off as a function of compressor pressure ratio for various turbine inlet temperatures, fan pressure ratio FPR = 1.5, and bypass ratio $\beta = 6$.

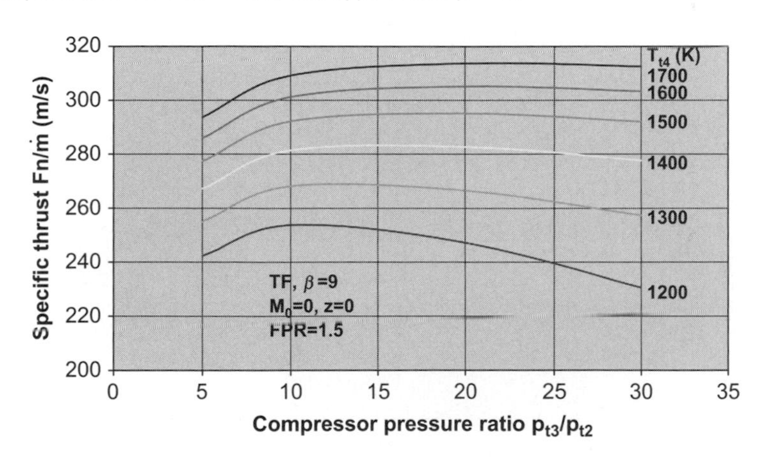


**FIGURE 3.26**

Specific thrust of a turbofan in take-off as a function of compressor pressure ratio for various turbine inlet temperatures, fan pressure ratio FPR = 1.5, and bypass ratio $\beta = 9$.

consumption of about 0.33 hr$^{-1}$ (Svoboda, 2000), which is consistent with results shown in Figure 3.29.

## 3.8.8 Real turbofan engine in take-off

Commercial jet airliners have a take-off thrust-to-weight ratio $(F_n/W)_{t\text{-}o}$ that is set by trying to satisfy the take-off and landing field requirements while meeting the general cruise zero-lift drag target. Generally, it lies in the range of $0.25 < (F_n/W)_{t\text{-}o} < 0.35$ (Torenbeek, 1982). The estimated take-off weight $W_{t\text{-}o}$ then fixes the number of engines and the take-off thrust required for each.

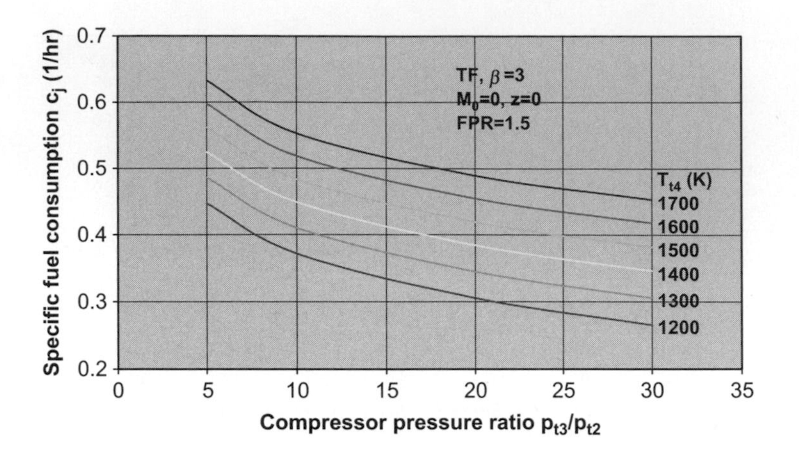


**FIGURE 3.27**

Specific fuel consumption of a turbofan in take-off as a function of compressor pressure ratio for various turbine inlet temperatures, fan pressure ratio FPR = 1.5, and $\beta = 3$.

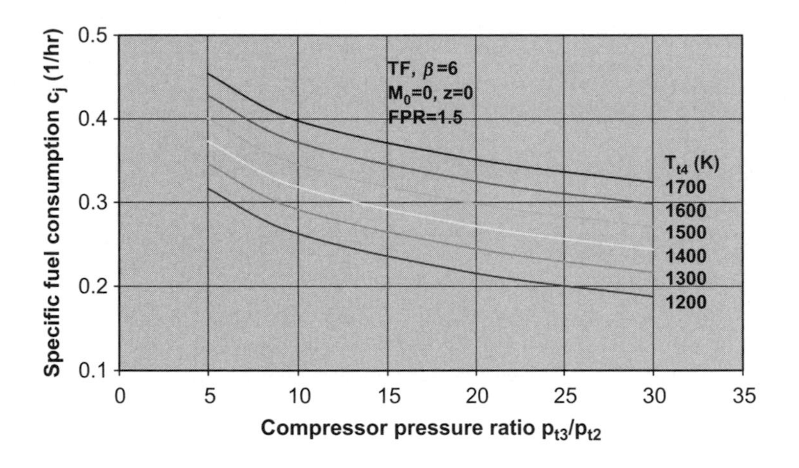


**FIGURE 3.28**

Specific fuel consumption of a turbofan in take-off as a function of compressor pressure ratio for various turbine inlet temperatures, fan pressure ratio FPR = 1.5, and $\beta = 6$.

For comparison, we may consider the TF39, a popular early military turbofan engine manufactured by General Electric; the commercial version was known as the CF6-50 and powered early wide-body commercial airliners such as the Douglas DC-10 and the Lockheed 1011. The CF6-50 had a thrust capability in the range of 227 to 240 kN, an overall compressor pressure ratio of 29.4, a bypass ratio $\beta = 4.35$, and a turbine inlet temperature of 1633K (*Aviation Week Space Technology*, August 2, 1971). Cross-plotting data from Figures 3.28 and 3.29, we estimate the take-off specific fuel consumption to be $c_j = 0.37\ \text{hr}^{-1}$, whereas the actual engine, at full static thrust, has $c_j = 0.38$ to $0.39\ \text{hr}^{-1}$. The

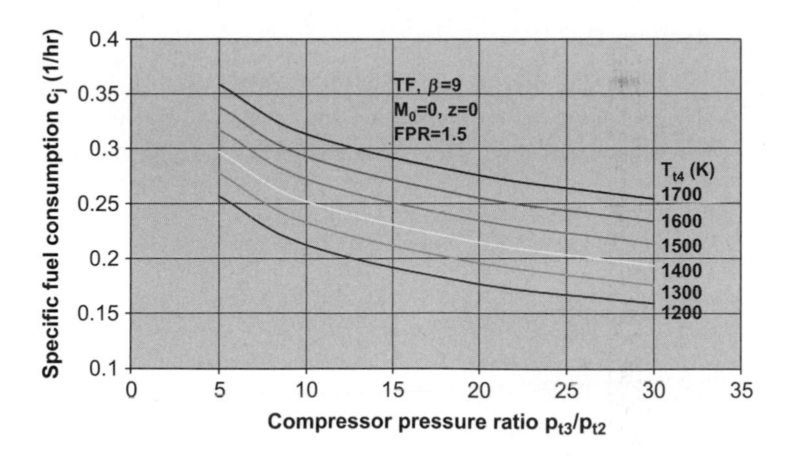


**FIGURE 3.29**

Specific fuel consumption of a turbofan in take-off as a function of compressor pressure ratio for various turbine inlet temperatures, fan pressure ratio FPR = 1.5, and $\beta = 9$.

estimated specific fuel consumption is about 5% less than that of the actual engine, making this a reasonably accurate prediction, given that this is a cycle analysis for an ideal engine. This result suggests that the ideal analysis provides about the right ratio of *f/a* compared to $F_n/\dot{m}_0$, but each of these two terms may not be predicted as accurately. For example, the specific thrust for this case would be estimated to be about 388 m/s, at the least, while reported engine data (*Aviation Week Space Technology*, August 2, 1971) indicate a value of about 341 m/s. Thus, the ideal analysis overpredicts the specific thrust by about 14%, an understandable discrepancy considering that the efficiencies of the various components of the engine have not been taken into account and affect the actual thrust levels achievable.

## 3.9 IDEAL TURBOFAN IN HIGH SUBSONIC CRUISE IN THE STRATOSPHERE

Later commercial jetliners, such as the wide-bodied Boeing B-747 and the Douglas DC-10, used larger diameter turbofan engines to cruise in the lower stratosphere, $z = 10$ km, at a high subsonic Mach number $M_0 = 0.80$. Although substantial effort was placed on keeping aerodynamic drag down at these almost sonic speeds, there was also a requirement for reducing the fuel consumption, and the turbofan engine filled that need. As mentioned in the previous section, current engines operate in the range $1200\text{K} < T_{t4} < 1700\text{K}$, although materials and cooling techniques permitting higher temperatures are being sought constantly. The basic cruise state is standard-day stratospheric conditions, that is, altitude $z = 10$ km, pressure $p_0 = 26.5$ kPa, temperature $T_0 = 223\text{K}$, and speed of sound $a_0 = 300$ m/s. The typical cruise flight condition is taken here as $M_0 = 0.80$ ($V_0 = 240\text{m/s}$), although current aircraft cruise in the range $0.75 < M_0 < 0.85$. We proceed through the cycle sequentially using various values of the turbine inlet stagnation temperature as in the case for take-off.

### 3.9.1 Inlet flow, stations 0-2

Because the flight Mach number is subsonic, the inlet of the ideal turbojet is considered to operate isentropically so that the stagnation values of pressure and temperature delivered to the compressor face are equal to those in the free stream, that is, $p_{t2} = p_{t0}$ and $T_{t2} = T_{t0}$. Again using Equations (3.7) and (3.8) for the given cruise conditions, the stagnation conditions at station 2 are calculated readily to be $p_{t2} = p_{t0} = 40.4$ kPa and $T_{t2} = T_{t0} = 252$K.

### 3.9.2 Compressor flow, stations 2-3

The compressor provides an increase in stagnation pressure from the inlet value determined by the compressor pressure ratio $p_{t3}/p_{t2}$. We continue to use typical values for comparison purposes: $p_{t3}/p_{t2} = 5$, 10, 20, and 30. The compressor is assumed to be ideal so that it provides isentropic compression and therefore permits calculation of the temperature ratio from the basic isentropic relation. Using the conditions determined thus far, we find the compressor exit stagnation temperature from Equation (3.9) to be given by

$$T_{t3} = 252\left(\frac{p_{t3}}{p_{t2}}\right)^{0.286}.$$

The stagnation pressure and temperature at the compressor exit for the various compressor pressure ratios are the same as those for the turbojet case given in Table 3.8 and are reproduced here as Table 3.35. Upon comparison of Tables 3.28 and 3.35, it is clear that the stagnation pressure being delivered to the combustor is much lower in the cruise state than at take-off because the ambient atmospheric pressure is only one-quarter that at sea level. The stagnation temperatures are also lower for the same reason. It should be noted that results in Table 3.35 are identical to those in Table 3.8 for the compressor of the pure turbojet, as we are considering exactly the same compressor pressure ratio in both cases.

### 3.9.3 Fan flow, stations 2-3F

The fan provides an increase in stagnation pressure from the inlet value as given by the fan pressure ratio $p_{t3F}/p_{t2}$. Current engines operate in the range $1.5 < p_{t3F}/p_{t2} < 3$, and again we use only the lower value of 1.5, which is typical of commercial engines, for comparison purposes. The fan is assumed to be ideal so that it provides isentropic compression and therefore permits calculation of the stagnation temperature ratio from the basic isentropic relation

**Table 3.35** Stagnation Pressure and Temperature at the Turbofan Compressor Exit in $M = 0.8$ Cruise at 10 km Altitude

| Pressure ratio | 5 | 10 | 20 | 30 |
|---|---|---|---|---|
| $p_{t3}$ (MPa) | 0.202 | 0.404 | 0.808 | 1.211 |
| $T_{t3}$ (K) | 399 | 486 | 593 | 666 |

$$\frac{T_{t3F}}{T_{t2}} = \left(\frac{p_{t3F}}{p_{t2}}\right)^{\frac{\gamma-1}{\gamma}}.$$

Then, using the conditions determined thus far, we find the compressor exit stagnation temperature, in degrees K, to be given by

$$T_{t3F} = 252\left(\frac{p_{t3F}}{p_{t2}}\right)^{0.286}.$$

The stagnation pressure and temperature at the fan exit for the various compressor pressure ratios are given in Table 3.35. The fan duct is generally a simple converging duct so that the fan exit Mach number $M_{3F} \leq 1$ with the upper limit indicative of a choked fan exit. In the case where the fan exit is subcritical, $M_{3F} < 1$, the exit pressure is equal to that of the local ambient pressure, which we consider to be the undisturbed free stream pressure $p_0$. However, if the fan exit is critical, $M_{3F} = 1$, then the exit pressure may be higher than the ambient pressure and is determined, using the isentropic relation of Equation (3.5), to be

$$\frac{p_{3F}}{p_{t3F}} = \left(1 + \frac{\gamma - 1}{2}M_{3F}^2\right)^{-\frac{\gamma}{\gamma-1}}.$$

For choked flow at the fan exit $M_{3F} = 1$ and for $\gamma = 1.4$, we have $p_{3F} = 0.5283p_{t3F}$. Note that in Table 3.35 the lowest fan pressure ratio produces a subcritical exit Mach number, while the two higher fan pressure ratios would produce a supercritical exit Mach number; therefore, the exit is instead considered to be choked, which results in an attendant increase in fan exit plane static pressure $p_{3F}$. The static temperature may be obtained from the isentropic relation of Equation (3.6) and is given by

$$\frac{T_{3F}}{T_{t3F}} = \left(1 + \frac{\gamma - 1}{2}M_{3F}^2\right)^{-1}.$$

The fan exit velocity may be found from the fan exit Mach number using

$$V_{3F} = M_{3F}\sqrt{\gamma R T_{3F}}.$$

Values for static pressure and temperature, as well as velocity, at the fan exit are also listed in Table 3.36.

### 3.9.4 Combustor flow, stations 3-4

Flow passing through the combustor is the same for the turbojet and turbofan cases because we are considering the same compressor pressure ratios and the same turbine entrance total temperatures. Therefore, results for the fuel-air ratio as a function of compressor pressure ratio and turbine inlet stagnation temperature, as given in Table 3.9, are the same for both turbojet and turbofan cases. The difference between the cases will be determined by the fan pressure ratio and the fan bypass ratio, as shown in the next section. For constant pressure combustion, note that $p_{t4} = p_{t3}$ and $p_{t3}$ may be found in Table 3.35.

**Table 3.36** Stagnation Pressure and Temperature, Mach Number, Velocity, and Static Pressure and Temperature at Fan Exit in $M = 0.8$ Cruise at 10 km for Various Fan Pressure Ratios

| $p_{t3F}/p_{t2F}$ | $p_{t3F}$ (kPa) | $T_{t3F}$ (K) | $M_{3F}$ | $p_{3F}$ (kPa) | $T_{t3F}$ (K) | $V_{3F}$ (m/s) |
|---|---|---|---|---|---|---|
| 1.5 | 61 | 282 | 1 | 32.0 | 234.9 | 307.2 |
| 2 | 81 | 306 | 1 | 42.7 | 255.0 | 320.1 |
| 2.5 | 101 | 326 | 1 | 53.4 | 271.8 | 330.5 |

## 3.9.5 Turbine flow, stations 4-5

The turbine is part of the turbofan engine solely because it is needed to drive the compressor and the fan and as such must produce just enough power to do so. Thus there is a matching condition to be met: the turbine power produced must equal the sum of the required compressor and fan power, which was shown to be represented by Equation (3.25). Note once again that the turbine must produce the required power using just the core mass flow. In terms of the bypass ratio $\beta$, this relation was given in Equation (3.27). With the ideal process imposed here and the conditions that $c_p$ is constant and $\gamma = 1.4$, we put the temperatures in terms of the pressures and arrived at Equation (3.27).

Because the bypass ratio, fan and compressor ratio, and turbine inlet temperature are all independent parameters, we solved for the required turbine pressure ratio in terms of these parameters in Equation (3.28), which is reproduced here for completeness:

$$\frac{p_{t4}}{p_{t5}} = \left[1 - \frac{T_{t2}}{T_{t4}}\left\{\beta\left(\frac{p_{t3F}}{p_{t2}}\right)^{0.286} + \left(\frac{p_{t3}}{p_{t2}}\right)^{0.286} - (\beta + 1)\right\}\right]^{-3.5}.$$

Therefore, for every selected value of the turbine inlet temperature, the required turbine pressure ratio depends on three independent parameters so that a large number of solutions are possible. Recall that for constant pressure combustion, we have $p_{t4} = p_{t3}$ and $p_{t3}$ may be found in Table 3.35. Using these values and Equation (3.28), we find the results for $p_{t5}$ as shown in Table 3.37 for the various limiting turbine inlet stagnation temperatures selected for various bypass and compressor pressure ratios. The turbine exit stagnation temperature $T_{t5}$ may then be found from the isentropic relation, Equation (3.16), and results for the different compressor pressure ratios and limiting turbine inlet stagnation temperatures are shown in Table 3.38.

## 3.9.6 Nozzle flow, stations 5-7

The nozzle for cruise is still considered to be of the simple converging duct type so that stations 6 and 7 are coincident. The maximum thrust for such a nozzle will be developed when the exit station is choked, that is, $M_7 = 1$. The nozzle is considered to operate ideally so that the stagnation pressure is constant throughout the nozzle, $p_{t5} = p_{t6} = p_{t7}$. Thus the pressure at the exit station $p_7$ may be found using the isentropic flow relation of Equation (3.5) with the condition $M_7 = 1$ and values for the turbine exit stagnation pressure from Table 3.37. Results for $p_7$ are shown in Table 3.39. Note that the ratio of exit pressure to free stream pressure $p_7/p_0 > 1$,

**Table 3.37** Turbine Exit Stagnation Pressures (kPa) for Various Bypass Ratios, Compressor Pressure Ratios, and Turbine Inlet Stagnation Temperatures in $M = 0.8$ Cruise at 10 km

| $\beta = 3$ | $p_{t3}/p_{t2}$ | 1200K | 1300K | 1400K | 1500K | 1600K | 1700K |
|---|---|---|---|---|---|---|---|
| | 5 | 93 | 99 | 105 | 110 | 114 | 119 |
| | 10 | 133 | 146 | 159 | 171 | 181 | 191 |
| | 20 | 168 | 195 | 221 | 245 | 267 | 288 |
| | 30 | 178 | 215 | 252 | 287 | 320 | 351 |
| $\beta = 6$ | $p_{t3}/p_{t2}$ | 1200K | 1300K | 1400K | 1500K | 1600K | 1700K |
| | 5 | 65 | 72 | 78 | 84 | 89 | 94 |
| | 10 | 90 | 103 | 116 | 128 | 139 | 150 |
| | 20 | 107 | 131 | 155 | 178 | 200 | 221 |
| | 30 | 108 | 139 | 171 | 203 | 234 | 265 |
| $\beta = 9$ | $p_{t3}/p_{t2}$ | 1200K | 1300K | 1400K | 1500K | 1600K | 1700K |
| | 5 | 44 | 50 | 57 | 63 | 68 | 74 |
| | 10 | 58 | 70 | 82 | 93 | 105 | 115 |
| | 20 | 64 | 84 | 105 | 125 | 146 | 166 |
| | 30 | 60 | 85 | 111 | 139 | 167 | 195 |

indicating that the jet exhaust is always underexpanded under cruise conditions. It has been pointed out previously that a variable area diverging section between stations 6 and 7 would permit the exhaust nozzle to get closer to matched conditions, and therefore greater thrust, but at the cost of mechanical complexity. In the same fashion, we may calculate the nozzle exit temperature $T_7$ using the isentropic flow relation of Equation (3.6) with the condition $M_7 = 1$ and values for the turbine exit stagnation temperature from Table 3.38. Results for $T_7$ are shown in Table 3.40. The exit velocity can then be found from the relation $V = aM$, where the sound speed is given by Equation (3.17):

$$a_7 = \sqrt{\gamma R T_7} = (20.04 m/s - K^{\frac{1}{2}})\sqrt{T_7}.$$

Here, $\gamma R = 1.4(287\ \text{m}^2/\text{s}^2\text{-K})$, and results for $V_7$ are shown in Table 3.41.

### 3.9.7 Turbofan thrust and fuel efficiency in cruise

The thrust of the turbofan is equal to the sum of the thrust of the fan and the nozzle, and the net thrust has already been shown in Equations (3.29) and (3.30). The specific net thrust is illustrated here for three bypass ratios, $\beta = 3$, 6, and 9, in Figures 3.30 through 3.32. Comparing these results to those for the pure turbojet in Figure 3.8, we see that the specific thrust of the turbofan is increasingly less as the bypass ratio increases. To produce a given amount of thrust with a turbofan requires a larger mass flow than a turbojet; therefore, the diameter of the turbofan will be increasingly larger than that of a turbojet as the bypass ratio increases. The specific fuel consumption was given in Equation (3.31) as

**Table 3.38** Turbine Exit Stagnation Temperatures (K) for Various Bypass Ratios, Compressor Pressure Ratios, and Turbine Inlet Stagnation Temperatures in $M = 0.8$ Cruise at 10 km

| $\beta = 3$ | $p_{t3}/p_{t2}$ | 1200K | 1300K | 1400K | 1500K | 1600K | 1700K |
|---|---|---|---|---|---|---|---|
| | 5 | 960 | 1060 | 1160 | 1260 | 1360 | 1460 |
| | 10 | 873 | 973 | 1073 | 1173 | 1273 | 1373 |
| | 20 | 766 | 866 | 966 | 1066 | 1166 | 1266 |
| | 30 | 693 | 793 | 893 | 993 | 1093 | 1193 |
| $\beta = 6$ | $p_{t3}/p_{t2}$ | 1200K | 1300K | 1400K | 1500K | 1600K | 1700K |
| | 5 | 867 | 967 | 1067 | 1167 | 1264 | 1367 |
| | 10 | 780 | 880 | 980 | 1080 | 1176 | 1280 |
| | 20 | 673 | 773 | 873 | 973 | 1069 | 1173 |
| | 30 | 601 | 700 | 800 | 900 | 995 | 1100 |
| $\beta = 9$ | $p_{t3}/p_{t2}$ | 1200K | 1300K | 1400K | 1500K | 1600K | 1700K |
| | 5 | 775 | 874 | 974 | 1074 | 1174 | 1274 |
| | 10 | 687 | 787 | 887 | 987 | 1087 | 1187 |
| | 20 | 581 | 680 | 780 | 880 | 980 | 1080 |
| | 30 | 508 | 607 | 707 | 807 | 907 | 1007 |

**Table 3.39** Choked Nozzle Exit Pressure (kPa) for Various Bypass Ratios, Compressor Pressure Ratios, and Turbine Inlet Stagnation Temperatures in $M = 0.8$ Cruise at 10 km

| $\beta = 3$ | $p_{t3}/p_{t2}$ | 1200K | 1300K | 1400K | 1500K | 1600K | 1700K |
|---|---|---|---|---|---|---|---|
| | 5 | 48.9 | 52.3 | 55.3 | 58.0 | 60.5 | 62.7 |
| | 10 | 70.1 | 77.4 | 84.1 | 90.2 | 95.8 | 101.0 |
| | 20 | 88.9 | 103.1 | 116.6 | 129.3 | 141.2 | 152.3 |
| | 30 | 94.0 | 113.7 | 133.0 | 151.4 | 169.0 | 185.6 |
| $\beta = 6$ | $p_{t3}/p_{t2}$ | 1200K | 1300K | 1400K | 1500K | 1600K | 1700K |
| | 5 | 34.3 | 37.9 | 41.3 | 44.4 | 47.2 | 49.8 |
| | 10 | 47.3 | 54.5 | 61.3 | 67.6 | 73.5 | 79.1 |
| | 20 | 56.6 | 69.4 | 81.9 | 94.0 | 105.6 | 116.7 |
| | 30 | 56.9 | 73.6 | 90.6 | 107.4 | 123.9 | 139.8 |
| $\beta = 9$ | $p_{t3}/p_{t2}$ | 1200K | 1300K | 1400K | 1500K | 1600K | 1700K |
| | 5 | 23.1 | 26.7 | 30.0 | 33.2 | 36.2 | 39.0 |
| | 10 | 30.4 | 36.9 | 43.3 | 49.4 | 55.2 | 60.8 |
| | 20 | 33.7 | 44.4 | 55.3 | 66.2 | 77.0 | 87.5 |
| | 30 | 31.7 | 44.8 | 58.9 | 73.4 | 88.1 | 102.8 |

**Table 3.40** Choked Nozzle Exit Temperature (K) for Various Bypass Ratios, Compressor Pressure Ratios, and Turbine Inlet Stagnation Temperatures in $M = 0.8$ Cruise at 10 km

| $\beta = 3$ | $p_{t3}/p_{t2}$ | 1200K | 1300K | 1400K | 1500K | 1600K | 1700K |
|---|---|---|---|---|---|---|---|
| | 5 | 800 | 883 | 967 | 1050 | 1133 | 1217 |
| | 10 | 727 | 810 | 894 | 977 | 1060 | 1144 |
| | 20 | 638 | 722 | 805 | 888 | 972 | 1055 |
| | 30 | 578 | 661 | 744 | 828 | 911 | 994 |
| $\beta = 6$ | $p_{t3}/p_{t2}$ | 1200K | 1300K | 1400K | 1500K | 1600K | 1700K |
| | 5 | 723 | 806 | 889 | 973 | 1053 | 1139 |
| | 10 | 650 | 733 | 816 | 900 | 980 | 1066 |
| | 20 | 561 | 644 | 728 | 811 | 891 | 978 |
| | 30 | 500 | 584 | 667 | 750 | 829 | 917 |
| $\beta = 9$ | $p_{t3}/p_{t2}$ | 1200K | 1300K | 1400K | 1500K | 1600K | 1700K |
| | 5 | 645 | 729 | 812 | 895 | 979 | 1062 |
| | 10 | 573 | 656 | 739 | 822 | 906 | 989 |
| | 20 | 484 | 567 | 650 | 734 | 817 | 900 |
| | 30 | 423 | 506 | 590 | 673 | 756 | 840 |

**Table 3.41** Choked Nozzle Exit Velocity (m/s) for Various Bypass Ratios, Compressor Pressure Ratios, and Turbine Inlet Stagnation Temperatures in $M = 0.8$ Cruise at 10 km

| $\beta = 3$ | $p_{t3}/p_{t2}$ | 1200K | 1300K | 1400K | 1500K | 1600K | 1700K |
|---|---|---|---|---|---|---|---|
| | 5 | 567.0 | 595.8 | 623.2 | 649.5 | 674.8 | 699.2 |
| | 10 | 540.6 | 570.7 | 599.3 | 626.6 | 652.8 | 677.9 |
| | 20 | 506.5 | 538.5 | 568.7 | 597.4 | 624.8 | 651.1 |
| | 30 | 481.8 | 515.3 | 546.8 | 576.6 | 605.0 | 632.0 |
| $\beta = 6$ | $p_{t3}/p_{t2}$ | 1200K | 1300K | 1400K | 1500K | 1600K | 1700K |
| | 5 | 538.9 | 569.1 | 597.8 | 625.1 | 650.6 | 676.6 |
| | 10 | 511.0 | 542.7 | 572.8 | 601.3 | 627.5 | 654.6 |
| | 20 | 474.8 | 508.8 | 540.7 | 570.8 | 598.2 | 626.7 |
| | 30 | 448.4 | 484.2 | 517.6 | 549.0 | 577.3 | 607.0 |
| $\beta = 9$ | $p_{t3}/p_{t2}$ | 1200K | 1300K | 1400K | 1500K | 1600K | 1700K |
| | 5 | 509.2 | 541.1 | 571.2 | 599.8 | 627.1 | 653.2 |
| | 10 | 479.7 | 513.3 | 544.9 | 574.8 | 603.3 | 630.4 |
| | 20 | 440.9 | 477.3 | 511.2 | 542.9 | 572.9 | 601.4 |
| | 30 | 412.3 | 451.0 | 486.7 | 520.0 | 551.2 | 580.8 |

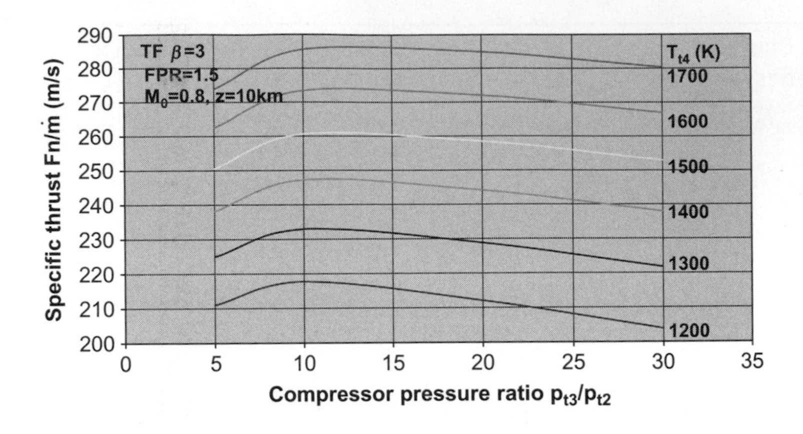


**FIGURE 3.30**

Specific thrust of a turbofan in cruise at $M_0 = 0.8$ at 10 km altitude as a function of compressor pressure ratio for various turbine inlet temperatures, fan pressure ratio FPR $= 1.5$, and bypass ratio $\beta = 3$.

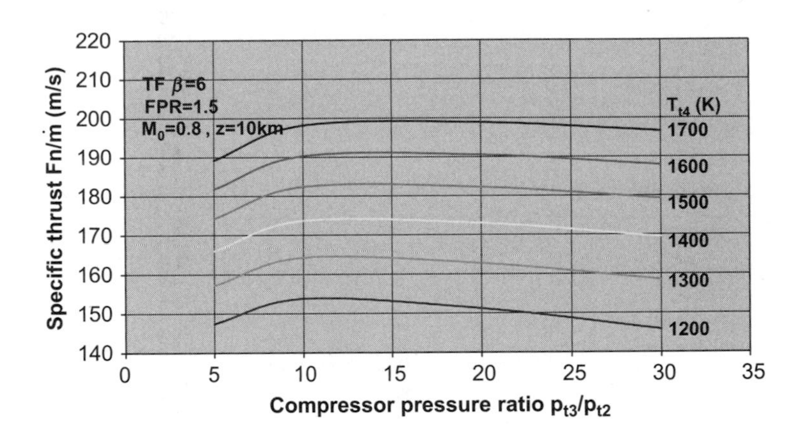


**FIGURE 3.31**

Specific thrust of a turbofan in cruise at $M_0 = 0.8$ at 10 km altitude as a function of compressor pressure ratio for various turbine inlet temperatures, fan pressure ratio FPR $= 1.5$, and bypass ratio $\beta = 6$.

$$c_j = \frac{\dot{m}_f g}{F_n} = \frac{\dot{m}_f}{\dot{m}_0}\frac{g}{(F_n/\dot{m}_0)} = \frac{1}{1+\beta}\frac{(f/a)}{(F_n/\dot{m}_0)}.$$

Note that this is equivalent to the turbojet result in Equation (3.21) divided by $1 + \beta$ so that when the bypass ratio is zero, the turbojet result is recovered. The specific fuel consumption is illustrated for three bypass ratios, $\beta = 3$, 6, and 9, in Figures 3.33 through 3.35. Comparing these results to those for the pure turbojet in Figure 3.7, we see that the specific fuel consumption of the turbofan is increasingly less as the bypass ratio increases. The turbofan operates more and more like a propeller, which, as recalled from Chapter 1, is more efficient than a pure jet engine.

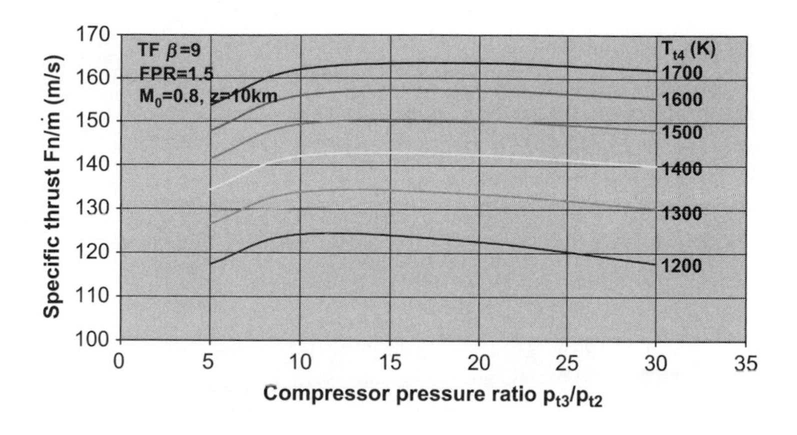


**FIGURE 3.32**

Specific thrust of a turbofan in cruise at $M_0 = 0.8$ at 10 km altitude as a function of compressor pressure ratio for various turbine inlet temperatures, fan pressure ratio FPR = 1.5, and bypass ratio $\beta = 9$.

### 3.9.8 Real turbofan in high subsonic cruise

We may consider the General Electric CF6-50 engine discussed in the previous section. Although not all engine operating conditions are known, we may estimate the specific fuel consumption by interpolating between Figures 3.33 ($\beta = 3$) and 3.34 ($\beta = 6$), and we find that $c_j \sim 0.68\ \text{hr}^{-1}$, which is equal to the actual value for maximum cruise thrust. Once again the ideal analysis yields a reasonable result for specific fuel consumption. This result suggests that the ideal analysis provides about the right ratio of *f/a* compared to $F_n/\dot{m}_0$ but each of these two terms may not be predicted as accurately. For example, the specific thrust for this case would be estimated to be about 233 m/s, at the least, whereas reported engine data (*Aviation Week*

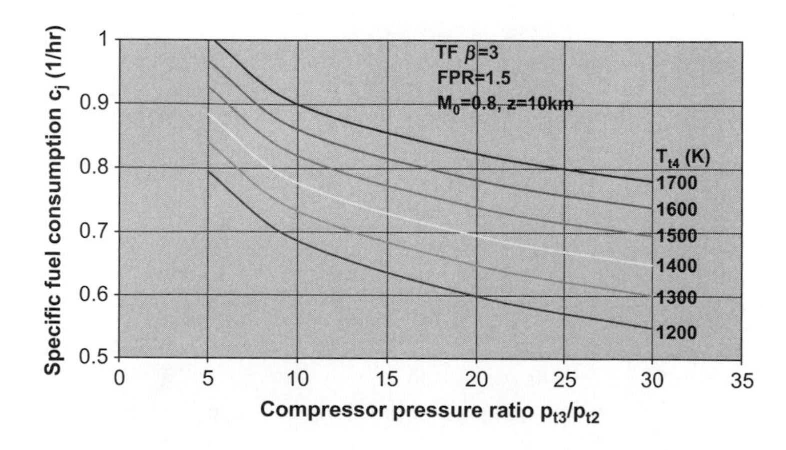


**FIGURE 3.33**

Specific fuel consumption of a turbofan in cruise at $M_0 = 0.8$ at 10 km altitude as a function of compressor pressure ratio for various turbine inlet temperatures, fan pressure ratio FPR = 1.5, and bypass ratio $\beta = 3$.

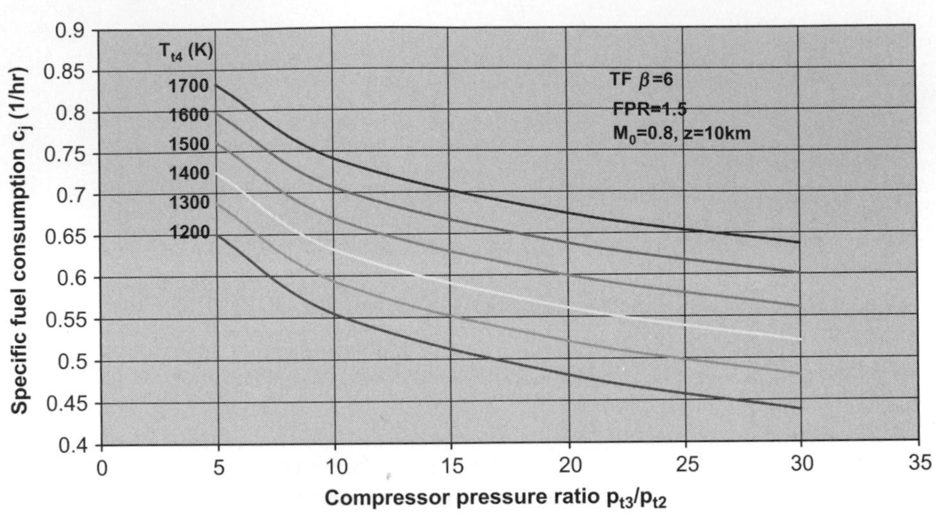


**FIGURE 3.34**

Specific fuel consumption of a turbofan in cruise at $M_0 = 0.8$ at 10 km altitude as a function of compressor pressure ratio for various turbine inlet temperatures, fan pressure ratio FPR = 1.5, and bypass ratio $\beta = 6$.

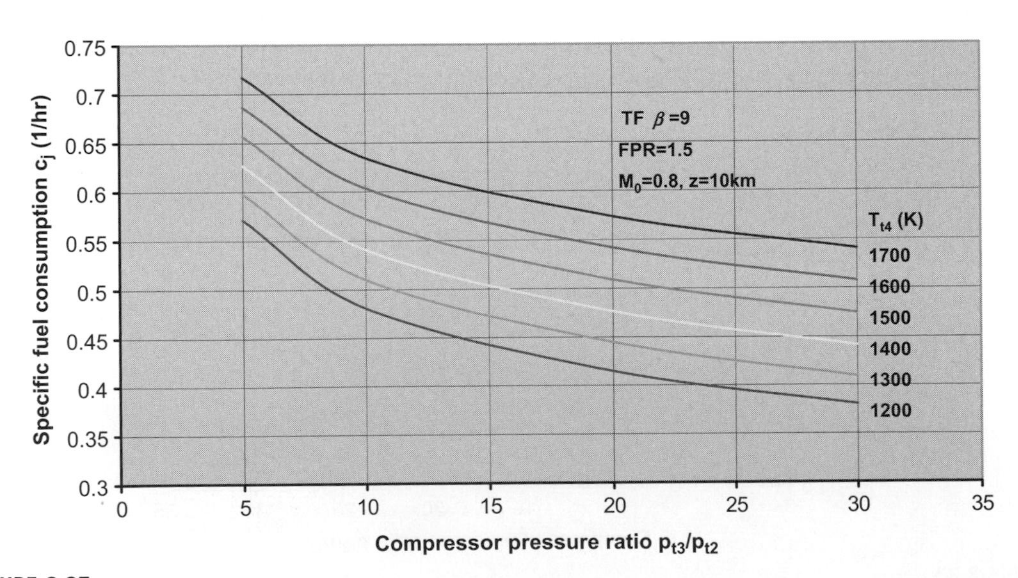


**FIGURE 3.35**

Specific fuel consumption of a turbofan in cruise at $M_0 = 0.8$ at 10 km altitude as a function of compressor pressure ratio for various turbine inlet temperatures, fan pressure ratio FPR = 1.5, and bypass ratio $\beta = 9$.

*Space Technology*, August 2, 1971) indicates a value of about 191 m/s. Thus, the ideal analysis overpredicts the specific thrust by about 22%, a typical result seen in previous sections. Thus, the efficiencies of the various components of the engine will have a substantial effect on the actual thrust levels achievable.

## 3.10 IDEAL INTERNAL TURBOFAN IN SUPERSONIC CRUISE IN THE STRATOSPHERE

The lure of improved fuel efficiency possible with turbofan engines led to the development of turbofans for high-performance military applications. However, turbofan engines for high-speed aircraft have low bypass engines compared to their commercial counterparts. This is a result of the requirement of low frontal area to avoid drag penalties in the supersonic regime. Because engines for high-speed aircraft must be amenable to afterburners, these engines must use a confined internal fan flow that mixes with the turbine exit flow prior to entrance into the single nozzle. The basic configuration, without afterburner, is shown in Figure 3.36.

As mentioned in previous sections, current engines operate in the range of turbine inlet temperatures $1200\text{K} < T_{t4} < 1700\text{K}$, although improved materials and cooling techniques are being sought constantly. The basic supersonic cruise state is standard-day stratospheric conditions, that is, altitude $z = 15$ km, pressure $p_0 = 12.1$ kPa, temperature $T_0 = 217\text{K}$, and speed of sound $a_0 = 295$ m/s. The typical cruise flight condition is taken here as $M_0 = 2$ ($V_0 = 590$ m/s). We will again proceed through the cycle sequentially using various values of the turbine inlet stagnation temperature as in the case for take-off and subsonic

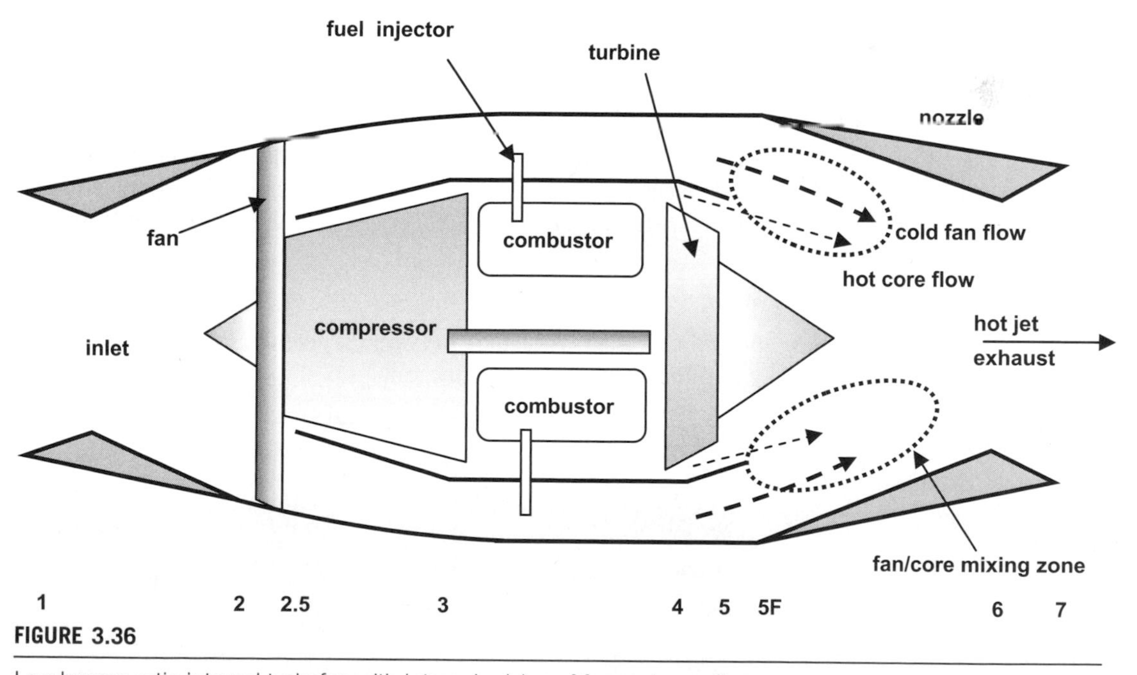


FIGURE 3.36

Low bypass ratio internal turbofan with internal mixing of fan and core flows.

cruise and then require the afterburner to raise the stagnation temperature in the jet exhaust back up to the selected turbine inlet stagnation temperatures. This provides an assessment of afterburning, as well as taking into account the material limitations represented by the selected turbine inlet temperature.

### 3.10.1 Inlet flow, stations 0-2

Although the flight Mach number is supersonic, the inlet of the ideal turbofan will be considered to operate isentropically so that the stagnation values of pressure and temperature delivered to the compressor face are equal to those in the free stream, that is, $p_{t2} = p_{t0}$ and $T_{t2} = T_{t0}$. Using Equations (3.7) and (3.8) for the given supersonic cruise conditions, the stagnation conditions at station 2 are calculated readily to be $p_{t2} = p_{t0} = 94.7$ kPa and $T_{t2} = T_{t0} = 391$K. We show, in Chapter 6, that a well-designed inlet can deliver a pressure recovery of more than 90% at $M_0 = 2$.

### 3.10.2 Compressor flow, stations 2-3

The compressor provides an increase in stagnation pressure from the inlet value determined by the compressor pressure ratio $p_{t3}/p_{t2}$. We will continue to use the typical values for comparison purposes: $p_{t3}/p_{t2} = 5$, 10, 20, and 30. The compressor is assumed to be ideal so that it provides isentropic compression and therefore permits calculation of the temperature ratio from the basic isentropic relation. Using the conditions determined thus far, we find the compressor exit stagnation temperature from Equation (3.9) to be given by

$$T_{t3} = 391\left(\frac{p_{t3}}{p_{t2}}\right)^{0.286}.$$

The stagnation pressure and temperature at the compressor exit for the various compressor pressure ratios have been given in Table 3.17 and are repeated here as Table 3.42. Upon comparison of Tables 3.42 and 3.35, it is clear that the stagnation pressure being delivered to the combustor is much higher in the supersonic cruise state than at subsonic cruise, even though the ambient atmospheric pressure at $z = 15$ km is only one-half that at $z = 10$ km. This is a consequence of the higher ram pressure in supersonic flight. The stagnation temperatures are also higher because of the ram compression.

### 3.10.3 Fan flow, stations 2-5F

The fan provides an increase in stagnation pressure from the inlet value as given by the fan pressure ratio $p_{t3F}/p_{t2}$. As mentioned previously, current engines operate in the range $1.5 < p_{t5F}/p_{t2} < 3$, but to provide

**Table 3.42** Stagnation Pressure and Temperature at Turbofan Compressor Exit in $M = 0.8$ Cruise at 10 km Altitude

| $p_{t3}/p_{t2}$ | $p_{t3}$ (MPa) | $T_{t3}$ (K) |
|---|---|---|
| 5 | 0.474 | 620 |
| 10 | 0.947 | 755 |
| 20 | 1.894 | 921 |
| 30 | 2.841 | 1034 |

the most effective mixing of the cold fan flow with the hot core flow from the turbine, we cannot arbitrarily specify both the fan pressure ratio and the bypass ratio. This is a result of two requirements: (i) matching the fan exit stagnation pressure to the turbine exit stagnation pressure for best mixing and (ii) ensuring that the power output of the turbine matches the power requirement of the compressor and fan.

The fan is assumed to be ideal so that it provides isentropic compression, and therefore the stagnation temperature ratio may be calculated from the basic isentropic relation

$$\frac{T_{t5F}}{T_{t2}} = \left(\frac{p_{t5F}}{p_{t2}}\right)^{\frac{\gamma-1}{\gamma}}.$$

Then, using the conditions determined thus far, we find the compressor exit stagnation temperature, in degrees K, to be given by

$$T_{t5F} = 391\left(\frac{p_{t5F}}{p_{t2}}\right)^{0.286}.$$

The stagnation pressure at the exit of the bypass duct $p_{t5F}$, being set by the fan pressure ratio, generally would be different from the stagnation pressure at the turbine exit. However, studies of turbulent mixing between two streams indicate that best mixing occurs when the stagnation pressure of both streams is equal, as described by Greitzer and colleagues (2004). In keeping with the present idealized analysis, we require that this equality between the stagnation pressure of the bypass exit flow and the turbine exit flow be equal, that is, $p_{t5F} = p_{t5}$.

### 3.10.4 Combustor flow, stations 3-4

Using Equation (3.12) again, we may determine the fuel–air ratio that will yield the selected limiting value of $T_{t4}$ from the suggested range of $1200K \leq T_{t4} \leq 1700$. Results are shown in Table 3.18 and are repeated here in Table 3.43 for convenience. Note that the higher the compressor ratio, the lower the fuel–air ratio required and that the stagnation temperature $T_{t3}$ in supersonic cruise is higher than in subsonic cruise. As a result, the fuel–air ratio in cruise is lower in supersonic cruise because less heat can be added to get to the limiting turbine inlet stagnation temperature.

### 3.10.5 Turbine flow, stations 4-5

For constant pressure combustion, we again take $p_{t4} = p_{t3}$, where $p_{t3}$ may be found using Table 3.17. With these values and the matching condition expressing the fact that the power provided by the turbine must

**Table 3.43** Fuel–Air Ratio for Various Compressor Pressure Ratios and Turbine Inlet Stagnation Temperatures in $M = 2$ Cruise at 15 km Altitude

| $p_{t3}/p_{t2}$ | 1200K | 1300K | 1400K | 1500K | 1600K | 1700K |
|---|---|---|---|---|---|---|
| 5 | 0.01375 | 0.01616 | 0.01858 | 0.02101 | 0.02345 | 0.02591 |
| 10 | 0.01054 | 0.01294 | 0.01535 | 0.01777 | 0.02020 | 0.02265 |
| 20 | 0.006610 | 0.009001 | 0.01140 | 0.01382 | 0.01624 | 0.01868 |
| 30 | 0.003927 | 0.006311 | 0.008707 | 0.01112 | 0.01353 | 0.01596 |

equal the power required by the fan and compressor, we are in a position to determine the turbine exit stagnation temperature. However, we must also enforce the condition that the fan exit stagnation pressure matches the turbine exit stagnation temperature. Equation (3.15) may be rearranged under this requirement to yield the following equation for the fan pressure ratio for a specified value of the bypass ratio $\beta$:

$$\frac{p_{t3F}}{p_{t2}} = \left[\frac{1+\beta+\dfrac{T_{t4}}{T_{t2}} - \left(\dfrac{p_{t3}}{p_{t2}}\right)^{0.286}}{\beta+\dfrac{T_{t4}}{T_{t2}}\left(\dfrac{p_{t3}}{p_{t2}}\right)^{-0.286}}\right]^{3.5}.$$

Conversely, if we wish to specify the fan pressure ratio, then the bypass ratio is determined by the fan-core stagnation pressure equality as follows:

$$\beta = \frac{\dfrac{T_{t4}}{T_{t2}}\left[1-\left(\dfrac{p_{t3F}}{p_{t2}}\dfrac{p_{t2}}{p_3}\right)^{0.286}\right]+\left[1-\left(\dfrac{p_{t3}}{p_{t2}}\right)^{0.286}\right]}{\left(\dfrac{p_{t3F}}{p_{t2}}\right)^{0.286}-1}.$$

It is convenient to specify the bypass ratio and proceed from there to calculate the associated fan pressure ratio. Results of fan pressure ratio as a function of compressor pressure ratio for three bypass ratios are shown in Figure 3.37. It is obvious that at a small bypass ratio such as $\beta = 0.5$, the required fan pressure ratio becomes rather large, while at larger values of $\beta$ the fan pressure ratio is relatively modest. The required fan pressure ratio as a function of bypass ratio for a compressor pressure ratio of 20 is shown in Figure 3.38 for all turbine inlet temperatures considered. Note that for $\beta \geq 1.5$ the curves of required fan pressure ratio become relatively flat.

The turbine exit pressure $p_{t5}$ may be found by recalling that one of the matching conditions is $p_{t5} = p_{t3F}$ so that

$$p_{t5} = p_{t2}\left(\frac{p_{t3F}}{p_{t2}}\right) = 94.7\left(\frac{p_{t3F}}{p_{t2}}\right).$$

Values for the turbine exit pressure for the cases considered here are given in Table 3.44.

The turbine exit temperatures shown in Table 3.45 are found by noting that $p_{t4} = p_{t3}$ and using the isentropic relation as follows:

$$T_{t5} = T_{t4}\left(\frac{T_{t5}}{T_{t4}}\right) = T_{t4}\left(\frac{p_{t5}}{p_{t4}}\right)^{\frac{\gamma-1}{\gamma}} = T_{t4}\left[\left(\frac{p_{t3F}}{p_{t2}}\right)\left(\frac{p_{t2}}{p_{t3}}\right)\right]^{\frac{\gamma-1}{\gamma}}.$$

### 3.10.6 Afterburner flow, stations 5-5AB

As discussed in Section 3.5, the exit flow from the turbine, station 5, is still reasonably rich in oxygen so that fuel may be injected into the duct leading to the nozzle throat, station 6, and burned at constant pressure. Thus the afterburner section, as illustrated in Figure 3.39, acts like a large constant pressure

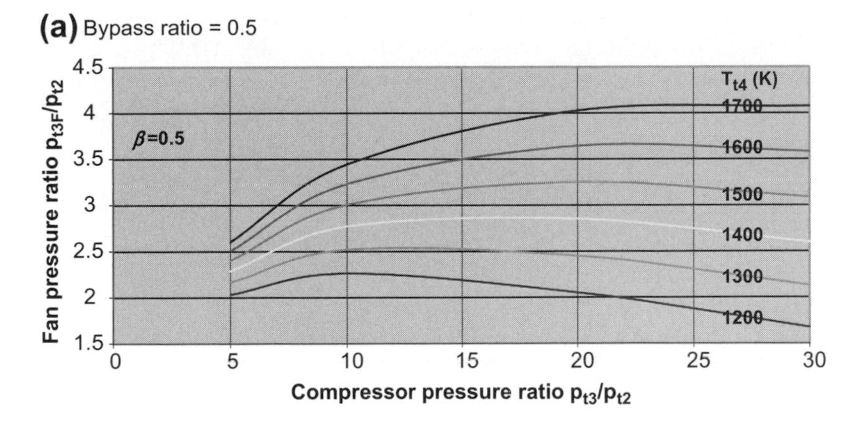


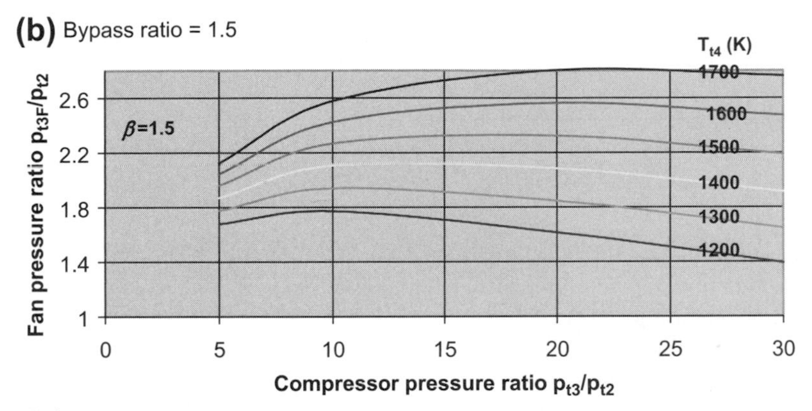


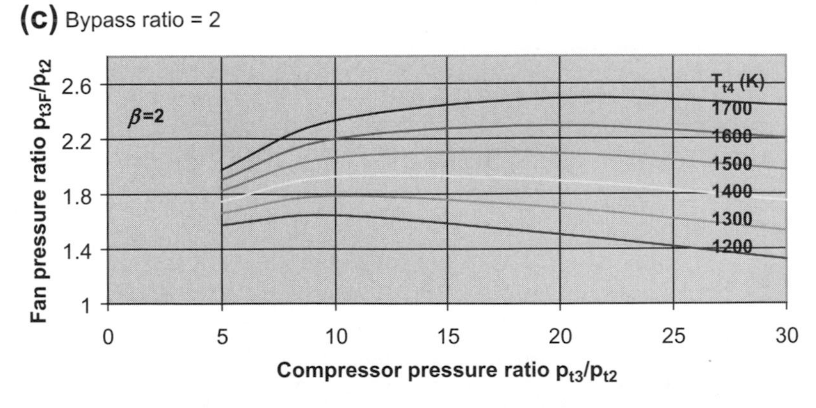


**FIGURE 3.37**

Fan pressure ratio as a function of compressor pressure ratio for various bypass ratios $\beta$ for flight at $M_0 = 2$ at 15 km altitude: (a) $\beta = 0.5$, (b) $\beta = 1.5$, and (c) $\beta = 2$.

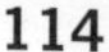

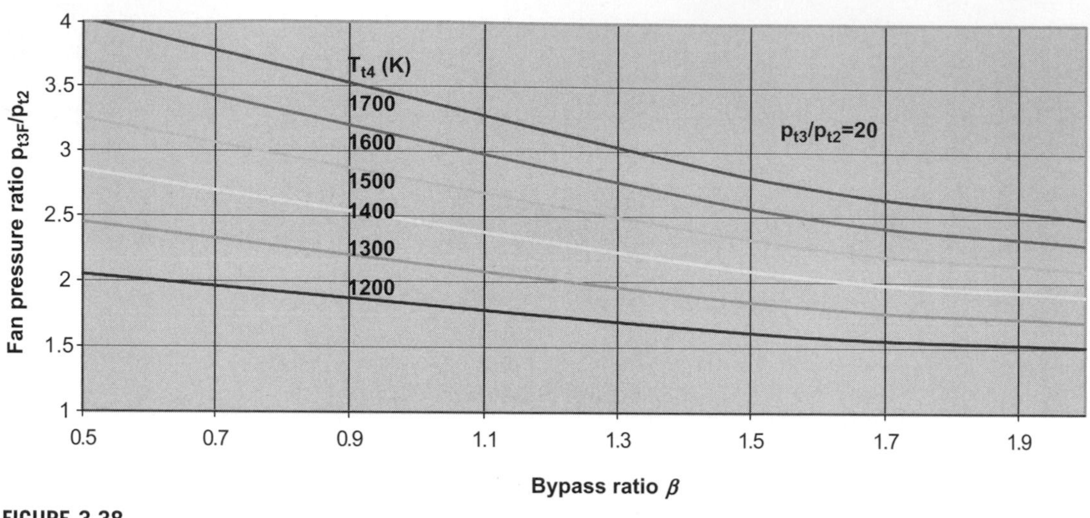


**FIGURE 3.38**

Required fan pressure ratio as a function of bypass ratio for various turbine inlet temperatures and a compressor ratio of 20.

combustor and the low temperatures coming from the turbine exit may be boosted to higher levels, $T_{t5b}$. The increased stagnation temperature permits the generation of higher exit velocity $V_7$, thereby increasing the thrust capability of the engine.

We may assume that combustion in the afterburner brings the final temperature $T_{t5b}$ back up to the limiting turbine inlet temperature so that the afterburner exit temperature $T_{t5} = T_{t4}$. This is the best that

**Table 3.44** Turbine Exit Pressures $p_{t5}$ (kPa) for Various Bypass Ratios, Compressor Pressure Ratios, and Turbine Inlet Temperatures

| $\beta$ | $p_{t3}/p_{t2}$ | 1200 | 1300 | 1400 | 1500 | 1600 | 1700 |
|---|---|---|---|---|---|---|---|
| 2 | 5 | 149.4 | 157.6 | 165.5 | 173.0 | 180.2 | 187.2 |
| 2 | 10 | 155.8 | 169.3 | 182.6 | 195.7 | 208.6 | 221.1 |
| 2 | 20 | 142.7 | 160.9 | 179.4 | 198.1 | 217.0 | 236.0 |
| 2 | 30 | 125.4 | 145.2 | 165.8 | 187.1 | 208.9 | 231.3 |
| 1.5 | 5 | 158.9 | 168.2 | 177.1 | 185.5 | 193.6 | 201.2 |
| 1.5 | 10 | 167.8 | 183.8 | 199.5 | 214.7 | 229.6 | 244.1 |
| 1.5 | 20 | 152.8 | 174.9 | 197.4 | 220.0 | 242.8 | 265.6 |
| 1.5 | 30 | 131.8 | 156.0 | 181.3 | 207.4 | 234.3 | 261.6 |
| 0.5 | 5 | 192.5 | 205.1 | 216.7 | 227.4 | 237.3 | 246.4 |
| 0.5 | 10 | 214.2 | 238.8 | 262.3 | 284.7 | 305.9 | 326.1 |
| 0.5 | 20 | 194.2 | 232.0 | 270.0 | 307.7 | 344.9 | 381.3 |
| 0.5 | 30 | 158.6 | 201.4 | 246.2 | 292.3 | 339.0 | 386.0 |

**Table 3.45** Turbine Exit Temperatures $T_{t5}$ (K) for Various Bypass Ratios, Compressor Pressure Ratios, and Turbine Inlet Temperatures

| $\beta$ | $p_{t3}/p_{t2}$ | 1200 | 1300 | 1400 | 1500 | 1600 | 1700 |
|---|---|---|---|---|---|---|---|
| 2 | 5 | 863 | 949 | 1036 | 1125 | 1214 | 1304 |
| 2 | 10 | 716 | 795 | 874 | 956 | 1038 | 1121 |
| 2 | 20 | 573 | 642 | 713 | 786 | 861 | 937 |
| 2 | 30 | 492 | 555 | 621 | 689 | 759 | 830 |
| 1.5 | 5 | 878 | 967 | 1057 | 1147 | 1239 | 1331 |
| 1.5 | 10 | 732 | 813 | 897 | 981 | 1067 | 1154 |
| 1.5 | 20 | 584 | 658 | 733 | 810 | 889 | 969 |
| 1.5 | 20 | 499 | 567 | 637 | 710 | 784 | 859 |
| 0.5 | 5 | 928 | 1023 | 1120 | 1216 | 1313 | 1410 |
| 0.5 | 10 | 784 | 877 | 970 | 1064 | 1158 | 1253 |
| 0.5 | 20 | 626 | 713 | 802 | 892 | 983 | 1075 |
| 0.5 | 30 | 526 | 610 | 696 | 783 | 871 | 961 |

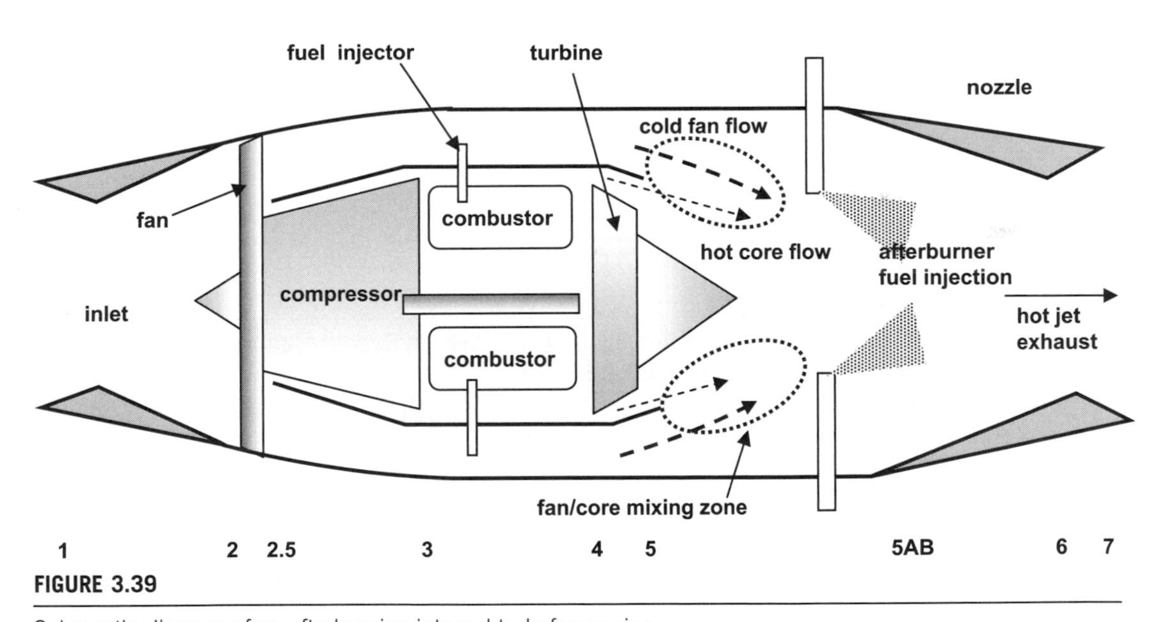


**FIGURE 3.39**

Schematic diagram of an afterburning internal turbofan engine.

can be achieved as we have assumed the turbine inlet temperature is set by the thermal capability of engine materials. This operation of the afterburner is described schematically in the cycle diagram shown in Figure 3.40, and the same use is made of the effective heating value *HV* as was done for the afterburning turbojet in Equations (3.25) and (3.26). The fuel to air ratios in the afterburner are given in Table 3.46.

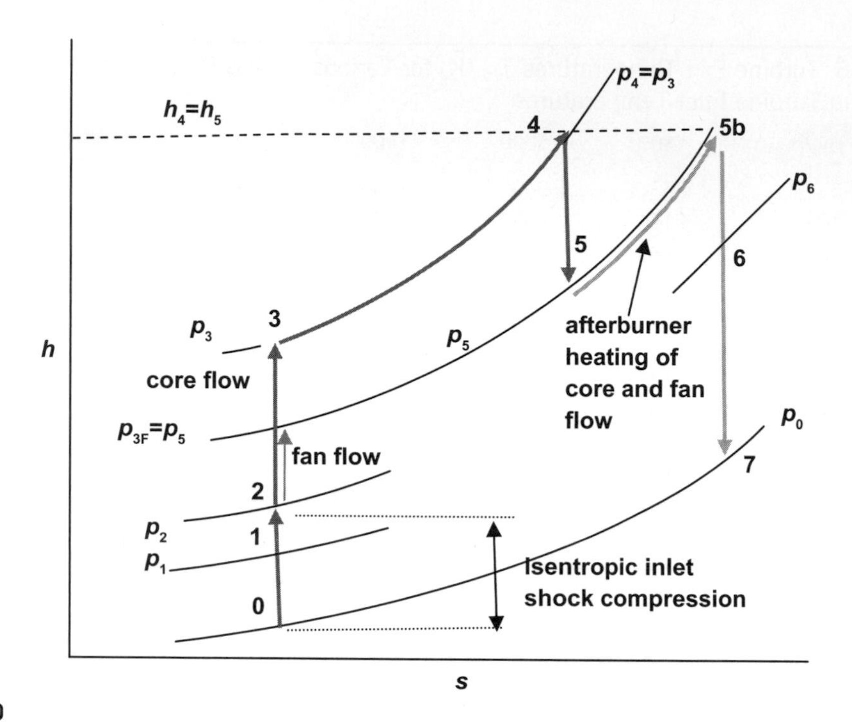


**FIGURE 3.40**

Afterburner operation shown on a turbofan cycle diagram.

**Table 3.46** Afterburner Fuel to Air Ratios $m_{fAB}/m_0$ for Various Bypass Ratios, Compressor Pressure Ratios, and Turbine Inlet Temperatures

| $\beta$ | $p_{t3}/p_{t2}$ | 1200 | 1300 | 1400 | 1500 | 1600 | 1700 |
|---|---|---|---|---|---|---|---|
| 2 | 5 | 0.04689 | 0.05275 | 0.05879 | 0.06503 | 0.07147 | 0.07811 |
| 2 | 10 | 0.04954 | 0.05535 | 0.06135 | 0.06753 | 0.07390 | 0.08048 |
| 2 | 20 | 0.05263 | 0.05838 | 0.06430 | 0.07041 | 0.07671 | 0.08320 |
| 2 | 30 | 0.05465 | 0.06036 | 0.06624 | 0.07229 | 0.07853 | 0.08497 |
| 1.5 | 5 | 0.03715 | 0.04175 | 0.04651 | 0.05146 | 0.05660 | 0.06193 |
| 1.5 | 10 | 0.03987 | 0.04443 | 0.04916 | 0.05406 | 0.05915 | 0.06443 |
| 1.5 | 20 | 0.04301 | 0.04752 | 0.05220 | 0.05704 | 0.06206 | 0.06727 |
| 1.5 | 20 | 0.04506 | 0.04954 | 0.05417 | 0.05897 | 0.06395 | 0.06911 |
| 0.5 | 5 | 0.01732 | 0.01925 | 0.02131 | 0.02349 | 0.02582 | 0.02832 |
| 0.5 | 10 | 0.02033 | 0.02229 | 0.02436 | 0.02657 | 0.02892 | 0.03143 |
| 0.5 | 20 | 0.02370 | 0.02568 | 0.02777 | 0.02999 | 0.03235 | 0.03486 |
| 0.5 | 30 | 0.02584 | 0.02783 | 0.02993 | 0.03215 | 0.03450 | 0.03701 |

### 3.10.7 Nozzle flow, stations 5AB-7

The nozzle considered for supersonic cruise must be of the variable-area converging–diverging duct type so that it may be considered to operate in the matched pressure mode $p_7 = p_0$, thereby assuring the maximum thrust achievable. Because the stagnation pressure is constant throughout the nozzle, $p_{t5b} = p_{t6} = p_{t7}$. Keeping $\gamma = 1.4$ for this ideal study, we may find $M_7$ using the isentropic relation of Equation (3.11); results are shown in Table 3.47.

### 3.10.8 Turbofan thrust and fuel efficiency in supersonic cruise

Using all the information gathered thus far, the specific thrust in the cruise mode for the ideal afterburning turbojet engine may be calculated using Equation (3.27); results are illustrated for three bypass ratios $\beta = 0.5$, 1.5, and 2 in Figures 3.41, 3.42, and 3.43, respectively. It is clear from Figure 3.16 that afterburning is crucial to good performance at supersonic Mach numbers. Note that the specific thrust performance for the afterburning turbojet at $M_0 = 2$ is about the same as that for the subsonic turbojet at $M_0 = 0.8$.

The corresponding results for specific fuel consumption for the afterburning turbofan at $M = 2$ and 15 km altitude are shown in Figures 3.44 to 3.46. Note that the specific fuel consumption for the afterburning turbofan is given by

$$c_j = \frac{3600g}{1+\beta} \frac{\left(\frac{f}{a}\right) + \left(\frac{f}{a}\right)_{AB}}{\frac{F_n}{\dot{m}_0}}.$$

**Table 3.47** Matched Nozzle Exit Mach Numbers for Various Compressor Pressure Ratios, Bypass Ratios, and Turbine Inlet Stagnation Temperatures in $M = 2$ Cruise at 15 km Altitude[a]

| $\beta$ | $p_{t3}/p_{t2}$ | 1200 | 1300 | 1400 | 1500 | 1600 | 1700 |
|---|---|---|---|---|---|---|---|
| 2 | 5 | 2.293 | 2.328 | 2.359 | 2.387 | 2.414 | 2.438 |
| 2 | 10 | 2.320 | 2.374 | 2.422 | 2.467 | 2.507 | 2.545 |
| 2 | 20 | 2.264 | 2.341 | 2.411 | 2.475 | 2.533 | 2.587 |
| 2 | 30 | 2.181 | 2.275 | 2.360 | 2.438 | 2.509 | 2.574 |
| 1.5 | 5 | 2.333 | 2.369 | 2.402 | 2.432 | 2.459 | 2.484 |
| 1.5 | 10 | 2.368 | 2.426 | 2.479 | 2.526 | 2.570 | 2.609 |
| 1.5 | 20 | 2.308 | 2.394 | 2.472 | 2.542 | 2.606 | 2.664 |
| 1.5 | 20 | 2.213 | 2.321 | 2.417 | 2.504 | 2.582 | 2.654 |
| 0.5 | 5 | 2.456 | 2.497 | 2.532 | 2.563 | 2.591 | 2.615 |
| 0.5 | 10 | 2.525 | 2.595 | 2.656 | 2.709 | 2.756 | 2.798 |
| 0.5 | 20 | 2.462 | 2.576 | 2.674 | 2.760 | 2.834 | 2.901 |
| 0.5 | 30 | 2.332 | 2.485 | 2.615 | 2.726 | 2.823 | 2.909 |

[a]*Note that exit Mach numbers lie in the range $2.29 < M_7 < 2.91$, which require area ratios in the range $2.2 < A_7/A_6 < 3.9$ in order to produced the matched exit flow. The exit velocity $V_7$ in the matched case may be found by first using the isentropic relation Equation (3.6) to find $T_7$ from $M_7$ and $T_{t7} = T_{t5b}$ and then using the Mach number definition $M_7 = V_7/a_7$*

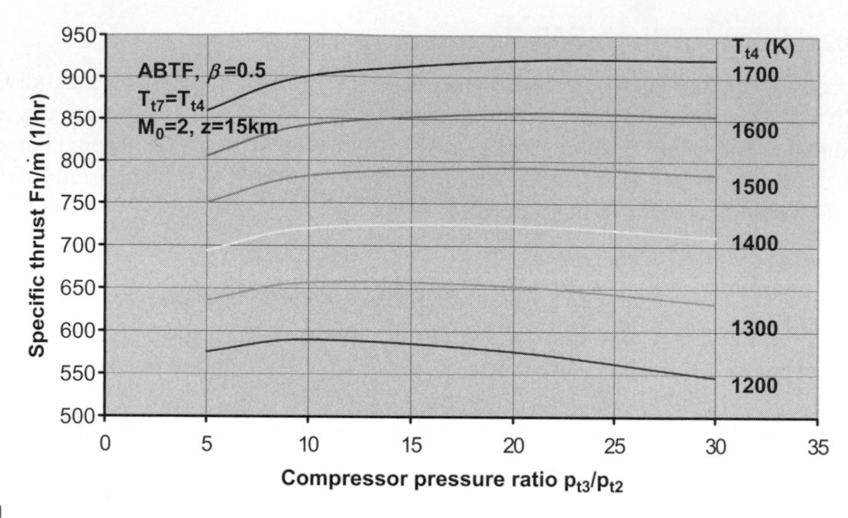


**FIGURE 3.41**

Specific thrust as a function of compressor pressure ratio and turbine inlet stagnation temperature for an afterburning internal turbofan with $\beta = 0.5$ and a matched nozzle in $M = 2$ cruise at 15 km altitude.

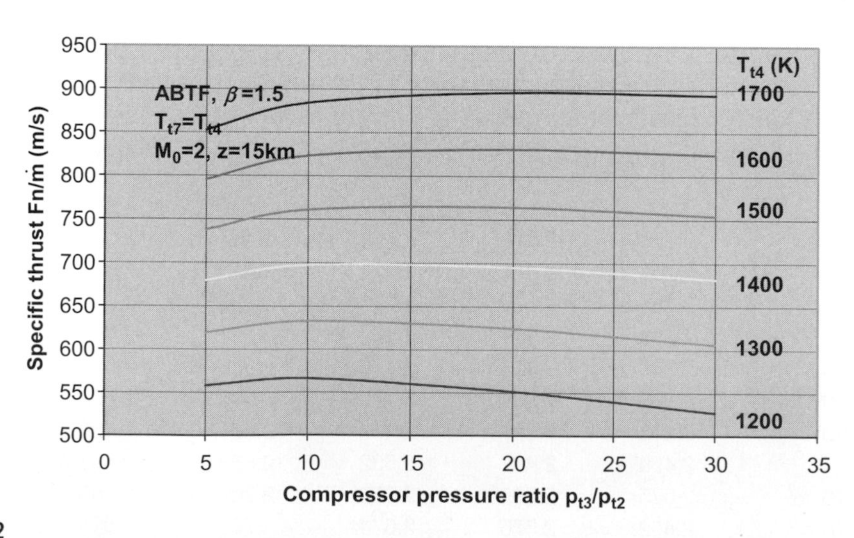


**FIGURE 3.42**

Specific thrust as a function of compressor pressure ratio and turbine inlet stagnation temperature for an afterburning internal turbofan with $\beta = 1.5$ and a matched nozzle in $M = 2$ cruise at 15 km altitude.

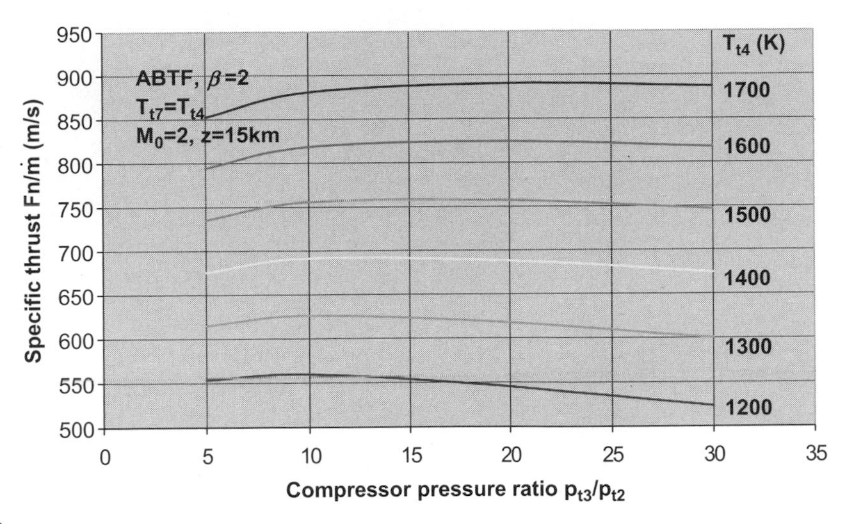


**FIGURE 3.43**

Specific thrust as a function of compressor pressure ratio and turbine inlet stagnation temperature for an afterburning internal turbofan with $\beta = 2$ and a matched nozzle in $M = 2$ cruise at 15 km altitude.

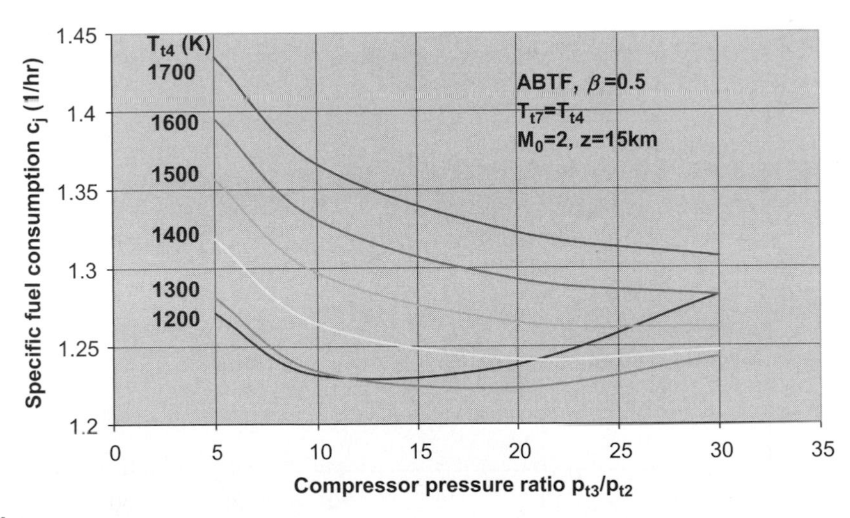


**FIGURE 3.44**

Specific fuel consumption as a function of compressor pressure ratio and turbine inlet stagnation temperature for an afterburning internal turbofan with $\beta = 0.5$ and a matched nozzle in $M = 2$ cruise at 15 km altitude.

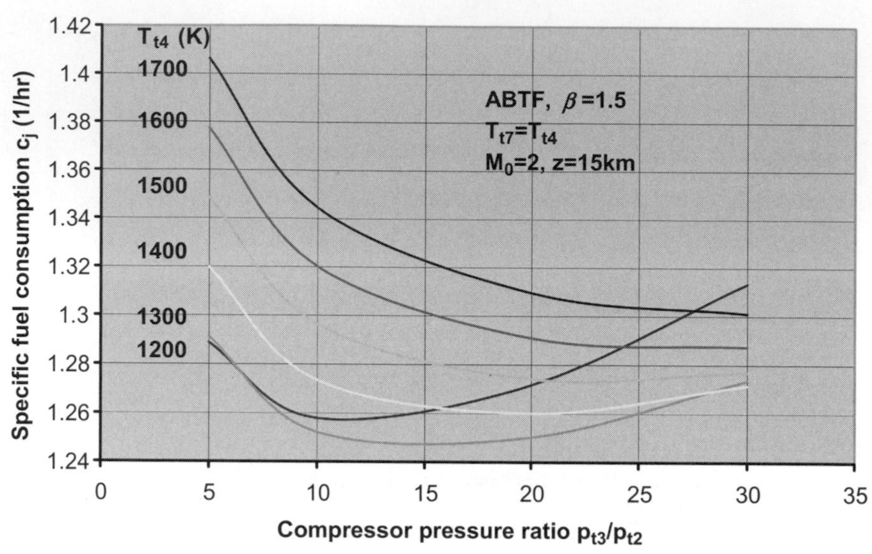


**FIGURE 3.45**

Specific fuel consumption as a function of compressor pressure ratio and turbine inlet stagnation temperature for an afterburning internal turbofan with $\beta = 1$ and a matched nozzle in $M = 2$ cruise at 15 km altitude.

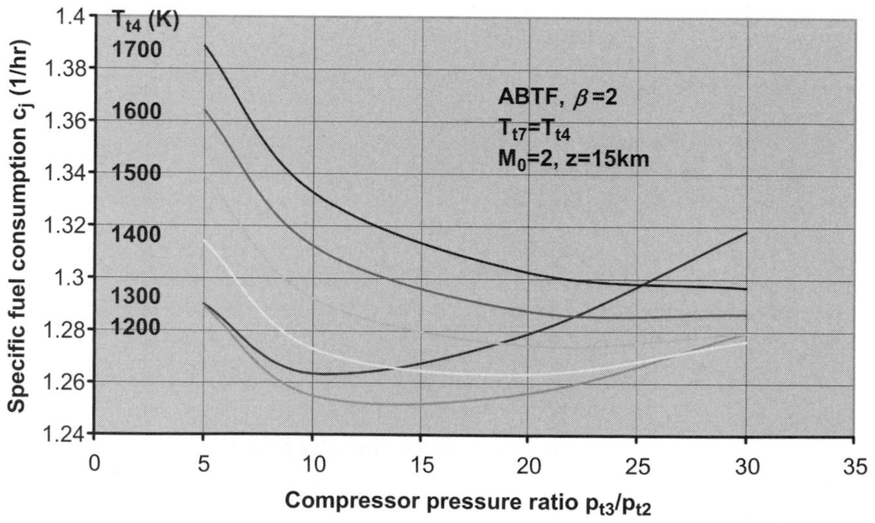


**FIGURE 3.46**

Specific fuel consumption as a function of compressor pressure ratio and turbine inlet stagnation temperature for an afterburning internal turbofan with $\beta = 2$ and a matched nozzle in $M = 2$ cruise at 15 km altitude.

The price for supersonic performance is fuel efficiency, and Figures 3.44 through 3.46 show that there is little advantage to internal bypass turbofans as compared to turbojets when afterburning is considered. In general, afterburning doubles the specific fuel consumption as compared to the standard turbojet in subsonic cruise.

### 3.10.9 Real internal turbofan in supersonic cruise

We considered cruising flight at $M_0 = 2$ in Section 3.5.8 and used the Olympus 593 as an example of an afterburning turbojet engine. The Pratt & Whitney TF30-P-412A ("Fact Sheet TF30-P-412A Turbofan Engine," Pratt & Whitney Aircraft S-4230, May 1973) was an afterburning internal turbofan that powered the Grumman F-14A Tomcat. At $M_0 = 2$ and 15 km altitude, the specific fuel consumption $c_j = 2.55$ hr$^{-1}$ with a net thrust $F_n = 71.2$ kN. The specific thrust produced was on the order of 650 m/s under these conditions in a standard atmosphere. The bypass ratio $\beta = 0.72$, the overall pressure ratio $p_{t3}/p_{t2} = 21.8$, and the turbine inlet temperature $T_{t4} = 1400$K. Results in Figure 3.41 can be used to get an estimate of the ideal specific thrust performance, which suggests $F_n/\dot{m} = 720$ m/s, an overprediction of about 9%. Using Figure 3.44, we see that the specific fuel consumption $c_j = 1.25$ hr$^{-1}$ for the case similar to that of the TF30. Here the ideal prediction greatly underestimates actual specific fuel consumption.

## 3.11 REAL ENGINE OPERATIONS

The preceding cases of jet engine performance under typical flight conditions were analyzed as ideal systems. As pointed out in the introduction, the jet engine is a complex system of components that must be suitably matched to ensure good performance over a broad flight envelope. These components are analyzed in detail in Chapters 4 through 7 to gain an understanding of their individual operations and limitations. In Chapter 8, several actual jet engines are reviewed in terms of component integration, and the efficiencies actually encountered are deduced. As an introduction to the efficiencies that may be encountered that impact the calculations presented thus far, a listing and brief discussion of suitable values is presented to close this chapter. Just as detailed calculations of ideal performance can be considered as sample problems, the reader can carry out additional calculations using some of the material presented here to get their own impression of the effects of efficiencies on performance. As pointed out in Chapter 1, good performance of jet engines depends mainly on maximizing the allowable stagnation temperature in the nozzle exhaust and minimizing stagnation pressure losses throughout the entire flow path.

### 3.11.1 Inlet operation

In inlet stagnation pressure, losses arise through friction and, for supersonic flight, shock waves. The details of inlet design influence the actual results, but to provide guidance for design efforts and for comparison purposes among different, perhaps competing, configurations, a military standard, "MIL-E-5008B: Military Specifications—Engines, Aircraft, Turbojet, Model Specification For," January, 1959, has been developed. The suggested values for the stagnation pressure ratio across the inlet are given in Table 3.48.

**Table 3.48** Standard Stagnation Pressure Recovery[a]

| $p_{t,2}/p_{t,0}$ | Mach number regime |
|---|---|
| 1 | $0 < M_0 < 1$ |
| $1 - 0.075(M_0 - 1)^{1.35}$ | $1 < M_0 < 5$ |
| $800/(M_0^4 + 935)$ | $M_0 > 5$ |

[a]From "MIL-E-5008B: Military Specifications—Engines, Aircraft, Turbojet, Model Specification For," January 1959

## 3.11.2 Compressor and fan operation

The compressor is subject to losses that diminish the actual pressure ratio delivered to that obtained ideally. The pressure ratio delivered by the compressor comes at the price of the efficiency of adiabatic compression, $\eta_{ad,c}$. Therefore, the temperature ratio actually produced is larger than that for isentropic compression and is given by

$$\frac{T_{t,3}}{T_{t,2}} = 1 + \frac{1}{\eta_{ad,c}}\left[\left(\frac{p_{t,3}}{p_{t,2}}\right)^{\frac{\gamma}{\gamma-1}} - 1\right].$$

Reasonable values for the efficiency of adiabatic compression of large-scale modern aircraft engine compressors are in the range of $0.85 < \eta_{ad,c} < 0.9$. Small-scale compressors that may be used in UAV applications are likely to have somewhat reduced values of efficiency due to the stronger influence of viscous effects, perhaps in the range of $0.75 < \eta_{ad,c} < 0.85$. Fan efficiency may be treated with the same range of efficiencies.

## 3.11.3 Combustor and afterburner operation

The actual combustor operation proceeds at burner efficiencies that effectively reduce the heating value of the fuel. This is apparent from the definition of burner efficiency in Chapter 1, and the equation for fuel to air ratio is actually

$$\frac{f}{a} = \frac{\dot{m}_f}{\dot{m}_0} = \frac{T_{t,4} - T_{t,3}}{\frac{\eta_b}{c_p}HV - T_{t,4}}.$$

Because the burner efficiency is proportional to combustor pressure, it falls off at higher altitudes. At design values of engine rpm and sea level conditions, it is in the range $0.90 < \eta_b < 0.96$ and falls off in the stratosphere to $0.85 < \eta_b < 0.90$. At lower engine rpm, the burner efficiency drops further from these values. There is also a stagnation pressure drop across the combustor due to the irreversible nature of the combustion process. This is generally around 5%.

Afterburner operation is not so efficient and would generally fall in the range of $0.6 < \eta_{b,AB} < 0.8$. Because the afterburner is used for short time intervals for acceleration, the efficiency of operation is not as important a criterion as it is for the main combustor. In the same fashion, the stagnation pressure losses connected with friction, strut drag, and heat addition combine to be about double that for the main combustor.

### 3.11.4 Turbine operations

The efficiency of adiabatic expansion $\eta_{ad,e}$ defines the ratio of the work extracted by the turbine to the ideal work extracted in an isentropic expansion. In a manner similar to that discussed for the compressor, we find that the stagnation pressure exiting the turbine and entering the nozzle is actually

$$\frac{T_{t,5}}{T_{t,4}} = 1 + \eta_{ad,e}\left[1 - \left(\frac{p_{t,5}}{p_{t,4}}\right)^{\frac{\gamma_4 - 1}{\gamma_4}}\right].$$

Therefore, for a given pressure drop across the turbine, a lower stagnation temperature is achieved than the ideal. Because the nozzle exit velocity is proportional to the square root of the stagnation temperature in the nozzle, this constitutes a loss in thrust. A reasonable value for the turbine efficiency would lie in the range $0.9 < \eta_{ad,e} < 0.95$. Once again, for small-scale turbines, these values should be reduced to, perhaps, $0.85 < \eta_{ad,e} < 0.9$.

### 3.11.5 Nozzle operations

Nozzles tend to be highly efficient and represent, like turbines, nearly perfect expansion devices. As a result, one would consider the nozzle efficiency of adiabatic expansion to be in the range of $0.92 < \eta_n < 0.96$. This is somewhat higher than the values for turbines because the nozzle is a much larger scale device.

## 3.12 NOMENCLATURE

| | |
|---|---|
| $A$ | cross-sectional area |
| $a$ | sound speed |
| $c_j$ | specific fuel consumption |
| $c_p$ | specific heat at constant pressure |
| $D$ | aircraft drag |
| $F_g$ | gross thrust |
| $F_n$ | net thrust |
| $f/a$ | fuel to air ratio |
| $g$ | acceleration of gravity |
| $h$ | enthalpy |
| $HV$ | heating value of the fuel |
| $I_{sp}$ | specific impulse |
| $L$ | aircraft lift |
| $M$ | Mach number |
| $\dot{m}$ | mass flow |
| $h$ | enthalpy |

| | |
|---|---|
| $p$ | pressure |
| $R$ | gas constant |
| $s$ | entropy |
| $T$ | temperature |
| $V$ | velocity |
| $W$ | aircraft weight or molecular weight |
| $\dot{w}$ | weight flow |
| $\beta$ | bypass ratio, $= \dot{m}_F/\dot{m}_C$ |
| $\gamma$ | ratio of specific heats |
| $\eta_{ad,c}$ | efficiency of adiabatic compression |
| $\eta_{ad,e}$ | efficiency of adiabatic expansion |
| $\eta_b$ | burner efficiency |

### 3.12.1 Subscripts

| | |
|---|---|
| $AB$ | afterburner |
| $a$ | air |
| $C$ | engine core |
| $e$ | conditions downstream of exit where $p = p_0$ |
| $eff$ | effective value |
| $F$ | engine fan |
| $f$ | fuel |
| $t$ | stagnation conditions |
| 0 | conditions in the free stream |
| 1 | conditions at the inlet entrance |
| 2 | conditions at the compressor entrance |
| 3 | conditions at the combustor entrance |
| $3F$ | conditions at the exit of the external fan |
| 4 | conditions at the turbine entrance |
| 5 | conditions at the nozzle entrance |
| $5b$ | conditions at the afterburner entrance |
| $5F$ | conditions at the exit of the internal fan |
| 6 | conditions at the nozzle throat |
| 7 | conditions at the nozzle exit |

## 3.13 EXERCISES

**3.1** A turbojet engine with a convergent nozzle is operating at 100% power at $M_0 = 0.9$ and an altitude of 11 km. The engine characteristics are such that the compressor, burner, and turbine efficiencies may be taken as 0.85, 0.96, and 0.90, respectively. The inlet pressure recovery may be taken as the MIL-E-5008B standard. The mass flow captured by the inlet is 29.17 kg/s, and the heating value of the fuel is 44 MJ/kg. The compressor pressure ratio is $p_{t,3}/p_{t,2} = 18$, the combustor pressure loss is $(p_{t,3} - p_{t,4}) = 0.02p_{t,3}$, and the turbine inlet temperature $T_{t,4} =$ 1330K. Calculate the properties of the flow at each station in the engine, the gross thrust, and the net thrust.

**3.2** A turbofan engine with the same component efficiencies as the turbojet engine in Exercise 3.1 is operating at the same flight conditions. The total mass flow, core flow plus fan flow, is 50 kg/s and the bypass ratio $\beta = 1$. The fan operates in a bypass duct without mixing with the core flow. The fan pressure ratio $p_{t,3F}/p_{t,2} = 2.2$, and its efficiency is 0.85. The nozzle of the fan bypass duct may be assumed to operate critically, that is, with an exit $M_{3F} = 1$. Calculate the properties of the flow at each station in the engine, the gross thrust, and the net thrust.

**3.3** Use the results of Exercises 3.1 and 3.2 to compare the performance of the two engines in terms of net thrust, specific fuel consumption, overall efficiency, and total mass flow. Discuss the significance of the comparison.

**3.4** Consider adding an afterburner to the turbojet engine of Exercise 3.1 with the additional information that the afterburner combustion efficiency is 0.85, and frictional losses in the afterburner yield $(p_{t,5} - p_{t,5AB}) = 0.05p_{t,5}$. In the afterburner, the fuel flow is $\dot{m}_f = 1.17$ kg/s. Calculate the properties of the flow at each station in the engine, the gross thrust, and the net thrust.

**3.5** Consider adding an afterburner to the turbofan engine of Exercise 3.2 with the additional information that the afterburner combustion efficiency is 0.85, and frictional losses in the afterburner yield $(p_{t,5} - p_{t,5AB}) = 0.05p_{t,5}$. In the afterburner, the fuel flow is $\dot{m}_f=1.00$ kg/s. The bypass duct burner flow is assumed to mix internally with the turbine exit flow with the resulting stagnation pressure being the average of the two. The combustion efficiency is 0.85, and frictional losses in the afterburner yield $(p_{t,5} - p_{t,5AB}) = 0.05p_{t,5F}$. Calculate the properties of the flow at each station in the engine, the gross thrust, and the net thrust.

**3.6** Use the results of Exercises 3.4 and 3.5 to compare the performance of the two engines in terms of net thrust, specific fuel consumption, overall efficiency, and thrust augmentation ratio. Discuss the significance of the comparison.

**3.7** The Garrett TFE 1042 low-bypass turbofan can operate with or without afterburning. The internal fan has a bypass ratio of 0.75 and a pressure ratio of 2.5; the overall pressure ratio for the engine is 20 and the inlet airflow rate is 75 lb/s. Assuming all efficiencies are 100%, $HV = 18{,}900$ Btu/lb, the turbine inlet temperature $T_{t4} = 2500$R, and the afterburning temperature $T_{t5} = 4000$R, find the net thrust and specific fuel consumption $c_j$ with and without afterburning for a static test, that is, $M_0 = 0$, under standard-day atmospheric conditions.

**3.8** Consider an ideal turbofan engine with a bypass ratio $\beta$ and fixed turbine entry stagnation temperature $T_{t4}$. Assuming that the working gas has constant thermal properties with $\gamma = 1.4$, determine the optimum value of $\beta$, that is, the $\beta$ for which the net thrust $F_n$ is maximum, under the following conditions: $T_0 = 216.7\text{K}$, $M_0 = 0.9$, fan pressure ratio $p_{t3F}/p_{t0} = 2.2$, compressor pressure ratio $p_{t3}/p_{t2} = 8$, and $T_{t4} = 1333\text{K}$. The compressor mass flow $m_c$ is to be held constant and may be assumed to be equal to the turbine mass flow (fuel addition is considered negligible). Determine the bypass ratio $\beta$ for which the normalized net thrust $F_n{}^* = F_n/\dot{m}_c V_0$ is maximum. Discuss and extend the results obtained to other cases.

## References


"Fact Sheet TF30-P-412A Turbofan Engine." Pratt & Whitney Aircraft S-4230, May 1973.

*Flight,* March 18, 1960, p. 385.

Greitzer, E. M., Tan, C. S., & Graf, M. B. (2004). *Internal Flow, Concepts and Applications.* New York: Cambridge University Press.

Hill, P. G., & Peterson, C. R. (1992). *Mechanics and Thermodynamics of Propulsion* (2nd Ed.). Reading, MA: Addison-Wesley Publishing Co.

"MIL-E-5008B: Military Specifications—Engines, Aircraft, Turbojet, Model Specification For," January 1959.

Segal, C. (2009). *Scramjet Engines.* New York: Cambridge University Press.

*Aviation Week Space Technology,* August 2, 1971.

Svoboda, C. (2000). Turbofan engine database as a preliminary design tool. In "Aircraft Design," Vol. 3, pp. 17–31.

Torenbeek, E. (1982). *Synthesis of Subsonic Airplane Design.* Dordrecht, The Netherlands: Kluwer Academic Publishers.

CHAPTER

# Combustion Chambers for Air-Breathing Engines

# 4

## CHAPTER OUTLINE



## 4.1 COMBUSTION CHAMBER ATTRIBUTES

The combustion chamber must direct large amounts of heat energy toward the turbine and/or nozzle at the proper temperature level and with an appropriate temperature distribution. A good combustion chamber must

- Provide full combustion with minimum pressure loss
- Operate without significant accumulation of deposits
- Ignite the fuel readily
- Give reliable service over long periods of time

Furthermore, these attributes must be evident across the entire operating range of the engine. Good combustion chamber design rests on satisfaction of the basic tenets of combustion processes:

- Proper mixture ratio and the three "Ts" of good combustion
- Temperature of the reactants
- Turbulence for good mixing
- Time for mixing and combustion to go to completion

Once again, it is expected that these fundamentals are achieved with minimal pressure loss across the combustor.

## 4.2 MODELING THE CHEMICAL ENERGY RELEASE

Let us consider the energy equation, as given in Equation (2.18), for the simplified case where no heat is transferred across the fluid boundaries and no work is added or removed from the fluid. Then the energy equation states that the total enthalpy of the fluid remains constant so that

$$dh_t = dh + \frac{1}{2}dV^2 = 0.$$

Using the definition of the change in enthalpy given in Equation (2.24), we may rewrite this equation as follows:

$$-\sum_{i=1}^{N} h_i dY_i = c_p dT + \frac{1}{2}dV^2. \tag{4.1}$$

We may introduce the Mach number and expand Equation (4.1) to

$$-\sum_{i=1}^{N} h_i dY_i = c_p\left(1 + \frac{\gamma - 1}{2}M^2\right)dT + c_p T\left(\frac{\gamma - 1}{2}\right)dM^2 + c_p T\left(\frac{\gamma - 1}{2}\right)M^2\left[\frac{d\gamma}{\gamma} - \frac{dW}{W}\right].$$

The stagnation temperature is

$$T_t = T\left(1 + \frac{\gamma - 1}{2}M^2\right).$$

The differential of the stagnation temperature is then

$$c_p dT_t = \left(1 + \frac{\gamma - 1}{2}M^2\right)c_p dT + c_p T\left(\frac{\gamma - 1}{2}\right)dM^2.$$

Then, using Equation (2.26), our volumetric source term representing the release of chemical energy into the fluid is

$$-dH = -\sum_{i=1}^{N} h_i dY_i = c_p dT_t + c_p T\left(\frac{\gamma - 1}{2}\right)M^2\left[\frac{d\gamma}{\gamma} - \frac{dW}{W}\right]. \tag{4.2}$$

Thus we see that a chemical energy release can almost be modeled as a simple increase in the stagnation temperature of the fluid, with the exception of the contribution of the second term on the right-hand side of Equation (4.2). The form of the independent variable for the heat release in our quasi-one-dimensional analysis becomes

$$-\frac{dH}{c_pT}=\frac{dT_t}{T}+\left(\frac{\gamma-1}{2}\right)M^2\left[\frac{d\gamma}{\gamma}-\frac{dW}{W}\right].$$

For cases in which we may reasonably assume that the isentropic exponent $\gamma$ and molecular weight $W$ are approximately constant, it is possible to consider the chemical heat release to be expressed solely as an increase in stagnation temperature. Note that this effectively requires that the specific heat remains constant as well. This is the simplest means for representing the heat release of a combustion process and is often used in practice with averaged values for $c_p$, $\gamma$, and $W$.

## 4.3 CONSTANT AREA COMBUSTORS

As a first approach to combustor design, we consider a combustor whose cross-sectional area is constant. Furthermore, we assume that the large energy release due to combustion represents the main driver for the combustor and that other effects, such as heat transfer and friction, are negligible in comparison. Using the table of influence coefficients for this flow yields

$$\begin{aligned}
\frac{dM^2}{M^2}&=-\left[\frac{1+\gamma M^2}{1-M^2}\right]\left[\frac{dH}{c_pT}+\frac{dW}{W}\right]-\frac{d\gamma}{\gamma}\\
\frac{d\rho}{\rho}&=\frac{1}{1-M^2}\left[\frac{dH}{c_pT}+\frac{dW}{W}\right]\\
\frac{dp}{p}&=\frac{\gamma M^2}{1-M^2}\left[\frac{dH}{c_pT}+\frac{dW}{W}\right]
\end{aligned}\tag{4.3}$$

Combining the first and third of Equations (4.3) yields

$$\frac{dp}{p}=-\frac{d(\gamma M^2)}{1+\gamma M^2}.\tag{4.4}$$

Now Equation (4.4) may be integrated to yield

$$\frac{p_3}{p_4}=\frac{1+\gamma_4M_4^2}{1+\gamma_3M_3^2}.\tag{4.5}$$

Combining the first and second equations in Equation (4.3) and integrating yields

$$\frac{\rho_3}{\rho_4}=\frac{\gamma_4M_4^2\,(1+\gamma_3M_3^2)}{\gamma_3M_3^2\,(1+\gamma_4M_4^2)}.\tag{4.6}$$

Then, from the equation of state we find that

$$\frac{T_4}{T_3} = \frac{\gamma_4 M_4^2}{\gamma_3 M_3^2} \frac{(1+\gamma_3 M_3^2)^2}{(1+\gamma_4 M_4^2)^2} \frac{W_4}{W_3}. \tag{4.7}$$

Note that because the area of the combustor is a constant, the mass flux $\rho V =$ constant and

$$\frac{\rho_3}{\rho_4} = \frac{V_4}{V_3}. \tag{4.8}$$

The adiabatic energy equation requires that

$$\frac{T_t}{T} = 1 + \frac{\gamma - 1}{2} M^2. \tag{4.9}$$

Therefore, the stagnation temperature ratio may be found as follows:

$$\frac{T_{t,4}}{T_{t,3}} = \frac{T_{t,4}}{T_4} \frac{T_4}{T_3} \frac{T_3}{T_{t,3}} = \frac{\gamma_4 M_4^2 W_4}{\gamma_3 M_3^2 W_3} \left[ \frac{1+\gamma_3 M_3^2}{1+\gamma_4 M_4^2} \right]^2 \left[ \frac{1+\frac{\gamma_4 - 1}{2} M_4^2}{1+\frac{\gamma_3 - 1}{2} M_3^2} \right]. \tag{4.10}$$

Using the same approach, that is, using

$$\frac{p_t}{p} = \left(1 + \frac{\gamma - 1}{2} M^2\right)^{\frac{\gamma}{\gamma - 1}}, \tag{4.11}$$

then the stagnation pressure ratio across the combustor is

$$\frac{p_{t,4}}{p_{t,3}} = \frac{1+\gamma_3 M_3^2}{1+\gamma_4 M_4^2} \frac{\left[1+\frac{\gamma_4 - 1}{2} M_4^2\right]^{\frac{\gamma_4}{\gamma_4 - 1}}}{\left[1+\frac{\gamma_3 - 1}{2} M_3^2\right]^{\frac{\gamma_3}{\gamma_3 - 1}}}. \tag{4.12}$$

The stagnation pressure loss in a constant area combustor as calculated directly from Equation (4.12) with a constant value of the isentropic exponent $\gamma = 1.33$ is shown in Figure 4.1. Note that combustors have a subsonic entry Mach number and adding heat will accelerate the flow and raise the Mach number until the flow chokes, that is, until $M_4 = 1$. This can be seen from the first of Equations (4.3) in which the release of heat makes all the differentials negative so that $dM^2 > 0$ if $M < 1$. It is clear that to increase the stagnation temperature to high values it is necessary to have a low initial Mach number or else the flow will promptly choke before the desired heat addition is accomplished. It is also seen that the stagnation pressure loss is reduced as the entry Mach number is reduced.

For constant values of specific heat ratio $\gamma$ and molecular weight $W$, the results reduce to the classical Rayleigh flow equations, that is, flow in a constant area duct with simple stagnation temperature change; see, for example, Anderson (1990).

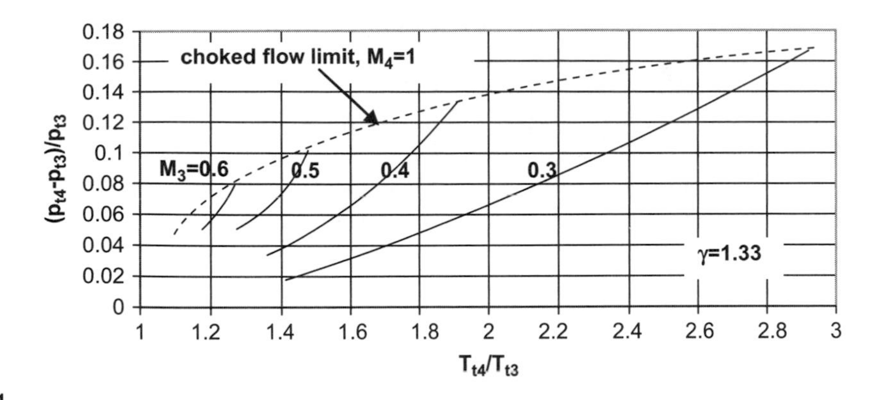


**FIGURE 4.1**

Fractional loss of stagnation pressure across a constant area combustor as a function of stagnation temperature ratio for a constant value of $\gamma = 1.33$.

## 4.4 EXAMPLE: CONSTANT AREA COMBUSTOR

Consider a ramjet-powered missile of the general type shown in a cross-sectional view in Figure 4.2. Fuel is added at mass flow rate $m_f$ while air enters the combustion chamber at rate $m_3$ with a stagnation temperature $T_{t,3} = 488$K and a Mach number of $M_3 = 0.35$. Assuming that the combustion chamber is of constant area throughout and the heating value of the fuel is 44 MJ/kg, find the maximum fuel to air ratio $m_f/m_3$ that can be sustained by the ramjet engine.

Assuming that the fuel mass flow is negligible compared to the air mass flow, the energy balance across the combustor is described by the following relation:

$$\dot{m}_3\left(c_{p,4}T_{t,4} - c_{p,3}T_{t,3}\right) \approx \dot{m}_f\eta_b HV \approx \dot{m}_3\bar{c}_p\left(T_{t,4} - T_{t,3}\right).$$

The stagnation temperature ratio across a constant area combustor is given by

$$\frac{T_{t,4}}{T_{t,3}} = \frac{T_{t,4}}{T_4}\frac{T_4}{T_3}\frac{T_3}{T_{t,3}} = \frac{\gamma_4 M_4^2 W_4}{\gamma_3 M_3^2 W_3}\left[\frac{1+\gamma_3 M_3^2}{1+\gamma_4 M_4^2}\right]^2\left[\frac{1+\frac{\gamma_4 - 1}{2}M_4^2}{1+\frac{\gamma_3 - 1}{2}M_3^2}\right]$$

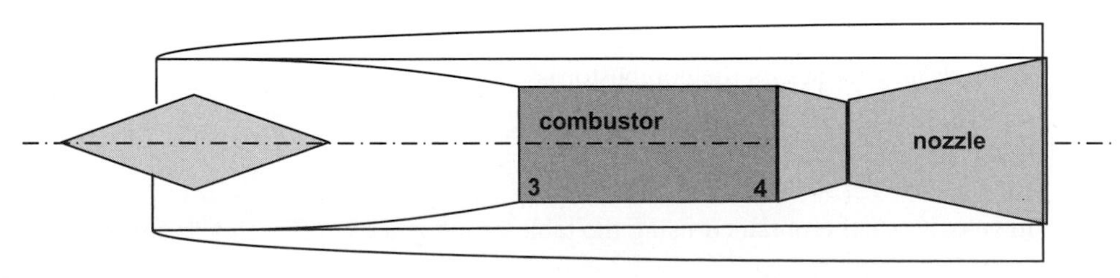


**FIGURE 4.2**

Cross-sectional view of a generic ramjet missile.

Assuming that $\gamma_3 = \gamma_4 = \gamma = 4/3$ leads to the result that

$$\frac{T_{t,4} - T_{t,3}}{T_{t,3}} = \frac{M_4^2}{M_3^2} \frac{\left(3 + 4M_3^2\right)^2}{\left(3 + 4M_4^2\right)^2} \frac{\left(6 + M_4^2\right)}{\left(6 + M_3^2\right)} - 1.$$

The maximum heat addition is that which chokes the combustor exit, that is, makes $M_4 = 1$. Setting $M_4 = 1$ and $M_3 = 0.35$ yields $\Delta T_{t,\max} = 1.327(488) = 648\text{K}$ and $T_{t,4\max} = 1136\text{K}$. Then

$$\Delta H_{\max} = \dot{m}_3 \overline{c}_p (T_{t,4\max} - T_{t,3}) = \dot{m}_f \eta_b HV$$

$$\left(\frac{\dot{m}_f}{\dot{m}_3}\right)_{\max} = \frac{\overline{c}_p (T_{t,4\max} - T_{t,3})}{\eta_b HV} = \frac{(1.15KJ/\gamma g - K)(648K)}{0.95(44MJ/\gamma g)} = 0.0178.$$

## 4.5 CONSTANT PRESSURE COMBUSTORS

It may be of interest to operate the combustor in a constant pressure mode. Using the table of influence coefficients (Table 2.1) shows that

$$\frac{dp}{p} = \frac{\gamma\gamma M^2}{\gamma - \gamma M^2}\left(\frac{dA}{A} + \frac{dH}{c_p T} + \frac{dW}{W}\right) = 0. \tag{4.13}$$

In the same fashion we find that

$$\frac{dM^2}{M^2} = -2\left[\frac{\gamma + \frac{\gamma - 1}{2}\gamma M^2}{\gamma - \gamma M^2}\right]\frac{dA}{A} - \left[\gamma \frac{1 + \gamma M^2}{\gamma - \gamma M^2}\right]\left[\frac{dH}{c_p T} + \frac{dW}{W}\right] - \frac{d\gamma}{\gamma}. \tag{4.14}$$

Combining Equations (4.13) and (4.14) yields

$$\frac{d\gamma M^2}{\gamma M^2} = -\frac{dA}{A}. \tag{4.15}$$

Therefore, the Mach number is given by the equation

$$\gamma M^2 = \frac{const}{A}. \tag{4.16}$$

The change in Mach number across the combustor is

$$\frac{\gamma_4 M_4^2}{\gamma_3 M_3^2} = \frac{A_3}{A_4}. \tag{4.17}$$

The change in velocity can be obtained using the table of influence coefficients as follows:

$$\frac{dV}{V} = -\frac{1}{1 - M^2}\left(\frac{dA}{A} - \frac{dH}{c_p T} - \frac{dW}{W}\right). \tag{4.18}$$

Comparing Equation (4.18) to Equation (4.13) for the nonzero Mach number leads to the result that $dV = 0$, or that $V =$ constant. This shouldn't be surprising, as the momentum equation for frictionless flow, from Section 2.8, is

$$\frac{dp}{p} + \frac{1}{2}\gamma M^2 \frac{dV^2}{V^2} = 0. \tag{4.19}$$

This clearly shows that if there is no pressure change, there is no velocity change, and the velocity is constant in a constant pressure combustor, $V_3 = V_4$. The variation of density may also be found using the table of influence coefficients along with the other thermodynamic variables:

$$\begin{aligned} \frac{\rho_3}{\rho_4} &= \frac{A_4}{A_3} \\ \frac{T_3}{T_4} &= \frac{A_3}{A_4}\frac{W_3}{W_4} \\ \frac{p_3}{p_4} &= 1 = \frac{V_3}{V_4} \\ \frac{M_3}{M_4} &= \sqrt{\frac{\gamma_4 A_4}{\gamma_3 A_3}} \end{aligned} \tag{4.20}$$

The stagnation pressure and temperature are given by

$$\frac{p_{t,3}}{p_{t,4}} = \frac{\left[1 + \frac{\gamma_3 - 1}{2}M_3^2\right]^{\frac{\gamma_3}{\gamma_3 - 1}}}{\left[1 + \frac{\gamma_4 - 1}{2}M_4^2\right]^{\frac{\gamma_4}{\gamma_4 - 1}}} \tag{4.21}$$

$$\frac{T_{t,3}}{T_{t,4}} = \frac{W_3}{W_4}\frac{\gamma_4 M_4^2}{\gamma_3 M_3^2}\frac{1 + \frac{\gamma_3 - 1}{2}M_3^2}{1 + \frac{\gamma_4 - 1}{2}M_4^2} \tag{4.22}$$

Thus all the flow variables at the end of the combustor are known if the inlet conditions are known, independent of the details of the heat release process.

The stagnation pressure loss in a constant pressure combustor as calculated directly from Equation (4.21) with a constant value of the specific heat ratio $\gamma = 1.33$ is shown in Figure 4.3. Note that to satisfy the constant pressure requirement for a subsonic flow with heat addition, Equation (4.13) shows that $dA > 0$, as the other terms will act to reduce the pressure. Then Equation (4.15) shows that the Mach number must decrease further when the combustor has a subsonic entry Mach number. Thus there is no difficulty with choking because the area increase needed to maintain constant pressure operation also maintains a subsonic Mach number. It is clear that to increase the stagnation temperature to high values it is necessary to have a low initial Mach number or else the flow will incur substantial stagnation pressure loss before the desired heat addition is accomplished. Stagnation pressure losses in the constant pressure combustor are smaller than those of the constant area

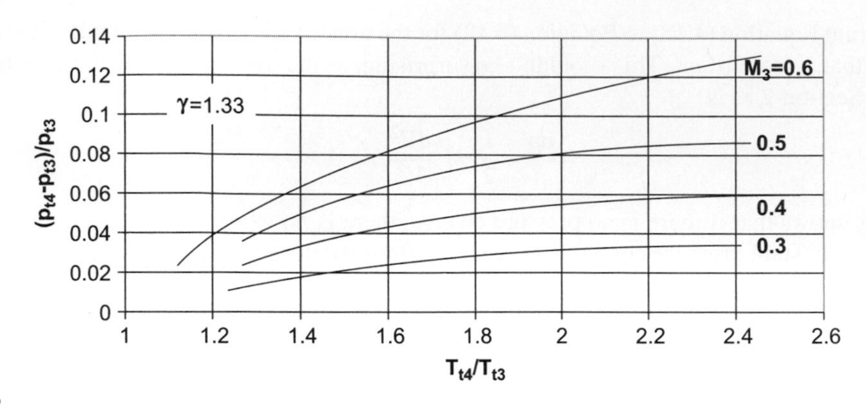


**FIGURE 4.3**

Fractional loss of stagnation pressure across a constant pressure combustor as a function of stagnation temperature ratio for a constant value of $\gamma = 1.33$.

combustor for the same stagnation temperature rise. For example, with $M_3 = 0.3$ and $T_{t4}/T_{t3} = 2$ the constant area combustor suffers a 7% loss, while the constant pressure combustor has only a 3% loss.

## 4.6 FUELS FOR AIR-BREATHING ENGINES

Jet fuels are primarily paraffin hydrocarbons with the chemical formula $C_nH_{2n+2}$. The stoichiometric chemical reaction between such fuels and air is given by

$$C_nH_{2n+2} + \alpha\left[O_2 + \frac{79}{21}N_2\right] \rightarrow ACO_2 + BH_2O + \alpha\frac{79}{21}N_2. \tag{4.23}$$

Conservation of atoms requires that

$$\begin{aligned} C &\rightarrow n = A \\ H &\rightarrow 2n + 2 = 2B \quad . \\ O &\rightarrow 2\alpha = 2(A + B) \end{aligned} \tag{4.24}$$

Then the molar fuel to air ratio $\alpha = \dfrac{3n+1}{2}$ and the stoichiometric reaction is written as

$$C_nH_{2n+2} + \frac{3n+1}{2}\left[O_2 + \frac{79}{21}N_2\right] \rightarrow nCO_2 + (n+1)H_2O + \frac{3n+1}{2}\left(\frac{79}{21}\right)N_2. \tag{4.25}$$

The fuel to air ratio by weight is then given by

$$\left(\frac{\dot{w}_f}{\dot{w}_a}\right)_{stoich} = \frac{\sum\limits_{fuel} W_iC_i}{\sum\limits_{air} W_iC_i} = \frac{(12)n + (1)(2n+2)}{\dfrac{3n+1}{2}\left[(32) + \dfrac{79}{21}(28)\right]} = \frac{14n+2}{68.67(3n+1)}. \tag{4.26}$$

Consider the case of octane, the paraffin hydrocarbon with $n = 8$:

$$C_8H_{18} + 12.5O_2 + 12.5\left(\frac{79}{21}\right)N_2 \rightarrow 8CO_2 + 9H_2O + 12.5\left(\frac{79}{21}\right)N_2 \tag{4.27}$$

The fuel to air ratio for stoichiometric combustion of octane is $\frac{f}{a} = \frac{\dot{w}_f}{\dot{w}_a} = 0.066$.

For hydrocarbon fuels of this type, the following characteristics are noted:

- Mixture ratios in the range $0.04 < f/a < 0.25$ can be burned
- Maximum reaction rate occurs for $f/a = 0.073$
- Complete combustion occurs for $f/a = 0.066$
- Combustion is slow and erratic for $f/a < 0.055$

The metric used to indicate the relationship of the actual mixture ratio to that required for stoichiometric combustion is the equivalence ratio:

$$\phi = \frac{\left(\frac{\dot{w}_f}{\dot{w}_a}\right)}{\left(\frac{\dot{w}_f}{\dot{w}_a}\right)_{stoich}}. \tag{4.28}$$

An equivalence ratio $\phi < 1$ signifies a fuel-lean mixture, whereas $\phi > 1$ signifies a fuel-rich mixture. In the former case, a fuel-lean mixture, there would be additional products on the right-hand side of Equation (4.23), including $CO_2$ and $O_2$, because there is insufficient fuel to completely use the available oxygen. Conversely, for a fuel-rich mixture, there would be additional products, such as CO and $H_2O$, because the abundance of fuel uses up all the available oxygen and the oxidation process can't be fully completed.

The energy content of the fuel is expressed by the heating value (*HV*) of the fuel, which is a measure of the enthalpy available in the fuel for increasing the temperature of the combustion products. It is equal in magnitude to the heat of reaction discussed earlier. There are typically two values given for the heating value: the lower *HV*, where water in the products is present in vapor form, and the higher *HV*, where the water is present as a liquid and the heat of vaporization is recovered. Because the temperatures of interest in propulsion applications are so high that liquid water would not be present in the combusted gases, we always use the lower heating value for the fuel.

An expression for the heating value of hydrocarbon fuels based on the specific gravity of the fuel was given by NBS (now NIST) as

$$HV\ (\text{Btu/lb}) = 19,960 + 1360(SG) - 3780(SG)^2. \tag{4.29}$$

With the conversion 1 Btu/lb $= 2.326 \times 10^{-3}$ MJ/kg, this becomes

$$HV(\text{MJ/kg}) = 46.43 + 3.163(SG) - 9.019(SG)^2.$$

Characteristics of some fuels for air-breathing engines taken from Goodger (1982) appear in Table 4.1.

It is worth noting that military aircraft use JP-8, a fuel almost identical to Jet A, which is used by commercial airliners, except for certain additives needed for military operations, such as rust inhibitors and icing inhibitors. The cloud point, the temperature at which wax crystals separate and cloud the

**Table 4.1** Characteristics of Some Fuels for Air-Breathing Engines

| Fuel | $(f/a)_{stoich}$ | Oxides | *HV* (MJ/kg)[a] | *HV'* (MJ/liter)[a] | $T_c$ (K)[b] |
|---|---|---|---|---|---|
| Jet A | 0.0678 | $CO_2$, $H_2O$ | 43.4 | 34.7 | 2.295 |
| Hydrogen | 0.0292 | $H_2O$ | 120.24 | 8.42 | 2.431 |
| Methane | 0.0582 | $CO_2$, $H_2O$ | 50 | 21.2 | 2.246 |
| Ethanol | 0.111 | $CO_2$, $H_2O$ | 27.23 | 21.62 | 2.214 |
| Ammonia | 0.165 | NO, $H_2O$ | 18.55 | 11.41 | 2.084 |

[a]*Lower heating value is implied*
[b]*Adiabatic flame temperature; see Section 4.12*

fuel, for JP-8 and Jet A is about −46°C (227K). The skin temperature of aircraft cruising at $M_0 = 0.8$ on a standard day is about −31°C (242K) and means for fuel heating must be considered for maintaining a margin of safety, particularly for long-range aircraft.

## 4.7 COMBUSTOR EFFICIENCY

Combustor efficiency is defined as the ratio of actual heat released by the reaction and the ideal amount of heat released according to the heating value of the fuel. It may be expressed as follows:

$$\eta_b = \frac{\dot{w}_4 h_{t,4} - \dot{w}_3 h_{t,3} - \dot{w}_f h_{t,f}}{\dot{w}_f HV} = \frac{\left(\dot{w}_3 + \dot{w}_f\right) h_{t,4} - \dot{w}_3 h_{t,3} - \dot{w}_f h_{t,f}}{\dot{w}_f HV}. \tag{4.30}$$

The fuel to air ratio for paraffin fuels, as described earlier, is on the order of 6% to 7%, and, as we'll soon see, the actual fuel to air ratio passing through the combustor will be only about one-third of that. The extra air is not used in the combustion process but rather is added to cool down the combustion products because of the temperature limitations of the materials available for combustors and the other downstream components of the flow path. Furthermore, the enthalpy of the fuel prior to injection into the combustor, denoted by the term $h_{t,f}$, is typically smaller than the total enthalpy of the entering air stream. Then we may rewrite Equation (4.30) for the condition $\dot{w}_f/\dot{w}_a = \dot{w}_f/\dot{w}_3 \ll 1$ (which also means that the quantity $\dot{w}_4/\dot{w}_3 \approx 1$) as follows:

$$\eta_b = \frac{\frac{\dot{w}_4}{\dot{w}_3} h_{t,4} - h_{t,3} - \frac{\dot{w}_f}{\dot{w}_3} h_{t,f}}{\frac{\dot{w}_f}{\dot{w}_3} HV} \approx \frac{h_{t,4} - h_{t,3}}{\frac{\dot{w}_f}{\dot{w}_3} HV} \approx \frac{\bar{c}_p\left(T_{t,4} - T_{t,3}\right)}{\frac{\dot{w}_f}{\dot{w}_3} HV}. \tag{4.31}$$

It is worth noting that although it is commonly assumed, for simplicity, that the fuel weight flow rate is merely neglected, the air weight flow rate bled from the compressor for various engine and accessory cooling purposes is about the same so that $\dot{w}_4/\dot{w}_3 \approx 1$ is a quite sound assumption Note that the approximate form in Equation (4.31) uses an average value for the specific heat at constant pressure across the combustor.

The parameter that best describes combustor efficiency is $p_{t,3}T_{t,3}/V_3$, which captures the attributes of a good combustor outlined at the start of this section, namely, high pressure, high temperature, and

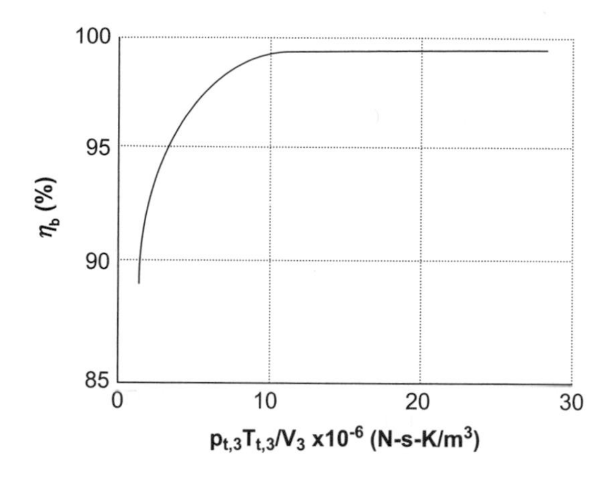


**FIGURE 4.4**

Typical variation of combustor efficiency $\eta_b$ with the parameter $(p_{t,3}T_{t,3}/V_3)$.

long residence time in the combustor. Achieving high values of this parameter involves having a high temperature of the reactants and sufficient time for complete mixing and reaction. The general trend of $\eta_b$ is shown in Figure 4.4.

The burner entrance pressure effectively limits altitude performance of the combustor because the burner efficiency drops off rapidly as the parameter $p_{t,3}T_{t,3}/V_3$ falls off. The compressor of the turbine engine, which is discussed in later chapters, provides an essentially fixed pressure ratio across it, as shown here:

$$p_{t,3} = \frac{p_{t,3}}{p_{t,2}}\frac{p_{t,2}}{p_{t,0}}\frac{p_{t,0}}{p_0}p_0 = \frac{p_{t,3}}{p_{t,2}}\frac{p_{t,2}}{p_{t,0}}p_0\left(1+\frac{\gamma-1}{2}M_0^2\right)^{\frac{\gamma}{\gamma-1}}. \tag{4.32}$$

Thus the burner entrance pressure depends on the compressor pressure ratio, inlet pressure recovery, flight Mach number, and free stream pressure, which is a function of altitude. The stagnation pressure $p_{t,3}$ drops by a factor of about 2.5 between take-off conditions and those at high subsonic cruise in the stratosphere. We will also see, in later sections, that the compressor pressure ratio depends on the rotational speed of the engine. The effect of different rotor speeds is illustrated schematically in Figure 4.5.

## 4.8 COMBUSTOR CONFIGURATION

The combustor is a flow duct that accepts the high-pressure air stream from the compressor and adds fuel to it in a manner that promotes good mixing. The fuel is typically in liquid form and must be vaporized to ensure the required mixing. This in turn requires the fuel to be injected through small orifices under high pressure, an approach that naturally destabilizes the fluid stream, causing it to break up into many fine droplets. The increased surface to volume ratio of the small moving fuel droplets enhances evaporation and promotes mixing with the interacting air stream. A detailed study of droplets and sprays is presented by Sirignano (2010).

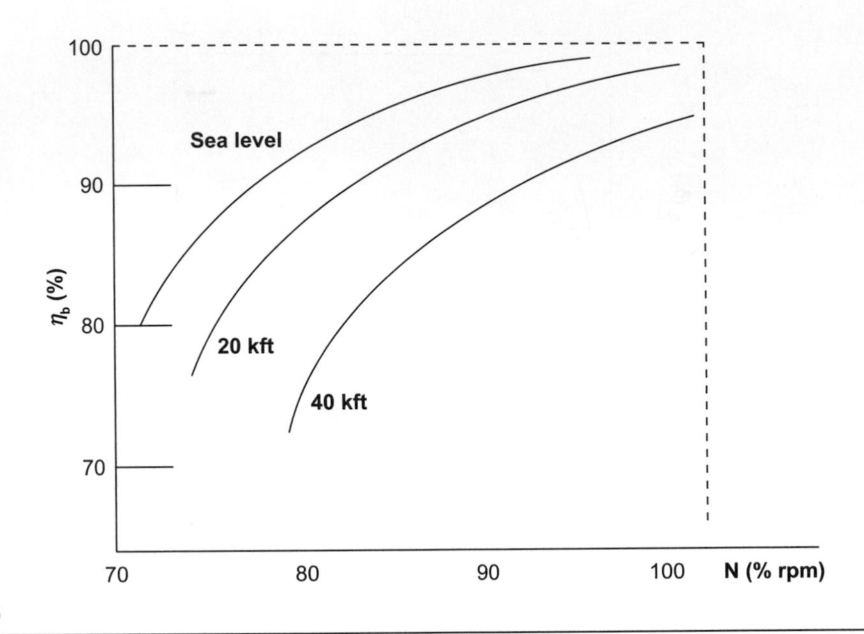


FIGURE 4.5

Effect of altitude and engine rpm on burner efficiency.

Air from the compressor enters a diffuser to slow it down and to split it into a primary and a secondary stream. The primary air mixes with the atomized fuel and reacts with it, raising the temperature to very high values, while the secondary stream is mixed with the combustion products to control the temperature of the gas leaving the combustor and entering the turbine. A schematic configuration of a simple can-type combustor is shown in Figure 4.6. As the name suggests, the

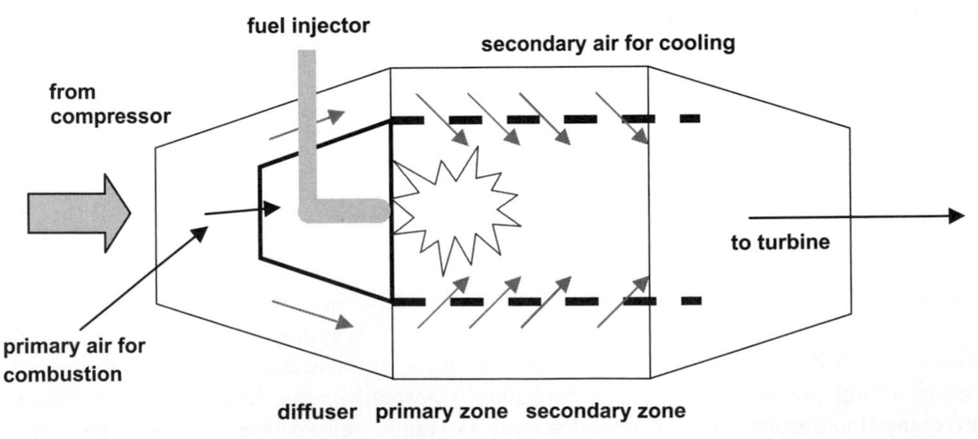


FIGURE 4.6

Schematic diagram of a combustor.

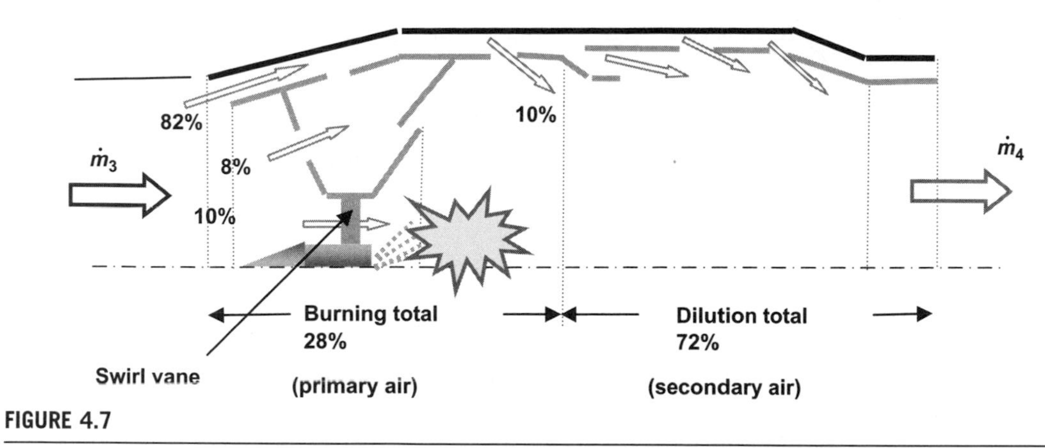


FIGURE 4.7

Apportionment of air in a typical combustor.

combustor is cylindrical in shape and has provision for the fuel injector to introduce fuel as a mist to mix with the primary air and burn locally at close to the stoichiometric mixture ratio. The primary zone serves as a flame holder where the low velocity ensures flame stability. Secondary air passes through the perforated or slotted liner, serving to keep it and the outer structure of the combustor at acceptable temperatures. This secondary air then mixes with the hot gas from the primary combustion zone and is directed toward the turbine station.

A more detailed description of the apportionment of the air between burning purposes (primary air) and cooling purposes is shown in Figure 4.7. About 82% of the entering air flow ($\dot{m}_3$) begins to bypass the combustion zone, while the other 18% continues axially. Of this 18%, 10% flows through a swirl vane directly toward the combustion zone where the fuel injector sprays in a fine mist of droplets that are mixed into the swirling flow and ignited. The other 8% diverts around the periphery of the combustion zone, along with about 10% from the originally diverted 82%. The remaining 72% of the incoming air is used primarily for cooling purposes and doesn't interact with the burning zone. The fully mixed combination of combustion products and cooling air leaves as a hot flow ($\dot{m}_4$) with a reasonably uniform temperature ($T_4$).

The temperature of the flow leaving the combustor is moderated by the degree to which cooling air is mixed with the products of combustion. An appreciation of the magnitude that may be expected is provided by Figure 4.8, which shows the combustor exit flow temperature ($T_4$) as a function of the percent cooling air mixed with the products of combustion under stoichiometric conditions. Also shown in Figure 4.8 is the range of usual operating conditions consistent with reliable performance.

Individual can combustors are arrayed around the engine such that the entrance and exit areas line up with the compressor outlet and the turbine inlet, respectively. Although this arrangement has advantages in terms of repair and replacement, an alternative configuration that has become standard is the annular combustor. This is simply an extension of the can combustor in which the cross section of the combustor, as shown in Figure 4.6, for example, is rotated around the axis of the engine, forming a single annular space. A typical practical annular combustor is shown in Figure 4.9.

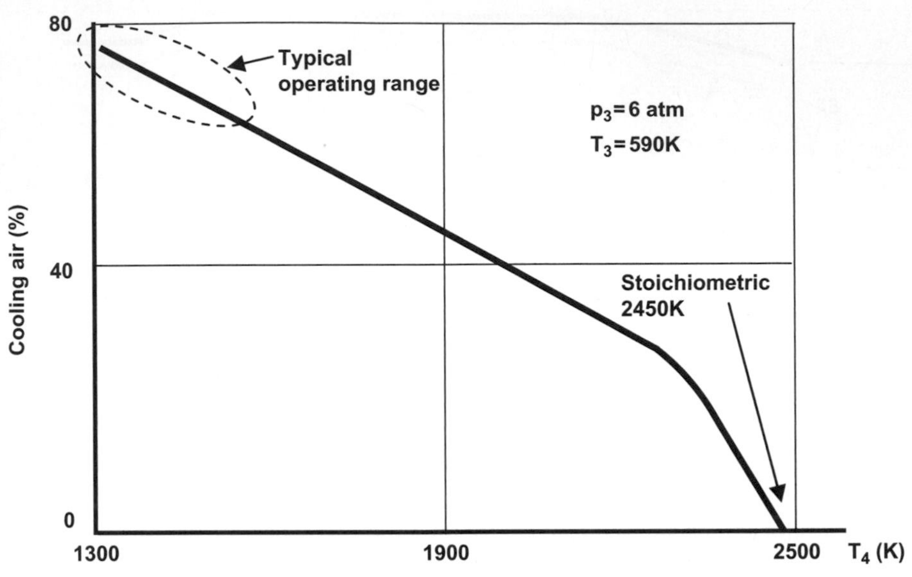


FIGURE 4.8

Variation of the combustor exit temperature as a function of the amount of cooling air provided to the products of a stoichiometric hydrocarbon–air reaction. Also shown is the typical operating range for maintaining reliable sustained performance.

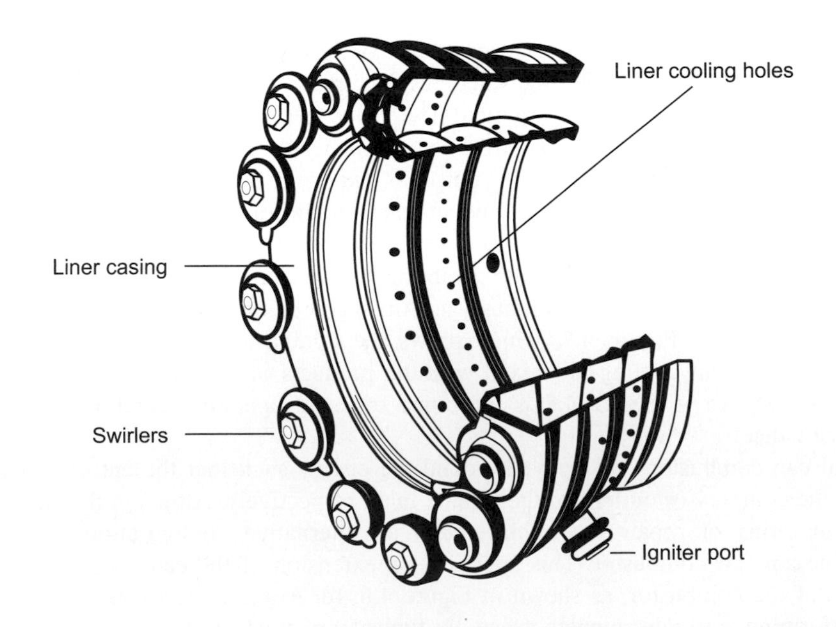


FIGURE 4.9

A cutaway view of an annular combustor.

## 4.9 EXAMPLE: SECONDARY AIR FOR COOLING

A constant pressure combustor takes on air from the compressor at a velocity $V_3 = 30$ m/s and a stagnation temperature $T_{t,3} = 690$K. The combustor is designed to burn fuel with a primary fuel to air ratio $(f/a)_p = 0.06$ with a burner efficiency $\eta_b = 90\%$. Jet A fuel has a heating value $HV = 43.4$ MJ/kg, and the ratio of secondary to primary air flow $\dot{m}_{3,s}/\dot{m}_3$ can be varied by inserting different perforated combustor liners. For values of secondary to primary air flow of 0, 0.2, 0.4, 0.6, and 0.8, determine the (a) stagnation temperature leaving the combustor $T_{t,4}$, (b) percentage stagnation pressure loss across the combustor, and (c) absolute velocity leaving the combustor $V_4$. Assume constant average values for $\gamma = 1.33$ and for $c_p = 1.19$ kJ/kg-K. Frictional effects may be neglected because of the low entrance velocity.

**(a)** Solving for $T_{t,4}$ from Equation (4.30) yields

$$T_{t,4} = \frac{\dfrac{\dot{m}_f}{\dot{m}_3}\eta_b HV + c_p T_{t,3}}{\left(1 + \dfrac{\dot{m}_f}{\dot{m}_3}\right)c_p}.$$

Here we have assumed that the contribution of the enthalpy of the injected fuel is negligible compared to that of the entering air flow. The air flow entering may be broken up into primary and secondary flows so that

$$\dot{m}_3 = \frac{\dot{m}_{3,p}}{\dot{m}_f}\dot{m}_f + \dot{m}_{3,s}.$$

Then we have

$$\frac{\dot{m}_f}{\dot{m}_3} = \left(1 - \frac{\dot{m}_{3,s}}{\dot{m}_3}\right)\frac{\dot{m}_f}{\dot{m}_{3,p}} = 0.6\left(1 - \frac{\dot{m}_{3,s}}{\dot{m}_3}\right).$$

Using this result in the equation for $T_{t,4}$ yields

$$T_{t,4} = \frac{0.06\left(1 - \dfrac{\dot{m}_{3,s}}{\dot{m}_3}\right)\eta_b HV + c_p T_{t,3}}{c_p\left[1 + 0.06\left(1 - \dfrac{\dot{m}_{3,s}}{\dot{m}_3}\right)\right]}.$$

This equation may be solved for the selected values of secondary to total air flow. Results for combustor exit temperature as a function of the cooling air ratio are shown in Figure 4.10.

**(b)** In order to apply the equations for flow in this constant pressure combustor we need the entering Mach number. At the entrance to the combustor, the energy equation may be written as

$$c_p T_3 + \frac{1}{2}V_3^2 = c_p T_{t,3}.$$

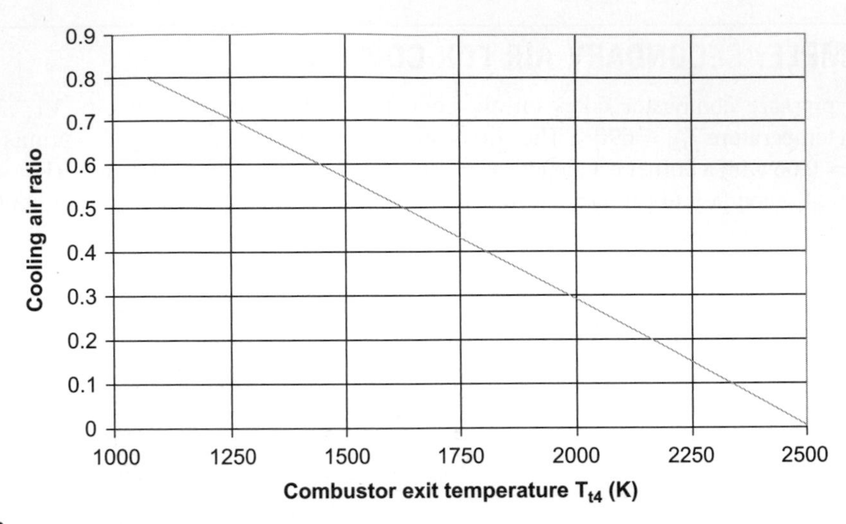


**FIGURE 4.10**

The reduction in combustor exit stagnation temperature as a function of the ratio of cooling secondary air to the total entering air flow.

With the values given, one may compute $T_3 = 689.6\,\text{K}$ so that $a_3 = 513.1$ m/s and $M_3 = 0.058$. Constant pressure combustors were shown to have decreasing Mach numbers along the combustor. Because the very low Mach number at the entrance means an even lower Mach number at the exit, we may approximate Equation (4.22) as follows:

$$\frac{T_{t,4}}{T_{t,3}} \approx \left(\frac{M_3}{M_4}\right)^2.$$

By the same reasoning it is clear from Equation (4.21) that

$$\frac{p_{t,4}}{p_{t,3}} \approx \left(\frac{M_4^2}{M_3^2}\right)^{\frac{\gamma}{\gamma-1}}.$$

Furthermore, the area ratio required for constant pressure operation may be calculated from Equation (4.20). Results are given in Table 4.2. It is clear that the stagnation pressure loss is only around 1% and that area ratios for the exit temperatures of interest are 2 or less. The low Mach number also means that the stagnation and static temperature values are essentially equal.

**(c)** The basic result for constant pressure combustors with negligible friction is that the velocity remains constant so $V_4 = 30$ m/s for all cooling ratios.

**Table 4.2** Constant Pressure Combustor Properties as a Function of the Ratio of Cooling Secondary Air to the Total Entering Air Flow

| Air Ratio | $T_{t,4}$ (K) | $p_{t,4}/p_{t,3}$ | $A_4/A_3$ |
|---|---|---|---|
| 0 | 2509 | 0.9984 | 3.514 |
| 0.2 | 2162 | 0.9985 | 3.028 |
| 0.4 | 1807 | 0.9986 | 2.530 |
| 0.6 | 1443 | 0.9989 | 2.021 |
| 0.8 | 1071 | 0.9993 | 1.500 |

## 4.10 CRITERIA FOR EQUILIBRIUM IN CHEMICAL REACTIONS

There are two basic conditions for establishing chemical equilibrium:

$dS = 0$ for equilibrium at constant $p$ and $H$
$dG = 0$ for equilibrium at constant $p$ and $T$

Chemical reactions in equilibrium chemistry involve dealing with the special case of stoichiometric reactions. Stoichiometric reactions have their reactants in exactly the correct proportion so as to yield the minimum number of products, as opposed to general reactions where there are more, and sometimes many more, product species than the minimum. Stoichiometric reactions are usually written as follows:

$$\sum_{j=1}^{n} \nu_j' M_j \rightarrow \sum_{j=1}^{n} \nu_j'' M_j, \tag{4.33}$$

where $\nu_j'$ and $\nu_j''$ are the stoichiometric coefficients of the reactants and products for the species $M_j$, and $j = 1, 2, 3, \ldots, n$, where $n$ is the total number of species involved. The difference in the Gibbs function is

$$\Delta G = \sum_{j=1}^{n} (\nu_j'' - \nu_j') G_j. \tag{4.34}$$

For ideal gases,

$$G = H - TS$$

$$\Delta G = \sum_{j=1}^{n} (\nu_j'' - \nu_j') G_j^0 + RT \left[ \sum_{j=1}^{n} \ln (p_j)^{\nu_j'' - \nu_j'} \right]. \tag{4.35}$$

Then we may define

$$\Delta G^0 = \sum_{j=1}^{n} (\nu_j'' - \nu_j') G_j^0.$$

Then Equation (4.35) becomes

$$\Delta G - \Delta G^o = RT \sum_{j=1}^{n} \ln \left( p_j \right)^{\nu_j'' - \nu_j'}. \tag{4.36}$$

For equilibrium at a fixed pressure, measured in atmospheres, and temperature $\Delta G = 0$, and

$$\Delta G^o = -RT \prod_{j=1}^{n} \left( p_{j,e} \right)^{\nu_j'' - \nu_j'} = -RT \ln K_p. \tag{4.37}$$

Here Equation (4.37) defines the equilibrium constant at constant pressure $K_p$ as

$$K_p = \prod_{j=1}^{n} (p_{j,e})^{\nu_j'' - \nu_j'}.$$

The utility of the equilibrium constant arises from the fact that $K_p = K_p(T)$ for any arbitrary constant pressure. Note that tables of equilibrium constants, like those in Appendix G, select certain species as being in the standard state for which $\ln K_p$, as given by Equation 4.37, will be zero. These species are then considered as the reactants from which other species are formed. From the definition of the Gibbs function,

$$\Delta H^o - T\Delta S^o = -RT \ln K_p. \tag{4.38}$$

Then, using the combined first and second laws of thermodynamics,

$$dH = TdS + Vdp.$$

We combine this with the definition of the standard Gibbs free energy function to find

$$\frac{d}{dT}\left(\frac{G^o}{T}\right) = -\frac{H^o}{T^2} \tag{4.39}$$

From Equation (4.37) we find

$$\frac{d}{dT}\left(\frac{\Delta G^o}{T}\right) = -R\frac{d}{dT}(\ln K_p) \tag{4.40}$$

Then, using Equations (4.39) and (4.40), we obtain

$$\frac{d}{d\left(\frac{1}{T}\right)}(\ln K_p) = -\frac{\Delta H^o}{R} \tag{4.41}$$

This is the Van't Hoff equation, which permits experimental determination of the equilibrium constant $K_p$ from the heat release in the reaction carried out at 1 atmosphere; the result is applicable to arbitrary constant pressure.

## 4.11 CALCULATION OF EQUILIBRIUM COMPOSITIONS

A general chemical reaction where $a$ and $b$ denote the number of moles in the various reactant and product species, respectively, may be written as follows:

$$\sum_{i=1}^{n} a_i A_{i,g} + \sum_{\Upsilon=1}^{n_\Upsilon} a_\Upsilon A_{\Upsilon,c} \leftrightarrow \sum_{j=1}^{m} b_j B_{j,g} + \sum_{q=1}^{m_q} b_q B_{q,c}, \tag{4.42}$$

where the symbols $g$ and $c$ refer to gaseous and condensed (solid or liquid) phases. A chemical reaction involving more than one phase is called a heterogeneous reaction. At a given temperature the vapor pressures of the condensed species are independent of the quantity of those species present. Then the vapor pressures are constant and will be small compared to the pressure of the gas mixture. The formal expression of the equilibrium constant for Equation (4.42) is as follows:

$$\tilde{K}_p = \frac{\prod_{j=1}^{m} p_{B_j}{}^{b_j} \prod_{q=1}^{m_q} p_{B_q}{}^{b_q}}{\prod_{i=1}^{n} p_{A_i}{}^{a_i} \prod_{\Upsilon-1}^{n_\Upsilon} p_{A_\Upsilon}{}^{a_\Upsilon}}. \tag{4.43}$$

Here we will concentrate on homogeneous reactions. For a more general discussion, see Penner (1957). The equilibrium constant for the gaseous species alone is

$$K_p = \frac{\prod_{j=1}^{m} \left(p_{B_{j,e}}\right)^{b_j}}{\prod_{i=1}^{n} \left(p_{A_{i,e}}\right)^{a_i}}. \tag{4.44}$$

Note that pressure is always measured in atmospheres in these calculations. The major requirement for determining the equilibrium composition is that all the subreactions are simultaneously in equilibrium. The subreactions considered are stoichiometric so that they involve the stoichiometric coefficients: $\nu_j''$ for the products and $\nu_j''$ for the reactants. Thus we will deal with the equilibrium constants for these stoichiometric subreactions as given by

$$K_p = \prod_{j=1}^{n} \left(p_{j,e}\right)^{\nu_j''-\nu_j'}. \tag{4.45}$$

In the reaction described by Equation (4.42), the total number of species present on both sides is $N$ (some species may not appear on both sides of a given reaction) and they are composed from a certain number of elements $L$. It can be shown (Wilkins, 1963) that given the individual number of moles of each reactant and the temperature and pressure, there are $L$ equations representing the conservation of elemental atoms, $N$-$L$ equations representing equilibrium among the subreaction, and one equation for the total number of moles of the products. The gas mixture thermodynamics review in Section 2.6.1 is of assistance in the analysis. The resulting system of equations rarely admits an analytic solution, and some computational approach must be applied to determine the composition of the products. Perhaps the most successful and long-lived approach was developed by NASA researchers, particularly Gordon and McBride (1994 and 1996). This approach remains applicable, has been developed and improved over the years, and is available for online calculation as part of the NASA Glenn Research Center Chemical Equilibrium Analysis program, which includes various applications to rockets, shock waves, etc. It is advantageous though to carry out solutions to representative problems in chemical equilibrium chemistry in order to develop an appreciation for the nature of the analysis that the large-scale codes perform for more complex problems. Several such fairly simple examples are carried out in the succeeding subsections.

## 4.12 EXAMPLE: HOMOGENEOUS REACTIONS WITH A DIRECT SOLUTION

Fluorine is a very energetic oxidizer, and we investigate it here in combination with hydrogen as the fuel. The homogeneous, or single phase, reaction of gaseous hydrogen and fluorine at a specified pressure and temperature may be described by the chemical equation

$$H_2 + F_2 \rightarrow a_1'' HF + a_2'' H_2 + a_3'' H + a_4'' F_2 + a_5'' F.$$

Here 1 mole of molecular hydrogen is reacted with 1 mole of molecular fluorine to form five species formed from two elements. Atom mass conservation yields

$$\begin{aligned} H : 2 &= a_1'' + 2a_2'' + a_3'' \\ F : 2 &= a_1'' + 2a_4'' + a_5'' \end{aligned} . \tag{4.46}$$

The total number of moles of products is

$$a'' = \sum_{j=1}^{5} a_j''. \tag{4.47}$$

We have six unknowns, the five $a_i''$ and the total number of moles $a''$ of the product species, and only three equations so we need three more equations. We may use the notion of simultaneous equilibrium among the subreactions forming the products by using the $K_{p,j}$ for three such subreactions. This means that among the five product species, the amounts of each must independently be consistent with their equilibrium formation. Dalton's law defines the total pressure of the mixture of product gases:

$$p = \sum_{j=1}^{n} p_j.$$

We also use the relation between mole fractions and partial pressures

$$X_j = \frac{p_j}{p} = \frac{a_j''}{a''}.$$

Among the product species assumed, we may consider that the following subreactions are continually maintaining equilibrium at the given pressure and temperature:

$$\begin{aligned} &H + F \rightarrow HF \\ &K_{p,HF} = \frac{p_1}{p_3 p_5} = \frac{a_1''}{a_3'' a_5''}\left(\frac{a''}{p}\right) \\ &2H \rightarrow H_2 \\ &K_{p,H_2} = \frac{p_2}{p_3^2} = \frac{a_2''}{a''^2_3}\left(\frac{a''}{p}\right) \\ &2F \rightarrow F_2 \\ &K_{p,F_2} = \frac{p_4}{p_5^2} = \frac{a_4''}{a_5''^2}\left(\frac{a''}{p}\right) \end{aligned} . \tag{4.48}$$

**Table 4.3** Equilibrium Constants for the $H_2$–$F_2$ System

| | 3000K | 4500K |
|---|---|---|
| $\text{Log}_{10} K_{p,HF}$ | 3.7832 | 0.3249 |
| $\text{Log}_{10} K_{p,H_2}$ | 1.6069 | −1.0731 |
| $\text{Log}_{10} K_{p,F_2}$ | −3.8207 | −4.8158 |

Now there are six equations, Equations (4.46), (4.47), and (4.48), and six unknowns, the five $a_j''$ and the total number of moles $a''$. This set may be solved for the equilibrium composition at a given pressure and temperature. We may seek to simplify the set by reducing the number of unknowns if we can estimate that some may be neglected. If we consider the range of combustion temperatures for the hydrogen–fluorine system, namely $3000\,\text{K} < T < 4500\,\text{K}$, we may examine the equilibrium constants (taken here, for example, from Gordon and McBride, 1996) in Table 4.3.

Because all the $K_p$ relations in Equation (4.48) have a common factor of $a''/p$ and $K_{p,F2} \ll K_{p,HF}$, $K_{pH2}$, we have $a_4'' \ll a_5''^2$ and therefore the number of moles of $F_2$ is negligible with respect to those of the other product species. We may eliminate the last equation in the set of Equation (4.48), leaving us with five equations and five unknowns. Eliminating $a_3''$ , $a_5''$, and $a''$ from this set leaves the following:

$$\begin{aligned} pK_{p,HF} &= \frac{a_1''(4 - a_1'' - a_2'')}{(2 - a_1'')(2 - a_1'' - 2a_2'')} \\ \frac{K_{p,HF}}{K_{p,H_2}} &= \frac{a_1''(2 - a_1'' - 2a_2'')}{a_2''(2 - a_1'')} \end{aligned} \quad (4.49)$$

For a specified pressure and temperature, Equation (4.49) constitutes a set of two equations in two unknowns, linear in $a_2''$, which may be solved by various techniques.

## 4.13 EXAMPLE: HOMOGENEOUS REACTIONS WITH TRIAL-AND-ERROR SOLUTION

A common rocket propellant combination consists of hydrogen fuel and oxygen. Consider a hydrogen–oxygen reaction given by

$$H_2 + O_2 \rightarrow a_1''O_2 + a_2''H_2 + a_3''O + a_4''H + a_5''OH + a_6''H_2O. \quad (4.50)$$

Here we have 1 mole of hydrogen fuel reacting with 1 mole of oxygen to form six species and two elements. Atom conservation requires that

$$\begin{aligned} H: 2 &= 2a_2'' + a_4'' + a_5'' + 2a_6'' \\ O: 2 &= 2a_1'' + a_3'' + a_5'' + a_6'' \end{aligned} \quad (4.51)$$

Equilibrium among the subreactions is given by the following:

$$\begin{aligned}
&\frac{1}{2}O_2 \rightarrow O \\
&K_{p,O} = \frac{p_3}{\sqrt{p_1}} \\
&\frac{1}{2}H_2 \rightarrow H \\
&K_{p,H} = \frac{p_4}{\sqrt{p_2}} \\
&\frac{1}{2}H_2 + \frac{1}{2}O_2 \rightarrow OH \\
&K_{p,OH} = \frac{p_5}{\sqrt{p_1 p_2}} \\
&H_2 + \frac{1}{2}O_2 \rightarrow H_2O \\
&K_{p,H_2O} = \frac{p_6}{p_2\sqrt{p_1}}
\end{aligned} \tag{4.52}$$

Using the equation of state, Dalton's law, and the relationship between mole fractions and the ratio of partial to total pressure of the mixture,

$$\begin{aligned}
pV &= a''RT \\
p &= \sum_{i=1}^{6} p_i \\
\frac{a_i''}{a''} &= \frac{p_i}{p}
\end{aligned}.$$

Introducing these relations permits rewriting the atom conservation relations as follows:

$$\begin{aligned}
2\frac{p}{a''} &= 2p_2 + p_4 + p_5 + 2p_6 \\
2\frac{p}{a''} &= 2p_1 + p_3 + p_5 + p_6
\end{aligned}. \tag{4.53}$$

We may recast the partial pressures of the products in terms of those products present in the reactants:

$$\begin{aligned}
p_3 &= K_{p,1}\sqrt{p_1} \\
p_4 &= K_{p,2}\sqrt{p_2} \\
p_5 &= K_{p,3}\sqrt{p_1 p_2} \\
p_6 &= K_{p,4}p_2\sqrt{p_1}
\end{aligned}. \tag{4.54}$$

Introducing these values into one of the atom conservation equations, Equation (4.53) yields

$$\sqrt{p_1} = \frac{\left[2\frac{p}{a''} - 2p_2 - K_{p,2}\sqrt{p_2}\right]}{2K_{p,4}p_2 + K_{p,3}\sqrt{p_2}}. \tag{4.55}$$

Combining Equation (4.54) and Equation (4.55) with the other atom conservation equation leads to

$$\begin{aligned} 2\frac{p}{a''} &= \left[2\frac{p}{a''} - 2p_2 - K_{p,2}\sqrt{p_2}\right]\frac{K_{p,4} + K_{p,3}\sqrt{p_2} + K_{p,1}}{2K_{p,4}p_2 + K_{p,3}\sqrt{p_2}} \\ &+2\left[\frac{2\frac{p}{a''} - 2p_2 - K_{p,2}\sqrt{p_2}}{2K_{p,4}p_2 + K_{p,3}\sqrt{p_2}}\right]^2 \end{aligned}. \tag{4.56}$$

Because there is an implicit unknown, $a''$, in this equation, its value must be guessed and the solution iterated upon until convergence is achieved. That is, the guess is used to calculate $p_2$ and then all the other pressures are calculated until finally a value for $a''$ is determined and compared to the initial guess. A good guess for $a''$ is usually the total number of moles for the stoichiometric reaction.

## 4.14 EXAMPLE: ESTIMATION OF IMPORTANCE OF NEGLECTED PRODUCT SPECIES

We have considered a couple of methods for calculating the equilibrium composition of reacting gas mixtures. In the examples shown, product species were specified. Unfortunately, one doesn't always know what species to include in the products and therefore must choose a set of products based on experience (or guesswork) or include every possible product species. There is a technique for determining whether some additional species should be present in the products of reaction, but the process depends on first making a choice of products and carrying one of the solution procedures through to the final results. The foundation for this procedure is that the selected products are assumed to be the major products from the point of view of energy contribution. It must be remembered that although the neglected species may be insignificant from an energy standpoint, they may be essential in terms of chemical kinetics. That is, they may be important in activating intermediate reactions in the complex chain of reactions that lead to the final products.

For example, consider the case of determining if an extra product is an important contributor to a given reaction, such as the homogeneous reaction given here:

$$4N_2 + O_2 \rightarrow a_1''N_2 + a_2''O_2 + a_3''NO + a_4''O.$$

The pressure is taken as $p = 10^{-2}$ atm and $T = 400$ K and is representative of air dissociation around a reentry vehicle in the atmosphere (the number of moles of $N_2$ would be 3.76 for a more accurate simulation of air). Then atom mass conservation requires that

$$\begin{aligned} N:8 &= 2a_1'' + a_3'' \\ O:2 &= 2a_2'' + a_3'' + a_4'' \end{aligned}.$$

The total number of moles in the products is

$$a'' = a_1'' + a_2'' + a_3'' + a_4''.$$

Equilibrium among the subreactions yields

$$\frac{1}{2}O_2 \rightarrow O$$

$$K_{pO} = \frac{p_4}{\sqrt{p_2}} = \sqrt{\frac{p}{a''}}\frac{a_4''}{\sqrt{a_2''}}$$

$$\frac{1}{2}N_2 + \frac{1}{2}O_2 \rightarrow NO$$

$$K_{pNO} = \frac{p_3}{\sqrt{p_1 p_2}} = \frac{a_3''}{\sqrt{a_1'' a_2''}}$$

Solving these five equations in five unknowns yields

$$a_1'' = 3.985$$
$$a_2'' = 2.77 \times 10^{-3}$$
$$a_3'' = 3.13 \times 10^{-2}$$
$$a_4'' = 1.963$$

It is clear that the amounts of $O_2$ and NO are much smaller than the amounts of $N_2$ and O, which suggests that the molecular oxygen has been mostly dissociated but leaves open the question: Would N have been important to include as one of the product species? It is only necessary to check the equilibrium subreaction between molecular and atomic nitrogen:

$$\frac{1}{2}N_2 \rightarrow N$$

$$K_{pN} = \frac{p_N}{\sqrt{p_1}} = \sqrt{\frac{p}{a''}}\frac{a_N''}{\sqrt{a_1''}}$$

At a pressure of 0.01 atm and temperature of 4000K the equilibrium constant is $K_{pN} = 0.0571$. Using the result that $a'' = 5.982$ in the aforementioned equation suggests that $a_N'' = 1.96$ and clearly atomic nitrogen cannot be neglected. However, it seems apparent that atomic oxygen could certainly be neglected, and perhaps NO as well, so that the overall complexity of the problem has remained about the same.

## 4.15 ADIABATIC FLAME TEMPERATURE

We have been considering the equilibrium composition for a given state, that is, a given temperature and pressure. It is often more important to find out the temperature that would result from the combustion of fuel and oxidizer under given pressure conditions in the combustor. We continue to make the following assumptions for the combustor:

- Velocities in the combustor are small enough so that the kinetic energy contribution may be neglected in the energy balance.
- The pressure in the combustor is constant, that is, the combustion is isobaric.
- The combustion products are gases and behave like perfect gases.
- The combustion process is adiabatic; all the heat released goes into the products.

The first law may be considered as $dQ = dH - Vdp$, which, for isobaric combustion, becomes $dQ = dH$, and because of our assumption of adiabaticity we have $dQ = dH = 0$. Because the enthalpy is a state function, we may write

$$\Delta H = \sum_{i=1}^{N} a_i'' H_i(T_c) - \sum a_i' H_i(T_j) = 0. \tag{4.57}$$

For a reaction of the type

$$\sum_{i=1}^{N} a_i' A_i \rightarrow \sum_{i=1}^{N} a_i'' A_i, \tag{4.58}$$

we may write Equation (4.57) as

$$\sum_{i=1}^{N} a_i'' \left[\Delta H_{f,i}^0 + \Delta H_i(T_c) - \Delta H_i(T_r)\right] - \sum_{i=1}^{N} a_i' \left[\Delta H_{f,i}^0 + \Delta H_i(T_j) - \Delta H_i(T_r)\right] = 0. \tag{4.59}$$

Here the quantities introduced are as follows:

$\Delta H_{f,i}^0$ is the standard molar heat of formation of species $i$ at $p = 1$ atm and $T = T_r$
$\Delta H_i(T) = H_i(T) - H_i(0)$
$T_r$ = reference temperature, 298.16 K
$T_c$ = adiabatic flame temperature in degrees K
$T_j$ = injection temperature of the reactants in degrees K

If liquid species are present in the injected reactants, one must include the latent heats of vaporization and the enthalpy change for the liquid phase in proceeding from the injection temperature to the vaporization temperature in Equation (4.59). For example, if the reactants are injected as liquids, then Equation (4.59) would become

$$\sum_{i=1}^{N} a_i'' \left[\Delta H_{f,i}^0 + \Delta H_i(T_c) - \Delta H_i(T_r)\right] - \sum_{i=1}^{N} a_i' \left[\Delta H_{f,i}^0 - L_i + \Delta H_i(T_{v,i}) - \Delta H_i(T_r) + \int_{T_{v,i}}^{T_j} C_{pi,l} dT\right] = 0 \tag{4.60}$$

In this equation, $L_i$ is the latent heat of vaporization of species $i$, $T_{v,i}$ is the vaporization temperature of species $i$, and $C_{p,i,l}$ is the specific heat of species $i$ in the liquid phase.

If we now confine ourselves to only homogeneous reactions, we can deal with two simplifications to Equation (4.60):

- If the injection temperature equals the reference temperature, $T_j = T_r$,

$$\sum_{i=1}^{N} a_i'' \left[\Delta H_{f,i}^0 + \Delta H(T_c) - \Delta H(T_r)\right] - \sum_{i=1}^{N} a_i' \Delta H_{f,i}^0 = 0. \tag{4.61}$$

- If the reactants are injected at $T_j = T_r$ and are in their reference state (recall that $\Delta H^0{}_{f,I} = 0$ for the reference state),

$$\sum_{i=1}^{N} a_i'' \left[\Delta H_{f,i}^0 + \Delta H(T_c) - \Delta H(T_r)\right] = 0. \tag{4.62}$$

The process of finding the adiabatic flame temperature for a specific reaction in a chamber at some specified pressure starts with choosing a temperature and calculating the equilibrium composition at that temperature and the specified pressure. Then the governing equation for the adiabatic flame temperature is entered with the state information and, if it is satisfied, the chosen temperature is the correct one. If not, the process is repeated until the correct temperature is found. A graphic representation of the adiabatic flame temperature equation is shown in Figure 4.11.

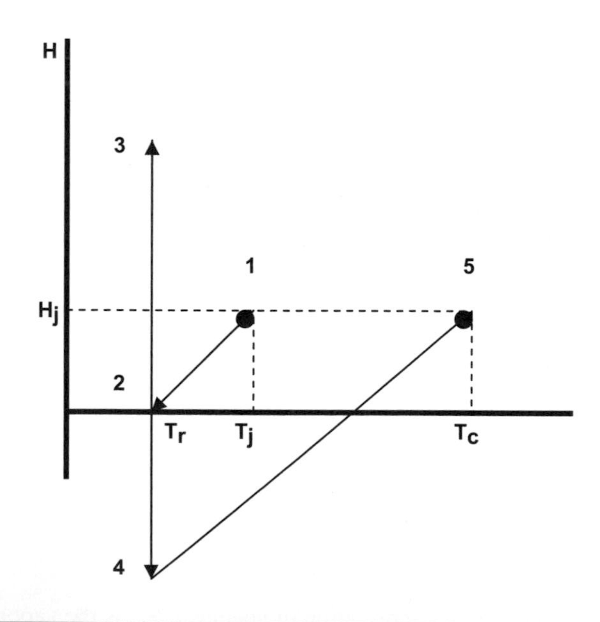


**FIGURE 4.11**

Graphical representation of the adiabatic flame temperature equation for homogeneous reactions.

The steps shown in Figure 4.11 may be explained as follows:

1. Propellants enter the combustion chamber at $T_j$ with $H_j = \sum_{i=1}^{N} a_i' \Delta H_i(T_j)$.
2. Propellants are brought to the reference state $H_j - \sum_{i=1}^{N} a_i' \Delta H_i(T_r)$, and this level of enthalpy is chosen arbitrarily as the abscissa in Figure 4.11.
3. Propellants are brought to the reference composition, that is, $\sum_{i=1}^{N} a_i' \Delta H_{f,i}^0$ is added to the results of step 2 at $T = T_r$.
4. Propellants are transformed into products of reaction at $T = T_r$, that is, the quantity $\sum a_i'' [\Delta H_{f,i}^0 - \Delta H_i(T_r)]$ is subtracted from the results of step 3.
5. Products are brought to the final temperature $T_c$ such that $\Delta H = H(5) - H(1) = 0$, that is, $\sum_{i=1}^{N} a_i'' H_i(T_c)$ is added to the result of step 4

For purposes of calculation, we may define the available heat due to chemical reaction as

$$Q_{avail} = \sum_{i=1}^{N} \left[ a_i' - a_i'' \right] \Delta H_{f,i}^0. \tag{4.63}$$

Then likewise define the heat of combustion at an assumed temperature $T_c$' as

$$Q_{comb}(T_c') = \sum_{i=1}^{N} a_i''(T_c') \left[ \Delta H_i(T_c') - \Delta H_i(T_r) \right] - H_j. \tag{4.64}$$

The quantity $H_j$ depends only on the reference temperature and the injection temperature and composition, which is generally given as

$$H_j = \sum_{i=1}^{N} a_i' \left[ \Delta H_i(T_j) - \Delta H_i(T_r) \right]. \tag{4.65}$$

Then the procedure for calculating the adiabatic flame temperature is as follows:

1. Assume a combustion temperature $T = T_c$' and calculate the equilibrium composition in the combustor at that temperature and the prescribed pressure.
2. Use the equilibrium composition of the products as calculated in the previous step to determine $Q_{avail}(T_c')$ from Equation (4.63).
3. Calculate the quantity $Q_{comb}(T_c')$ from Equation (4.64).
4. If $Q_{avail}(T_c') > Q_{comb}(T_c')$, then the adiabatic flame temperature $T_{ad} > T_c'$ and vice versa. When $Q_{avail}(T_c') = Q_{comb}(T_c')$, then $T_c' = T_{ad}$.

A reasonable guess for the initial value of the adiabatic flame temperature and a linear interpolation for the final result is generally sufficient.

## 4.16 EXAMPLE: ADIABATIC FLAME TEMPERATURE FOR STOICHIOMETRIC $H_2$–$O_2$ MIXTURE

Consider the adiabatic flame temperature for a stoichiometric mixture of hydrogen and oxygen:

$$H_2 + \frac{1}{2} O_2 \rightarrow H_2O.$$

Table 4.4 Evaluation of Equation (4.66)

| $T$ (K) | $\Delta H_f^0$ ($H_2O$) (kJ/mol) | a" ($H_2O$) (mol) | $\Delta H(T)$-$\Delta H(298.16)$ (kJ/mol) | $Q_c$–$Q_a$ (kJ) |
|---|---|---|---|---|
| 4000 | −241.988 | 1 | 184.7 | −57.3 |
| 4500 | −241.988 | 1 | 214.5 | −27.4 |
| 5000 | −241.988 | 1 | 244.9 | 2.90 |

The adiabatic flame temperature for this reaction would be the highest temperature achievable with these reactants. More than 2 moles of hydrogen to each mole of oxygen would mean excess hydrogen would absorb some of the heat produced, reducing the final temperature. Less than that proportion would mean having excess oxygen that would again absorb some of the heat released in forming water. This case is simplified from the standpoint of chemical composition, as the only product would be water. Then if we assume that hydrogen and oxygen are introduced in their standard state at the reference temperature, that is, molecular hydrogen and oxygen injected at 298.16 K, Equation (4.63) becomes

$$Q_{avail} = (1mole)\Delta H^0_{f,H_2} + \left(\frac{1}{2}mole\right)\Delta H^0_{f,O_2} - (1mole)\Delta H^0_{f,H_2O}.$$

Because the standard heat of formation of molecular hydrogen and molecular oxygen is zero,

$$Q_{avail} = -\Delta H^0_{f,H_2O}.$$

With $T_j = T_r = 298.16$ K and the only species in the products being water so that $a'' = 1$ mole, Equation (4.64) becomes

$$Q_{comb}(T'_c) = \Delta H_i(T'_c) - \Delta H_i(298.6K).$$

We are searching for the value of $T'_c$ that satisfies the following equation:

$$Q_{comb} - Q_{avail} = \Delta H(T'_c) - \Delta H(298.16) + \Delta H^0_{f,H_2O}. \qquad (4.66)$$

Using tabulated thermodynamic values, e.g., those presented McBride and colleagues (1963) or those in Appendix G, we may construct Table 4.4. Interpolation leads to $T_c = 4920$K.

## 4.17 NOMENCLATURE

| | |
|---|---|
| $A$ | cross-sectional area |
| $a$ | sound speed |
| $c_p$ | specific heat at constant pressure |
| $f/a$ | fuel to air ratio, $\dot{m}_f/\dot{m}_a$ |
| $G$ | Gibbs free energy |
| $g$ | acceleration of gravity |
| $H$ | heat release due to chemical reaction or molar enthalpy |
| $HV$ | lower heating value of fuel |
| $h$ | enthalpy |

| | |
|---|---|
| $I_{sp}$ | specific impulse |
| $K_p$ | equilibrium constant |
| $k$ | isentropic exponent |
| $M$ | Mach number |
| $\dot{m}$ | mass flow |
| $p$ | pressure |
| $R$ | gas constant |
| $S$ | molar entropy |
| $SG$ | specific gravity |
| $T$ | temperature |
| $V$ | velocity |
| $W$ | molecular weight |
| $\dot{w}$ | weight flow rate $= mg$ |
| $X$ | mole fraction |
| $Y$ | mass fraction |
| $\alpha$ | molar fuel to air ratio |
| $\gamma$ | ratio of specific heats |
| $\phi$ | equivalence ratio |
| $\rho$ | gas density |
| $\eta$ | efficiency |
| $\nu$ | stoichiometric number of moles |

### 4.17.1 Subscripts

| | |
|---|---|
| $a$ | air |
| $b$ | combustor |
| $c$ | adiabatic flame condition |
| $e$ | exit conditions |
| $f$ | fuel |
| $i$ | species index |
| $j$ | species index |
| lim | limiting conditions |
| max | maximum conditions |
| stoich | stoichiometric conditions |
| $t$ | stagnation conditions |
| 0 | free stream conditions |
| 2 | compressor entrance conditions |
| 3 | combustor entrance conditions |
| 4 | combustor exit conditions |

### 4.17.2 Superscripts

| | |
|---|---|
| * | critical conditions, $M = 1$ |
| $(\bar{\ })$ | average value |
| ()' | reactant species |
| ()'' | product species |
| $()^0$ | standard state, 1 atmosphere and 298.16K |

## 4.18 EXERCISES

**4.1** Consider a combustion chamber designed for constant static temperature operation. (a) Develop quasi-one-dimensional equations to describe the thermodynamic and flow properties at the exit station under the assumptions of heat addition by stagnation temperature change, variable cross-sectional area, negligible friction, and constant chemical composition. (b) Show a plot of the variation of Mach number $M_4$ with stagnation temperature ratio $T_{t,4}/T_{t,3}$ for $M_3 = 0.4, 0.8$, and 2. (c) Determine if this combustion chamber is limited by thermal choking.

**4.2** The SCORPION (Statoréacteur Cruciforme comme ORgane Portant Intégré à Orifices Nasaux) was a prototype of a long-range, ground-to-air missile (M = 6600 km); 220-mm diameter liquid-fueled ramjet (demonstrator) designed by ONERA of France. Successful ground tests of the fully integrated vehicle were carried out in their S4 MA wind tunnel (schematic diagram shown in Figure E4.1). Assume that ramjet combustion has a constant cross-sectional area and that air from the intakes enters it at a speed of 100 fps (30.5 m/s) with a stagnation temperature $T_{t,3}$ = 780F (416°C). The combustor is designed to burn a mixture of fuel and primary air in the ratio of 0.06:1 with a burner efficiency $\eta_b = 90\%$. The kerosene fuel has a heating value of 18,900 Btu/lb (44 MJ/kg), and the ratio of secondary air flow to total airflow $\dot{w}_{3,s}/\dot{w}_3$ can be varied using different combustor liners.

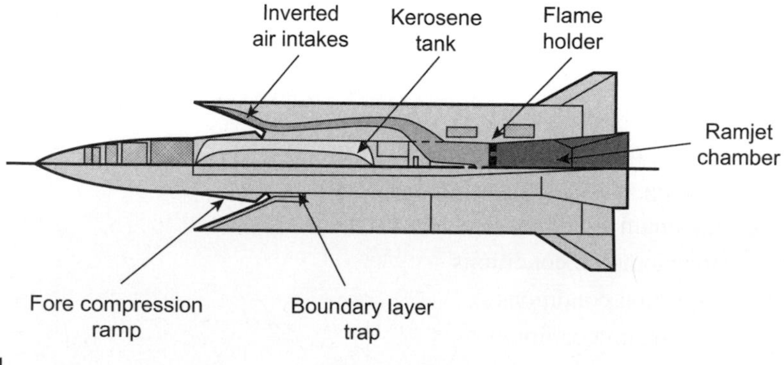


**FIGURE E4.1**

Schematic diagram of the SCORPION ramjet powered missile.

For values of $\dot{w}_{3,s}/\dot{w}_3 = 0$, 0.2, 0.4, and 0.6, determine the stagnation temperature ($T_{t,4}$) leaving the combustor, percentage loss in stagnation pressure across the combustor, and velocity leaving the combustor, $V_4$. Show results on graphs using $\dot{w}_{3,s}/\dot{w}_3$ as the abscissa and briefly explain the behavior of the flow. Assume that friction and fuel injector strut drag are negligible and that the average $c_p = 0.276$ Btu/lb-R (1.16 kJ/kg-K) and the average $\gamma =$ 1.33.

**4.3** An S-shaped exponential curve may be used to model the spatial distribution of heat released in the combustion chamber. One such model is as follows:

$$\Delta H = H_{eff} \exp\left(-B_1 \varsigma^{B_2}\right)$$

$$\zeta = \left|\frac{z}{z_1} - 1\right|,$$

where $z$ is the axial location in the combustor and $z_l$ is the axial location where combustion is complete. Quantities $B_1$ and $B_2$ provide the shape considered to best characterize the heat release as a function of the axial coordinate. The quantity $\Delta H$ represents the heat released up to a given axial station in the combustor, while $H_{eff}$ represents the total heat release per unit mass during combustion and is proportional to the heating value of the fuel. Using this information, (a) describe the significance of the term $z_1$ and how it influences the design of a combustor for a practical aerospace vehicle, (b) relate the term $z_1$ to the Damkohler similarity parameter, (c) consider a combustor typical of the Pratt & Whitney TF30 engine used on the Grumman F-14A Tomcat in which the combustor length $L = 50$ cm, the combustor height $d = 20$ cm, and $H_{eff}$ can be taken as 40 MJ/kg. Assuming that $z_1 = 50$ cm and $B_1 = 4.605$, plot the heat release as a function of axial distance for $B_2 = 2$, 3, and 4. (d) Define a combustion zone length $L_c$ based on your results and show a plot of $L_c$ as a function of $B_2$ (you may wish to carry out calculations for additional values of $B_2$ to clarify the functional relationship sought). (e) Discuss how results for $L_c$ are related to the general concept of the chemical production rate of the combustion reactions taking place in the combustor.

**4.4** A combustor is to accept air from a compressor at a pressure $p_3 = 1$ MPa, a speed $V_3 = 150$ m/s, and a temperature $T_3 = 500°C$. It is to operate with a stoichiometric primary zone fuel to air ratio $(f/a)_p = \dot{m}_f/\dot{m}_{a,p} = 0.066$ and a burner efficiency $\eta_b = 96\%$. The hydrocarbon fuel has a heating value $HV = 44$ MJ/kg, and secondary air mass flow can be supplied to the combustor through an appropriate combustor liner design over the range $0.8 > \dot{m}_{a,s}/\dot{m}_a > 0$. Two designs are being considered: a constant area combustor design and a constant pressure combustor design. Show the variation of (a) temperature leaving the combustor and entering the turbine $T_4$, (b) stagnation pressure loss across the combustor, and (c) velocity leaving the combustor and entering the turbine $V_4$ as functions of the cooling secondary air flow ratio $m_{a,s}/m_a$. Compare the performance of the two combustor designs for this range of cooling flow and discuss the results. For the purposes of this problem, the thermodynamic properties of the combustion products may be assumed to be constant with average values $c_p = 1.37$ kJ/kg-K, $\gamma = 1.27$, and $W = 28.96$. Furthermore, you may assume that the mass flow of fuel is much less than that of the air so that $\dot{m} = (\dot{m}_f + \dot{m}_{a,p} + \dot{m}_{a,s}) \approx (\dot{m}_{a,p} + \dot{m}_{a,s}) \approx \dot{m}$

**4.5** Hydrogen and fluorine make a powerful propellant combination. The space shuttle main engines use an $H_2$–$O_2$ propellant mix operating at a combustion chamber pressure of 200 atmospheres at a temperature of about 3500K. At these chamber conditions, determine the combustion products of a mixture of $H_2$ and $F_2$ using 1 mole of each. Denote the products as $b_{HF}$, $b_H$, $b_O$, etc. and clearly indicate the unknown quantities and the equation set being used to solve for them.

**4.6** Turbo pumps are used on the Space Shuttle Main Engines (SSME) to inject the hydrogen and oxygen propellants into the main combustion chamber. The fuel and oxidizer pumps are powered by turbines driven by combustion of some of the hydrogen and oxygen in a preburner. Find the adiabatic flame temperature in the preburner assuming that molecular hydrogen and oxygen are injected at a temperature $T_j = 298.16$K and a pressure $p_c = 20$ atm and that they react stoichiometrically.

**4.7** For the propellant system composed of $C_nH_{2n}$ as fuel and $O_2$ as oxidizer:

$$a_1 C_nH_{2n} + a_2 O_2 \rightarrow products$$

Determine (a) the stoichiometric fuel to oxidizer ratio $(F/O)_s$ and (b) the equilibrium composition of the products of reaction for the following conditions: $p = 20$ atm, $T = 3000$K, equivalence ratio $\phi = 1.2$, and total mass of reactants equal to 1 kg.

**4.8** Consider the gaseous propellant system described by the reaction

$$4H_2 + O_2 \rightarrow products$$

Determine (a) equivalence ratio $\phi$ and (b) adiabatic flame temperature $T_{ad}$. Note that this reaction is fuel rich so it should be possible to reduce the number of product species accordingly so as to simplify the calculation, if desired. Use $p = 20$ atm for the chamber pressure.

**4.9** Consider the following reaction between solid carbon and gaseous oxygen in which all elemental carbon is consumed:

$$C(c) + O_2(g) \rightarrow b_1 O_2(g) + b_2 CO(g) + b_3 CO_2(g) + b_4 O(g).$$

Determine the adiabatic flame temperature for this reaction assuming the reactants are introduced at the reference temperature, 298.16K, and a pressure of 20 atmospheres.

**4.10** Consider the reaction of 1 mole of bromine ($Br_2$, gas) reacting with varying numbers of moles of hydrogen ($H_2$, gas) injected at a temperature $T_j = 298.16$K. Show a graph of the variation of the adiabatic flame temperature $T_{ad}$ with the weight fraction of hydrogen.

**4.11** A turbojet engine combustor is fed with air from a compressor at a Mach number $M_3 = 0.3$ and a stagnation temperature $T_{t,3} = 315$°C and discharges the combustion products to the turbine at a stagnation temperature $T_{t,4} = 838$°C.

**(a)** For a burner efficiency $\eta_b = 90\%$ and fuel heating value $HV = 43.3$ MJ/kg, find the overall fuel–air ratio *f/a*, assuming $\gamma = 1.33$ and $R = 287$ J/kg.

**(b)** If the combustor operates at constant pressure, determine the exit Mach number $M_4$.

**4.12** Consider the high-performance propellant that uses hydrogen as the fuel and bromine as the oxidizer. The reaction may be taken in the following form:

$$H_2 + Br_2 \rightarrow b_1 H_2 + b_2 H + b_3 HBr + b_4 Br_2.$$

Determine the equilibrium composition at a temperature of 1500K and a pressure of 20 atm. Some pertinent equilibrium constants at 1500K are given here:

$$K_{p,2} = \frac{p_H}{\sqrt{p_{H_2}}} = 1.76 \times 10^{-5}$$

$$K_{p,15} = \frac{p_{Br}}{\sqrt{p_{Br_2}}} = 0.298$$

$$K_{p,19} = \frac{p_{HBr}}{\sqrt{p_{H_2} p_{Br_2}}} = 163.0$$

**4.13** Compute the adiabatic flame temperature arising from the combustion of octane ($C_8H_{18}$) with air at a fuel to air ratio of 0.02 and where the initial temperature of the reactants is 811K. The lower heating value of octane is $HV = 44.79$ MJ/kg at 25°C.

## References


Anderson, J. A. (1990). *Modern Compressible Flow*. New York: McGraw-Hill.

Goodger, E. M. (1982). Comparative Properties of Conventional and Alternative Fuel. In: *Alternative Fuel Technology Series, Vol 2*. Cranfield, U.K: Cranfield Press.

Sirignano, W. A. (2010). *Fluid Dynamics and Transport of Droplets and Sprays* (2nd ed.). New York: Cambridge University Press

Penner, S. S. (1957). *Chemistry Problems in Jet Propulsion*. New York: Pergamon.

Wilkins, R. (1963). *Theoretical Evaluation of Chemical Propellants*. New Jersey: Prentice-Hall.

Gordon, S., & McBride, B. (October, 1994). Computer Program for Complex Chemical Equilibrium Calculations and Applications – I. Analysis. *NASA RP-1311*.

Gordon, S., & McBride, B. (June, 1996). Computer Program for Complex Chemical Equilibrium Calculations and Applications – II. User's Manual and Program Description. *NASA RP-1311–P2*.

McBride, B., Heimel, S., Ehlers, J., & Gordon, S. (1963). Thermodynamic Properties to 6000 K for 210 Substances Involving the first 18 Elements. *NASA SP-3001*.

CHAPTER

# Nozzles

5

## CHAPTER OUTLINE



## 5.1 NOZZLE CHARACTERISTICS AND SIMPLIFYING ASSUMPTIONS

As pointed out in Chapter 1, the nozzle is the primary component of a jet propulsion system. It transforms random internal energy into ordered kinetic energy to increase the momentum of the flow, thereby producing thrust. The basic nozzle is a very simple and passive piece of equipment, although the demands of a broad range of operating conditions often add substantial complexity to it in practical engines. Although the fundamental analysis of nozzles applies to both air-breathing and rocket engines, this chapter emphasizes nozzles for jet aircraft; a discussion of the special characteristics of rocket nozzles is presented in Chapter 11. The basic equations for ideal flow in nozzle are developed, and the fundamental concepts of nozzle performance, such as mass flow capability, shock wave effects, and nozzle geometry requirements, are addressed.

The release of heat energy in the combustor serves to raise the internal energy of the combustion products. In order to create thrust, it is necessary to convert that energy into kinetic energy and thereby increase the velocity of the flow when it exits the propulsion device. A simple device for accelerating a fluid is the nozzle, a duct whose area is varied in such a fashion as to increase the velocity of the flow through it. Recall that our mass conservation equation for quasi-one-dimensional flow, in logarithmic differential form, may be written as

$$\frac{dV}{V} = -\frac{d\rho}{\rho} - \frac{dA}{A}. \tag{5.1}$$

For an incompressible flow it is clear that the velocity varies inversely as the area, so in order to speed up such a flow we need only provide a nozzle with decreasing area. However, in a compressible flow, the effect of area change on the density must be considered and we will find that this is the controlling factor in such flows.

The momentum equation in logarithmic differential form can be obtained from the table of influence coefficients, Table 2.1, as

$$\begin{aligned}\frac{dp}{p} = {} & \left[\frac{\gamma M^2}{1-M^2}\right]\frac{dA}{A} + \left[-\frac{\gamma M^2}{1-M^2}\right]\frac{dQ - dW_k - dH}{c_p T} \\ & + \left[-M^2\frac{\gamma + (\gamma-1)kM^2}{2(1-M^2)}\right]\left(4c_f\frac{dz}{D} + \frac{dF_z}{\frac{1}{2}\gamma p M^2 A}\right) + \left[\frac{\gamma M^2}{1-M^2}\right]\frac{dW}{W}.\end{aligned} \tag{5.2}$$

### 5.1.1 Frictional effects

Because the nozzle is meant to be an aerodynamically efficient duct, it is reasonable to consider the nature of the frictional and drag forces on the flow passing through it. The frictional force term in Equation (5.2) may be written as

$$4c_f\frac{dz}{D} = 4c_f\frac{L}{D}d\left(\frac{z}{L}\right).$$

Clearly the magnitude of the frictional force on the nozzle walls is proportional to the skin friction coefficient and the ratio of nozzle length to hydraulic diameter. When $L/D \gg 1$ the duct is better classified as a pipe, and the local skin friction coefficient may be obtained from a Moody chart for pipe flow, as described in White (2006). If $L/D = O(1)$, the local skin friction coefficient to use is the one for a flat plate flow, as in such a case the effect of curvature is negligible; flat plate skin friction coefficient values are also available in White (2006). Nozzles of jet aircraft typically have $L/D \sim 1$, and although the skin friction coefficient is a function of Reynolds number, Mach number, and wall roughness, for the conditions of interest here its magnitude is typically in the range $1.5 \times 10^{-3} < c_f < 4.5 \times 10^{-3}$. Thus the coefficient of $d(z/L)$ is typically on the order of $10^{-2}$ and therefore frictional terms are not extremely important as they would be in a pipe flow where $L/D$ is large. Thus frictional effects may be safely considered negligible in preliminary analyses. In a subsequent section, the effect of friction on nozzle performance is considered in more detail, and an expression for the flat plate skin friction coefficient for compressible flat plate flow is given.

### 5.1.2 Drag effects

The drag force $F_z$ of a body in the nozzle flow field is defined in terms of the dynamic pressure, $q$, a characteristic body area, $S$, and a drag coefficient, $c_d$, based on that area so that $F_z = c_d qS$. For plate-like bodies, like wings or fins, reference area $S$ is usually taken as the planform area, which is essentially the area of the projection of the body normal to its major (plate-like) plane. However, for axisymmetric bodies, reference area $S$ is usually considered to be the maximum cross-sectional area in the plane normal to the axis of symmetry and is often called the maximum frontal area. The drag force term in Equation (5.2) comprises the total drag force exerted by bodies immersed in the flow up to any given station in the axial direction and we may write this as

$$\frac{F_z}{\frac{1}{2}\gamma pAM^2}\frac{dF_z}{F_z} = \frac{\sum_{i=1}^{n} c_{d,i}S_i q}{qA}\frac{dF_z}{F_z} = \frac{\sum_{i=1}^{n} c_{d,i}S_i}{A}\frac{dF_z}{F_z}. \quad (5.3)$$

Because drag coefficients for streamlined objects are typically on the order of $10^{-1}$ and for bluff bodies are no larger than unity, the maximum for the ratio in Equation (5.3) is effectively set by the magnitude of $S_i/A$, the ratio of the body reference area to the cross-sectional flow area. For illustration purposes, assume an appropriate average value for the drag coefficient $c_{d,avg}$ and a total reference area $S$ equal to the sum of all the areas $S_i$ so that we may write

$$\frac{F_z}{\frac{1}{2}\gamma pM^2 A}\frac{dF_z}{F_z} = c_{d,avg}\frac{S}{A}\frac{dF_z}{F_z}.$$

Thus, for a few, small streamlined objects in the nozzle, such as afterburner fuel injectors or support struts, the coefficient of the drag term will be on the order of $S/A$ or less. It is reasonable to expect that $S/A$ will be less than $10^{-2}$ (in wind tunnel experimentation, blockage becomes a serious concern when the cross-sectional area of bodies starts to approach 5% of the nozzle cross-sectional area). Therefore, it is often reasonable to neglect the effect of drag forces on the flow for initial design analyses of nozzles. This subject is addressed again in a subsequent section dealing with practical nozzle performance.

### 5.1.3 Energy transfer effects

There is negligible heat transfer to the walls of the nozzle once transient start-up heating raises the wall temperature up to the adiabatic wall temperature, which is generally close to the stagnation temperature of the flow, which is not to say that the wall of the nozzle is not affected by high temperatures as far as structural considerations are concerned but rather that nozzle flow may be considered adiabatic under steady conditions, so that in Equation (5.2), for example, $dQ = 0$. Nozzles have no work extraction devices in them so there is no work term to account for and therefore $dW_k = 0$. The only other energy term to consider is the heat release due to chemical reaction, $dH$. The objective of the nozzle is to accelerate the flow to high speed, and therefore fluid particles have short residence times within the nozzle. For preliminary design studies, it is reasonable to assume that the Damkohler similarity parameter $\tau_f/\tau_c = 0$, so that the chemical composition is fixed throughout the flow, or that the flow through the nozzle is chemically "frozen," and therefore the molecular weight change through the nozzle is zero and thus in the table of influence coefficients the independent variable $dW/W = 0$.

A final assumption is that the isentropic exponent is constant though the flow and is equal to the ratio of specific heats, $\gamma =$ constant. Because the nozzle increases flow speed at the expense of flow temperature and $\gamma = \gamma(T)$, is likely that $\gamma$ will increase down the nozzle as the temperature decreases. Because simplifications arising from assuming $\gamma =$ constant are substantial, it is common to assume a constant average value for $\gamma$ that will reasonably accommodate the effects of its actual variation through the nozzle.

## 5.2 FLOW IN A NOZZLE WITH SIMPLE AREA CHANGE

The approximations described earlier reduce the number of independent variables that seriously affect the flow in a nozzle to just one: variable cross-sectional area. Consider a nozzle with the numbering system typical of an air-breathing jet engine as shown in Figure 5.1. Under these ideal conditions, pressure and temperature variations are formed from the table of influence coefficients as follow:

$$\begin{aligned} \frac{dp}{p} &= \frac{\gamma M^2}{1 - M^2}\frac{dA}{A} \\ \frac{dT}{T} &= \frac{(\gamma - 1)M^2}{1 - M^2}\frac{dA}{A} \end{aligned}. \tag{5.4}$$

Note that we are using $\gamma = c_p/c_v =$ constant as described in the previous section. A relationship for entropy involving only state variables can be written using the first and second laws of thermodynamics as shown here:

$$\frac{ds}{c_p} = \frac{dT}{T} - \frac{\gamma - 1}{\gamma}\frac{dp}{p}. \tag{5.5}$$

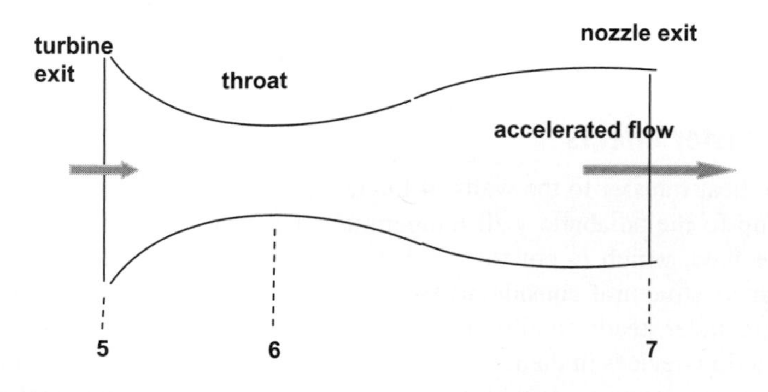


FIGURE 5.1

Schematic diagram of a jet nozzle with the usual numbering system for axial stations.

Using Equation (5.4) in Equation (5.5) reveals the fact that $ds = 0$ so that the flow, under the current assumptions, is isentropic.
The Mach number variation according to the table of influence coefficients is

$$\frac{dM^2}{M^2} = -2\frac{1+\frac{\gamma-1}{2}M^2}{1-M^2}\frac{dA}{A}. \tag{5.6}$$

Combining this equation with the previous one for pressure yields

$$\frac{dp}{p} = -\frac{\gamma}{2\left(1+\frac{\gamma-1}{2}M^2\right)}dM^2. \tag{5.7}$$

This equation may be integrated, to yield, for the illustrative case of the pressure ratio across the nozzle, the following:

$$\frac{p_7}{p_5} = \left[\frac{1+\frac{\gamma-1}{2}M_7^2}{1+\frac{\gamma-1}{2}M_5^2}\right]^{-\frac{\gamma}{\gamma-1}}. \tag{5.8}$$

The energy equation along a streamline in adiabatic flow is

$$c_pT + \frac{1}{2}V^2 = c_pT_t$$

$$\frac{T_t}{T} = 1+\frac{\gamma-1}{2}M^2. \tag{5.9}$$

Here again note that we assume a constant value for $c_p$, and therefore of $\gamma$, and appropriate average values over the temperature range involved should be used in any calculations. At very high Mach numbers, for example, the temperatures involved can be in substantial error if inappropriate choices are made for $c_p$ and $\gamma$.

Making use of the isentropic relation ($p \sim \rho^\gamma$) and the perfect gas law permits determination of the temperature ratio

$$\frac{T_7}{T_5} = \left[\frac{1+\frac{\gamma-1}{2}M_7^2}{1+\frac{\gamma-1}{2}M_5^2}\right]^{-1}. \tag{5.10}$$

Then, because total temperature $T_t$ is constant, the total pressure $p_t \sim T_t^{\frac{\gamma-1}{\gamma}}$ is also constant, which of course is expected for an isentropic flow. With this result we may note the relationship for the static to stagnation pressure ratio as

$$\frac{p}{p_t} = \left(1+\frac{\gamma-1}{2}M^2\right)^{-\frac{\gamma}{\gamma-1}}. \tag{5.11}$$

Similarly, we may determine the ratio of the pressure at any station in the nozzle to the pressure at the critical station:

$$\frac{p}{p*} = \left[\frac{2}{\gamma+1}\left(1+\frac{\gamma-1}{2}M^2\right)\right]^{-\frac{\gamma}{\gamma-1}}. \tag{5.12}$$

## 5.3 MASS FLOW IN AN ISENTROPIC NOZZLE

The mass flow rate through the nozzle may be written as

$$\dot{m} = \rho AV = \frac{p}{RT}AaM. \tag{5.13}$$

This equation is in terms of variables solved for thus far. Noting that the total pressure and temperature are constant and that pressure and temperature are functions of Mach number and that the sound speed, $a$, depends on the temperature permit us to rewrite the mass flow equation in the following form:

$$\dot{m} = \frac{p_t A}{\sqrt{T_t}}\sqrt{\frac{\gamma}{R}}M\left(1+\frac{\gamma-1}{2}M^2\right)^{-\frac{\gamma+1}{2(\gamma-1)}}. \tag{5.14}$$

For given stagnation conditions and gas composition, this is in the form $\dot{m} = KMf^{\Gamma}$so that the derivative may be expressed as

$$\frac{d\dot{m}}{dM} = Kf^{\Gamma}\left[1+(\gamma-1)M^2\frac{\Gamma}{f}\right] = Kf^{\Gamma}\left[\frac{1-M^2}{1+\frac{\gamma-1}{2}M^2}\right]. \tag{5.15}$$

The mass flow rate is stationary where $M = 1$, and we may determine if this is a maximum or minimum by examining the second derivative, which is given by

$$\frac{d^2\dot{m}}{dM^2} = \frac{d}{dM}\left[\frac{\dot{m}}{M}+(\gamma-1)\frac{\dot{m}}{M}\frac{M^2\Gamma}{f}\right]. \tag{5.16}$$

Carrying out the derivative and evaluating it at the stationary point $M = 1$ yields

$$\left[\frac{d^2\dot{m}}{dM^2}\right]_{M=1} = -\dot{m}\frac{2}{\gamma+1} < 0. \tag{5.17}$$

Therefore, for fixed stagnation conditions, the mass flow rate is a maximum at $M = 1$, which is termed the critical Mach number. Because the gas composition and the stagnation conditions are independent of the nozzle, the only other factor to be explored is the relationship between the Mach number and the flow area, which, according to the table of influence coefficients, is given by

$$\frac{dM^2}{M^2} = -2\frac{1+\frac{\gamma-1}{2}M^2}{1-M^2}\frac{dA}{A}. \tag{5.18}$$

At the critical Mach number, $M = 1$, the area distribution must be stationary, that is, $dA = 0$, in order for Equation (5.17) to be valid. This equation can be integrated to yield

$$A = C\frac{1}{M}\left(1 + \frac{\gamma - 1}{2}M^2\right)^{\frac{\gamma+1}{2(\gamma-1)}}. \tag{5.19}$$

To determine the change of Mach number with area we calculate

$$\frac{dA}{dM} = -\frac{A}{M}\left[1 - \frac{(\gamma+1)M^2}{2 + (\gamma - 1)M^2}\right]. \tag{5.20}$$

This area has a stationary point at the critical Mach number, $M = 1$, and the second derivative is

$$\frac{d^2A}{dM^2} = \frac{A}{M^2\left(1 + \frac{\gamma - 1}{2}M^2\right)}\left[1 + \frac{(\gamma - 1)M^2(1 - M^2)}{1 + \frac{\gamma - 1}{2}M^2}\right]. \tag{5.21}$$

Evaluation of this equation at the critical point yields

$$\left[\frac{d^2A}{dM^2}\right]_{M=1} = \frac{2A}{\gamma + 1} > 0. \tag{5.22}$$

This result shows that the area is a minimum at the critical Mach number, $M = 1$, where the mass flow is a maximum, and we term this critical area $A^*$. For fixed gas composition and stagnation conditions, the maximum mass flow possible occurs at the throat, where $M = 1$, and is given by

$$\dot{m}_{\max} = \dot{m}^* = \frac{p_t A^*}{\sqrt{T_t}}\sqrt{\frac{\gamma}{R}}\left(\frac{\gamma + 1}{2}\right)^{-\frac{\gamma+1}{2(\gamma-1)}}. \tag{5.23}$$

When the nozzle is passing the maximum mass flow it is said to be "choked." In order to increase the mass flow of a given gas it is necessary to increase the stagnation pressure or the throat area or to reduce the stagnation temperature. For various values of the constant $\gamma$ the mass flow may be reasonably approximated by

$$\dot{m}^* \approx \frac{p_t A^*}{\sqrt{RT_t}}(0.192\gamma + 0.417). \tag{5.24}$$

Note that the effect of reducing $\gamma$ is to reduce the maximum mass flow rate through the nozzle when the stagnation conditions are fixed. For hydrocarbon-fueled, air-breathing engines the nozzle inlet temperature, which is also the turbine exit temperature, is on the order of 1000K and $\gamma = 1.33$ is usually taken as a suitable average value.

The relationship between the flow area and the critical area follows from Equation (5.19). Note that the critical area is a defined function and need not actually occur in a nozzle. That is, the minimum geometric area, or throat, of a nozzle only becomes the critical area when $M = 1$ is established there:

$$\frac{A}{A^*} = \frac{1}{M}\left[\frac{1 + \frac{\gamma - 1}{2}M^2}{\frac{\gamma + 1}{2}}\right]^{\frac{\gamma+1}{2(\gamma-1)}}. \tag{5.25}$$

Combining Equations (5.14) and (5.25) leads to the following general relationship for mass flow based on stagnation properties and the critical flow area:

$$\dot{m} = \frac{p_t A^*}{\sqrt{T_t}} \sqrt{\frac{\gamma}{R}} \left( \frac{\gamma + 1}{2} \right)^{-\frac{(\gamma+1)}{2(\gamma-1)}}.$$

If one instead substitutes the isentropic relations from Equations (5.9) and (5.10) into Equation (5.19), an expression for mass flow in terms of the static properties and the local flow area is obtained as follows:

$$\dot{m} = \frac{pAM}{\sqrt{T}} \sqrt{\frac{\gamma}{R}}.$$

## 5.4 NOZZLE OPERATION

For a given area distribution that satisfies the basic geometric requirements for a quasi-one-dimensional isentropic flow, one may calculate the distribution of all the other properties of the flow. However, this determination must be done indirectly, as all the flow variables are given in terms of Mach number. This requires developing a table of Mach numbers for which the corresponding area ratios and flow properties are calculated. This table is used to generate the axial distribution of the desired properties, including the area distribution. Another common method is to specify a desired Mach number distribution, $M = M(z)$, and determine the area distribution required, along with the corresponding distribution of flow variables. Of course, this requires some care in evaluating the resulting area distribution so that it satisfies the general requirements for quasi-one-dimensional flow, that is, a smooth wall contour with no rapid changes in slope. It is then possible to use other techniques, such as the method of characteristics or Euler solvers, to evaluate the final shape of the nozzle and ensure there is no tendency to form shock waves within the nozzle. The general nature of the possible flow states in a jet nozzle is summarized in the pressure versus Mach number plot shown in Figure 5.2.

There are available tables of isentropic flow properties as a function of Mach number for a given $\gamma$, such as NACA Ames Research Staff (1953)($\gamma = 1.4$), part of which appears as Appendix A. One may use such a table to analyze a given nozzle, assuming it meets the basic shape requirements. The results will be applicable to flows that are continuously isentropic throughout the nozzle, a situation that is not always possible. Indeed the operation of the nozzle will be continuously isentropic only for exit pressure ratios $p/p_t$ above and below a certain band of values. These values are determined by the two solution branches, which represent the bifurcation originating at the critical point of the nozzle where $M = 1$. For a given exit area ratio $A_7/A^*$ there is only one supersonic exit Mach number, which is the one for which the exit pressure matches the pressure of the ambient into which the nozzles exhausts, that is, $p_7/p_{t,5} = p_e/p_{t,5}$, and the exit Mach number is that which satisfies Equation (5.11). For exit pressures higher than that value, a shock may appear inside the nozzle in order to raise the ultimate pressure at station 7 to that of the ambient.

To illustrate the nature of the flow in the different regions of Figure 5.2 we consider the case of a nozzle designed for air ($\gamma = 1.4$) and an exit Mach number $M_7 = 3$ under different back pressure

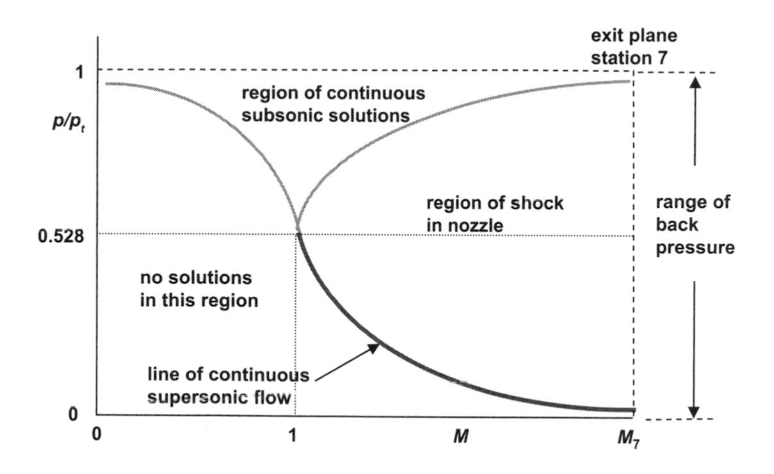


**FIGURE 5.2**

Schematic diagram of the variation of pressure ratio with Mach number in an exhaust nozzle illustrating the possible states of flow.

ratios, as shown in Figure 5.3. The different back pressure cases may be treated sequentially, as follows:

**a)** $p_a/p_t = 1$: The back pressure is equal to the supply stagnation pressure and there is no flow in the nozzle.

**b)** $p_a/p_t = 0.987$: The back pressure is low enough to have the flow accelerate and choke the throat (i.e., $M_6 = 1$) and then decelerate to a low subsonic Mach number at the exit station. This is a critical case and is generally not achievable in practice. However, all values in the range $1 > p_a/p_t > 0.987$ will produce a totally subsonic flow throughout the nozzle, with the maximum (subsonic) Mach number occurring at the throat, $A_6$. This is the case of the so-called Venturi tube.

**c)** $p_a/p_t = 0.70$: Here the back pressure is low enough to ensure starting the nozzle ($M_6 = 1$) but not low enough to permit supersonic flow throughout the nozzle. The supersonic flow must "shock down," that is, a normal shock must appear to bring down the Mach number (and the stagnation pressure) to an appropriate subsonic level so that a higher static pressure is produced behind the shock. Further diffusion through the rest of the increasing area duct serves to bring the pressure at the exit up to the appropriate level.

**d)** $p_a/p_t = 0.28$: In this instance, the back pressure is low enough to ensure supersonic flow throughout the nozzle but still higher than the pressure at the end of the nozzle, $p_7$. In order to adjust this pressure, the flow must once again "shock down" and form a normal shock exactly at the exit station such that the pressure behind the shock is exactly equal to $p_a$.

**e)** $0.28 > p_a/p_t > 0.027$: In this range of back pressure the flow is continuously supersonic throughout the nozzle and the adjustment of the exhaust pressure takes place through shock waves occurring outside the nozzle proper. Because the back pressure is higher than the exit pressure, the nozzle is said to be "overexpanded."

**f)** $p_a/p_t = 0.027$: Here the exit pressure is exactly equal to the back pressure, the flow is continuously supersonic throughout, and the exhaust stream is perfectly adapted to the surrounding ambient pressure. This case is called the perfectly expanded nozzle.

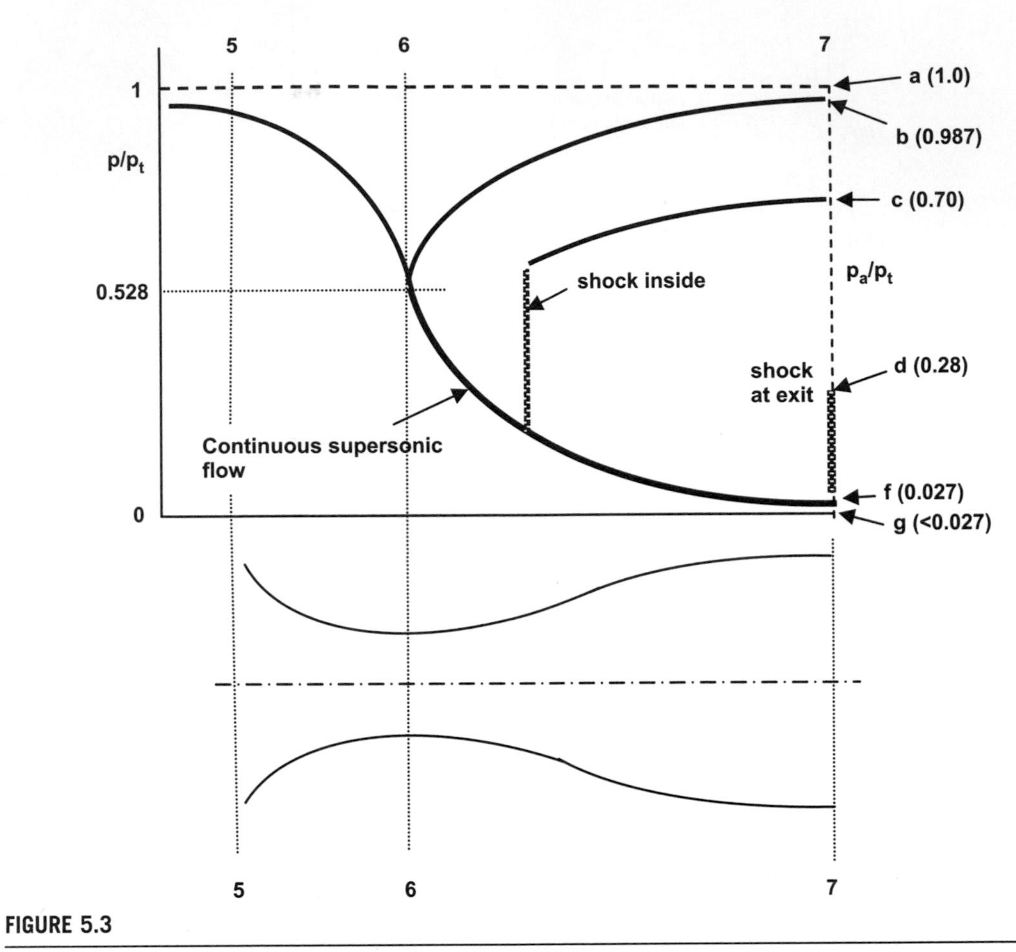


**FIGURE 5.3**

Schematic diagram of possible flow states in a jet nozzle designed for an exit Mach number $M_7 = 3$ shown in terms of the pressure distribution as a function of downstream distance, $z$. Values of the ratio of ambient pressure to upstream stagnation pressure corresponding to different possible flow states are denoted by letters with the appropriate ratios shown in parentheses.

**g)** $p_a/p_t < 0.027$: For all back pressures below the perfectly expanded value the flow through the nozzle is supersonic and the adjustment to the low back pressure takes place outside the nozzle. This case is called the "underexpanded" nozzle.

## 5.5 NORMAL SHOCK INSIDE THE NOZZLE

Processes occurring inside the nozzle are illustrated in Figure 5.4. An ideal isentropic expansion is shown proceeding through the nozzle along the $s_5$ = constant line from the initial enthalpy and

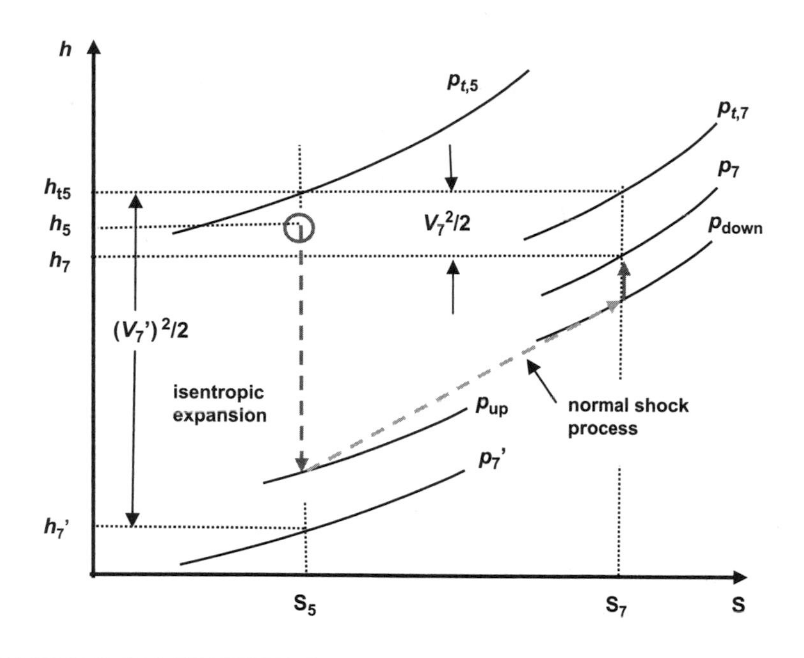


**FIGURE 5.4**

Process diagram for a normal shock somewhere in the nozzle. Flow starts at the turbine exit conditions denoted by the circle and then proceeds isentropically to $p_{up}$, at which point a shock appears and raises the pressure to $p_{down}$. Flow then continues isentropically (and subsonically), but at increased entropy, up to the exit pressure $p_7$.

pressure point ($s_5$, $h_{t,5}$) to the nozzle exit plane pressure $p_7$' defined by the point ($s_5$, $h_7$'), where the prime denotes an isentropic process. If ambient pressure $p_7$ at the nozzle exit is substantially higher than $p_7$', a normal shock is produced somewhere along the nozzle such that the ambient exit pressure can be matched. The shock is a dissipative, although adiabatic, process so that the stagnation enthalpy $h_{t,5}$ remains constant but the entropy jumps from $s_5$ to $s_7$, resulting in a corresponding abrupt drop in stagnation pressure from $p_{t,5}$ to $p_{t,7}$. Thus the pressure starts from $p_5$ and drops to the value $p_{up}$, the pressure just upstream of the shock. The effect of the shock is to cause the pressure to jump to $p_{down}$, the pressure just downstream of the shock, but now with the reduced stagnation pressure $p_{t7}$, which remains constant through the remainder of the nozzle. The Mach number is supersonic ahead of the shock and subsonic behind it. The flow thus continues as a subsonic flow in a duct of increasing area so the flow decelerates to lower Mach numbers and higher pressure, ultimately reaching the ambient pressure, that is, $p_7 = p_e$, which is the pressure of the surroundings at the nozzle exit.

When there is a shock in the nozzle, the mass flow upstream and downstream of the shock will still be equal, as will the stagnation temperature, although the stagnation pressure will be lower on the downstream side due to the entropy increase across the adiabatic shock. A schematic diagram of the general case of a normal shock within the nozzle is illustrated in Figure 5.5.

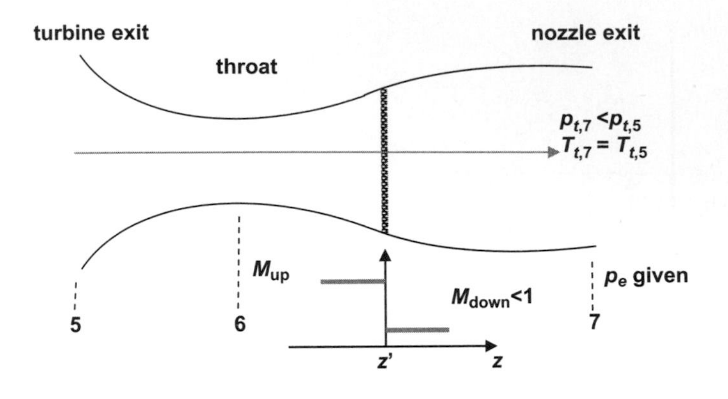


**FIGURE 5.5**

Schematic diagram of a normal shock in the nozzle and the associated notation.

The mass conservation principle requires that

$$m_{\max} = p_{t,up}A^*_{up}\sqrt{\frac{\gamma}{RT_t}}\left(\frac{\gamma+1}{2}\right)^{-\frac{\gamma+1}{2(\gamma-1)}} = p_{t,down}A^*_{down}\sqrt{\frac{\gamma}{RT_t}}\left(\frac{\gamma+1}{2}\right)^{-\frac{\gamma+1}{2(\gamma-1)}} \tag{5.26}$$

Because $T_t$ and $\gamma$ are constant, we have

$$p_{t,up}A^*_{up} = p_{t,down}A^*_{down}$$

Because a shock wave separates the states just upstream and just downstream of it, the ratio of the stagnation pressures may be determined from the following relation:

$$\frac{p_{t,7}}{p_{t,5}} = \frac{p_e}{p_{t,5}}\left\{\frac{1}{2}\left[1 \pm \sqrt{1+\left(\frac{2}{\Gamma}\frac{A_5^*}{A_e}\left[\frac{p_{t,5}}{p_e}\right]^{\frac{1}{\gamma}}\right)^2}\right]\right\}^{\frac{\gamma}{\gamma-1}}. \tag{5.27}$$

Here the quantity

$$\Gamma = \sqrt{\frac{2}{\gamma-1}}\left(\frac{\gamma+1}{2}\right)^{\frac{\gamma+1}{2(\gamma-1)}}. \tag{5.28}$$

To locate a shock in a nozzle knowing the upstream stagnation pressure $p_{t,5}$, the pressure of the surroundings at the nozzle exit $p_e$, and the geometry of the nozzle, first use Equation (5.27) to determine the stagnation pressure ratio across the shock; note that $A_5{}^* = A_6$. This result may be used to find the Mach number just ahead of the shock. Knowing the geometry of the nozzle, one may determine where along the nozzle that Mach number is reached, which is the shock location.

## 5.6 EXAMPLE: SHOCK IN NOZZLE

Consider an ideal nozzle operating with a stagnation temperature $T_{t,5} = 333$ K and a stagnation pressure $p_{t,5} = 276$ kPa. The throat area is $A_6 = 6.45$ cm$^2$, and the gas may be considered to have $\gamma = 1.4$. A shock is situated between the throat and the exit of the nozzle at an axial station where the Mach number $M = 2$. (a) At the throat, calculate Mach number $M_6$, pressure $p_6$, and temperature $T_6$. (b) Using subscripts $x$ and $y$ to denote conditions just upstream and just downstream of the shock, respectively, calculate $p_x$, $T_x$, $p_{t,x}$, $A_x$, $p_y$, $T_y$, $p_{t,y}$. (c) Assuming $A_7/A_6 = 2$, calculate the exit conditions $p_7$, $T_7$, $M_7$, $T_{t,7}$.

**(a)** If there is a shock at the nozzle, the throat must be sonic, that is, $M_6 = 1$, because a minimum section is the only location where the flow can transition from a subsonic to a supersonic Mach number. The isentropic flow relations or (since $\gamma = 1.4$) the supersonic flow tables in Appendix A may be used to find that at $M = 1$ $p/p_t = 0.5283$ and $T/T_t = 0.8333$ so that $p_6 = 0.5283(276 \text{ kPa}) = 146$ Pa and $T_6 = 0.8333(333\text{K}) = 277\text{K}$.

**(b)** In front of the shock $M = M_x = 2$ and in an ideal nozzle, the flow is isentropic from station 5 right up to station $x$ so that isentropic relations or the tables in Appendix A may be used to find that at station $p/p_t = 0.1278$, $T/T_t = .5556$, and $A/A^* = 1.6875$ so that $p_x = 35.3$ kPa, $T_x = 185$K, and $A_x = (A/A^*)A_6 = 10.88$ cm$^2$. To cross the shock, we may use the shock relations or the tables in Appendix A to find that $p_y/p_x = 4.50$, $T_y/T_x = 1.6875$, $p_{t,y}/p_{t,x} = 0.7209$, and $M_y = 0.5774$. Then $p_y = 159$ kPa, $T_y = 312$K, and $p_{t,y} = 199$ kPa.

**(c)** With $M_y = 0.5774$, enter the subsonic compressible flow tables in Appendix A and interpolate, as necessary, to find $A_y/A_y^* = 1.217$. Across the shock $p_{t,x}A_x^* = p_{t,y}A_y^*$ and at the shock $A_x = A_y$ so $A_y^* = (p_{t,x}/p_{t,y})(A_x^*/A_x)A_x = (0.7209)^{-1}(1.6875)^{-1}(10.88) = 8.94$ cm$^2$. Then $A_7/A_y^* = 2(6.45)/8.94 = 1.443$. Entering the subsonic compressible flow tables with this value of $A/A^*$ yields $M_7 = 0.45$ with the associated values $p_7/p_{t,7} = 0.8703$ and $T_7/T_{t,7} = 0.9611$. Because the flow is assumed to be adiabatic $T_t = 333\text{K} =$ constant throughout the nozzle, the exit temperature is $T_7 = (0.9611)(333) = 320$K. The flow is isentropic on either side of the shock, although the entropy is a different value on each side. Then downstream of the shock the stagnation pressure is constant so $p_{t,7} = p_{t,y} = 199$ kPa and $p_7 = (0.8703)(199) = 173$ kPa.

## 5.7 TWO-DIMENSIONAL CONSIDERATIONS IN NOZZLE FLOWS

The quasi-one-dimensional considerations discussed so far have not considered any two-dimensional effects, and these enter the discussion for evaluation of the effects in regions $c$, $d$, and $e$ in Figure 5.3. In a compressible flow, infinitesimal pressure disturbances are carried by acoustic waves, which propagate at the local speed of sound, which is finite, as opposed to the infinite sound speed in an incompressible flow. Thus if one imagines a sound source moving at some constant speed, information about the presence of the disturbance is propagated on a front that is the locus of points tangent to the spherical sound waves, moving at sound speed $a$, as illustrated in Figure 5.6.

Thus, ahead of the front the presence of the sound source cannot be perceived; that region is called the zone of silence. The front is a Mach wave and is inclined at angle $\mu$ with respect to the direction of motion of the sound source. From the trigonometry of the figure it is clear that

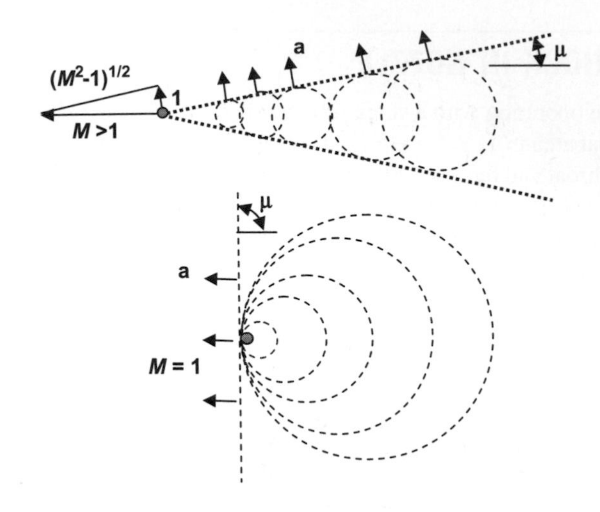


FIGURE 5.6

Propagation of sound waves from a source moving to the left supersonically (upper diagram) or sonically (lower diagram).

$$\mu = \arcsin \frac{1}{M}. \tag{5.29}$$

Then information about pressure disturbances is propagated along Mach waves, each of which represents an infinitesimal pressure rise. Pressure disturbances caused by boundary conditions, such as flow deflections, are transmitted in this manner. If there are no changes in pressure, as for example in the constant area section of a supersonic nozzle, then the Mach number is constant, as are the Mach angles, as illustrated in Figure 5.7. The last Mach waves from the last point on the nozzle wall, that is, the nozzle lip, define the boundary of uniform flow. Any changes in boundary conditions outside the nozzle cannot penetrate into the uniform region shown. Therefore, flow conditions in the exit plane of the nozzle do not depend on changes in back pressure in the region *d-c-e* in Figure 5.3, but instead are set by the area ratio of the nozzle.

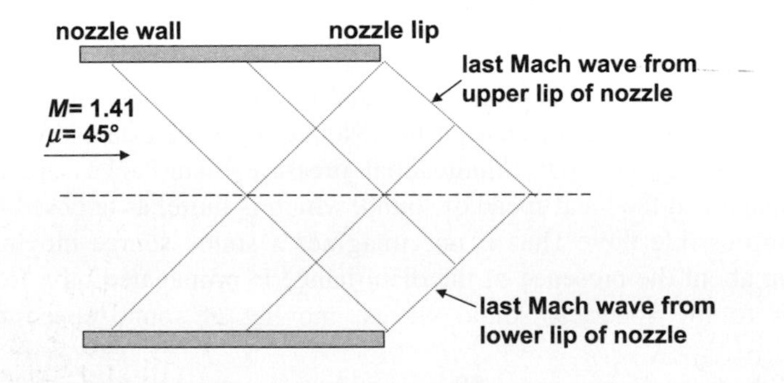


FIGURE 5.7

Uniform supersonic flow in the nozzle and uniform Mach waves in the nozzle. Flow field remains uniform up to the last Mach waves indicated.

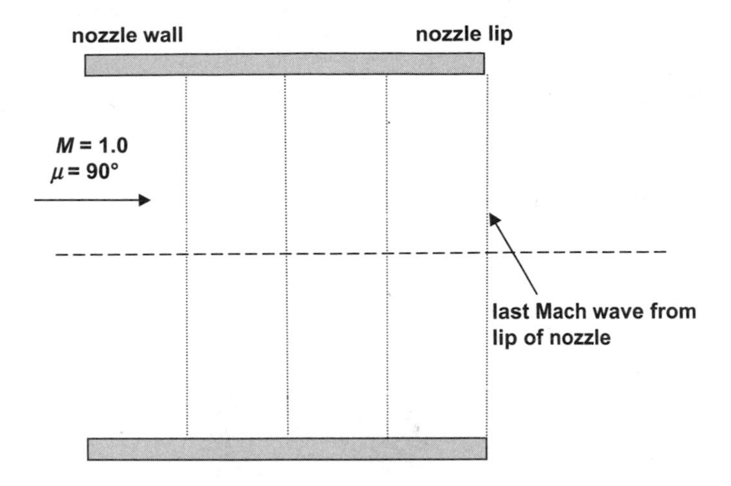


FIGURE 5.8

Uniform supersonic flow in the nozzle and uniform Mach waves in the nozzle. Flow field remains uniform up to the last Mach waves indicated.

If the flow is just sonic in the nozzle shown in Figure 5.7, the Mach waves will be vertical lines and the last one, which emanates from the lip, defines the boundary of uniform sonic flow in the nozzle, as shown in Figure 5.8. Once again, any changes on back pressure will not affect the flow in the nozzle, providing that the back pressure remains below $p^*$, the critical pressure.

Two general cases of interest will be treated in subsequent sections. One case is for the region *c-d* in Figure 5.3 where the back pressure is higher than the perfectly expanded pressure at point *c*, called the overexpanded nozzle, and one case is for the region *c-e* where the back pressure is lower than the perfectly expanded pressure at point *c*, called the underexpanded nozzle. A detailed study of jet flows exhausting into quiescent and moving surroundings at various pressure levels may be found in Love and colleagues (1959).

## 5.8 EXAMPLE: OVEREXPANDED NOZZLES

When the back pressure is higher than the design exit pressure $p_7$, but lower than that causing a normal shock at the exit plane, the flow is continuously supersonic throughout the nozzle and adaptation to the high outside pressure occurs by means of a system of oblique shocks and expansion waves. The example case of Figure 5.3 is once again considered in the case where $p_a/p_{t,5} = 0.056$ and the design exit pressure (the pressure at point *c* in Figure 5.3) is $p_7/p_{t,5} = 0.02722$. The nature of the flow field under these conditions is illustrated in Figure 5.9. Information about the higher pressure outside the exit cannot be felt in the nozzle for the reasons described in the previous section and illustrated in Figure 5.7.

Here we will consider the nozzle to be two dimensional so that plane flow relations can be applied readily. Somehow the pressure in the uniform flow region of the nozzle must be raised so that streamlines leaving the lip of the nozzle have a pressure exactly equal to that of the ambient atmosphere. To raise the pressure of the supersonic flow, streamlines at the lip must be deflected 10° toward the axis so as to reduce the streamtube area. This finite deflection of the flow causes an oblique shock wave to form at an

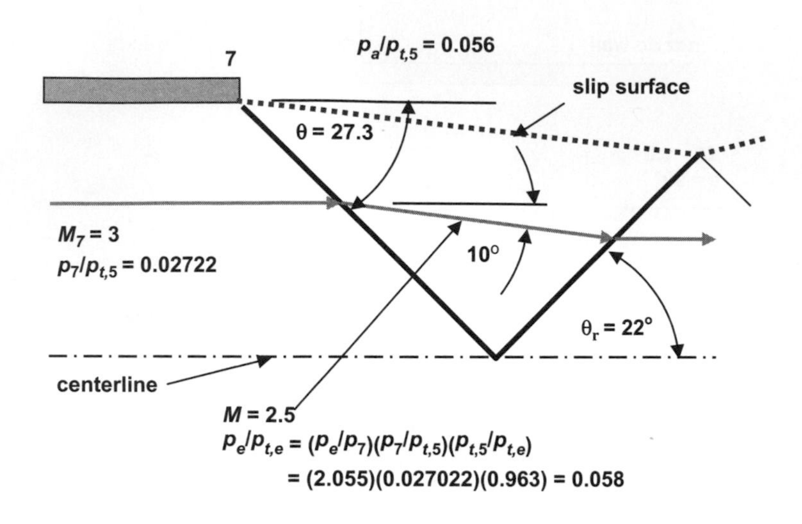


**FIGURE 5.9**

Flow field for an overexpanded nozzle designed for an exit Mach number $M_7 = 3$.

angle of 27.3° with respect to the undisturbed flow direction. Now the flow is moving toward the axis, but the flow at the centerline must remain straight, so the shock wave reflects off the axis by an angle of 22°, which straightens the flow out, making it all now parallel to the axis. Therefore, in the triangular region shown in Figure 5.9, the flow has the correct pressure, but to straighten out the flow direction the flow passes through another, weaker, shock, which now will make the pressure of the jet too high. The flow will have to expand, that is, diverge away from the axis; this is achieved by a Prandtl–Meyer expansion from the slip surface. Thus the process repeats itself in order to maintain pressure-balanced flow across the slip surface. The first few such cells are well described by these inviscid flow calculations, but ultimately frictional effects act to further equilibrate the flow, as described in White (2006).

## 5.9 EXAMPLE: UNDEREXPANDED NOZZLES

When the back pressure is lower than design exit pressure $p_7$, the flow is continuously supersonic throughout the nozzle, and adaptation to the low outside pressure occurs by means of a system of expansion waves and oblique shocks. The example case of Figure 5.3 is once again considered in the case where $p_a/p_{t,5} = 0.01$ and the exit pressure is $p_7/p_{t,5} = 0.02722$. The flow field under these conditions is illustrated in Figure 5.10. Once again, it should be clear that information about the lower pressure outside the exit cannot be felt in the nozzle for the reasons described in the previous section and illustrated in Figure 5.7.

We again consider the nozzle to be two dimensional so that plane flow relations can be applied readily. Somehow the pressure in the uniform flow region of the nozzle must be decreased so that streamlines leaving the lip of the nozzle have a pressure exactly equal to that of the ambient atmosphere. To reduce the pressure of the supersonic flow, streamlines at the lip must be deflected 11.69° away from the axis so as to increase the streamtube area. This finite deflection of the flow causes a Prandtl–Meyer expansion fan to form starting at the Prandtl–Meyer angle of $\nu = 49.76°$ appropriate

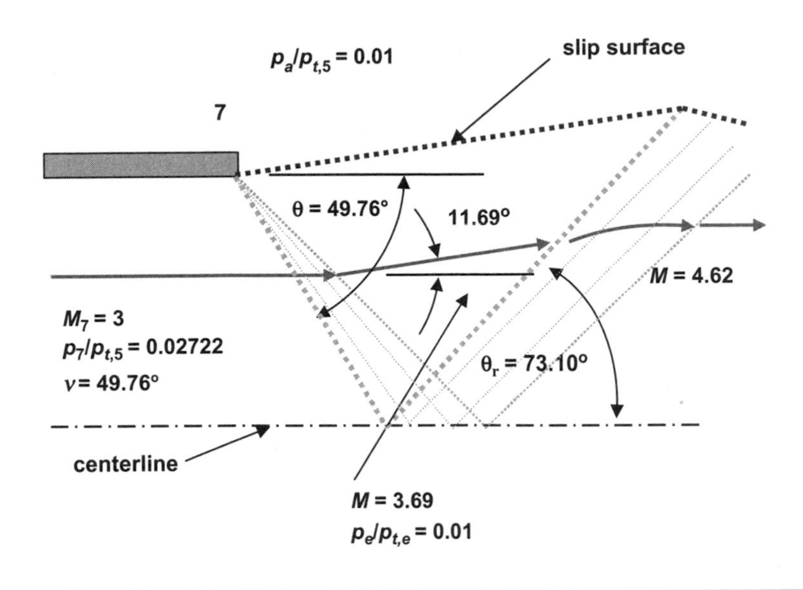


**FIGURE 5.10**

Flow field for an underexpanded nozzle designed for an exit Mach number $M_7 = 3$.

to $M = 3$ flow. The required 11.69° of expansion equilibrates the pressure but increases the flow Mach number to $M = 3.69$ and directs the streamline away from the centerline. Because, as in the previous case, the centerline of the flow must remain straight, the expansion fan reflects off the axis by an amount appropriate to straighten the flow out, making it all now parallel to the axis. Therefore, in the smallest triangular region shown in Figure 5.10 the flow has the correct pressure, but to straighten out the flow direction the flow passes through the reflected expansion fan shock, which now will make the pressure of the jet too low. The flow will have to compress, that is, converge toward the axis; this is achieved by a shock wave formed by coalescence of the expansion waves reflecting from the slip surface, which is not shown in Figure 5.10. As before in the overexpanded nozzle case, the process repeats itself in order to maintain pressure-balanced flow across the slip surface. The first few such cells are well described by these inviscid flow calculations, but frictional effects ultimately act to further equilibrate the flow, as described in White (2006).

## 5.10 AFTERBURNING FOR INCREASED THRUST

The turbine outlet temperature $T_{t,5}$ is limited by the structural integrity of the turbine blades and considerations of reliability, component lifetime, and maintenance requirements. However, the gross thrust of the engine is dependent on the nozzle exhaust velocity, which was shown in Chapter 1 to be given by

$$V_7 = \sqrt{2c_pT_{t,7}\left[1 - \left(\frac{p_7}{p_{t,7}}\right)^{\frac{\gamma_7}{\gamma_7-1}}\right]}.$$

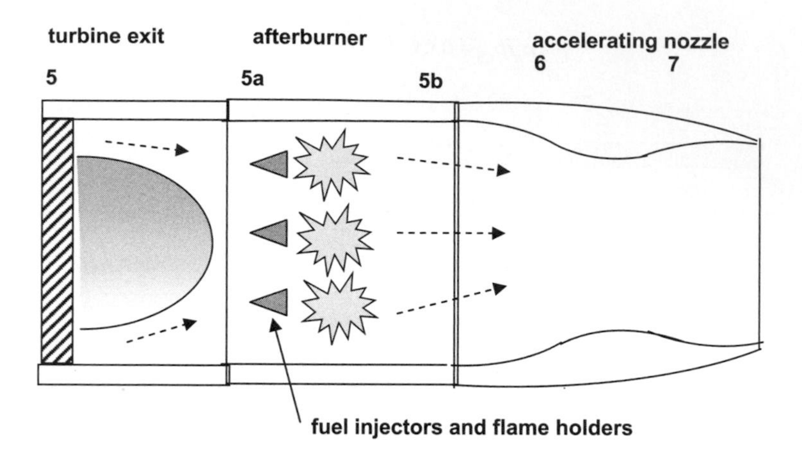


**FIGURE 5.11**

Afterburner section added between turbine exit and accelerating nozzle.

We generally assume that the nozzle flow is at least adiabatic, so $T_{t,5} = T_{t,7}$ and therefore $V_7 \sim T_{t,7}^{1/2}$. Because of the secondary air cooling requirements of the combustor and turbine components, hot gas flowing into the nozzle is oxygen rich and can easily support additional combustion if additional fuel is added to the stream. The portion of the nozzle between the turbine exit, station 5, and the minimum section, station 6, can be extended and fuel injectors placed there to provide a second combustion chamber, as shown in Figure 5.11. Here burning takes place after the conventional combustor; this section is therefore called the afterburner. In this afterburner section, the techniques used for conventional combustors, as described in Chapter 3, may be applied. The energy balance, assuming an average value for the mixture specific heat, is given by

$$\dot{m}_5 \bar{c}_p \left(T_{t,5b} - T_{t,5}\right) = \eta_{ab} \dot{m}_{f,ab} HV.$$

Then the stagnation temperature in the nozzle is

$$T_{t,5b} = T_{t.5} + \frac{\eta_{ab}}{\bar{c}_p} \frac{\dot{m}_{f,ab}}{\dot{m}_5} HV.$$

Because we assume that flow in the nozzle is adiabatic, we have $T_{t,5b} = T_{t,7}$ and the usual nozzle calculations may be applied at this new, higher energy state of the exhaust gas. Because $HV = 44$ MJ/kg and $\bar{c}_p$=1.16 kJ/kg/K for conventional fuels and exhaust conditions, the temperature rise may be approximated by

$$T_{t,7} \approx T_{t,5} + 37,930 \eta_{ab} \frac{\dot{m}_{f,ab}}{\dot{m}_5}.$$

Thus for each percent of fuel fraction added to the afterburner, there is a net return of around 360K added to the nozzle stagnation temperature.

Typical exhaust conditions for transonic flight of a military fighter aircraft are shown in Figure 5.12 (Fuhs, 1972). The stagnation temperature is increased greatly and the ratio of specific heats is decreased as power is increased from military power to partial afterburner operation to full afterburner operation.

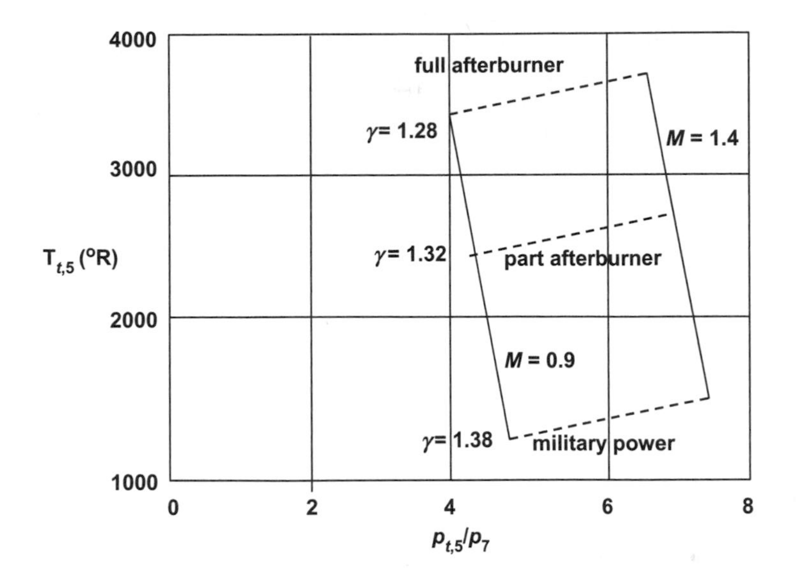


**FIGURE 5.12**

Range of nozzle exhaust conditions for different power settings for flight of fighter aircraft at transonic Mach numbers, $0.9 < M < 1.4$.

Because the exit velocity is proportional to $T_{t,5}^{1/2}$, we see that this velocity, and therefore the thrust, will increase 40 to 70% by operating in the afterburning mode. Of course, this is accomplished at the cost of increased fuel consumption, but afterburning is necessary for most fighter aircraft in order to have sufficient thrust to overcome the substantial drag rise in the transonic flight regime. The Lockheed F-22 is one of the new fighter aircraft that can achieve so-called supercruise, which denotes the ability to achieve supersonic flight without recourse to afterburning.

## 5.11 NOZZLE CONFIGURATIONS

### 5.11.1 Geometry requirements

Nozzles for subsonic flight are simple converging nozzles, whereas converging–diverging nozzles are required for supersonic flight. Two generic nozzles are shown in Figure 5.13, both having the same critical area $A_6$, so that both would pass the same mass flow for given $p_{t,5}$ and $T_{t,5}$ and the same gas properties. Obviously, if afterburning is used, the minimum section would choke because of the increase in stagnation temperature produced by afterburning. Recall that the maximum mass flow through the nozzle has the following form:

$$\dot{m}_{\max} \sim \frac{p_{t,5}A_6}{\sqrt{T_{t,5}}}.$$

Therefore, the increase in stagnation temperature would increase the exhaust velocity $V_7$, but at the expense of reduced mass flow, which would offset the possible thrust gains. Thus the use of

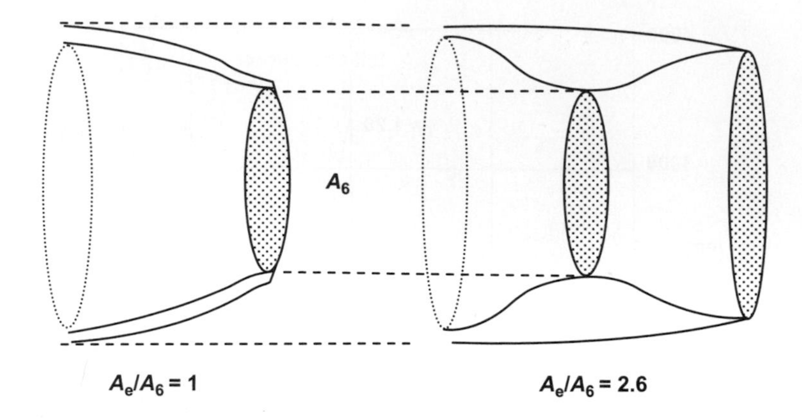


**FIGURE 5.13**

Two nozzles with the same critical area, $A_6$. The converging nozzle has an exit $M = 1$, while the converging–diverging nozzle has a supersonic Mach number exit.

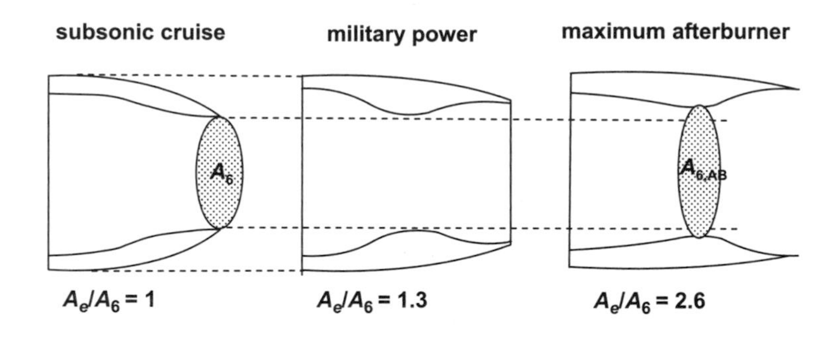


**FIGURE 5.14**

Changes in nozzle geometry required in going from subsonic cruise ($M_0 < 0.9$) to military power for transonic acceleration ($0.9 < M_0 < 1.2$) to full afterburning for high Mach number flight ($2 < M_0 < 2.4$).

afterburning requires a variable area nozzle, one in which the minimum area must be capable of being increased. Typical changes of the nozzle geometry for different power settings are shown in Figure 5.14.

The variation of the ratio of nozzle exit stagnation pressure $p_{t,7}$ to ambient pressure at altitude $p_0$ for the flight conditions illustrated in Figure 5.12 is shown in Figure 5.15 (Aulehla and Lotter, 1972). There are various means of producing variable area in a nozzle, but these all entail additional complexity, weight, and therefore cost. Nozzles also require cooling, which became an important aspect of nozzle design.

### 5.11.2 Simple ejector theory

Early jet aircraft nozzles were merely circular ducts with convergent exits and were made of high-temperature steel. As performance improvements were made, particularly in the area of increased

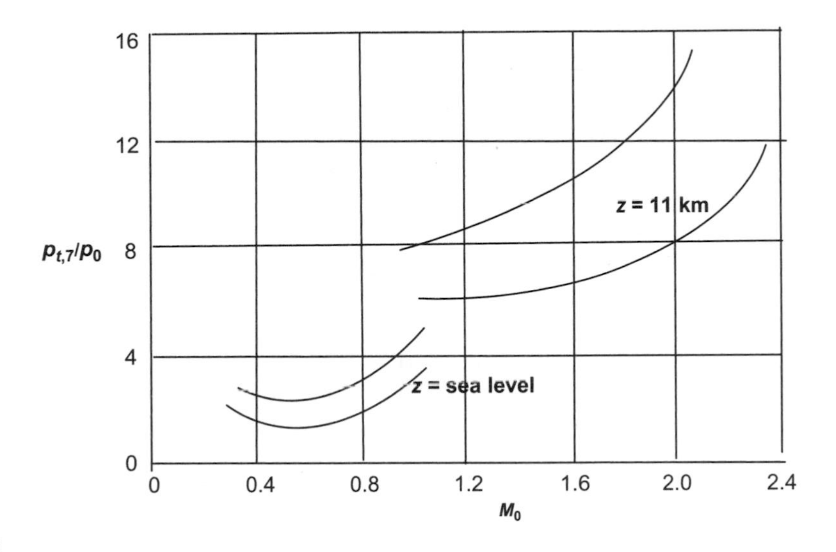


**FIGURE 5.15**

Variation of nozzle stagnation pressure with altitude $z$ and $M_0$.

exhaust temperature, the nozzle structural strength became increasingly compromised and simple, lightweight cooling methods were sought. Obviously, blowing cooler air over the tailpipe would be effective, but the weight and cost of pumping seemed prohibitive. However, one means of pumping was actually already available in the nozzle exhaust: the jet pump. High-velocity gas flowing out of a pipe naturally entrains the surrounding air and induces movement of that air. In essence, the jet drags the surrounding air along with it and this effect can be refined for cooling the nozzle. A nozzle designed for substantial entrainment is called an ejector. The induced motion can even enhance the total thrust of the nozzle. A simple case provides a simple illustration of the concept and is illustrated in Figure 5.16.

The conservation of mass and momentum equations may be applied between the primary nozzle exit station and the ejector nozzle exit station. We assume that the primary nozzle exit has velocity $V_j$, pressure $p_j$, and area $A_j$. At the primary nozzle exit station the ejector nozzle velocity is $U$ and the pressure is constant and equal to $p_j$. The ejector nozzle continues as a circular duct of cross-sectional area $A$ and has sufficient length to ensure that mixing between primary and ejector flows is complete and that the exit velocity is uniform and equal to $V$. The flow everywhere is assumed incompressible and frictionless so the exit station of the ejector nozzle is at the ambient pressure $p$. Because the air entrained through the ejector nozzle originates in the quiescent ambient the stagnation pressure of the entrained flow up to the primary nozzle exit station is $p_{t,0} = p_0$. The mass conservation equation for flow in the control volume that extends from the primary nozzle exit to the ejector nozzle exit is

$$-\rho A_j V_j - \rho\left(A - A_j\right)U + \rho A V = 0.$$

The conservation of momentum for the flow in the control volume is

$$-\rho V_j^2 A_j - \rho U^2\left(A - A_j\right) + \rho V^2 A = (p_j - p_0)A.$$

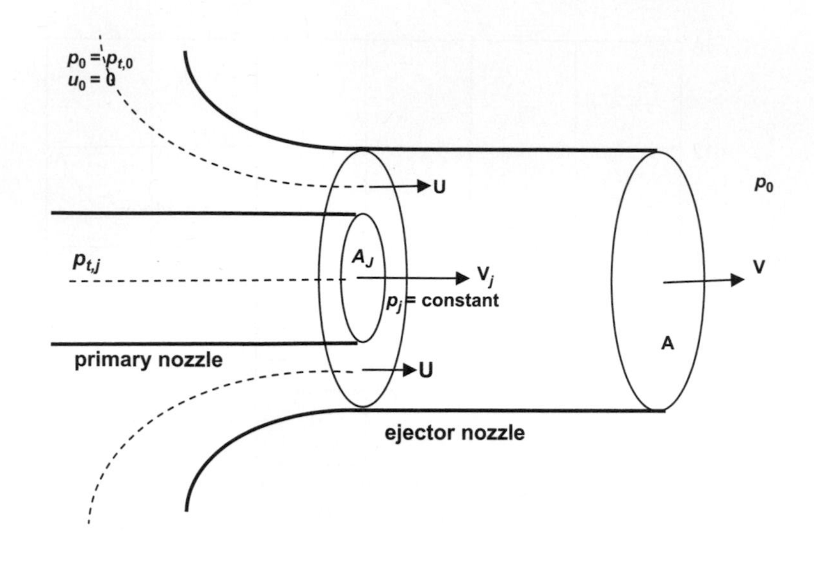


**FIGURE 5.16**

Primary nozzle exhausting into ejector nozzle in a quiescent ambient.

Bernoulli's equation along the streamline starting out in the quiescent ambient and entering the control volume through the ejector nozzle provides the following information:

$$p_j + \frac{1}{2}\rho U^2 = p_{t,o} = p_0.$$

Introducing the area ratio $\alpha = A_j/A$, the conservation of mass equation yields

$$U = \frac{V - \alpha V_j}{1 - \alpha}.$$

Then the conservation of momentum equation becomes

$$-\alpha V_j^2 - \left(\frac{V - \alpha V_j}{1 - \alpha}\right)^2 (1 - \alpha) + V^2 = -\frac{1}{2}\left(\frac{V - \alpha V_j}{1 - \alpha}\right)^2.$$

This equation may be written in terms of $\overline{V} = V/V_j$ as follows:

$$\left(\frac{\overline{V} - \alpha}{1 - \alpha}\right)^2 \left(\frac{1}{2} - \alpha\right) - \overline{V}^2 + \alpha = 0.$$

This quadratic equation may be solved and the results used to calculate two important quantities for nozzle operation: mass flow entrained by the ejector that can provide cooling and the primary nozzle thrust augmentation capability of the ejector. The ratio of mass flow entrained through the ejector nozzle to that passing through the primary jet nozzle, $\dot{m}_e/\dot{m}_j$, and the ratio of the total thrust compared to the primary nozzle jet thrust, $F/F_j$, are shown in Figure 5.17 as functions of the ratio of the diameters of the primary jet nozzle diameter to the ejector nozzle, $D_j/D$.

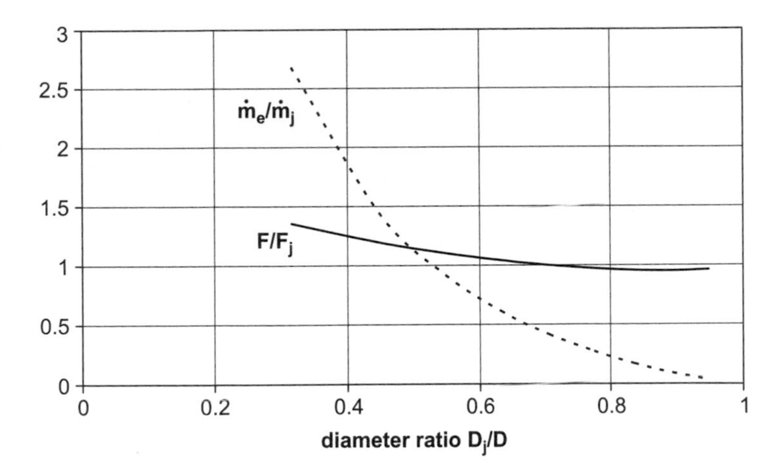


**FIGURE 5.17**

Ratio of ejector to primary nozzle mass flow, $\dot{m}_e/\dot{m}_j$, and ratio of total thrust to primary nozzle jet thrust, $F/F_j$, as functions of the diameter ratio, $D_j/D$.

It is clear that there can be a substantial benefit to using a simple ejector nozzle to pump air over the primary nozzle for cooling as well as the possibility of enhancing the thrust produced by the primary nozzle alone. The momentum flux of the mixed ejector and nozzle flow leaving the ejector grows faster than the area ratio for small area ratios but slower for larger area ratios so the thrust enhancement falls off. Of course this analysis doesn't account for friction or compressibility, but the same effects are observed under more practical conditions.

### 5.11.3 Ejector application to high-performance aircraft

For subsonic flight, a simple fixed convergent nozzle is generally suitable for a broad range of flight conditions. However, in supersonic flight, the nozzle requirements are more severe and variable area capability is necessary. In addition, high temperatures associated with afterburning force the use of cooling air in the nozzle. There is an additional condition that must be considered, which is the need to minimize the additional drag attendant upon variable geometry nozzles. The manner in which the geometry of the aft regions of the aircraft fuselage is altered to accommodate nozzle geometry variations gives rise to varying degrees of so-called boat-tail drag. Three ejector nozzles are shown schematically in Figure 5.18. The first, in Figure 5.18a, is a simple ejector with some basic degree of geometry variation in that the area of the convergent nozzle can be increased to accommodate the mass flow in the afterburning mode. Bypass air is drawn over the nozzle by the jet pump action of the ejector. The variable flap ejector (VFE) nozzle sketched in Figure 5.18b has a variable geometry capability that provides for convergent nozzle operation at subsonic speed and convergent divergent nozzle operation with greater throat area for supersonic speeds. This type of nozzle is often called a long flap ejector nozzle, and variants of it have been used on the J79 and J93 engines used on the B-58, XB-70, F-4, and F-104 aircraft. The long flaps provide for small boat-tail angle in the supersonic configuration, which is important in reducing fuselage drag while still providing for supersonic expansion within the nozzle. Bypass air again provides cooling for the nozzle.

The variable flap ejector nozzle is necessarily of higher weight and complexity, and the longer flaps are susceptible to aeroelastic problems, which can lead to destruction of the ejector. Optimization for high-pressure ratios for supersonic afterburning dictates long flaps, which have reduced performance at low pressure ratios where shorter flaps would be superior. However, the requirements for supersonic flight were met by this type of ejector nozzle and were satisfactory for supersonic dash capability, that is, for intermittent use.

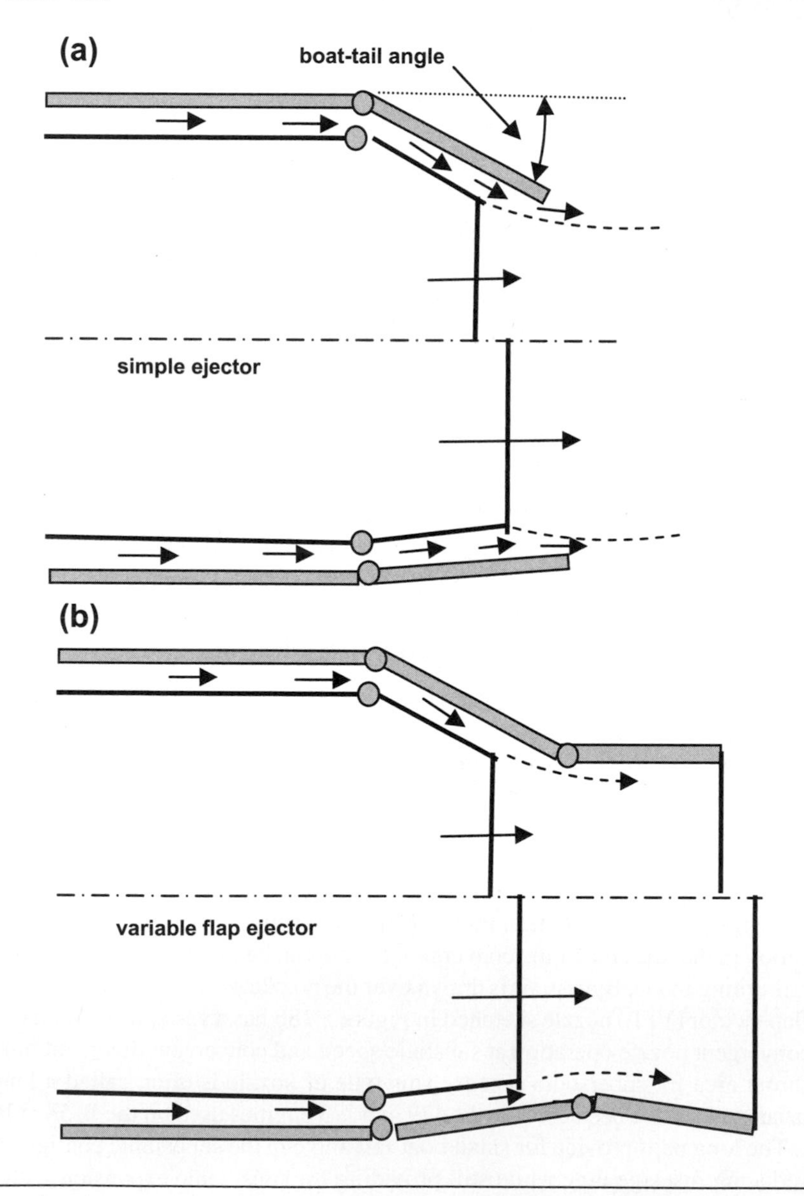


**FIGURE 5.18**

Three ejector nozzles showing subsonic (upper view) and supersonic (lower view) operation.

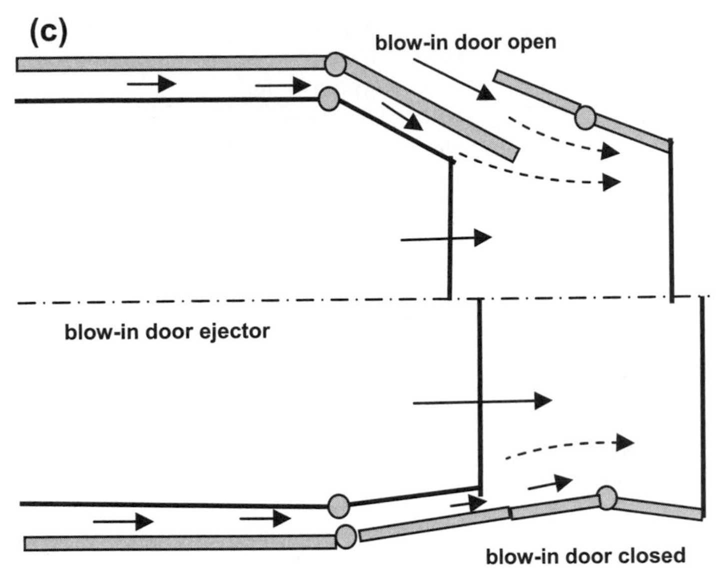


FIGURE 5.18 *(Continued)*

The blow-in door ejector (BIDE) nozzle is shown in Figure 5.18c. At supersonic speeds, where nozzle pressures are high, the blow-in door is closed and flaps provide variable exit area appropriate to the nozzle conditions. The bypass cooling air is required to cool the flaps, and typically the flow rate is around 4% of the nozzle flow. At lower flight Mach numbers, the nozzle pressure ratio decreases and therefore the force on the flaps decreases and they are permitted to float inward. As the Mach number decreases further, the blow-in doors are opened to admit external air into the nozzle to prevent overexpansion of the nozzle, and the attendant loss in thrust, at the lower pressures.

The structural and mechanical construction of the BIDE ejector nozzle is relatively simple. The flaps are loaded from within by nozzle pressure so that they can rely on simple mechanical stops to limit their maximum opening. They can be permitted to float to intermediate positions dictated by the local conditions and have little unbalanced load to withstand. Of course, any floating control surfaces must be selected properly to avoid oscillatory instability. In terms of weight and complexity, the BIDE falls between the simple ejector and the VFE. The BIDE has been used on the SR-71, the F-111, and the Concorde supersonic transport. The use of convergent–divergent iris nozzles, such as those on the F-14, has largely supplanted the VFE and BIDE nozzles in high-performance applications.

### 5.11.4 Convergent–divergent iris nozzle

One successful type of variable area nozzle is similar in operation to the iris on a camera: the convergent iris nozzle and the convergent–divergent iris nozzle, both of which are shown in Figure 5.19.

The curved iris segments are extended or retracted by actuators and slide over one another to accommodate the changing geometry. Smooth low boat-tail angles are achieved with this type of nozzle, improving drag performance of the aircraft fuselage, which accounts for a large fraction of the

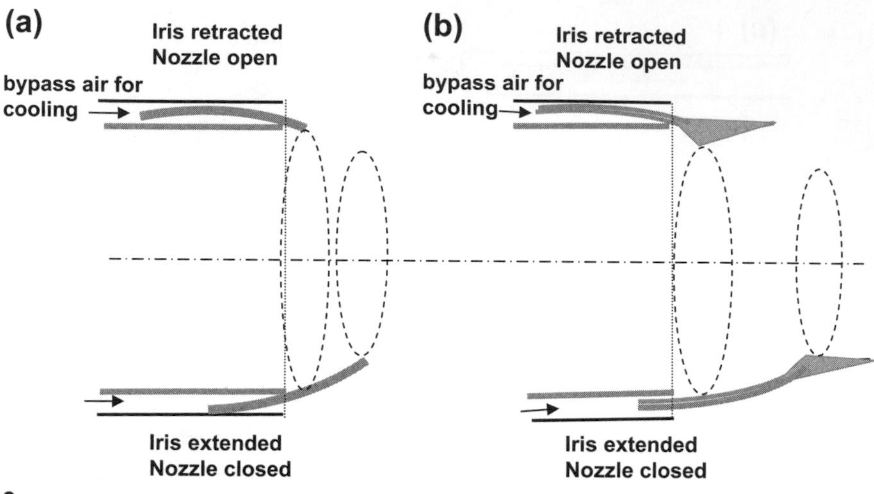


**FIGURE 5.19**

Iris-type variable area nozzles: (a) convergent and (b) convergent–divergent. The upper half of the diagrams represents the high throat and/or exit area setting, while the lower half depicts the smaller throat and/or exit area setting.

drag in supersonic flight. Grumman's F-14 aircraft pioneered use of the convergent–divergent iris variable area nozzle, which was used on later aircraft such as the F-15, F-16, and the B1-B.

### 5.11.5 Thrust-vectoring nozzles

The ability to move the thrust force vector can be of great benefit to military aircraft in terms of enhancing maneuverability, particularly at low speeds where control surfaces have less authority. Like rockets, which have been using gimbaled motors for many years to change the direction of the thrust vector for trajectory control, aircraft are now using similar mechanisms. It has generally been considered preferable to move the entire nozzle rather than reconfigure the arrangement of the interior of the nozzle to affect the thrust vector. Thrust vectoring has been applied to axisymmetric nozzles using overlapping flaps in an iris-like arrangement or by a gimbal mechanism on the nozzle. A two-dimensional nozzle has individually actuated upper and lower flaps, which permit varying throat and exit area, as well as deflecting in concert so as to provide thrust vectoring. This type of nozzle is shown in Figure 5.20.

There are a variety of techniques for achieving thrust vector control of nozzle flows, and a discussion of several has been given by Wing and colleagues (1997). The latest advanced fighter aircraft, such as the F-22 and the F-35, incorporate thrust-vectoring nozzles.

## 5.12 NOZZLE PERFORMANCE

The concept of quasi-one-dimensional flow requires that the flow properties are considered average properties across a given cross section. Main departures from the averaged values tend to arise in the boundary layer on the nozzle wall, as well as any localized disturbances in the main flow, such as the drag of structures or the heat release from flames. These are all losses reflected by an entropy increase

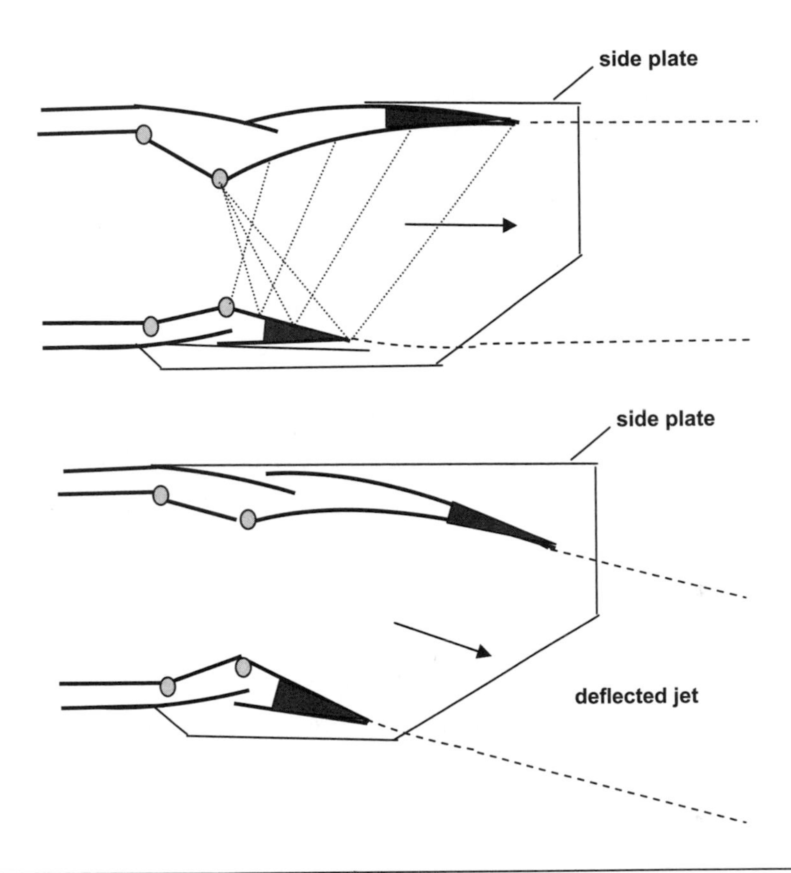


**FIGURE 5.20**

Two-dimensional thrust vectoring nozzle showing flow deflection caused by variable geometry flaps. The side plate bounds either side of the nozzle.

in the flow. A comparison of an ideal isentropic expansion and a real expansion in a nozzle can be easily appreciated on an *h,s* diagram such as that in Figure 5.21.

The nozzle cannot realize the full isentropic potential because friction and other losses have used up some of the internal energy that might have gone into increasing the kinetic energy, and therefore the velocity, of the gas. The efficiency of expansion, which is addressed again in Chapter 7 when expansion in a turbine is studied, is called here the nozzle efficiency and is expressed by the ratio of actual enthalpy change to ideal enthalpy change as follows:

$$\eta_n = \frac{h_{t,5} - h_7}{h_{t,5} - h_7'} = \left(\frac{V_7}{V_7'}\right)^2. \tag{5.30}$$

To be considered a good nozzle, nozzle efficiency should lie in the range $0.9 < \eta_n < 0.96$. It is often more usual to define a nozzle velocity coefficient as

$$c_V = \frac{V_7}{V_7'}. \tag{5.31}$$

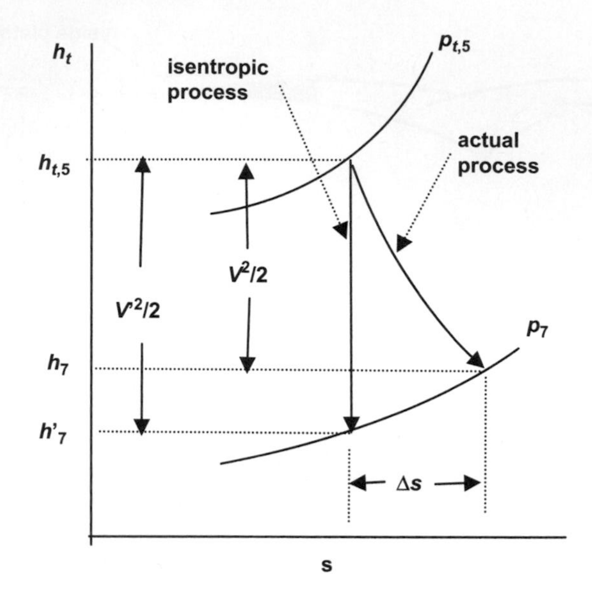


**FIGURE 5.21**

Isentropic and actual expansion process in nozzle.

Obviously, good nozzle performance means that this coefficient would lie in the range $0.95 < c_V < 0.98$. Liepmann and Roshko (1957) show that the change in stagnation pressure is a simple indicator of an entropy change and that the relationship between the two, using the notation of Figure 5.16, may be written as follows:

$$\frac{\Delta s}{R} = \frac{s_7 - s_5}{R} = -\ln \frac{p_{t,7}}{p_{t,5}}.$$

The change in stagnation pressure in the nozzle flow may be written as

$$\frac{dp_t}{p_t} = \frac{1}{p_t} d\left\{ p\left[1 + \frac{\gamma - 1}{2} M^2\right]^{\frac{\gamma}{\gamma - 1}} \right\} = \frac{dp}{p} + \frac{1}{2}\gamma \frac{dM^2}{1 + \left(\frac{\gamma - 1}{2}\right) M^2}.$$

Using the influence coefficients in Table 2.1 under the assumptions stated and neglecting any drag forces in the flow leads to

$$\frac{dp_t}{p_t} = -\frac{\gamma M^2}{2}\left[4c_f \frac{L}{D} d\left(\frac{z}{L}\right)\right]. \tag{5.32}$$

The skin friction coefficient is a function of Mach number and Reynolds number. In particular, for compressible turbulent constant pressure boundary layers, the local skin friction coefficient is reasonably approximated by the flat plate reference enthalpy method as given by

$$c_f = 0.0592\left[1 + 0.72 \, \mathrm{Pr}^{1/3} \left(\frac{\gamma - 1}{2}\right) M^2\right]^{-0.6734} \mathrm{Re}^{-1/5}. \tag{5.33}$$

Thus the skin friction has a relatively complicated dependence on detailed development of the flow. The Prandtl number $\Pr = \mu c_p/k$ and the Reynolds number $\mathrm{Re} = \rho V z/\mu$ are well-known similarity parameters and obviously vary with the thermodynamic state of the gas. Therefore, Equation (5.32) is coupled to the equations for the other flow variables through the Mach number and thermodynamic state, and integration is not possible analytically, although solutions could be carried out numerically. To gain some insight to the effect of friction on the stagnation pressure, assume that the skin friction coefficient and Mach number in Equation (5.32) are replaced by representative average values so that

$$\frac{p_{t,7}}{p_{t,5}} = \exp\left(-2\gamma M_{avg}^2 c_{f,avg}\frac{L}{D}\right).$$

As mentioned previously, for the conditions of interest here, the argument of the exponential term is $O(10^{-2})$, so it may be expanded to

$$\frac{p_{t,7}}{p_{t,5}} \approx 1 - 2\gamma M_{avg}^2 c_{f,avg}\frac{L}{D}.$$

For reasonable nozzle conditions, such as $\gamma = 1.33$, $M_{avg} = 1.8$, $c_{f,avg} = 0.0027$, and $L/D = 1$, we find $p_{t,7}/p_{t,5} \cong 0.975$. Thus, as predicted, the effect of friction on stagnation pressure is small. In the case where an afterburner section is considered, the effect of frictional losses would be even smaller because even though $L/D$ might be doubled, the Mach number in the afterburner would be subsonic. For example, taking $\gamma = 1.33$, $M_{avg} = 0.7$, $c_{f,avg} = 0.0033$, and $L/D = 2$, we find $p_{t,5ab}/p_{t,5} \cong 0.99$. Then, across the afterburner–nozzle combination, the effect of friction on stagnation pressure ratio is $p_{t,7}/p_{t,5} \cong 0.967$. Of course, when the afterburner is in operation the major effect on stagnation pressure arises through the heat addition caused by combustion. We may characterize the afterburner as a constant area combustor and use the result shown in Equation (4.12), assuming a constant value of $\gamma = 1.33$ and using the notation of Figure 5.11:

$$\frac{p_{t,5b}}{p_{t,5}} = \frac{1 + 1.33M_5^2}{1 + 1.33M_{5b}^2}\left[\frac{1 + 0.165M_{5b}^2}{1 + 0.165M_5^2}\right]^{4.03}.$$

The maximum value for $M_{5b} = 1$ and we may assume $M_5 = 0.5$, which leads to a value of $p_{t,7}/p_{t,5} \cong 0.90$. This is a substantial loss in stagnation pressure, and the drag associated with the fuel injectors would raise that a bit more. Thus, in afterburner mode, the stagnation pressure ratio across the whole exhaust system would be $p_{t,7}/p_{t,5} \cong 0.877$.

Carrying out detailed calculations can be counterproductive during the early stages of the design cycle where many different nozzle configurations are to be considered and time constraints are strict. It is sometimes useful to consider the bulk effects of nozzle efficiency by making use of nozzle efficiency as a parameter. A basic assumption is that the flow is steady and adiabatic so that the stagnation temperature is a constant throughout. In addition, because we assumed that the thermodynamic properties of the gas flowing through the nozzle can be assumed to be constant, Equation (5.30) may be rewritten as

$$\eta_n = \frac{1-\left(\frac{T_7}{T_{t,5}}\right)}{1-\left(\frac{T_7'}{T_{t,5}}\right)}.$$

Then the ideal temperature at the exit is that which would have been reached in an isentropic process between stations 5 and 7, which may be written as

$$\frac{T_7'}{T_{t5}} = 1 - \frac{1}{\eta_n}\left(1 - \frac{T_7}{T_{t,5}}\right) = \left(\frac{p_7}{p_{t,5}}\right)^{\frac{\gamma-1}{\gamma}}. \tag{5.34}$$

Of course, for an adiabatic flow where $T_{t,5} = T_{t,7}$, we may use Equation (5.9) in Equation (5.34) and solve for the pressure at the exit as

$$\frac{p_7}{p_{t,5}} = \left[\frac{\eta_n - 1 + \left(1 + \frac{\gamma-1}{2}M_7^2\right)^{-1}}{\eta_n}\right]^{\frac{\gamma}{\gamma-1}}. \tag{5.35}$$

Because we typically specify the stagnation pressure upstream of the nozzle and can measure the pressure at any other station, such as the exit, Equation (5.35) provides a means to estimate the exit Mach number using the nozzle efficiency as a parameter. A plot of the nozzle pressure ratio for $\eta_n =$ 1, 0.95, and 0.9 is shown in Figure 5.22 for the typical exhaust nozzle with $\gamma = 1.33$. The other variables may be put into a similar form by using the mass conservation equation along with the appropriate influence coefficients in Table 2.1.

The action of friction on the nozzle walls is to reduce the effective flow area and momentum flux because of growth of the boundary layer displacement and momentum thicknesses, respectively. Because the Reynolds number in nozzle flows is high, the boundary layers are thin and a reasonable

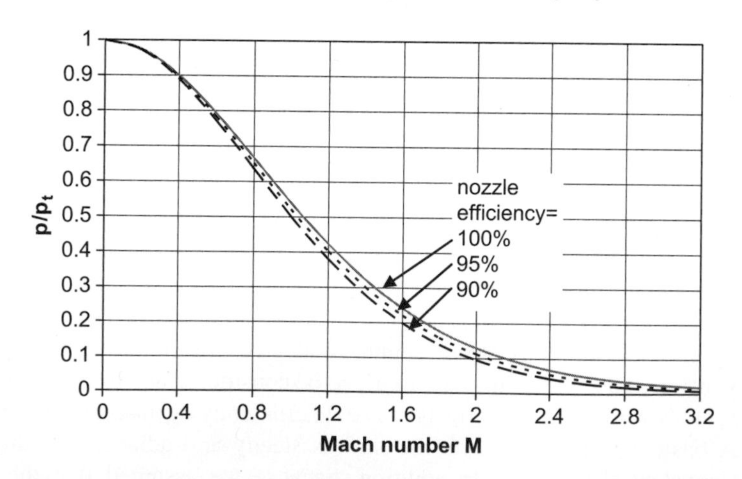


**FIGURE 5.22**

Nozzle pressure ratios for various values of nozzle efficiency and $\gamma = 1.33$.

design approach is to develop the nozzle configuration considering purely isentropic flow with area variation and then carry out corrections to the geometry using boundary layer analysis. After the design has reached this point, a more accurate computational analysis can productively be employed to fine-tune the design. A discussion of nozzle design techniques is given by Korte (1992).

## 5.13 NOMENCLATURE

| | |
|---|---|
| $A$ | cross-sectional area |
| $a$ | sound speed |
| $C$ | constant of integration |
| $c_d$ | drag coefficient |
| $c_f$ | skin friction coefficient |
| $c_p$ | specific heat at constant pressure |
| $c_V$ | nozzle velocity coefficient |
| $D$ | hydraulic diameter |
| $H$ | heat release due to chemical reaction |
| $HV$ | heating value of fuel |
| $F$ | force |
| $g$ | acceleration of gravity |
| $h$ | enthalpy |
| $k$ | isentropic exponent or gas thermal conductivity |
| $L$ | length |
| $M$ | Mach number |
| $\dot{m}$ | mass flow |
| $h$ | enthalpy |
| Pr | Prandtl number $= \mu c_p/k$ |
| $p$ | pressure |
| $Q$ | heat added to the fluid |
| $q$ | dynamic pressure |
| $R$ | gas constant |
| $S$ | reference area for drag |
| $s$ | entropy |
| $T$ | temperature |
| $V$ | velocity |
| $W$ | molecular weight |
| $W_k$ | work done on the fluid |
| $z$ | axial coordinate or altitude |
| $\Gamma$ | function of $\gamma$, Equation (5.28) |

| | |
|---|---|
| $\gamma$ | ratio of specific heats |
| $\rho$ | gas density |
| $\mu$ | Mach angle, $\sin^{-1}(1/M)$ or gas viscosity |
| $\eta_{\uparrow n}$ | nozzle efficiency |

### 5.13.1 Subscripts

| | |
|---|---|
| *a* | ambient |
| *ab* | afterburner |
| *avg* | average |
| *down* | downstream of normal shock |
| *e* | exit |
| *f* | fuel |
| max | maximum conditions |
| *t* | stagnation conditions |
| *up* | upstream of normal shock |
| *z* | axial direction |
| 5 | nozzle entrance conditions |
| 6 | nozzle minimum section conditions |
| 7 | nozzle exit conditions |

### 5.13.2 Superscripts

| | |
|---|---|
| $()*$ | critical conditions, $M = 1$ |
| $(\bar{\ })$ | average value |
| $()'$ | isentropic value |

## 5.14 EXERCISES

**5.1** A simple two-dimensional nozzle is shown in Figure E5.1. The area of minimum section is $A_{\text{min}} = 0.1\ \text{m}^2$, and the area of the exit station is $A_{\text{e}} = 0.5\ \text{m}^2$; the area varies linearly with axial distance $z$ between the two locations. During a test, measurements indicate that total pressure at the nozzle inlet $p_{ti} = 172.3$ kPa, while the ambient pressure outside the nozzle $p_a = 101.3$ kPa. The test engineer claims that a normal shock is standing midway between the minimum and the exit stations, that is, at $z/L = 0.5$. Assuming $\gamma = 1.4$, (a) determine if a normal shock is actually located at the station claimed. (b) If the shock is not located at that station, is it upstream or downstream of the location stated by the test engineer? (c) Determine the exact location of the shock, if one is present at all.

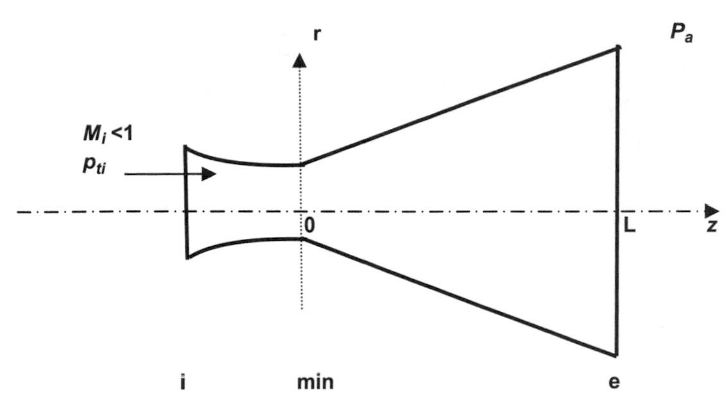


FIGURE E5.1

Nozzle for turbojet engine.

**5.2** The contoured (Laval) nozzle shown in Figure E5.2 has a ratio of exit area to minimum area $A_e/A_{\min} = 2$ and exhausts into an ambient atmosphere where the static pressure is $p_a = p_t/30$, that is, one-thirtieth of the stagnation pressure $p_t$ in the nozzle flow. For $\gamma = 1.4$, determine the direction $\alpha$ of the free streamline leaving the exit plane of the nozzle.

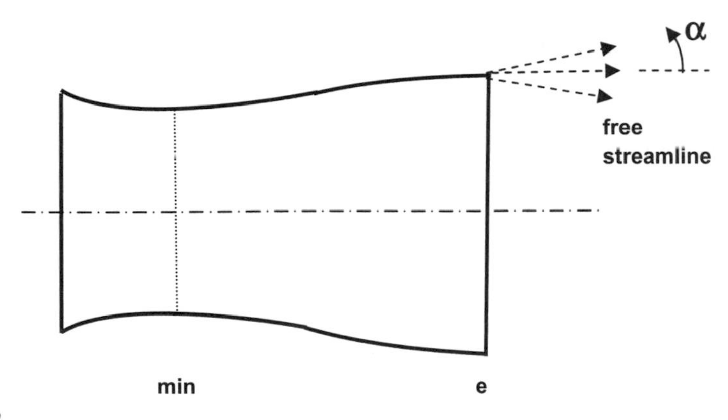


FIGURE E5.2

Contoured nozzle with unknown exit streamline deflection $\alpha$.

**5.3** The NASA F-18 Hornet Systems Research Aircraft (SRA) is currently being flown by NASA's Dryden Flight Research Center, Edwards, California, in a multiyear, joint NASA/DOD/industry program; the F/A-18 former Navy fighter has been modified into a unique SRA to investigate a host of new technologies in the area of flight controls. A convergent nozzle like that shown in Figure E5.3 is operated at standard sea level conditions with an exit pressure ratio $p_7/p_{t,5} = 0.55$ and a nozzle entrance total temperature

**FIGURE E5.3**

The NASA F-18 Hornet Systems Research Aircraft (SRA).

$T_{t,5} = 1450\text{R}$. If the exit area is 1 ft$^2$ and the flow in the nozzle flow is ideal, calculate the following using $\gamma = 1.33$: (a) exit velocity $V_7$, (b) exit temperature $T_7$, (c) exit weight flow rate, (d) gross thrust $F_g$, (e) gross thrust if the nozzle entrance total temperature is raised to $T_{t,5} = 3000\text{R}$, (f) gross thrust if $T_{t,5}$ remains at 3000R and the exit area is increased to 1.3 ft$^2$.

**5.4** Two SR-71 aircraft have been used by NASA as testbeds for high-speed and high-altitude aeronautical research. The aircraft, an SR-71A and an SR-71B pilot trainer aircraft, have been based at NASA's Dryden Flight Research Center, Edwards, California. They were transferred to NASA after the U.S. Air Force program was cancelled. The shock diamonds in the exhaust are caused by supersonic flow being generated by the exhaust nozzles (Figure E5.4). In a supersonic nozzle that has a converging diverging shape, the flow can be accelerated to supersonic speed by passing through the critical point ($M = 1$) at the minimum area formed by the converging section ($A_6 = A^*$) and then through the diverging section where the area increases to the exit area $A_7$. For such an ideal nozzle with $\gamma = 1.33$, a minimum area $A_6 = 4$ ft$^2$ and an expansion out to an exit area $A_7$ at which the pressure $p_7$ matches the ambient pressure $p_0 = 5$ psia (altitude = 27,000 ft), calculate (a) velocity,

**FIGURE E5.4**

An SR-71 aircraft used by NASA as a testbed for high-speed and high-altitude aeronautical research.

temperature, pressure, Mach number, and density at the minimum area station and then (b) repeat the calculation for the exit station and (c) the gross thrust developed.

**5.5** Design a converging diverging nozzle for a missile ramjet engine under the constraint of maximum thrust. The combustor delivers hot gas (assume $\gamma = 1.33 =$ constant) to the nozzle at a stagnation pressure $p_{t,5} = 81$ kPa through a minimum section of fixed area $A_6 = 0.0845\ m^2$. The missile is to fly at $M_0 = 2.2$ at altitudes between 12.2 and 18.3 km. Provide an illustration of the nozzle contour, drawn to scale and showing all pertinent features, and explain these features in accompanying text. (a) Explain how your design ensures maximum thrust over the mission specification envelope. (b) Show a plot of the variation of gross thrust with altitude for your nozzle design. (c) How did you decide on the length of the nozzle and its contour? (d) Present a simplified design for a nozzle that will approximate the maximum gross thrust constraint. (d) Compare the gross thrust produced by your simplified design with that of your ideal design over the flight envelope.

**5.6** Consider a jet engine that exhausts hydrocarbon–air combustion products (the overall fuel to air ratio here is $f/a = 0.02$) through a nozzle with stagnation conditions $T_{t,5} = 1500$F and $p_{t,5} = 30$ psia. Use the one-dimensional adiabatic energy equation and the assumption of an isentropic expansion through the nozzle to develop equations used to generate a plot of the (dimensional) static and stagnation temperature and pressure variation as a function of Mach number through the nozzle for Mach numbers in the range $0 < M < 3$. Clearly state any assumptions you made in producing the plot.

**5.7** A turbojet engine with an exit area of $0.116\ m^2$ is operated at an altitude of 6.1 km where $p_0 = 46$ kPa and $T_0 = 249$K. Assume that the nozzle is choked ($M_7 = 1$), exit pressure is equal to the ambient pressure of the undisturbed atmosphere at the operating altitude, and exhaust gases are characterized by $\gamma = 1.33$ and $R = 287\ m^2/s^2$-K. Determine the following:

**(a)** Gross thrust produced

**(b)** Exit mass flow rate, if the exit temperature $T_7 = 811$K

**(c)** Ram drag at flight speed $V_0 = 168$ m/s (assuming an overall fuel to air ratio of 0.02)

**(d)** Specific fuel consumption

**5.8** A turbojet engine with a nozzle exit radius of 19.23 cm is operated with an exit pressure $p_7 = 62$ kPa at an altitude of 6.1 km where $p_0 = 46$ kPa and $T_0 = -24$°C. Assume the nozzle is choked ($M_7 = 1$) and exhaust gases are characterized by $\gamma = 1.33$ and $R = 287$ J/kg-K. Determine the following:

**(a)** Gross thrust $F_g$ produced

**(b)** Exit mass flow rate, if the exit temperature $T_7 = 538$°C

**(c)** Ram drag $F_r$ at the flight speed $V_0 = 168$ m/s (assuming an overall fuel to air ratio f/a $= 0.02$)

**(d)** Specific fuel consumption $c_j$

**5.9** A turbojet engine has a converging diverging nozzle that operates in the matched pressure mode and accepts the gas flow from the turbine at a stagnation temperature $T_{t,5} = 838°C$ and a stagnation pressure $p_{t,5} = 275.6$ kPa. The nozzle throat radius $r_6 = 10$ cm and $\gamma = 1.33$ throughout the nozzle.

**(a)** Calculate the exit Mach number $M_7$.

**(b)** Find the nozzle exit area $A_7$ required to attain the maximum gross thrust $F_g$ possible at an altitude of $z = 23$ km, where $p_0 = 4.35$ kPa.

**(c)** Calculate the mass exit flow.

**5.10** At the entrance to a convergent turbojet nozzle, the stagnation values of pressure and temperature are 138 kPa and 560°C and $\gamma = 1.33$.

**(a)** For a back pressure of 101.3 kPa, calculate the exit velocity for isentropic flow.

**(b)** If the exit area of the nozzle is 0.11 $m^2$, determine the mass flow rate and the thrust produced.

**(c)** If the back pressure in part (a) were reduced to the critical value, find the values of exit pressure, exit velocity, mass flow rate, and thrust.

**(d)** If the back pressure were reduced further to 34.5 kPa, why would it be necessary to add a divergent segment to the nozzle to maintain isentropic flow conditions?

**(e)** For this condition, determine exit velocity and mass flow rate.

**(f)** Calculate pressure, temperature, and Mach number at the throat formed by the juncture of the converging and diverging sections.

**5.11** A blowdown supersonic tunnel, that is, one that operates by exhausting air from one receiver at high pressure through a contoured nozzle to another receiver at lower pressure, utilizes standard sea level air from the atmosphere as the high-pressure receiver and a large vacuum vessel as the low-pressure receiver.

**(a)** For isentropic flow through the nozzle, calculate pressure, temperature, velocity, Mach number, and density at the nozzle throat.

**(b)** If the cross-sectional area of the throat is 25.81 $cm^2$ and the exit pressure is 20.7 kPa, calculate temperature, velocity, Mach number, density, and area at the nozzle exit.

**(c)** Determine mass flow through the nozzle.

**5.12** A contoured nozzle is to be designed to accelerate air from stagnation conditions of 345 kPa and 889 K to a Mach number of 2.0 at the nozzle exit at a flow rate of 14.6 kg/s.

**(a)** Calculate pressure, temperature, velocity, area, and density at the nozzle throat.

**(b)** Calculate pressure, temperature, velocity, area, and density at the nozzle exit.

**(c)** Calculate the axial force on the nozzle due to the acceleration of the air.

**5.13** A contoured nozzle with a throat area of 323 $cm^2$ is operated with stagnation conditions of 276 kPa and 1110 K. It is designed to have the exit pressure match the ambient pressure of 34.5 kPa. If the adiabatic expansion efficiency of the nozzle is 90% and the ratio of actual to theoretical gross thrust is 0.93

**(a)** Calculate pressure, temperature, velocity, Mach number, and density at the throat and exit stations.

**(b)** Determine the exit area.

**(c)** Calculate the thrust developed.

**(d)** When the ambient back pressure is raised to 179 kPa, a normal shock occurs within the nozzle. Find pressure, temperature, velocity, and Mach number ahead of the shock, behind the shock, and at the exit station.

**(e)** Calculate the percent reduction in thrust caused by the shock wave in the nozzle.

**5.14** A convergent nozzle exhausts to standard sea level conditions with a pressure ratio $p_6/p_{t5} = 0.55$ and an inlet stagnation temperature $T_{t5} = 806\text{K}$. If the exit diameter is 0.344 m and the nozzle efficiency is 100%, assume $\gamma = 1.30$ and calculate the following: (a) exit velocity $V_6$, (b) exit temperature $T_6$, (c) mass flow rate, (d) gross thrust $F_g$, (e) exit Mach number $M_6$, (f) gross thrust if $T_{t5}$ is increased to 1667K, and (g) gross thrust if the exit diameter is also increased to 0.392 m.

**5.15** Consider air flowing through a smoothly contoured Laval nozzle as shown in Figure E5.5. Pressure in the very large reservoir, region S, is $p_0 = 1$ MPa. Pressure in region $A$, the unconfined ambient into which the nozzle issues, is $p_E = 100$ kPa. The area of the narrowest section is $A_1 = 0.15\ \text{m}^2$ and that of the nozzle exit is $A_2 = 0.6\ \text{m}^2$. Determine the following:

**(a)** Mach number and pressure at station 1, $M_1$ and $p_1$, respectively

**(b)** Mach number and pressure at station 2, $M_2$ and $p_2$, respectively

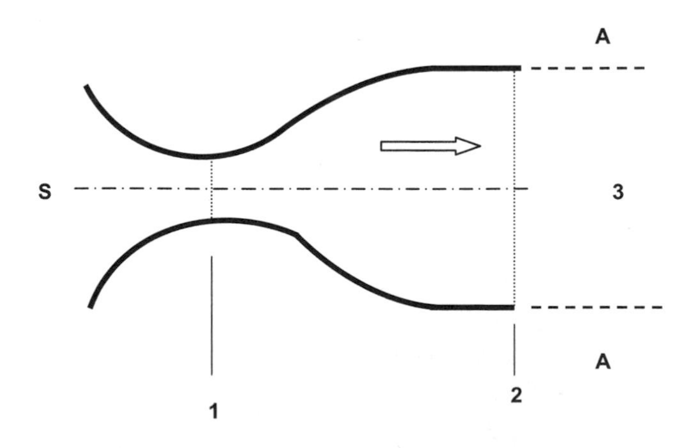


**FIGURE E5.5**

Schematic diagram of nozzle.

**(c)** Mach number and pressure in nozzle exit region 3, $M_3$ and $p_3$, respectively, and sketch the flow pattern there

(The universal gas constant is R = 8.413 kJ/kmol-K, and the molecular weight and ratio of specific heats for air are 29 kg/kg-mol and 1.4, respectively)

## References

Aulehla, F., & Lotter, K. (1972). Nozzle/Airframe Interference and Integration. AGARD Lecture Series, No. 53.

Fuhs, A.E. (1972). Engine Integration and Thrust/Drag Definition. AGARD Lecture Series, No. 53.

Korte, J.J. (1992). Aerodynamic design of axisymmetric hypersonic wind tunnel nozzles using least-squares/parabolized navier stokes procedure. *Journal Spacecraft Rockets*, *Vol. 29*, Sept.–Oct. pp. 685–691. Errata, *Vol. 29*, Nov.–Dec. 1992, pp. 870–871.

Liepmann, H. W., & Roshko, A. (1957). *Elements of Gasdynamics.* New York: John Wiley & Sons.

Love, E.S., Grigsby, C.E., Lee, L.P., & Woodling, M.J. (1959). Experimental and Theoretical Studies of Axisymmetric Free Jets. NASA Technical Report R-6.

NACA Ames Research Staff (1953). Equations, Tables, and Charts for Compressible Flow. NACA Report 1135.

White, F. (2006). *Viscous Fluid Flow,* (3rd ed.). New York: McGraw-Hill.

Wing, D.J., Mills, C.T.L., & Mason, M.L. (1997). Static Investigation of a Multiaxis Thrust-Vectoring Nozzle with Variable Internal Contouring Ability. NASA Technical Paper 3628, June.

CHAPTER

# Inlets

6

## CHAPTER OUTLINE



## 6.1 INLET OPERATION

The inlet is the first component on the engine and interacts directly with the ambient atmosphere. The major attributes of the inlet may be listed as follows:

- Handle a wide range of mass flow
- Duct air to the engine with low total pressure loss and low drag
- Diffuse the flow over its length to high pressure and low Mach number
- Minimize distortions in the flow field exiting the inlet
- Low weight, small size, and mechanically simple

The discussion about general performance in an idealized flow machine demonstrated the impact of inlet total pressure recovery on the net thrust developed by the engine and illustrated the possibility of an added drag increment due to the nature of the induction of the oncoming air into the inlet. The inlet should provide a means for ensuring that the flow is increased continually in pressure and decelerated so as to present the best flow state for downstream components, such as the compressor. At the same

Theory of Aerospace Propulsion.

time, the diffusion process within the inlet should take place in a fashion that reduces the distortion of the flow leaving the inlet. That is, the distribution of flow and state variables across the inlet exit area should be as uniform as possible. This capability should extend over the operating range of the inlet, including effects of angle of attack variations and gust encounters. The design of the inlet should also provide for the capability of handling a wide range of mass flow rates required by the engine over its performance envelope. In keeping with the idea of a practical inlet with good operability and reliability, the inlet size and weight should be kept as low as possible and the mechanisms involved should be as simple as possible.

## 6.2 INLET MASS FLOW PERFORMANCE

The mass flow equations given in Section 5.3 permit writing the mass flow equation in terms of free stream static conditions where

$$\dot{m}_0 = \frac{p_0 A_0 M_0}{\sqrt{T_0}}\sqrt{\frac{\gamma}{R}}.$$

For standard atmospheric conditions (see Appendix C), we may write the mass flow in the inlet as

$$\dot{m}_0 = A_0 M_0 f(z).$$

Then it is sufficient to deal with $A_0$ as the indicator for mass flow captured by the inlet at a given flight condition of altitude and Mach number. A simple approximation for the mass flow in kg/s and area $A_0$ in $m^2$ is as follows:

$$\dot{m}_0 = \left\{ \begin{matrix} 430e^{-z/8.2} \\ 630e^{-z/6.3} \end{matrix} \right\} A_0 M_0.$$

The upper function of $z$ covers the range $0 < z < 10$ km and the lower 10 km $< z <$ 30 km, both with an error within $\pm 3\%$. This altitude range covers the usual range of altitudes for supersonic air-breathing vehicles.

Consider the simple inlet shown in Figure 6.1. The quasi-one-dimensional results for the relationship between the streamtube area and the Mach number obtained in Chapter 2 may be used to express the area ratio between any two stations in the streamtube in terms of Mach numbers at those stations. In particular, we seek the maximum value of the ratio of capture streamtube area to inlet entrance area, which is given as

$$\left(\frac{A_0}{A_1}\right)_{max} = \frac{M_1}{M_0}\left(\frac{1+\frac{\gamma-1}{2}M_0^2}{1+\frac{\gamma-1}{2}M_1^2}\right)^{\frac{\gamma+1}{2(\gamma-1)}} = \frac{1}{M_0}\left[\frac{2}{\gamma+1}\left(1+\frac{\gamma-1}{2}M_0^2\right)\right]^{\frac{\gamma+1}{2(\gamma-1)}}. \tag{6.1}$$

The Mach number at station 1 of the inlet has been set to $M_1 = 1$ denoting the condition of maximum mass flow. For the case of air, where $\gamma = 1.4$, Equation (6.1) becomes

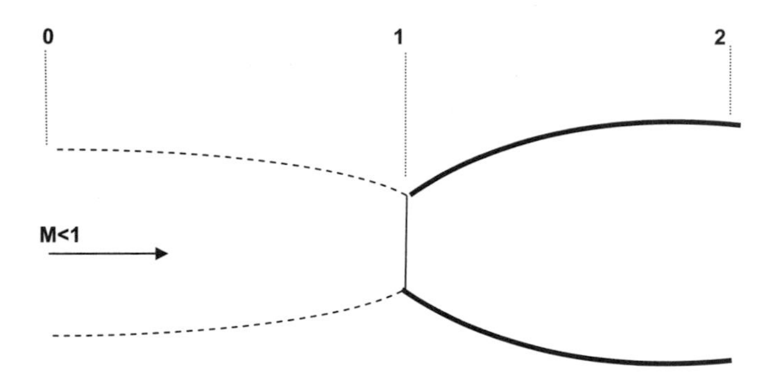


FIGURE 6.1

Schematic diagram of a simple inlet showing the streamtube captured by the inlet as a dashed line.

$$\left(\frac{A_0}{A_1}\right)_{max} = \frac{1}{M_0}\left[\frac{5}{6}\left(1+\frac{M_0^2}{5}\right)\right]^3. \tag{6.2}$$

The maximum mass flow of air that can pass through the inlet, as determined by Equation (6.2), is shown in Figure 6.2.

The captured streamtube thus diminishes in area as the Mach number approaches unity. This behavior of the captured streamtube is illustrated in Figure 6.3. Also shown in the figure is the behavior of the captured streamtube under static or very low-speed operation, such as during taxi or the beginning of take-off. Under such low-speed conditions the inlet acts like a sink, drawing fluid upstream over the nacelle in a manner that can easily cause separation from the inlet lip as the flow completes an almost 180° turn upon entry.

This poses a design constraint because for high-speed flight the inlet should be relatively sharp and this exacerbates the problem of inlet lip separation at very low speeds. To alleviate this problem, aircraft may be equipped with an inflatable cuff around the inlet lip, which increases the radius of curvature or blow-in doors behind the lip that permits the intake of air

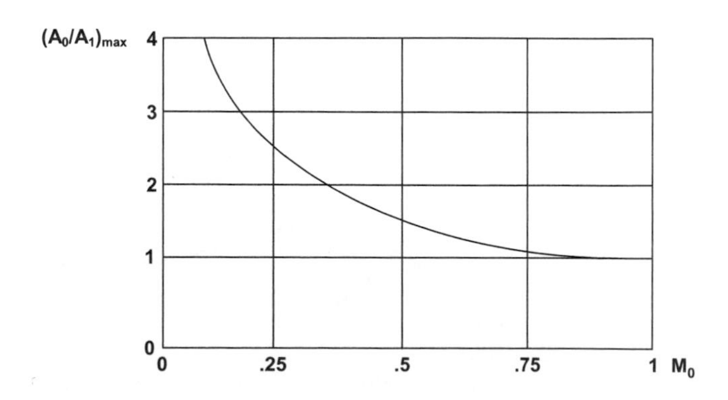


FIGURE 6.2

Maximum mass flow of air passed by the inlet as a function of flight Mach number.

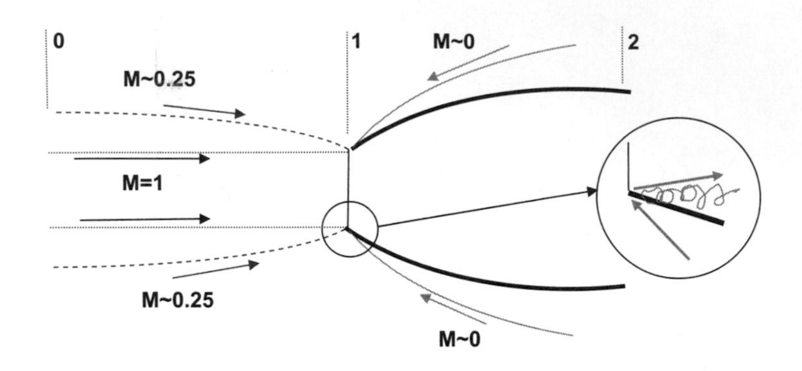


**FIGURE 6.3**

Captured streamtube for maximum mass flow through an inlet at various forward speeds. Detail for low forward speed shows the development of separation at the lip due to rapid turning of the flow.

normal to the nacelle surface, thereby reducing the severity of the turn the flow must make inside the inlet.

In the case of supersonic flow, a shock will appear upstream of the inlet, as shown in Figure 6.4. Because the flow upstream of the shock wave has no knowledge of the presence of the shock wave, the captured streamtube is of constant area there.

The shock wave generated by the inlet is curved, and therefore normal only at the centerline of the flow. Thus the streamline entering the inlet must curve toward the inlet lip and $A_0/A_1 < 1$. The maximum mass flow that can pass through the inlet in supersonic flight occurs when the ratio $A_0/A_1 = 1$, which means that the shock wave must somehow be drawn back toward the lip of the inlet

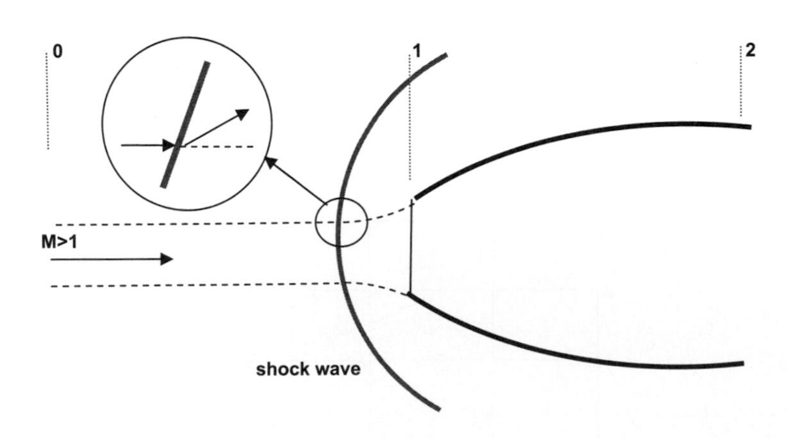


**FIGURE 6.4**

Captured streamtube for an inlet in supersonic flight. Because the shock wave is curved, it is normal only at the centerline of the flow. The detail in the inset illustrates that the streamline downstream of the shock, which is locally oblique, must deflect toward the shock.

**FIGURE 6.5**

Location of the bow shock wave ahead of a normal shock inlet as the back pressure is increased from the value for maximum mass flow (top) to higher values. Here $M_0 = 1.14$ and the area ratio $A_0/A_1$ decreases from approximately 1 to 0.9 to 0.8 from top to bottom (from Olstad, 1956).

until it sits as a normal shock just at the inlet lip. The behavior of the shock wave ahead of a normal shock inlet is illustrated in the flow visualization photographs shown in Figure 6.5.

Note that the shock wave moves further and further upstream of the inlet lip as the back pressure is increased. Although the total pressure recovery is the same for the different back pressures, the mass flow processed by the inlet varies. The operational factors for these simple normal shock inlets under supersonic flight conditions are discussed in a subsequent section.

## 6.3 INLET PRESSURE PERFORMANCE

Inlet performance is based primarily on three criteria involving the stagnation pressure at its exit: mean stagnation pressure, distortion in stagnation pressure, and turbulence of the stagnation pressure. Because the stagnation pressure at station 2 is, in general, a function of space and time, $p_t = p_t(r,\theta,t)$ and we may separate it into a time-averaged stagnation pressure component and a fluctuating stagnation pressure component as follows:

$$p_t(r,\theta,t) = \bar{p}_t(r,\theta) + p_t'(r,\theta,t).$$

The time-averaged stagnation pressure is defined as follows:

$$\bar{p}_t(r,\theta) = \frac{1}{\tau}\int_0^{\tau} p_t(r,\theta,t)dt.$$

Thus, at each location $r,\theta$ at station 2 the stagnation pressure may be measured as a function of time and the result integrated over a sufficiently long integration time $\tau$ . The integration time should be long enough so that the time-averaged stagnation pressure doesn't change if $\tau$ is increased further. This assures us that the measurement is made under flow conditions that are steady in the mean. Then the time-averaged stagnation pressure may be integrated over the area of station 2 to yield the mean stagnation pressure

$$p_{t,2} = \iint_{A_2} \bar{p}_t(r,\theta)rdrd\theta.$$

When stagnation pressures are discussed, we are always referring to mean stagnation pressures unless otherwise noted. The mean stagnation pressure may have a value that is quite high and therefore seemingly very good. However, variations in local values of the time-averaged stagnation pressure may be detrimental to compressor performance. These variations are referred to as inlet distortions in which some regions at station 2 have high values of time-averaged stagnation pressure, whereas others have low values, even though the space-averaged quantity seems acceptable, as illustrated in Figure 6.6. These spatial gradients, or distortions, in time-averaged stagnation pressure should be minimized to ensure acceptable inlet performance over the range of flight conditions expected. The turbulence level of the stagnation pressure may be described by the root-mean-square deviation of the stagnation pressure from its mean value, as illustrated in Figure 6.6 and given by

$$\sqrt{\overline{p_t'^2}} = \left\{\frac{1}{\tau}\int_0^{t} [p_t(r,\theta,t) - \bar{p}_t(r,\theta)]dt\right\}^{\frac{1}{2}}.$$

The higher the turbulence level, the less effective the inlet will be. Inlets are expected to perform well in terms of mean stagnation pressure recovery, low distortion, and low turbulence level over the flight envelope. Factors that influence these capabilities are variations in angles of attack caused by maneuvers or crosswinds, changes in acceleration or altitude, and free stream turbulence in the ambient atmosphere. This suggests that these requirements are particularly important for fighter

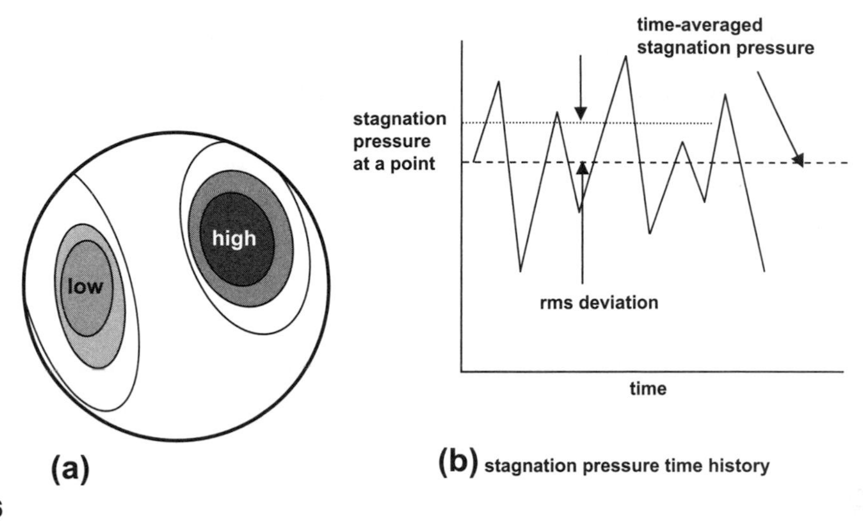


FIGURE 6.6

(a) Inlet distortion shown as contours of time-averaged stagnation pressure $p_{t,2}$ and (b) time variation of stagnation pressure $p_{t,2}(r,\theta)$ at a given point showing the time averaged value of $p_{t,2}$ and rms deviation from the average.

aircraft and less so for commercial airliners. This chapter confines attention to just the first performance determinant: pressure recovery.

## 6.4 SUBSONIC INLETS

The fundamental consideration in subsonic intakes is the inlet lip design. Two conflicting requirements are suggested by the general flow structure around the inlet shown in Figure 6.3 and in greater detail in Figure 6.7.

- At cruise speeds, $M_0 \sim 0.8$, the capture streamtube is about the size of the inlet entrance so the mass ratio $A_0/A_1 \sim 1$. A large inlet radius would reduce the critical Mach number for the inlet outer surface, that is, the value of $M_0$ where the surface speed over some point of the inlet becomes sonic. This overspeed, like that on the airfoil of a wing, occurs in a region above the surface where the Mach number is supersonic. The flow must decelerate through a normal shock wave, leading to subsequent boundary layer flow separation with the attendant increase in drag. The larger the nose radius, the more extensive the supersonic region produced and the greater the drag penalty. In this speed range, drag considerations favor a small inlet lip radius.
- At low speeds, where high thrust is required, such as take-off and climb, $0 < M_0 < 0.3$, the mass flow demand is large and $A_0/A_1 > 1$. Flow entering the inlet must negotiate a large turn to enter the inlet, and therefore separation at the lip can diminish mass flow entering the inlet. Cross-flow winds during take-off and the angle of attack increase caused by aircraft rotation at lift-off can also exacerbate the separation problem. In this speed range, mass flow considerations favor a large inlet lip radius.

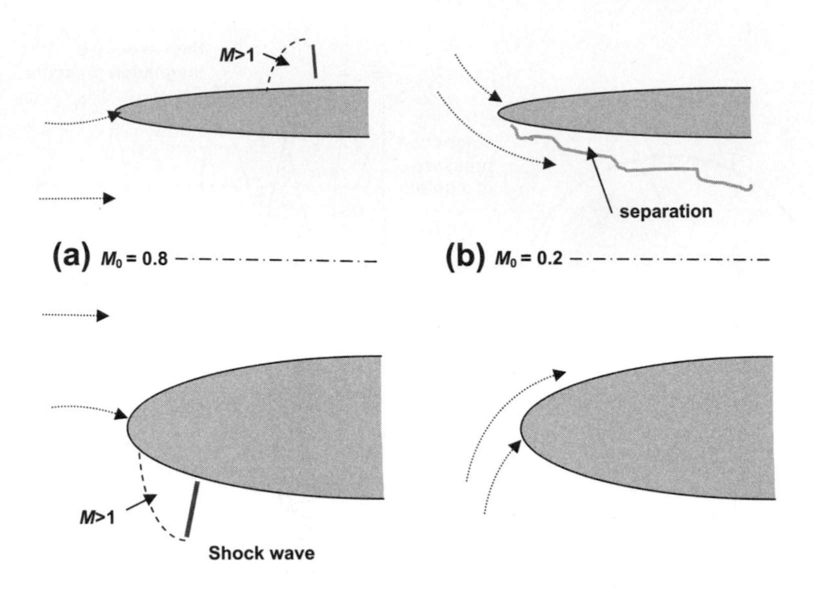


FIGURE 6.7

Flow structure around small and large nose radius inlet lips for (a) cruise, $M_0 = 0.8$, and (b) take-off, $M_0 = 0.2$.

There are a number of approaches for accommodating this conflicting requirement on nose radius, including variable angle lip section, movable leading edge flap, blow-in doors, and other devices that all lead to some penalty in weight and complexity. A high-performance installation with fixed geometry is achievable for high subsonic cruise speeds typical of commercial airliners. Acoustic treatment of nacelles for noise abatement makes variable geometry systems difficult to implement. Supersonic aircraft require active air flow control throughout the flight regime and are equipped with variable geometry features that accommodate the low-speed requirements discussed earlier, which are discussed in Chapter 9.

The stagnation pressure recovery in a subsonic inlet is generally very high because losses accrue through friction on the interior walls of the inlet and turns in the inlet path. Most inlets are straight and short, with lengths being restricted usually to several inlet duct diameters. Longer and more serpentine ducts, such as those on stealth aircraft with deeply buried engines, may require more careful attention. Frictional losses are treated readily by the quasi-one-dimensional approach of Chapter 2. For example, the equation for stagnation pressure change with purely frictional losses in a variable area duct may be found to be given by

$$\frac{dp_t}{p_t} = \left(-\frac{\gamma}{2}M^2\right)4C_f\frac{dz}{D}.$$

To obtain an estimate of the pressure recovery magnitude expected of well-designed subsonic inlet flows, we may assume that $M$ and $C_f$ are taken as average properties in the duct from the entrance ($z = 0$) to the exit ($z = L$) of the inlet and then integrating over the length of the duct to obtain

$$\frac{p_{t,2}}{p_{t,1}} \simeq \exp\left(-2\gamma M_{avg}^2 C_{f,avg}\frac{L}{D}\right).$$

The pressure recovery of an inlet is defined as the ratio of the stagnation pressure at the exit of the inlet to that in the free stream, $p_{t,2}/p_{t,0}$. Because stagnation pressure in the subsonic flow doesn't vary between the free stream and the inlet lip, $p_{t,1} = p_{t,0}$. The friction coefficient $C_f$ is typically in the vicinity of $4 \times 10^{-3}$ for practical flight conditions so that in supersonic flight where inside the inlet we can assume $M_{avg} = 0.6$ and $L/D = 4$ the exponential term is on the order of $10^{-2}$ so that we may expand it to yield the pressure recovery as

$$\frac{p_{t,2}}{p_{t,0}} \simeq 1 - 2\gamma M^2 C_f \frac{L}{D} \simeq 0.98.$$

This result would be typical of most straight inlet diffusers with modest expansion angles, that is, with local slopes of less than 10° so as to avoid separation of the boundary layer flow on the walls of the inlet. Bends in the duct would, of course, lead to greater reductions in pressure recovery and should be avoided. Typical commercial airliner inlets generally have $L/D \sim 1$ so that their pressure recovery is high, whereas aircraft with buried engines, such as some fighter and stealthy aircraft (B-2, F-22) and even older airliners (McDonnell Douglas MD-10 and Boeing 727), can pay a greater penalty in pressure recovery without the addition of flow control devices to maintain attached flow in the ducts feeding the engines.

## 6.5 NORMAL SHOCK INLETS IN SUPERSONIC FLIGHT

In order to have the maximum possible mass flow enter the inlet under supersonic flight conditions, we have just seen that it is necessary that the shock be situated at the inlet lip. The shock wave boosts the pressure by adiabatic compression that drops the Mach number to subsonic levels; as the flow proceeds through the expanding area of the duct, the pressure increases further through isentropic expansion. This is shown schematically in Figure 6.8.

Although it increases the static pressure, the effect of the shock wave also decreases the stagnation pressure, a direct indication of the entropy increase caused by the shock wave. The stagnation pressure loss in the inlet is crucial to overall engine performance, which was shown in Chapter 1. The stagnation pressure recovery for a normal shock inlet operating in the fashion we have considered is shown in Figure 6.9.

Let us consider the performance of such an inlet by first recalling from Section 5.3 that the mass flow in a duct, such as an inlet, is directly proportional to the stagnation pressure at station 2, $p_{t,2}$. In addition, in Section 1.6, the exhaust velocity, and therefore the net thrust per unit mass flow, is also proportional to $p_{t,2}$. The normalized change in net thrust due to inlet losses may be characterized (Hesse and Mumford, 1964) as follows:

$$\frac{\Delta F_n}{F_n} = -\left(1 - \frac{p_{t,2}}{p_{t,0}}\right) C_R.$$

The correction factor $C_R$ is in the range of 1.1 and 1.6 according to Antonatas and colleagues (1972). The general behavior shown in Figure 6.10 is for an average value of $C_R = 1.35$ in order to provide a reasonable estimate of the performance loss attributable to using a normal shock inlet in supersonic flight. It is apparent from Figure 6.10 that the normal shock inlet is satisfactory at Mach numbers up to about $M = 1.5$, especially since it is a very simple and robust device.

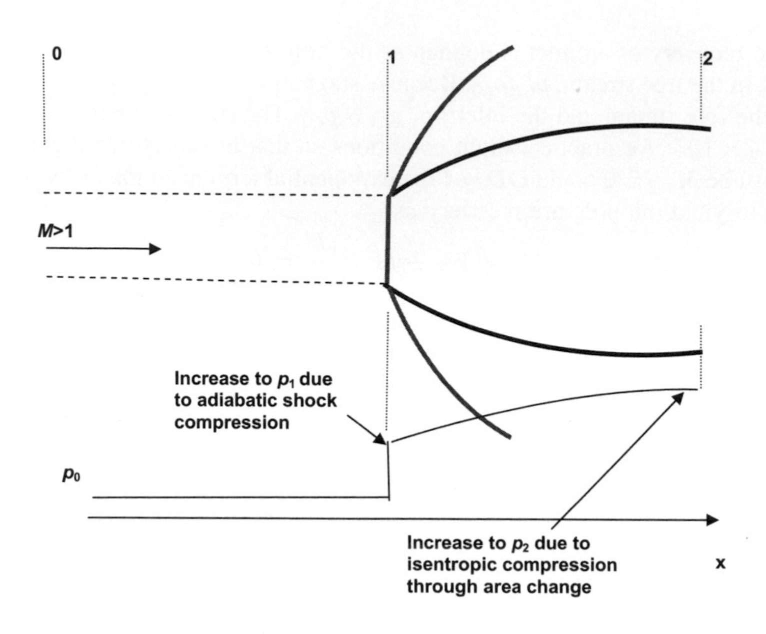


**FIGURE 6.8**

Normal shock inlet in supersonic flight showing shock placement for maximum inlet mass flow. The adiabatic shock wave compresses the flow to $p_1$ and then the flow is compressed further to $p_2$ through the effect of an area increase in a subsonic flow.

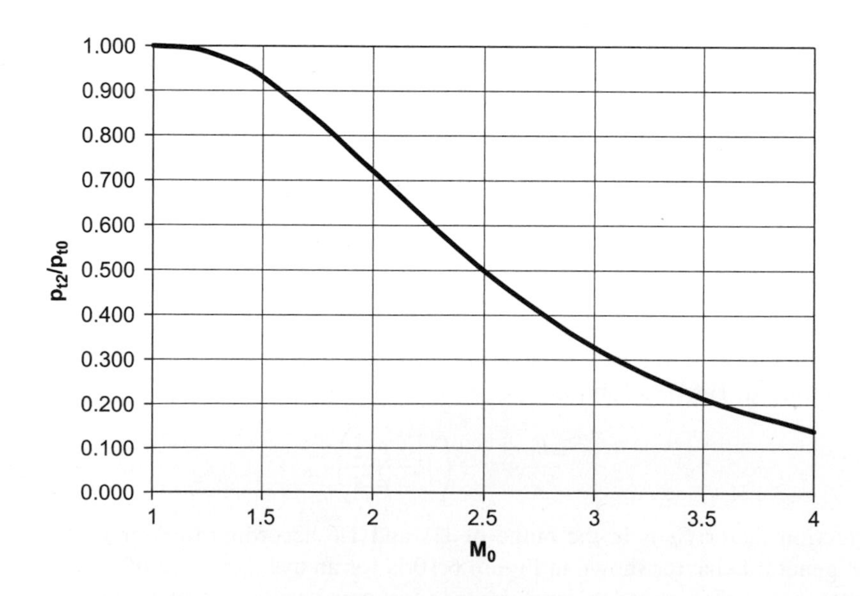


**FIGURE 6.9**

Stagnation pressure recovery for normal shock inlets (normal shock at the inlet lip) as a function of flight Mach number.

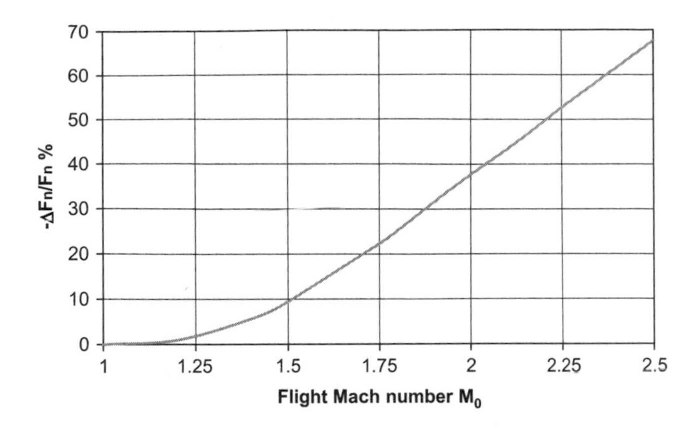


**FIGURE 6.10**

Nominal reduction in net thrust as a function of Mach number for normal shock inlets.

However, beyond that speed the losses are too great to justify its use and a more efficient inlet must be designed. Increased efficiencies may be obtained by reducing the Mach number ahead of the (necessary) normal shock. Such reductions may be achieved by inlets that reduce the Mach number, that is, compress the flow, within the inlet duct (internal compression inlets) or outside the inlet duct (external compression inlets). These higher performance inlets are discussed subsequently.

## 6.6 INTERNAL COMPRESSION INLETS

The simplest way to improve the total pressure recovery of an inlet in supersonic flight is to reduce the Mach number before the normal shock to values below that in the free stream. If the inlet was designed as a converging–diverging duct, such as that in a nozzle, and the shock wave at the lip of the inlet could be drawn into the inlet, then, for an inviscid flow, the decreasing area of the internal duct would serve to isentropically decelerate the Mach number from $M_0$ to some lower value. If the flow were then permitted to further decelerate to a subsonic Mach number, through a normal shock within the duct, total pressure recovery would be improved.

Consider such an inlet equipped with an ideal valve at station 2, as shown conceptually in Figure 6.11. Beyond the valve there is a reservoir at very low pressure. If the valve is closed, the inlet functions as a pitot tube with no net gas motion inside the inlet. Under such conditions the shock wave stands out in front of the inlet and the pressure inside the inlet is equal to the total pressure behind the normal shock so that the pressure recovery is that for a normal shock at the flight Mach number. Now if the valve is opened so as to allow flow through the inlet, then the shock wave pattern won't change much but will move closer to the inlet lip, as shown in Figure 6.12. The captured streamtube area $A_0$ is still smaller than the inlet lip area $A_1$, but now there is flow through the inlet. Further opening of the valve, permitting more flow to pass through the inlet, continues to draw the shock wave pattern closer and closer to the inlet lip until a normal shock sits there.

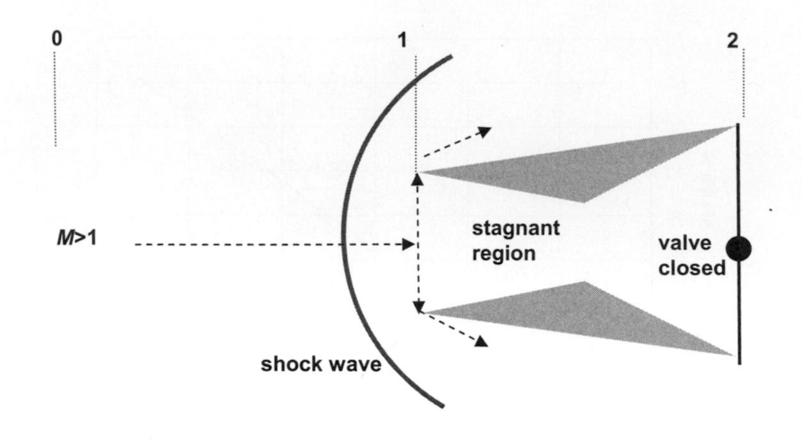


FIGURE 6.11

Schematic diagram of a converging diverging inlet in a supersonic stream with flow through the inlet blocked by an ideal valve at station 2. Flow streamlines are indicated by dashed lines.

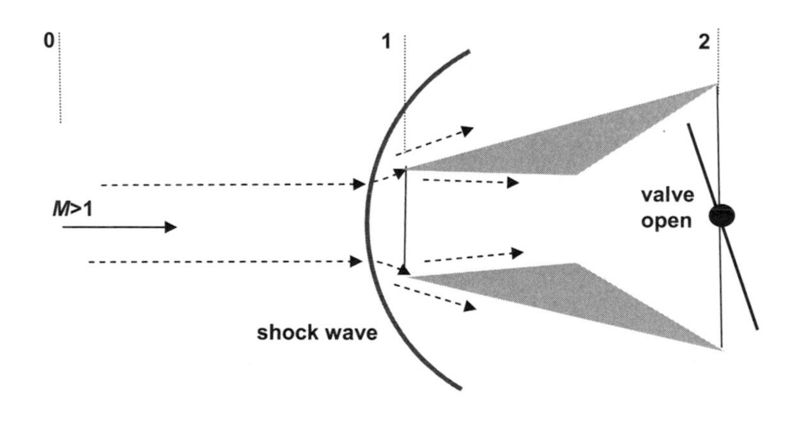


FIGURE 6.12

Schematic diagram of a converging diverging inlet in a supersonic stream with flow through the inlet due to opening of the ideal valve at station 2. Flow streamlines are indicated by dashed lines.

This state of affairs is depicted schematically in Figure 6.13, and at the given supersonic flight Mach number, the mass flow passing through the inlet is a maximum and the ratio $A_0/A_1 = 1$. It must be remembered that because the flow behind the shock wave in the duct is subsonic, it is affected instantly by changes in pressure downstream of it. Therefore, if the valve is opened more, the reduction in pressure draws the shock further into the converging section, as shown in Figure 6.14.

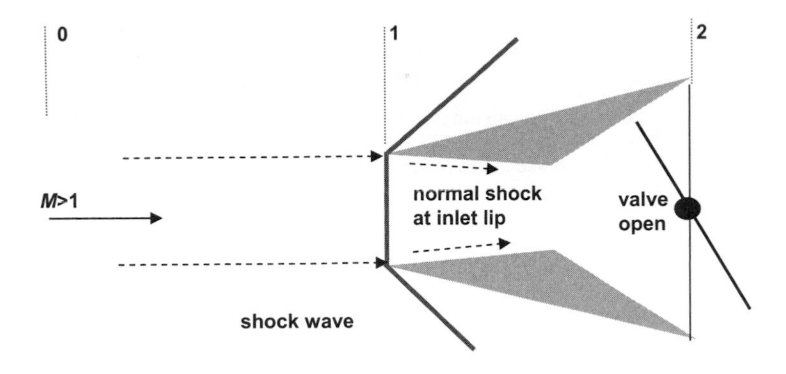


**FIGURE 6.13**

Schematic diagram of a converging diverging inlet in a supersonic stream with flow through the inlet due to opening of the ideal valve at station 2. Here the valve is opened just enough to position the shock at the lip as a normal shock. Flow streamlines are indicated by dashed lines.

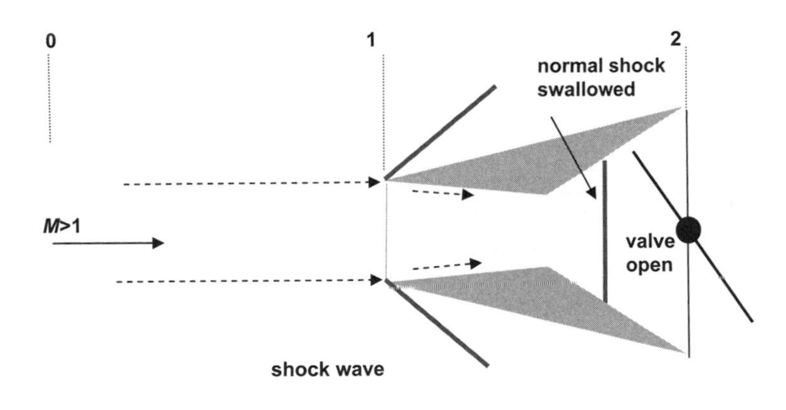


**FIGURE 6.14**

Schematic diagram of a converging diverging inlet in a supersonic stream with flow through the inlet due to opening of the ideal valve at station 2. Here the valve is opened enough to draw the shock into the duct. Flow streamlines are indicated by dashed lines.

The decreasing area ahead of the shock provides isentropic compression, reducing the Mach number and thereby improving the total pressure recovery of the inlet. This all supposes of course that the area decrease is gentle enough to preclude the formation of oblique shock waves, as well as to avoid separation of the boundary layers on the duct walls, which are advancing through an adverse pressure gradient. The variation of $A_0/A_1$ with the ratio of total pressure recovery at $M_0$ to that achievable with the shock swallowed, $(p_{t,2}/p_{t,0})_{M0}/(p_{t,2}/p_{t,0})$, is shown in Figure 6.15. Of course the maximum pressure recovery is achieved when the shock is positioned at the minimum section of the inlet.

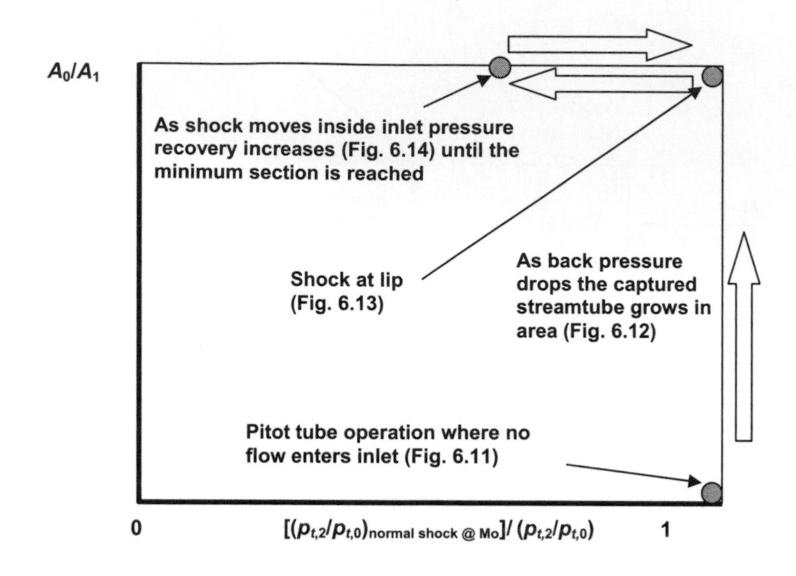


**FIGURE 6.15**

Variation of total pressure recovery with captured streamtube area for the internal compression inlet operation shown in Figures 6.11 to 6.14.

## 6.7 INTERNAL COMPRESSION INLET OPERATION

Consider an internal compression inlet with a given contraction ratio, $A_{min}/A_1$. To start the inlet, that is, to have the normal shock placed at the entrance of the inlet and minimum station of the inlet be critical, $M_{\min} = 1$, we must have a specific area ratio for the inlet that will permit this situation to exist. The flight Mach number for which this situation exists is called the starting Mach number, $M_s$. The required area ratio at this Mach number, noting that under the condition of maximum captured streamtube area $A_1 = A_0$, is given as follows:

$$\frac{A_1}{A_{\min}} = \left(\frac{A_1}{A_0^*}\right)_{M=M_s}\left(\frac{A_0^*}{A_{\min}}\right) = \left(\frac{A_0}{A_0^*}\right)_{M=M_s}\left(\frac{A_0^*}{A_{\min}}\right). \tag{6.3}$$

The steps in operating the inlet in supersonic flight may be visualized with the help of Figure 6.16. Below the starting Mach number (Figure 6.16a), the captured streamtube area is less than that of the inlet, $A_1$, as the shock is outside the inlet. At the starting Mach number, the shock is at the lip of the inlet and the minimum section is sonic (Figure 6.16b). Reducing the back pressure, $p_2$, draws the shock in and permits it to be swallowed (Figure 6.16c). The back pressure is then increased slowly until the shock is brought into position at the minimum section (Figure 6.16d) and now the pressure recovery will be a maximum.

Equation (6.3) requires the ratio of the critical area in the free stream, $A_0$*, to the critical area within the inlet, $A_{\min}$. The relationship between the critical area in free stream $A_0$* and the critical area in inlet

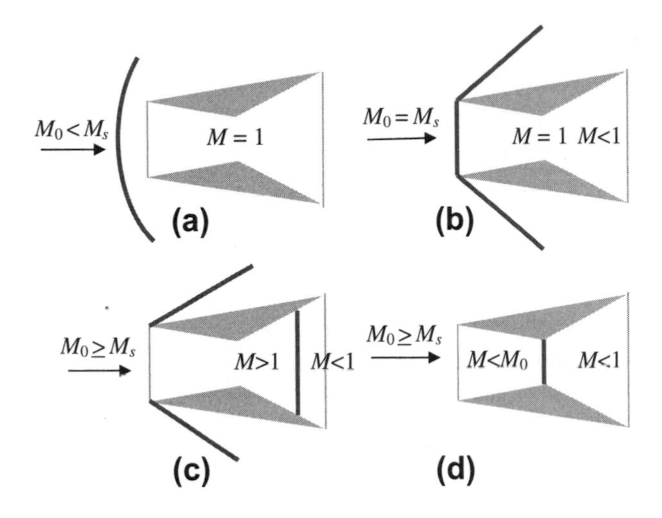


**FIGURE 6.16**

Stages of operation of internal compression inlet under maximum mass flow conditions: (a) shock expelled, (b) inlet started at $M_0 = M_s$, (c) shock swallowed, and (d) shock adjusted.

$A_{min}$ may be obtained from the condition for mass conservation across a normal shock wave, as developed in Chapter 5:

$$\dot{m} = \frac{p_{t0}A^*}{\sqrt{RT_{t0}}}f(\gamma) = \frac{p_{ty}A_{min}}{\sqrt{RT_{ty}}}f(\gamma),$$

where subscript $y$ denotes conditions behind normal shock. Recalling that the stagnation temperature is constant across the (adiabatic) shock, as are the gas constant $R$ and the ratio of specific heats $\gamma$, reduces the mass conservation to the simple expression

$$A_o^* p_{t,0} = A_{min} p_{t,y}. \tag{6.4}$$

Note that because the stagnation pressure across a shock wave always decreases, the critical area behind the shock is always larger than that upstream of the shock. With this information, the expression for the inlet geometry at the starting Mach number becomes

$$\frac{A_1}{A_{min}} = \frac{p_{t,y}}{p_{t,0}}\left(\frac{A}{A^*}\right)_{M=M_s}. \tag{6.5}$$

Consideration of the streamtube capture area for all operating regimes can be determined as follows. For flight at subsonic Mach number, $M_0 < 1$, the maximum value of the ratio of the streamtube capture area to the inlet entrance area is

$$\left(\frac{A_0}{A_1}\right)_{max} = \frac{A_0}{A^*}\frac{A_0^*}{A_1}. \tag{6.6}$$

But for maximum mass flow at this subsonic flight condition, the inlet minimum, or throat, area $A_{min} = A_0{*}$ so that

$$\left(\frac{A_0}{A_1}\right)_{max} = \frac{A_0}{A_0^*}\frac{A_{min}}{A_1}. \tag{6.7}$$

For $M_0 = 0.5$ and $M_s = 2$, for example,

$$\frac{A_0}{A_1} = \frac{A_0}{A^*}\frac{A^*}{A_1} = \frac{1.34}{1.22} = 1.10.$$

For flight at supersonic speed but at a Mach number less than the starting Mach number, that is, $M_s > M_0 > 1$, we have the following expression for maximum captured streamtube area

$$\left(\frac{A_0}{A_1}\right)_{max} = \frac{A_0}{A_0^*}\frac{A_0^*}{A_1} = \frac{A_0}{A_0^*}\frac{p_{t,y}}{p_{t,0}}\frac{A_{min}}{A_1}. \tag{6.8}$$

Here the Mach number behind the normal shock is higher than that necessary to achieve $M = 1$ at the minimum section if one considered the shock to be on the lip. Therefore, the shock moves out in front of the inlet, that is, the shock-expelled case of Figure 6.16a. For example, in the case where $M_s = 2$ and $M_s > M_0 = 1.5$, Equation (6.8) yields the result $(A_0/A_1)_{max} = 0.895$, which occurs, of course, with $M_{min} = 1$. For flight at the starting Mach number or greater, $M_0 \geq M_s$, the maximum capture streamtube area is always equal to the inlet entrance area, that is, $(A_0/A_1)_{max} = 1$.

We may construct Table 6.1 to collect some of the pertinent information discussed in the preceding material and determine the contraction ratios required at different starting Mach numbers. The first three columns refer to flight Mach number and associated area ratio with normal shock pressure recovery denoted by the subscript NS. The next column is the required contraction for $M_s = M_0$, and the next column is the area ratio at the minimum section based on Equation (6.5). The next column is the Mach number ahead of the normal shock when it is positioned properly at the minimum section, and the final column is the pressure recovery at that condition. In Table 6.1, the contraction required $A_1/A_{min}$ is tending toward a limiting value of 1.666 as $M_0$ gets larger. For any larger contraction, the inlet cannot be started no matter how large $M_0$ becomes.

Let us consider the case of an internal compression inlet designed for a starting Mach number of $M_s = 2$. The captured streamtube area as a function of flight Mach number may be calculated using Table 6.1 and Equation (6.8); the result is shown in Figure 6.17. The maximum pressure recovery possible for internal compression inlets with $M_s = M_0$ is shown in Figure 6.17. Note that this type of inlet extends useful operation to about $M_0 = 2$.

**Table 6.1** Contraction Ratios for Internal Compression Inlets

| $M_0$ | $A_0/A^*$ | $(p_{t2}/p_{t0})_{NS}$ | $A_1/A_{min}$ | $A_{min}/A^*$ | $M_{min}$ | $(p_{t2}/p_{t0})_{IC}$ |
|---|---|---|---|---|---|---|
| 1.00 | 1 | 1.000 | 1.00 | 1.00 | 1.00 | 1.000 |
| 1.20 | 1.03 | 0.993 | 1.02 | 1.01 | 1.10 | 0.999 |
| 1.40 | 1.11 | 0.958 | 1.07 | 1.04 | 1.24 | 0.988 |
| 1.50 | 1.18 | 0.930 | 1.09 | 1.08 | 1.31 | 0.978 |
| 1.75 | 1.39 | 0.835 | 1.16 | 1.20 | 1.53 | 0.920 |
| 1.83 | 1.47 | 0.799 | 1.18 | 1.25 | 1.60 | 0.895 |
| 2.00 | 1.69 | 0.721 | 1.22 | 1.39 | 1.75 | 0.835 |
| 2.50 | 2.64 | 0.499 | 1.32 | 2.00 | 2.20 | 0.628 |
| 3.00 | 4.23 | 0.328 | 1.39 | 3.05 | 2.65 | 0.442 |
| 3.50 | 6.79 | 0.213 | 1.45 | 4.70 | 3.11 | 0.299 |
| 4.00 | 10.72 | 0.139 | 1.49 | 7.21 | 3.56 | 0.202 |

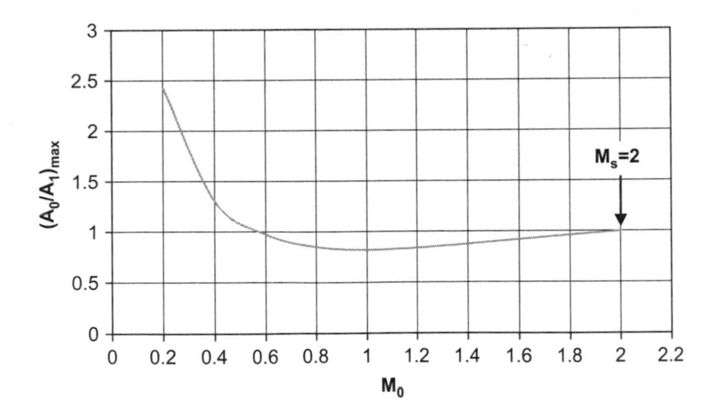


**FIGURE 6.17**

Variation of captured streamtube area as a function of flight Mach number for an internal compression inlet with a starting Mach $M_S = 2$.

A military standard (MIL = E-5008B) inlet pressure recovery is prescribed for the Mach number range $1 < M_0 < 3$ in "MIL-E-5008B: Military Specifications—Engines, Aircraft, Turbojet, Model, Specification For" (January, 1959) and is also shown in Figure 6.18. The standard pressure recovery is given by

$$\frac{p_{t,2}}{p_{t,0}} = 1 - 0.0075(M_0^2 - 1)^{1.35}. \tag{6.9}$$

Note that this pressure recovery is considerably better than the maximum internal compression inlet at Mach numbers greater than 2. It will be shown that a more complicated shock pattern must be developed by an inlet in order to achieve the higher pressure recoveries indicated by the standard.

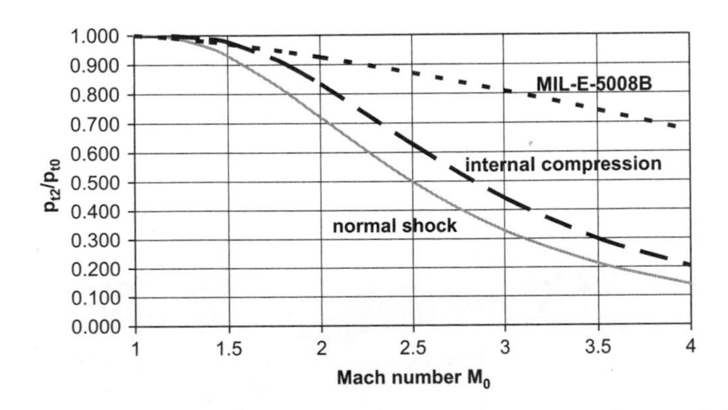


**FIGURE 6.18**

Variation of total pressure recovery as a function of flight Mach number for internal compression inlets with a starting Mach number equal to the flight Mach number.

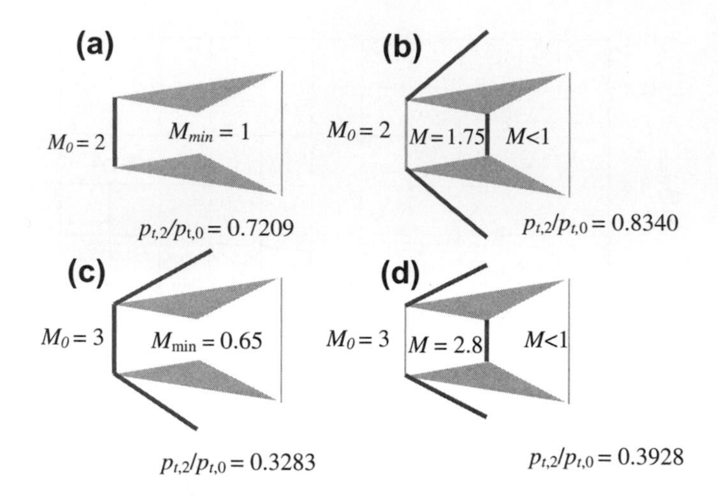


**FIGURE 6.19**

Conditions in an internal compression inlet designed for a starting Mach number $M_S = 2$: (a) inlet started, (b) shock positioned at the minimum section, (c) shock at lip for $M_0 = 3$, and (d) shock positioned at the minimum section at $M_0 = 3$.

Consider an internal compression inlet designed for a given starting Mach number $M_s$ and flying at $M_0 = M_s$. It is possible to cause the inlet to swallow the normal shock and then position it at or near the minimum section by regulating the exhaust nozzle minimum area and/or burner temperature. The maximum possible inlet pressure recovery occurs, as described previously, when the shock is precisely at the minimum area section of the inlet. Take the case of an inlet designed for $M_s = 2$, for which the required contraction ratio $A_1/A_{min} = 1.217$, as shown in Table 6.1. Conditions in the inlet at different flight Mach numbers at or above the starting Mach number are summarized in Figure 6.19, while for $M < M_s$ they are summarized in Figure 6.20.

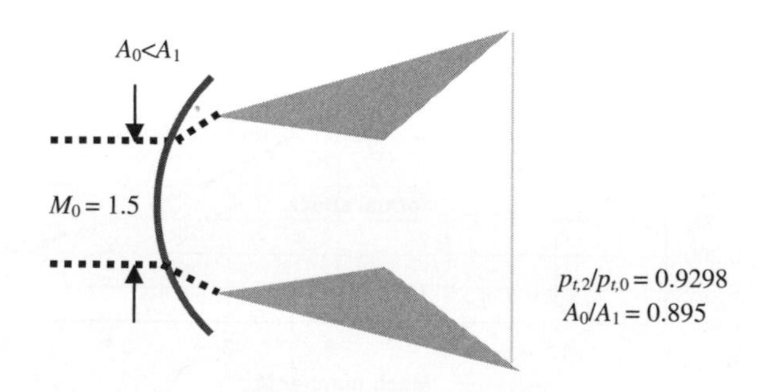


**FIGURE 6.20**

Conditions in an internal compression inlet designed for a starting Mach number $M_S = 2$ at a flight Mach number $M_0 = 1.5$.

For subsonic Mach numbers, the area ratio is greater than unity, for example, at $M_0 = 0.5$ for this case of an inlet with $M_s = 2$ we have

$$\frac{A_0}{A_1} = \frac{A_0}{A^*}\frac{A^*}{A_1} = \frac{1.340}{1.217} = 1.10.$$

## 6.8 EXAMPLE: INTERNAL COMPRESSION INLET

Consider the axisymmetric internal compression inlet shown in Figure 6.21 for which the inlet diameter $d_1 = 1$ m and the area ratios $A_2/A_1 = 2$ and $A_{min}/A_1 = 0.85$. (a) Determine the starting Mach number $M_s$ for the inlet. (b) Assuming that the pressure and temperature in the free stream at the altitude of flight (around 3100 m) are $p_0 = 68.9$ kPa and $T_0 = -5°C$, respectively, find the corresponding pressure $p_2$ and the capture streamtube area ratio $A_0/A_1$ for the maximum stagnation pressure recovery when the flight Mach number is equal to the starting Mach number, $M_0 = M_s$. (c) Using the same conditions as those in the previous item, calculate the mass flow and stagnation pressure recovery for the following pressures at station 2, $p_2 = 330.8$, 310.1, and 224 kPa; and (d) the variation of $(A_0/A_1)_{max}$ as a function of $M_0$ for the range $0.2 < M_0 < M_s$.

**(a)** The starting Mach number is that value of $M_0$ for which the normal shock wave is located at station 1 and the Mach number at the minimum section is $M_{min} = 1$. Under those conditions, the area ratio $A_1/A_{min} = (0.85)^{-1} = 1.176 = A/A^*$. The tables in Appendix A show that for this area ratio $M = 0.61$. Therefore, the Mach number behind the normal shock standing at station 1 is 0.61, as shown in Figure 6.22.

The free stream Mach number that would produce that Mach number behind it may also be found in the same tables by finding the 0.61 in the column headed $M_2$ (Hesse and Mumford (1964) use subscript 1 for conditions just ahead of a shock and subscript 2 for conditions just behind a shock) in the supersonic section of the tables. The corresponding value for $M_1$ is found under the column headed $M_1$ as 1.83. Therefore, $M_0 = M_s = 1.83$. The total pressure recovery may be found in the same table as $p_{t2}/p_{t0} = 0.799$.

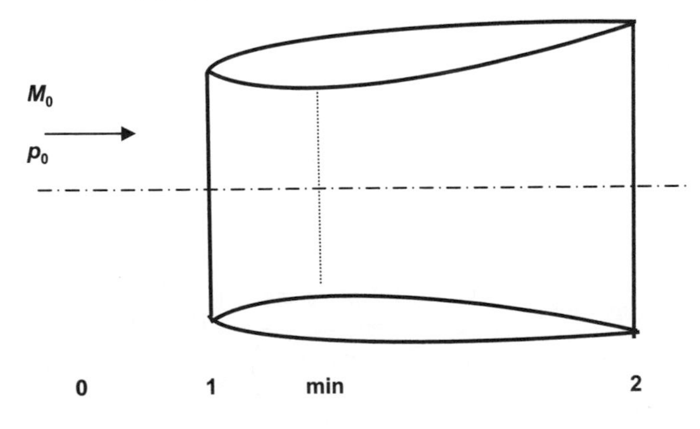


**FIGURE 6.21**

Schematic diagram of an axisymmetric internal compression inlet.

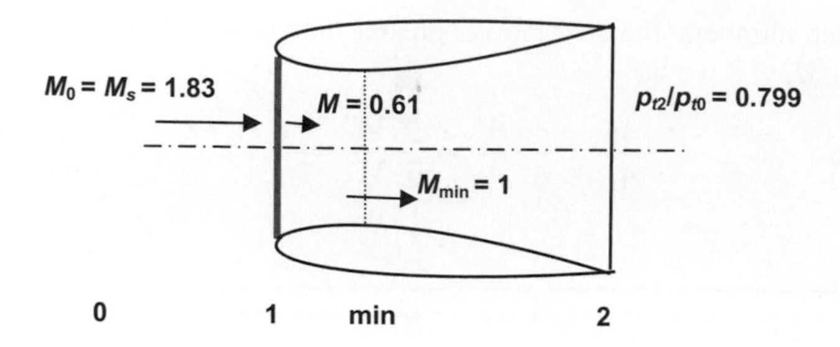


**FIGURE 6.22**

Flow conditions in the started inlet.

**(b)** To find the maximum pressure recovery at $M = M_s$, we must determine conditions when the shock is placed exactly at the minimum section. The area ratio $A_0/A_1 = 1$ for the conditions considered is the maximum value achievable. Because $M_0 = M_s = 1.83$, the corresponding value of $A_1/A^*$ may be found from the tables as $A/A^* = 1.472$. The ratio of area at the minimum station to the critical area is

$$A_{min}/A^* = (A_{min}/A_1)(A_1/A^*) = (0.85)(1.472) = 1.251.$$

From the tables, this area ratio corresponds to a Mach number $M_{min} = 1.60$, and a Mach number behind the normal shock at the minimum station is found to be $M = 0.668$ and the total pressure recovery across the shock is $p_{t2}/p_{t,0} = 0.895$ (recall that the stagnation pressure is constant at $p_{t1} = p_{t0}$ upstream of the shock and constant at $p_{t2}$ downstream of the shock). At $M = 0.668$ behind the shock, the area ratio $A_{min}/A^* = 1.1179$ as may be found in the subsonic section of the tables in Appendix A. Because the flow is isentropic downstream of the shock, the critical area $A^*$ is constant and we may find the area ratio at station 2 as follows:

$$A_2/A^* = (A_2/A_{min})(A_{min}/A^*) = (A_2/A_1)(A_1/A_{min})(A_{min}/A^*) = (2)(1.176)(1.1179) = 2.63.$$

Conditions appropriate to the inlet having the shock positioned optimally are shown in Figure 6.23.

The corresponding Mach number and static to stagnation pressure ratio at station 2 are $M_2 = 0.227$ and $p_2/p_{t2} = 0.965$. The pressure at station 2 may be found from

$$p_2 = (p_2/p_{t,2})(p_{t,2}/p_{t,0})(p_{t,0}/p_0)p_0 = (0.965)(0.895)(6.017)(68.9\text{ kPa}) = 358\text{ kPa}.$$

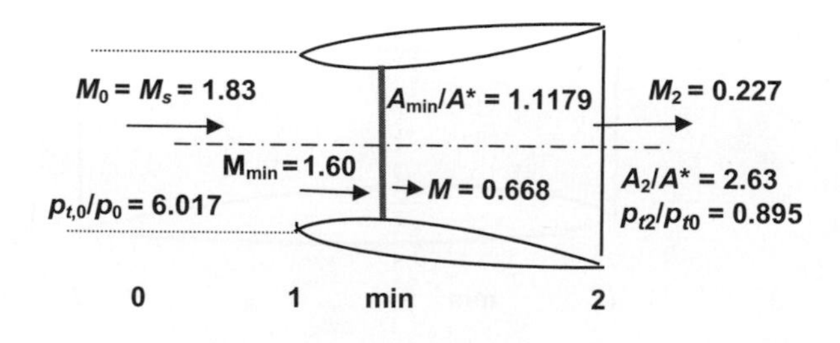


**FIGURE 6.23**

Inlet with the shock positioned properly.

**(c)** Decreasing the back pressure will cause the shock to move from the throat toward station 2. We may examine the variation in total pressure recovery as a function of the back pressure, which is equivalent to location of the shock between the throat and station 2, by picking shock locations and finding the corresponding back pressure $p_2$. Choose, for example, shock locations at $A_y/A_{min} = 1.25$, 1.5, and 1.75 (we know the result at $A_y/A_{min} = 1$), where the subscript $y$ denotes a general station between the throat and station 2. This area ratio may be written as $A_y/A_{min} = (A_y/A^*)(A^*/A_{min})$ and we have found $A_{min}/A^* = 1.251$ for the started inlet. Then the local area ratio is simply $A_y/A^* = 1.251(A_y/A_{min})$. Using this value in the compressible flow tables and following the approach used in part (b) leads to the results for pressure recovery as a function of back pressure in Table 6.2.

A plot of back pressure $p_2$ as a function of inlet pressure recovery is shown in Figure 6.22. If the back pressure exceeds 358 kPa, the shock will be expelled and the pressure recovery will drop to the normal shock value. If the back pressure is less than 315 kPa, the pressure recovery will be too low. The useful range of back pressure is shown as the shaded area in Figure 6.24.

Thus the inlet with the shock positioned at the throat yields the greatest pressure recovery and the highest pressure delivered to the compressor face at station 2. The mass flow depends

**Table 6.2** Stagnation Pressure Recovery as a Function of Back Pressure

| $A_y/A_{min}$ | $A_y/A^*$ | $M_y$ | $p_{t,2}/p_{t,0}$ | $p_2/p_{t,2}$ | $p_2$ (kPa) |
|---|---|---|---|---|---|
| 1.0 | 1.25 | 1.60 | 0.895 | 0.964 | 358 |
| 1.25 | 1.56 | 1.90 | 0.767 | 0.951 | 302 |
| 1.50 | 1.875 | 2.12 | 0.665 | 0.933 | 257 |
| 1.75 | 2.185 | 2.30 | 0.583 | 0.910 | 220 |

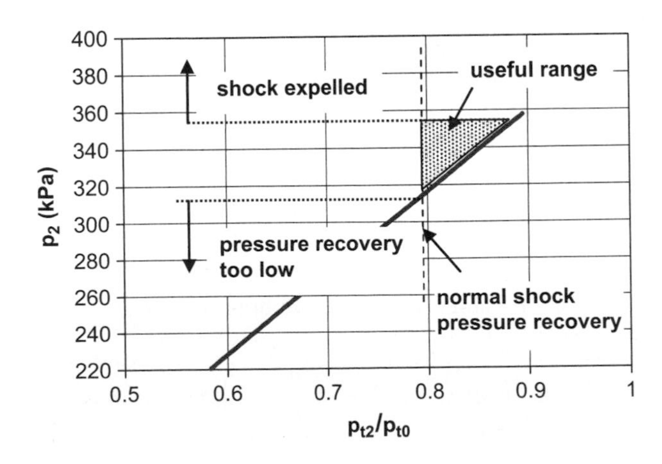


**FIGURE 6.24**

Graph showing stagnation pressure recovery in the inlet as a function of back pressure.

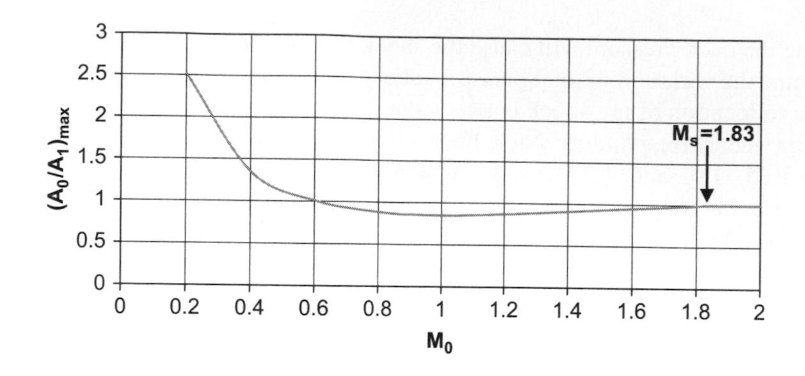


**FIGURE 6.25**

Maximum streamtube capture area ratio as a function of flight Mach number.

on the free stream stagnation temperature, which, for $\gamma = 1.4$, is $T_{t0} = T_0(1 + M^2/5)$ or $T_{t0} = 268\text{K}(1 + 1.83^2/5) = 448\text{K}$. Likewise, the stagnation pressure $p_{t0} = p_0(1 + M^2/5)^{3.5}$ or $p_{t0} = 68.9 \text{ kPa}(1 + 1.83^2/5)^{3.5} = 415 \text{ kPa}$. For flight in the atmosphere, mass flow into the engine is $m = 0.0404\ p_{t0}A_0/T_{t0}^{1/2}$ in kg/s. Now, at the flight Mach number $M_0 = M_s = 1.83$, we know that $A_0 = A_1 = \pi d_1^2/4 = 0.785 \text{ m}^2$; therefore $m = 0.622$ kg/s.

**(d)** We may find the maximum streamtube capture area ratio $(A_0/A_1)_{max}$ by using Equation (6.8) in the regime $M_0 < M_s$ and noting that for $M_0 \geq M_s$ the area ratio $(A_0/A_1)_{max} = 1$. Results are shown in Figure 6.25.

## 6.9 ADDITIVE DRAG

As discussed in Chapter 1, there may be a component of force acting on the segment of the streamtube captured by the inlet. This component, $F_1$, is called the inlet additive drag and is of substantial significance only under supersonic flight conditions when an expelled shock wave forces the capture streamtube area to be smaller than the inlet entrance area as in Figure 6.26. We may examine the nature of the additive drag under these conditions by considering momentum conservation for the region between stations 0 and 1, which leads to the following equation:

$$-(\rho_0 A_0 V_0)V_0 + (\rho_1 A_1 V_1)V_1 = p_0 A_0 - p_1 A_1 + \overline{p}(A_1 - A_0).$$

This results in the following expression for the additive drag force $F_1$:

$$F_1 = \int_{A_0}^{A_1} (p - p_0)dA = \left(\frac{\overline{p}}{p_0} - 1\right) p_0 (A_1 - A_0). \tag{6.10}$$

In this equation, we have assumed that the pressure, which generally varies along the capture streamtube segment under consideration, as illustrated in Figure 6.26, may be replaced by an average

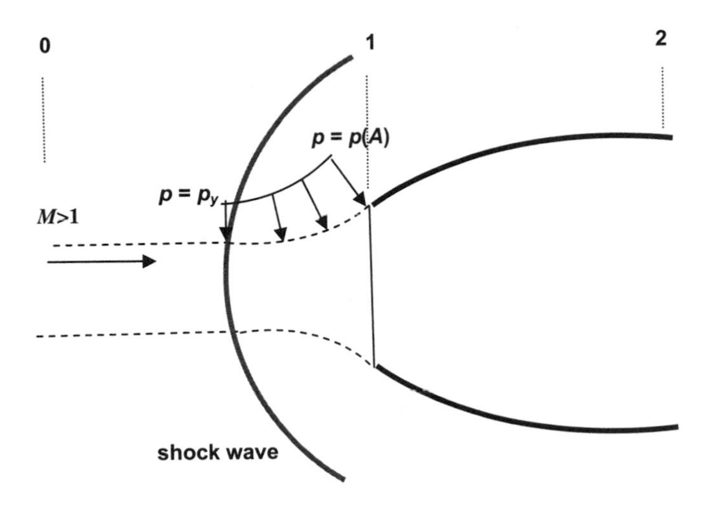


**FIGURE 6.26**

Schematic diagram of an expelled shock case showing pressure $p(A)$ acting on the segment of the captured streamtube between the shock and the inlet lip.

pressure, denoted here by $\overline{p}$. Forming a drag coefficient increment by normalizing $F_1$ with the product of dynamic pressure and inlet entrance area $A_1$ leads to

$$\Delta c_D = \frac{F_1}{\frac{1}{2}\gamma p_0 M_0^2 A_1} = \left(\frac{\overline{p}}{p_0} - 1\right)\left(\frac{1 - \frac{A_0}{A_1}}{\frac{1}{2}\gamma M_0^2}\right). \tag{6.11}$$

An estimate of the average pressure on the streamline may be made by considering it to be equal to the pressure on the streamline immediately behind the shock wave, $p_y$ (see Figure 6.26). Because the streamtube will be expanding and the flow within it is subsonic, the pressure will always larger than $p_y$, rendering a somewhat optimistic drag estimate. Using the normal shock relation for the pressure ratio across a normal shock provides the following result:

$$\frac{\overline{p}}{p_0} - 1 \approx \frac{2\gamma M_0^2 - (\gamma - 1)}{\gamma + 1} - 1 = \frac{7(M_0^2 - 1)}{6}. \tag{6.12}$$

Substitution of this estimate for the pressure into the equation for the additive drag coefficient leads to

$$\Delta c_D = \frac{5}{3}\left(1 - \frac{1}{M_0^2}\right)\left(1 - \frac{A_0}{A_1}\right) \approx O(10^{-1}). \tag{6.13}$$

The estimate of the order of magnitude of the additive drag coefficient is based on a flight Mach number $M_0 \sim 1.5$ and an area ratio $A_0/A_1 \sim 0.9$. Similar results may be obtained for external compression inlets that use oblique shocks to reduce the Mach number ahead of the terminal normal shock. Additional analysis of additive, or spillage, drag is discussed by Seddon and Goldsmith (1985).

## 6.10 EXTERNAL COMPRESSION INLETS

Using internal inlet contraction to decelerate the flow isentropically and then a normal shock to complete the process and bring the flow to a subsonic Mach number and increased static pressure at the end of the inlet has practical difficulties. Because the boundary layer on the inlet walls can tolerate only a very modest adverse pressure gradient before separating, the area decrease must be slow, which leads to long and therefore heavy inlets. Because the objective is to reduce the flow Mach number and raise the static pressure without incurring too great a penalty in total pressure recovery, one may instead turn to generating oblique shocks *outside* the inlet proper to accomplish this task. For a weak oblique shock wave where the normal component of the Mach number ahead of the shock is $M_{n,x} = M_x \sin\theta$, is only slightly greater than unity, the total pressure ratio across it is given approximately by

$$\frac{p_{t,y}}{p_{t,x}} = e^{-\frac{\Delta s}{R}} = 1 - \frac{2\gamma}{3(\gamma+1)^2}\left(M_x^2 \sin^2\theta - 1\right)^3 + \ldots, \tag{6.14}$$

where subscripts $x$ and $y$ denote conditions just upstream and downstream of the shock, respectively, $\theta$ denotes the angle the shock makes with the upstream flow direction, $s$ is the entropy, and $R$ is the gas constant. The reason for introducing the analytic form for the pressure recovery in Equation (6.14) is that it shows clearly that total pressure recovery can be quite good, as long as the component of the Mach number normal to the shock wave is close to unity. The shock wave angle $\theta$ is related to the deflection angle $\delta$ of the surface, causing the shock by the following equation (NASA Report 1135, 1953):

$$\tan\delta = \frac{2\cot\theta}{2 + M_x^2(\gamma + \cos 2\theta)}\left(M_x^2 \sin^2\theta - 1\right). \tag{6.15}$$

Note that the tangent of the deflection angle is proportional to the normal component of the Mach number upstream of the shock $M_x$. Substituting moderate shock wave angles and supersonic Mach numbers into the fraction on the right-hand side of Equation (6.15) will show that it is on the order of unity. Thus the normal component of the Mach number will be close to unity, that is, $M_x^2 \sin^2\theta - 1 \ll 1$, when the deflection angle is small. It is reasonable then to conclude that small flow deflections may be used to decelerate a supersonic flow in a process that comes close to being isentropic. Detailed discussion of oblique shock waves is presented in Appendix A.

External compression inlets are those that make use of a system of weak oblique shock waves external to the inlet, reducing the Mach number and culminating in a normal shock at the cowl lip. Upon passing through the terminal normal shock, the flow becomes subsonic and passes through the remainder of the inlet wherein subsonic diffusion raises the static pressure to a higher value at the exit station of the inlet (station 2). This arrangement is shown schematically in Figure 6.27 for the simplest (two shock) external compression inlet and in Figure 6.28 for a more complicated (three shock) external compression inlet. Oswatitsch (1947) first demonstrated the effectiveness of external compression inlets, and Ferri and Nucci (1954) first presented analytical and experimental studies of such inlets. Discussion of these inlets is presented by Seddon and Goldsmith (1985). Design aspects of various inlets were analyzed in some detail by Connors and Meyer (1956). An interactive design tool for external compression inlets is described by Benson (1994).

It should be noted that deceleration comes about from turning the flow into the mainstream direction. As a result, the flow is captured by the inlet cowl, which must be deflected into the flow and

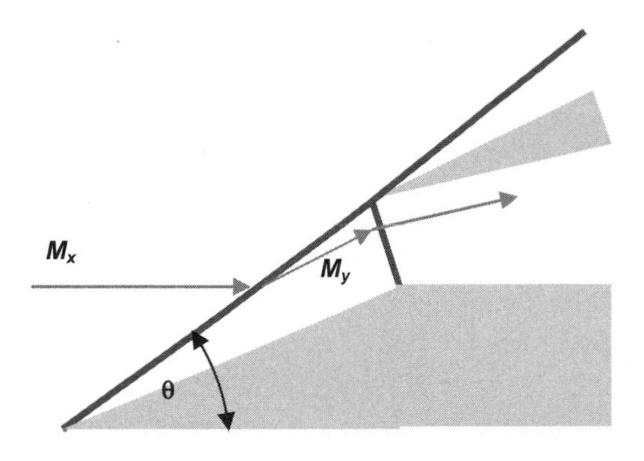


**FIGURE 6.27**

Schematic diagram of an external compression inlet incorporating two shock waves.

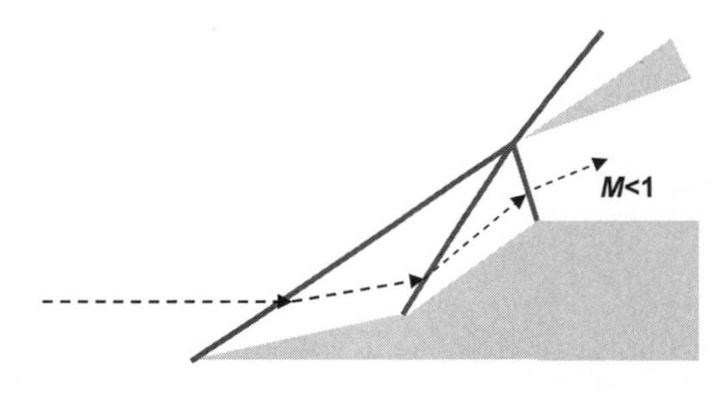


**FIGURE 6.28**

Schematic diagram of an external compression inlet incorporating three shock waves.

therefore gives rise to an additional source of drag that must be taken into account. Of course these inlets are also prone to performance limitations due to additive drag, as well as the drag associated with the relatively large cowl lip angles required to capture the flow that has been turned away from the mainstream direction. These factors are illustrated schematically in Figure 6.29 and are shown in flow visualization photographs in Figure 6.30.

The problem of determining the optimum shock pattern for an external compression inlet with a given number $n$ shocks, $n - 1$ oblique shocks followed by one terminal normal shock, was solved by Oswatitsch (1947), as discussed in Seddon and Goldsmith (1985). He showed that the condition for optimum performance is that the oblique shocks be all of equal strength, which means that the normal component of the Mach number must be the same for all oblique shocks. The general character of the pressure recovery for such optimal systems is shown in Figure 6.31.

The limiting case of an external compression inlet is an isentropic compression inlet where the several weak oblique shock waves caused by small finite flow turns, as shown in Figure 6.32, are taken to the limit of an infinite number of infinitesimal turns, leading to a smooth shockless isentropic compression, as shown in Figure 6.33. In practice, this would lead to an excessively long inlet with the

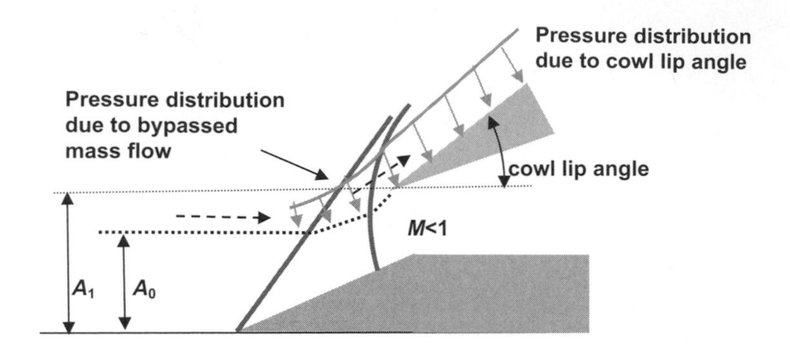


**FIGURE 6.29**

Schematic diagram illustrating sources of drag in an external compression inlet.

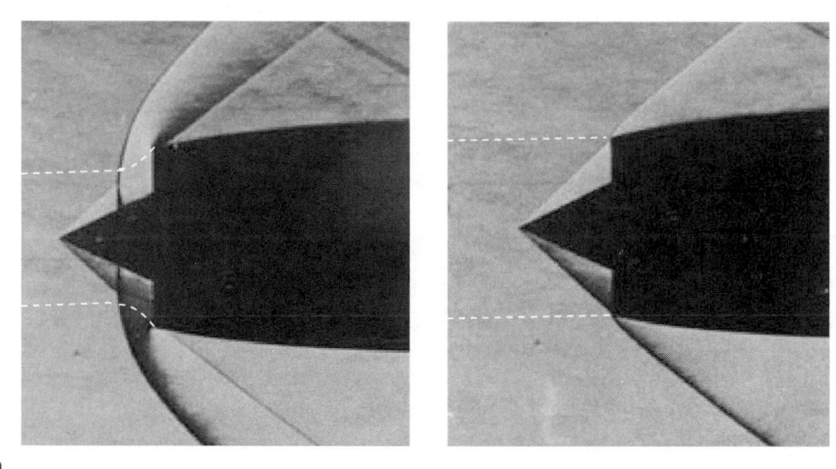

**FIGURE 6.30**

Schlieren flow visualization photographs showing subcritical (left) and supercritical (right) operations of an external compression inlet at $M_0 = 2$ (Anonymous, 1955).

attendant penalties arising from viscous boundary layer effects. This is particularly aggravated in axisymmetric compression surfaces, as the transverse curvature is very large, which exacerbates boundary layer growth.

Another problem that can arise in external compression inlets is a pulsating pressure effect called inlet buzz. This instability in inlet operation may arise when the inlet is operating in a subcritical mode under reduced mass flow conditions. The main effect is rapid movement of the shock wave train in and out of the inlet itself, subjecting it to rapidly varying pressure and acoustic loads, which can compromise the structural integrity of the inlet. A detailed discussion of this phenomenon is given by Seddon and Goldsmith (1985).

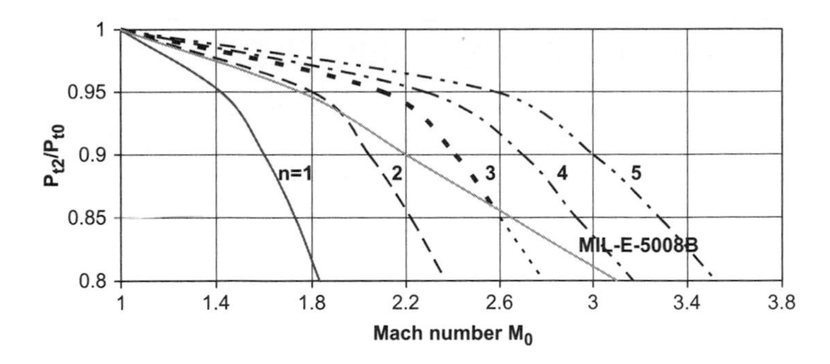


**FIGURE 6.31**

Optimum pressure recovery for external compression inlets with $n$ shocks, $n-1$ oblique shocks followed by a terminal normal shock (Seddon and Goldsmith 1985).

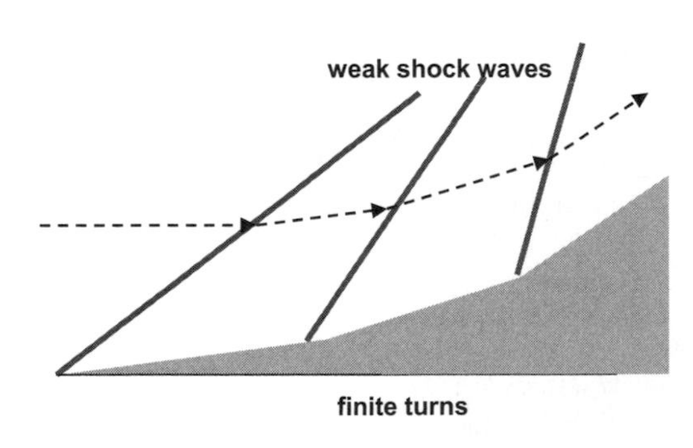


**FIGURE 6.32**

Schematic diagram of a series of weak shock waves in an external compression inlet.

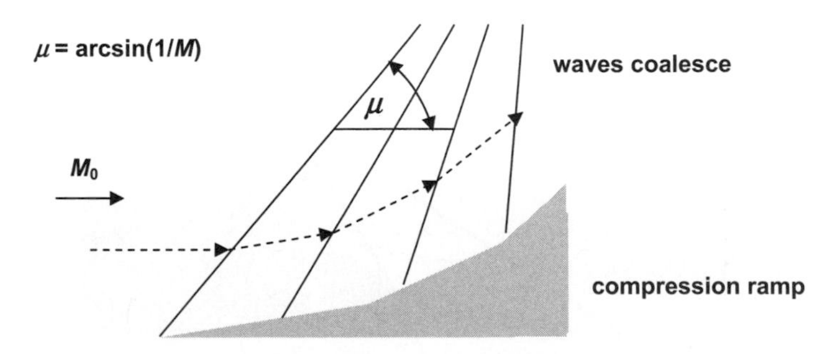


**FIGURE 6.33**

Limiting case of an isentropic compression ramp as an external compression inlet.

## 6.11 EXAMPLE: EXTERNAL COMPRESSION INLET

Consider the three shock external compression inlet shown in Figure 6.34. Assume the following flight conditions: an altitude of $z = 35$ kft where the free stream static pressure $p_0 = 3.44$ psia, the free stream static temperature $T_0 = 390$R, and a flight Mach number $M_0 = 2.5$. (a) Determine the pressure recovery of the inlet. (b) If instead of a three shock inlet this were a two shock inlet with one 20° turn rather than two 10° turns, what would the pressure recovery be and how would the inlet length differ, assuming the same inlet area $A_1$? (c) What would the pressure recovery be for a normal shock inlet under the same conditions?

**(a)** Under the given flight conditions, $p_{t,0} = p_0\,(1 + M_0^2/5)^{3.5} = 58.79$ psia and $T_{t,0} = T_0\,(1 + M_0^2/5) = 877.5$ R. From the oblique shock chart in Appendix A (or from the oblique shock equations there), find $\theta_a = 32°$. Then the Mach number in region a may be found using

$$M_a^2 \sin^2(\theta_a - \delta_a) = \frac{M_0^2 \sin^2\theta_a + 5}{7M_0^2 \sin^2\theta_a - 1} \tag{6.16}$$

where $M_a = 2.066$ and the normal component of the free stream Mach number is $M_{0,n} = M_0 \sin\theta_a = 1.32$. Using normal shock tables or the normal shock equations, we find

$$p_{t,a}/p_{t,0} = 0.975, \; p_{t,a} = 57.32 \text{ psia}.$$

Because $M_a = 2.066$, we can find the ratio of static to stagnation pressure

$$p_a/p_{t,a} = (1 + M_a^2/5)^{-3.5} = 0.115. \tag{6.17}$$

Thus the static pressure is $p_a = 0.115p_{t,a} = 6.59$ psia. Again from the oblique shock chart (or equations), $\theta_a = 38°$ and $M_a \sin\theta_b = 1.272$. Then, from Equation (6.16) $M_b = 1.71$ and from Equation (6.17) $p_b/p_{t,b} = 0.201$. Because $M_a \sin\theta_b = 1.27$, then $p_{t,b}/p_{t,a} = 0.985$ and $p_{t,b}/p_{t,0} = 0.961$. Thus $p_b = 11.36$ psia.
Across the normal shock between **b** and **c** using $M_b = 1.71$ in the normal shock relations yields $p_{t,c}/p_{t,b} = 0.852$ and therefore the pressure recovery $p_{t,c}/p_{t,0} = 0.852(0.961) = 0.818$ and $p_{t,c} = 48.09$

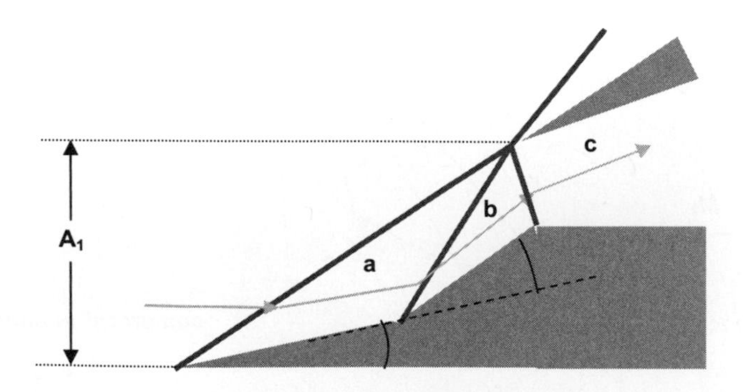


**FIGURE 6.34**

Schematic diagram of a three shock inlet with three flow regions.

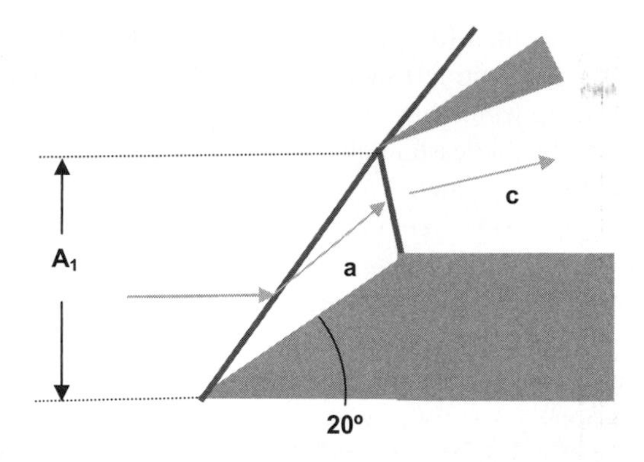


**FIGURE 6.35**

Schematic diagram of a two shock inlet with two flow regions.

psia. Downstream of the terminal shock the Mach number $M_c = 0.638$ and therefore $p_c = 0.76 p_{t,c} = 36.05$ psia.

**(b)** The two shock inlet is shown in Figure 6.35, and it is clear from comparison to Figure 6.34 that for the same inlet area $A_1$ the two shock inlet is shorter in length than the three shock inlet, although the cowl angle is the same. Thus this inlet should be lighter in weight and have no more cowl drag than the three shock inlet.From the oblique shock chart in Appendix A (or from the oblique shock equations there), find $\theta_a = 42°$. Then the Mach number in region **a** may be found using Equation (6.16). Here $M_a = 1.73$, and the normal component of the free stream Mach number is $M_{0,n} = M_0 \sin\theta_a = 1.67$. Then, using normal shock tables or the normal shock equations, we find

$$p_{t,a}/p_{t,0} = 0.868, \; p_{t,a} = 51.03 \text{ psia.}$$

Because $M_a = 1.729$, we can find the ratio of static to stagnation pressure:

$$p_a/p_{t,a} = (1 + M_a^2/5)^{-3.5} = 0.194.$$

Thus, static pressure is $p_a = 0.194 p_{t,a} = 11.4$ psia. Across the normal shock between **a** and **c** using $M_a = 1.73$ in the normal shock relations yields $p_{t,c}/p_{t,a} = 0.843$ and therefore the pressure recovery $p_{t,c}/p_{t,0} = 0.843(0.868) = 0.732$ and $p_{t,c} = 43.03$ psia. Downstream of the terminal shock the Mach number $M_c = 0.633$ and therefore $p_c = 0.762 p_{t,c} = 32.8$ psia.

The two shock inlet has almost 11% less stagnation pressure recovery than the three shock inlet, and the latter inlet would likely be worth the extra weight it would require.

**(c)** A normal shock inlet would have given a much smaller pressure recovery, that is, $p_{t,c}/p_{t,0} = 0.499$.

## 6.12 MIXED COMPRESSION INLETS

As the flight Mach number increases, the shock waves in an external compression inlet must deflect the flow further from the flight direction in order to decrease the Mach number before the terminal normal shock, thereby achieving optimal pressure recovery. For example, at a flight Mach number $M_0 = 3$, a four shock external compression inlet ($n = 4$ in Figure 6.31) with an inlet area $A_1$ requires turns of 11.1, 13.1, and 15.4° before the last (normal) shock is reached. This is a total turn of 39.6° up from the horizontal. Therefore, the cowl lip would have to be at an angle of at least that amount to capture the flow, as shown in Figure 6.36. This increase in effective frontal area leads to increased cowl drag and diminishes the net performance of the inlet.

However, if some of the compression takes place downstream of the cowl lip with the interior of the cowl providing suitable flow deflections, the angle of the cowl lip is much reduced, as is the associated cowl drag, as shown in Figure 6.37. This inlet is considered at the same $M_0$ and inlet area as the external compression inlet.

In Figure 6.37, one shock compression is seen to take place external to the cowl lip, whereas the other three shock compressions take place within the inlet. Therefore, this is called a mixed compression inlet. The first external shock turns the flow up 11.1°, but the succeeding internal shocks

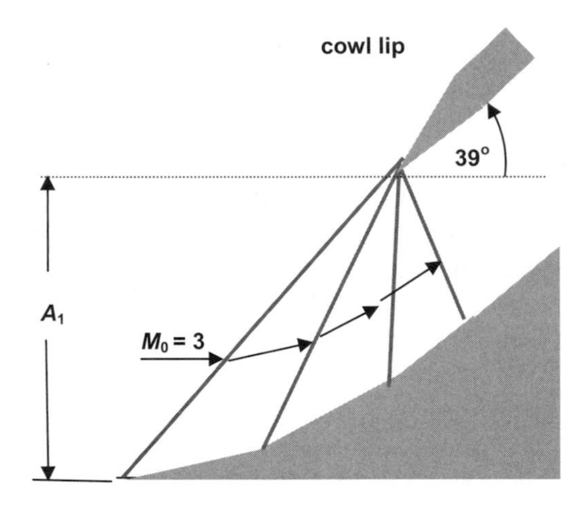


**FIGURE 6.36**

Schematic diagram illustrating necessary flow deflection for an external compression inlet.

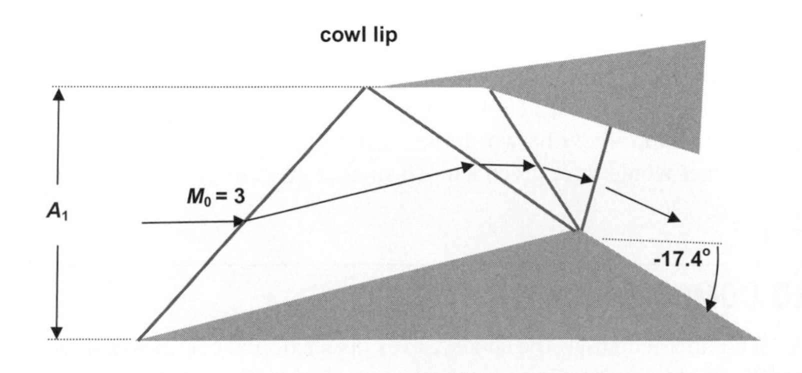


**FIGURE 6.37**

Schematic diagram illustrating flow deflections for a mixed compression inlet.

turn the flow back down by –13.1–15.4° = –28.5° so that the net turn is then –17.4°, which is the angle of inclination of the subsonic diffuser. The mixed compression inlet trades cowl angle for inlet length, and weight and boundary layer problems are incurred.

Comparisons of maximum range achievable in supersonic cruise using inlets with various degrees of internal compression from 0% (all external compression) to 80% internal compression show a decided edge toward those with a high degree of internal compression, according to Bowditch and colleagues (1971). However, such inlets are sensitive to flow disturbances caused by gusts and maneuvers and are susceptible to inlet unstarts, where the swallowed terminal (normal) shock moves out of the inlet severely, reducing the pressure recovery of the inlet, as described in Ferri and Nucci (1954). Throat bypass bleed systems can reduce this sensitivity but add weight and complexity to the engine. Lower performance inlets, like all external compression inlets, constitute a viable alternative to the airplane designer. A design approach for fixed geometry, mixed compression inlets has been proposed by Creta and Valorani (2002).

## 6.13 HYPERSONIC FLIGHT CONSIDERATIONS

The compression processes discussed in this section are of particular importance as flight Mach numbers in the hypersonic regime, approximately $M_0 > 5$, are considered by Segal (2009). At such flight conditions, high temperatures begin to be encountered as may be inferred from the equation for the free stream stagnation temperature:

$$T_{to} = T_0\left(1 + \frac{\gamma - 1}{2}M_0^2\right).$$

If $M_0 = 5$, then with $\gamma = 1.4$ and the temperature in the stratosphere $T_0 = 216$K, the stagnation temperature $T_{t0} \sim 1300$K or over 1000°C. At $M_0 = 10$, the stagnation temperature would appear to be about 4500K! Of course, under increasingly high temperature, the air would dissociate and thermodynamic properties such as $\gamma$ and $c_p$ would vary with the temperature. As a consequence, it is more convenient to avoid the complications due to chemistry and instead examine inlet performance on an enthalpy–entropy plot, as shown in Figure 6.31. The free stream stagnation enthalpy $h_{t0}$ is constant for an adiabatic process, expressing the balance between internal and kinetic energy as follows:

$$h_{t0} = h_0 + \frac{1}{2}V_0^2.$$

The real inlet brings the total pressure from the free stream value of $p_{t0}$ to a value of $p_{t2}$ at the end of the inlet, but this is at the expense of an entropy increase due to the irreversibility of the shock wave system employed in decelerating the flow. An ideal inlet would act isentropically, leaving the total pressure unchanged. Using Figure 6.38, the inlet kinetic energy efficiency may be defined as follows:

$$\eta_{ke} = \frac{h_{t,0} - h_2'}{h_{t,0} - h_0} = \frac{V'2}{V_0^2}. \tag{6.18}$$

The numerator in Equation (6.18) represents the kinetic energy that could be recovered from the inlet after isentropic expansion from the actual stagnation pressure $p_{t0}$ to the ambient pressure $p_0$. Because the denominator represents the kinetic energy in the free stream, the kinetic energy efficiency

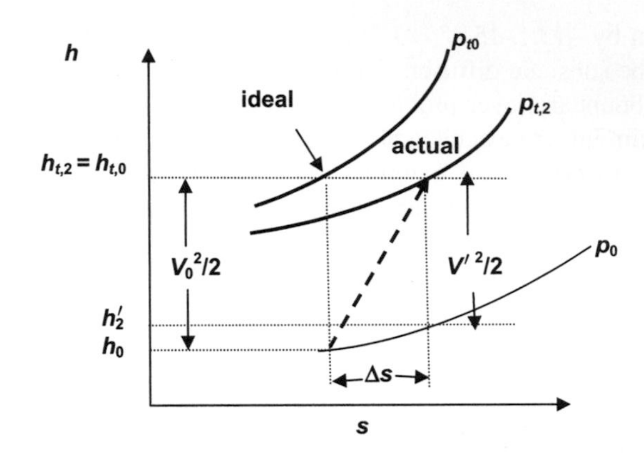


**FIGURE 6.38**

Inlet compression process on an enthalpy–entropy diagram for an adiabatic flow. The ideal inlet would have $\Delta s = 0$ and $p_{t,2} = p_{t,0}$. The entropy increment due to shock losses is reflected in a lower value of $p_{t,2}$.

$\eta_{ke}$ provides a measure of the effectiveness of the inlet in capturing the kinetic energy from the free stream for the adiabatic compression process. It should be clear from Figure 6.32 that at a given altitude and velocity there is a fixed value of $h_{t0}$ and the higher the efficiency, that is, the smaller the entropy rise, the better the stagnation pressure recovery. If we consider the case of constant thermal properties, which is appropriate for supersonic Mach numbers, we may rewrite the inlet kinetic energy efficiency as

$$\eta_{ke} = 1 - \frac{2}{(\gamma - 1)M_0^2}\left[\left(\frac{p_{t2}}{p_{t0}}\right)^{\frac{1-\gamma}{\gamma}} - 1\right].$$

For the case of atmospheric air with $\gamma = 1.4$, this equation becomes

$$\eta_{ke} = 1 - \frac{5}{M_0^2}\left[\left(\frac{p_{t2}}{p_{t0}}\right)^{-\frac{2}{7}} - 1\right].$$

## 6.14 NOMENCLATURE

| | |
|---|---|
| $A$ | cross-sectional area |
| $ar$ | sound speed |
| $C_f$ | friction coefficient |
| $C_R$ | correction factor for inlet thrust losses |
| $c_D$ | drag coefficient |

| | |
|---|---|
| $c_p$ | specific heat at constant pressure |
| $D$ | hydraulic diameter of a duct or aircraft drag |
| $F_I$ | additive drag force |
| $F_n$ | net thrust |
| $g$ | acceleration of gravity |
| $h$ | enthalpy |
| $L$ | length of duct or aircraft lift |
| $M$ | Mach number |
| $M_s$ | starting Mach number |
| $\dot{m}$ | mass flow |
| $n$ | number of shocks |
| $p$ | pressure |
| $R$ | gas constant |
| $s$ | entropy |
| $T$ | temperature |
| $t$ | time |
| $V$ | velocity |
| $w$ | weight flow |
| $z$ | altitude |
| $\gamma$ | ratio of specific heats |
| $\delta$ | flow deflection angle |
| $\theta$ | shock wave angle or general angular coordinate |
| $\eta_{ke}$ | inlet kinetic energy efficiency |
| $\tau$ | integration time for time averaging |

### 6.14.1 Subscripts

| | |
|---|---|
| IC | internal compression inlet conditions |
| max | maximum condition |
| NS | normal shock inlet conditions |
| $s$ | inlet starting condition |
| $t$ | stagnation conditions |
| $x$ | condition just upstream of a normal shock |
| $y$ | condition just downstream of a normal shock |
| 0 | conditions in the free stream |
| 1 | conditions at the inlet entrance |
| 2 | conditions at the compressor entrance |

### 6.14.2 Superscripts

| | |
|---|---|
| $*$ | critical condition where $M = 1$ |
| $\overline{(\,)}$ | time-averaged property |
| $(\,)'$ | time-fluctuating property |

## 6.15 EXERCISES

**6.1** Consider a converging–diverging internal compression inlet as shown in Figure E6.1:

**(a)** For an area ratio $A_1/A_{min} = 1.30$, determine the starting Mach number $M_s$ and the corresponding values of $p_{t2}/p_{t1}$ and $(A_0/A_1)_{max}$.

**(b)** For a flight Mach number $M_0 = 3$, find the corresponding values of $p_{t2}/p_{t1}$ and $(A_0/A_1)_{max}$.

**(c)** For a flight Mach number $M_0 = 2$, find the corresponding values of $p_{t2}/p_{t1}$ and $(A_0/A_1)_{max}$.

**(d)** For a flight Mach number $M_0 = 0.2$, find the corresponding values of $p_{t2}/p_{t1}$ and $(A_0/A_1)_{max}$.

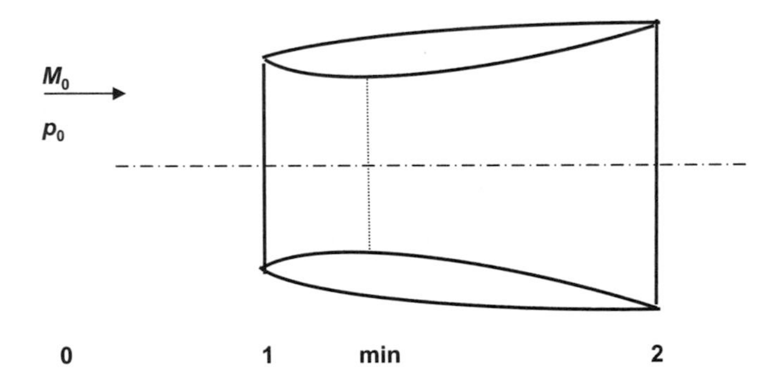


**FIGURE E6.1**

Converging–diverging internal compression inlet.

**6.2** Consider the internal compression inlet shown in Figure E6.1:

**(a)** For an area ratio $A_1/A_{min} = 1.1265$, determine the starting Mach number $M_s$ and the corresponding values of $p_{t2}/p_{t1}$ and $(A_0/A_1)_{max}$.

**(b)** For a flight Mach number $M_0 = 2$, find the corresponding values of $p_{t2}/p_{t1}$ and $(A_0/A_1)_{max}$.

**(c)** For a flight Mach number $M_0 = 1.25$, find the corresponding values of $p_{t2}/p_{t1}$ and $(A_0/A_1)_{max}$.

**(d)** For a flight Mach number $M_0 = 0.75$, find the corresponding values of $p_{t2}/p_{t1}$ and $(A_0/A_1)_{max}$.

**6.3** Consider the internal compression inlet shown in Figure E6.1:

**(a)** For an area ratio $A_1/A_{min} = 1.23$, determine the starting Mach number $M_s$ and the corresponding values of $p_{t2}/p_{t1}$ and $(A_0/A_1)_{max}$.

**(b)** For a flight Mach number $M_0 = 2.5$, find the corresponding values of $p_{t2}/p_{t1}$ and $(A_0/A_1)_{max}$.

**(c)** For a flight Mach number $M_0 = 1.5$, find the corresponding values of $p_{t2}/p_{t1}$ and $(A_0/A_1)_{max}$.

**6.4** Consider the internal compression inlet shown in Figure E6.1:

**(a)** For an area ratio $A_1/A_{min} = 1.26$, determine the starting Mach number $M_s$ and the corresponding values of $p_{t2}/p_{t1}$ and $(A_0/A_1)_{max}$.

**(b)** For a flight Mach number $M_0 = 2.5$, find the corresponding values of $p_{t2}/p_{t1}$ and $(A_0/A_1)_{max}$.

**(c)** For a flight Mach number $M_0 = 0.25$, find the corresponding values of $p_{t2}/p_{t1}$ and $(A_0/A_1)_{max}$.

**6.5** The North American Aviation F-100 Super Sabre had a wing area of 400 ft$^2$, a wing sweepback angle of 45°, a taper ratio of 0.25, and an airfoil-section thickness ratio of about 7%. Like the wing of its ancestor the F-86, the leading edge was equipped with automatic slats for stall control and the trailing edge incorporated plain flaps. Location of the ailerons mounted a short distance inboard of the tip reduced adverse wing twisting due to aileron deflection. (Under conditions of high dynamic pressure, adverse wing twist can become so large that, on some aircraft, roll takes place in a direction opposite to that intended. This condition is known as aileron reversal.) The low-mounted horizontal tail of the F-100 is clearly shown in Figure E6.2. This tail position assists in preventing pitch-up. The oblong nose inlet shown in Figure E6.2 provides an immediate recognition feature of the F-100 series of aircraft. As compared with a circular inlet, the oblong design ($A_1 = 6$ ft$^2$) provides better pilot visibility over the nose and, because the duct passes under the pilot's seat, the vertical dimension of the fuselage is reduced at this location. The photograph in Figure E6.2 shows the variable-area nozzle necessary for efficient operation of the afterburning engine; the nozzle is in the nonafterburning configuration. The petals of the nozzle open to a larger diameter for afterburning. The Pratt & Whitney J57-P-21A engine produces 16,000 pounds thrust with afterburner. The maximum speed of the F-100 was $M = 1.39$ at 35,000 ft altitude. Under this condition, and assuming that 1976 U.S. standard atmosphere properties apply, determine the (a) free stream stagnation pressure, $p_{t,0}$; (b) free stream stagnation temperature, $T_{t,0}$; (c) maximum pressure recovery of the inlet, $p_{t,2}/p_{t,0}$; (d) static pressure at the end of the inlet, $p_2$, assuming that the inlet duct expands smoothly to $A_2/A_1 = 5/3$; and (e) weight flow through the inlet, (f) Assuming the gross thrust to be the full afterburning thrust quoted, determine the drag coefficient $c_D = D/qS$, where $q$ is the dynamic pressure and $S$ is the wing area.

**FIGURE E6.2**

Photograph of F-100 Super Sabre.

**FIGURE E6.3**

Photograph of F4 Phantom.

**6.6** The McDonnell F4 Phantom, a supersonic all-weather fighter-bomber had a maximum speed of Mach 2.23 at 40,000 ft altitude. It was powered by two GE J79-GE-17 turbojets producing 1788 lbs of thrust each. The inlet used on the Phantom is a "D-type" inlet of area 6.5 ft$^2$ mounted on either side of the fuselage and containing variable angle ramps and a boundary layer splitter plate to divert low-speed boundary layer flow around the inlet rather than into it. Some of these details can be seen in the photograph in Figure E6.3. The inlet can be considered to have nominally two-dimensional flow entering it for the supersonic flight condition shown in the diagram given. The configuration shown has one ramp deflected up 10° in order to position the oblique shock generated by the ramp such that it impinges directly on the cowl lip. The back pressure is adjusted so that the supersonic flow in region "a" shocks down through a normal shock wave, as shown on the diagram, and passes into region "b" and then on through the duct to exit station 2. For conditions upstream of the inlet shown in Figure E6.4, calculate the following: (a) angle of the oblique shock wave $\theta$; (b) Mach number, pressure, stagnation pressure, temperature, and stagnation temperature in region "a"; (c) weight flow rate through the inlet; (d) Mach number, pressure, stagnation pressure, temperature, and stagnation temperature in region "b"; (e) stagnation pressure at inlet station 2 and pressure recovery of the inlet; (f) ratio of the area at station 2 to the area at the location of the normal shock that would bring the Mach number at station 2 to $M_2 = 0.35$; and (g) ram drag of the inlet, $F_r$.

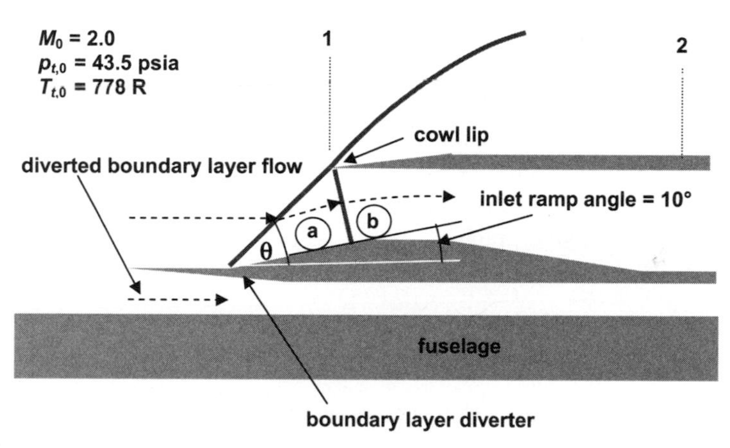


**FIGURE E6.4**

Schematic diagram of F4 inlet.

**6.7** The fuselage-mounted, two-shock D-type inlet shown in Figure E6.5 has a height $h = 0.193$ m and a width $w = 0.577$ m. It is operated at $M_0 = 2.5$ at an altitude where $p_0 = 68.91$ kPa and $T_0 = 250$K. The pressure recovery inside the inlet from station $y$ to station 2 is $p_{t2}/p_{ty} = 0.9$. For the case of maximum mass flow through the inlet and

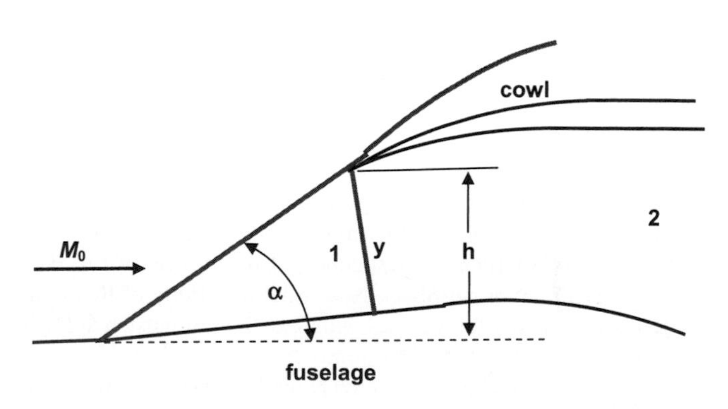


**FIGURE E6.5**

Fuselage-mounted D-type inlet with two-shock compression.

assuming two-dimensional flow, calculate (a) $\alpha$, (b) $M_1$, (c) $p_1$, (d) $T_1$, (e) mass flow rate, (f) $A_0/A_1$, (g) $M_y$, (h) $p_y$, (i) $T_y$, (j) $p_{t2}$, and (k) ram drag $F_r$.

**6.8** Assuming constant thermal properties, derive the expression for the kinetic energy efficiency of a high speed given by

$$\eta_{k.e.} = 1 - \frac{\left(\frac{p_{t2}}{p_{t0}}\right)^{-\frac{\gamma-1}{\gamma}} - 1}{\frac{\gamma-1}{2} M_0^2}.$$

It may be useful to consider Equation (6.14) in carrying out this exercise.

**6.9** Derive an expression for the contraction ratio $A_{min}/A_1$ for a started internal contraction inlet as a function of Mach number $M_0$ and present a plot of the result. Prove that as $M_0$ grows without bound, the limiting value of the contraction ratio is 1.666 for air where $\gamma = 1.4$. Indicate this value on your graph.

## References


Anonymous (1955). NASA Glenn Research Center Image Gallery No. C-1955-37520.

Antonatas, P., Surber, L., & Stava, D. (1972). Inlet/Airplane Interference and Integration. AGARD Lecture Series No. 53, Engine/Airframe Integration.

Benson, T. J. (1994). An Interactive Design and Educational Tool for Supersonic External Compression Inlets. NASA TM 1066581.

Bowditch, D. N., Coltrin, R. E., Sanders, B. W., Sorenson, N. E., & Wasserbauer, J. F. (1971). Supersonic cruise inlets. In Aircraft Propulsion (pp. 283–312). NASA SP-259.

Connors, J. F., & Meyer, R. C. (1956). Design Criteria for Axisymmetric and Two-Dimensional Supersonic Inlets and Exits. NACA TN 3589.

Creta, F., & Valorani, M. (2002). Optimal Shape Design of Supersonic, Mixed-Compression, Fixed-Geometry Air Intakes for SSTO Mission Profiles. AIAA 2002-4133, 38th Joint Propulsion Conference, Indianapolis, IN.

Ferri, A., & Nucci, L. M. (1954). Theoretical and Experimental Analyses of Low-Drag Supersonic Inlets Having a Circular Cross Section and a Central Body at Mach Numbers of 3.30, 2.75, and 2.45. NASA Technical Report 1189.

Hesse, W. J., & Mumford, N. V. S. (1964). *Jet Propulsion for Aerospace Applications* (2nd ed.). New York: Pitman

MIL-E-5008B: Military Specifications—Engines, Aircraft, Turbojet, Model, Specification For, January 1959.

NACA Report 1135 (1953). Equations Tables and Charts for Compressible Flow.

Olstad, W. B. (1956). Transonic-Wind-Tunnel Investigation of the Effects of Lip Bluntness and Shape on the Drag and Pressure Recovery of a Normal-Shock Nose Inlet in a Body of Revolution. NACA-RM-L56C28.

Oswatitsch, K. (1947). Pressure Recovery for Missiles with Reaction Propulsion at High Supersonic Speeds: The Efficiency of Shock Diffusers. NACA TM 1140, June.

Seddon, J., & Goldsmith, E. L. (1985). Intake Aerodynamics. American Institute of Aeronautics and Astronautics, New York.

Segal, C. (2009). *Scramjet Engines.* New York: Cambridge University Press

CHAPTER

# Turbomachinery

7

## CHAPTER OUTLINE



**Theory of Aerospace Propulsion.**

## 7.1 THERMODYNAMIC ANALYSIS OF A COMPRESSOR AND A TURBINE

Turbomachines are flow processors that depend on rotation to affect changes in pressure in the flow that passes through them. Those turbomachines that add work to the flow and increase the output pressure are compressors, and those that extract work from the flow and decrease output pressures are turbines. Consider an idealized turbomachine as shown in Figure 7.1.

An energy balance may be formulated for the machine under the assumption that potential energy effects for this gas compressor are negligible, that is,

$$h_2 + \frac{1}{2}V_2^2 + Q + W_c = h_3 + \frac{1}{2}V_3^2, \tag{7.1}$$

where $W_c$ is work done on the fluid passing through the machine and $Q_{ext}$ is the heat transfer to the fluid from the surroundings. Further assuming that the transfer of heat from the surroundings is likewise negligible, work done on the fluid by the turbomachine is

$$W_c = \left(h_3 + \frac{1}{2}V_3^2\right) - \left(h_2 - \frac{1}{2}V_2^2\right) = h_{t,3} - h_{t,2}. \tag{7.2}$$

In this case, because the numbering system suggests that the turbomachine under consideration is a compressor, $W_c$ is positive and represents work put into the fluid. A similar approach for the turbine, where station 4 denotes the entrance to the turbine and station 5 denotes its exit, would yield

$$W_t = h_{t,4} - h_{t,5}. \tag{7.3}$$

Here we have reversed the terms on the right-hand side in order to preserve a positive sign for $W_t$, which is work removed from the fluid by the turbine.

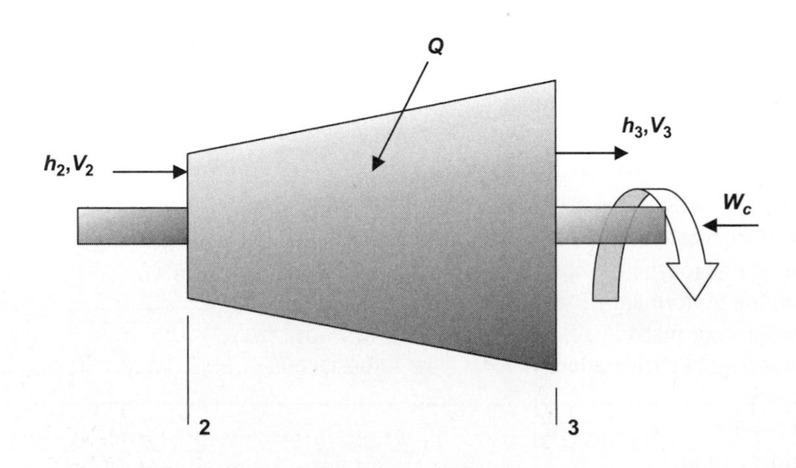


**FIGURE 7.1**

Schematic diagram of a compressor wheel.

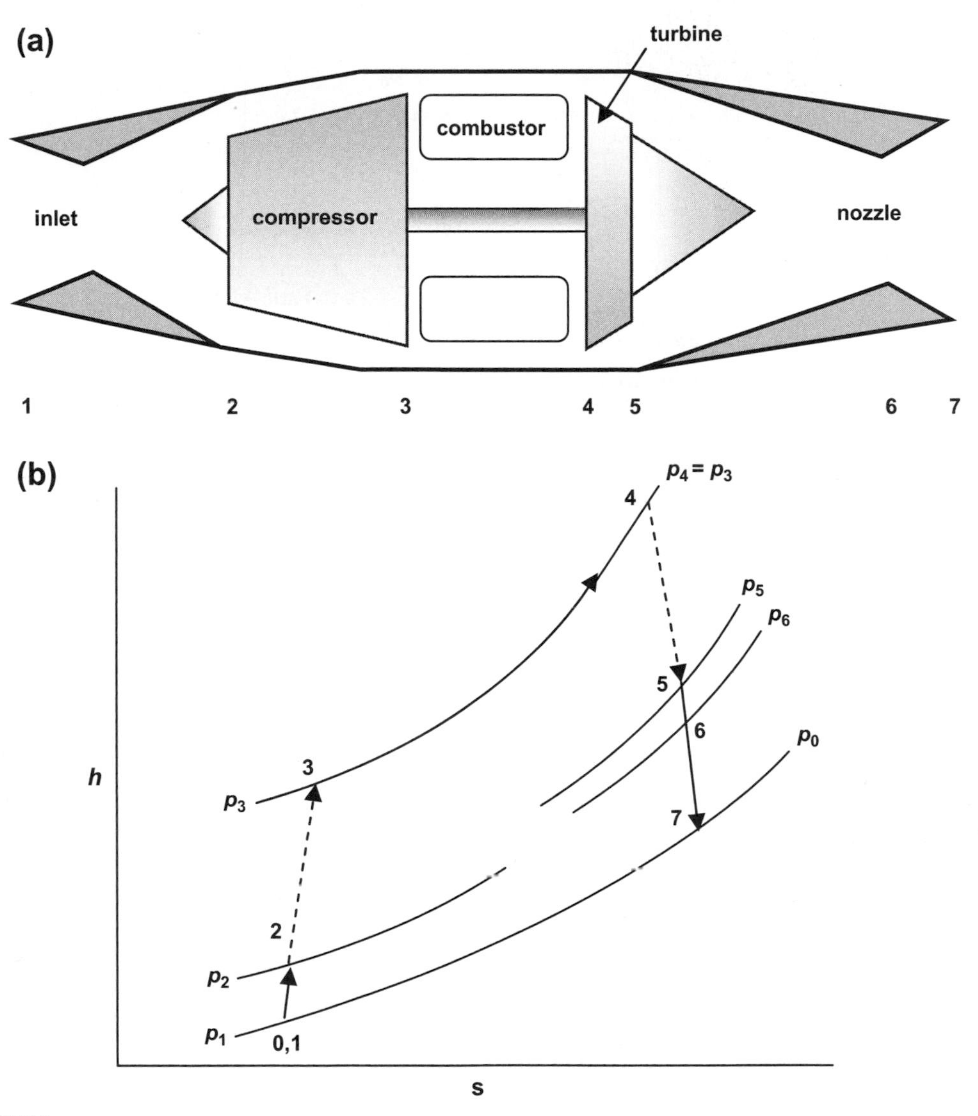


**FIGURE 7.2**

(a) Schematic of turbojet engine with typical numbering system and (b) thermodynamic cycle on *h-s* plot showing compressor and turbine paths as dashed lines.

A generic gas turbine engine is shown in Figure 7.2, along with the associated thermodynamic cycle. The compressor and turbine paths are shown as dashed lines, and the basic operation of both is covered in this chapter. Because of the similarity in the operation of compressors, they will be sometimes be discussed simultaneously where those similarities are clear. However, there are aspects of operation where it is preferable to treat each independently.

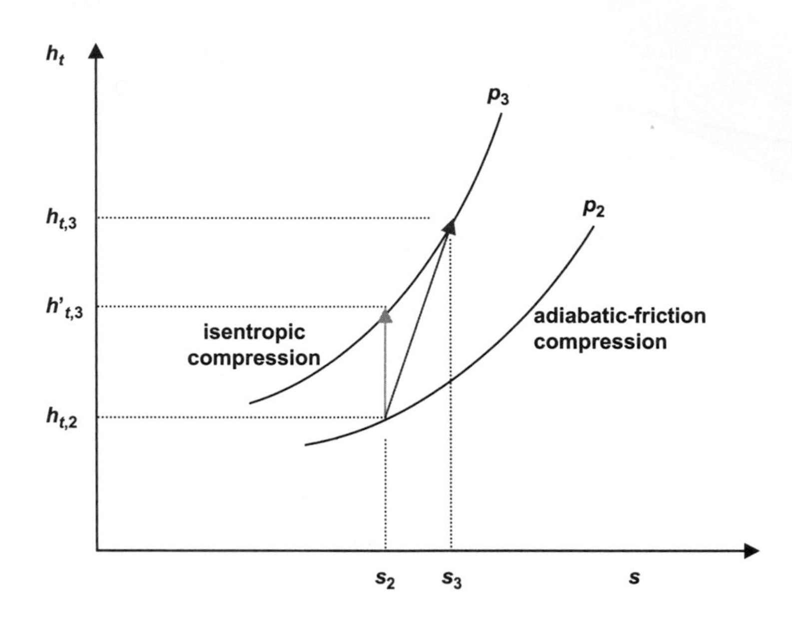


**FIGURE 7.3**

Ideal and real compression processes on an $h,s$ plot.

### 7.1.1 Compressor thermodynamics

Assuming that the specific heat of the fluid is constant through the compressor, the work required is

$$W_c = c_p\left(T_{t,3} - T_{t,2}\right). \tag{7.4}$$

This result assumes that all the work put into the fluid is involved in raising the total enthalpy, which would imply an adiabatic compression. Although we have assumed that the compression process is adiabatic, we haven't assumed a lossless system, and the efficiency of adiabatic compression is given as

$$\eta_{ad,c} = \frac{h'_{t,3} - h_{t,2}}{h_{t,3} - h_{t,2}} = \frac{T'_{t,3} - T_{t,2}}{T_{t,3} - T_{t,2}}. \tag{7.5}$$

This process is shown on an $h$-$s$ plot in Figure 7.3, where the primed quantity $h_t'$ (and $T_t'$) arises from an isentropic process, while the unprimed quantity $h_t$ (and $T_t$) is associated with the dissipative effects that lead to an increase in entropy and therefore represents less than ideal efficiency.

Note that the first and second laws of thermodynamics provide the following relation

$$Tds = dh - \frac{dp}{\rho}.$$

Then

$$\frac{dh}{ds} = T + \frac{1}{\rho}\frac{dp}{ds}.$$

For a constant pressure process with constant specific heat, the slope

$$\frac{dh}{ds} = \frac{h}{c_p}. \tag{7.6}$$

Thus, along a constant pressure line, the slope increases with the magnitude of the enthalpy and constant pressure lines diverge.

In an ideal, or isentropic, compression, the stagnation temperature is related to the stagnation pressure by the relation

$$\frac{T'_{t,3}}{T_{t,2}} = \left(\frac{p_{t,3}}{p_{t,2}}\right)^{\frac{\gamma_2-1}{\gamma_2}}. \tag{7.7}$$

The ratio of specific heats $\gamma_2$ has the subscript 2 to denote that it is the constant average value of $\gamma$ associated with the compressor, stations 2 to 3. The efficiency of adiabatic compression may then be expressed in terms of actual thermodynamic state variables as follows:

$$\eta_{ad,c} = \frac{\left(\frac{p_{t,3}}{p_{t,2}}\right)^{\frac{\gamma_2-1}{\gamma_2}} - 1}{\frac{T_{t,3}}{T_{t,2}} - 1}. \tag{7.8}$$

The work required to operate the compressor is then given by

$$W_c = \frac{c_{p,2}T_{t,2}}{\eta_{ad,c}}\left[\left(\frac{p_{t,3}}{p_{t,2}}\right)^{\frac{\gamma_2-1}{\gamma_2}} - 1\right]. \tag{7.9}$$

### 7.1.2 Turbine thermodynamics

In the same fashion, we may arrive at an expression for the adiabatic efficiency of expansion through the turbine, as operation of the turbine is similar to that of the compressor. For an expansion process going down to $p_{t,5}$ from $p_{t,4}$, the actual work extracted by the turbine is less than that which would be extracted if the expansion were isentropic. The efficiency of adiabatic expansion is defined as

$$\eta_{ad,e} = \frac{h_{t,4} - h_{t,5}}{h_{t,4} - h'_{t,5}}. \tag{7.10}$$

If it is assumed that the specific heat is constant through the expansion, Equation (7.10) becomes

$$\eta_{ad,e} = \frac{1 - \frac{T_{t,5}}{T_{t,4}}}{1 - \frac{T'_{t,5}}{T_{t,4}}}. \tag{7.11}$$

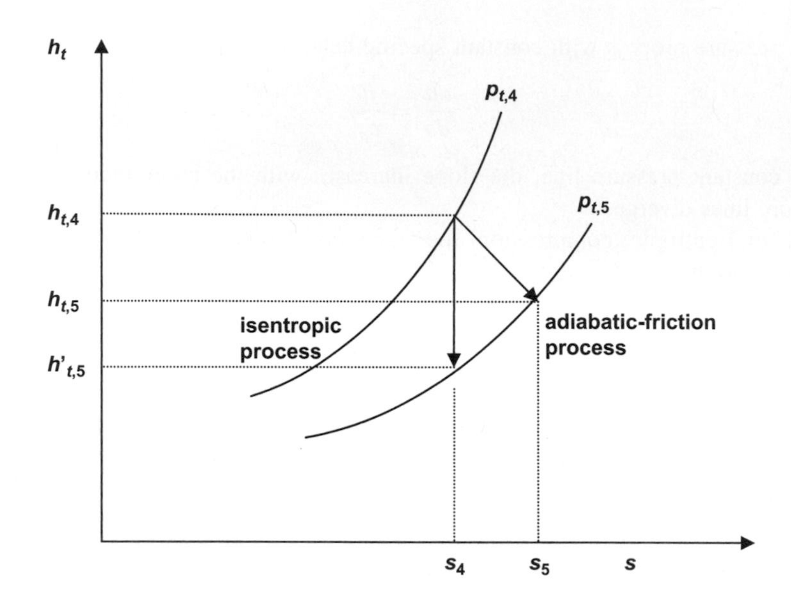


**FIGURE 7.4**

Ideal and real expansion processes on an *h-s* plot.

Using the isentropic relation, the adiabatic efficiency of expansion is

$$\eta_{ad,e} = \frac{1 - \dfrac{T_{t,5}}{T_{t,4}}}{1 - \left(\dfrac{p_{t,5}}{p_{t,4}}\right)^{\frac{\gamma_4 - 1}{\gamma_4}}}. \tag{7.12}$$

The process is shown in Figure 7.4. The work delivered by the turbine is then

$$W_t = \eta_{ad,e} c_{p,4} T_{t,4} \left[1 - \left(\frac{p_{t,5}}{p_{t,4}}\right)^{\frac{\gamma_4 - 1}{\gamma_4}}\right]. \tag{7.13}$$

The use of subscript 4 on $c_p$ and $\gamma$ is to denote that constant (average) values typical of turbine hot gas composition and temperature are to be used.

### 7.1.3 Units used in compressors and turbines

Although the units used in different applications are sometimes confusing, it is necessary to keep in mind the concept behind the units, which is particularly true for flowing systems. The use of units varies from one industry to another, and it takes some thought to keep them straight. We denote compressor and turbine work, denoted by $W_c$ and $W_t$, respectively, in terms of specific units, that is, work per unit mass ([N-m/kg] = [J/kg] in the SI system, which is mass based) or per unit weight ([ft-lb/lb] = [ft] in the English system, which is force based). In flowing systems, we generally see units of mass flow rate ([kg/s] in the SI system) or weight flow rate ([lbs/s] in the English system). Then the power, which has units of work per unit time, refers to the work per unit mass multiplied by the flow rate in mass per unit time, giving us power ([J/kg][kg/s] = [W] in the SI system), or else refers to the work per unit weight multiplied by the flow rate in weight per unit time ([ft-lbs/lb][lbs/s] = [ft-lb/s] = [hp/550]).

For example, the energy equation for flowing systems under adiabatic and frictionless conditions is $h + u^2/2 = \text{constant}$ and is written on a per unit mass basis. Obviously the units must match: for example, in the SI system, [kJ/kg] for $h$ and [$m^2/s^2$] for $u^2/2$. Note that [kJ/kg] = [(kN-m)/kg] = [(kN/kg)-m] = [$10^3$(m/$s^2$)-m] = [$10^3 m^2/s^2$] so that the units do indeed match. In the English system, the units of $h$ and $u^2/2$ are [Btu/lb] and [$ft^2/s^2$], respectively. The specific enthalpy in [Btu/lb] = [778ft-lb/lb] = [ft] is a per weight unit, which must be multiplied by $g$ [32.2 ft/$s^2$] in order to get the enthalpy in the correct units, that is, [(Btu/lb)(32.2ft/$s^2$)] = [(778ft-lb/lb)(32.2ft/$s^2$)] = [$2.51 \times 10^4 ft^2/s^2$].

The units for $W_c$ or $W_t$ had historically been cast in units of the English system as work per unit weight of fluid being processed. In other words, $W = P/mg$, that is, power divided by the weight flow rate for which $P/mg$ has the units [hp/(lbs/s)] = [(550 ft-lb/s)/(lb/s)] = [550 ft]. Historically, this approach was used during the development of hydraulic pumps and turbines where the flow was incompressible and work done per pound of liquid being pumped could be expressed as a "head" or height of that fluid. Conveniently, that height of fluid would exert a pressure equal to the product of the weight density of the fluid and its head, or height. Now in SI units we would have [W/(kg/s-m/$s^2$)] = [(N-m)/s]/[(kg/s-m/$s^2$)] = [(N/kg)-$s^2$] = [(m/$s^2$)-$s^2$] = [m], and again the work per unit weight of fluid has the units of length. If one considers the power per unit mass flow, then the units are those of velocity squared, that is, [W/(kg/s)] = [N-m/kg] = [$m^2/s^2$] or [hp/(lb-s/ft)] = [(550 ft-lb/s)(32.2 ft/$s^2$)/(lb/s)] = [$2.51 \times 10^4 ft^2/s^2$]. Symbols and units are not often uniform in practice, which requires each case to be evaluated individually and to understand the concepts involved in applying them.

## 7.2 ENERGY TRANSFER BETWEEN A FLUID AND A ROTOR

The rotor of a generalized turbomachine is depicted in Figure 7.5; here the laboratory-frame velocity is denoted by $c$. Its components are shown at the entrance to a spinning rotor, for example, a compressor, hence the subscript 2; for a turbine, the subscript is 4. The linear speed of rotation of the rotor and the speed of the fluid relative to the rotor are denoted by $u$ and $w$, respectively. There will be a corresponding set of velocity components leaving the rotor, and this station will be denoted by the subscript 3; for the turbine, the subscript is 5. Subscripts $a$, $r$, and $u$ denote the axial, radial, and tangential components of velocity, respectively.

The relationship between the linear speed of rotation and the absolute velocity is one that introduces a relative wind, the velocity seen by an observer fixed to a spinning rotor blade, as shown in Figure 7.6.

Forces acting on the fluid in the various component directions may be obtained by applying the momentum conservation principle to a control volume enclosing the fluid and entering and leaving the rotor through some as yet unspecified passages. Using the symbol $\dot{m}$ to denote the mass flow rate, the axial force, for compressors and turbines, respectively, is given by

$$\begin{aligned} F_{a,c} &= \dot{m}\left(c_{3,a} - c_{2,a}\right) \\ F_{a,t} &= \dot{m}\left(c_{5,a} - c_{4,a}\right) \end{aligned}. \tag{7.14}$$

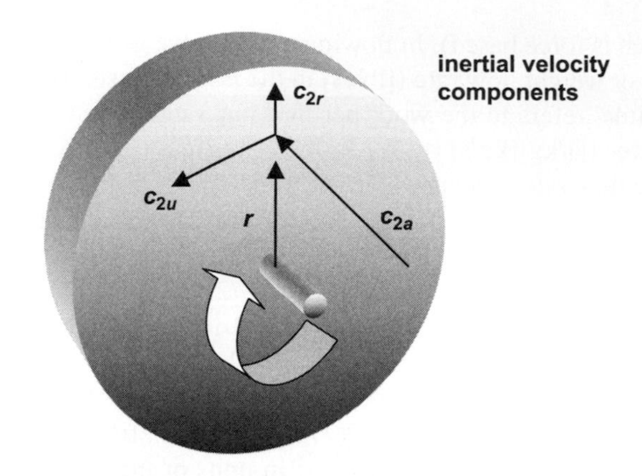


FIGURE 7.5

Laboratory frame velocity components entering a spinning rotor. Subscripts shown denote entry of compressor as an example.

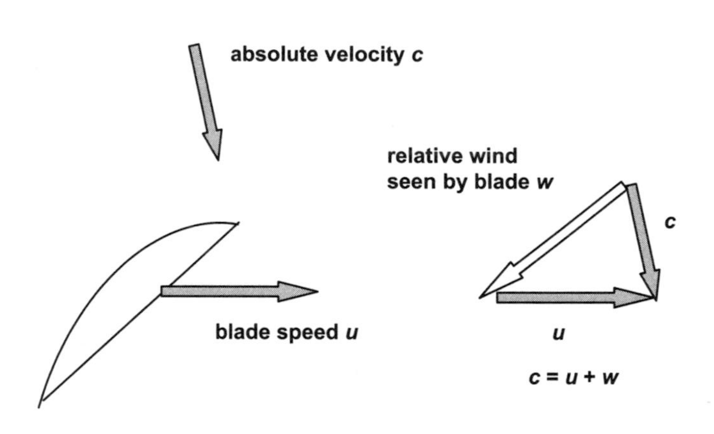


FIGURE 7.6

Relative velocity seen by an observer on the rotor in terms of absolute velocity of the gas and the linear speed of rotation of a blade on the rotor.

The axial forces must be taken up by thrust bearings in an appropriate structure. Although the main interest in jet engines is the propulsive thrust produced, it is important to recognize that substantial axial loads are placed on the different components of the engine, as will be discussed subsequently. The tangential force in compressors and turbines is found to be

$$\begin{aligned} F_{u,c} &= \dot{m}(c_{3,u} - c_{2,u}) \\ F_{u,t} &= \dot{m}(c_{5,u} - c_{4,u}) \end{aligned} . \tag{7.15}$$

This does not take into account frictional forces in the system and represents the ideal net turning force required by a compressor or supplied by a turbine. The radial force in compressor and turbines must be absorbed by journal bearings in appropriate structures and is given by

$$\begin{aligned} F_{r,c} &= \dot{m}\left(c_{3,r} - c_{2,r}\right) \\ F_{r,t} &= \dot{m}\left(c_{5,r} - c_{4,r}\right) \end{aligned}. \tag{7.16}$$

The conservation of angular momentum requires that the change in the moment of momentum of the fluid passing through the control volume, that is, the flow passage through the rotor, is equal to the net external torque applied to the fluid, which may be expressed as follows:

$$\begin{aligned} T_{q,c} &= \dot{m}\left(r_3 c_{3,u} - r_2 c_{2,u}\right) \\ T_{q,t} &= \dot{m}\left(r_5 c_{5,u} - r_4 c_{4,u}\right) \end{aligned}. \tag{7.17}$$

Quantity $rc_u$ is called the *whirl*, so that the term $r_3c_{3,u} - r_2c_{2,u}$ is called the change in whirl in a compressor. Similarly, $r_5c_{5,u} - r_4c_{4,u}$ is the change in whirl for a turbine. Keep in mind that $F_u$ and $T_q$ are the tangential force and torque acting on the fluid, respectively.

Consider the rate at which energy must be supplied to the fluid by the rotor, that is, the power imparted to the fluid, which may be written for the compressor as

$$P_c = T_q\omega = \dot{m}\omega\left(r_3c_{3,u} - r_2c_{2,u}\right).$$

But the linear speed of rotation $u = \omega r$ and $\omega = (2\pi/60)N$, where $N$ is the rotational speed of the rotor in revolutions per minute (rpm). Therefore, the compressor power becomes

$$P_c = \dot{m}\left(u_3c_{3,u} - u_2c_{2,u}\right). \tag{7.18}$$

And the work done by the compressor on the fluid per unit mass of fluid is

$$W_c = \frac{P_c}{\dot{m}} = \left(u_3c_{3,u} - u_2c_{2,u}\right). \tag{7.19}$$

For a compressor, work done on the fluid is therefore positive, so $W_c > 0$. The power extracted by the turbine is given by

$$P_t = T_{q,t}\omega = \dot{m}\omega\left(r_5c_{5,u} - r_4c_{4,u}\right).$$

As was done for the compressor, $r\omega = u$ is substituted into the power equation to yield

$$P_t = \dot{m}\left(u_4c_{4,u} - u_5c_{5,u}\right). \tag{7.20}$$

It is important to note here that the order of the entrance and exit conditions has been reversed. This is to keep the power a positive quantity for convenience by avoiding carrying the negative sign throughout. Work done by the turbine on the fluid would be negative because our convention is that extracting work from the fluid is negative. It should not cause any confusion because when one speaks of compressor work or turbine work it is implicit that the compressor does work on the fluid and the turbine extracts work from the fluid. The turbine work would then be written as

$$W_t = \left(u_4c_{4,u} - u_5c_{5,u}\right). \tag{7.21}$$

### 7.2.1 Velocity components and work in turbomachines

The absolute velocity components may be related to those involving the relative wind and the linear speed of rotation by considering the velocity triangle, as shown in Figure 7.7. The vector relation between the velocity components is

$$\vec{u} + \vec{w} = \vec{c}.$$

From trigonometry, we find from the diagram that the following relations apply:

$$\begin{aligned} c^2 &= c_u^2 + c_a^2 \\ w^2 &= (u - c_u)^2 + c_a^2 \end{aligned} \quad (7.22)$$

Combining these equations leads to the result that

$$c^2 - w^2 = -u^2 + 2uc_u. \quad (7.23)$$

This may be recast as follows:

$$uc_u = \frac{1}{2}(c^2 + u^2 - w^2). \quad (7.24)$$

Then, considering the case of the compressor, we may substitute this result into Equation (7.19) for compressor work to obtain

$$W_c = \frac{1}{2}\left[\left(c_3^2 - c_2^2\right) + \left(u_3^2 - u_2^2\right) + \left(w_2^2 - w_3^2\right)\right]. \quad (7.25)$$

This equation is often referred to as the general ideal total head equation. Terms in the square brackets of Equation (7.25) may be described as follows:

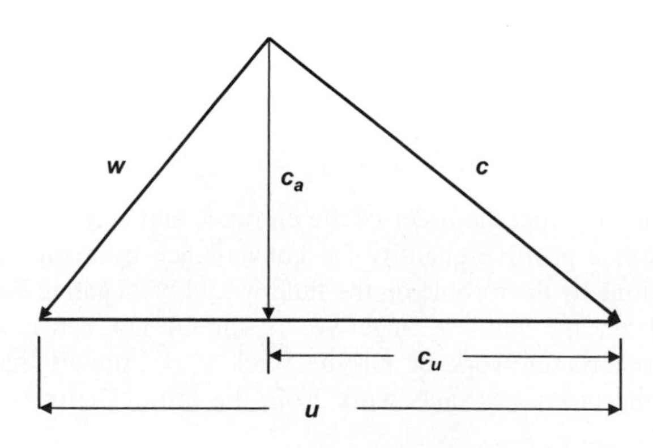


**FIGURE 7.7**

Velocity triangle illustrating the relationship between velocity components.

- The first term, $(c_3^2 - c_2^2)/2$, is called the *external effect* and represents the increase in absolute velocity due to the rotor's work on the fluid, which may be converted to a pressure increase by some external means, such as a diffuser.
- The second term, $(u_3^2 - u_2^2)/2$, represents the *centrifugal effect* in which a fluid particle may enter the rotor at a radius less than that at which it leaves and thereby increase in tangential speed from $u_2$ to $u_3$ as it proceeds through the rotor. Even for the simple case of a fluid rotating as a solid body, the Euler equation for the radial direction is

$$\rho\omega^2 r = \frac{\partial p}{\partial r}. \tag{7.26}$$

  In such a case, integration yields

$$p_3 - p_2 = \frac{1}{2}\rho\omega^2\left(r_3^2 - r_2^2\right) = \frac{1}{2}\rho\left(u_3^2 - u_2^2\right).$$

  Thus the centrifugal effect arises from the generation of a radial pressure force necessary to balance the product of fluid mass and its angular acceleration.
- The third term, $(w_2^2 - w_3^2)/2$, is called the *internal effect* and represents the increase in pressure that may be achieved by diffusion of velocity relative to the rotating passage through which the fluid travels. Because this diffusion occurs inside the rotating passage it is called an internal effect. Note that even if the internal diffusion effect is not exploited, and $w_3 > w_2$, the rotating machine can still be a compressor because the other terms may be positive. The case of the turbine is essentially the reverse of the aforementioned discussion for the compressor.

If we now consider the case of a turbine, we find that substituting Equation (7.24) into Equation (7.21) for turbine work yields

$$W_t = \frac{1}{2}\left[\left(c_4^2 - c_5^2\right) + \left(u_4^2 - u_5^2\right) + \left(w_5^2 - w_4^2\right)\right]. \tag{7.27}$$

Terms in the square brackets of Equation (7.27) may be described as follows:

- The first term, $(c_4^2 - c_5^2)/2$, is still an *external effect* and may be considered the decrease in kinetic energy due to the rotor's extraction of work from the fluid. In the turbine, the remaining kinetic energy is channeled into the nozzle, a component of the engine external to the rotor.
- The second term, $(u_4^2 - u_5^2)/2$, again represents a *centrifugal effect* in which a fluid particle enters the rotor at a radius greater than that at which it leaves and thereby loses tangential speed from $u_4$ to $u_5$ as it proceeds through the rotor. Using Equation (7.26) and integrating yields

$$p_4 - p_5 = \frac{1}{2}\rho\omega^2\left(r_4^2 - r_5^2\right) = \frac{1}{2}\rho\left(u_4^2 - u_5^2\right).$$

  Thus the centrifugal effect may be exploited to provide an expansion process by extracting kinetic energy from the rotating fluid.
- The third term, $(w_5^2 - w_4^2)/2$, is an *internal effect* and represents the decrease in pressure that may be achieved by acceleration of the velocity relative to the rotating passage through which the fluid travels. Because this acceleration occurs inside the rotating passage it is called an internal effect. The internal passages of the rotor are often called nozzles because of this accelerating effect.

## 7.3 THE CENTRIFUGAL COMPRESSOR

As its name implies, the centrifugal compressor exploits the centrifugal effect in the generalized ideal total head [Equation (7.25)]. This is accomplished by directing the flow outward in a generally radial direction, as shown in Figure 7.8. The general velocity components $c$, $w$, and $u$ are shown, as well as the radial and tangential components of the inertial velocity $c$ at station 3; a similar decomposition of the inertial velocity at station 2 is obvious. The blade width at a general radial distance is denoted by $b$. Radial component $c_r$ is important because it is the component normal to the flow area and therefore is proportional to the mass flow passing through any station according to the relation

$$\dot{m} = \rho(2\pi rb)c_r. \tag{7.28}$$

In Section 7.2 it was shown, Equation (7.17), that torque on the fluid is given by

$$T_q = \dot{m}(r_3c_{3u} - r_{2c_{2u}}).$$

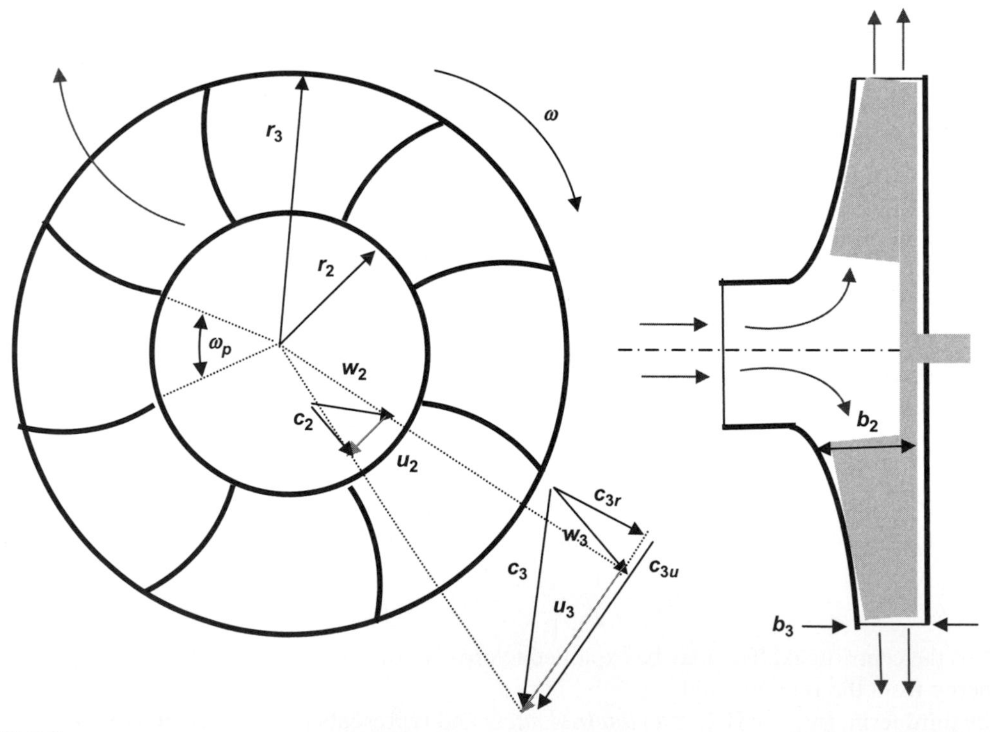


**FIGURE 7.8**

A centrifugal compressor showing velocity components and the general fluid motion within it. The radial and tangential components for the absolute velocity $c$ are only shown at station 3 to maintain clarity; also illustrated is the typical shroud or casing enclosing the spinning blades.

Similarly, the power required to drive the compressor was shown in Equation (7.18) to be

$$P = \omega T_q = \dot{m}\omega(r_3 c_{3u} - r_2 c_{2u}) = \dot{m}(u_3 c_{3u} - u_2 c_{2u}).$$

The energy balance across the compressor may be written as

$$\dot{m}\left(h_2 + \frac{1}{2}c_2^2\right) + \dot{m}W_c = \dot{m}\left(h_3 + \frac{1}{2}c_3^2\right). \tag{7.29}$$

Recall that $W_c$ is the work done on the fluid per unit mass of fluid. A specific case can be made for an incompressible fluid by introducing the thermodynamic relation $h = e + p/\rho$ so that Equation (7.29) becomes

$$e_2 + \frac{p_2}{\rho_2} + \frac{1}{2}c_2^2 + W_c = e_3 + \frac{p_3}{\rho_3} + \frac{1}{2}c_3^2. \tag{7.30}$$

Because the internal energy of an incompressible fluid is unchanged due to the pressure increase and if no work is done on the fluid, Equation (7.30) becomes

$$p_{t3} - p_{t2} = (p_3 - p_2) + \frac{1}{2}\rho\left(c_3^2 - c_2^2\right). \tag{7.31}$$

This, of course, leaves us with Bernoulli's equation, which states that the stagnation pressure of an incompressible frictionless fluid is constant in the absence of any work addition. Then Equation (7.30) shows that the change in the stagnation pressure for an incompressible frictionless fluid is equal to the work done per unit volume:

$$p_{t3} - p_{t2} = \rho W_c. \tag{7.32}$$

It is common for pressure to be measured in terms of the height of a column of fluid, for example, $\Delta p = \rho g z$, so the work per unit weight of fluid is often expressed in terms of the height, or head, of a column of the working fluid, or $W_c/g = z$. We have shown already in Section 7.2 that the work done per unit mass of fluid, measured as a kinetic energy, may be written as

$$W_c = \frac{1}{2}\left[\left(c_3^2 - c_2^2\right) + \left(u_3^2 - u_2^2\right) + \left(w_2^2 - w_3^2\right)\right]. \tag{7.33}$$

Substituting Equation (7.33) into Equation (7.32) and comparing it to Equation (7.31) show that the static pressure rise in incompressible frictionless flow is

$$p_3 - p_2 = \frac{1}{2}\rho\left(u_3^2 - u_2^2\right) + \frac{1}{2}\rho\left(w_2^2 - w_3^2\right). \tag{7.34}$$

It is useful to consider the mass flow once again, noting that

$$\dot{m} = \rho_2 c_{2r}(2\pi r_2 b_2) = \rho_3 c_{3r}(2\pi r_3 b_3). \tag{7.35}$$

Then the ratio of the radial component of inertial velocity at the exit to that at the entrance is given, for incompressible flow, by

$$\frac{c_{3r}}{c_{2r}} = \frac{\rho_2}{\rho_3}\frac{rb_2}{r_3 b_3} = \frac{r_2 b_2}{r_3 b_3}, \tag{7.36}$$

where $b_2$ and $b_3$ denote the effective height of the blade passage at the entrance and exit stations, respectively, as illustrated in Figure 7.8. Thus a reduction in blade passage height may offset the increase in radius between entrance and exit so that the radial components of velocity may be relatively close in magnitude.

### 7.3.1 Axial entry centrifugal compressor

A typical, yet relatively simple, centrifugal compressor configuration is shown in Figure 7.8. The compressor has purely axial flow at the inlet, that is, the absolute velocity $c$ enters axially with the following conditions:

$$\begin{aligned} c_2 &= c_{2a} \\ c_{2u} &= c_{2r} = 0 \end{aligned} . \tag{7.37}$$

When there is a tangential component of the entering velocity, the flow is said to have prewhirl, signifying that guide vanes imparted the swirling motion prior to entry. Because axial entry centrifugal compressors are the rule for aircraft turbine engines, we will focus on such machines. Under these assumptions, the work per unit mass and therefore the total pressure rise are given by

$$W_c = \frac{p_{t,3} - p_{t,2}}{\rho} = u_3 c_{3u} - u_2 c_{2u} = u_3 c_{3u}. \tag{7.38}$$

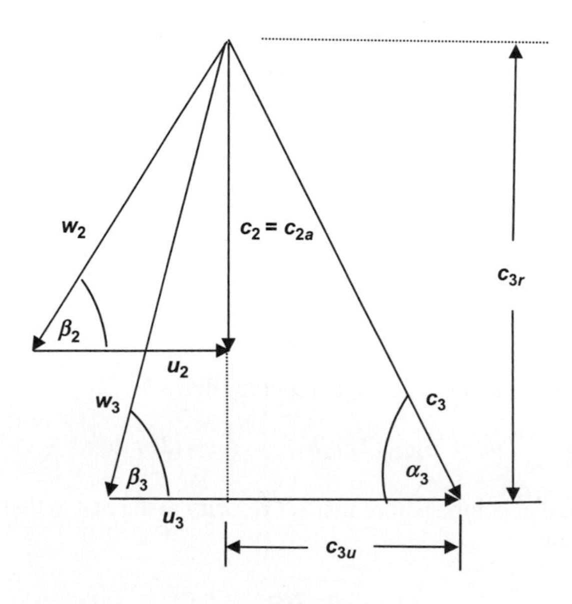


FIGURE 7.9

Combined velocity triangles for an axial entry centrifugal compressor.

This result may be written in the nondimensional form of a turning, or deflection, coefficient $\tau$ as follows:

$$\tau = \frac{p_{t,3} - p_{t,2}}{\rho u_3^2} = \frac{c_{3u}}{u_3}.$$

Clearly the pressure rise is proportional to the degree to which the flow is turned, that is, how much the exit absolute velocity is turned into the direction of rotation. This component can be seen in Figure 7.8 where the effect of prewhirl, which is being neglected in the current application, is also illustrated. It is useful at this point to examine the velocity triangles for the case at hand, which are shown in Figure 7.9.

The deflection coefficient can also be put in terms of the trailing edge angles of the blades using the velocity diagram in Figure 7.9 as follows:

$$\tau = \frac{c_{3u}}{u_3} = \frac{\tan \beta_3}{\tan \beta_3 + \tan \alpha_3}. \tag{7.39}$$

From Figure 7.9 we also see that

$$\tau = \frac{u_3 - w_{3u}}{u_3} = 1 - \frac{w_{3u}}{u_3}.$$

For a given mass flow through a particular compressor the values of $\tau$ are dependent on how the trailing portions of the rotor are shaped, which can be seen in Figure 7.10. We may also define the degree of reaction, that is, the fraction of pressure rise that takes place in the rotor compared to the total pressure rise. The degree of reaction may be written as

$$r = \frac{p_3 - p_2}{p_{t,3} - p_{t,2}} = \frac{(u_3^2 - u_2^2) + (w_2^2 - w_3^2)}{(u_3^2 - u_2^2) + (w_2^2 - w_3^2) + (c_3^2 - c_2^2)}.$$

From Figure 7.9, for the case of radial entry $c_2^2 = w_2^2 + u_2^2$ and using Equation (7.38), the degree of reaction becomes

$$r = \frac{c_2^2 + u_3^2 - w_3^2}{2u_3c_{3u}}.$$

Then using Equation (7.36) and the discussion accompanying it, we can assume that $c_{2r} = c_{3r}$ and note that

$$r = \frac{u_3^2 - w_3^2 + c_{3r}^2}{2u_3c_{3u}} = 1 - \frac{c_{3u}}{2u_3} = 1 - \frac{1}{2}\tau. \tag{7.40}$$

It is clear from Equations (7.39) and (7.41) that the degree of reaction, like $\tau$, is dependent on the details of the blade shape near the trailing edge. This information can be put into perspective by examining Figure 7.10 where three rotors of the same size and turning in the same direction at the same rotational speed are depicted. The velocity triangles at the exit are shown for each, and the blade angle $\beta$ defines the geometry of the blade-trailing edge. The backward curved blades have a high degree of reaction but a low pressure rise and a blade angle less than 90°. Forward curved blades are the opposite: a low degree of reaction but a high pressure rise and a blade angle greater than 90°. Radial-tipped blades fall between the two with a degree of reaction of one-half and a moderate

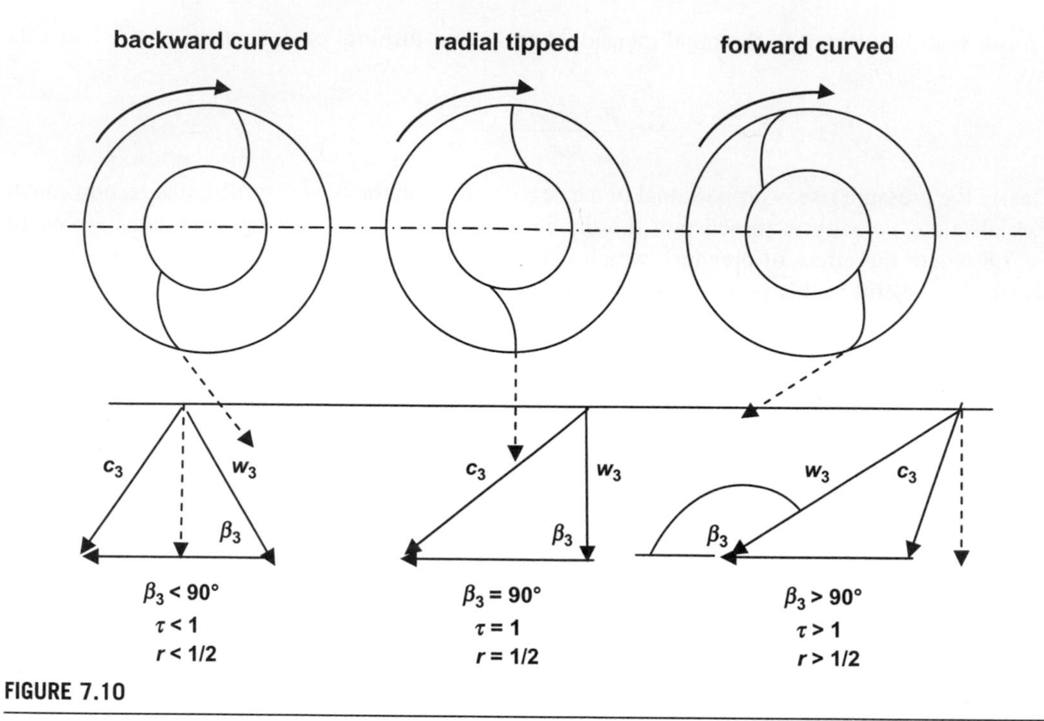


FIGURE 7.10

Different blade configurations for a centrifugal compressor showing blade angles $\beta$, deflection coefficients $\tau$, and degrees of reaction $r$.

pressure rise with a blade angle of exactly 90°. Radial-tipped blades are encountered more commonly, primarily due to manufacturing considerations.

## 7.3.2 Example: Centrifugal compressor

Consider a centrifugal compressor running on a test stand under standard sea level conditions. The tip speed $u_3 = 460$ m/s at which the efficiency of adiabatic compression $\eta_{ad,c} = 0.85$. If the compressor has no prewhirl and $c_{3u} = 0.95\ u_3$, determine the following: (a) pressure ratio, (a) work required per kilogram of air, (b) pressure ratio, and (c) power required for a mass flow of 29 kg/s.

**(a)** No prewhirl means that $c_{2u} = 0$, and therefore the work done per unit mass is

$$W_c = u_3 c_{3u} - u_2 c_{2u} = u_3 c_{3u} = u_3(0.95 u_3) = 2.01 \times 10^5 m^2/s^2 = 201 kJ/kg.$$

**(b)** The pressure ratio may be found from Equation (7.9) as follows:

$$\frac{p_{t,3}}{p_{t,2}} = \left[1 + \frac{\eta_{ad,c} W_c}{c_{p,2} T_{t,2}}\right]^{\frac{\gamma_2}{\gamma_2 - 1}}.$$

Under the conditions stated, we may assume that $\gamma = 1.4$, and because the compressor is on a test stand, $V_0 = 0$. Then $T_{t,2} = T_{t,0} = T_0$, and for a standard sea level day $T_0 = 288$K. Furthermore, for air at these conditions, $c_{p,2} = 1$ kJ/kg-K. Then the pressure ratio is

$$\frac{p_{t,3}}{p_{t,2}} = \left[1 + \frac{0.85(201kJ/kg)}{(1kJ/kg - K)(288K)}\right]^{3.5} = 5.10.$$

**(c)** The power required may be obtained from Equation (7.19):

$$P = \dot{m}W_c = (29kg/s)201kJ/kg = 5,830kW.$$

If the compressor had negative prewhirl, the work required would be greater. However, for a fixed amount of work, the introduction of negative prewhirl permits a reduction in $u_3$. For a fixed rotational speed, this reduction in linear speed of rotation at the exit means that a smaller diameter impeller could be used. This is attractive for aircraft applications because it reduces frontal area and thus aircraft drag. Negative prewhirl may be introduced merely by installing appropriately shaped guide vanes at the entrance to the compressor.

### 7.3.3 Pressure coefficient

A nondimensional stagnation pressure coefficient $\Psi$ may be introduced as follows:

$$\Psi = \frac{p_{t,3} - p_{t,2}}{\frac{1}{2}\rho u_3^2} = \frac{\rho u_3^2 \tau}{\frac{1}{2}\rho u_3^2} = 2\tau.$$

Static pressure may be treated in the same way:

$$\Psi_{static} = \frac{p_3 - p_2}{\frac{1}{2}\rho u_3^2} = r\Psi = 2\tau\left(1 - \frac{1}{2}\tau\right).$$

All this information is combined in Figure 7.11 where the pressure coefficients and the degree of reaction are shown as functions of the deflection parameter $\tau$.

Although attention has focused on the pressure rise produced by the compressor, the mass flow processed by the compressor is of interest. In particular, it is important to know the variation of the pressure rise produced as a function of the mass flow passed by the compressor. From Equation (7.38), total pressure rise depends on the tangential component of absolute exit velocity:

$$c_{3u} = u_3 - w_{3u} = u_3 - \frac{c_{3r}}{\tan\beta_3}.$$

Then Equation (7.38) may be written as

$$p_{t,3} - p_{t,2} = \rho u_3^2 - \frac{\rho u_3 c_{3r}}{\tan\beta_3}.$$

The mass flow is given by

$$\dot{m} = \rho c_{3r}\pi d_3 b_3.$$

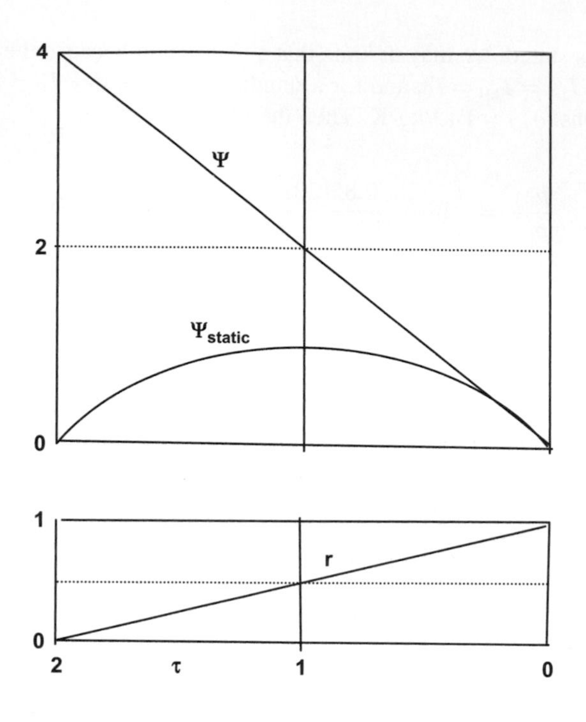


**FIGURE 7.11**

Pressure coefficients and degree of reaction are shown as a function of the deflection parameter $\tau$.

Then the total pressure rise is

$$p_{t,3} - p_{t,2} = \rho u_3^2 - \frac{u_3}{\tan\beta_3}\frac{1}{\pi d_3 b_3}\dot{m}.$$

Thus the pressure rise is linearly proportional to the mass flow through the compressor. Furthermore, the magnitude of $b_3$ will determine whether the pressure rise increases or decreases with mass flow. In addition, the power required is

$$P = \frac{p_{t,3} - p_{t,2}}{\rho}\dot{m} = u_3^2\dot{m} + \frac{u_3}{\tan\beta_3}\frac{1}{\pi d_3 b_3 \rho}\dot{m}^2. \tag{7.41}$$

Note that the power varies as the square of the mass flow. Pressure rise and power required trends are depicted in Figure 7.12.

After considering the pressure rise and power characteristics it is useful to continue the analysis of a particular blade configuration. Therefore, in addition to the assumption of axial entry, we further assume that the compressor rotor has a radial exit, that is, blade tips are aligned radially so that the relative velocity leaves the blade in the radial direction (as shown in Figure 7.13). In this case,

$$w_3 = w_{3r}$$

$$c_{3u} = u_3$$

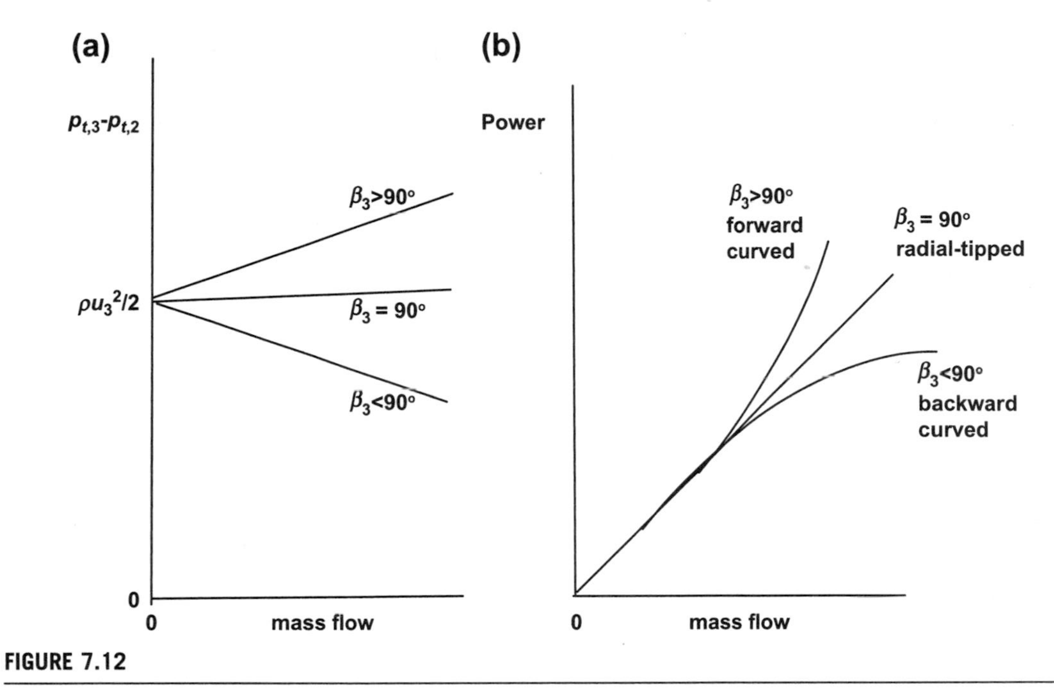


**FIGURE 7.12**

The variation of (a) total pressure rise and (b) power required as a function of mass flow for different blade configurations.

These two design constraints cause the power required, as given by Equation (7.41), to be simplified, becoming

$$P = \dot{m}u_3^2. \tag{7.42}$$

In the same fashion, work done on the fluid, per unit mass of fluid, becomes

$$W_c = \frac{P}{\dot{m}} = u_3^2.$$

Thus the power required is easily seen to be directly proportional to the mass flow processed and the square of the linear speed of rotation, $u_3^2 = (\omega r_3)^2$. With this same reasoning, the work done per unit mass of fluid, which is indicative of the pressure rise produced by the machine, depends on the squares of the rotational speed and the tip radius.

### 7.3.4 Effects due to number and shape of blades

The purpose of the blades is to guide the flow into the required turn. In the idealized case of an infinite number of zero-thickness blades processing a frictionless fluid, the exit relative velocity $w_3$ would follow the blade exactly and the velocity field would be fixed by the blade geometry alone. However, complicating factors serve to make the flow depart from this idealized situation. Most obvious is the fact that there are actually a finite number of blades with finite thickness processing a fluid that generates friction at the blade surface. Even in the inviscid case, the streamline

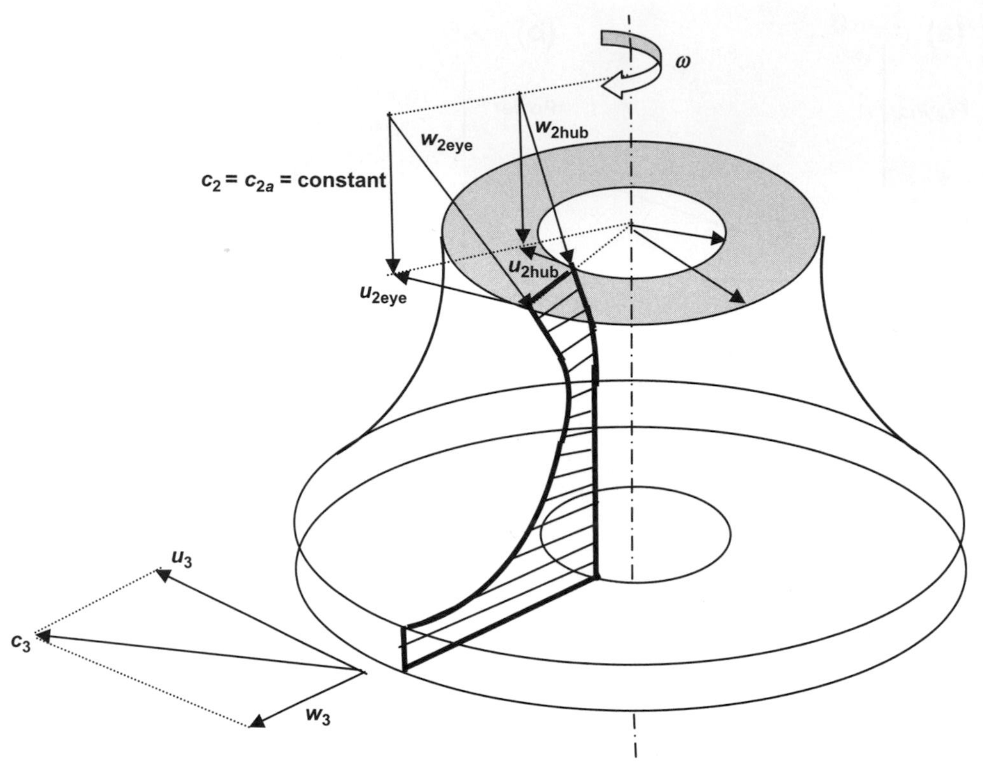


**FIGURE 7.13**

A typical centrifugal compressor showing an axial inlet flow and a radial exit flow. The associated twisting of the blade between the hub and the eye of the compressor at station 2 is illustrated, as is the radial exit flow with $c_{3u} = u_3$.

from the trailing edge of the blade leaves with a higher pressure on one side than the other, with the difference causing a tendency for the flow to bend toward the low-pressure region, as shown in Figure 7.14.

Note that the effect of the movement of the flow away from the direction of rotation serves to reduce $c_{3u}$ and thereby reduce the power imparted to the flow and likewise the pressure rise achievable. This effect is exacerbated by the generation of secondary flows in the blade passages as a result of rotation. In the case of an inviscid fluid there will be no means of generating solid body rotation of the fluid in the passages, and a secondary motion instead will be generated in the direction opposite to that of rotation of the wheel. An example is the simple case of a fluid-filled, two-dimensional circular cylinder divided internally by walls into identical compartments and rotating about its center, as shown in Figure 7.15. The reduction in pressure rise due to the finite number of blades is typically around 85%.

The effects of the pressure difference-induced deviation near the blade tips and the secondary flow produced by the rotation give rise to relative velocities opposing the direction of rotation. As a result, the whirl velocity $c_{3u}$ is decreased and the pressure rise achievable is reduced from the ideal value. The

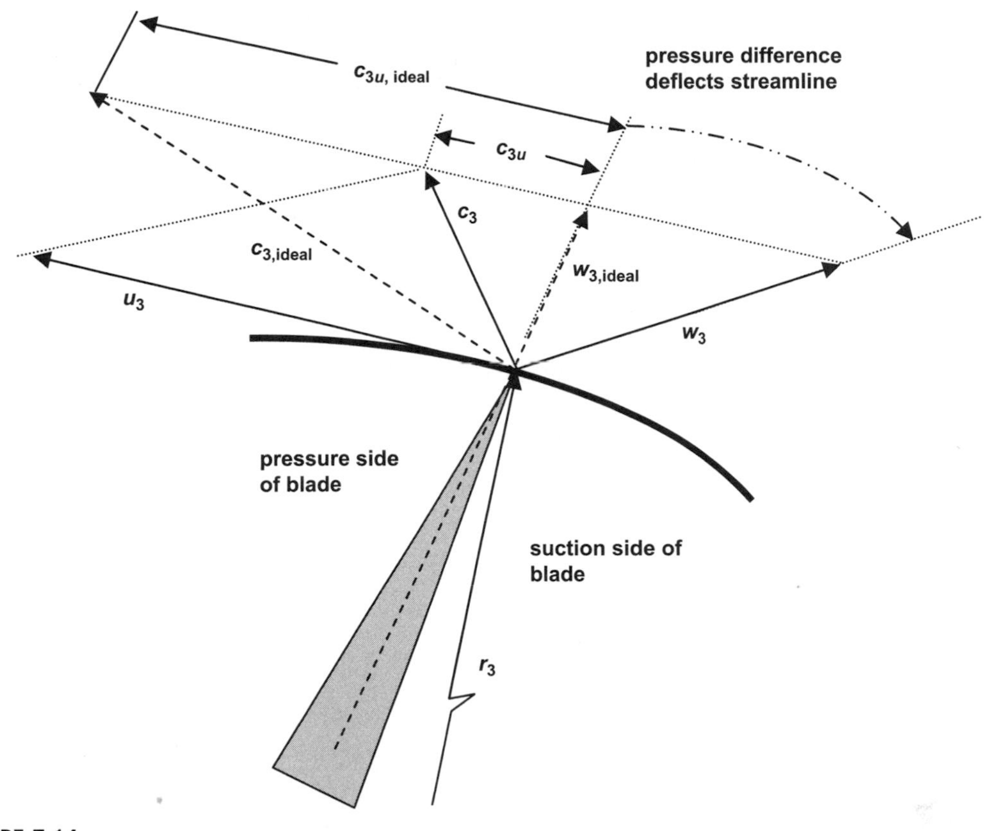


**FIGURE 7.14**

Pressure difference across blade gives rise to deflection of departing streamline away from the ideal direction.

influence of general blade tip configurations for an axial entry compressor may be discussed with the aid of Figure 7.16.

The power required for this more general case becomes

$$P = mc_{3u}u_3 = mu_3^2 \frac{c_{3u}}{u_3}. \tag{7.43}$$

The ratio of the whirl velocity $c_{3u}$ to the linear speed of rotation $u_3$ is called the slip or deflection coefficient:

$$\tau = \frac{c_{3u}}{u_3}. \tag{7.44}$$

The slip coefficient is related to the velocity triangle configuration by the following:

$$\tau = \frac{\tan \beta_3}{\tan \alpha_3 + \tan \beta_3}.$$

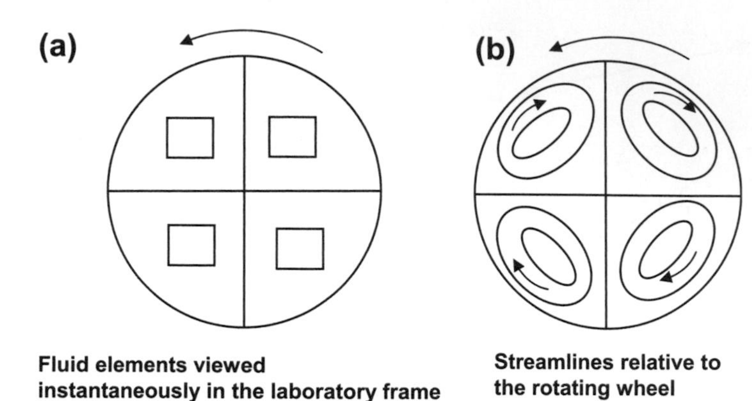


FIGURE 7.15

Irrotational flow in the laboratory frame shown in (a) results in nonrotating fluid elements, while in the rotating frame, fluid motion is shown by the relative streamlines (b).

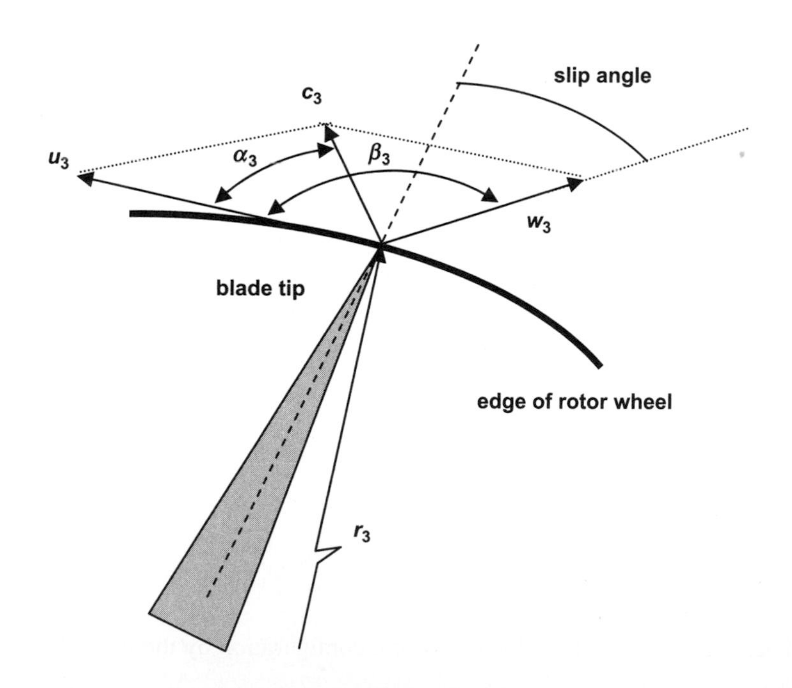


FIGURE 7.16

A general blade tip configuration showing velocity components and the slip angle caused by flow deviation from the blade tip direction.

The work done per unit mass of fluid is

$$W_c = \frac{P}{\dot{m}} = u_3^2 \frac{c_{3u}}{u_3} = u_3^2 \tau.$$

The effect of the slip coefficient on the flow field is best illustrated by Figure 7.17.

The comparison of the effect of the slip coefficient, which arises from the geometry of the blade tip, is based on a constant mass flow ($c_{3r}$ = constant) and a constant tip speed $u_3$. Note that as $\beta_3$ increases from values less than 90° the absolute velocity $c_3$ decreases, as does its tangential component $c_{3u}$. The blade shape proceeds from a forward-curved blade with $\tau > 1$ (shown in Figure 7.17) to a radial tip blade with $\tau = 1$ at $\beta_3 = 90°$ to a backward-curved blade with $\tau < 1$ at still higher values of $\beta_3$. Equations (7.47) and (7.48) showed that for the same linear speed of rotation and mass flow $P \sim \tau$ and $W_c \sim \tau$ so that a forward-curved blade requires high power but can supply high pressure, while a backward-curved blade requires less power but will provide lower pressure. The radial tip blade will perform somewhere between the other two. Manufacturing considerations often turn the table in favor of radial tip blades for centrifugal compressors.

The effects of practical operation on a radial-tipped compressor are shown in Figure 7.18.

The finite number of blades in the real machine reduces performance from that of the ideal case of an infinite number of blades providing perfect guidance of the flow. Very many blades will guide the flow well but will also increase the frictional loss because of the added surface area of the blades. There also will be some reduction in flow area because the blades do provide some physical obstruction to the flow, typically reducing flow area by about 5%. Very few blades relieve those problems but offer

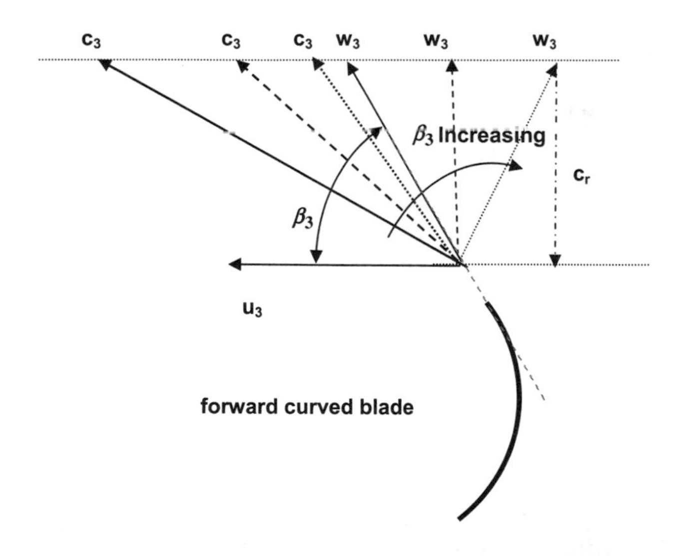


**FIGURE 7.17**

The change in slip coefficient in terms of the blade tip angle $\beta_3$. Arrows of the same shade denote the same $\beta_3$ case, with dashed arrows denoting the relative velocity $w_3$ and solid arrows denoting the absolute velocity $c_3$. One blade shape for $\beta_3 < 90°$ is shown, drawn in the shade appropriate to the velocity triangle to which it applies.

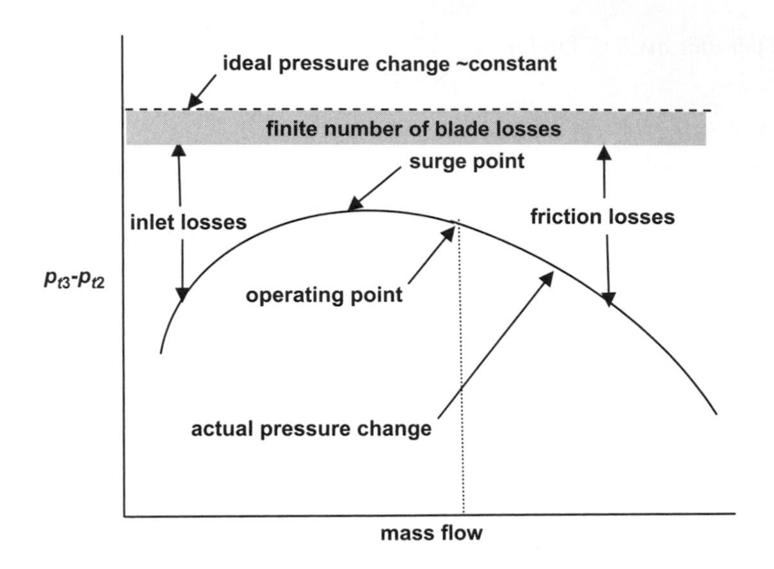


**FIGURE 7.18**

Pressure rise as a function of mass flow for a radial-tipped centrifugal compressor showing effects of practical operating conditions.

reduced guidance for the flow, and separation effects will be more likely. The number of blades is a design issue that involves structural, manufacturing, and cost considerations. A rough rule of thumb for the number of blades to use is given by

$$n_b \approx 8.5 \frac{\sin \beta_3}{1 - \frac{r_2}{r_3}}.$$

After taking into account the effect of a finite number of blades, one may consider frictional effects, which grow as the square of the mass flow (because mass flow is proportional to speed and frictional drag is proportional to the square of the speed), and reduce performance as the mass flow increases. Because the blade orientation is fixed but the flow angles vary with mass flow, there are losses in performance due to separation of the flow from the blade surfaces, as shown in Figure 7.19. At the design point, Figure 7.19b, the velocity relative to the blade at the entrance, $w_2$, aligns properly with the leading edge of the blade and flows smoothly around it. However, at high or low mass flows, Figures 7.19a and 7.19c, the relative velocity $w_2$ is at too low or too high an angle, respectively, to the blade-leading edge and separation bubbles appear, causing performance to drop. The point of maximum pressure rise is called the surge point and is analogous to stall of an airfoil.

### 7.3.5 Guide vanes, diffusers, and volutes

The flow leaving the rotor has a radial component of absolute velocity $c_{2r}$ that represents the velocity in the mass conservation equation

$$\dot{m} = \rho c_{3r} 2\pi r_3 b_3 = const.$$

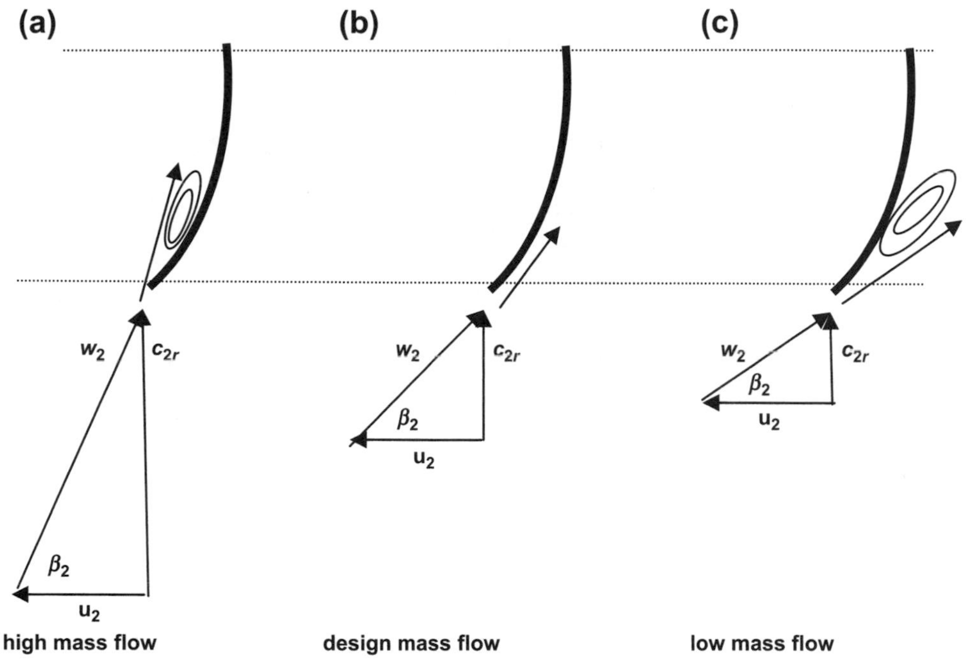


**FIGURE 7.19**

Entrance flow at various mass flows: (a) high, (b) design, and (c) low.

Once the flow leaves the rotor, its angular momentum must be conserved in the absence of any external torques, and in the frictionless case this means that

$$\dot{m} r c_u = const.$$

In the housing receiving the flow as it leaves the rotor, the angle $\alpha$ between the absolute velocity c and any circle $r = $ constant is given by

$$\tan \alpha = \frac{c_{3r}}{c_u}.$$

Then, because the mass flow is constant, $rc_u = $ constant and, if the width $b$ is constant, then $rc_r = $ constant and

$$\tan \alpha = \frac{c_{3r}}{c_{3u}} = \frac{c_r}{c_u} = const.$$

Then the flow in the housing is characterized by a constant angle $a$, as shown in Figure 7.20.

The equation of the streamline in the flow beyond the rotor may be obtained by considering the equation for $\alpha$ in polar coordinates, which may be written as

$$\tan \alpha = \frac{dr}{rd\theta}.$$

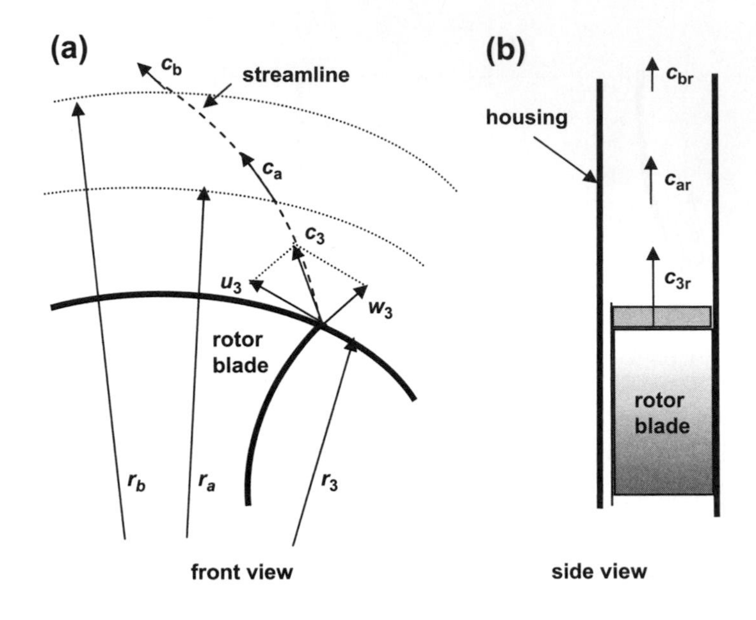


**FIGURE 7.20**

Streamline path of flow beyond rotor exit.

Because $\alpha$ is a constant, this may be integrated to yield

$$\ln\left(\frac{r}{r_3}\right) = \theta \tan \alpha.$$

This is the equation of the logarithmic spiral, and therefore the housing may be designed with this in mind. It is clear from Figure 7.20 that flow external to the rotor is naturally decelerating as it moves further from the rotor and therefore the pressure increases. This is the external effect that was described in Section 7.2. The flow in the housing may include guide blades, the passages between which act as diffusers. Use of the logarithmic spiral is confined to the leading edges of any diffuser blades because the turn it provides is too great and separation may occur. The portion beyond the entrance region of the blades may be designed as a straightened diffuser where the area change is such that a straight channel diffuser with the same area variation would have an included angle of less than 12°. For low-pressure rise compressors, such as blowers or fans, the blades may be done away with completely and just a spiral volute housing is used.

## 7.4 CENTRIFUGAL COMPRESSORS, RADIAL TURBINES, AND JET ENGINES

Although centrifugal compressors are used on jet engines, they are confined to relatively small size and thrust applications. Because they depend primarily on centrifugal effects to produce a pressure

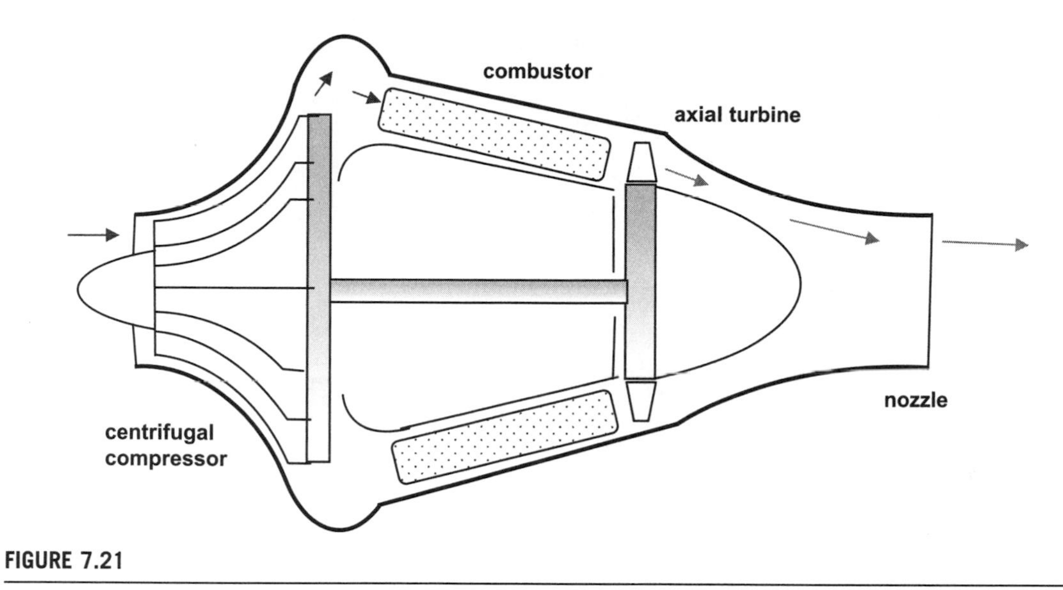


FIGURE 7.21

This schematic diagram shows a typical application of a centrifugal compressor in an aircraft engine. Note that though the engine employs a centrifugal compressor, an axial turbine is used.

rise, they are capable of delivering a high-pressure rise in just a single stage. Of course, the advantage of using the centrifugal effect carries a disadvantage in that it is difficult to use more than one stage because of the excessive flow turning involved and the attendant pressure losses. Centrifugal compressors also suffer from the inability to be adapted readily for use as turbofan engines. Still one more deficiency arising from using the centrifugal effect is the requirement for a comparatively large maximum diameter in centrifugal compressors, and the large frontal area leads to high drag penalties.

There is a turbine counterpart to the centrifugal compressor called the radial turbine. As in the case of axial flow machines, it is essentially a compressor with flow moving in the opposite direction with flow entering at the outer periphery of the rotor and exiting close to the axis of rotation. Such turbines are very popular in hydroelectric power-generating stations and have been developed to a high degree of efficiency over the years. However, the long curving passages of centrifugal turbomachines are at a structural disadvantage when operating with high-temperature exhaust gases. However, for small-scale turbines, as in the case of turbines used in turbochargers for internal combustion engines, the radial flow turbine is widely used. When centrifugal compressors are used on jet engines, the turbines used are almost invariably of the axial flow type. Such an application is shown in Figure 7.21.

## 7.5 AXIAL FLOW COMPRESSOR

Unlike centrifugal flow turbomachines, axial flow turbomachines make little, if any, use of the centrifugal effect and derive their performance from the internal and external effects shown in

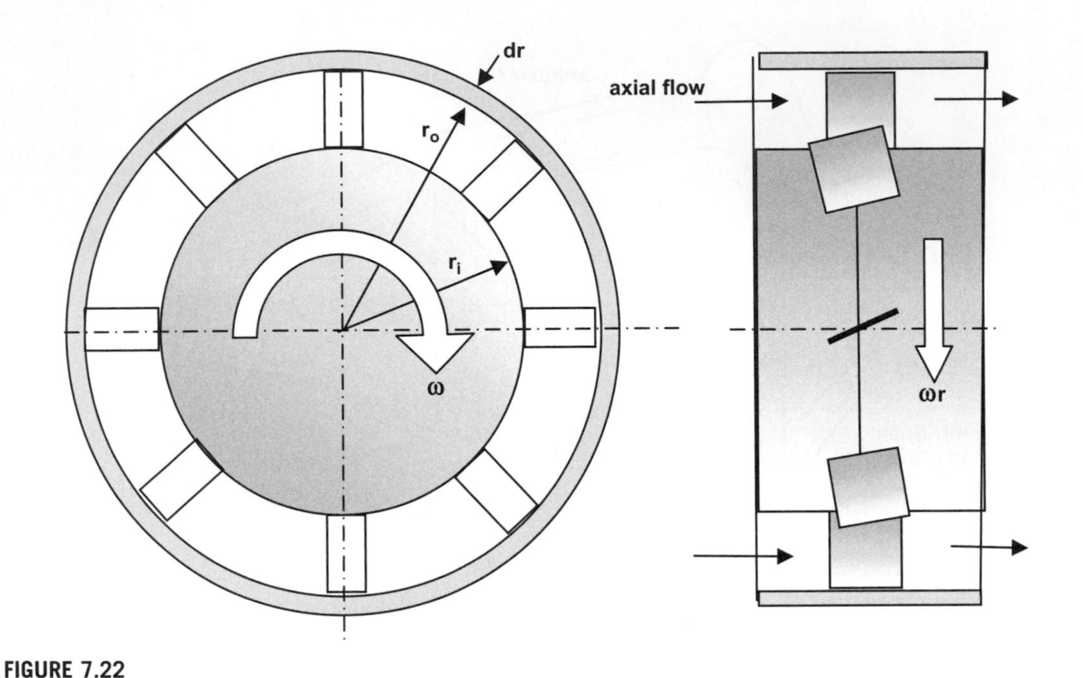


FIGURE 7.22

Schematic diagram of an axial flow turbomachine rotating at angular velocity $\omega$ showing the general annular region through which the flow passes.

Equation (7.25). In axial flow turbomachines, as the name implies, the flow is mainly parallel to the axis of rotation, and the flow is guided between inner and outer radii of the blade passage, as illustrated in Figure 7.22. As a result of the flow being guided between essentially parallel walls the changes in tangential velocity $u$ are minimized and often ignored altogether until the final design stages are reached.

## 7.5.1 Velocity diagrams

We may visualize the blade shapes as defining the flow passages in the rotor as well as the stator, with the stationary passages delivering flow to and accepting flow from the rotor. In Figure 7.23, the so-called velocity triangle introduced in Figure 7.7 is drawn on the turbomachine from the perspective of an observer looking down at it. When looking at the side view of the axial flow turbomachine in Figure 7.22, one would perceive the linear speed of rotation $u = \omega r$ vertically, as shown by the arrow there, whereas in Figure 7.23 the speed $u$ appears horizontal because of the viewpoint chosen.

The velocity diagrams entering and leaving the blade passage may be combined into one diagram, as shown in Figure 7.24. The airfoil sections are designed to align the flow so as to produce a particular streamline pattern. Separation is to be avoided so the angles of attack of the airfoils, that is, the angle between the airfoil chord and the appropriate relative wind, must be maintained at relatively small values. The curvature of the blades arises from the requirement to accept the flow

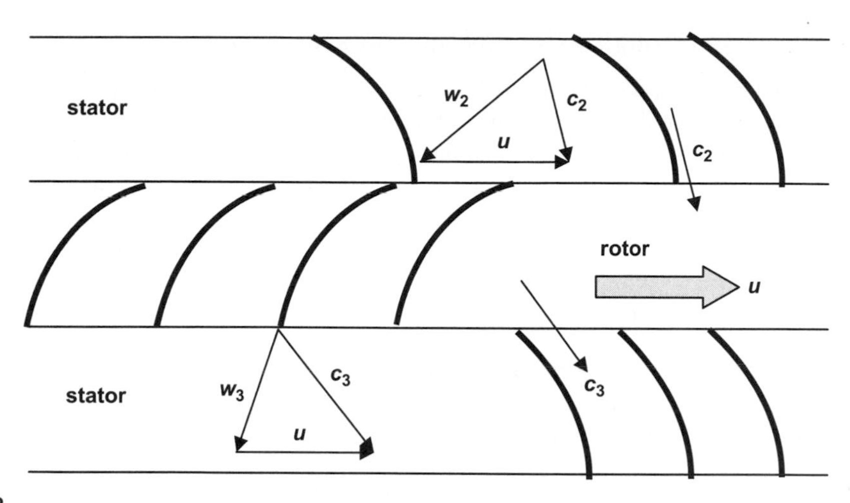


**FIGURE 7.23**

Blades shown as moving with linear speed of rotation $u$ and the associated velocity triangles entering and leaving a rotor.

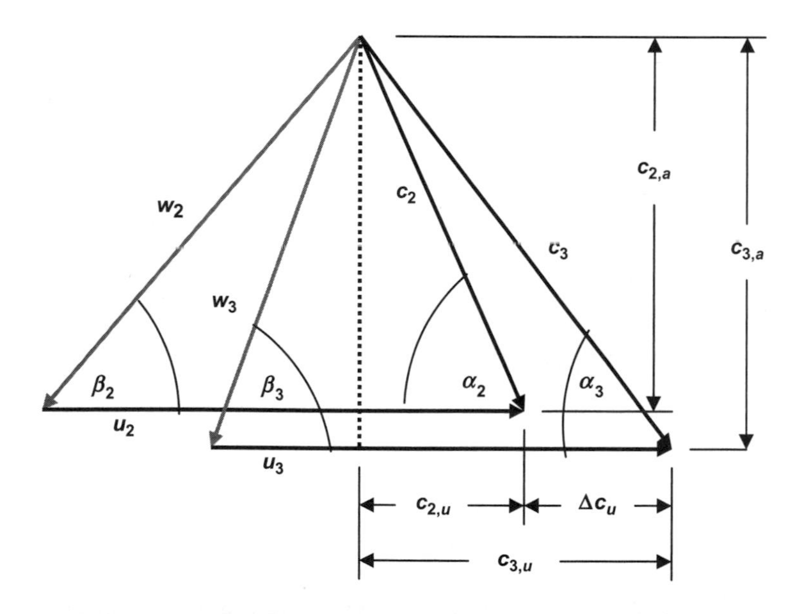


**FIGURE 7.24**

A combined velocity diagram for a compressor stage.

facing it with small angles of attack and yet to turn the flow appreciably, that is, increase the change in whirl. The curvature of the airfoil sections of the rotor and stator depends on the degree of reaction, or the extent to which the pressure rises in each segment of the stage. This aspect is discussed in Chapter 8.

### 7.5.2 Pressure rise through axial flow compressor stages

In axial flow turbomachines, the height of the blade is generally much smaller than the maximum radius of the machine so that the change in the linear speed of rotation $u = \omega r$ across the height of the blade is also small with respect to $u$ itself, and therefore $u$ is generally taken as a constant across the blade passage. We may consider rows of rotating and stationary blades, called rotors and stators, respectively, arranged in stages, as shown in Figure 7.25. The $i$th stage may be considered to occupy the space between station $2_i$ and station $2_{i+1}$ with the rotor passage lying between stations $2_i$ and $3_i$ and the stator passage between stations $3_i$ and $2_{i+1}$.

Across the $i$th rotor the pressure rise is given by

$$p_{3,i} - p_{2,i} = \frac{1}{2}\rho\left(w_{2,i}^2 - w_{3,i}^2\right). \tag{7.45}$$

Similarly, across the $i$th stator, the pressure rise is given by

$$p_{2,i+1} - p_{3,i} = \frac{1}{2}\rho\left(c_{3,i}^2 - c_{2,i+1}^2\right). \tag{7.46}$$

The sum of the two equations is

$$p_{2,i+1} - p_{2,i} = \frac{1}{2}\rho\left[\left(c_{3,i}^2 - c_{2,i+1}^2\right) + \left(w_{2i}^2 - w_{3i}^2\right)\right].$$

This is the static pressure rise across the $i$th stage. Assuming that the flow across the stator passage is frictionless, the stagnation pressure is constant through it and is given by

$$p_{t3,i} = p_{t2,i+1} = p_{2,i+1} + \frac{1}{2}\rho c_{2,i+1}^2. \tag{7.47}$$

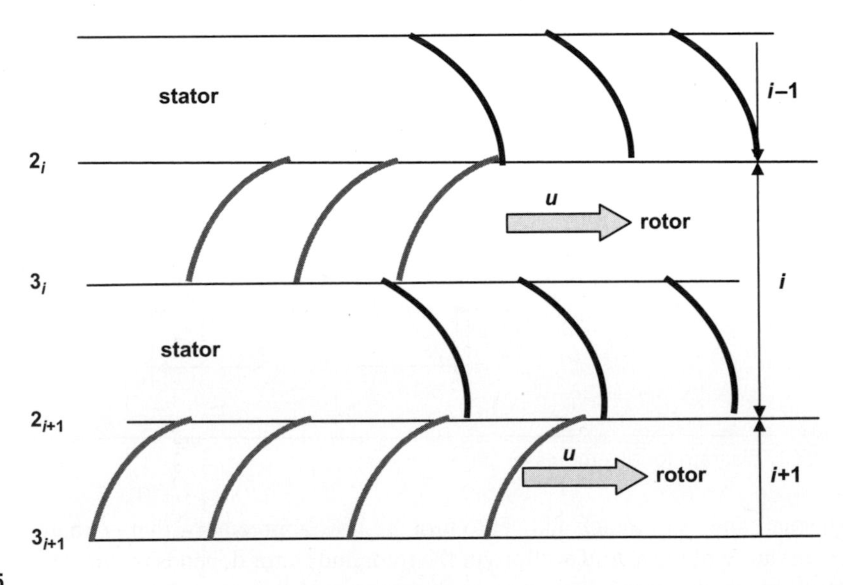


**FIGURE 7.25**

A representative stage in a compressor.

The total pressure rise occurs solely across the rotor where work is put into the flow, which may be expressed as follows:

$$p_{t3,i} - p_{t2,i} = \frac{1}{2}\rho\left[\left(c_{3,i}^2 - c_{2,i}^2\right) + \left(w_{2,i}^2 - w_{3,i}^2\right)\right]. \tag{7.48}$$

The static pressure rise may be written in terms of the stagnation pressure rise as

$$p_{2,i+1} - p_{2,i} = p_{t3,i} - p_{t2,i} - \frac{1}{2}\rho\left(c_{2,i+1}^2 - c_{2,i}^2\right). \tag{7.49}$$

Thus we see that if all stages were identical, then the static pressure rise is equal to the total pressure rise. Note that in the case of incompressible flow the total pressure rise is related to the work required per unit mass according to the following equation:

$$W_c = \frac{1}{\rho}(p_{t,3} - p_{t,2}) = \frac{p_{t2}}{\rho}\left(\frac{p_{t3}}{p_{t2}} - 1\right). \tag{7.50}$$

In most cases of high-performance turbomachinery, the incompressible flow approximation is not valid and the energy equation must be used, as at the start of this chapter. Recall that the stagnation pressure ratio across the stage may be determined from the energy equation as

$$\frac{p_{t3}}{p_{t2}} = \left(1 + \frac{\eta_{ad,c} W_c}{c_p T_{t2}}\right)^{\frac{\gamma}{\gamma-1}}. \tag{7.51}$$

One must be careful in determining the pressure rise across a stage because of the likelihood that a compressible flow relation may have to be employed.

However, we shall see that axial flow compressor stages are relatively lightly loaded so that some simplifications can be employed. For example, for relatively small stagnation pressure rises across the stage, the stagnation pressure ratio may be expanded as follows:

$$\left(\frac{p_{t3}}{p_{t2}}\right)^{\frac{\gamma-1}{\gamma}} = \left(1 + \frac{\Delta p_t}{p_{t2}}\right)^{\frac{\gamma-1}{\gamma}} = 1 + \frac{\gamma-1}{\gamma}\left(\frac{\Delta p_t}{p_{t2}}\right) + \frac{\gamma-1}{\gamma}\frac{1}{2}\left(\frac{\gamma-1}{\gamma} - 1\right)\left(\frac{\Delta p_t}{p_{t2}}\right)^2 + \ldots \tag{7.52}$$

If we neglect the nonlinear terms in Equation (7.52) because $\Delta p_t/p_{t,2} \ll 1$ and substitute the result into the general compressor work equation, Equation (7.9), we obtain

$$W_c = \frac{c_p T_{t2}}{\eta_c}\left[\left(\frac{p_{t3}}{p_{t2}}\right)^{\frac{\gamma-1}{\gamma}} - 1\right] \approx \frac{c_p T_{t2}}{\eta_c}\left[\frac{\gamma-1}{\gamma}\left(\frac{\Delta p_t}{p_{t2}}\right)\right]. \tag{7.53}$$

Rearranging this equation yields

$$W_c \approx \frac{c_p T_{t2}}{\eta_c}\frac{R}{c_p}\left(\frac{p_{t3}}{p_{t2}} - 1\right) = \frac{p_{t2}}{\eta_c \rho_{t2}}\left(\frac{p_{t3}}{p_{t2}} - 1\right). \tag{7.54}$$

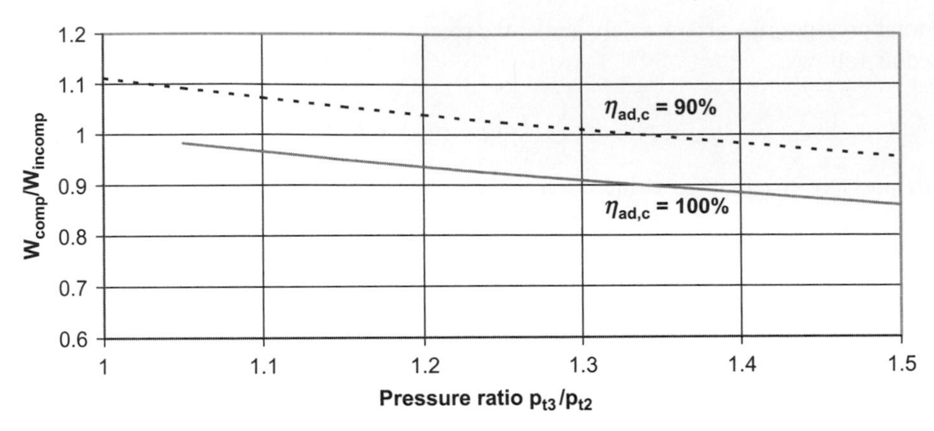


**FIGURE 7.26**

The ratio of work required for a compressible flow to that for an incompressible flow for the same stagnation pressure ratio. Here $\eta = 1.4$.

This equation is analogous to the compressor work required for an incompressible fluid given in Equation (7.50) with $\rho$ replaced by an average density $\eta_c \rho_{t2}$. Thus the incompressible flow approach yields useful results if the density is considered to be an average density fairly near the stagnation density at the stage entrance.

The ratio of work required for a compressible flow, Equation (7.9), to that for an incompressible flow, Equation (7.50), for the same stagnation pressure ratio is shown in Figure 7.26 where the gas is assumed to have $\gamma = 1.4$ and the incompressible fluid density is equal to the stagnation density of the entering gas. This illustrates the idea that for small pressure ratios across a stage the incompressible flow approximation will provide a reasonably accurate representation of the flow if the state variables are treated appropriately. For example, it is apparent from the 90% compressor efficiency line shown that the incompressible fluid density must be taken as the product of the compressor efficiency and the stagnation density of the gas in order for the approximation to be reasonable.

### 7.5.3 Types of compressor stages

The degree of reaction is defined as the ratio of the static pressure rise in the stage to the stagnation pressure rise in the stage or

$$\%r = \frac{\Delta p_{rotor}}{\Delta p_{rotor} + \Delta p_{stator}} \times 100. \tag{7.55}$$

This may be expressed in terms of enthalpy change or velocity components as follows:

$$\%r = \frac{\Delta h_{rotor}}{\Delta h_{rotor} + \Delta h_{stator}} = \frac{\left(w_2^2 - w_3^2\right)}{\left(w_2^2 - w_3^2\right) + \left(c_3^2 - c_2^2\right)}. \tag{7.56}$$

First, we may consider a 100% reaction stage where the entire pressure rise takes place in the rotor and none in the stator. Here the absolute velocity is unchanged, $c_2 = c_3$, while the relative velocity does change, $w_3 < w_2$, giving a pressure rise due entirely to the internal effect. The velocity diagram for

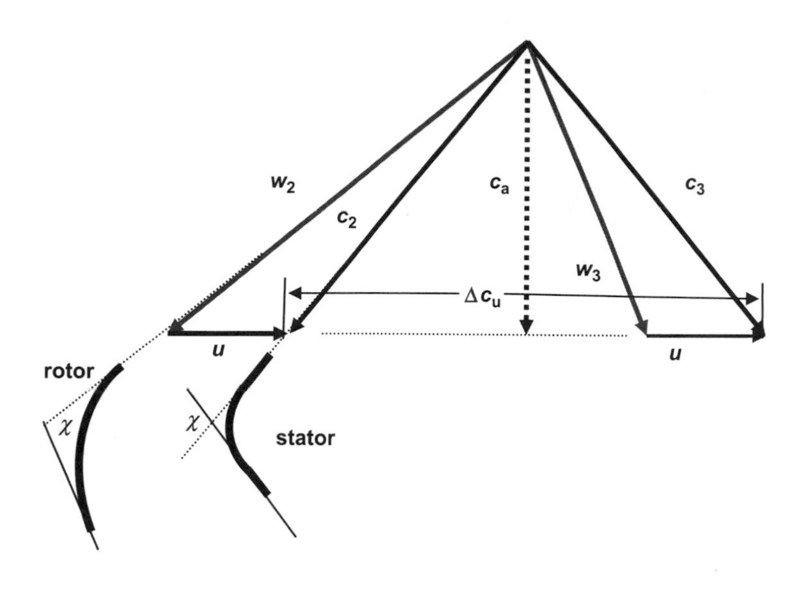


**FIGURE 7.27**

Velocity diagram for 100% reaction stage: $c_2 = c_3$. Camber lines of rotor and stator blades are outlined.

such a stage is shown in Figure 7.27. The general shape of the camber lines of the rotor and stator blades are also shown in Figure 7.27. Note that now the stator blade is symmetrical about the midpoint of its chord. The flow is turned symmetrically and there is no change in the enthalpy or temperature in the stator as a result. The rotor blade is less curved with the angle $\chi$ between the tangents to the leading and trailing edges larger than that for the stator.

Next we may consider a 50% reaction stage where half the pressure rise takes place in the rotor and half in the stator. Here the entering absolute velocity $c_2 = w_3$ and the exiting absolute velocity $c_3 = w_2$ providing for an equal share of the pressure rise due to the internal effect and to the external effect. The velocity diagram for this type of stage, often called a symmetric stage because of the shape of the velocity diagram, is shown in Figure 7.28. The general shape of the camber lines of the rotor and stator blades is also shown in Figure 7.28. In addition, the rotor and stator blades have the same cross-sectional shape. That is, if the stator blade is rotated 180° about the axis A-A in Figure 7.28 it will be identical in shape to the rotor blade. Note that the rotor blade has the same included angle $\chi$ between the tangents to the leading and trailing edges larger as the stator. This stage is used most commonly in aircraft turbine engines for reasons that will be discussed subsequently.

We may also consider a 0% reaction stage where none of the pressure rise takes place in the rotor and the entire pressure rise occurs in the stator. Here the relative velocity is unchanged, $w_2 = w_3$ while the absolute velocity does change, giving a pressure rise due entirely to the external effect. The velocity diagram for such a stage, also called an impulse stage, is shown in Figure 7.29. Now the rotor blade is symmetrical about the midpoint of its chord. The flow is turned symmetrically and there is no change in the enthalpy or temperature in the rotor as a result. The stator blade is less curved with the angle $\chi$ between the tangents to the leading and trailing edges larger than that for the rotor.

A final case often encountered in practice is that of an axial exit stage, which has a degree of reaction greater than 100%. Such a stage is often the last stage in an axial flow compressor and is intended to

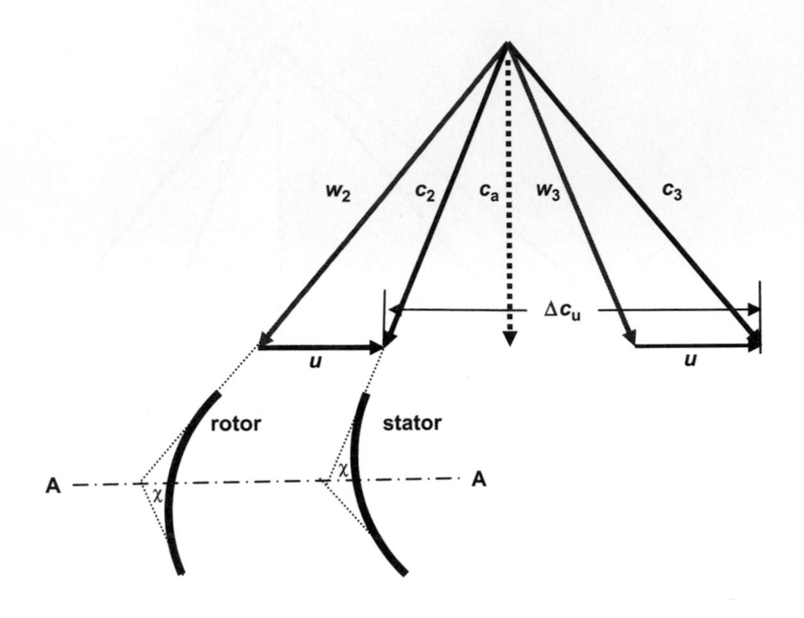


**FIGURE 7.28**

Velocity diagram for 50% reaction stage, $w_2 = c_3$ and $w_3 = c_2$; also called a symmetric stage. Camber lines of rotor and stator blades are outlined.

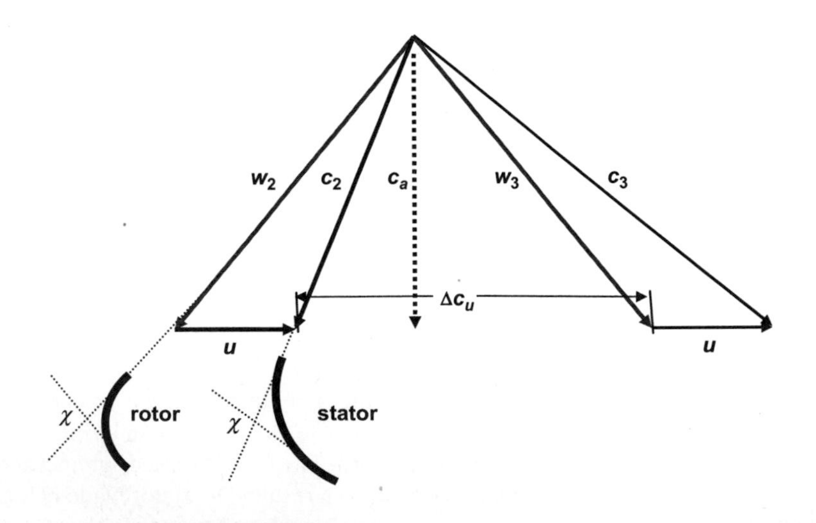


**FIGURE 7.29**

Velocity diagram for 0% reaction stage, $w_2 = w_3$; also called an impulse stage. Camber lines of rotor and stator blades are outlined.

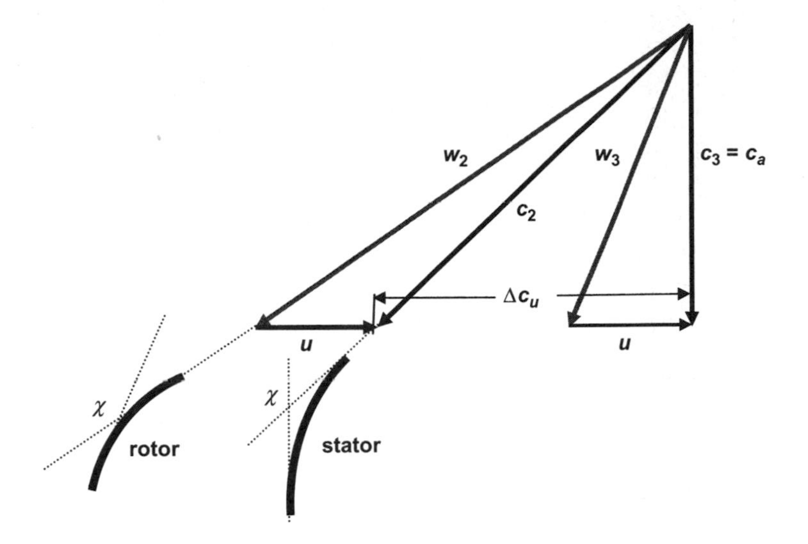


**FIGURE 7.30**

Velocity diagram for axial outlet stage, $c_3 = c_a$. The degree of reaction is greater than 100% ($c_2 > c_3$). Camber lines of rotor and stator blades are outlined.

provide a nonswirling flow to the entrance of the combustor. Here the absolute velocity decreases, that is, $c_2 > c_3$, as does the relative velocity. The velocity diagram and blade shapes are shown in Figure 7.30.

We may construct a diagram of pressure as a function of axial distance through the stage as shown in Figure 7.31.

It is evident from Figure 7.31 that the axial pressure gradient is least adverse for the symmetric (50% reaction) stage. If we wish to design a compressor with the specification that the adverse pressure gradient in any stage does not exceed some maximum value, we may obtain an idea of the number of stages needed to achieve a desired overall pressure ratio. This may be visualized by examining the axial pressure variation through a number of stages, as presented in Figure 7.32.

## 7.5.4 Compressor stages

Consider now the effects of staging, that is, of building a turbomachine with a set of rotor and stator stages connected serially. The pressure ratio across such a compressor may be written as follows:

$$\frac{p_{t3}}{p_{t2}} = \frac{p_{t3}^{(1)}}{p_{t2}^{(1)}} \frac{p_{t3}^{(2)}}{p_{t2}^{(2)}} \cdots \frac{p_{t3}^{(n)}}{p_{t2}^{(n)}} = \prod_{i=1}^{n} \frac{p_{t3}^{(i)}}{p_{t2}^{(i)}}. \tag{7.57}$$

In this notation, the following conventions are observed:

$$\begin{aligned} p_{t3}^{(i)} &= p_{t2}^{(i+1)} \\ p_{t2}^{(1)} &= p_{t2} \\ p_{t3}^{n} &= p_{t3} \end{aligned} \tag{7.58}$$

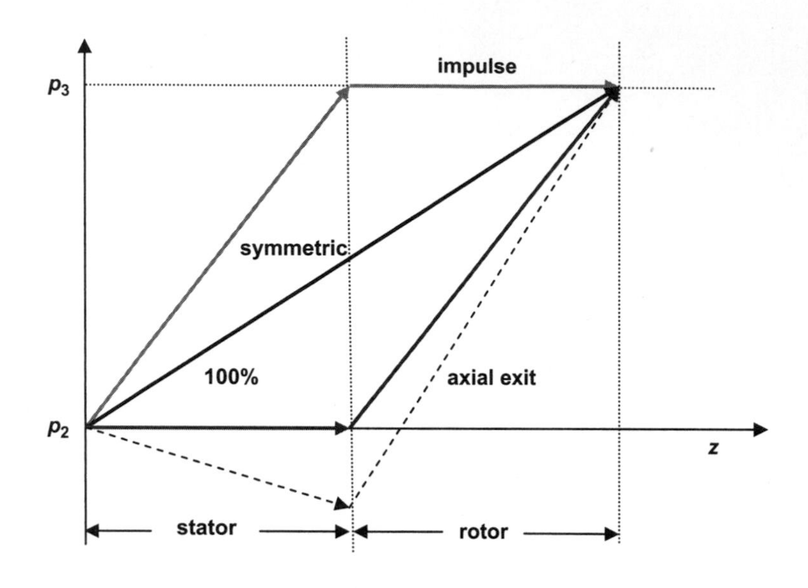


**FIGURE 7.31**

Axial pressure variation through various compressor stages.

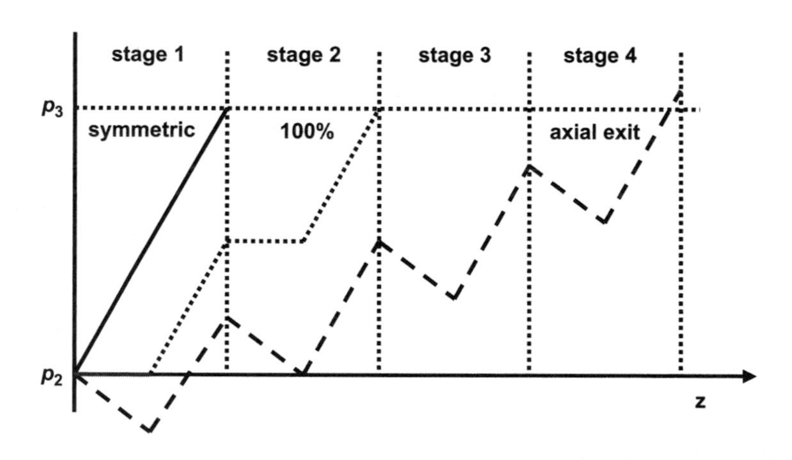


**FIGURE 7.32**

Variation of pressure through a series of different stages for a fixed pressure rise anywhere in the turbomachine.

In the special case where we may consider that the pressure rise across each stage is a given constant, we would have, for $n$ stages, the following stagnation pressure ratio across the entire turbomachine:

$$\frac{p_{t3}}{p_{t2}} = \left(\frac{p_{t3}^{(i)}}{p_{t2}^{(i)}}\right)^n. \tag{7.59}$$

We may define the efficiency of adiabatic compression for each individual stage as we did previously, that is,

$$\eta_{ad,c}^{(i)} = \frac{\left(\dfrac{p_{t3}^{(i)}}{p_{t2}^{(i)}}\right)^{\frac{\gamma-1}{\gamma}} - 1}{\dfrac{T_{t3}^{(i)}}{T_{t2}^{(i)}} - 1}. \tag{7.60}$$

Solving for the stagnation temperature ratio yields

$$\frac{T_{t3}^{(i)}}{T_{t2}^{(i)}} = 1 + \frac{1}{\eta_{ad,c}^{(i)}}\left[\left(\frac{p_{t3}^{(i)}}{p_{t2}^{(i)}}\right)^{\frac{\gamma-1}{\gamma}} - 1\right]. \tag{7.61}$$

The stagnation temperature ratio across the entire turbomachine may be obtained by applying the chain rule again, yielding

$$\frac{T_{t3}}{T_{t2}} = \prod_{i=1}^{n}\left[1 + \frac{1}{\eta_{ad,c}^{i}}\left\{\left(\frac{p_{t3}^{(i)}}{p_{t2}^{(i)}}\right)^{\frac{\gamma-1}{\gamma}} - 1\right\}\right]. \tag{7.62}$$

Then the efficiency of adiabatic compression for the entire machine is

$$\eta_{ad,c} = \frac{\left(\dfrac{p_{t3}}{p_{t2}}\right)^{\frac{\gamma-1}{\gamma}} - 1}{\left(\displaystyle\prod_{i=1}^{n}\left[1 + \frac{1}{\eta_{ad,c}^{(i)}}\left\{\left(\frac{p_{t3}^{(i)}}{p_{t2}^{(i)}}\right)^{\frac{\gamma-1}{\gamma}} - 1\right\}\right]\right) - 1}. \tag{7.63}$$

For the case of constant pressure rise across each stage we have

$$\frac{p_{t3}^{(i)}}{p_{t2}^{(i)}} = \left(\frac{p_{t3}}{p_{t2}}\right)^{\frac{1}{n}}. \tag{7.64}$$

In this special case, the efficiency of adiabatic compression is

$$\eta_{ad,c} = \frac{\left(\dfrac{p_{t3}}{p_{t2}}\right)^{\frac{\gamma-1}{\gamma}} - 1}{\left(1 + \dfrac{1}{\eta_{ad,c}^{(i)}}\left[\left(\dfrac{p_{t3}}{p_{t2}}\right)^{\frac{\gamma-1}{n\gamma}} - 1\right]\right)^{n} - 1}. \tag{7.65}$$

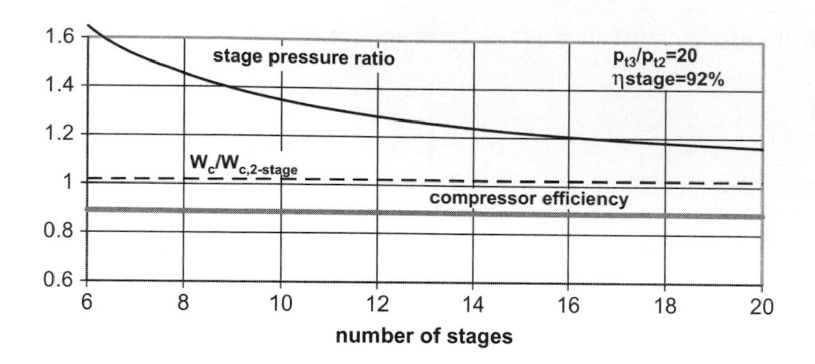


**FIGURE 7.33**

Variation of compressor work and efficiency as a function of the number of stages for constant stage efficiency and given overall pressure ratio.

The behavior of the work done by the compressor and its efficiency is illustrated in Figure 7.33. The work done in the case where the stages are identical is

$$W_c = c_{p,2}T_{t,2}\left[\left\{1+\frac{1}{\eta^i_{ad,c}}\left[\left(\frac{p_{t3}}{p_{t2}}\right)^{\frac{\gamma-1}{n\gamma}}-1\right]\right\}^n-1\right]. \tag{7.66}$$

## 7.5.5 Example: Axial compressor stages

Consider a compressor with $n$ stages, each of which has a stage efficiency of adiabatic compression $\eta^{(i)}_{ad,c} = 0.92$, and an overall pressure ratio $p_{t,3}/p_{t,2} = 20$. Find the overall efficiency of adiabatic compression $\eta_{ad,c}$, the stage pressure ratio required, and the work input as a function of the number of stages. What is the effect of a 1% increase in stage efficiency?

The efficiency of adiabatic compression can be determined from Equation (7.65), the stage pressure ratio required from Equation (7.64), and the work input from Equation (7.66). Results are shown in Figure 7.33. Here it is seen that the required stage pressure ratio clearly decreases as the number of stages increases. It was mentioned previously, in connection with Figure 7.31, that the stage pressure rise in axial flow compressors is limited by boundary layer separation considerations. Thus, as higher overall pressure ratios are sought, the number of stages must rise, or the stage efficiency of adiabatic compression must rise. Typically the number of stages for aircraft axial flow compressors lies somewhere between 10 and 20 with a trade-off between part count and high stage loading. Performance is always tempered by the need for long engine life, high reliability, and capital and maintenance costs. The Pratt & Whitney TF30 engine for the F-14A had a pressure ratio of 19.8, close to that considered in this problem, and used 16 stages in the compressor, 9 in the low-pressure compressor (including three fan stages), and 7 in the high-pressure compressor.

The change in overall efficiency, which drops from 90.3% for 2 stages to 88.4% for 20 stages, is difficult to see on the scale of Figure 7.33. Similarly, work required for the compressor is shown normalized to the work output required for two stages, $W_c/W_{c,2\text{-stage}}$, so as to maintain the same scale in

Figure 7.33. The work required rises about 2.2%, again an amount not readily discernible on the plot. However, it must be recalled that if this were the compressor on a high output engine, such as the Pratt & Whitney TF30 engine mentioned earlier, which has about the same pressure ratio and can pass more than 100 kg/s, the full static sea level power required is about 50 MW or about 68,000 hp. Then a 2% increase in power required represents about 1 MW more power demanded from the turbine, an increase that can impact performance significantly. One of the reasons that seemingly small percentage increases in efficiency are highly sought after is because they can represent large absolute benefits in performance. For example, a 1% increase in stage efficiency to 92.92% provides more than a 1.4% increase in overall efficiency of adiabatic compression. This increase in efficiency provides a reduction of more than 1.4% in work required by the compressor. Note that in the range $10 < n < 20$ the stage loading, or pressure rise, is $1.15 < (p_{t,3}/p_{t,2})^i < 1.35$, and a value around 1.25 is common.

### 7.5.6 Polytropic efficiency of adiabatic expansion

Another approach has been used to characterize the efficiency of adiabatic compression. This is based on considering an infinite number of stages, each providing an infinitesimal adiabatic compression. Then a so-called polytropic, or small-stage, efficiency can be defined as the ratio of the ideal isentropic temperature rise to the actual adiabatic temperature rise, as in Equation (7.5), but here the increases are infinitesimal so that

$$\eta_{pc} = \frac{dT_t'}{dT_t} = \frac{\frac{dT_t'}{T_t}}{\frac{dT_t}{T_t}}.$$

The isentropic stagnation temperature rise during the infinitesimal stagnation pressure rise is then given by

$$\frac{T_t + dT_t'}{T_t} = \left[\frac{p_t + dp_t}{p_t}\right]^{\frac{\gamma-1}{\gamma}} = 1 + \frac{\gamma - 1}{\gamma}\frac{dp_t}{p_t} + \ldots$$

Then the actual temperature rise during the same stagnation pressure rise, neglecting quadratic and higher terms in $dp_t$, is

$$\frac{dT_t}{T_t} = \frac{1}{\eta_{pc}}\frac{dT_t'}{T_t} = \frac{\gamma - 1}{\eta_{pc}\gamma}\frac{dp_t}{p_t}.$$

Integrating over an infinite number of stages taking the stagnation temperature from $T_{t2}$ to $T_{t3}$ across the whole compressor yields

$$\frac{T_{t,3}}{T_{t,2}} = \left(\frac{p_{t,3}}{p_{t,2}}\right)^{\frac{\gamma-1}{\eta_{pc}\gamma}}.$$

Substituting this result into the definition of overall efficiency of adiabatic compression given by Equation (7.8) leads to

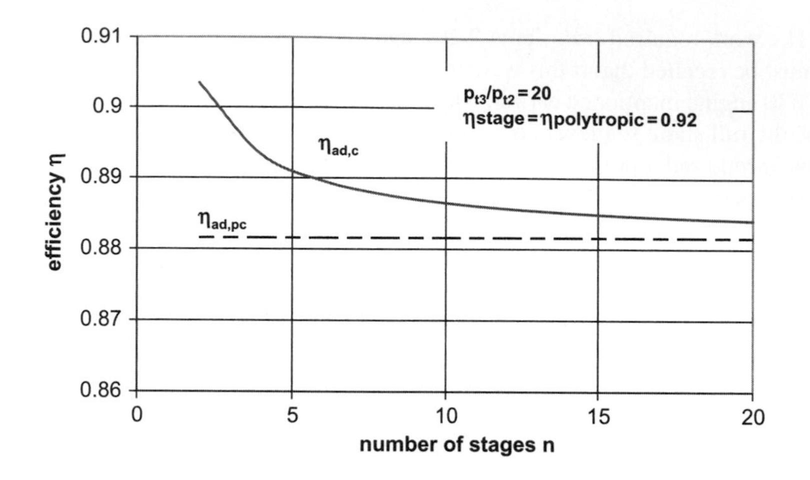


FIGURE 7.34

Comparison of overall efficiency of adiabatic compression based on finite stage efficiency ($\eta_{ad,c}$) with that based on the polytropic efficiency ($\eta_{ad,pc}$).

$$\eta_{ad,pc} = \frac{\left(\frac{p_{t,3}}{p_{t,2}}\right)^{\frac{\gamma-1}{\gamma}} - 1}{\left(\frac{p_{t,3}}{p_{t,2}}\right)^{\frac{\gamma-1}{\eta_{pc}\gamma}} - 1}.$$

This result for the overall efficiency of adiabatic compression based on constant infinitesimal-stage polytropic efficiency $\eta_{pc}$ is not equivalent to that using constant finite-stage efficiencies [Equation (7.65)]. The difference between the two is shown in Figure 7.34 for the conditions dealt with earlier and with polytropic efficiency equal to the stage efficiency of 92%. Note that polytropic efficiency is less than overall efficiency as computed using polytropic efficiency, that is, $\eta_{pc} < \eta_{ad,pc}$. However, the more accurate value of overall efficiency computed with Equation (7.65) using individual stage efficiency yields a larger value than that computed on the basis of polytropic efficiency $\eta_{ad,c} > \eta_{ad,pc}$ and that the former approaches the latter as the number of stages increases.

## 7.6 AXIAL FLOW TURBINE

### 7.6.1 Velocity diagrams

We may visualize the blade shapes as defining the flow passages in the rotor as well as the stator, and the stationary passages delivering flow to and accepting flow from the rotor. In Figure 7.35, the so-called velocity diagram introduced in Figure 7.7 is drawn on the turbomachine from the perspective of

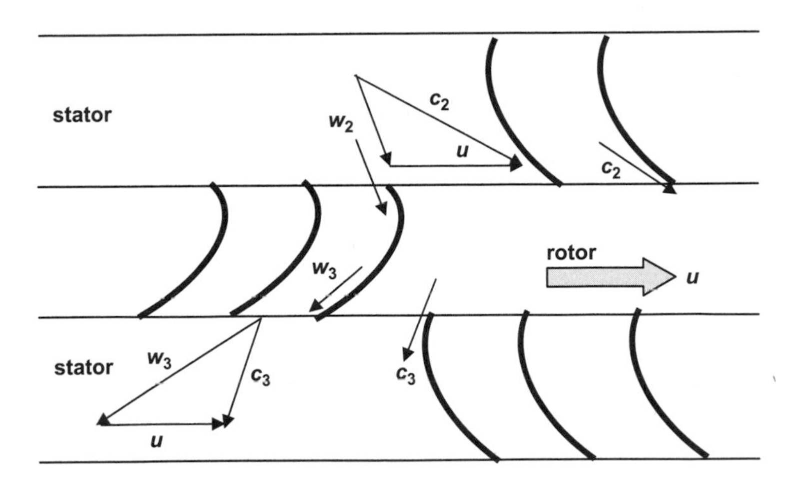


**FIGURE 7.35**

Turbine blades shown as moving with the linear speed of rotation $u$ and the associated velocity triangles entering and leaving a turbine rotor.

an observer looking at it from the side. For example, when looking at the side view of the axial flow turbomachine in Figure 7.22, one would perceive the linear speed of rotation $u = \omega r$ as shown by the arrow there or as shown in Figure 7.35.

The velocity diagrams entering and leaving the blade passage may be combined into one diagram, as shown in Figure 7.36. The airfoil sections are designed to align the flow so as to produce a particular streamline pattern. Separation is to be avoided so the angles of attack of the airfoils, that is, the angle between the airfoil chord and the appropriate relative wind, must be maintained at relatively small values. Thus one may sketch the mean camber line of the airfoil by having the leading and trailing edges aligned with the entering and exiting velocity vectors, as will be shown subsequently. The magnitude of the power extracted is proportional to the amount the flow is turned, that is, the degree to which the whirl is increased.

## 7.6.2 Pressure drop through axial flow turbine stages

In axial flow turbomachines, we have noted that it is reasonable to assume that the linear speed of rotation $u$ may be taken as a constant equal to its value at the middle of the blade passage. In the case of a compressor, we considered many rows of rotating and stationary blades, called rotors and stators, respectively, arranged in stages, as was shown in Figure 7.25. The need for a large number of stages in compression (10 to 20) is a result of the inability of the viscous boundary layer to resist separation except at small values of adverse pressure gradient in any given stage. In turbines, however, the pressure is falling continually so that limitation doesn't exist. Therefore, turbines in aircraft engines have few stages, typically two to four. The major limitation on turbine stages is choking, that is, the attainment of critical, $M = 1$, conditions in the turbine, thereby limiting the mass flow passing through the stage.

The fact that compressor stages had relatively small pressure rises permitted the application of incompressible flow analysis in many cases, as described in Section 7.5.2. With the large pressure drops

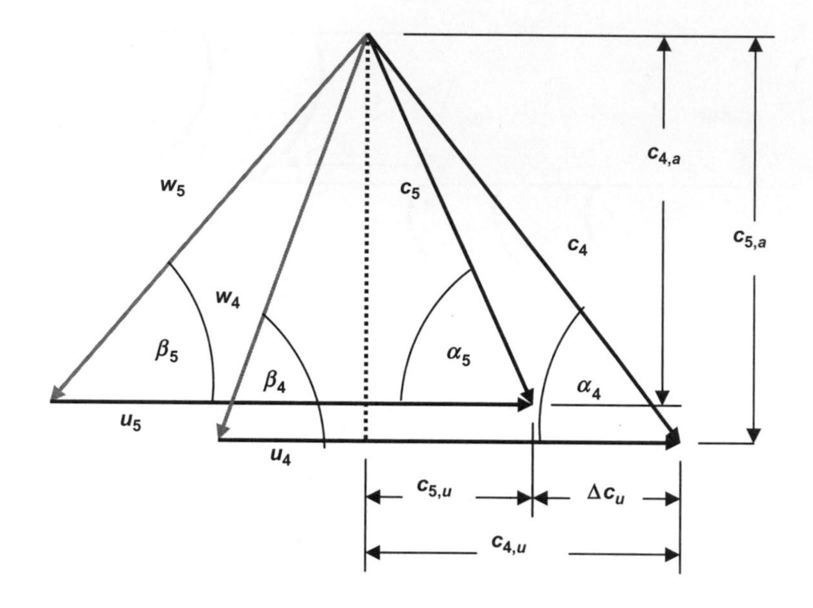


**FIGURE 7.36**

A combined velocity diagram for a turbine stage.

allowable in turbine stages, the compressibility of the gas must be considered, and therefore it is often more convenient to deal with enthalpy, or temperature, changes in the adiabatic expansion process.

The work output of the turbine is used primarily to supply the work input required by the compressor to achieve a sufficiently high level of stagnation pressure for delivery to the combustor. Considering this point we may examine the stagnation pressure of the flow exiting the turbine using Equation (7.13) and find that

$$\frac{p_{t,5}}{p_{t,4}} = \left(1 - \frac{W_t}{\eta_{ad,e} c_{p,4} T_{t,4}}\right)^{\frac{\gamma_4}{\gamma_4 - 1}}.$$

The turbine exit stagnation pressure is given by

$$\frac{p_{t,5}}{p_5} = \left(1 + \frac{\gamma_5 - 1}{2} M_5^2\right)^{\frac{\gamma_5}{\gamma_5 - 1}}.$$

This is the stagnation pressure entering the nozzle and to get the static conditions one may use the mass flow, the flow area, and the stagnation pressure and temperature to find the Mach number.

### 7.6.3 Example: Turbine pressure drop

An axial flow turbine must develop 5830 kW from a mass flow of 29 kg/s. If the turbine inlet temperature is 1200K and the efficiency of adiabatic expansion is 0.9, determine (a) the pressure ratio across the turbine and (b) the exit temperature.

**(a)** The work is related to the power by the mass flow, as in Equation (7.19), so

$$W_t = \frac{P}{\dot{m}} = \frac{5830kW}{29kg/s} = \frac{5830kJ/s}{29kg/s} = 201kJ/kg.$$

The pressure ratio across the turbine may be found from Equation (7.13) as follows:

$$\frac{p_{t,5}}{p_{t,4}} = \left(1 - \frac{W_t}{\eta_{ad,e} c_{p,4} T_{t,4}}\right)^{\frac{\gamma_4}{\gamma_4 - 1}}$$

Because gas flowing through the turbine is hot ($T_{t,4} = 1200$K), we may assume $\gamma = 1.33$ and $c_{p,4} = 1.16$ kJ/kg-K. Then the pressure ratio is

$$\frac{p_{t,5}}{p_{t,4}} = \left[1 - \frac{201kJ/kg}{0.9(1.16kJ/kg - K)1200K}\right]^4 = 0.49.$$

We may invert this to get $p_{t4}/p_{t,5} = 2.04$. In the example problem in Section 7.3.2, a centrifugal compressor that handled the same mass flow and needed the same amount of power as developed by this turbine had to pump through a pressure ratio of $p_{t,3}/p_{t,2} = 5.1$. This is a result of the divergence of the constant pressure lines on an *h,s* cycle diagram.

**(b)** The turbine exit temperature is given by the work equation:

$$T_{t,5} = T_{t,4} - \frac{W_t}{c_{p,4}} = 1200K - \frac{201kJ/kg}{1.16kJ/kg - K} = 1026K.$$

Here we have assumed that the specific heat is constant through the turbine and equal to that appropriate to the entry conditions. As shown in Appendix B, specific heat does vary with temperature and for more accurate calculations it is recommended that one deals with enthalpy rather than temperature.

### 7.6.4 Types of turbine stages

In the previous discussion of the different degrees of reaction in compressor stages, the 50% reaction, or symmetric, stage was shown to be the preferred one. This was the case because the limiting operational factor in a compressor stage is the magnitude of the adverse pressure gradient, which is a natural consequence of compressor operation. Too large a gradient induces boundary layer separation in the stage and a substantial loss in performance. The 50% or symmetric stage has the lowest possible pressure gradient and is therefore generally the favored choice. However, in a turbine stage, the pressure gradient is always favorable so it has no special influence in the choice of the degree of reaction in a turbine stage.

In the turbine, the work extracted per unit mass is given by

$$W_t = \left(u_4 c_{4,u} - u_5 c_{5,u}\right) = h_{t,4} - h_{t,5}. \tag{7.67}$$

It has been pointed out that in axial flow machines, where the height of the blades is much smaller than the radius of the turbomachine wheel, it is reasonable to take a mean value for $u$ so that we may simplify Equation (7.14) to

$$W_t = u\left(c_{4,u} - c_{5,u}\right) = u\Delta c_u = h_{t,4} - h_{t,5}. \tag{7.68}$$

We may also write this as

$$W_t = \eta_{ad,e} c_{p,4} T_{t,4} \left[ 1 - \left( \frac{p_{t,5}}{p_{t,4}} \right)^{\frac{\gamma_4 - 1}{\gamma_4}} \right]. \quad (7.69)$$

Here $\eta_{ad,e}$ is the efficiency of adiabatic expansion, and $\gamma_4$ is assumed to be an average ratio of specific heats in the turbine and is often taken as 1.33 for combustion of hydrocarbon fuels. Because the turbine has a continual pressure drop across each stage, and the effects of separation are no longer an issue, more work can be extracted per stage than can be put in, per stage, in a compressor. Indeed the major limitation of flow in turbine stages is choking, which can limit the mass flow passing through the system. In the turbine, the flow must be considered compressible, and the possibility of shock waves being present cannot be ignored and must be guarded against.

In the rotor of a turbine stage, velocity is defined with respect to the rotor passage so that within the rotor the stagnation enthalpy may be taken as constant, assuming the flow in the rotor is adiabatic. Therefore, the enthalpy change is given by the change in kinetic energy of the fluid in the rotor. Then the energy equation in the rotor is

$$h_4 - h_5 = \frac{1}{2}\left(w_5^2 - w_4^2\right). \quad (7.70)$$

The energy equation in the absolute, or laboratory, frame is

$$h_4 - h_5 = \frac{1}{2}\left(c_5^2 - c_4^2\right) + W_t. \quad (7.71)$$

Then, subtracting Equation (7.71) from Equation (7.70) yields

$$\frac{1}{2}\left(w_5^2 - w_4^2\right) + \frac{1}{2}\left(c_4^2 - c_5^2\right) = W_t. \quad (7.72)$$

In a turbine stage, the degree of reaction may be expressed as the ratio of enthalpy change in the rotor to the total enthalpy change across the complete stage. Using Equations (7.15), (7.70), and (7.72) in this definition leads to the following equation for degree of reaction:

$$\%r = \frac{\Delta h_{rotor}}{\Delta h_{stage}} = \frac{\left(w_5^2 - w_4^2\right)}{\left(w_5^2 - w_4^2\right) + \left(c_4^2 - c_5^2\right)}. \quad (7.73)$$

First, we may consider a 0% reaction, or impulse, stage where the entire expansion, and therefore the entire enthalpy drop, takes place in the stator and none in the rotor. Thus the relative velocity is unchanged, $w_4 = w_5$, while the relative velocity does change, $c_5 < c_4$, giving an enthalpy, and therefore temperature, drop due entirely to the external effect. The velocity diagram for such a stage is shown in Figure 7.37. The general shapes of the camber lines of the rotor and stator blades are also shown in Figure 7.37. Note that the rotor blade is symmetrical about the midpoint of its chord. The flow is turned symmetrically and there is no change in the enthalpy or temperature in the rotor as a result. The stator blade is less curved, with the angle $\chi$ between the tangents to the leading and trailing edges larger than that for the rotor. Use of this type of blade dates back to early stationary steam power plant turbines.

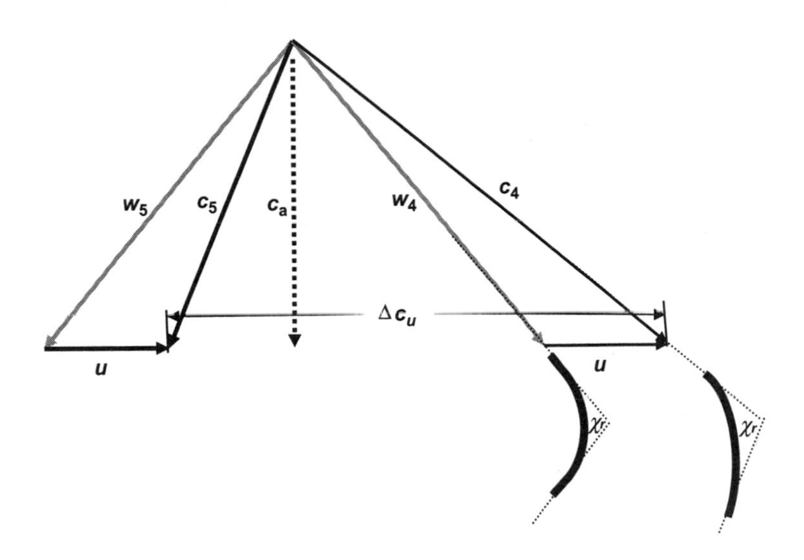


**FIGURE 7.37**

Velocity diagram for turbine stage with 0% reaction, also called an impulse stage, $w_2 = w_3$. Camber lines of rotor and stator blades are outlined.

Next, we may consider a 100% reaction stage where the entire expansion, and therefore the entire enthalpy, and temperature, drop takes place in the rotor and none in the stator. Thus the absolute velocity is unchanged, $c_4 = c_5$, while the relative velocity does change, $w_5 > w_4$, giving an enthalpy, and therefore temperature, drop due entirely to the internal effect. The velocity diagram for such a stage is shown in Figure 7.38. The general shapes of the camber lines of the rotor and stator blades are also shown in Figure 7.38. Note that now the stator blade is symmetrical about the midpoint of its chord. The flow is turned symmetrically, and there is no change in the enthalpy or temperature in the stator as a result. The rotor blade is less curved, with the angle $\chi$ between the tangents to the leading and trailing edges larger than that for the stator.

Next we may consider a 50% reaction stage where half the enthalpy drop takes place in the rotor and half in the stator. Here the entering absolute velocity $c_4 = w_5$ and the leaving absolute velocity $c_5 = w_4$, providing for an equal share of the pressure change due to the internal effect and to the external effect. The velocity diagram for this type of stage, often called a symmetric stage because of the shape of the velocity diagram, is shown in Figure 7.39. In addition, the rotor and stator blades have the same cross-sectional shape. That is, if the stator blade is rotated 180° about the axis A-A in Figure 7.39, it will be identical in shape to the rotor blade. This stage is used most commonly in aircraft turbine engines.

As a final case often encountered in practice, we may consider a greater than 100% reaction stage where the intent is to have an axial exit velocity for entrance into the nozzle. Here the absolute velocity decreases, $c_5 < c_4$, as does the relative velocity, $w_5 < w_4$. The velocity diagram for such a stage is shown in Figure 7.40.

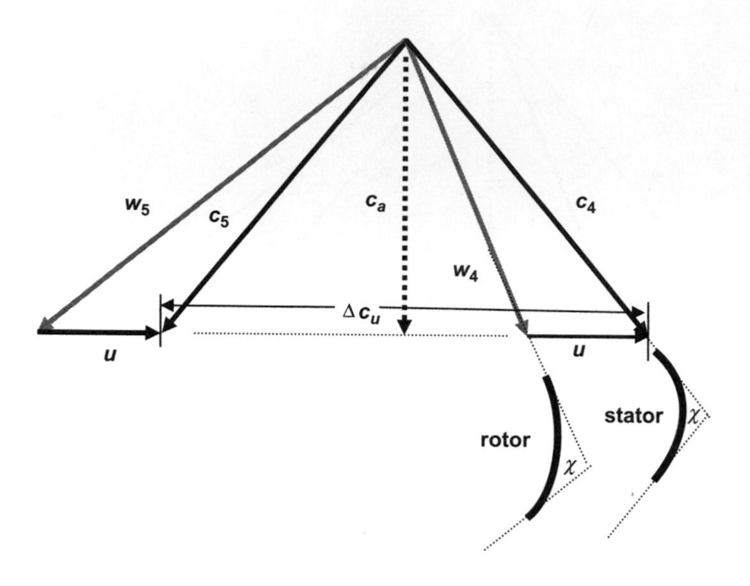


**FIGURE 7.38**

Velocity diagram for turbine stage with 100% reaction: $c_4 = c_5$. Camber lines of rotor and stator blades are outlined.

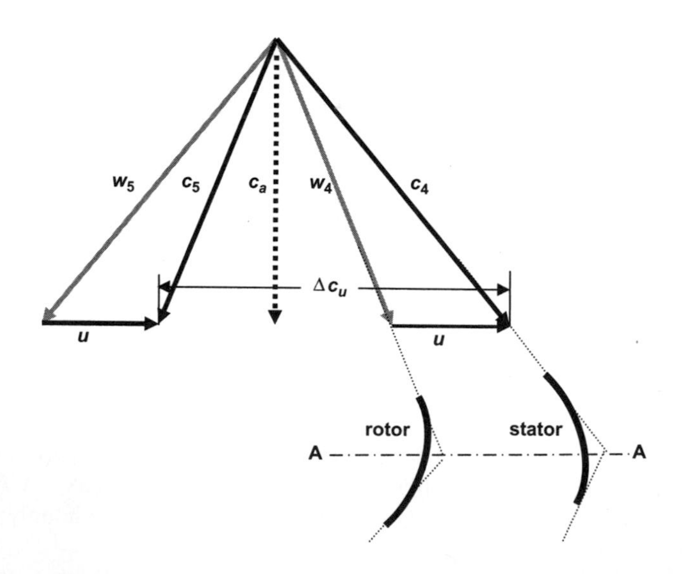


**FIGURE 7.39**

Velocity diagram for turbine stage with 50% reaction, also called a symmetric stage, $w_2 = c_3$ and $w_3 = c_2$. Camber lines of rotor and stator blades are outlined. Note that if the stator blade is rotated 180° around axis A-A it will be identical in shape to the rotor blade.

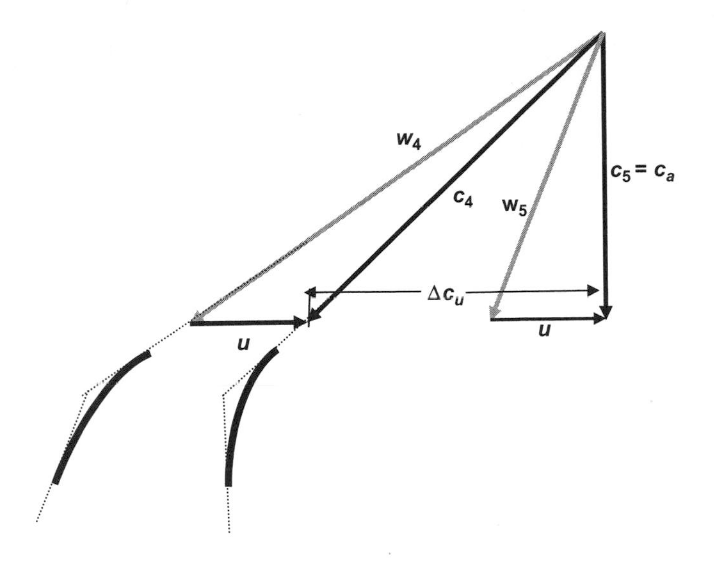


**FIGURE 7.40**

Velocity diagram for axial outlet stage, $c_3 = c_a$. Because $c_2 > c_3$, the degree of reaction is $\%r > 100$. Camber lines of rotor and stator blades are outlined.

## 7.7 AXIAL FLOW COMPRESSOR AND TURBINE PERFORMANCE MAPS

### 7.7.1 General aerodynamic considerations

We may use the blade element theory to determine the performance characteristics of turbomachines. We may consider compressors and turbines with the same approach since one is essentially the reverse of the other. Consider the force field acting on a turbine or compressor blade as shown in Figure 7.41. The turning force required from the blade by the compressor or supplied by the blade to the turbine is given by

$$F_u = L\sin\beta_2 + D\cos\beta_2 = L\sin(\pi - \beta_4) + D\cos(\pi - \beta_4). \quad (7.74)$$

The lift and drag of the blade may be written in terms of the lift and drag coefficients of the airfoil cross section of the blade, $c_L$ and $c_D$, the dynamic pressure of the flow relative to the blade, $q = \rho w^2/2$, and the planform area of the blade, $S_b$. The lift and drag coefficients are dependent on the airfoil geometry, the Mach number relative to the blade, and the Reynolds number based on the chord length of the blade $\mathrm{Re}_c = \rho wc/\mu$. The expressions for lift and drag of a blade are as follows:

$$L = c_L \frac{1}{2}\rho w^2 S_b$$

$$D = c_D \frac{1}{2}\rho w^2 S_b.$$

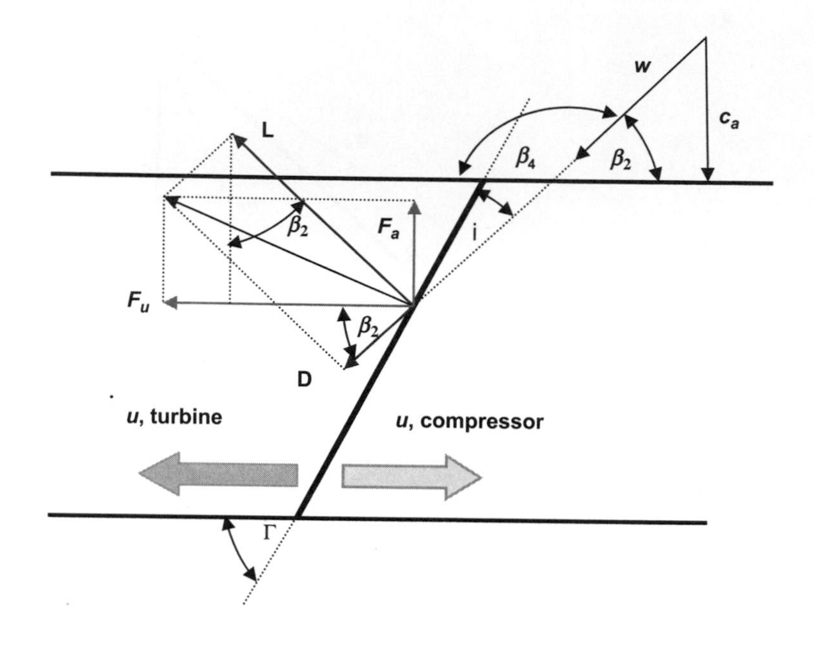


FIGURE 7.41

Generalized force structure and flow field on an axial flow turbomachine blade. Subscript 2 refers to compressor blade entrance conditions, and subscript 4 refers to turbine blade entrance conditions.

With this substitution, the turning force becomes

$$F_u = \frac{1}{2} c_L \rho S_b w^2 (\sin \beta_2 + \varepsilon \cos \beta_2) = \frac{1}{2} c_L \rho S_b w^2 (\sin \beta_4 - \varepsilon \cos \beta_4). \tag{7.75}$$

Here we have also introduced the inverse lift to drag ratio, $\varepsilon = c_D/c_L$, which defines the aerodynamic efficiency of the blade. The drag to lift ratio $\varepsilon$ is typically quite small for well-designed compressor or turbine blades, with values around 0.05 not unusual.

Recall that Equation (7.74) provides an aerodynamic approach to calculating the work put in by a compressor blade or extracted by a turbine blade. Attention will be focused on turbine blades here with the result for compressor blades following directly by analogy. Our previous fluid dynamic analysis for turbine work led to the work per unit mass generated by a turbine blade as given here:

$$W_t = u\left(c_{5,u} - c_{4,u}\right) = \frac{uF_u}{\dot{m}} = \frac{\omega r F_u}{\dot{m}}. \tag{7.76}$$

We are assuming that the turbine is of axial flow type with blade height $\Delta r$ much smaller than the maximum radius of the machine, $\Delta r \ll r_{\max}$, so that we may assume a single, average, radius $r$ for the flow path. Introducing the expression for $F_u$ from Equation (7.75), which applies to a single blade and multiplying by the number of blades $n_b$, we obtain the following result for the work output of the entire turbine:

$$W_t = \frac{\omega r}{\dot{m}} c_L \frac{1}{2} \rho_4 (n_b S_b) w_4^2 (\sin \beta_4 - \varepsilon \cos \beta_4). \tag{7.77}$$

From Figure 7.34 we may note that the relative velocity seen by the turbine blade may be written as

$$w_4 = \frac{c_a}{\sin \beta_4}. \quad (7.78)$$

Then the mass flow entering the turbine through station 4 is

$$\dot{m} = \rho_4 c_a A_4. \quad (7.79)$$

We may replace the rotational speed $\omega$ (1/s) by $N$ (rpm) using the relation

$$\omega = \frac{2\pi N}{60},$$

which permits us to write the turbine work as follows:

$$W_t = \frac{2\pi N}{60} \frac{r}{\rho_4 A_4^2} (\dot{m}) \frac{1}{2} c_L (n_b S_b) \frac{1 - \varepsilon \cot \beta_4}{\sin \beta_4}. \quad (7.80)$$

We introduce the following nondimensional variables

$$\begin{aligned} \theta &= \frac{T_t}{T_{ref}} \\ \delta &= \frac{p_t}{p_{ref}} \end{aligned}, \quad (7.81)$$

where $p_{ref} = 101.3$ kPa and $T_{ref} = 288$K. Now Equation (7.80) can be put into the form

$$W_t = \frac{\pi r n_b S_b}{60 A_4^2} \frac{\delta_4}{\rho_4} \left( \frac{N}{\sqrt{\theta_4}} \right) \left( \frac{\dot{m}\sqrt{\theta_4}}{\delta_4} \right) \frac{c_L}{\sin \beta_4} (1 - \varepsilon \cot \beta_4). \quad (7.82)$$

To eliminate the density in this equation, we use the perfect gas equation

$$\rho_4 = \frac{p_4}{RT_4} = \frac{p_{ref}}{RT_{ref}} \left( \frac{p_4}{p_{t,4}} \right) \left( \frac{T_4}{T_{t,4}} \right)^{-1} \frac{\delta_4}{\theta_4}. \quad (7.83)$$

Regrouping terms and substituting $r = d/2$ yields

$$\frac{W_t}{\theta_4} = \left[ \frac{\pi d n_b S_b}{120 A_4^2} \frac{RT_{ref}}{p_{ref}} \left( \frac{T_4}{T_{t,4}} \right) \left( \frac{p_4}{p_{t,4}} \right)^{-1} \right] \frac{N}{\sqrt{\theta_4}} \frac{\dot{m}\sqrt{\theta_4}}{\delta_4} \frac{c_L}{\sin \beta_4} (1 - \varepsilon \cot \beta_4). \quad (7.84)$$

Carrying out the same process for the compressor yields the following analogous result:

$$\frac{W_c}{\theta_2} = \left[\frac{\pi d n_b S_b}{120 A_2^2} \frac{R T_{ref}}{p_{ref}} \left(\frac{T_2}{T_{t,2}}\right) \left(\frac{p_2}{p_{t,2}}\right)\right]^{-1} \frac{N}{\sqrt{\theta_2}} \frac{\dot{m}\sqrt{\theta_2}}{\delta_2} \frac{c_L}{\sin\beta_2}(1 + \varepsilon \cot\beta_2). \tag{7.85}$$

## 7.7.2 Turbine performance maps

The term in the square brackets in Equation (7.84) involves the machine size and configuration, as well as the conditions at the entrance to the turbomachine, and may be considered a constant $K$ for purposes of this discussion. Furthermore, because the aerodynamic efficiency of the turbomachine blades is typically quite high, we may take $\varepsilon \sim 0$ without loss in generality for the purpose of discussing the major aspects of turbomachine operation. Under these approximations we may rewrite Equation (7.84), which is for the turbine, as

$$\frac{W_t}{\theta_4} = K_t \frac{N}{\sqrt{\theta_4}} \frac{\dot{m}\sqrt{\theta_4}}{\delta_4} \frac{c_L}{\sin\beta_4}. \tag{7.86}$$

We see that for a fixed corrected rpm $N/\sqrt{\theta_4}$ the corrected turbine work $\dot{m}\sqrt{\theta_4}/\delta_4$ is linearly proportional to the corrected mass flow.

The velocity diagram in Figure 7.42 shows that the angle $\beta$ is related to the angle of incidence (or angle of attack) $i$ by the following equation:

$$\beta_4 = \pi - \Gamma + i. \tag{7.87}$$

Note that $\Gamma$ is the stagger angle of the blade, that is, the angle between the blade chord line and the tangential velocity direction, as defined in Figure 7.41. The lift coefficient is linearly proportional to the angle of incidence over much of the range of small incidence angles, starting from $i_0$ until the

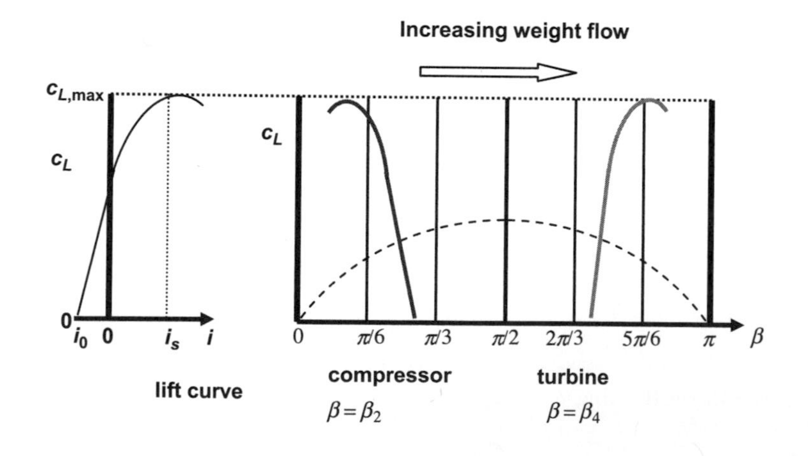


**FIGURE 7.42**

Illustration of the variation of blade lift coefficient with $\beta$ and $i$ along with $\sin\beta$ and the direction of increasing weight flow rate.

maximum, or stall, angle for the blade $i_s$ is approached, as illustrated in Figure 7.42; typically the range of $i$ is $i_s–i_0 \sim 20°$. The lift coefficient $c_L$ in Equation (7.86) will increase, and $\sin\beta_4$ will decrease as $\beta_4$ grows along with the corrected mass flow, for a fixed corrected rpm. Thus the corrected work in Equation (7.86) will grow more rapidly than linearly with corrected mass flow as the corrected mass flow approaches the maximum value at choking of the turbine inlet, $M_4 = 1$. A concise description of turbine performance may be obtained by plotting corrected turbine work as a function of corrected weight flow for given values of corrected rpm. Also shown in Figure 7.42 is the behavior of $c_L$ in the first quadrant of $\beta$ where the lift coefficient variation for a compressor blade is shown. The fact that it is reversed from that for a turbine blade is discussed in the next section.

One way to illustrate this behavior of the last term in Equation (7.86) is to show the lift curve as it would appear as a function of $\beta$. The regions of $\beta$ in which the compressor ($\beta = \beta_2$) and turbine ($\beta = \beta_4$) operate are shown in Figure 7.42. Superimposed on this plot are the relevant orientations of the lift curve, along with the variation of $\sin\beta$, thereby illustrating the effect of increasing mass flow rate on the blade-turning force. Note that $c_L \sim \sin\beta_4$ for much of the operating range of the blade so that the ratio of $c_L$ to $\sin\beta_4$ is constant over that range and the corrected work is approximately linear with the corrected weight flow over a fairly broad range for a given corrected rpm. When the airfoil $c_L$ moves out of the linear range there are deviations from this simple behavior. The general behavior of the corrected work as a function of corrected mass flow for different values of corrected rpm is shown in Figure 7.43.

The mass flow rate through a passage of area $A_4$ is given by

$$\dot{m} = \frac{p_{t,4}A_4}{\sqrt{RT_{t,4}}} f(\gamma_4, M_4) = \frac{p_{ref}}{\sqrt{RT_{ref}}} \frac{\delta_4}{\sqrt{\theta_4}} f(\gamma_4, M_4). \tag{7.88}$$

Writing out Equation (7.88) in full, using, for example, Equation (5.14), yields the following corrected mass flow form:

$$\frac{\dot{m}\sqrt{\theta_4}}{\delta_4} = \frac{p_{ref}A_4}{\sqrt{RT_{ref}}}\sqrt{\gamma}M_4\left(1 + \frac{\gamma_4 - 1}{2}M_4^2\right)^{-\frac{\gamma+1}{2(\gamma-1)}}. \tag{7.89}$$

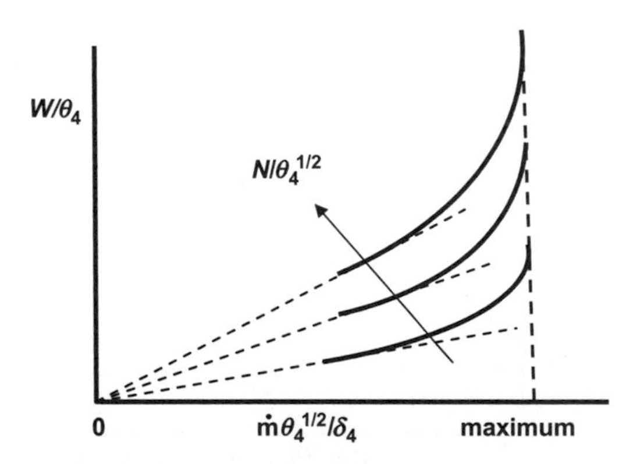


**FIGURE 7.43**

Turbine performance map showing the corrected turbine work varying with corrected mass flow for various corrected rpm. At lower mass flows the behavior is shown as approximately linear.

Maximum mass flow occurs where $M_4 = w_4/a_4 = 1$, that is, when the entrance station to the turbine is choked. The expression for the turbine work given by Equation (7.13) may be put in terms of the corrected work:

$$\frac{W_t}{\theta_4} = c_{p,4} T_{ref} \eta_{ad,e} \left[ 1 - \left( \frac{p_{t,5}}{p_{t,4}} \right)^{\frac{\gamma-1}{\gamma}} \right]. \tag{7.90}$$

The turbine work performance map has been shown in Figure 7.43 as the corrected turbine work as a function of corrected mass flow with corrected rpm as a parameter. Note that all the curves rise rapidly near a maximum value of the corrected weight flow. This condition represents choked flow at the turbine inlet described by Equation (7.89) for $M_4 = 1$. Using Equation (7.86) in Equation (7.90), the stagnation pressure across the turbine may be written as

$$\frac{p_{t,4}}{p_{t,5}} \approx \left[ 1 - \frac{K_t}{\eta_{ad,t} c_{p,4} T_{ref}} \frac{N}{\sqrt{\theta_4}} \left( \frac{c_L}{\sin \beta_4} \right) \frac{\dot{m}\sqrt{\theta_4}}{\delta_4} \right]^{\frac{-\gamma}{\gamma-1}}. \tag{7.91}$$

In the turbine, because the hot gas value of $\gamma$ is around 1.33, the exponent in Equation (7.91) is $-4$ and the pressure ratio will start to rise rapidly as the maximum mass flow condition is approached. The pressure ratio will have weak rpm dependence, especially when the turbine entrance is close to choked. The turbine performance map in terms of stagnation pressure ratio is shown in Figure 7.44.

Well-designed axial turbines show high values of efficiency, and presentation of the variation on a performance map tends to be rather crowded and therefore not very clear. Such information is generally presented on a separate plot, as depicted in Figure 7.45. Because the adiabatic expansion

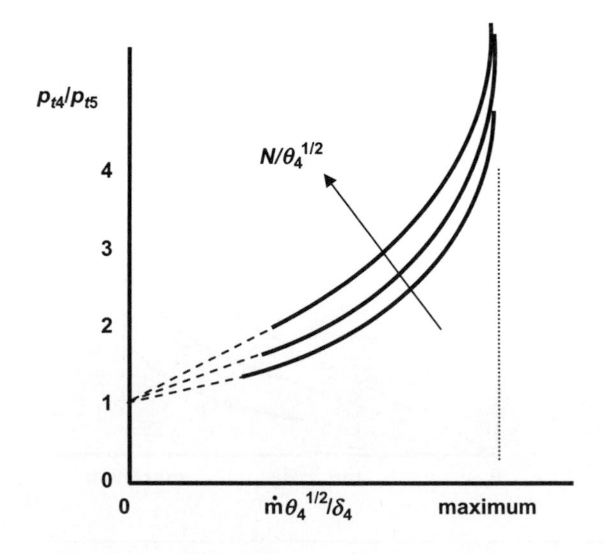


**FIGURE 7.44**

Turbine performance map in terms of stagnation pressure ratio.

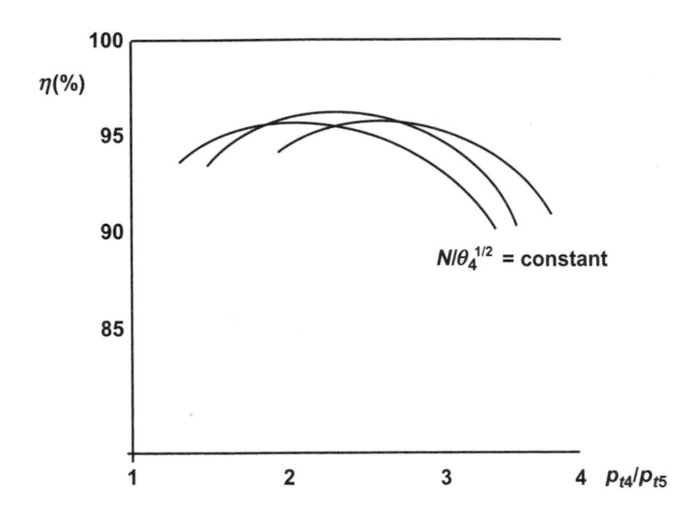


**FIGURE 7.45**

Efficiency of adiabatic expansion as a function of turbine pressure ratio for various values of corrected rpm.

efficiency is relatively constant in aircraft engines, it is often accepted to treat it as a constant for a given turbine.

### 7.7.3 Compressor maps

The compressor work required follows the same analysis as for turbine work, starting from Equation (7.85). The corrected compressor work may be written as

$$\frac{W_c}{\theta_2} = K_c \frac{N}{\sqrt{\theta_2}} \frac{\dot{m}\sqrt{\theta_2}}{\delta_2} \frac{c_L}{\sin \beta_2}(1 + \varepsilon \cot \beta_2). \tag{7.92}$$

Here the constant $K_c$ is the analog to $K_t$ for the turbine and expresses information on machine size and compressor inlet conditions. The compressor work equation in Equation (7.92) may be regrouped as follows:

$$\frac{W_c}{\theta_2} = K_c \left[\frac{N}{\sqrt{\theta_2}}\right] \left[\frac{\dot{m}\sqrt{\theta_2}}{\delta_2} \frac{1}{\sin \beta_2}\right] [c_L][1 + \varepsilon \cot \beta_2]. \tag{7.93}$$

The first term in square brackets is the corrected rpm, whereas the second term involves the corrected mass flow and the angle of the flow relative to the linear speed of rotation $u$. The corrected mass flow is proportional to the axial velocity component $c_a$ and, referring to the entering velocity diagram shown in Figure 7.46, we may write $c_a$ as

$$c_a = c_2 \sin \alpha_2 = w_2 \sin \beta_2. \tag{7.94}$$

Therefore, the corrected mass flow

$$\frac{\dot{m}\sqrt{\theta_2}}{\delta_2} \sim w_2 \sin \beta_2. \tag{7.95}$$

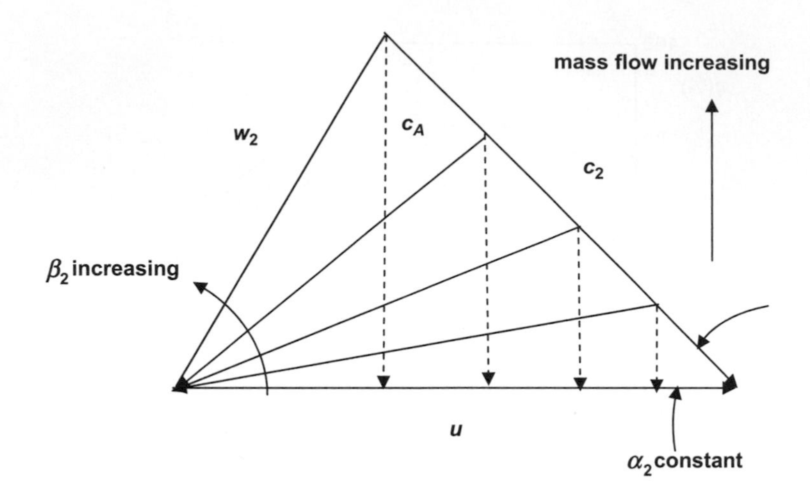


**FIGURE 7.46**

Velocity diagram for flow entering the rotor illustrating the increase in the angle $\beta_2$ as mass flow is increased. The stator is fixed so that $\alpha_2$ remains constant as the mass flow increases.

Thus the second term in square brackets is not explicitly dependent on the relative entry angle $\beta_2$ or

$$\frac{\dot{m}\sqrt{\theta_2}}{\delta_2}\frac{1}{\sin\beta_2} \neq f(\beta_2). \tag{7.96}$$

However, the term in Equation (7.96) does depend on the relative velocity $w_2$, which is seen to decrease as the mass flow increases. Thus the second term in the square brackets in Equation (7.93) decreases as mass flow increases. The lift coefficient in Equation (7.93) may be expressed in terms of the angle of attack as follows:

$$c_L = c_{L,\max}\sin\left(\frac{\pi}{2}\frac{i - i_0}{i_s - i_0}\right), \tag{7.97}$$

where $i_0$ is the zero-lift angle of attack of the rotor blade, while $i_s$ is the stall angle of attack of the rotor blade. A generic representation of the lift characteristics of a rotor blade with $c_{L,\max} = 1.2$ and $i_s = 18°$ is shown in Figure 7.47.

In Figure 7.48 we have expressed the lift curve in terms of $\beta_2 = \gamma - i$ as follows:

$$c_L = c_{L,\max}\sin\left(\frac{\pi}{2}\frac{\beta_2 - (\gamma - i_0)}{\beta_{2,s} - (\gamma - i_0)}\right). \tag{7.98}$$

Then the lift curve as a function of $\beta_2$ is effectively rotated 180° from its appearance in Figure 7.47 as illustrated in Figure 7.48. This is the reason for the reversal of the $c_L$–$\beta$ curve for the compressor blade as compared to that for the turbine blade in Figure 7.42.

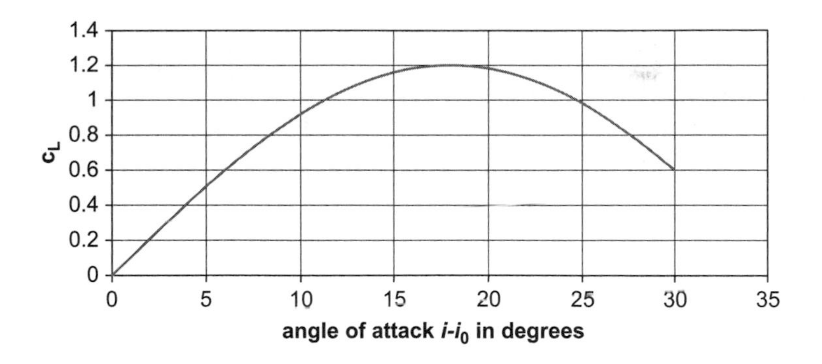


**FIGURE 7.47**

Representative variation of the lift coefficient as a function of angle of attack for a rotor blade.

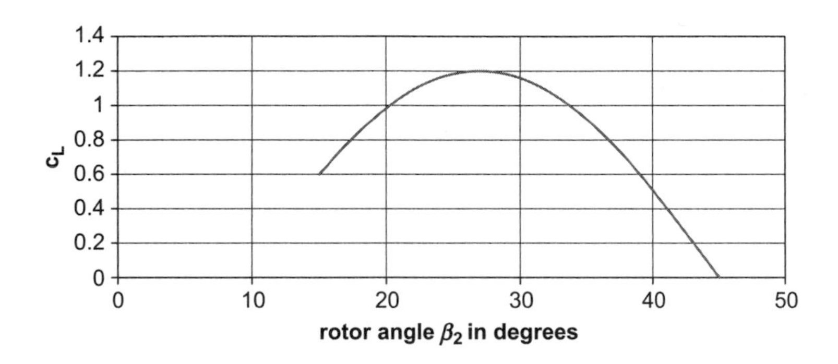


**FIGURE 7.48**

Variation of the lift coefficient as a function of the rotor entrance angle $\beta_2$. Note that increasing $\beta_2$ implies increasing mass flow.

The drag factor is the fourth term in square brackets in Equation (7.93), and its behavior as a function of angle of attack is shown in Figure 7.49 and as a function of rotor entrance angle in Figure 7.50 for a nominal drag to lift ratio $\varepsilon = 0.02$. Note that the drag factor decreases as the mass flow increases.

Now that all the terms in Equation (7.93) have been considered in some detail, we may examine the variation of the corrected work as a function of corrected weight flow for a given value of corrected rpm, for which we have

$$\frac{W_c}{\theta_2} = K_c\left[\frac{w'\sqrt{\theta_2}}{\delta 2}\frac{1}{\sin\beta_2}\right]c_L(1+\varepsilon\cot\beta_2). \tag{7.99}$$

The first term in the square brackets tends to decrease with mass flow, the second term increases and then decreases with mass flow (Figure 7.48), and the last term decreases with mass flow (Figure 7.50). The net effect is to have a curve of corrected work versus corrected weight flow that resembles the $c_L$ vs $\beta_2$ curve in Figure 7.48. Then for each increasing value of corrected rpm the corrected work curves

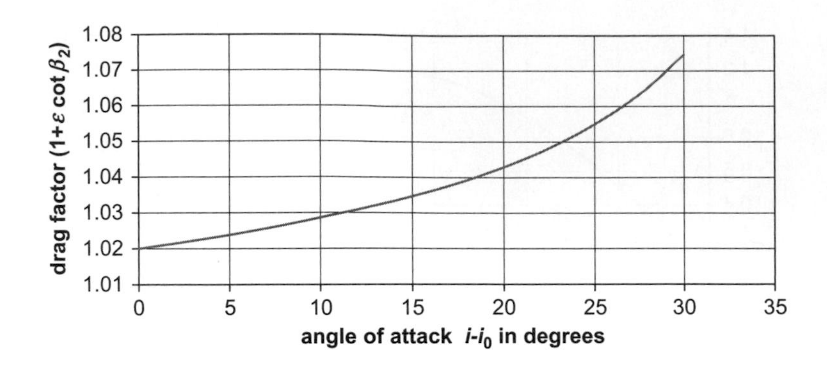


**FIGURE 7.49**

Variation of the drag factor as a function of the angle of attack of the rotor blade.

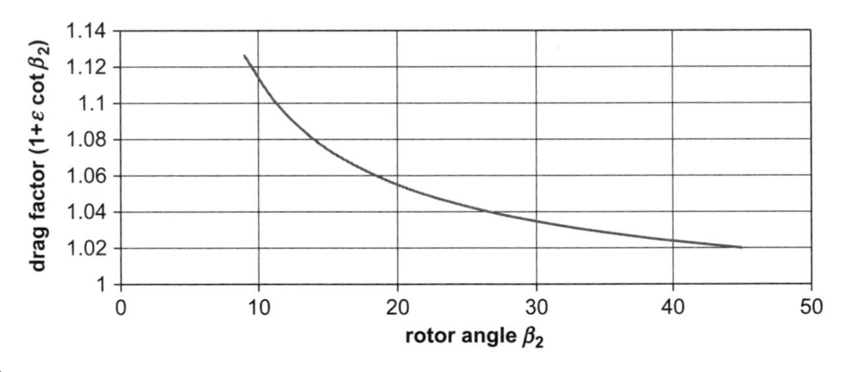


**FIGURE 7.50**

Variation of the drag factor as a function of the rotor entrance angle $\beta_2$. Note that increasing $\beta_2$ implies increasing mass flow.

will be repeated as shown in Figure 7.51. The maximum corrected work values correspond to the $c_{L,\max}$ values of the individual curves for given corrected rpm, and the locus of these points is shown in Figure 7.51. Note that for mass flows lower than that indicated by the locus of maximum lift coefficient, the compressor is operating in the stall region. That is, the angles of attack experienced by the rotor are large enough to provoke flow separation resulting in (usually rapidly) declining lift coefficients. Operation in this region is to be avoided, and the locus of maximum lift coefficient is also called the stall, or surge, line. The performance map may also be put in terms of the compressor pressure ratio $p_{t3}/p_{t2}$, as shown in Figure 7.52; this is the more usual form for the compressor map.

It should be pointed out that the curves for the corrected compressor work or the compressor pressure ratio at constant corrected rpm are more peaked for axial flow machines than for centrifugal machines, as depicted in Figure 7.52. For axial flow machines the curves are very much like the lift curves for airfoils, but reversed as discussed previously. For centrifugal compressors the curves are smoother and broader.

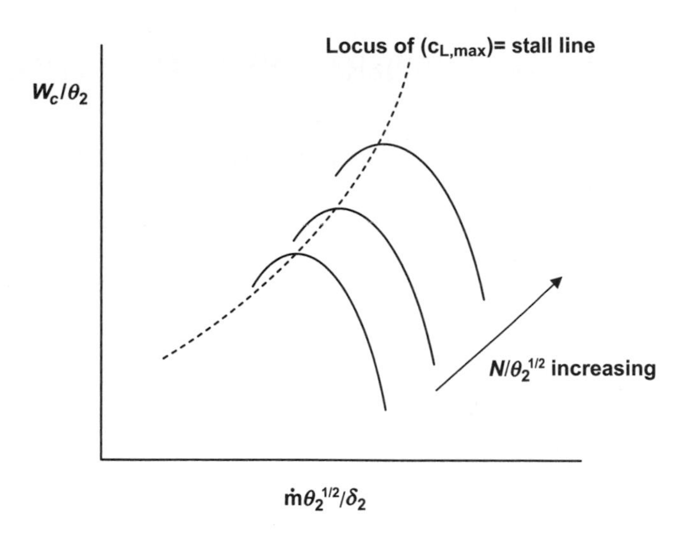


**FIGURE 7.51**

Typical compressor performance map in terms of corrected work and corrected weight flow.

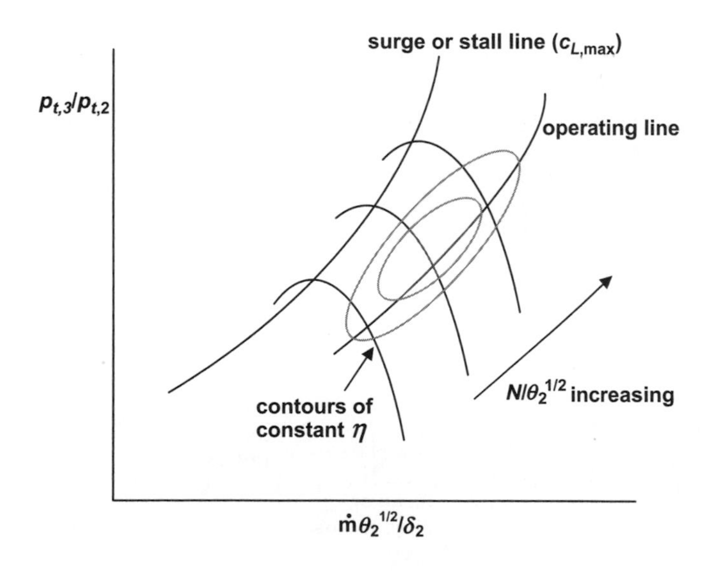


**FIGURE 7.52**

Typical compressor performance map in terms of pressure ratio and corrected mass flow.

## 7.8 THREE-DIMENSIONAL CONSIDERATIONS IN AXIAL FLOW TURBOMACHINES

An important assumption made throughout this chapter is that unlike centrifugal machines, the radial variations of flow properties are negligible. It is useful to examine this assertion in somewhat greater detail. Consider the perspective view of a sector of several blades on an axial turbomachine, as shown in Figure 7.53. The blades are attached to the wheel at their root and extend a height $h$ to the tip. The pitch circle is defined by a mean or pitch radius $r_p$, the average of the tip and root radii. The spacing $s$ between the blades on the pitch circle is called the pitch of the blades. The blades are defined by a camber line that is the locus of points midway between the upper and lower surfaces of the blade and a chord line that extends a length $l$ from the leading to trailing edge of the blade. Lightly loaded compressors and turbines may have a camber that has little deviation from the chord line, whereas heavily loaded ones may display substantial curvature. Remember that the basis for generating a force on the fluid is making the flow turn through some angle. The larger the force, the greater the turn required.

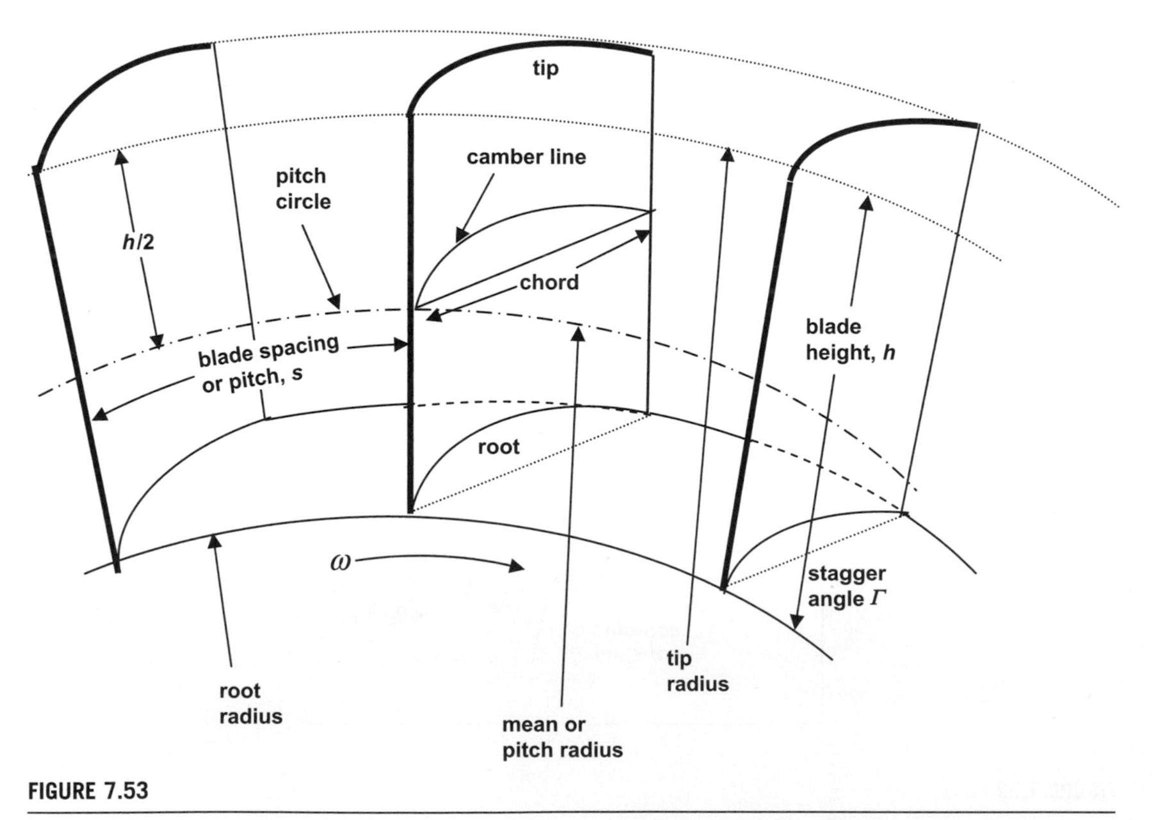


**FIGURE 7.53**

Schematic diagram of blades and passages they form and definition of their various features.

The pitch of the blades is related to the number of blades by the equation $s = 2\pi r_p/n_b$. Then each blade passage, that is, the channel formed by the blades, has an internal surface area approximately equal to $2h(l + s)$ and an entrance area equal to $sh$, and the ratio of the two is $2(1 + l/s)$. The ratio $l/s = \sigma$, called the solidity, and it represents the blockage to the flow caused by the presence of the blades. Larger values of the solidity mean greater passage surface area on which friction can act to retard the flow. The ratio $h/l$ may be considered the aspect ratio of the blades, and the larger this ratio, the less the effects of losses at the tip of the blade due to leakage across the gap between the tip of the blade and the outer casing. The influence of these factors is discussed more fully in Chapter 8.

When the blade height is considerably smaller than the pitch radius $h/r_p \ll 1$ there is little room for any substantial radial motion of the flow, because the wheel surface and the casing surface provide strong guidance for the flow to keep it mainly in the axial direction. In addition, the linear speed of rotation $u = \omega r$ so that small changes in radius lead to proportionally small changes in $u$. However, the change in $u$, although relatively small, does have an effect, which must be considered. This is best seen by examining Figure 7.54, which shows the velocity triangles at the root, pitch, and tip radius locations on a single blade. The absolute velocity is constant in the radial direction because of the guidance

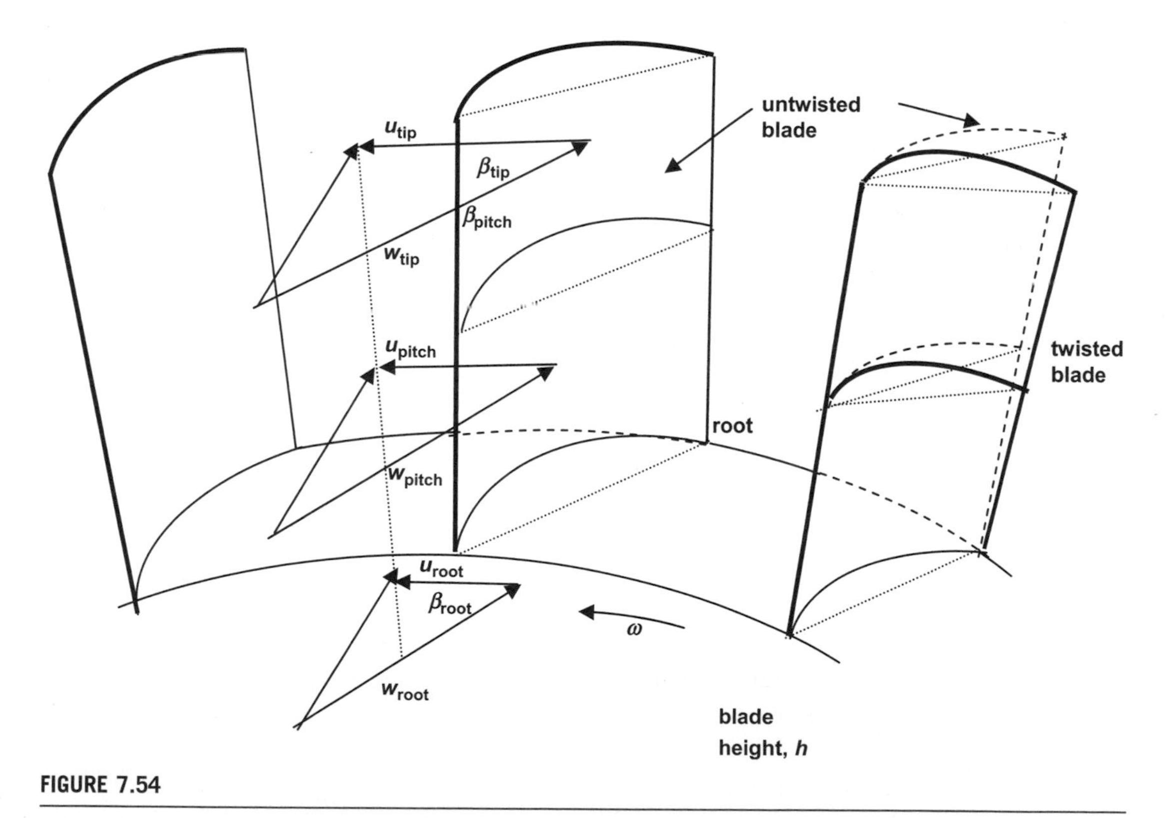


**FIGURE 7.54**

Schematic diagram of blades showing effect of increasing radius on blade angle $\beta$. To compensate for decreasing blade angle $\beta$, the blade must be twisted as shown.

offered by the wheel and the casing, as described previously. However, the linear speed of rotation increases as the radius increases, leading to a reduction of the blade angle $\beta$ in the same direction. Thus, if one wishes to maintain the optimum angle of incidence for the blade along its entire length, the blade must be twisted. Then, because the design conditions have been developed for the pitch radius, they will be met at all radii if the blade has been twisted appropriately. This effect is seen more easily in propellers, where the large change in radius from root to tip requires an appreciable change in blade angle over that span.

## 7.9 NOMENCLATURE

| | |
|---|---|
| $A$ | cross-sectional area |
| $b$ | blade width of centrifugal compressor |
| $c$ | absolute velocity |
| $c_L$ | lift coefficient |
| $c_p$ | specific heat at constant pressure |
| $D$ | drag |
| $e$ | internal energy |
| $F$ | force |
| $g$ | acceleration of gravity |
| $h$ | enthalpy or blade height |
| $i$ | blade incidence |
| $i_0$ | blade incidence at zero lift |
| $i_s$ | blade incidence at stall |
| $K_c$ | constant in a compressor work equation |
| $K_t$ | constant in a turbine work equation |
| $L$ | lift |
| $M$ | Mach number |
| $\dot{m}$ | mass flow |
| $N$ | engine rotational speed in rpm |
| $n$ | number of stages |
| $n_b$ | number of blades |
| $p$ | pressure |
| $P$ | power |
| $Q$ | heat transfer into turbomachine |
| $r$ | radius or degree of reaction in a stage |
| $R$ | gas constant |
| $s$ | entropy or blade spacing |
| $S_b$ | blade area |
| $T$ | temperature |

| | |
|---|---|
| $T_q$ | torque |
| $u$ | tangential velocity |
| $V$ | velocity |
| $W_c$ | work required by the compressor |
| $W_t$ | work delivered by the turbine |
| $w$ | velocity relative to blade |
| z | axial coordinate |
| $\alpha$ | angle between $u$ and $c$ velocity components |
| $\beta$ | angle between $u$ and $w$ velocity components |
| $\Gamma$ | blade stagger angle |
| $\gamma$ | ratio of specific heats |
| $\delta$ | $p_t/p_{ref}$ |
| $\varepsilon$ | drag to lift ratio, $D/L$ |
| $\rho$ | density |
| $\omega$ | angular speed of rotation |
| $\theta$ | $T_t/T_{ref}$ |
| $\eta$ | efficiency |
| $\chi$ | angle included between tangents to leading and trailing edges of a blade |
| $\sigma$ | solidity, $l/s$ |
| $\tau$ | slip coefficient $c_u/u$ |

### 7.9.1 Subscripts

| | |
|---|---|
| *a* | axial direction |
| *ad* | adiabatic |
| *c* | compressor |
| *e* | expansion |
| *ext* | external |
| max | maximum condition |
| *i* | stage designation |
| *p* | conditions on the pitch circle |
| *pc* | polytropic compression |
| *r* | radial direction |
| *%r* | percent reaction |
| *ref* | standard sea level reference condition |
| *t* | stagnation conditions |
| *u* | tangential direction |
| *z* | axial direction |

| | |
|---|---|
| 2 | conditions at the compressor entrance |
| 3 | conditions at the combustor entrance |
| 4 | conditions at the turbine entrance |
| 5 | conditions at the nozzle entrance |
| $\alpha$ | angle between $u$ and $c$ velocity components |
| $\beta$ | angle between $u$ and $w$ velocity components |

### 7.9.2 Superscripts

| | |
|---|---|
| $()'$ | isentropic conditions |
| $()^i$ | $i$th stage |

## 7.10 EXERCISES

**7.1** An axial flow compressor for a jet engine is operating on a test stand under standard sea level atmospheric conditions. The pressure ratio provided by the compressor is $p_3/p_1 = 10$ and processes a mass flow rate of 45.3 kg/s. The dimensions of the compressor are shown in Figure E7.1. Assuming that all processes are isentropic, determine the force $F$ experienced by the load cell on the test stand, the power $P$ required to drive the compressor, and the temperature leaving the compressor $T_3$.

**7.2** A turbojet engine is operated under standard sea level conditions and the following measurements are made at the compressor exit: stagnation temperature equals 216°C and stagnation pressure equals 482 kPa. The airflow through the compressor is determined to be 27.2 kg/s. Determine the (a) work input required per unit mass of air, (b) power required,

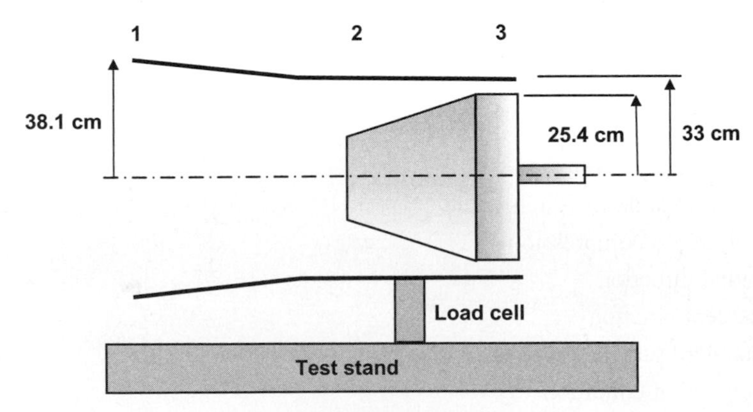


**FIGURE E7.1**

Schematic diagram of compressor installation.

(c) adiabatic compression efficiency of the compressor, and (d) torque exerted on the fluid at a rotational speed of 12,000 rpm.

**7.3** A compressor operates with a stagnation pressure ratio of 4 and an inlet stagnation temperature of 5°C. If the exit stagnation temperature is 171°C, what is the adiabatic efficiency of compression?

**7.4** Axial flow compressors are characterized by flow passages whose radial coordinates change little over the length of the machine, whereas centrifugal compressors have blade passages that start near the axis of rotation and increase substantially in radius by the time the exit of the machine is reached. Some idea of the difference is afforded by Figure E7.2 from NASA showing examples of two such machines side by side. Consider standard sea level air entering a centrifugal flow compressor with the following conditions: $r_2 = 6$ in., $c_2 = 300$ ft/s, and $\alpha_2 = 70°$. The air leaves the rotor with $r_3 = 18$ in., $c_3 = 1200$ ft/s, and $\alpha_3 = 25°$. The rotational speed of the compressor is 9000 rpm and it processes 32.2 lb/s of air. Carry out the following: (a) draw the combined entrance and exit velocity diagrams to scale, (b) determine the entrance and exit blade angles $\beta_2$ and $\beta_3$, (c) determine the torque required to drive the compressor, (d) determine the horsepower required, (e) find the ideal pressure head, (f) determine the contribution to the pressure rise provided by the external, internal, and centrifugal effects, (g) determine the whirl velocity $\Delta c_u$, and (h) determine the ideal pressure ratio of the compressor.

**7.5** Repeat Exercise 4, changing only the entrance radius to $r_2 = r_3 = 18$ in. so that the machine now acts as an axial flow compressor.

**FIGURE E7.2**

Photograph of typical axial flow compressor wheel (left) and centrifugal flow compressor wheel (right).

**7.6** A centrifugal compressor rotor is operated at 10,000 rpm on a test stand where the atmospheric conditions are those of a standard sea level day (Figure E7.3). The compressor has an axial inlet with a hub diameter of 13.35 cm and an eye diameter of 25.40 cm. The compressor has a radial exhaust, and the rotor tip diameter is 50.80 cm. If the average inlet Mach number is 0.7, determine the (a) mass flow through the compressor, (b) power required to drive the compressor, (c) total temperature at the compressor exit, (d) portion of the required power due to the centrifugal effect, and (e) overall pressure ratio of the machine, if the adiabatic efficiency of compression $\eta_c = 85\%$.

**FIGURE E7.3**

Photograph of the entry region of a centrifugal flow compressor wheel showing the hub bolted around the shaft and the eye, defined here by the locus of the sharp blade tips.

**7.7** Carry out Exercise 6 for the case where instead of the engine being operated on a test stand it is operating in an aircraft flying at $M_0 = 0.7$ at sea level where the atmospheric conditions are those of a standard day.

**7.8** A radial bladed centrifugal compressor designed with zero prewhirl (axial entry) for a turbojet engine is shown in Figure E7.4 and has the following data:

slip angle = 14°
impeller tip radius = 26.3 cm
impeller eye radius = 15 cm
impeller hub radius = 7 cm
impeller tip width = 4.4 cm
number of vanes = 29
diffuser inner radius = 30.5 cm
diffuser outer radius = 40.6 cm
N = 16,750 rpm

The compressor operates with an axial inlet Mach number of 0.8 at maximum rpm at the impeller inlet as the aircraft moves through the atmosphere at standard sea level conditions.

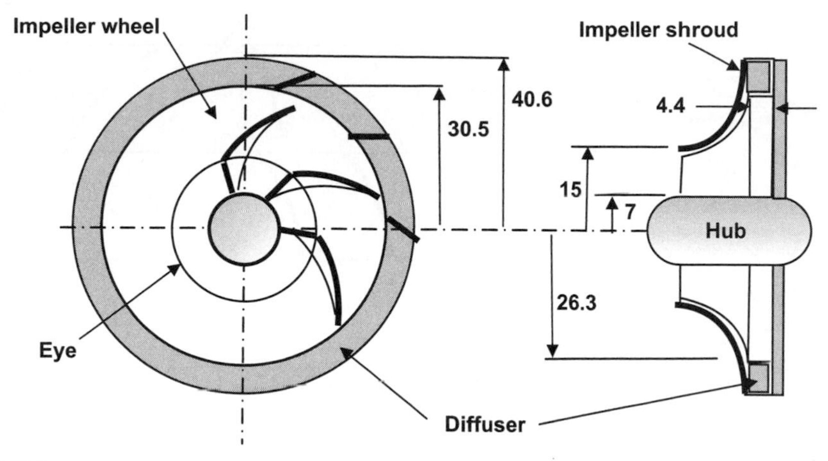


**FIGURE E7.4**

Sketch of a centrifugal compressor (all dimensions in cm).

**(a)** Determine the linear speed of rotation at the impeller tip, eye, and hub.

**(b)** Determine the inlet flow area allowing for 5% blockage by vane thickness.

**(c)** Sketch the inlet blade design showing proper diameters and blade angles at the hub, eye, and at the station midway between the two.

**(d)** Calculate the airflow rate.

**(e)** If $\tau = 0.94$, draw the exit velocity diagram to scale and indicate all magnitudes.

**(f)** If $\eta_c = 80\%$, determine $p_3$, $p_{t3}$, $T_3$, and $T_{t3}$.

**(g)** What is the entrance angle for the fixed diffuser vanes? (Note that angular momentum of the fluid is conserved in the annulus between the impeller tip and the diffuser entrance.)

**(h)** What is the diffuser entrance Mach number?

**(i)** What is the pressure ratio of the compressor $p_{t3}/p_{t2}$?

**(j)** What is the torque and power required to drive the impeller?

**7.9** Redo Exercise 8 at a reduced rotational speed N = 6750 rpm. Discuss major differences between the two cases.

**7.10** Consider a staged axial flow compressor with the general configuration shown in Figure E7.5. Only the rotor of the first stage is shown and it accepts air from the inlet guide vanes with the following conditions: $p_2 = 96.2$ kPa and $T_2 = 289$K. The rotor mean line has $r = 30.5$ cm, $\alpha_2 = 60°$, $\beta_2 = 80°$, and $u_2 = 427$ m/s.

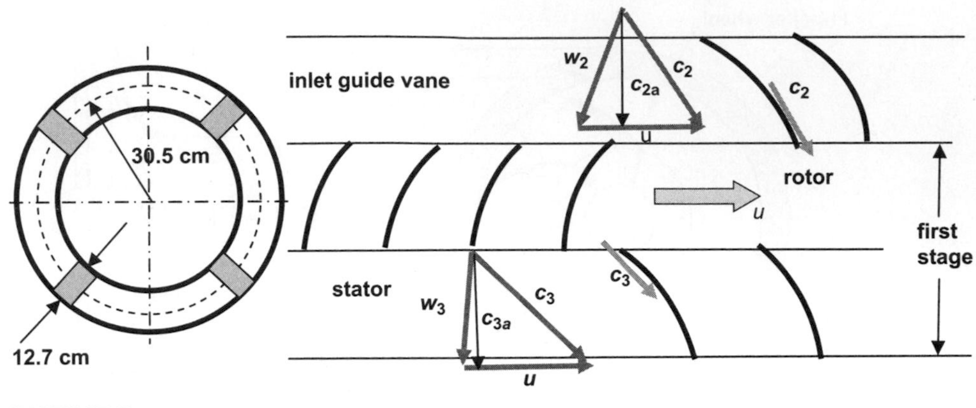


**FIGURE E7.5**

Sketch of a stage of an axial flow compressor.

**(a)** Draw the velocity diagram of the entry to the rotor and find $w_2$, $c_2$, and $c_{2a}$ and then find $p_{t2}$, $T_{t2}$, and $M_2$, where the last three items are based on the absolute velocity $c_2$.

**(b)** Determine N, the rotational speed of the machine in rpm.

**(c)** If the stagnation pressure ratio across the rotor is 1.15 and $\eta_c = 85\%$, find $c_{3u} - c_{2u}$.

**(d)** Draw the exit velocity diagram to scale and find $w_3$, $c_3$, $\beta_3$, and $\alpha_3$, assuming $c_{2a} = c_{3a}$.

**(e)** Find the blade angle of the first stator passage at the median radius $r$ of the rotor.

**(f)** Determine the airflow rate for the compressor if the blade height is 12.7 cm, assuming that 5% of the flow passage is blocked by blade-leading edges.

**(g)** Calculate the required torque, power, and work per unit mass for this first stage.

**(h)** Determine the power required to drive a compressor using 13 stages like the first stage.

**(i)** Calculate the static pressure rise achieved in the one stage shown.

**(j)** Determine the percent reaction of the one stage shown.

**7.11** A 50% reaction stage of an axial flow turbine shown in Figure E7.6 has blades 12.7 cm in height and a linear speed of rotation $u = 366$ m/s at the mean line of the rotor where the diameter is $d = 76.2$ cm. The stagnation temperature and stagnation pressure entering the stator from the combustor are $T_t = 1144$K and $p_t = 687$ kPa. The angle for the stator exit absolute velocity is $\alpha_4 = 25°$.

**(a)** Draw two possible combined velocity diagrams for this stage. The diagrams should be neat, executed with a straight edge or by computer, of reasonable size, clearly marked, and shown to scale. The combined velocity diagram is one that shows the inlet and

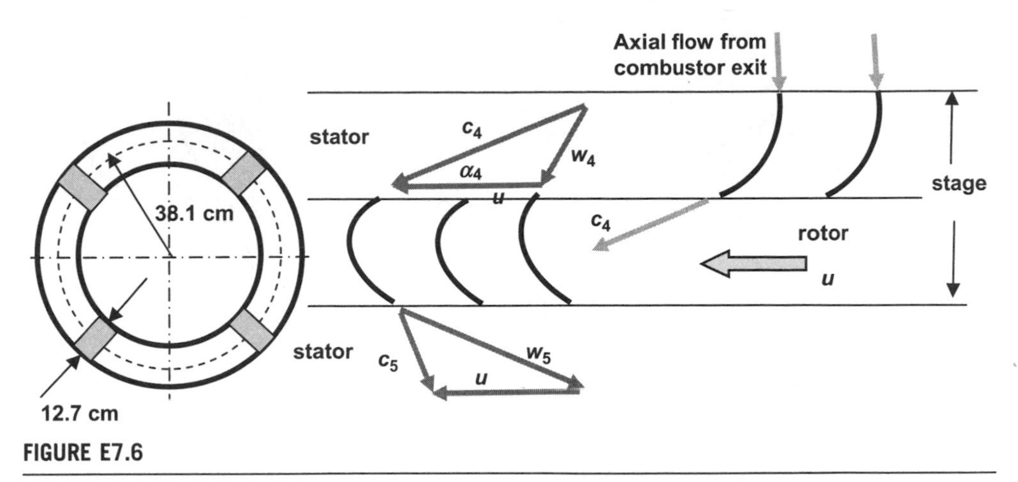


**FIGURE E7.6**

Sketch of an axial flow turbine.

outlet velocities referenced to a single $u$ velocity. For example, the general velocity diagrams shown in Figure E7.6 may be superimposed to form a combined velocity diagram.

**(b)** Calculate the work per unit mass $W$ for the case where the turbine is extracting the maximum work per unit of entering kinetic energy $W/(c_4^2/2)$.

**(c)** Determine the blade efficiency $\eta_b$, which is defined as $\eta_b = u/c_4$, for the case of maximum work described in part (b) and draw the corresponding combined velocity diagram.

**(d)** Plot the distribution of stagnation and static values of pressure and temperature, as well as the Mach number, through the stage as a function of $x$, starting from the entrance to the stator blades and ending at the exit of the rotor blades.

**(e)** If the blades block 5% of the flow area, determine the mass flow rate through the turbine.

**(f)** Determine the power developed by the turbine.

**7.12** A turbine blade receives hot gas from a stator with an outlet angle of 70° as shown in Figure E7.7. The blade is designed to have 3° incidence at the root when the linear speed of rotation at the root is 213.4 m/s. The absolute speed leaving the stator is 548.6 m/s and the radius of the turbine disc at the root.

**(a)** Find the rotor blade angle $\beta_4$ for a blade that is untwisted over its entire length.

**(b)** Find the incidence at the tip for this untwisted blade.

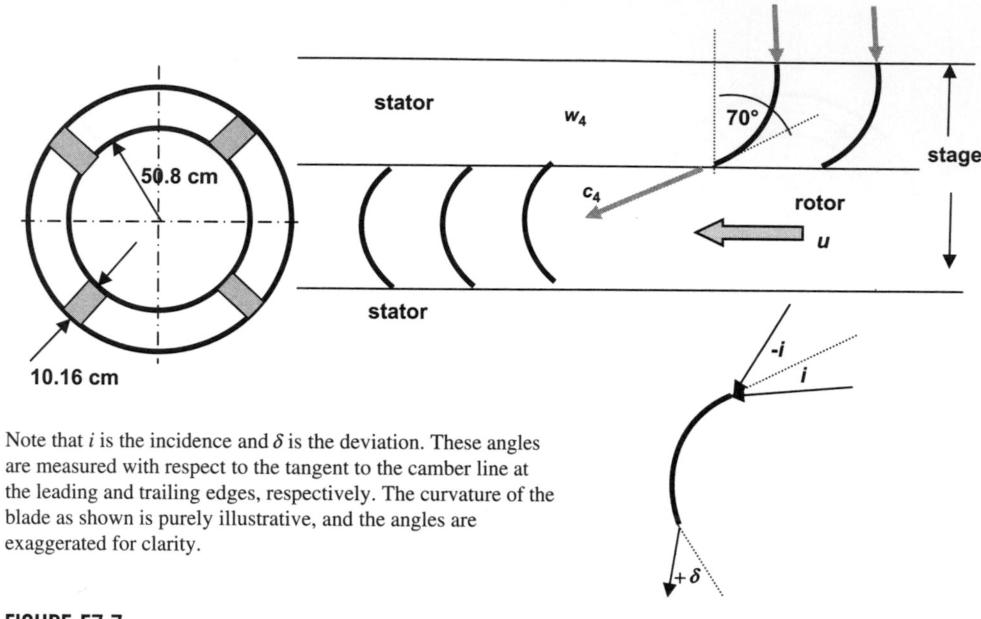


**FIGURE E7.7**

Sketch of an axial flow turbine stage.

**(c)** Find the stator outlet angle $\alpha_4$ and the rotor blade inlet angle $\beta_4$ at the root and at the tip under the following set of assumptions:

**(i)** The original conditions pertain only to the root station.

**(ii)** There is free vortex flow in the gap between the stator and the rotor, which implies that $uc_u =$ constant through the gap and, likewise, the axial component $c_a =$ constant.

**(iii)** The incidence at all points along the leading edge of the blade is 3°.

**(iv)** The rotor outlet absolute velocity $c_4$ is 231.6 m/s and is in the axial direction.

**(v)** The deviation is 5° at all points along the trailing edge of the stator blade.

**7.13** An axial flow turbine blade 10.2 cm in height is attached to a 50.8-cm radius turbine wheel as shown in Figure E7.8. The blade is designed to receive air from a stator that has an outlet angle $\alpha_4 = 20°$. When the linear speed of rotation at the root of the blade (i.e., at the point where the blade is attached to the wheel) is 213 m/s, the angle of incidence of the chord line of the blade there is $i = 3°$. The absolute velocity from the stator outlet is constant at $c_4 = 549$ m/s. Consulting Figure E7.8 and assuming incompressible flow,

**(a)** Find the stagger angle $\gamma$ of the blade chord at the root.

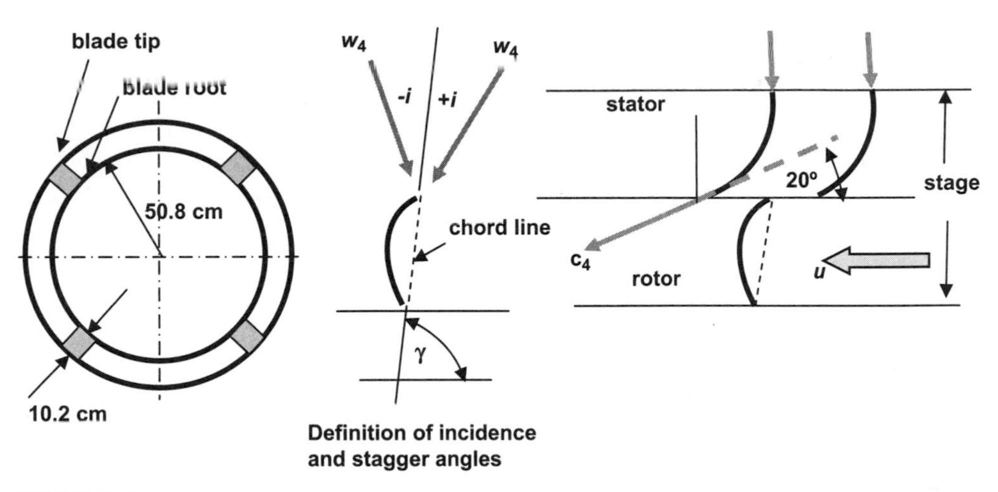


**FIGURE E7.8**

Sketch of an axial flow turbine blade.

**(b)** If the blade is untwisted throughout its height, find the angle of incidence at the tip of the blade under the given conditions.

**(c)** Find the work performed per unit mass of fluid processed if the rotor blade provides an exit absolute velocity that is purely axial.

**7.14** An axial flow compressor stage is designed for 50% reaction, a rotational speed of 13,760 rpm, and a mean radius of 25.4 cm. Carry out all analyses at the mean radius location and assume the flow is incompressible in this problem.

**(a)** If the rotor blade relative velocity exits at an angle $\beta_3 = 70°$ and $c_a = 152$ m/s, draw a velocity diagram for the entrance and the exit of the rotor at the mean radius approximately to scale and determine all appropriate angles and velocities and note them directly on the diagram.

**(b)** If the airflow rate is 36.2 kg/s, find the torque and the power required and the ideal total head produced.

CHAPTER

# 8 Blade Element Theory for Axial Flow Turbomachines

## CHAPTER OUTLINE



## 8.1 CASCADES

The flow through a rotor described in previous sections is characterized by motion through a number of identical flow passages distributed periodically around the periphery of the device. However, the motion may also be considered as that passing around a number of identical vanes distributed periodically around the periphery of the device. These periodic structures constitute a cascade of vanes,

Theory of Aerospace Propulsion.

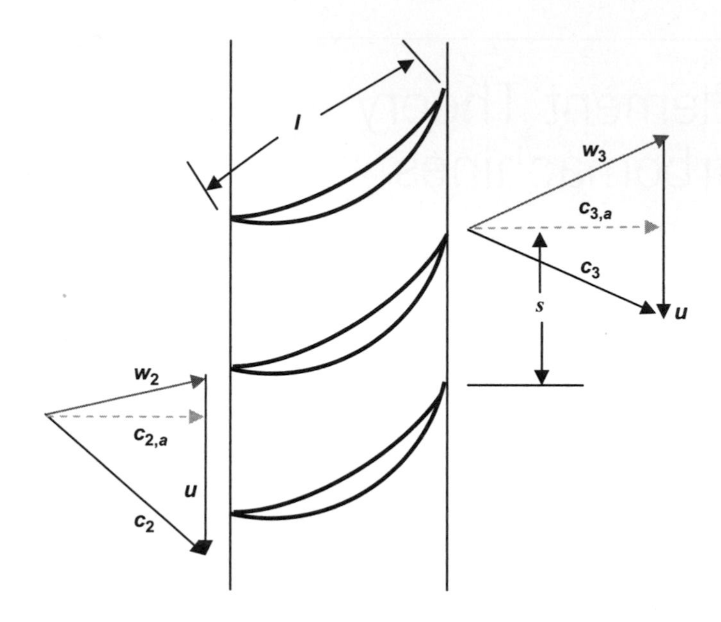


**FIGURE 8.1**

The straight, or two-dimensional, cascade showing velocity components entering and leaving.

which are generally idealized in two forms. The compressor numbering system will be used here, although turbines may be analyzed in the same manner. One is the straight, or two-dimensional, cascade shown in Figure 8.1, and the other is the circular cascade shown in Figure 8.2.

The flow in a cascade repeats itself as one travels a linear distance equal to the pitch distance $s$ along the cascade or an angular distance equal to the pitch angle $\theta_p$. If the cascade is moving, one may choose a coordinate system fixed to the cascade, and in that system the cascade will be stationary with flow passing through it. Note that in the case of a circular cascade, which looks like a section of a centrifugal turbomachine, such a reference frame is rotating and the flow with respect to that frame cannot be a potential flow. In a frame of reference rotating at angular velocity $\omega$, the vorticity vector $\Omega$ is given by

$$\vec{\Omega} = \nabla \times \vec{V} - 2\vec{\omega}. \tag{8.1}$$

Thus even if the curl of the velocity vector is zero, the vorticity is nonzero in a rotating frame, so the simplification possible by dealing with a nonrotating cascade is limited to the straight cascade, which is a reasonable representation of flow in an axial flow turbomachine.

## 8.2 STRAIGHT CASCADES

Circulation is an integral around a closed contour $C$ expressed as follows:

$$\Gamma = \oint \vec{V} \cdot dC. \tag{8.2}$$

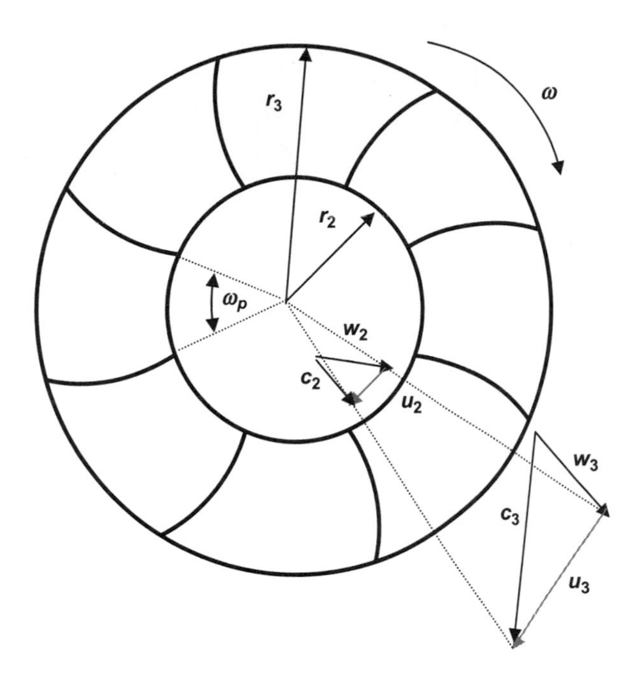


**FIGURE 8.2**

A circular cascade showing velocity components entering and leaving.

Circulation is considered positive when taken in the clockwise sense. In a straight cascade, circulation about an individual vane may be determined with the aid of the diagram in Figure 8.3. Because we are considering a stationary straight cascade, whether it is actually stationary like a stator cascade or it is a moving cascade, which is brought to rest by the addition of $-u$ to the flow, we will use the notation for absolute velocity, namely $c^2 = c_a^2 + c_u^2$. The contour is taken to be that delineated by *ABCD*, and integration is taken in the counterclockwise sense so that circulation may be written as shown in Equation (8.3):

$$-\Gamma = \int_C \vec{V}\cdot d\vec{C} = c_{3u}s + \int_B^C \vec{V}\cdot d\vec{C} - c_{2u}s + \int_D^A \vec{V}\cdot d\vec{C}. \tag{8.3}$$

The contribution from segment *AB* is calculated as illustrated in Equation (8.4):

$$\int_A^B \vec{V}\cdot d\vec{C} = \int_A^B (c_{3u}\hat{j}\cdot dy\hat{j}) = c_{3u}s. \tag{8.4}$$

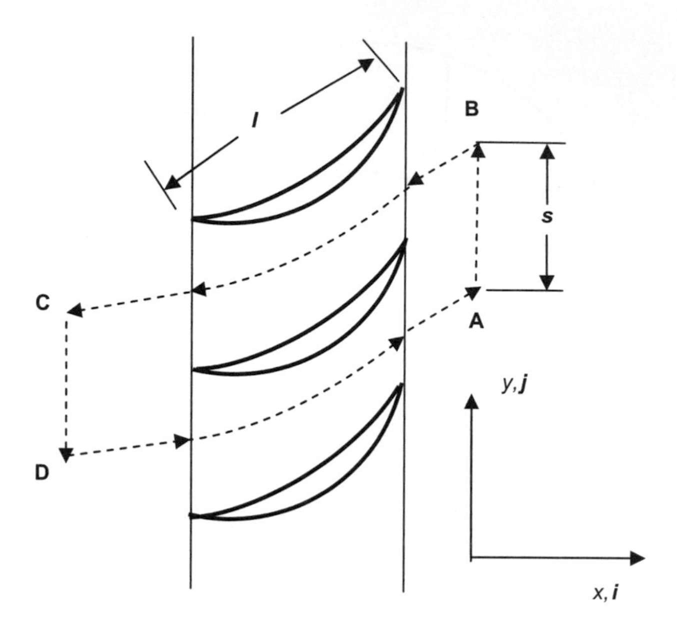


**FIGURE 8.3**

The contour taken around an individual vane.

Similarly, the contribution from the segment *CD* is calculated as illustrated in Equation (8.5):

$$\int_C^D \vec{V}\cdot d\vec{C} = \int_C^D (c_{2u}\hat{j}\cdot(-dy\hat{j})) = -c_{2u}s. \tag{8.5}$$

Referring again to Figure 8.3, it is clear that the periodicity of the flow as described previously ensures that the contribution from segment *BC* is equal and opposite to that from segment *DA* so that the circulation reduces to

$$-\Gamma = (c_{3u} - c_{2u})s. \tag{8.6}$$

As shown previously in Equation (7.15), the tangential, or turning, force

$$F_u = \dot{m}(c_{3u} - c_{2u}).$$

We may now express this force in terms of the circulation as follows:

$$F_u = -\frac{\dot{m}\Gamma}{s}. \tag{8.7}$$

The mass flow passing through the region of the contour may be written as

$$\dot{m} = \rho c_a sh, \tag{8.8}$$

where the flow area is the product of the spacing $s$ and the height of the cascade passage $h$.

The tangential force on the fluid may then be written as

$$F_u = -\rho c_a h \Gamma. \tag{8.9}$$

In the same fashion, we recall that the axial force on the fluid is

$$F_a = \dot{m}(c_{3a} - c_{2a}). \tag{8.10}$$

Therefore, because the pitch is fixed and the height of the passage is uniform, mass conservation requires that

$$\rho_2 c_{a2} = \rho_3 c_{a3}.$$

For incompressible flows, this means that the axial component of velocity $c_a =$ constant and the net external axial force acting on the fluid $F_a = 0$. Note that in compressors for typical aerospace applications, the fluid can rarely be considered incompressible.

The definition of an incompressible fluid is that the density is independent of the pressure so that as pressure rises in a compressor processing an incompressible fluid, there is no corresponding change in density. For incompressible fluids, the compressor is generally spoken of as a pump moving a mass flow against a pressure gradient, for example, vertically to higher levels. In a gas compressor, the ideal process is an isentropic compression where $p \sim \rho^\gamma$ so that

$$\frac{c_{3a}}{c_{2a}} = \frac{\rho_2}{\rho_3} = \left(\frac{p_2}{p_3}\right)^{1/\gamma} = \left(\frac{p_3}{p_2}\right)^{-1/\gamma}. \tag{8.11}$$

Assuming, as an illustration, a compression pressure ratio of 1.2 across the cascade row and $\gamma = 1.4$ leads to an axial velocity ratio $c_{3a}/c_{2a} = 0.88$, demonstrating that the pressure rise causes a density rise, in turn requiring a 12% decrease in the axial velocity component.

Now consider the control volume bounded by the contour C and the forces acting on it, as shown in Figure 8.4. The net force on the fluid in the tangential, or turning, direction is

$$F_u = F'_u + \int_A -p(x,y)\hat{n}\cdot\hat{j}dA = F'_u.$$

Likewise, the net force on the fluid in the axial direction is

$$F_a = F'_a + \int_A -p(x,y)\hat{n}\cdot\hat{i}dA = F'_a + (p_3 - p_2)sh. \tag{8.12}$$

The integrals in Equations (8.11) and (8.12) are taken over the contour ABCD, and because the curved portions $BC$ and $DA$ are assumed to be streamlines, no mass crosses them. However, because the flow is spatially periodic and repeats itself over each interval of the blade pitch $s$, the pressure forces generated on the BC and CD boundaries are equal and opposite in each case so that the integrals on these boundaries cancel themselves in both the tangential and the axial directions.

Focusing attention on the forces the fluid exerts on the blade and assuming frictionless flow, the resultant force on the blade may be considered a lift force and thus

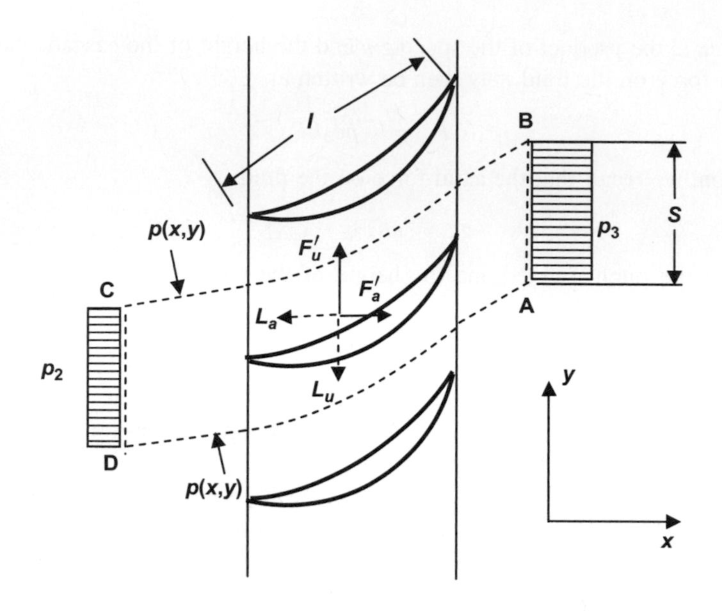


**FIGURE 8.4**

Control volume bounded by contour C showing the constant pressure acting on the entrance and exit faces and the varying pressure acting on the curved boundary enclosing the vane. Forces $F_u'$ and $F_a'$ represent force components on the fluid produced by the boundary of the vane, and $L_u$ and $L_a$ represent equal and opposite force components on blade boundaries produced by the fluid.

$$\begin{aligned} L_u &= -F_u' = -\dot{m}(c_{3u} - c_{2u}) \\ L_a &= -F_a' = (p_2 - p_3)sh - \dot{m}c_{3a}\left(1 - \frac{\rho_3}{\rho_2}\right). \end{aligned} \tag{8.13}$$

For an incompressible frictionless fluid flowing through the cascade, we may write Bernoulli's equation along any streamline within the control volume as follows:

$$p_2 + \frac{1}{2}\rho\left(c_{2u}^2 + c_{2a}^2\right) = p_3 + \frac{1}{2}\rho\left(c_{3u}^2 + c_{3a}^2\right). \tag{8.14}$$

As a consequence of the incompressibility of the fluid, that is $c_{2a} = c_{3a}$, Equation (8.14) becomes

$$p_2 - p_3 = \frac{1}{2}\rho\left(c_{3u}^2 - c_{2u}^2\right). \tag{8.15}$$

Now, using Equations (8.9) and (8.13), the lift component in the tangential direction is

$$L_u = -\dot{m}(c_{3u} - c_{2u}) = \rho c_a h \Gamma. \tag{8.16}$$

Similarly, using Equations (8.13) and (8.15), the lift component in the axial direction becomes

$$L_a = (p_3 - p_2)sh = \frac{1}{2}\rho\left(c_{3u}^2 - c_{2u}^2\right)sh. \tag{8.17}$$

Noting that $c_{3u}^2 - c_{2u}^2 = (c_{3u} + c_{2u})(c_{3u} - c_{2u})$ and inserting this result into Equation (8.17) yields

$$L_a = -\frac{1}{2}\rho(c_{3u} + c_{2u})h\Gamma = -\rho\bar{c}_u h\Gamma. \tag{8.18}$$

The previous equation introduces the average tangential velocity component through the cascade

$$\bar{c}_u = \frac{1}{2}(c_{3u} + c_{2u}). \tag{8.19}$$

The resultant, or lift, force on the blade is

$$L = \sqrt{L_u^2 + L_a^2} = \rho\bar{c}\Gamma h. \tag{8.20}$$

In Equation (8.20), the velocity

$$\bar{c} = \sqrt{\bar{c}_u^2 + c_a^2}. \tag{8.21}$$

The force field acting on the blade and the corresponding equivalent velocity field are shown in Figure 8.5. Note that the angle $\alpha$ may be written as follows:

$$\tan\alpha = \frac{L_u}{L_a} = \frac{\rho c_a \Gamma h}{\rho c_u \Gamma h} = -\frac{c_a}{\bar{c}_u}. \tag{8.22}$$

This shows that the resultant force, the lift, acts normal to $\bar{c}$ and therefore, in the sense of the Kutta–Joukowski law for isolated airfoils with circulation, the relative wind is essentially $\bar{c}$. Recall that this is a straight, or two-dimensional, cascade and that all the results pertain to purely two-dimensional flow. The foregoing results could also have been obtained using the Blasius theorem for irrotational flow about a two-dimensional cylinder. In that instance, the force in the axial ($x$) direction is given by

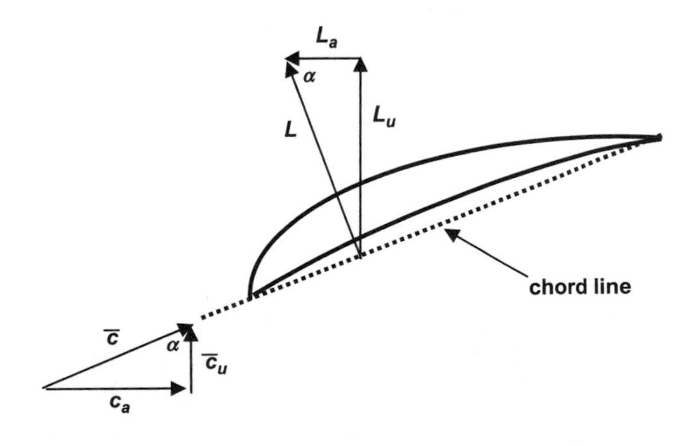


**FIGURE 8.5**

Force field acting on the blade and the corresponding equivalent velocity field.

$-\rho V_y \Gamma$ and the force in the tangential ($y$) direction is given by $\rho V_x \Gamma$, where $V_x$ and $V_y$ denote the $x$ and $y$ components of velocity, respectively. The forces, in this form, are independent of the airfoil shape. The moment acting on the airfoil is also given by the Blasius theorem, but because this is dependent on the airfoil shape, it will not be given here.

In the case of compressible flow, the Bernoulli equation in the form shown in Equation (8.14) is not applicable. The compressible form of the Bernoulli equation is given by

$$\frac{p_2}{\rho_2} + \frac{\gamma - 1}{\gamma}\frac{c_2^2}{2} = \frac{p_3}{\rho_3} + \frac{\gamma - 1}{\gamma}\frac{c_3^2}{2}. \tag{8.23}$$

However, to use the compressible form of the Bernoulli equation leads to tedious calculations. Instead, for the small pressure rises in typical compressor stages we may use the result from Equation (7.54) that the density across the stage may be considered constant and equal to $\overline{\rho} \approx \eta_{ad,c}\rho_{t,2}$, at least for initial design studies. In that case the result of Equation (8.20) holds with

$$L_u = -\overline{\rho} c h \Gamma. \tag{8.24}$$

If we now give our stationary cascade a constant translational speed $u$ in the positive $y$ direction, as shown in Figure 8.6, we may consider a coordinate system attached to the blade such that the velocity relative to the blade is

$$\overline{w} = \overline{c} - \overline{u}. \tag{8.25}$$

To move the cascade in the direction shown requires an input of power

$$P = -uL_u = u[\rho c_a s h(c_{3u} - c_{2u})]. \tag{8.26}$$

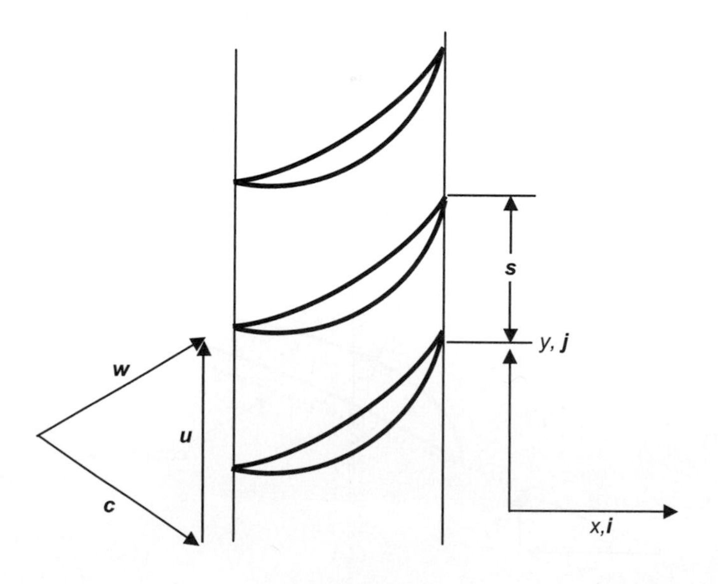


**FIGURE 8.6**

A straight cascade moving vertically with constant speed $u$ showing the relative velocity $w$, which faces an apparently stationary blade.

We have developed equations for the cascade at rest in this section and we may now apply them to this moving cascade by recalling that as far as the relative velocity $w$ is concerned, the cascade is stationary. Then, for the case of incompressible flow the change in total pressure across the stage is

$$p_{t3} - p_{t2} = p_3 - p_2 + \frac{1}{2}\left(c_3^2 - c_2^2\right). \tag{8.27}$$

However, we have already determined that the pressure difference in the stationary blade passage involves velocities relative to the blade passage, as in Equation (8.14), and this may be described in terms of the velocity $w$ relative to the blade:

$$p_3 - p_2 = \frac{1}{2}\rho\left(w_2^2 - w_3^2\right). \tag{8.28}$$

Thus, in incompressible flow, the total pressure change across the steadily moving cascade is

$$p_{t3} - p_{t2} = \frac{1}{2}\rho\left[\left(c_3^2 - c_2^2\right) + \left(w_2^2 - w_3^2\right)\right]. \tag{8.29}$$

This is the result obtained in Section 7.2 by other means. Recall that the relation between the velocity components is

$$c^2 + u^2 - w^2 = 2uc_u. \tag{8.30}$$

Then, because the linear speed $u$ is constant, Equation 8.29 becomes

$$p_{t3} - p_{t2} = \rho u\left(c_{3u} - c_{2u}\right). \tag{8.31}$$

## 8.3 ELEMENTAL BLADE FORCES

In axial flow compressors, the centrifugal effect is minimal because the radial component of velocity is suppressed and the flow is basically quasi-two-dimensional, a function primarily of the axial and azimuthal coordinates $z$ and $\theta$. Thus $c_r \sim 0$ and $w_r \sim 0$, and $c = c(z,\theta)$ and $w = w(z,\theta)$. In this sense, we may imagine the flow through the rotor to be composed of a number of infinitesimal annuli, each essentially independent of the other, in which the flow interaction with the element of blade height $dr$ occurs. This scheme is shown schematically in Figure 8.7.

Consider the flow entering the rotor at station 2 according to the diagram in Figure 8.8. Here $dF$ is the resultant aerodynamic force acting on the blade element of height $dr$, and $dF_a$ and $dF_u$ are the force components in the axial and tangential direction, respectively. The angle $i_m$ is the mean angle of incidence of the blade and has a somewhat different definition from that usually employed in the study of an isolated airfoil in a uniform stream. Recall that in the discussion of linear cascades in the previous section we found that the lift force due to frictionless flow over a blade in the cascade was determined in terms of its circulation as follows:

$$L = \sqrt{L_u^2 + L_a^2} = \rho\bar{c}\Gamma h\,. \tag{8.32}$$

The lift acts normal to the velocity given by

$$\bar{c} = \sqrt{\bar{c}_u^2 + c_a^2}. \tag{8.33}$$

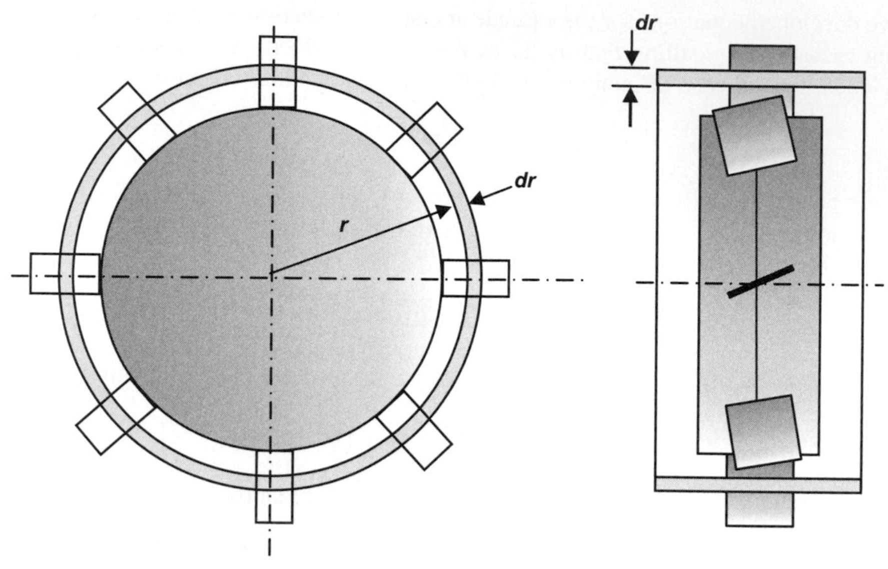


**FIGURE 8.7**

Schematic diagram of axial flow turbomachine showing the general annulus of thickness *dr* in which blade element analysis is carried out.

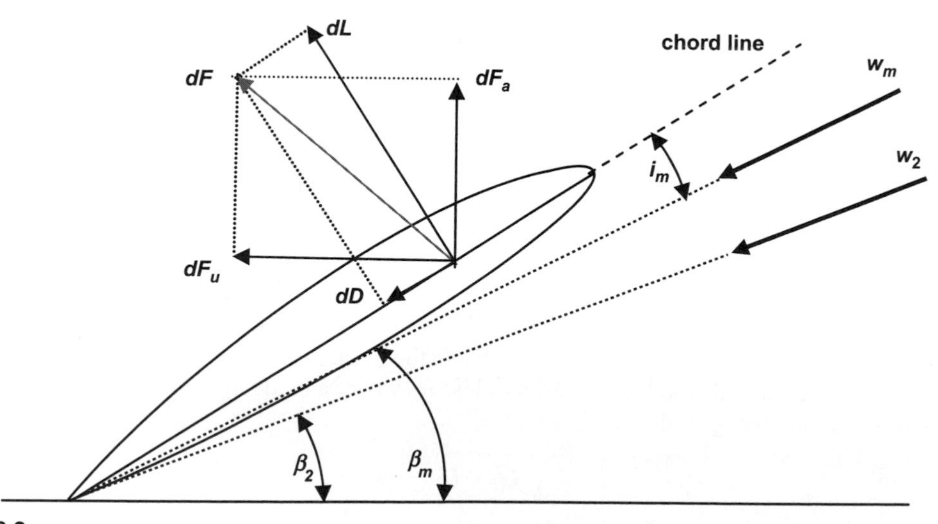


**FIGURE 8.8**

Schematic diagram of flow past blade element of height *dr* showing the forces acting and the incoming velocity relative to the blade element.

In this velocity, quantity $c_a$ is the axial component of velocity passing through the cascade, while the other quantity

$$\bar{c}_u = \frac{1}{2}(c_{3u} + c_{2u}). \tag{8.34}$$

This term is the average value of the tangential component of velocity entering and leaving the blade passage. If the cascade is given a uniform velocity $u$, we have the equivalent case of a rotor blade rotating at an angular velocity $\omega = u/r$ and the blade therefore sees a relative velocity $w$. Thus, the appropriate velocity upon which the blade lift should be based will be analogous to that for the stationary cascade, that is,

$$w_m = \sqrt{w_{mu}^2 + w_a^2}. \tag{8.35}$$

Here the components involved are defined as follows:

$$\begin{aligned} w_{mu} &= \frac{1}{2}(w_{2u} + w_{3u}). \\ w_a &= c_a \end{aligned} \tag{8.36}$$

The combined velocity diagram for the flow passing the blade element of Figure 8.7 is shown in Figure 8.9 where the quantity $\beta_m$ is defined along with the different $u$ components of the relative velocities. Then the differential turning and axial forces are given in terms of the differential lift and drag forces as follows:

$$\begin{aligned} dF_u &= dL \sin \beta_m + dD \cos \beta_m \\ dF_a &= dL \cos \beta_m - dD \sin \beta_m. \end{aligned} \tag{8.37}$$

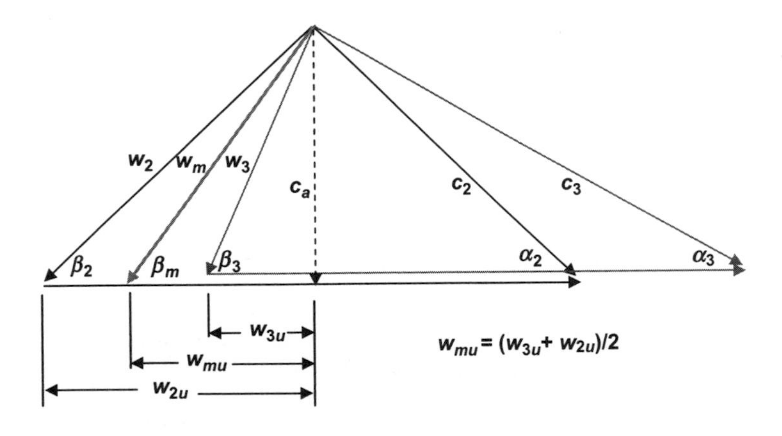


**FIGURE 8.9**

Combined velocity diagram for flow past a blade element showing $w_m$, the appropriate velocity upon which to base the lift force.

The differential lift and drag forces in turn are defined in terms of blade properties:

$$\begin{aligned} dL &= c_l\left(\frac{1}{2}\rho w_m^2\right) l dr \\ dD &= c_d\left(\frac{1}{2}\rho w_m^2\right) l dr. \end{aligned} \tag{8.38}$$

Combining Equations (8.37) and (8.38) yields

$$\begin{aligned} dF_u &= c_l\left(\frac{1}{2}\rho w_m^2\right) l dr\left[\sin\beta_m + \frac{c_d}{c_l}\cos\beta_m\right] \\ dF_a &= c_l\left(\frac{1}{2}\rho w_m^2\right) l dr\left[\cos\beta_m - \frac{c_d}{c_l}\sin\beta_m\right]. \end{aligned} \tag{8.39}$$

## 8.4 ELEMENTAL BLADE POWER

Recalling from the study of cascades [Equation (8.13)] that the turning force is proportional to the change in $c_u$ and that the axial force is proportional to the change in pressure, we may write, still assuming constant average density across the stage, the following:

$$\begin{aligned} dF_u &= (c_{3u} - c_{2u})\rho c_a s dr \\ dF_a &= (p_3 - p_2) s dr. \end{aligned} \tag{8.40}$$

Equating the results from Equations (8.39) and (8.40) yields the following results for the change in tangential component of absolute velocity and the change in pressure across the stage:

$$\begin{aligned} (c_{3u} - c_{2u}) &= \frac{c_l\left(\frac{1}{2}\rho w_m^2\right) l}{\rho c_a s}[\sin\beta_m + \varepsilon\cos\beta_m] \\ (p_3 - p_2) &= \frac{c_l\left(\frac{1}{2}\rho w_m^2\right) l}{s}[\cos\beta_m - \varepsilon\sin\beta_m]. \end{aligned} \tag{8.41}$$

The differential power necessary to drive the blade element is

$$dP = \dot{m}u(c_{3u} - c_{2u}) = uc_l\left(\frac{1}{2}\rho w_m^2\right) l dr[\sin\beta_m + \varepsilon\cos\beta_m]. \tag{8.42}$$

The total pressure rise is

$$p_{t3} - p_{t2} = \rho u(c_{3u} - c_{2u}). \tag{8.43}$$

Expressing the total pressure rise in terms of blade element properties yields

$$p_{t3} - p_{t2} = c_l\frac{l}{s}\frac{u}{c_a}\left(\frac{1}{2}\rho w_m^2\right)[\sin\beta_m + \varepsilon\cos\beta_m]. \tag{8.44}$$

The degree of reaction expresses the ratio of the pressure rise in the rotor to the total pressure rise across the stage and may be expressed as follows:

$$
\begin{aligned}
r &= \frac{p_3 - p_2}{p_{t3} - p_{t2}} = \frac{c_a}{u}\frac{\cos\beta_m - \varepsilon\sin\beta_m}{\sin\beta_m + \varepsilon\cos\beta_m} \\
r &= \phi\cot\beta_m\frac{1 - \varepsilon\tan\beta_m}{1 + \varepsilon\cot\beta_m}.
\end{aligned}
\tag{8.45}
$$

Here we have introduced the flow rate coefficient

$$\phi = \frac{c_a}{u}. \tag{8.46}$$

This parameter provides a nondimensional measure of the mass flow passing through the stage. From Figure 8.9 it may be shown that

$$
\begin{aligned}
\cot\beta_m &= \frac{w_{mu}}{c_a} = \frac{\frac{1}{2}(w_{3u} + w_{2u})}{c_a} = \frac{\frac{1}{2}(w_{3u} + (u - c_{2u}))}{c_a} \\
\cot\beta_m &= \frac{1}{2}\left(\frac{1}{\phi} + \cot\beta_3 - \cot\alpha_2\right).
\end{aligned}
\tag{8.47}
$$

## 8.5 DEGREE OF REACTION AND PRESSURE COEFFICIENT

Introducing the expression for the mean rotor blade angle $\beta_m$ from Equation (8.47) into the expression for the degree of reaction $r$ in Equation (8.45) leads to the following result:

$$r = \frac{1}{2}[1 + \phi(\cot\beta_3 \quad \cot\alpha_2)]. \tag{8.48}$$

Thus, the degree of reaction depends on the mass flow passing through the stage and the exit angles from the rotor and the stator, which are fixed by the geometry of the blade installation, assuming attached flow at or near the rotor blade and stator blade-trailing edges, as shown in Figure 8.10.

The degree of reaction may then be expressed simply as follows:

$$r = \frac{1}{2}(1 + \phi f), \tag{8.49}$$

where the quantity $f$ is given by

$$f = \cot\beta_3 - \cot\alpha_2. \tag{8.50}$$

Within our constant density approximation, the stagnation pressure change across the stage given by Equation (8.43) may be written as

$$p_{t3} - p_{t2} = \bar{\rho}u(c_{3u} - c_{2u}) = \bar{\rho}u[(u - c_a\cot\beta_3) - (u - c_a\cot\alpha_2)]. \tag{8.51}$$

Gathering terms in Equation (8.51) yields

$$p_{t,3} - p_{t,2} = \bar{\rho}u^2(1 - \phi g). \tag{8.52}$$

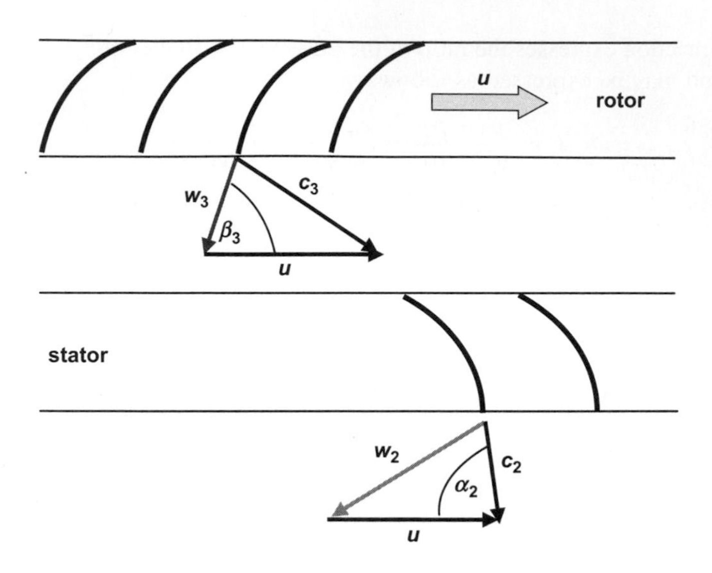


FIGURE 8.10

Illustration of the important trailing edge angles in evaluation of flow properties in the stage.

The quantity $g$ is given by

$$g = \cot \beta_3 + \cot \alpha_2. \tag{8.53}$$

We may introduce a nondimensional ideal pressure coefficient:

$$\psi = \frac{p_{t3} - p_{t2}}{\frac{1}{2}\bar{\rho}u^2} = 2(1 - \phi g). \tag{8.54}$$

Thus the function $g$ in Equation (8.53) and the function $f$ in Equation (8.50) are determined by the exit angles from the rotor and stator blades and are constant in a frictionless flow where the fluid always remains attached to the blade contour. The nondimensional ideal pressure rise coefficient $\psi$ is shown as a function of the nondimensional flow rate coefficient $\phi$ in Figure 8.11. The general area of operation for axial compressors is shown in color in Figure 8.11 and occupies the region $0.5 < \psi < 1$. In the absence of friction and other losses, the pressure coefficient is linear in the mass flow coefficient. The line for $g > 0$ applies to axial flow compressors and lightly loaded centrifugal compressors, like those with backward curved blades. The line with $g < 0$ applies to highly loaded centrifugal compressors, like those with forward curved blades, and the line $g = 0$ refers to the special intermediate case of a centrifugal compressor with radial-tipped blades.

In practice, losses occurring in the flow are due to viscous effects, which grow on either side of the design mass flow rate. In axial flow compressors, which are of particular interest here, mass flow rates much larger or smaller than the design value are particularly damaging to performance. It is reasonable to expect frictional losses to grow with $\phi^2$ because they are proportional to the velocity over the blades; the higher the flow rate, the higher the fluid velocity. At flow rates less than design, where the pressure

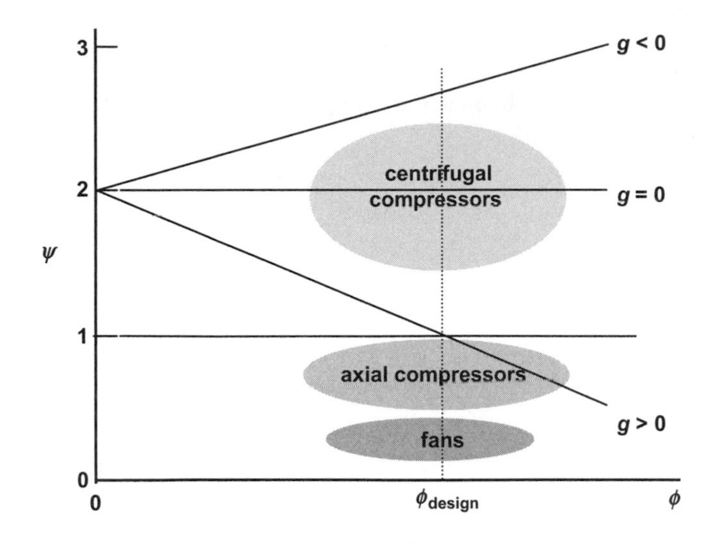


**FIGURE 8.11**

Variation of the nondimensional pressure coefficient $\Psi$ as a function of the nondimensional flow rate coefficient $\phi$. The general area of operation for compressors and fans is illustrated by shaded areas.

rise is higher, the boundary layers are prone to separate, and therefore the flow deviates from the angles at the trailing edges of the blades and turning is not as complete, thus reducing the pressure rise from the ideal value. At much smaller flow rates the angle of incidence of the flow approaching the blades becomes too large and leading edge separation occurs, resulting in a large drop in pressure compared to the ideal. This is the situation commonly called blade stall. A general description of how these viscous effects alter the ideal pressure coefficient $\Psi$ curve is provided by Figure 8.12. The design point is shown as being located a sufficient distance from the stall point to provide good performance with a reasonable margin of safety.

## 8.6 NONDIMENSIONAL COMBINED VELOCITY DIAGRAM

We may use the linear speed of rotation $u$ to normalize the velocities in the combined velocity diagram, as shown in Figure 8.12. Several parameters appear in that diagram: $\phi$, $\tau$, and $r_i$. We have already introduced the flow rate parameter $\phi$. The slip coefficient $\tau$, which was introduced for centrifugal compressors in Section 7.3, is a measure of the change of the $u$ component of velocity in the rotor or the stator. The parameter $r_i$ is the ideal degree of reaction. The relationship between the linear speed of rotation $u$ and the $u$ components of the absolute and relative velocities is given by

$$\begin{aligned} c_{2u} + w_{2u} &= u \\ c_{3u} + w_{3u} &= u. \end{aligned} \tag{8.55}$$

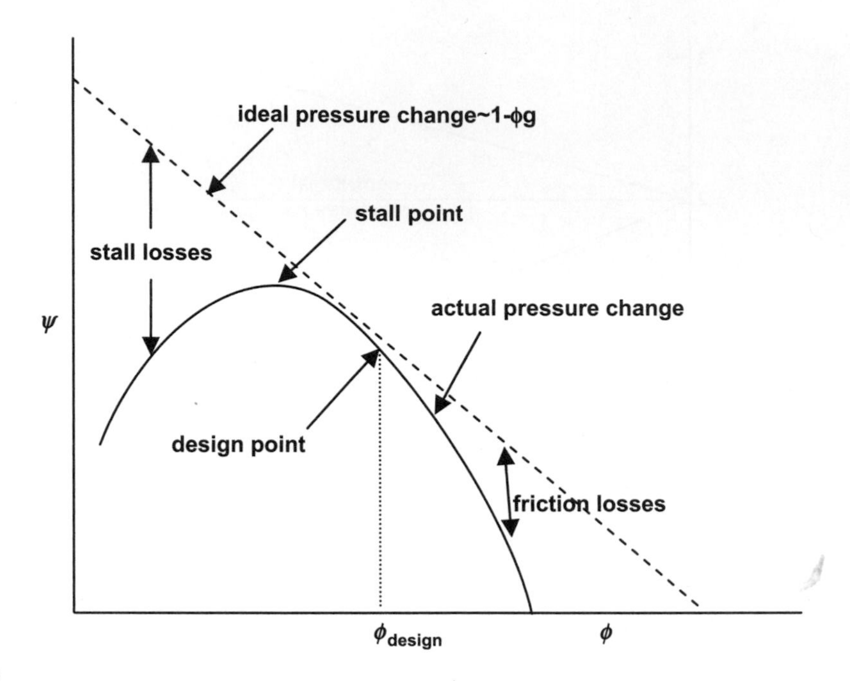


FIGURE 8.12

Ideal pressure coefficient as a function of mass flow parameter (dashed line) showing effect of viscous losses (solid line).

We may subtract the first equation from the second to obtain

$$c_{3u} - c_{2u} = w_{2u} - w_{3u}. \tag{8.56}$$

Then the slip coefficient is

$$\tau = \frac{c_{3u} - c_{2u}}{u} = \frac{w_{2u} - w_{3u}}{u}. \tag{8.57}$$

The ideal degree of reaction, Equation (7.56), may be rewritten in terms of tangential components of velocity, under the continuing assumption that the density is approximately constant, as follows:

$$r_i = \frac{w_2^2 - w_3^2}{w_2^2 - w_3^2 + c_3^2 - c_2^2} = \frac{(w_{2u} + w_{3u})(w_{2u} - w_{3u})}{(w_{2u} + w_{3u})(w_{2u} - w_{3u}) + (c_{3u} + c_u)(c_{3u} - c_{2u})}. \tag{8.58}$$

Equation (8.55) also admits the relation

$$(c_{3u} + c_{2u}) + (w_{3u} + w_{2u}) = 2u. \tag{8.59}$$

Then the ideal degree of reaction may be written as

$$r_i = \frac{1}{2}\frac{w_{2u} + w_{3u}}{u}. \tag{8.60}$$

Thus the ideal degree of reaction is the ratio of the average tangential component of relative velocity to the linear speed of rotation. Applying momentum conservation in the $u$ direction for the rotor and the stator to obtain the differential force components on each as in Equation (8.13) leads to the following:

$$\begin{aligned} dF_{r,u} &= (\rho c_a s_r dr)(w_{3u} - w_{2u}) = \rho u^2 \tau\phi(s_r dr) \\ dF_{s,u} &= (\rho c_a s_s dr)(c_{3u} - c_{2u}) = \rho u^2 \tau\phi(s_s dr). \end{aligned} \tag{8.61}$$

Here the differential mass flow through the rotor and stator passages is given by

$$\begin{aligned} d\dot{m}_r &= \rho c_a s_r dr \\ d\dot{m}_s &= \rho c_a s_s dr \end{aligned}$$

The quantities $s_r$ and $s_s$ denote blade spacing in the rotor and the stator, respectively. Applying the same approach to the axial components of force leads to

$$\begin{aligned} dF_{r,a} &= \Delta p_r\,(s_r dr) \\ dF_{s,a} &= \Delta p_s\,(s_s dr). \end{aligned} \tag{8.62}$$

Using Figure 8.13, we may show that

$$\begin{aligned} dF_{r,a} &= dF_{r,u}\cot(\beta_m + \varepsilon_r) \\ dF_{s,a} &= dF_{s,a}\cot(\alpha_m + \varepsilon_s). \end{aligned} \tag{8.63}$$

For efficient blades with low drag to lift ratios $\varepsilon_s$ and $\varepsilon_r \ll 1$, Equation (8.63) may be written as

$$\begin{aligned} dF_{r,a} &= dF_{r,u}\frac{1 - \varepsilon_r \tan\beta_m}{\varepsilon_r + \tan\beta_m} \\ dF_{s,a} &= dF_{s,u}\frac{1 - \varepsilon_s \tan\alpha_m}{\varepsilon_s + \tan\alpha_m}. \end{aligned} \tag{8.64}$$

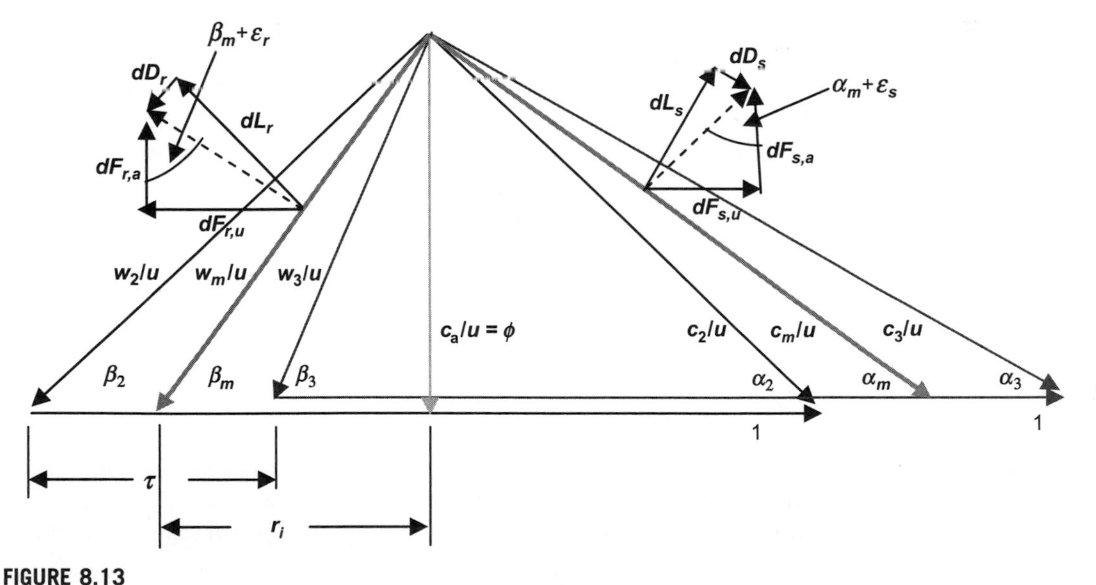


**FIGURE 8.13**

Combined velocity diagram in nondimensional form. Differential forces on the rotor and stator blades are shown along with the angle between the mean velocity and the axial force component.

From the velocity diagram in Figure 8.13, we may also show that

$$\begin{aligned} \tan \beta_m &= \frac{\phi}{r_i} \\ \tan \alpha_m &= \frac{\phi}{1 - r_i}. \end{aligned} \tag{8.65}$$

Combining Equations (8.61), (8.62), (8.64), and (8.65) permits the pressure rise across the rotor and the stator to be expressed as follows:

$$\begin{aligned} \Delta p_r &= \rho u^2 \tau \phi \left( \frac{r_i - \varepsilon_r \phi}{\phi + \varepsilon_r r_i} \right) \\ \Delta p_s &= \rho u^2 \tau \phi \left[ \frac{1 - r_i - \varepsilon_s}{\phi + \varepsilon_s (1 - r_i)} \right]. \end{aligned} \tag{8.66}$$

Thus the total pressure rise across the stage is

$$\Delta p_t = \Delta p_r + \Delta p_s = \rho u^2 \tau \phi \left[ \frac{r_i - \varepsilon_r \phi}{\phi + \varepsilon_r r_i} + \frac{1 - r_i - \varepsilon_s}{\phi + \varepsilon_s (1 - r_i)} \right]. \tag{8.67}$$

If the flow through the stage were frictionless,

$$\begin{aligned} \Delta p_{r,i} &= \rho u^2 \tau r_i \\ \Delta p_{s,i} &= \rho u^2 \tau (1 - r_i) \\ \Delta p_{t,i} &= \rho u^2 \tau \\ r_i &= \frac{\Delta p_{r,i}}{\Delta p_{t,i}} \end{aligned} \tag{8.68}$$

The theoretical degree of reaction may be found directly from the combined velocity diagram, while the actual value depends on the lift to drag ratio of the rotor and stator blades.

Assuming two-dimensional flow in the rotor passage, that is, no variations in the $r$ direction over the height $h$ of the passage, the power necessary to move one blade at the peripheral speed $u$ is

$$P = uF_{r,u} = \rho u^3 \tau \phi s_r h. \tag{8.69}$$

The general assumption about the compression process is that it is adiabatic. The power transferred to the flow must be manifested as an increase in total pressure if the density is assumed to be approximately constant. Because the energy increase per unit mass for a constant density flow is $\Delta p / \rho$, the power required becomes

$$P = \left( \frac{\Delta p}{\rho} \right) \dot{m}_r = \left( \frac{\Delta p}{\rho} \right) \rho u \phi s_r h. \tag{8.70}$$

Comparing Equations (8.69) and (8.70) shows that if all the power goes into the total pressure rise the process must be frictionless, and here $\Delta p$ would be the frictionless pressure rise $\Delta p_{t,i}$. Therefore, if we desired to produce a total pressure rise $\Delta p$ by an ideal frictionless process the power required would be

$$P_i = \left( \frac{\Delta p}{\rho} \right) \dot{m}_r. \tag{8.71}$$

## 8.7 ADIABATIC EFFICIENCY

We may examine the efficiency of adiabatic compression by considering a constant combined velocity diagram and determining the theoretical power required to achieve the pressure rise as given by Equation (8.71) and comparing it to the pressure rise that would be achieved by a frictionless process with the same mass flow. This may be expressed as follows:

$$\eta_{ad,c} = \frac{P_i}{P} = \frac{\Delta p_t}{\Delta p_{t,i}} = \frac{\Delta p}{\rho u^2 \tau}. \tag{8.72}$$

Using Equation (8.67) in Equation (8.72) we find that

$$\eta_{ad,c} = \phi\left[\frac{r_i - \phi\varepsilon_r}{\phi + \varepsilon_r r_i} + \frac{1 - r_i - \phi\varepsilon_s}{\phi + \varepsilon_s(1 - r_i)}\right]. \tag{8.73}$$

Obviously, this equation for efficiency is a function of the flow rate parameter $\phi$ and it may have an optimal value for particular values of $r_i$ and given values of $\varepsilon_r$ and $\varepsilon_s$. Taking the partial derivative of $\eta_c$ with respect to $r_i$ and setting it equal to zero permits one to find the maximum value of the theoretical reaction. Neglecting the squares of $\varepsilon_r$ and $\varepsilon_s$ leads to the following simple result for the optimal theoretical degree of reaction:

$$r_{i,opt} \approx \left(1 + \frac{\varepsilon_r}{\varepsilon_s}\right)^{-1}. \tag{8.74}$$

Thus, to first order in the aerodynamic efficiency of the blades ($\varepsilon_r$ and $\varepsilon_s$), the optimum value for the theoretical degree of reaction is independent of the flow rate parameter $\phi$. It is reasonable to assume that the aerodynamic efficiency of the rotor is comparable to that of the stator ($\varepsilon_r = \varepsilon_s = \varepsilon$). Substituting this assumption into Equations (8.73) and (8.74) leads to an efficiency given by

$$\eta_{ad,c} = 2\phi\frac{1 - 2\phi\varepsilon}{\varepsilon + 2\phi}. \tag{8.75}$$

Under this assumption, we find that $r_{i,opt} = 1/2$, which represents a symmetrical stage, a type common in aircraft gas turbine engines. From Equation (8.75) we may determine that the value of $\phi$ which maximizes the efficiency is

$$\phi_{opt} = \frac{1}{2}\left(\sqrt{1 + \varepsilon^2} - \varepsilon\right) \approx \frac{1 - \varepsilon}{2}. \tag{8.76}$$

Therefore, we see that the optimum values for $\phi$ and $r_i$ are just about one-half. Using the optimum values yields

$$\eta_{ad,c,opt} \approx 1 - 2\varepsilon + 2\varepsilon^2. \tag{8.77}$$

For the typical range $0.04 < \varepsilon < 0.07$, we have efficiencies of about 92 to 86%.

## 8.8 SECONDARY FLOW LOSSES IN BLADE PASSAGES

Three-dimensional effects in the blade passages make the analysis of turbomachines very complicated. The leakage flow from the pressure side of the blade to the suction side of the blade through the blade

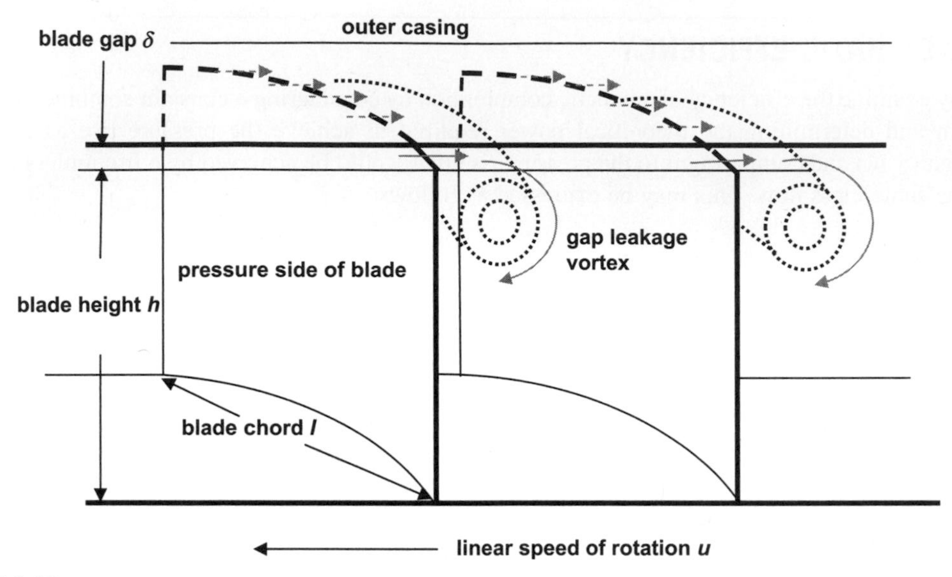


FIGURE 8.14

Schematic diagram of leakage flow from the pressure side to the suction side through the blade gap and formation of the gap leakage vortex.

end gap is one of the most important of these effects. The blade end gap is the distance between the blade tip and the outer casing enclosing the rotating wheel. A schematic diagram of this leakage flow is shown in Figure 8.14. This gap leakage vortex is similar to the trailing vortex from the tip of an airplane wing. The leakage reduces the lift force produced by the blade and represents a reduction in performance of the turbomachine.

Another secondary motion is generated by the curvature of the streamlines as they flow through the curved blade passage. In the central region of the blade passage between the root and tip of the blades there is the least effect of friction, as this region is furthest from the passage walls. The essentially inviscid flow there is subjected to a pressure gradient given by

$$\frac{\partial p}{\partial n} = \frac{\rho V^2}{R}. \tag{8.78}$$

A schematic diagram of the flow is shown in Figure 8.15. The coordinate $n$ is measured normal to the blade surface and in the same sense as the radius of curvature of the inviscid streamline passing through the blade passage. The pressure gradient is positive in the direction $n$, although the magnitude decreases as $n$ grows because $R$, the radius of curvature of the streamline, also grows. Thus the pressure increases as distance increases across the blade passage from the suction side of one blade to the pressure side of the next blade.

Because the boundary layers on the upper and lower surfaces of the blade passage have no pressure variation in the direction normal to those surfaces, the low momentum boundary layer flow is less able to resist the imposed centrifugal pressure gradient. This results in streamlines near the walls departing from the direction taken by the inviscid central flow. Because the velocities in the viscous regions are lower,

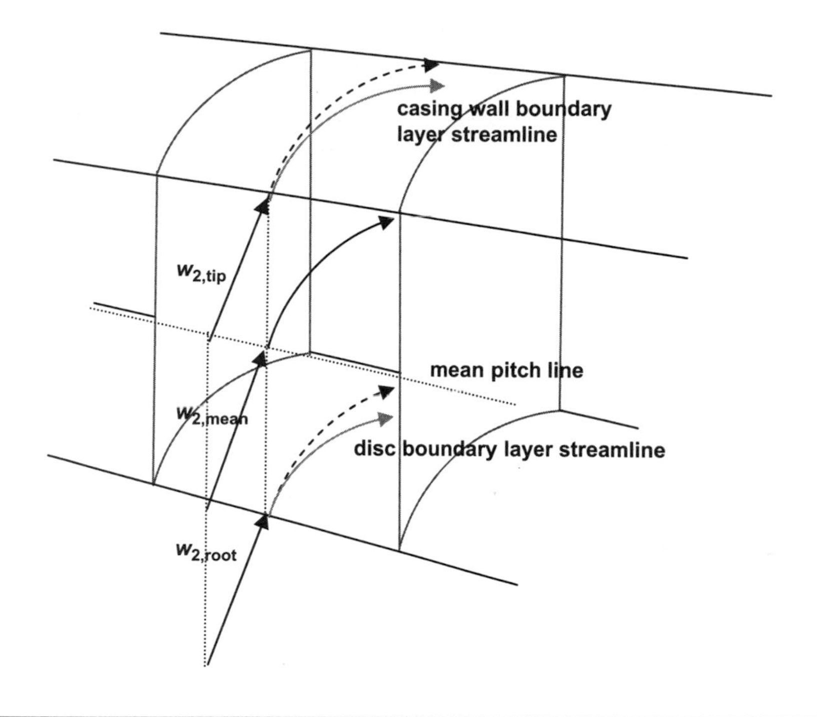


**FIGURE 8.15**

Streamlines through a curved blade passage.

the radius of curvature of the streamlines there must be smaller so as to properly satisfy the pressure gradient. The result is a secondary flow in which the flow near the walls moves the fluid in the direction of decreasing $n$. Then the requirements of continuity force fluid elsewhere to move in a fashion that results in the formation of secondary flow vortices as depicted schematically in Figure 8.16.

Both the gap leakage vortex and the secondary flow vortices are present simultaneously and they generate a counterpart to the gap leakage vortex near the suction side blade root as shown schematically in Figure 8.17.

Calculation of losses due to gap leakage and secondary flow is difficult, and empirical data are often used to account for them. The simplest of such rules of thumb for estimating the loss due to the induced secondary flows is given as an additional, induced, drag coefficient by

$$c_{D,i} = 0.018c_L^2. \tag{8.79}$$

Note that the simple induced drag coefficient given in Equation (8.79) does not depend explicitly on the actual tip clearance, or gap height, $\delta$, so it should only be applied when the gap height is considered to be optimal. Clearly though, the tip clearance must be of importance in determining the tip loss portion of the secondary flow loss estimate. An approximation for secondary flow losses that does account for the tip clearance is

$$c_{D,i} = 0.04\, c_L^2\frac{l}{h} + 0.25c_L^2\sigma\frac{\delta}{h}\frac{1}{\sin\beta_3}. \tag{8.80}$$

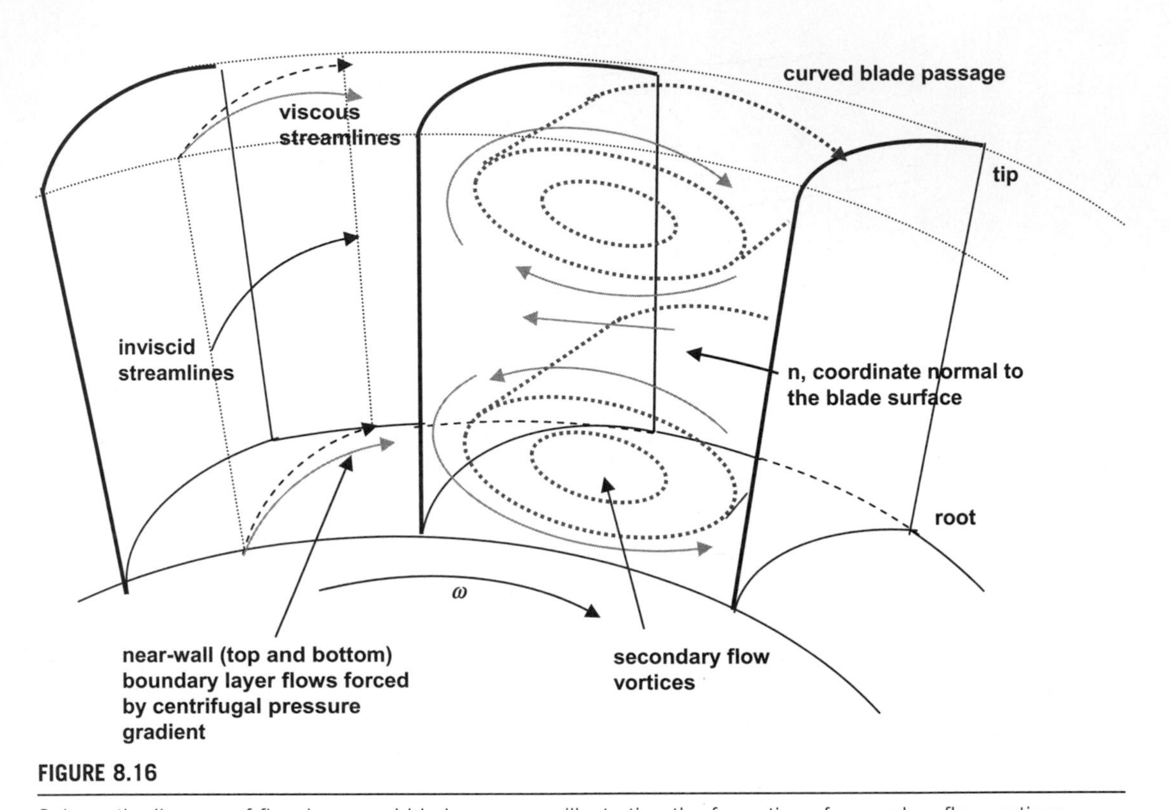


FIGURE 8.16

Schematic diagram of flow in curved blade passages illustrating the formation of secondary flow vortices.

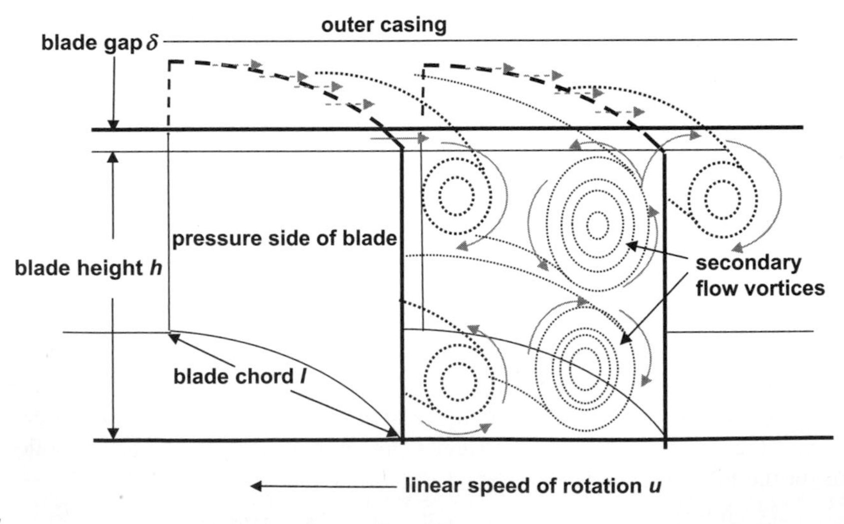


FIGURE 8.17

Schematic diagram of gap leakage vortex and associated secondary flows generated in the curved blade passages of turbomachines.

The first term accounts for the secondary flow effects, whereas the second term accounts for gap leakage. The secondary flow effect is seen to depend on the ratio of blade chord to blade height, indicating that the longer the curved blade passages, the greater the losses. The gap leakage loss is seen to be proportional to the ratio of the gap height to the blade height, as would be expected. The secondary flow loss and the gap leakage loss are both proportional to the square of the lift coefficient, which shows the nonlinear effect of higher blade loading. We may rewrite Equation (8.80) as

$$c_{D,i} = \left[0.04 + 0.25\frac{\delta}{s(\sin\beta_3)}\right]\frac{c_L^2}{\left(\frac{h}{l}\right)}. \tag{8.81}$$

In this form, we see that the additional drag coefficient is proportional to the square of the lift coefficient and inversely proportional to the aspect ratio of the blade $h/l$, in keeping with the functional form of the induced drag coefficient in wing theory. In the second term in the square brackets, $s(\sin\beta_3)$ effectively represents the perpendicular distance between adjacent blades. Small values of $\beta_3$ mean closer blades for a given spacing, for which the effects of leakage will be relatively more important.

Similarly, an additional drag coefficient representing the drag on the annulus wall surfaces is

$$c_{D,S} = 0.020\frac{s}{h}, \tag{8.82}$$

where the quantity $s/h$ is the ratio of blade pitch to blade height and suggests that the drag coefficient would increase continually as the blade spacing increases, independent of the magnitude of the solidity. This would appear to overemphasize the annulus boundary layer while neglecting the interaction between the blade and wall boundary layers. A more appropriate expression might be

$$c_{D,s} = 0.018\sigma\frac{s}{h}. \tag{8.83}$$

These drag coefficients must be added on to the profile drag coefficients introduced in the blade element analysis of axial flow turbomachines so as to account for these three-dimensional flow effects.

## 8.9 BLADE LOADING AND SEPARATION

The detailed aerodynamic behavior of turbomachines is controlled to a large extent by boundary layer effects. Because the geometry of the flow passages necessarily generates three-dimensional flow effects, accurate prediction of such flow is difficult and many problems arise that pose substantial challenges to the designer. It is convenient to separate the flow in turbomachines into three distinct regimes of boundary layer analysis:

- Flow on turbomachine walls or annuli
- Flow through interconnecting ducts
- Flow over blade profiles

The first two areas pose significant problems for theoretical analyses and may be treated, for preliminary design purposes at least, by ad hoc empirical or experimental means. However, the third regime is now reasonably amenable to analytic treatment and will be discussed in some detail. The state of the flow over a typical turbomachine blade is illustrated in Figure 8.18.

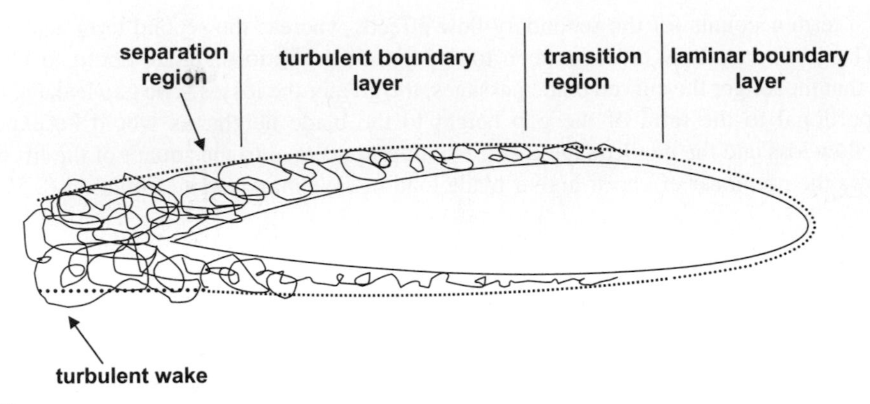


**FIGURE 8.18**

General state of the boundary layer flow over a turbomachine blade.

In analysis of the lift and drag in cascades, we developed the following relationship among lift and drag coefficients, solidity, and geometry of the velocity diagram:

$$\sigma c_L = 2\sin\beta_m(\cot\beta_2 - \cot\beta_3)(1 - \varepsilon\cot\beta_m). \tag{8.84}$$

Recall that

$$\cot\beta_2 - \cot\beta_3 = \frac{\tau}{\phi}, \tag{8.85}$$

where $\phi$ is the mass flow coefficient and $\tau$ is the slip, or deflection, coefficient, indicating the degree to which the flow is turned. Obviously the flow cannot be turned through any arbitrary angle because turning of the flow is accompanied by a deceleration process in the blade passage and is always subject to the possibility of separation, which invariably degrades the performance. Therefore, it is important to have some criterion by which one may limit the blade loading, $\sigma c_L$, and the simplest thing to suggest is that

$$\sigma c_L \le (\sigma c_L)_{\max}. \tag{8.86}$$

Unfortunately, experiments show that there is too much variability in maximum blade loading to make this a viable criterion. Correlation with cascade experiment results can be used to provide a blade-loading criterion. For example, it has been found that for blades with a typical maximum blade thickness to chord length ratio $(t/l)_{\max} = 10\%$, the following correlation for maximum blade loading is reasonable:

$$c_L\left(\frac{\sin\beta_3}{\sin\beta_m}\right)^2\frac{3}{6-\sigma} \le 0.675. \tag{8.87}$$

Substituting the value for $c_L$ from Equation (8.84) yields

$$(\cot\beta_2 - \cot\beta_3)\frac{\sin^2\beta_3}{\sin\beta_m} \le 0.1125\sigma(6-\sigma). \tag{8.88}$$

Permissible flow deflections $\Delta\beta = \beta_2 - \beta_3$ determined from experiment show that this correlation is consistent, although perhaps a bit too conservative. Note that the right-hand side of Equation (8.88) has a maximum value for a solidity $\sigma = 3$ and shouldn't be used for values of $\sigma > 5$. Because the correlation shown has relatively wide applicability, it should have some basis in the physics of the flow field.

## 8.10 CHARACTERISTICS OF BLADE PRESSURE FIELD

The pressure distribution on a typical compressor blade is shown in Figure 8.19 in terms of the pressure coefficients on the pressure and suction sides of the blades as given by

$$S_p = \frac{p_l - p_3}{\frac{1}{2}\rho w_3^2} \tag{8.89}$$

$$S_s = \frac{p_u - p_3}{\frac{1}{2}\rho w_3^2} \tag{8.90}$$

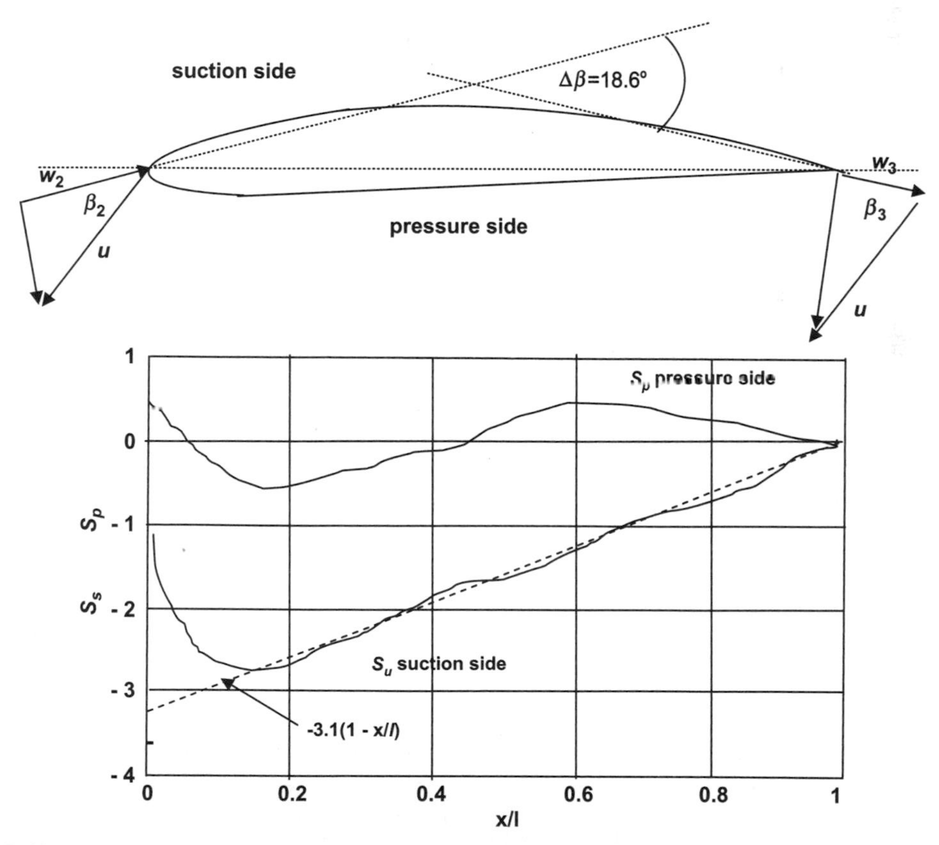


**FIGURE 8.19**

Pressure distribution on a typical compressor blade as a function of distance along the chord. Here $\beta_2 = 30°$, $\beta_3 = 48.6°$, $\sigma = 1$.

These pressure coefficients are a bit different from the usual form in that they are referred to trailing edge conditions instead of upstream conditions. The lift, which is only weakly affected by friction in attached flows, may be determined using Blasius' theorem:

$$L = h\int_0^l (p_l - p_u)dx. \tag{8.91}$$

Substituting the pressure coefficients into Equation (8.91) yields

$$L = \frac{1}{2}\rho w_3^2 hl\int_0^1 (S_p - S_s)d\left(\frac{x}{l}\right). \tag{8.92}$$

Noting from Figure 8.19 that the integral of the pressure side pressure coefficient will be approximately zero, the suction side pressure coefficient may be approximated by

$$S_s \approx K\left(1 - \frac{x}{l}\right). \tag{8.93}$$

The approximation in Equation (8.93) will be in error to some extent in the forward portion of the blade, ahead of the location of the suction peak, but this will not change the general behavior to any appreciable extent. Then the lift may be considered to be given by

$$L \approx \frac{1}{2}\rho w_3^2 hl\int_0^1 K\left(1 - \frac{x}{l}\right)d\left(\frac{x}{l}\right) = \frac{1}{2}\rho w_3^2 hl\left(-\frac{1}{2}K\right). \tag{8.94}$$

Then the lift coefficient, which is defined in terms of $w_m$, becomes

$$c_L = \frac{L}{\frac{1}{2}\rho w_m^2 hl} = -\frac{1}{2}K\left(\frac{w_3}{w_m}\right)^2. \tag{8.95}$$

Using this result in Equation (8.84) for $c_L$ and neglecting drag effects ($\varepsilon \ll 1$) results in

$$(\cot\beta_2 - \cot\beta_3)\frac{\sin^2\beta_3}{\sin\beta_m} = -\frac{1}{4}\sigma K. \tag{8.96}$$

For the case shown in Figure 8.19, we find that

$$(\cot\beta_2 - \cot\beta_3)\frac{\sin^2\beta_3}{\sin\beta_m} = 0.775. \tag{8.97}$$

The empirical result of Equation (8.88) yields

$$(\cot\beta_2 - \cot\beta_3)\frac{\sin^2\beta_3}{\sin\beta_m} = 0.5625. \tag{8.98}$$

The empirical result is quite different from the derived result, and cascade tests for the blades in question find that

$$(\cot\beta_2 - \cot\beta_3)\frac{\sin^2\beta_3}{\sin\beta_m} = 0.781. \tag{8.99}$$

It is clear that experimental results bear out the current analysis. The experimental drag to lift ratio for this case is shown in Figure 8.20 as a function of the turning angle $\Delta\beta = (\beta_3 - \beta_2)$.

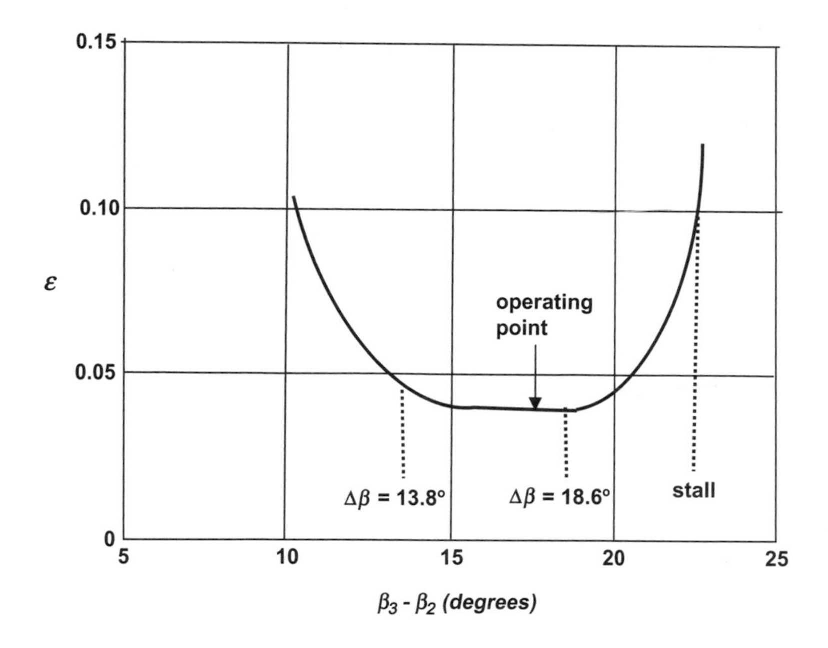


**FIGURE 8.20**

Behavior of drag to lift ratio as a function of the flow turning angle for the blade shown in Figure 8.19, that is, for $\Delta\beta = 18.6°$.

It is seen that stall may be said to occur at $\Delta\beta = 22°$ where the drag rises rapidly. Note that the operating point for the case under study, $\Delta\beta = 18.6°$, is relatively close to the stall point. The usual operating rule is to be at a turning angle that is 80% of that for which stall occurs, which in this case would be at 17.6°, as shown. The more conservative blade-loading estimate of Equation (8.88) would suggest $\Delta\beta = 13.8°$, which is quite far from the stall point and would represent a less efficient operating point. It is desirable to get somewhere in the middle of the region of minimum drag so as to keep high efficiency while providing a safety margin to guard against stall. A simple correlation that provides results consistent with this analysis suggests a maximum turn angle, in degrees, given by

$$(\Delta\beta)_{max} = 0.625\beta_3 - 12.5. \tag{8.100}$$

## 8.11 CRITICAL MACH NUMBER

The inviscid flow over an airfoil accelerates to speeds above that in the free stream, thereby reducing the pressure over the airfoil surface to values below that in the free stream. Variation of the surface velocity for a family of NACA 6-series airfoils is shown in Figure 8.21 (Abbott and von Doenhoff, 1959). Acceleration to higher speeds on the upper surface of the airfoil than on the lower surface of the airfoil results in a net force in the direction normal to the free stream, that is, the lift force. The decreased pressure is accompanied by a decrease in temperature; for example, in an isentropic flow $T \sim p^{(\gamma-1)/\gamma}$. Therefore, the Mach number on the surface of the airfoil also increases above that in the

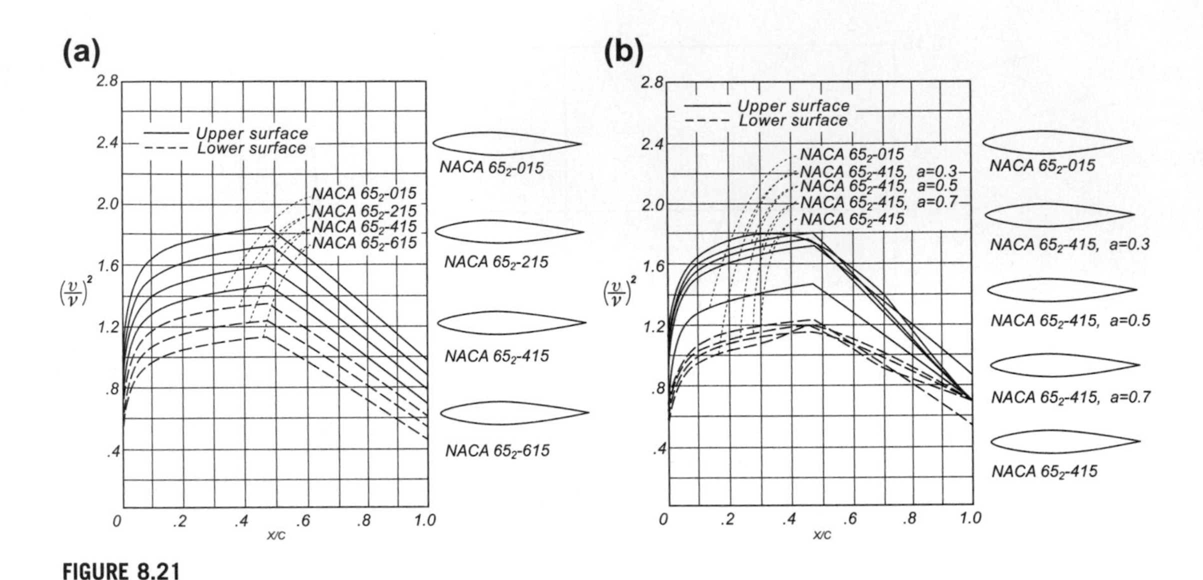


**FIGURE 8.21**

The square of the velocity ratio $v/V$ is shown as a function of axial distance along the chord of 15% thick NACA 6-series airfoils for both upper and lower surfaces: (a) effects of changes in the amount of camber and (b) changes in the type of camber (Abbott and von Doenhoff, 1959).

free stream as the flow accelerates over the airfoil. Then the Mach number at some point on the surface of the airfoil can reach unity before the free stream Mach number does. The free stream Mach number at which this occurs is called the critical Mach number, $M_{cr}$.

The variation of the critical Mach number with lift coefficient for a typical NACA airfoil section is shown in Figure 8.22. Note that at the usual operating lift coefficients the critical Mach number is

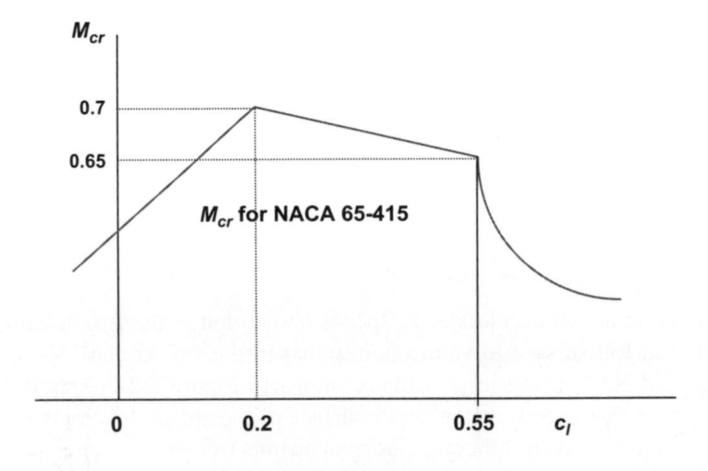


**FIGURE 8.22**

Typical critical Mach number variation with section lift coefficient.

around 0.7, suggesting that operation beyond $M_m = 0.7$ would soon involve shock waves and drastic losses in the turbomachine. As the entrance Mach number increases beyond $M_{cr}$, the drag will soon begin to increase rapidly. A drag-divergence Mach number $M_{dd}$ is defined arbitrarily as that Mach number at which the drag coefficient rises to some increment, typically 10%, above that at $M = M_{cr}$.

## 8.12 LINEARIZED SUBSONIC COMPRESSIBLE FLOW

Considering the flow relative to the rotating blades focuses attention on the relative velocity $w$. Assuming two-dimensional inviscid compressible subsonic flow in the blade passages permits us to write the Euler equations as follows:

$$\begin{aligned} w_x \frac{\partial w_x}{\partial x} + w_y \frac{\partial w_x}{\partial y} &= -\frac{1}{\rho}\frac{\partial p}{\partial x} \\ w_x \frac{\partial w_y}{\partial x} + w_y \frac{\partial w_y}{\partial y} &= -\frac{1}{\rho}\frac{\partial p}{\partial y} \end{aligned}. \tag{8.101}$$

The corresponding continuity equation is

$$\frac{\partial}{\partial x}(\rho w_x) + \frac{\partial}{\partial y}(\rho w_y) = 0. \tag{8.102}$$

The sound speed in the gas is

$$a^2 = \gamma \frac{p}{\rho}. \tag{8.103}$$

If we assume further that the flow is isentropic, then $p = C\rho^\gamma$ and therefore the pressure gradient term may be written as follows:

$$\begin{aligned} \frac{\partial p}{\partial x} &= \frac{\partial}{\partial x}(C\rho^\gamma) = C\gamma\rho^{\gamma-1}\frac{\partial \rho}{\partial x} = a^2\frac{\partial \rho}{\partial x} \\ \frac{\partial p}{\partial y} &= a^2 \frac{\partial \rho}{\partial y}. \end{aligned} \tag{8.104}$$

Introducing this description into Equation (8.101) and combining the result with Equation (8.102) yields

$$w_x^2 \frac{\partial w_x}{\partial x} + w_x w_y \left(\frac{\partial w_x}{\partial x} + \frac{\partial w_y}{\partial y}\right) + w_y^2 \frac{\partial w}{\partial y} = \frac{a^2}{\rho}\left(\frac{\partial w_x}{\partial x} + \frac{\partial w_y}{\partial y}\right). \tag{8.105}$$

This nonlinear equation may be addressed under the small-perturbation assumption that the velocity everywhere is close to the relative wind speed $w_m$ approaching the blade element, as shown in Figure 8.23. This assumption leads to the following definitions:

$$\begin{aligned} w_x &= w_m + w'_x \\ w_y &= w'_y \\ w'_x, w'_y &\ll w_m. \end{aligned} \tag{8.106}$$

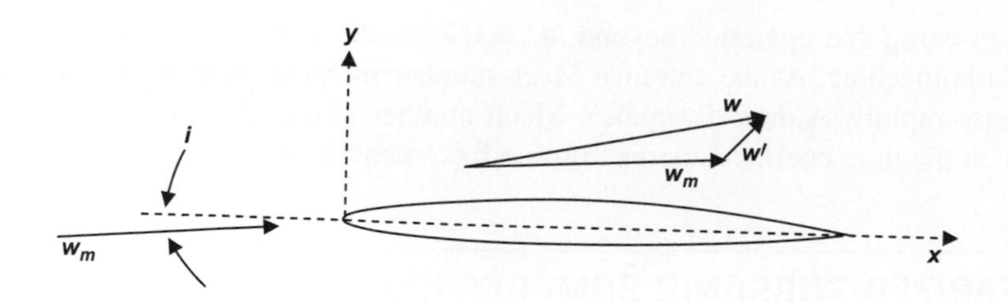


FIGURE 8.23

A slender airfoil at small angle of attack such that the velocity everywhere is very close to the oncoming velocity, that is, $w' \ll w_m$.

Furthermore, we assume that the density changes only slightly from the upstream value:

$$\begin{aligned} \rho &= \rho_2 + \rho' \\ \rho' &\ll \rho_2. \end{aligned} \tag{8.107}$$

Using the small perturbation assumptions of Equations (8.106) and (8.107) in the Euler [Equation (8.101)] and continuity [Equation (8.102)] equations and neglecting all nonlinear terms leads to the following linearized equations for the perturbation velocity:

$$(1 - M_m^2)\frac{\partial w'_x}{\partial x} + \frac{\partial w'_y}{\partial y} = 0 \tag{8.108}$$

$$\frac{\partial w'_x}{\partial x} + \frac{\partial w'_y}{\partial y} = 0. \tag{8.109}$$

Here the quantity $M_m = w_m/a$, the upstream Mach number approaching the blade element. We may introduce the perturbation potential to satisfy the continuity equation as follows:

$$\begin{aligned} w'_x &= \frac{\partial \phi}{\partial x} \\ w'_y &= \frac{\partial \phi}{\partial y} \end{aligned}. \tag{8.110}$$

Then the resulting single equation for the perturbation potential becomes

$$(1 - M_m^2)\frac{\partial^2 \phi}{\partial x^2} + \frac{\partial^2 \phi}{\partial y^2} = 0. \tag{8.111}$$

The coordinates may be transformed according to

$$\begin{aligned} x_i &= x \\ y_i &= y\sqrt{1 - M_m^2} \end{aligned}. \tag{8.112}$$

Then, for generality, we introduce $\phi_i = \Lambda\phi$, where $\Lambda$ is a constant, and the perturbation equation becomes

$$\frac{\partial^2 \phi_i}{\partial x_i^2} + \frac{\partial^2 \phi_i}{\partial y_i^2} = 0. \tag{8.113}$$

This is Laplace's equation, which describes the velocity potential in inviscid incompressible two-dimensional flow. Then the relationship between the compressible flow field and the transformed incompressible flow field is given by

$$\begin{aligned} w'_{x,i} &= \Lambda w'_x \\ w'_{y,i} &= \frac{\Lambda}{\sqrt{1-M_m^2}} w'_y. \end{aligned} \tag{8.114}$$

For small angles of incidence $i$ such that $i \sim \tan i$ we have

$$i_i \approx \tan i_i = \frac{w'_y}{w_m + w'_x} \approx \frac{w'_y}{w_m} = \frac{\Lambda}{\sqrt{1-M_m^2}} i. \tag{8.115}$$

Because we are considering geometrically similar airfoils, it is necessary that

$$\begin{aligned} i &= i_i \\ \Lambda &= \sqrt{1-M_m^2}. \end{aligned} \tag{8.116}$$

This means that perturbation velocities are given by

$$\begin{aligned} w'_x &= \frac{1}{\sqrt{1-M_m^2}} w'_{x,i} \\ w'_y &= w'_{y,i} \end{aligned}. \tag{8.117}$$

In small perturbation theory, the pressure coefficient is

$$C_p = \frac{p - p_m}{\frac{1}{2}\rho_m w_m^2} \approx -2\frac{w'_x}{w_m}. \tag{8.118}$$

Therefore, in incompressible flow,

$$C_{p,i} = -2\frac{w'_{x,i}}{w_m}. \tag{8.119}$$

Then, in linearized compressible flow,

$$C_p = \frac{1}{\sqrt{1-M_m^2}} C_{p,i}. \tag{8.120}$$

This is the Prandtl–Glauert correction for compressible flow and is applicable where small perturbations to the oncoming flow are experienced. Because the lift coefficient is determined by integration of the pressure coefficient around the airfoil, it is subject to the same correction and is given by

$$c_l = \frac{1}{\sqrt{1-M_m^2}} c_{l,i}. \tag{8.121}$$

This coordinate deformation in the $y$ direction in Equation (8.112) implies that the incompressible counterpart airfoil cascade has closer blade spacing, and it is given by

$$s = \frac{1}{\sqrt{1-M_m^2}} s_i. \tag{8.122}$$

Because there is no stretching of the axial coordinate, the chord length remains the same in the compressible flow and its incompressible counterpart so that the solidity is less in the compressible flow, or

$$\sigma = \frac{s}{l} = \frac{1}{\sqrt{1-M_m^2}}\frac{s_i}{l_i} = \frac{1}{\sqrt{1-M_m^2}}\sigma_i. \tag{8.123}$$

## 8.13 PLANE COMPRESSIBLE FLOW

It is obvious that the Prandtl–Glauert correction to the lift coefficient for linearized subsonic flow will fail as $M_m$ approaches unity. This suggests that there are essential nonlinearities in the description of flows near sonic velocity. The full combined Euler and continuity equations yielded Equation (8.105), which draws on no assumptions about the magnitude of perturbations. A potential function that satisfies the continuity equation may be substituted into Equation (8.105) to yield the general potential equation for plane subsonic inviscid flow and is given as follows:

$$\left[a^2 - \left(\frac{\partial\phi}{\partial x}\right)^2\right]\frac{\partial^2\phi}{\partial x^2} - 2\left(\frac{\partial\phi}{\partial x}\right)\left(\frac{\partial\phi}{\partial y}\right)\frac{\partial^2\phi}{\partial x\partial y} + \left[a^2 - \left(\frac{\partial\phi}{\partial y}\right)^2\right]\frac{\partial^2\phi}{\partial y^2} = 0. \tag{8.124}$$

The hodograph transformation, which changes the independent variables from $x$ and $y$ to the velocity vector, or $V$ and $\theta$, the speed and direction of the velocity vector may be employed to pursue a solution. Shapiro (1953) describes in detail how Karman and Tsien used this approach by incorporating a simplified approximation to the equation of state given by

$$p - p_m = \rho_m a_m^2\left(\frac{1}{\rho} - \frac{1}{\rho_m}\right). \tag{8.125}$$

The subscript $m$ refers to the upstream values as used previously in describing the oncoming velocity $w_m$. Note that this approximation is limited to only small changes in pressure and that although the equations are not linearized, the requirement for their solution is that $M_m^2\tau \ll 1$, where $\tau$ is a characteristic slope for the airfoil, such as the thickness to chord ratio. The major result of the analysis is the Karman–Tsien correction for the pressure coefficient given by

$$C_p = C_{p,i}\left[\sqrt{1-M_m^2} + \frac{M_m^2}{1+\sqrt{1-M_m^2}}\frac{C_{p,i}}{2}\right]^{-1}. \tag{8.126}$$

It is useful to note that the Karman–Tsien correction is always less than the Prandtl–Glauert correction.

## 8.14 TURBINE BLADE HEAT TRANSFER

In previous sections, analysis was geared to compressors because they are limited by viscous effects connected with boundary layer separation. The following sections concentrate on boundary layer

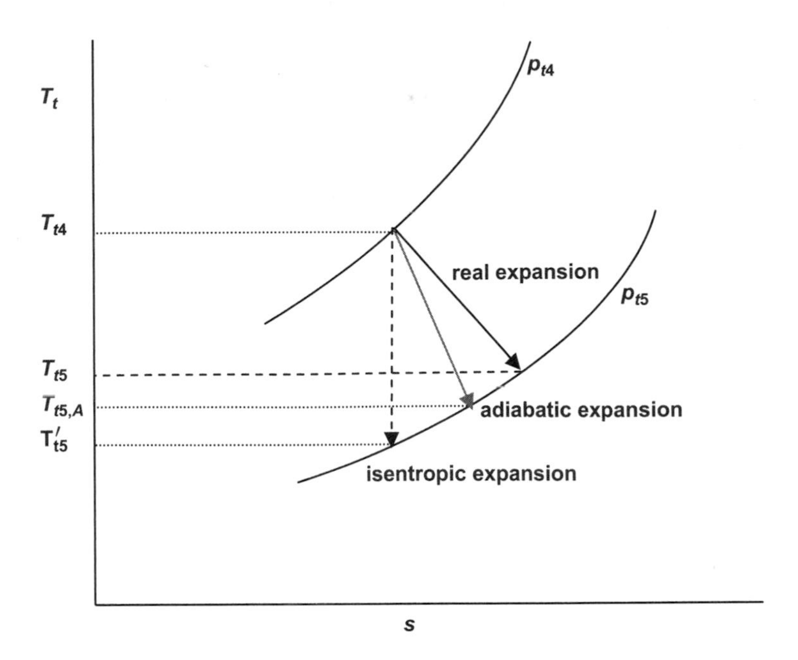


**FIGURE 8.24**

Expansion processes in a turbine.

heat transfer, and therefore we will deal with turbine blades because they tend to be limited by thermal constraints. Expansion processes in a turbine are shown in Figure 8.24 under three conditions: (i) isentropic flow, (ii) adiabatic flow with friction, and (iii) real flow with heat transfer and friction.

For a given work output necessary to drive the compressor, a certain temperature difference across the turbine is required, $T_{t4}$-$T_{t5}$. Then, if the stagnation temperature entering the turbine $T_{t4}$ is increased, there will be a corresponding increase in the turbine exit temperature, $T_{t5}$. We have shown that the exit velocity increases with the square root of the stagnation temperature in the nozzle. Thus the thrust will also increase. However, increased temperatures in the turbine place structural integrity and operational reliability at greater risk. To take advantage of the benefits of higher turbine operating temperatures, research has been expended on developing more highly heat-resistant materials along with designing more effective, even if relatively complex, systems for cooling turbine components.

### 8.14.1 Boundary layer over the turbine blade

We may consider the flow in a typical blade passage such as that shown in Figure 8.25 under entrance conditions that provide a high enough Reynolds number so that thin boundary layers develop on both the suction and the pressure surfaces of the blades forming the walls of the passage. For reference, we

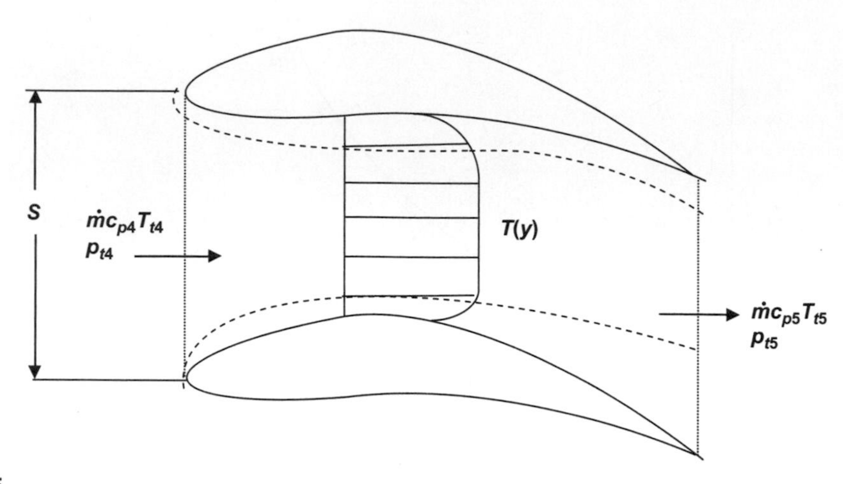


**FIGURE 8.25**

Flow field in a typical turbine blade passage. Dashed lines represent boundary layer edges. Scale exaggerated for clarity.

may refer to the boundary layer thickness on a flat plate, which for laminar and turbulent flows, respectively, is given by (Schlichting, 1968)

$$\begin{aligned}\frac{\delta_{lam}}{x} &= \frac{3.2}{(\mathrm{Re}_x)^{\frac{1}{2}}}, \\ \frac{\delta_{turb}}{x} &= \frac{0.37}{(\mathrm{Re}_x)^{\frac{1}{5}}}\end{aligned} \tag{8.127}$$

where $\mathrm{Re}_x$ is the Reynolds number based on distance $x$ measured from the leading edge and on the conditions in the inviscid flow entering the blade passage relative to the rotating blade as follows:

$$\mathrm{Re}_x = \frac{\rho_4 w_4 x}{\mu_4}. \tag{8.128}$$

With the help of the perfect gas law and the isentropic relations, we may write the Reynolds number as follows:

$$\mathrm{Re}_x = p_{t4}\sqrt{\frac{\gamma_4}{RT_{t4}}}\left(1+\frac{\gamma_4-1}{2}M_4^2\right)^{-\frac{\gamma+1}{2(\gamma-1)}}\frac{M_4 x}{\mu_4}, \tag{8.129}$$

where pressure is in N/m$^2$, temperature is in K, and viscosity is in N-s/m$^2$. Note that, using mass flow [Equation (5.14)], the Reynolds number may also be recast in terms of mass flow $\dot{m}$ through the engine as follows:

$$\mathrm{Re}_x = \frac{\dot{m}}{A}\frac{x}{\mu_4}, \tag{8.130}$$

where $A$ is the flow area of the turbine passages $A = n_b h s$, $n_b$ is the number of blades, $h$ is the height of the blade passage, and $s$ is the blade spacing. To get a reasonable estimate of the order of magnitude of the Reynolds number entering the turbine, let us assume our standard hot gas value of $\gamma_4 = 1.33$ with $R_4 = 287$ m²/s²-K and a nominal value of $M_4 = 0.5$. Then, introducing our reference pressure of one atmosphere, or $p_{ref} = 101.3$ kPa, the equation for the Reynolds number becomes

$$\frac{\mathrm{Re}_x}{\left(\frac{p_{t4}}{p_{ref}}\right)x} = \frac{2989}{\mu_4\sqrt{T_{t4}}}. \tag{8.131}$$

With this form, we keep the right-hand side dependent on temperature only, as viscosity is not a function of pressure in the range of interest. Furthermore, the total pressure ratio shown is merely the actual stagnation pressure measured in atmospheres. The usual combustion chamber pressures are O(10) atmospheres and the blade chord lengths are $O(10^{-1})$ meters so the right-hand side of Equation (8.131) is effectively the Reynolds number based on the blade chord length and is shown as a function of $T_{t4}$ in Figure 8.26.

Although the exact value may be calculated, it is clear that the Reynolds number based on blade chord length is $O(10^6)$ and, therefore, according to Equation (8.127), the boundary layer thickness will be very small compared to the chord length.

Because of the relatively high turbulence level in internal flows in turbine passages, on the order of 2% to 10%, transition from laminar to turbulent flow will occur at lower Reynolds numbers than would occur in low turbulence (less than 0.5%) external flows such as those over wings. Therefore, although laminar flow may be in evidence in the forward part of the blade where $x$ is small, thereafter it will

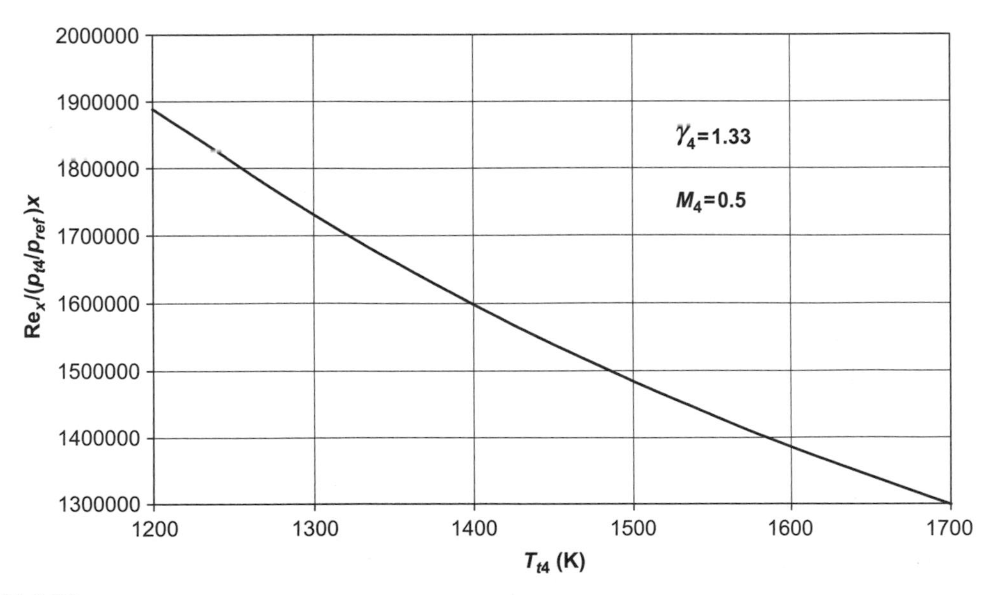


**FIGURE 8.26**

Reynolds number variation with turbine entry stagnation temperature for nominal turbine entry Mach number $M_4 = 0.5$ and $\gamma_4 = 1.33$.

likely be turbulent over a good part of a typical turbine blade. As can be inferred from Figure 8.26, higher pressure turbines will operate at higher Reynolds numbers than lower pressure turbines and therefore have a greater extent of turbulent flow over the turbine blade. Calculating the location of the onset of transition and the flow field during the transition to fully turbulent flow is not highly accurate, especially in the complicated flow fields over turbine blades. Furthermore, turbine blade-cooling systems often introduce coolant flow through the blade surface, making calculations even more problematical. The assumption of fully developed turbulent flow over a blade provides a conservative estimate, which is often helpful in design and sizing exercises.

### 8.14.2 General heat transfer effects in the blade passage

The distribution of flow variables across the blade passage is assumed to be uniform except within the thin boundary layers over the blade surfaces wherein steep gradients occur. The no-slip condition requires the velocity to be zero at the surface, while the temperature at the surface is determined by heat transfer considerations. The heat transfer rate at the wall is given by Fourier's law of heat conduction, which for heat transfer in the normal direction alone is

$$\dot{q}_w = -k_w \left( \frac{\partial T}{\partial y} \right)_w, \tag{8.132}$$

where the subscript $w$ refers to conditions in the gas at the wall ($y = 0$). The general behavior of the temperature profile in the boundary layer is shown in Figure 8.27.

We may consider first the simplest case where there is no heat or work transfer across the boundaries of the blade passage. The energy equation under these conditions requires the stagnation enthalpy to be constant. For quasi-one-dimensional flow with constant specific heat, this leads to the following result:

$$T + \frac{1}{2c_p} w^2 = T_e + \frac{1}{2c_p} w_e^2. \tag{8.133}$$

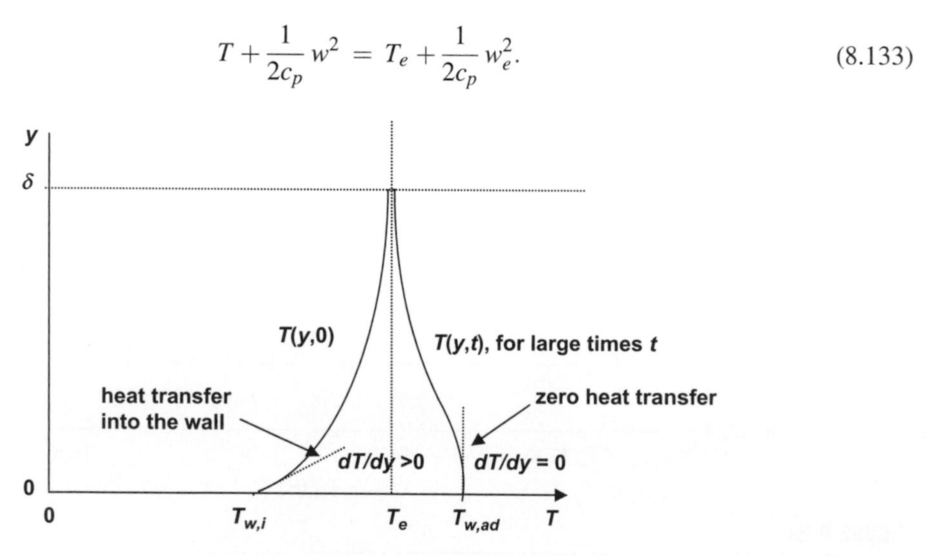


**FIGURE 8.27**

Generic temperature distributions in the boundary layer for flows with and without heat transfer.

However, in a turbine passage, we expect to extract work from the hot gas and that there will be significant heat transfer to the blade surfaces. The quasi-one-dimensional energy equation then becomes

$$\dot{m}dQ = \dot{m}dh_t + dP.$$

Assuming that there are no chemical reactions occurring in the passage, we may rewrite the energy equation as

$$\dot{m}dQ \approx \dot{m}c_p dT_t + dP.$$

Integration around the control volume boundaries yields the power extracted from the flow by the turbine passage per unit span of the passage as follows:

$$P_{turbine} = \int_0^S \rho_4 c_{a4} c_{p4} T_{t4} dy - \int_0^S \rho_5 c_{a5} c_{p5} T_{t5} dy - \int_0^l \dot{q}_{w,c,pressure} dx - \int_0^l \dot{q}_{w,c,suction} dx. \quad (8.134)$$

Because all blade passages on the turbine wheel are identical, we may shift the control volume to surround one blade, as shown in Figure 8.28, with no loss in generality, as it is assumed there is no heat

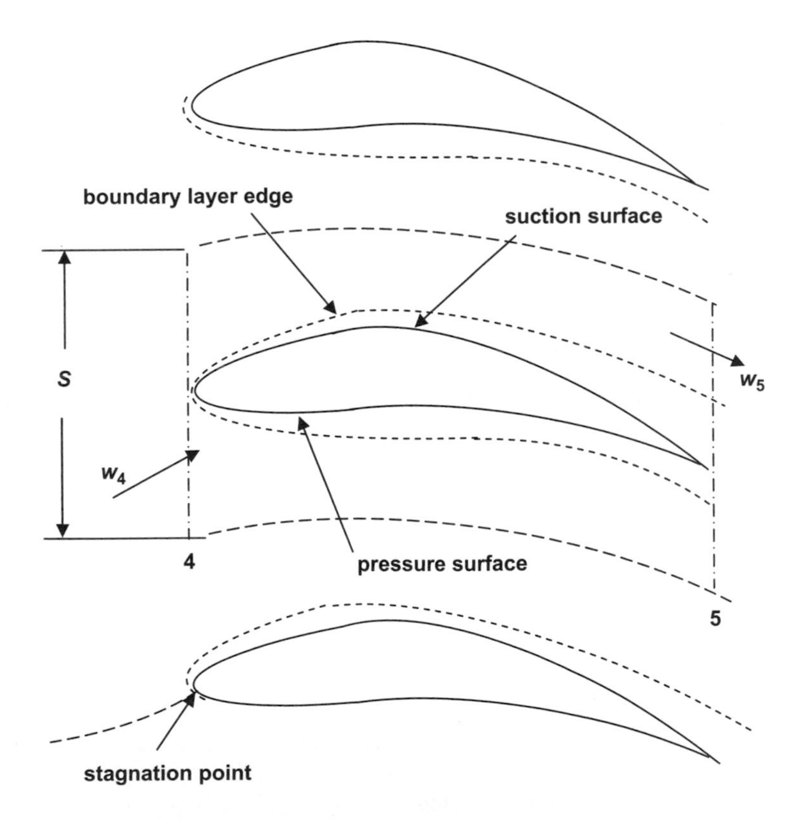


**FIGURE 8.28**

Control volume drawn around a blade rather than around the blade passage. Boundary layers are depicted by dotted lines and their thickness is exaggerated for clarity. Streamlines are shown as dashed lines.

transfer across the streamlines defining the control volume contour. Thus we are considering the power output of one blade. Here the elemental length $dx$ is taken along the surface contour over the entire chord length $l$. We see from Equation (8.134) that heat loss to the blade represents a loss in blade power generation. At the same time, heat flow into the blade will continue to raise the temperature of the blade until an equilibration temperature is reached, which occurs when the blade surface supports no further heat flow or when the blade surface is at an adiabatic condition. The blade temperature may be reduced by supplying a cooling system within the blade that can carry heat away to be rejected somewhere other than in the blade material. Such systems can be quite complicated, as a coolant must be circulated through a highly stressed, rapidly rotating structure. Of course, the higher the temperature the blade can withstand, the less heat lost and the more power produced. In general, elevated temperatures diminish the strength of the blade, as well as reducing the life expectancy of the blade. Therefore, a trade-off between the durability of the blade at high temperatures and the complexity of the blade cooling system required to manage an appropriate blade temperature must be considered.

### 8.14.3 Similarity parameters in heat transfer

There are several important similarity parameters in boundary layer heat transfer. The first is the Stanton number, which normalizes the convective wall heat transfer by the external flow enthalpy flux based on conditions at the wall and is given by

$$St = \frac{\dot{q}_{c,w}}{\rho_e u_e (h_w - h_{aw})}. \tag{8.135}$$

The second is the Nusselt number, which normalizes the convective wall heat transfer by an effective heat transfer based on a distance $x$ in the streamwise direction and is given by

$$Nu = \frac{\dot{q}_{c,w}}{\frac{k_e}{c_{p,e}} \frac{(h_w - h_{aw})}{x}}. \tag{8.136}$$

The two nondimensional similarity parameters are related by the following expression:

$$St = \frac{\mathrm{Pr}_e}{\mathrm{Re}_{x,e}} Nu. \tag{8.137}$$

The Prandtl number, Pr, is a similarity parameter that measures the relative importance of frictional heating to heat conduction in the same manner that the Reynolds number, Re, measures the relative importance of inertia forces to frictional forces. In Equations (8.135)–(8.137), the subscript $e$ denotes evaluation at the edge of the boundary layer. The Prandtl number is defined as

$$\mathrm{Pr}_e = \frac{\mu_e c_{pe}}{k_e}. \tag{8.138}$$

In gases, $\mathrm{Pr} = O(1)$ and the thermal and viscous boundary layers are approximately equal in thickness. In Equation (8.137), the Reynolds number is based on the streamwise distance $x$ measured from the leading edge of a blade and is evaluated at the edge of the boundary layer; it is defined as

$$\mathrm{Re}_{x,e} = \frac{\rho_e w_e x}{\mu_e}. \tag{8.139}$$

In Equations (8.135) and (8.136), the wall enthalpy $h_w$ is unambiguous but it remains to explain the adiabatic wall enthalpy, $h_{aw}$. If there were no heat transfer to the wall, that is, if it were adiabatic, the wall would reach a certain equilibrium temperature and the gas at the wall would take on the enthalpy appropriate to that temperature. The no-slip condition requires that the velocity at the surface be zero, but because the flow is not brought to rest adiabatically through the boundary layer because of viscous dissipation, the enthalpy at the surface is not the stagnation enthalpy but a lower value, the so-called adiabatic wall enthalpy, $h_{aw}$. We may equally speak of an adiabatic wall temperature $T_{aw} = h_{aw}/c_p$. The ratio of the adiabatic wall enthalpy and the stagnation enthalpy is called the recovery factor and is given by

$$\frac{h_{aw} - h_w}{h_{s,e} - h_w} = \frac{\left(h_e - r\frac{1}{2}u_e^2\right) - h_w}{\left(h_e - \frac{1}{2}u_e^2\right) - h_w} = r. \tag{8.140}$$

The recovery factor $r$ represents the fraction of ordered kinetic energy that is lost to heat by friction. It has been found that the recovery factor is well correlated by a function of the Prandtl number of the gas, as shown here:

$$h_{aw} = h_e + \mathrm{Pr}_e^m \frac{u_e^2}{2}. \tag{8.141}$$

The quantity $m = 1/2$ for laminar flow and 1/3 for turbulent flow. Here $h_e$ and $h_w$ are the enthalpies at the edge of the boundary layer and at the wall, respectively, and $\mathrm{Pr}_e$ is the Prandtl number evaluated at the edge of the boundary layer.

It is often suggested that rather than evaluating the flow properties denoted by subscript $e$ at the temperature $T_e$, they should be evaluated at an intermediate temperature, called the reference temperature, $T^*$. The associated reference enthalpy is defined as follows:

$$h^* = 0.28h_e + 0.5h_w + 0.22h_{aw}. \tag{8.142}$$

This approach is more appropriate in flows where the external wall and adiabatic wall temperatures are very different, which is more the case for supersonic external flows than in the present case of turbine flows. For high subsonic flows, one may show that the $T^* \cong (T_w + T_e)/2$ and for hot flow over turbine blades this means $T^* \cong T_e$. The Nusselt number for constant pressure, or flat plate, flow is correlated by

$$Nu = A\left(\frac{\rho^*}{\rho_e}\right)^a\left(\frac{\mu^*}{\rho_e}\right)^b \mathrm{Re}_x^c. \tag{8.143}$$

The coefficients for the general case of Equation (8.155) are explained in Table 8.1. If we choose to consider $T^* \cong T_e$, then the coefficients $a = b = 0$.

**Table 8.1** Coefficients for Nusselt Number Correlation in Equation (8.143)

| Type of Flow | A | a | b | c |
|---|---|---|---|---|
| Laminar | $0.332\ \mathrm{Pr}^{1/3}$ | 0.5 | 0.5 | 0.5 |
| Turbulent | $0.0296\ \mathrm{Pr}^{1/3}$ | 0.8 | 0.2 | 0.8 |

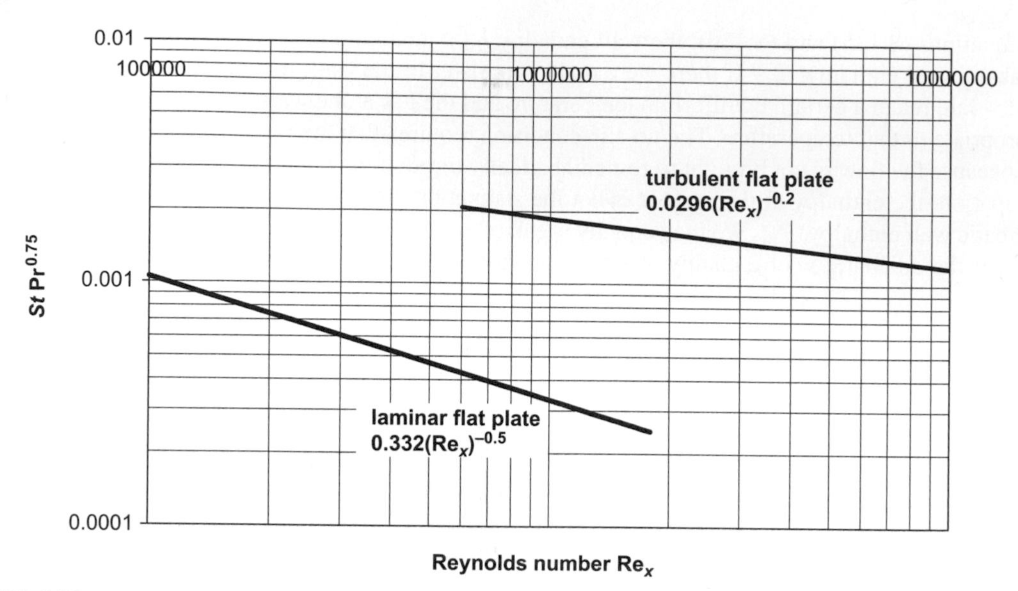


**FIGURE 8.29**

Variation of the Stanton number with Reynolds number for flat plate flows with properties evaluated at the temperature at the boundary layer edge.

Thus in the case of flow over a flat surface where the pressure is everywhere constant, the external conditions could be used in Equation (8.143) to calculate the Nusselt number at any location $x$ measured from the leading edge. For a given constant wall temperature $T_w$, one may use Equation (8.136) to compute the heat transfer at the wall. Note that even for this case of constant pressure flow where all the variables at the edge of the boundary layer are constant, the Nusselt number is inversely proportional to the streamwise distance $x$ and therefore the wall heat transfer $q_w$ will have similar behavior. We may use Equations (8.137) and (8.143) to develop a simple relationship between the Stanton number and the Reynolds number for the case where we evaluate all properties at the temperature at the edge of the boundary layer rather than at the reference temperature $T^*$. In this case, we have

$$St\,\mathrm{Pr}^{0.75} = 0.332\,\mathrm{Re}_x^{-0.5} \text{ laminar flow}$$

$$St\,\mathrm{Pr}^{0.75} = 0.0296\,\mathrm{Re}_x^{-0.2} \text{ turbulent flow}$$

The nature of these relations is shown in Figure 8.29. Laminar flow will usually be confined to $\mathrm{Re}_x < 10^6$, while turbulent flow will be in evidence at values of $\mathrm{Re}_x > 10^6$ because of the high turbulence level common to turbine flows.

## 8.14.4 Flat plate blade heat transfer

For illustrative purposes, let us consider the turbine blade to be a flat plate aligned with the relative velocity entering the blade passage $w_4$ as shown in Figure 8.30. We may use the Nusselt number, which

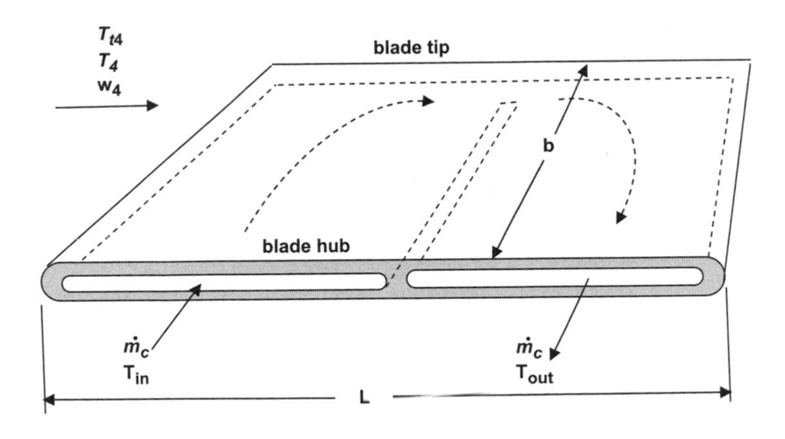


**FIGURE 8.30**

Schematic diagram of a flat plate turbine blade aligned with the relative velocity $w_4$. A coolant passage is shown with the inlet on the left (upstream side) and exit on the right (downstream) side.

normalizes the convective wall heat transfer by an effective heat transfer based on streamwise distance $x$, to evaluate the convective heat flux into the wall as follows:

$$\dot{q}_{c,w} = N_u \frac{k_e}{c_{p,e}} \frac{h_w - h_{aw}}{x}. \tag{8.144}$$

Recall that the recovery factor $r$ represents the fraction of ordered kinetic energy that is lost to heat by friction. It has been found that the recovery factor is well correlated by a function of the Prandtl number of the gas so that the adiabatic wall enthalpy is given by

$$h_{aw} = h_e + \mathrm{Pr}_e^m \frac{u_e^2}{2}. \tag{8.145}$$

The quantity $m = 1/2$ for laminar flow and 1/3 for turbulent flow. Here, $h_e$ and $h_w$ are the enthalpies at the edge of the boundary layer and at the wall, respectively, $\mathrm{Pr}_e$ is the Prandtl number evaluated at the edge of the boundary layer, and $u_e$ is the velocity at the edge of the boundary layer, $w_4$. As noted previously in the discussion of heat transfer similarity variables, rather than evaluating the flow properties denoted by the subscript $e$, a more accurate approach is to evaluate them at an intermediate temperature, called the reference temperature, $T^*$. The associated reference enthalpy is defined as follows:

$$h^* = 0.28h_e + 0.5h_w + 0.22h_{aw} \tag{8.146}$$

The Nusselt number for constant pressure flow is correlated by Equation (8.143). Considering the case of laminar flow over a flat plate, the heat flux, measured in W/m$^2$, is then

$$\dot{q}_{c,w} = 0.332\mathrm{Pr}_e^{\frac{1}{3}} \sqrt{\frac{\rho^* \mu^*}{\rho_e \mu_e} \mathrm{Re}_x} \frac{k_e}{c_{p,e}} \frac{h_w - h_{aw}}{x}. \tag{8.147}$$

In the case of flow in a turbine passage, the Mach number $M < 1$ everywhere, and under such conditions there isn't a significant difference between static and stagnation temperatures, at least from the standpoint of evaluating thermodynamic properties. We will assume that all those properties are constant and equal to the values corresponding to the static temperature at the edge of the boundary layer $T_e$ so that Equation (8.147) may be simplified to

$$\dot{q}_{c,w} = 0.332\,\mathrm{Pr}^{\frac{1}{3}}\sqrt{\frac{\rho_e w_e}{\mu_e x}}\,k_e(T_w - T_{aw}). \tag{8.148}$$

For thermal conditions in the turbine blade passage, the Prandtl number is typically around 0.77 so that the fractional powers of the Prandtl number are practically unity; it will simplify matters to assume that $\mathrm{Pr}^{1/2} \sim \mathrm{Pr}^{1/3} \sim 1$. Then $T_{aw} \sim T_{t4}$ and the local heat flux to the wall given by Equation (8.148) becomes

$$\dot{q}_{c,w} = 0.332\sqrt{\frac{\rho_e w_e}{\mu_e x}}\,k_e(T_w - T_{t4}). \tag{8.149}$$

Assuming the flat plate is of chord length $L$ and width $b$ and that both sides of the plate are exposed to the hot gas flow, we may determine the rate of heat transfer to it by integrating Equation (8.149) over the chord to obtain

$$\dot{Q} = 2\int_0^L \dot{q}_{c,w}\,b\,dx = 0.664\int_0^L \sqrt{\frac{\rho_e w_e}{\mu_e}}\,k_e(T_w - T_{t4})\,b\,\frac{dx}{\sqrt{x}}. \tag{8.150}$$

This may be solved readily to yield

$$\dot{Q} = 1.328 k_e(T_{t4} - T_w)\,b\sqrt{\mathrm{Re}_L}. \tag{8.151}$$

Note that the heat transfer is measured here in watts. Of course, as the wall temperature increases, the heat transfer decreases until the adiabatic wall condition is reached where $T_w = T_{t4} \sim T_{aw}$. Reliability and longevity of the turbine blades are enhanced as the wall temperature is decreased, but that increases $\dot{Q}$. Then, in order to achieve that blade temperature reduction, some heat must be removed by an external cooling arrangement. One may consider passing a coolant at a mass flow $m_c$ through internal passages within the blade such that it enters with an inlet temperature $T_{in}$ and leaves with a final outlet temperature $T_{out}$ so that the rate of heat removal by the coolant is given by

$$\dot{Q}' = \dot{m}_c c_{p,c}(T_{in} - T_{out}). \tag{8.152}$$

Setting the heat transfer from the gas into the blade equal to the heat transferred to the coolant, that is, $\dot{Q} = \dot{Q}'$, permits solving for the wall temperature as follows:

$$T_w = T_{t4} - 0.753\,\frac{\dot{m}_c c_{p,c}}{b\sqrt{\mathrm{Re}_L}}\,\frac{T_{out} - T_{in}}{k_e}. \tag{8.153}$$

Obviously, if there is no coolant flow ($\dot{m}_c = 0$), no heat is removed and the blade heats up to the adiabatic wall temperature, and here $T_{aw} \sim T_{t4}$. Although we would like the coolant to have a high specific heat, the coolant on an aircraft gas turbine is generally air bled from the compressor so that $c_{p,c} \sim c_{p,e}$ and therefore we may put $c_{p,c}/k_e \sim \mathrm{Pr}/\mu_e \sim 1/\mu_e$. To get a large change in coolant temperature it would be advantageous to have a long flow path for the coolant within the blade. However, the longer the coolant duct, the greater the pressure drop required to force a given flow rate of coolant through it so there is a design trade-off to be considered in this regard.

Carrying out the same approach for turbulent flow on the plate yields a heat flux of

$$\dot{q}_{c,w} = 0.0296\left(\frac{\rho_e w_e}{\mu_e x^{1/4}}\right)^{4/5} k_e(T_w - T_{t4}). \tag{8.154}$$

The heat transfer rate for turbulent flow is then

$$\dot{Q} = 0.00592 k_e (T_{t4} - T_w) b \,\mathrm{Re}_L^{4/5}.$$

The corresponding wall temperature would be described by

$$T_w = T_{t4} - 168.9 \frac{\dot{m}_c c_{p,c}}{b\mathrm{Re}_L^{4/5}} \frac{T_{out} - T_{in}}{k_e}.$$

### 8.14.5 Heat transfer mechanisms in turbine passages

The simple analysis carried out for a flat plate blade aligned with the relative velocity entering the passage provides an introduction for understanding the fundamental nature of the heating problem. The practical case of turbine blade heat transfer involves a number of complicating issues, such as

- Thermal and flow conditions entering the blade passage
- Details of the flow field over blades and vanes of practical shape
- Methods for predicting the heat transfer over blade and vane surfaces

The hot gas entering the turbine passages of high-performance aviation turbines range up to, and can exceed, 1800K. The thermodynamic properties of the flow are, of course, dependent on temperature and, to a lesser extent, on the composition of the combustion products. On the way to the turbine, the gas has passed through many stages of compression, combustors with fuel injection, atomization, and elaborate cooling air mixing arrangements, and redirection by turbine inlet guide vanes so as to present a high turbulence level (2 to 10%) flow to the rotating turbine blades. In addition, the movement of turbine blades relative to guide vanes introduces a periodic unsteadiness into the flow, and additional turbine stages only contribute to the level of disturbance in the flow. Therefore, the variable gas properties, the high level of turbulence, and the periodic unsteadiness of the flow all contribute to the practical difficulties that must be dealt with in order to produce a reliable and efficient gas turbine engine.

In addition to these complexities of flow entering the turbine passage, there are those connected with the geometry of the blade cross section. The flow is subjected to local acceleration and deceleration rather than the simple constant pressure and velocity flow over a flat plate considered in the previous sections. The streamlines are curved and the flow is within a rotating passage, both of which contribute secondary flow effects, which can be important. The turbine blade is subject to high levels of heat transfer at the stagnation point on the leading edge, while upstream disturbances influence the location of the transition to turbulence and the corresponding increase in heat transfer accompanying the turbulent boundary layer flow. Details of boundary layer development along the blade may lead to separation on the suction surface, which can substantially deteriorate performance. Additional considerations are that the turbine blade has finite length and that there is a gap between the tip of the moving blade and the surrounding casing. Tip effects are important in maintaining efficient performance and including them brings three-dimensional considerations into the picture. The elaborate cooling schemes developed for turbine blades and guide vanes can have considerable effect on altering the flow field over the blades and add to the difficulties in predicting the flow field.

Based on the preceding discussion, it should be clear that calculating the details of flow over turbine blades is quite difficult. An appreciation for these may be found in Goldstein (2001). The correlation techniques described previously have some utility in determining the level of heat transfer expected based on gross characteristics of the flow entering the blade passage but cannot provide detailed information useful for design purposes. Computational fluid dynamics (CFD) methods have been developed that can accommodate several of the flow complexities mentioned previously, but often at substantial computational expense. Obviously, experimental investigations are valuable in complementing CFD studies, but these constitute considerable expenditures and are used primarily for verification of design developments.

### 8.14.6 Turbine blade cooling

The simplified case of a cooled flat plate blade was presented to depict the nature of the cooling problem. Elaborate cooling schemes have been developed for practical turbine blades and guide vanes operating at high temperatures. An external view of a typical cooled turbine blade is shown in Figure 8.31. Cooling air is bled from the compressor and forced through the turbine disc and up through the root of the blade. The cooling air is directed through internal passages and exits at different locations on the blade,

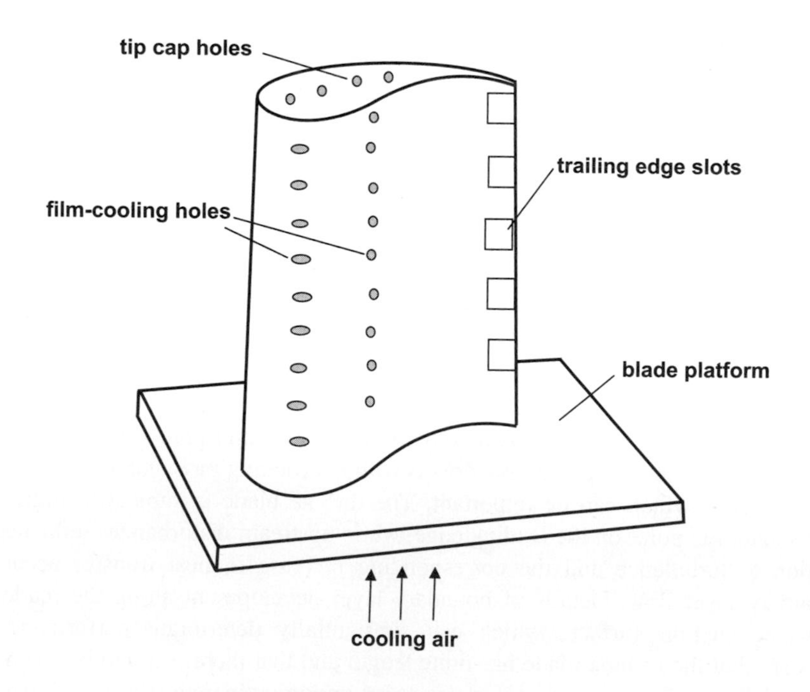


**FIGURE 8.31**

Typical external configuration of a turbine blade-cooling system.

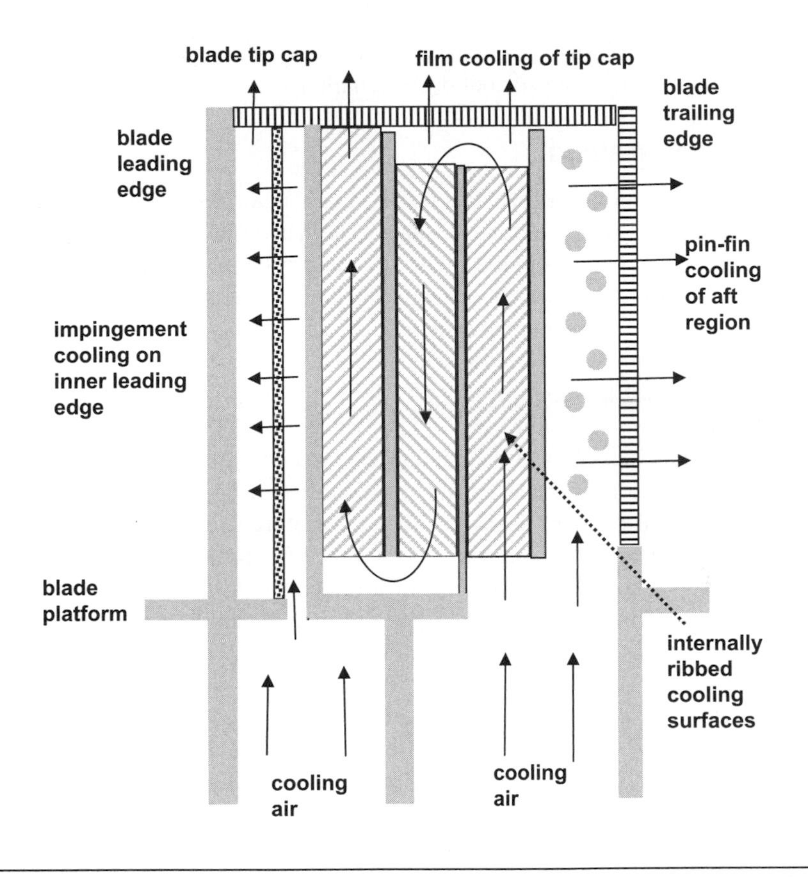


**FIGURE 8.32**

Typical internal configuration of a turbine blade-cooling system.

providing different advantages. The coolant removes some heat as it flows through the internal passages on its way to different exit points on the blade. The film-cooling holes exhaust some of the coolant at discrete points, which subsequently spreads out over the surface of the blade, forming a "film" that mixes with, and cools, the hot boundary layer flow. Another portion of the coolant flows out through the tip cap holes, cooling off that part of the blade. The final portion of the coolant is directed out through the trailing edge slots, removing heat from the aft regions of the blade on its way out of the blade.

The cooling air passes through a labyrinth of internal passages as shown in Figure 8.32. Different cooling mechanisms are exploited to achieve suitable blade temperatures. Near the leading edge the cooling air is injected so as to impinge directly on the inner surface of the leading edge and then exits through holes in the blade tip cap, providing cooling there as well. In the midsection of the blade, the cooling air makes several passes through passages that are ribbed, enhancing turbulence and increasing available surface area, thereby removing heat from the main surfaces of the blade before exiting through the blade tip cap. Cooling air is also directed to the aft portion of the blade where it passes over pin-fins, short cylindrical elements bridging the gap between suction and pressure surfaces of the blade. This arrangement also serves to increase turbulence and surface area for improved heat transfer.

It should be evident that the cooling system thus described, operating in a small highly stressed blade rotating at high rpm, represents a substantial design challenge.

### 8.14.7 Turbine blade materials

The development of high-temperature materials was always a major research thrust in the jet propulsion community. With temperatures in the turbine passages as high as 1800K, special attention has to be paid to the material used in the hot sections of the engine. Cooling systems for blades have been shown to be complicated and therefore costly, as well as requiring the bypass of some of the compressor flow, leading to performance reductions. Therefore, increasing the thermal capabilities of turbine blades and guide vanes has obvious advantages. Because standard vacuum cast blades made of high-temperature, nickel-based superalloys start to soften at around 1500K, cooling is necessary in first-stage turbine blades. The cooled blades must be investment cast under vacuum conditions to permit accurate rendition of the labyrinthine passages of the cooling system and minimize the possibility of incurring defects due to impurities. Ceramic thermal barrier coatings are then applied to the surface of the blade to provide an additional several hundred degrees to the temperature resistance of the blade.

Conventional casting techniques result in a crystalline structure with varying orientation of the crystals, often called grains, and this lack of coherent alignment contributes to the strength and lifetime limitations of the material. This weakness poses performance penalties because the allowable turbine entry temperature must be reduced to provide sufficient reliability. Directional solidification techniques pioneered in the mid-1960s at Pratt & Whitney led to the production of blades with the crystals aligned in one direction so that the blade would have much increased strength in that direction. Although this improved performance and blade life, research was continued with the aim of eliminating crystal boundaries: development of a blade composed of a single crystal. Manufacture of single crystal blades was started by Pratt & Whitney in the 1970s and has been continued and improved by them and other companies. The blades were introduced on the JT9D engine, variants of which are used on Boeing 767 and Airbus A310 aircraft, and on the J58 engine that powered the SR-71 Blackbird aircraft. These aircraft engine blades have a length around 8 cm, but now single crystal blades are even being fashioned for high-power industrial turbines with blades around five times longer, a much more difficult manufacturing process.

## 8.15 NOMENCLATURE

| | |
|---|---|
| $A$ | cross-sectional area |
| $a$ | sound speed |
| $C$ | closed contour path |
| $C_p$ | pressure coefficient |
| $c$ | absolute velocity |
| $c_D$ | drag coefficient |
| $c_{D,i}$ | induced drag coefficient |
| $c_L$ | lift coefficient |
| $c_p$ | specific heat |

| | |
|---|---|
| $D$ | drag |
| $F$ | force |
| $g$ | acceleration of gravity |
| $h$ | blade height or enthalpy |
| $i$ | angle of incidence |
| $K$ | constant coefficient |
| $k$ | thermal conductivity |
| $L$ | lift |
| $l$ | blade chord length |
| $M$ | Mach number |
| $\dot{m}$ | mass flow |
| $N$ | Nusselt number |
| $n$ | coordinate normal to blade surface |
| $n_b$ | number of blades |
| $h$ | enthalpy |
| $P$ | power |
| Pr | Prandtl number |
| $p$ | pressure |
| $\dot{Q}$ | heat transfer rate |
| $\dot{q}_c$ | convective heat flux |
| $R$ | gas constant or radius of curvature |
| Re | Reynolds number |
| $r$ | degree of reaction or radial coordinate |
| $S$ | surface pressure coefficient |
| St | Stanton number |
| $s$ | blade spacing |
| $T$ | temperature |
| $u$ | tangential velocity |
| $V$ | velocity |
| $w$ | velocity relative to the blade |
| $x$ | distance along the blade surface |
| $y$ | distance normal to the blade surface |
| $\alpha$ | angle between absolute velocity and tangential velocity |
| $\beta$ | angle between relative velocity and tangential velocity |
| $\Gamma$ | circulation |
| $\gamma$ | ratio of specific heats |
| $\delta$ | tip clearance or boundary layer thickness |

$\varepsilon$ drag to lift ratio
$\phi$ ratio of axial to tangential velocity $c_a/u$ or potential function
$\psi$ nondimensional pressure coefficient
$\rho$ gas density
$\mu$ viscosity
$\theta$ azimuthal coordinate
$\sigma$ solidity factor, $l/s$
$\tau$ slip or deflection parameter
$\eta$ efficiency
$\Omega$ vorticity
$\omega$ angular velocity

## 8.15.1 Subscripts

$a$ axial direction
$ad,c$ adiabatic compression
$cr$ critical
$e$ external to boundary layer
$i$ incompressible or ideal
$m$ mean
max maximum conditions
$opt$ optimum conditions
$l$ lower surface
$p$ pressure side
$s$ suction side
$t$ stagnation conditions
$u$ tangential direction or upper surface of blade
$w$ wall conditions
0 free stream conditions
2 compressor entrance conditions
3 compressor exit conditions
4 combustor exit conditions

## 8.15.2 Superscripts

$(\;)$ arithmetical average
$(\;)'$ perturbation quantity

## 8.16 EXERCISES

**8.1** (a) Derive the Kutta–Joukowski law for subsonic compressible flow. (b) Determine the velocity components and show the appearance of the Prandtl–Glauert factor $(1\text{-M}^2)^{1/2}$ in the results. (b) Explain how the lifting characteristics of a blade are affected by the compressibility effect.

**8.2** Consider an infinite circular cylinder rotating at constant angular speed $\omega$. The cylinder is divided into sectors of included angle $2\alpha$, which are filled with an inviscid fluid. (a) Show that the fluid in each sector circulates in the opposite sense to that of $\omega$. (b) Calculate the flow in a sector of included angle 60° and show streamlines and velocity vectors.

**8.3** Using the equation for the efficiency of adiabatic compression, Equation (8.79), with the assumption that $\epsilon = \epsilon_s = \epsilon_r$, show that contours of constant efficiency may be depicted as circles on a $\phi, r_i$ plot. Show the contours for $\epsilon = 0.04$ and efficiencies of 88%, 90%, and 92%, and indicate the value and location of the maximum efficiency on the same plot.

**8.4** A compressor stage processing air under standard sea level conditions is designed for ideal degree of reaction $r^* = 50\%$. The rotor speed is $u = 460$ m/s, the axial velocity component $c_a = 200$ m/s, and the rotor blade entrance angle is 35°. The blade chord length $l = 5$ cm, the blade pitch distance $s = 7$ cm, the blade height is $h = 12.7$ cm, and the hub radius $r_i = 30$ cm. The aerodynamic efficiency of the blade in the design operating range is $\varepsilon = 0.008$.

**(a)** Draw the combined entrance and exit velocity diagram to scale.

**(b)** Find the slip coefficient $\tau$ and the flow rate coefficient $\phi$.

**(c)** Find the force components $F_u$, $F_a$, and the resultant force $F$.

**(d)** Find the lift and drag components $L$ and $D$.

**(e)** Draw a scaled vector diagram showing the forces found in parts (b) and (c).

**(f)** Determine the solidity of the compressor.

**(g)** Find the torque and power required per blade.

**(h)** Determine the pressure rise in the rotor.

**8.5** Compressor characteristics may be known for a specific design point, typically sea level static conditions, and performance is desired for other, off-design, flight conditions. A semiempirical equation has been developed that provides quite accurate estimates of the compressor pressure ratio $p_{t3}/p_{t2}$ at a given altitude and speed when the value is known at some other flight condition, both being evaluated at the same rpm $N$. The empiricism arises from the following ad hoc definition of the compressor pressure ratio equation:

$$\frac{p_{t,3}}{p_{t,2}} = \left[1 + \frac{u(c_{3,u} - c_{2,u})}{\eta_{ad,c} c_p \overline{T}_2}\right].$$

Note that in this empirical definition the temperature in the denominator in the equation is not $T_{t,2}$ as arises in the proper formulation of the energy equation but instead the average stagnation temperature given by

$$\overline{T}_2 = \frac{1}{2}(T_{t,2} + T_{t,3}) = T_{t2}\left\{1 + \frac{1}{2\eta_{ad,c}}\left[\left(\frac{p_{t,3}}{p_{t,2}}\right)^{\frac{\gamma-1}{\gamma}} - 1\right]\right\}.$$

The basis for the remainder of the development is that the kinematic work term $u(c_{3,u} - c_{2,u})$ is independent of the compressor inlet temperature $T_{t,2}$, and of course $T_{t,2}$ varies with altitude and Mach number. Consider two conditions: subscript $a$, where the pressure ratio and temperature are specified, and $b$, where the flight conditions are specified but the compression ratio must be found.

**(a)** Use the equations given, the correct basic equation for the stagnation temperature ratio $(T_{t,3}/T_{t,2})$, and the condition that $u(c_{3,u} - c_{2,u})$ is a constant to find a relation for $(p_{t3}/p_{t2})_b$.

**(b)** For the case where $\eta_{ad,c} = 100\%$, plot a graph that shows the variation of $(p_{t3}/p_{t2})_b$ as a function of $(p_{t3}/p_{t2})_a$ with the temperature ratio $(T_{t,2b}/T_{t,2a})$ as a parameter. Consider case $a$ to be the known case, which is evaluated at standard sea level static conditions, and case $b$ to cover flight in the atmosphere between sea level and 11 km and Mach numbers between 0 and 1.

**(c)** For a compressor with a stagnation pressure ratio of 20 under standard sea level static conditions, determine its value at an altitude of 11 km and $M_0 = 0.85$ for $\eta_{ad,c} = 100\%$ and 85% and discuss the difference between the two answers.

**8.6** Another consideration for off-design performance calculations concerns the flow rate through the engine at flight conditions different from those at the design point. Considering the compressor to act like a constant volume pump over the range of flight conditions as long as the engine rpm is N is maintained constant has been found to be a quite accurate assumption. Mass flow entering the compressor may be written in terms of the axial component of the axial velocity and the flow area as follows:

$$\dot{m} = \rho_2 c_{2,a} A_2 = \rho_2 \dot{v}.$$

The assumption of constant volume flow rate is seen to be equivalent to requiring $c_{2,a} = \text{constant}$. Then the mass flow ratio

$$G = \frac{\dot{m}_b}{\dot{m}_a} = \frac{\rho_{2,b}}{\rho_{2,a}} = \frac{p_{2,b}}{p_{2,a}}\frac{T_{2,a}}{T_{2,b}}.$$

Assume, as in the previous exercise, that case $a$ is the design condition that is evaluated at standard sea level static conditions and case $b$ covers flight in the atmosphere between sea level and 11 km and Mach numbers between 0 and 1.

**(a)** Derive an equation to estimate the mass flow ratio $G$ in terms of the case $b$ flight conditions and describe any further assumptions made in arriving at the final result.

**(b)** Plot a graph showing the variation of $G$ as a function of flight Mach number with altitude as a parameter.

**(c)** For a compressor processing a mass flow of 82 kg/s at sea level static conditions, determine the mass flow at an altitude of 11 km and $M_0 = 0.85$.

## References

Goldstein, R.J. (2001). *Heat Transfer in Gas Turbine Systems.* New York Academy of Sciences, New York.

Abbott, I.H. and von Doenhoff, A.E. (1959), *Theory of Wing Sections*, Dover, New York

Schlichting, H. (1968). *Boundary Layer Theory.* New York: McGraw-Hill.

Shapiro, A.H. (1953), *The Dynamics and Thermodynamics of Compressible Flow*, vol. 1, The Ronald press, New York

CHAPTER

# Turbine Engine Performance and Component Integration

# 9

## CHAPTER OUTLINE



## 9.1 TURBOJET AND TURBOFAN ENGINE CONFIGURATIONS

Turbojet and turbofan engines are composed of many thermodynamic systems that must be matched properly so that they can work together effectively. As a consequence, there are many possible configurations for combining such systems depending on the design constraints considered. The complexity of practical aircraft engines is substantial, and cutaway drawings of existing engines are often so detailed as to lack clarity in emphasizing the basic design features. To gain an appreciation for the general arrangement of major systems in a complete engine, we will consider some specific turbojet and turbofan engines that have seen extensive use. Cross-sectional diagrams of a simple schematic form are shown in Figures 9.1 through 9.5, respectively. The axial variation of the stagnation pressure and temperature through the different components of the engine is also shown on these figures, along with other pertinent performance-related information for the case of static ($V_0 = 0$) operation. The availability of much of this data makes it possible to estimate the efficiency of the various components. Recall that in Chapter 3 engine performance was developed for ideal components with 100% efficiency. In this section we will get a good sense of the magnitude of component efficiencies achievable in practice and these may be used by the reader to redo the cycle analysis presented in Chapter 3. Detailed values for the various thermodynamic and flow properties are presented in Table 9.1. The published thrust levels are compared to the values obtained by applying the methods of Chapter 3 with practical rather than ideal values for the efficiency of the various components. The models illustrated all employ axial flow compressors driven by axial flow turbines; this is typical for aircraft engines, except perhaps at the smallest sizes where centrifugal compressors may sometimes be employed. Note that all data used were presented originally in English units and were converted to SI for purposes of this study.

### 9.1.1 Single-shaft turbojet

Figure 9.1 shows a simple turbojet configuration with a compressor and its driving turbine riding on a single driveshaft. This was the configuration of choice for early turbojet engines. The engine shown represents the turbojet used as an illustration in Figure 2.5 of "The Jet Engine" (3rd ed. 1969, Rolls-Royce Ltd., Derby, UK) from which the pressures and temperatures were deduced. Here $T_{t4} = 900°C$ and the exit stagnation temperature $T_{t7}$ appears on the figure as about 600°C. The exit static to stagnation pressure ratio is taken such that $M_7 = 1$, which leads to an underexpanded sonic nozzle condition. The exit velocity is difficult to read on Figure 2.5 of "The Jet Engine" because the slope of the line on the graph shown is steep. However, the derived value of 1753 ft/s for a choked nozzle is consistent with that line. The arrangement, thrust level, and compressor pressure ratio are typical of the Rolls-Royce Avon engine family. According to *Flight* (March 1959, pp. 394–395), the Avon Ra.29 Mk 527 is a single-shaft engine with a 16-stage compressor and a 3-stage turbine that provides 11,700 lbs static thrust, with a mass flow of 173 lb/s and a compressor pressure ratio of 9.63. This engine was used commercially in the de Havilland Comet and Sud Aviation Caravelle commercial airliners, while military versions powered aircraft such as English Electric's Lightning fighter and Canberra bomber. The published values of weight flow and pressure ratio given in *Flight* (March 1959, pp. 394–395) were chosen for the current analysis. The stagnation pressures and temperatures were used to calculate the compressor and turbine efficiencies and the stagnation pressure drop across the combustor. The value for specific fuel consumption of $c_j = 0.75$ lbs of fuel per hour per pound of thrust was taken from *Flight* (March 1959, pp. 394–395) and was used to estimate the fuel to air ratio so that the burner efficiency $\eta_b$ could be calculated. It was pointed out in *Flight* (March 1959, pp. 394–395) that the design flame temperature was kept so low so as to afford very low specific fuel consumption. All data and

**Table 9.1** Detailed Properties of Engines Considered

| Property | Avon | JTC-3 | JT8D | JT3D | JT9D | J57-P-23 |
|---|---|---|---|---|---|---|
| $p_{t0}$ (kPa) | 101 | 101 | 101 | 101 | 101 | 101 |
| $p_{t2}$ (kPa) | 101 | 101 | 101 | 101 | 101 | 101 |
| $p_t$ FAN (kPa) | | | 286 | 179 | 155 | |
| $p_{t3}$ (kPa) | 1003 | 1264 | 1601 | 1374 | 2171 | 1147 |
| $p_{t4}$ (kPa) | 931 | 1202 | 1512 | 1305 | 2075 | 1086 |
| $p_{t5}$ (kPa) | 264 | 247 | 199 | 192 | 144 | 247 |
| $p_{t6}$ (kPa) | 264 | 247 | 199 | 192 | 144 | 247 |
| $p_{t7}$ AB (kPa) | | | | | | 219 |
| $T_{t0}$ (K) | 288 | 288 | 288 | 288 | 288 | 288 |
| $T_{t2}$ (K) | 288 | 288 | 288 | 288 | 288 | 288 |
| $T_t$ FAN (K) | | | 361 | 349 | 327 | |
| $T_{t3}$ (K) | 673 | 622 | 699 | 652 | 744 | 617 |
| $T_{t4}$ (K) | 1173 | 1127 | 1211 | 1144 | 1349 | 1127 |
| $T_{t5}$ (K) | 873 | 818 | 749 | 749 | 727 | 818 |
| $T_{t6}$ (K) | 873 | 818 | 749 | 749 | 727 | 818 |
| $T_{t7}$ AB (K) | | | | | | 1667 |
| $V$ FAN (m/s) | | | | 302 | 270 | |
| $V_6$ (m/s) | 534 | 517 | 442 | 475 | 363 | |
| $V_7$ AB (m/s) | | | | | | 820 |
| $m'$ FAN (kg/s) | 0 | 0 | 74.7 | 120.0 | 565.3 | 0 |
| $m'$ core (kg/s) | 78.4 | 81.5 | 68.0 | 88.3 | 111.9 | 89.2 |
| $\beta$ | 0 | 0 | 1.1 | 1.4 | 5.3 | 0 |
| $\eta_c$ (%) | 68 | 91 | 84 | 88 | 89 | 87 |
| $\eta_t$ (%) | 95 | 84 | 96 | 91 | 95 | 89 |
| $\eta_b$ (%) | 97 | 86 | 89 | 90 | 89 | |
| $\eta_b$ (%) AB | | | | | | 63 |
| $\Delta p_t/p_{t3}$ (%) | 5 | 5 | 6 | 5 | 4 | |
| $\Delta p_t/p_{t5}$ (%) AB | | | | | | 11 |
| $F_n$ (N) listed | 49640 | 49818 | 62272 | 80064 | 193488 | 71168 |
| $F_n$ (N) calc. | 51094 | 49924 | 63424 | 78485 | 193270 | 73250 |
| Error % | 2.93 | 0.21 | 1.85 | −1.97 | −0.11 | 2.84 |

derived results for this engine are listed in Table 9.1. The calculated thrust is found to be about 3% higher than the published value.

## 9.1.2 Dual-shaft turbojet

In Figure 9.2, a common turbojet configuration is shown in which a low-pressure compressor and its driving turbine ride on an inner driveshaft, while a high-pressure compressor rides on a concentric outer driveshaft connected to a separate high-pressure compressor. In this fashion,

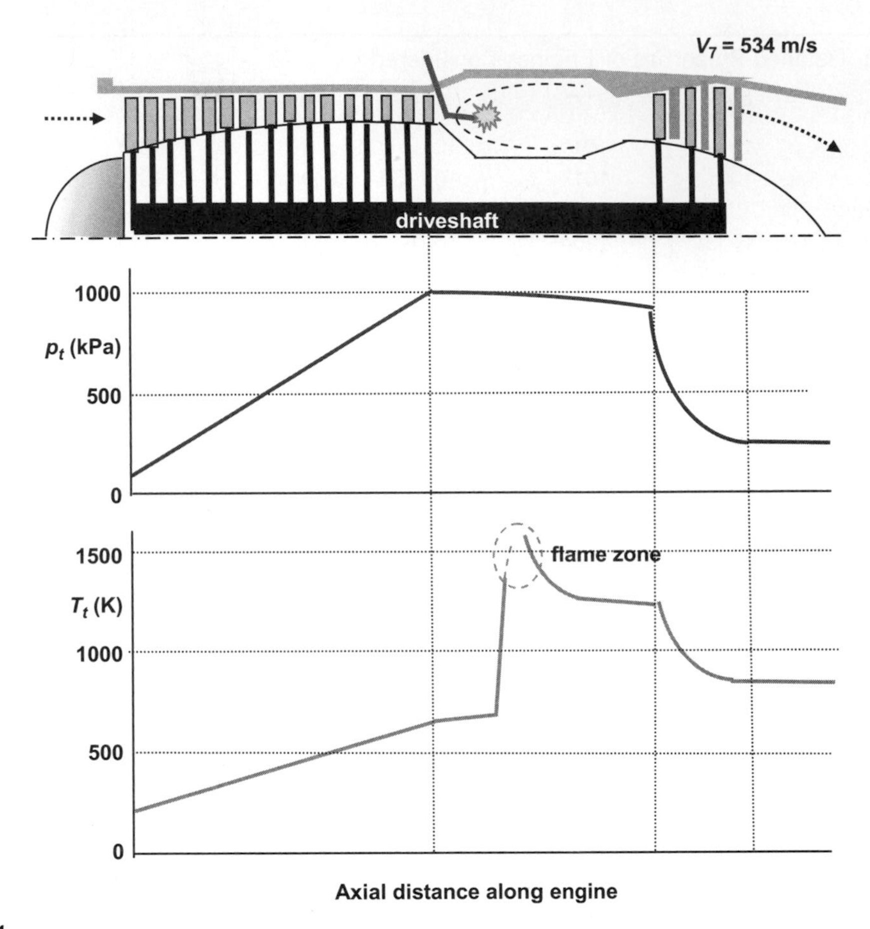


**FIGURE 9.1**

Schematic diagram of a single-shaft turbojet engine similar to the Rolls Royce Avon showing the axial variation of the major flow properties through the engine.

each compressor–turbine set is free to rotate at a different rotational speed appropriate to the desired performance characteristics. The Pratt & Whitney (P&W) J57 is a two-shaft engine with a nine-stage low-pressure compressor, a seven-stage high-pressure compressor, a single-stage high-pressure turbine, and a two-stage low-pressure turbine. The military versions of this engine powered such aircraft as the North American F-100 Super Sabre and Vought F8U Crusader fighter aircraft, while the civil versions were flown by the Boeing 707 and Douglas DC-8 commercial airliners. The schematic diagram in Figure 9.2 represents the P&W J57-P-43WB, the military version of the P&W JT3C-6 turbojet.

Pressures, temperatures, and exit velocity are taken from "The Gas Turbine Engine and Its Operation" (PWA OI 200 Pratt & Whitney Aircraft, May 1974, E. Hartford, CT) and Mattingly (1996). Those data were given for the J57 with no additional identification, but it seems as if it were for the J57-P-23, an afterburning turbojet, which is discussed subsequently in Section 9.12.6.

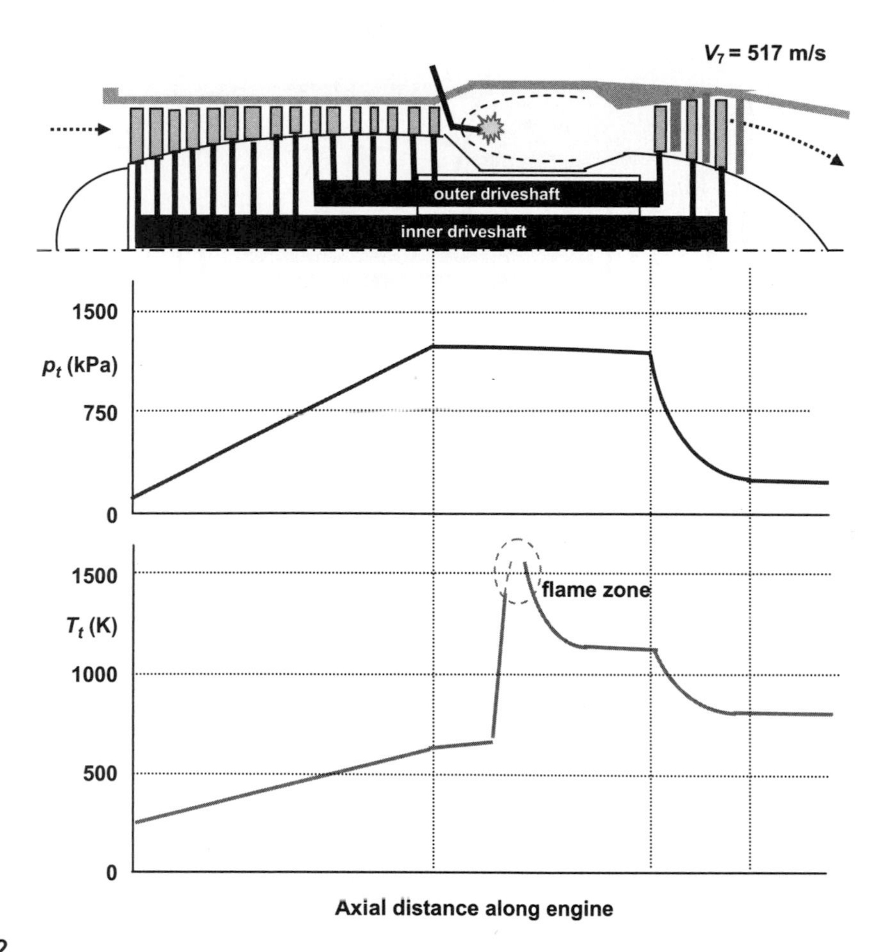


**FIGURE 9.2**

Schematic diagram of a dual-shaft turbojet engine similar to the Pratt & Whitney JT3C showing the axial variation of the major flow properties through the engine.

Weight flow, thrust, and compressor pressure ratio are taken from *Flight* (March 18 1960, p. 385) and are for the JT3C-6, a commercial variant. The calculated nozzle area for the given weight flow checks with that for choked flow. The temperatures are those quoted for the J57-P-23 without the afterburner section.

### 9.1.3 Dual-shaft internally mixed turbofan

Figure 9.3 shows a turbofan engine with internally mixed fan flow. The compressor and its driving turbine are divided into two specialized subsystems: a low-pressure compressor with an integral fan driven by a turbine mounted on the same driveshaft and a high-pressure compressor driven by another turbine.

The JT8D is a two-shaft engine with a four-stage low-pressure compressor and integral two-stage internally bypassed fan, a seven-stage high-pressure compressor, a single-stage high-pressure turbine,

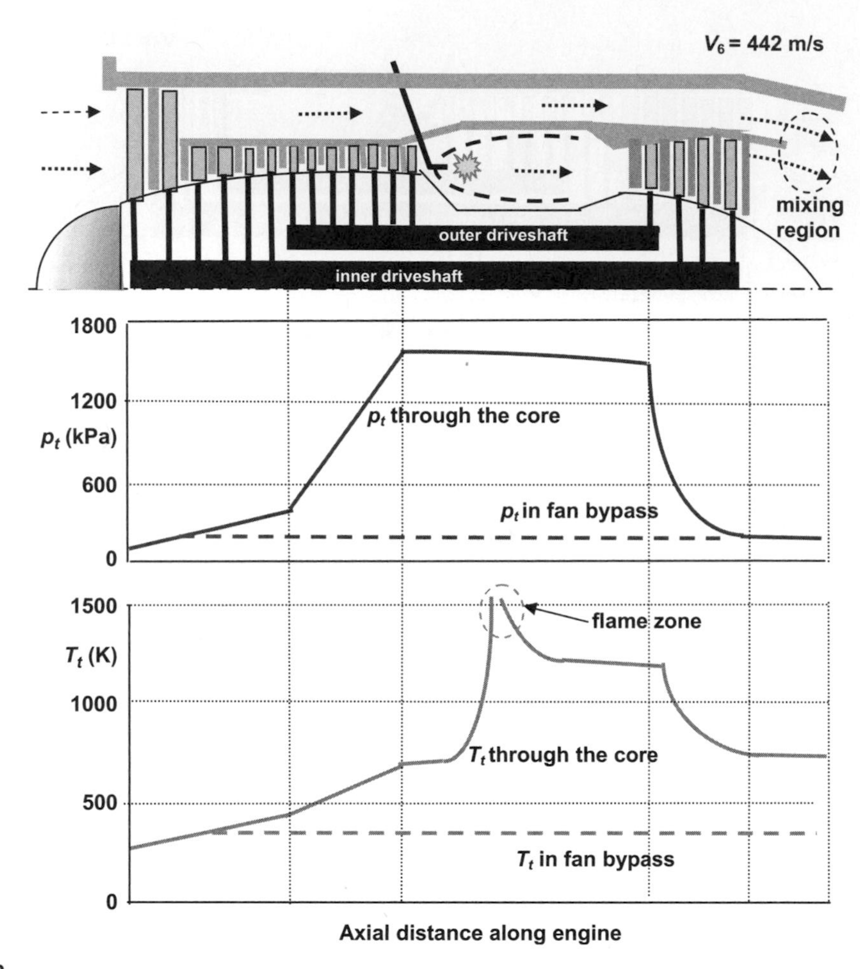


**FIGURE 9.3**

Schematic diagram of a dual-shaft, internally mixed, low bypass turbofan engine similar to the Pratt & Whitney JT8D showing the axial variation of the major flow properties through the engine.

and a three-stage low-pressure turbine. The bypass ratio is low, with a value of $\beta = 1.1$. It is an engine used commonly on narrow-body airliners such as the Boeing 727, Boeing 737, and the McDonnell-Douglas MD80. The schematic diagram in Figure 9.3 represents the P&W JT8D-3B low bypass ratio ($\beta = 1.1$) turbofan with a single nozzle for the mixed fan and the jet flow. Stagnation pressures and temperatures, as well as fan and jet exit velocities, are taken from "The Gas Turbine Engine and Its Operation" (PWA OI 200 Pratt & Whitney Aircraft, May 1974, E. Hartford, CT) and Mattingly (1996). The nozzle area is based on the weight flow given and calculation of an exit Mach number $M_7 = 1$ according to the stagnation pressures given. Thus the nozzle is choked, and the nozzle area is determined for that condition. The mixing region for the fan and nozzle flow is identified in Figure 9.3. As pointed out in Chapter 3, the most effective mixing takes place when the stagnation pressures of the two streams are equal. It is clear from the curves presented in Figure 9.3 that the stagnation pressure of the fan and the nozzle are about the same.

### 9.1.4 Dual-shaft low bypass turbofan

In the turbofan engine shown in Figure 9.4 we see a similar arrangement, but in this case the fan flow is not bypassed internally to mix with thc exhaust from the turbine, but instead flows around the outside of the nacelle.

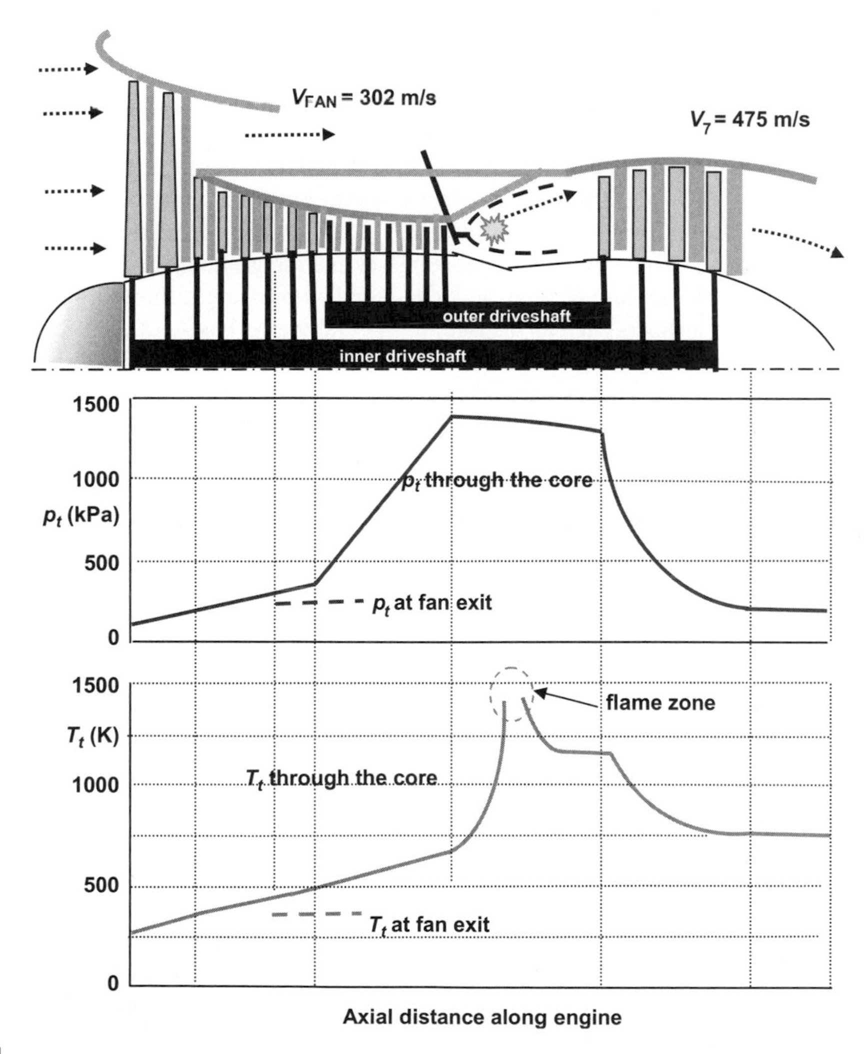


**FIGURE 9.4**

Schematic diagram of a dual-shaft, externally mixed, low bypass turbofan engine similar to the Pratt & Whitney JT3D showing the axial variation of the major flow properties through the engine.

The JT3D is a two-shaft engine with a six-stage low-pressure compressor and integral two-stage fan, a seven-stage high-pressure compressor, a single-stage high-pressure turbine, and a three-stage low-pressure turbine. It is an engine used commonly on wide-body airliners such as the Boeing 747 and the Airbus A300. Figure 9.4 represents the P&W JT3D-3B low bypass ratio ($\beta = 1.4$) turbofan with separate nozzles for the fan and the jet. Pressures and fan and jet exit velocities are taken from *Flight* (March 1959, pp. 394–395) and "The Gas Turbine Engine and Its Operation" (PWA OI 200 Pratt & Whitney Aircraft, May 1974, E. Hartford, CT). Those data were given for the JT3D-3B. Weight flow, thrust, and compressor pressure ratio are taken from the data set mentioned. The nozzle area is based on the weight flow given and calculation of an exit Mach number $M_7 = 1$ according to the stagnation pressures given. Thus the nozzle is choked, and the nozzle area is determined for that condition.

### 9.1.5 Dual-shaft high bypass turbofan

In Figure 9.5, a high bypass turbofan, characterized by a large fan diameter, is illustrated. Some high bypass turbofans carry this specialization of compressor–turbine pairs further so as to have three concentric driveshafts with the very inner one dedicated to the fan and its driving turbine. The JT9D is a two-shaft engine with a 3-stage low-pressure compressor and integral fan, an 11-stage high-pressure compressor, a 2-stage high-pressure turbine, and a 4-stage low-pressure turbine. It is an engine used commonly on wide-body airliners such as the Boeing 747 and the Airbus A300. Figure 9.5 represents the P&W JT9D high bypass ratio ($\beta = 5$) turbofan with separate nozzles for the fan and the jet. Pressures and fan and jet exit velocities are taken from *Flight* (March 1959, pp. 394–395) and "The Gas Turbine Engine and Its Operation" (PWA OI 200 Pratt & Whitney Aircraft, May 1974, E. Hartford, CT). Those data were given for the JT9D with no additional identification. Weight flow, thrust, and compressor pressure ratio are taken from the data set mentioned. The nozzle area is based on the weight flow given and calculation of an exit Mach number $M_7 = 0.744$ according to the stagnation pressures given. Thus the nozzle is not choked, and the nozzle area is determined for that condition.

### 9.1.6 Dual-shaft afterburning turbojet

Figure 9.6 shows an afterburning two-shaft turbojet configuration. The P&W J57 has been described previously as a two-shaft engine with a nine-stage low-pressure compressor, a seven-stage high-pressure compressor, a single-stage high-pressure turbine, and a two-stage low-pressure turbine. The military versions of this engine, which featured afterburning, powered such aircraft as the North American F-100 Super Sabre and Vought F8U Crusader fighter aircraft. The schematic diagram in Figure 9.6 represents the P&W J57-P-23, a military version of the P&W JT3C-6 turbojet. Pressures, temperatures, and exit velocity are taken from "The Gas Turbine Engine and Its Operation" (PWA OI 200 Pratt & Whitney Aircraft, May 1974, E. Hartford, CT) and Mattingly (1996). Those data were given for the J57 with no additional identification, but it seems as if it were for the J57-P-23. Weight flow, thrust (no water injection), and compressor pressure ratio are taken from Mattingly (1996) and are for the JT3C-6, a commercial variant.

It is worth noting that the efficiencies achieved in the compressor, turbine, and burner are quite high, while the stagnation pressure losses in the combustor are reasonably low. The afterburner efficiency, however, is low and the associated stagnation pressure losses are considerable. The thrust

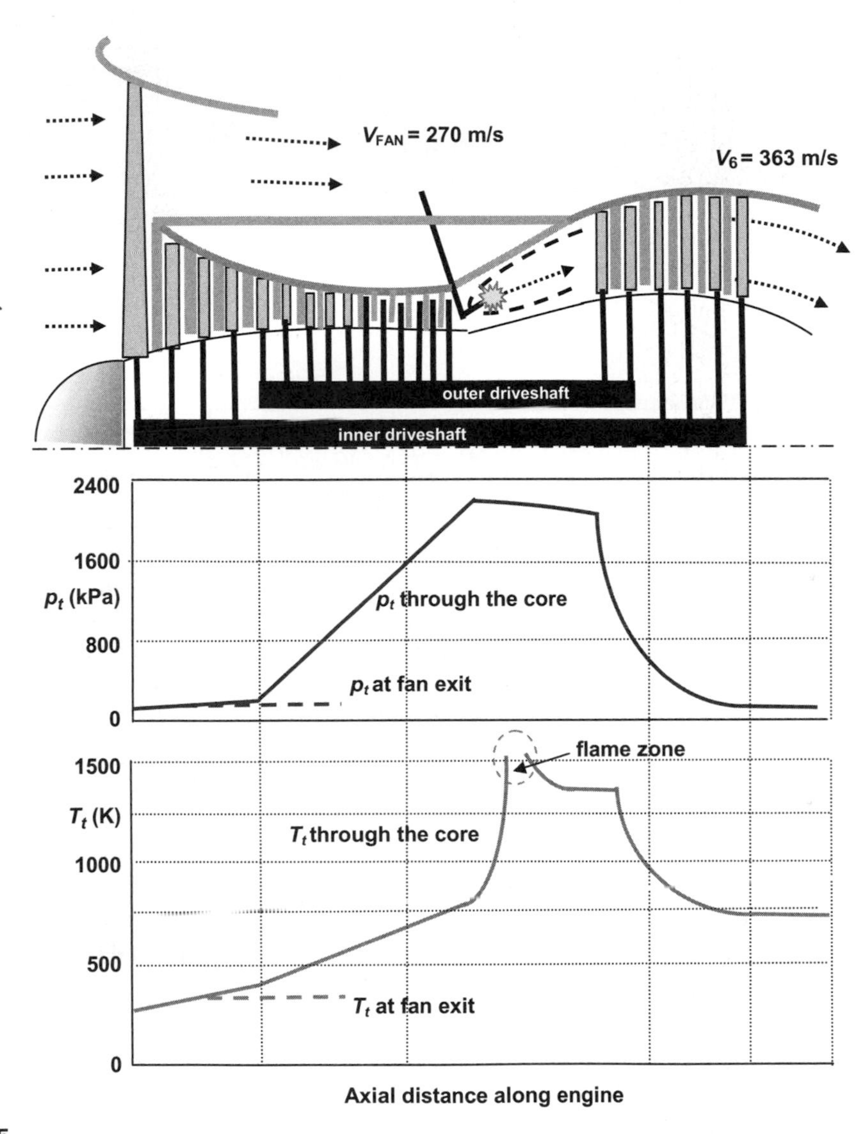


**FIGURE 9.5**

Schematic diagram of a dual-shaft, externally mixed, high bypass turbofan engine similar to the Pratt & Whitney JT9D showing the axial variation of the major flow properties through the engine.

calculations were carried out neglecting the weight flow of fuel added to the flow. This is not unreasonable because all engines have air bled from some stage in the compressor to carry out tasks such as providing cooling air for the hot components of the engine. This subtraction of air from the engine flow rate is large enough to offset the fuel weight added. Indeed, the engine-bleed air requirements can be larger than the fuel flow added. The specific heat used in the determination of

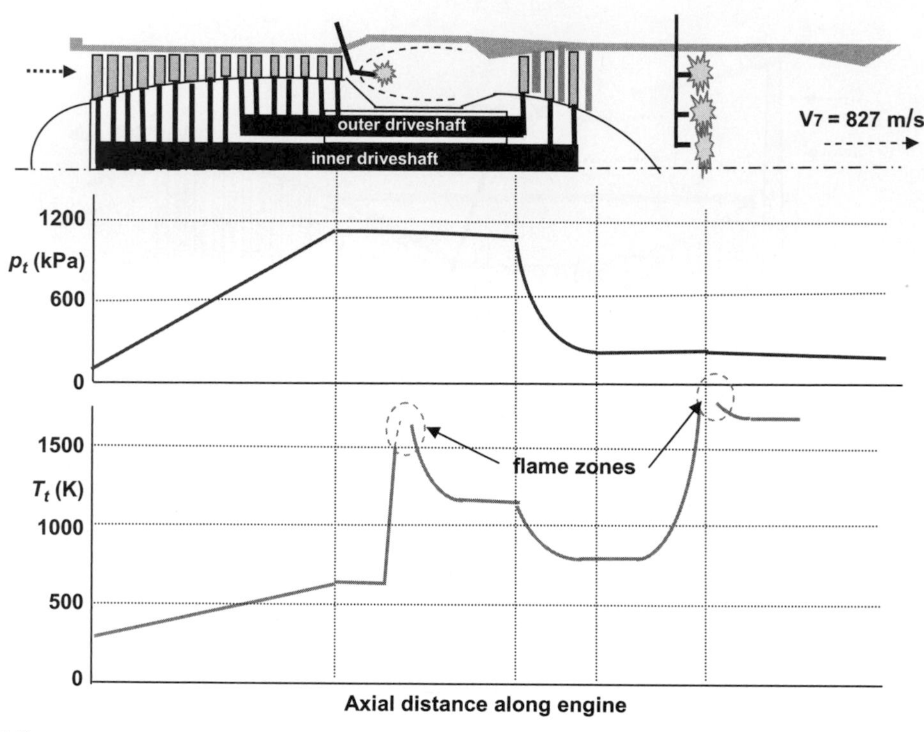


FIGURE 9.6

Schematic diagram of a dual-shaft afterburning turbojet engine similar to the Pratt & Whitney J57-P-23 showing the axial variation of the major flow properties through the engine.

burner efficiency was an average value based on the combustor inlet and exit conditions using specific heat data given in Appendix B.

## 9.2 OPERATIONAL REQUIREMENTS

The engine is required to provide appropriate thrust levels with safe and efficient operation throughout the design operational envelope of the aircraft. This means that all the components of the engine—the inlet, compressor, combustor, turbine, and nozzle—must work as a unit in that design envelope. A flight envelope for the McDonnell F4-C Phantom is shown in Figure 9.7.

Generally the flight altitude and Mach number are stipulated, and some properties of the various engine components are given with the intent of determining the remaining properties of those and other engine components pertinent to matched operation at that flight condition. The analysis would be repeated to develop the engine characteristics required throughout the flight envelope. The compressor and turbine maps for the engine must be used to ensure that the turbine and compressor are operating at the same rpm and that the power extracted by the turbine is equal to that required by the compressor.

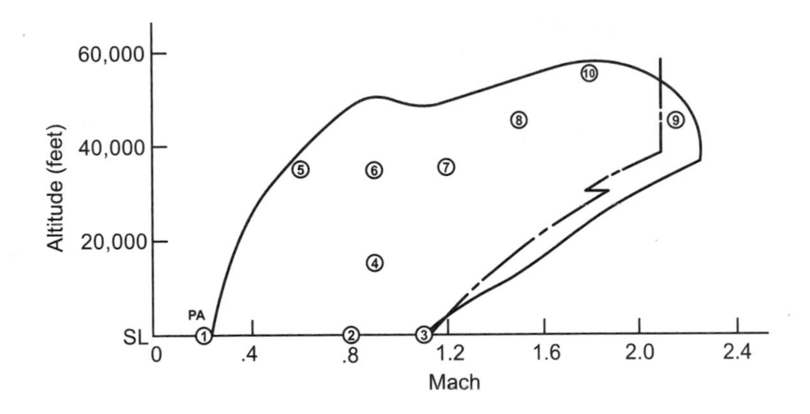


**FIGURE 9.7**

Level flight envelope for the F-4C Phantom in its nominal configuration. The broken line denotes speed restrictions (from Hefley and Jewell, 1972). (Numbers on flight envelope refer to points where handling qualities were evaluated in that reference).

This is the matching condition that must be met for the turbomachinery in the engine to operate successfully.

## 9.3 COMPRESSOR–TURBINE MATCHING—CASE 1: NOZZLE MINIMUM AREA AND COMBUSTOR EXIT STAGNATION TEMPERATURE SPECIFIED

Here the flight conditions $z$ and $M_0$ are given along with the inlet pressure recovery $p_{t2}/p_{t0}$, and the nozzle minimum area $A_6$ and the combustor exit stagnation temperature $T_{t4}$ are stipulated. It is desired to determine the operating rpm $N$, the captured streamtube area $A_0$, and the nozzle exit area $A_7$ under these conditions. The nozzle flow is assumed to be isentropic so that $p_t$ and $T_t$ are both constant throughout. For this reason, the corrected weight flow is also constant throughout. Assuming that the throat ($A_6$) is always operating in the choked condition and therefore passing the maximum mass flow for the stagnation conditions at the turbine exit, the actual mass flow is

$$\dot{m} = \frac{\dot{w}}{g} = p_{t5}A_6\sqrt{\frac{\gamma_5}{RT_{t5}}}\left(\frac{\gamma_5+1}{2}\right)^{\frac{\gamma_5+1}{2(1-\gamma_5)}}. \tag{9.1}$$

It is convenient to use the throat area $A_6$ for the flow area; therefore, in Equation (9.1) we may use $p_{t6}=p_{t5}$ and $T_{t6}=T_{t5}$. Recall that the nondimensional stagnation temperature and stagnation pressure are defined by

$$\begin{aligned}\theta &= \frac{T_t}{T_{ref}} \\ \delta &= \frac{p_t}{p_{ref}}\end{aligned}. \tag{9.2}$$

Using these variables in Equation (9.1) leads to the corrected mass flow at station 5 as given by

$$\frac{\dot{m}\sqrt{\theta_5}}{\delta_5} = p_{ref}A_6\sqrt{\frac{\gamma_5}{RT_{ref}}}\left(\frac{\gamma+1}{2}\right)^{\frac{\gamma+1}{2(1-\gamma)}}. \tag{9.3}$$

Turbine performance maps are given both in terms of pressure ratio $p_{t4}/p_{t5}$ as a function of corrected mass flow and in terms of corrected work per unit weight as a function of corrected mass flow, where the corrected values are referred to station 4. Note that in practice, corrected weight flow is often used instead of corrected mass flow and sometimes the terms are used interchangeably; therefore, one must be careful to check the units being used. Thus, to use that data, a correlation between the corrected mass flow at station 4 and station 5 must be constructed. We note that because the actual mass flow is constant, the corrected mass flow at station 5 is related to that at station 4 as follows:

$$\frac{\dot{m}\sqrt{\theta_5}}{\delta_5} = \frac{\dot{m}\sqrt{\theta_4}}{\delta_4}\frac{p_{t4}}{p_{t5}}\sqrt{\frac{T_{t5}}{T_{t4}}}. \tag{9.4}$$

The basic turbine work equation may be manipulated, assuming constant $c_p = c_{p4}$ through the turbine, to yield the stagnation temperature ratio as

$$\frac{T_{t5}}{T_{t4}} = 1 - \frac{W_t}{c_{p4}T_{ref}\theta_4}. \tag{9.5}$$

The turbine performance map is given most usually in terms of pressure ratio and efficiency as functions of corrected mass (or weight) flow, as shown in the representative turbine map in Figure 9.8. Therefore, the temperature ratio in Equation (9.5) may also be written in terms of the turbine pressure ratio and the efficiency as follows:

$$\frac{W_t}{c_{p4}T_{ref}\theta_4} = \eta_t\left[1 - \left(\frac{p_{t5}}{p_{t4}}\right)^{\frac{\gamma_4-1}{\gamma_4}}\right]. \tag{9.6}$$

Now the corrected mass flow may be written as

$$\frac{\dot{m}\sqrt{\theta_5}}{\delta_5} = \frac{\dot{m}\sqrt{\theta_4}}{\delta_4}\frac{p_{t4}}{p_{t5}}\sqrt{1 - \eta_t\left[1 - \left(\frac{p_{t5}}{p_{t4}}\right)^{\frac{\gamma_4-1}{\gamma_4}}\right]}. \tag{9.7}$$

We may now use the turbine performance map shown in Figure 9.8 to complete the relationship of Equation (9.7). For a given value of corrected rpm $N/\theta_4^{1/2}$ we may enter a value of corrected mass flow $\dot{m}\sqrt{\theta_4}/\delta_4$ and determine the associated corrected work and the corresponding pressure ratio from Figure 9.8. This information permits calculation of a curve of $\dot{m}\sqrt{\theta_5}/\delta_5$ as a function of $\dot{m}\sqrt{\theta_4}/\delta_4$ as shown in Figure 9.9.

The value of $\dot{m}\sqrt{\theta_5}/\delta_5$ is known from Equation (9.3), and a corresponding value of $\dot{m}\sqrt{\theta_4}/\delta_4$ is found from the correlation graph in Figure 9.9. This value of corrected mass flow, taken at a particular corrected rpm, can be entered into the turbine performance map of Figure 9.8 to obtain the appropriate

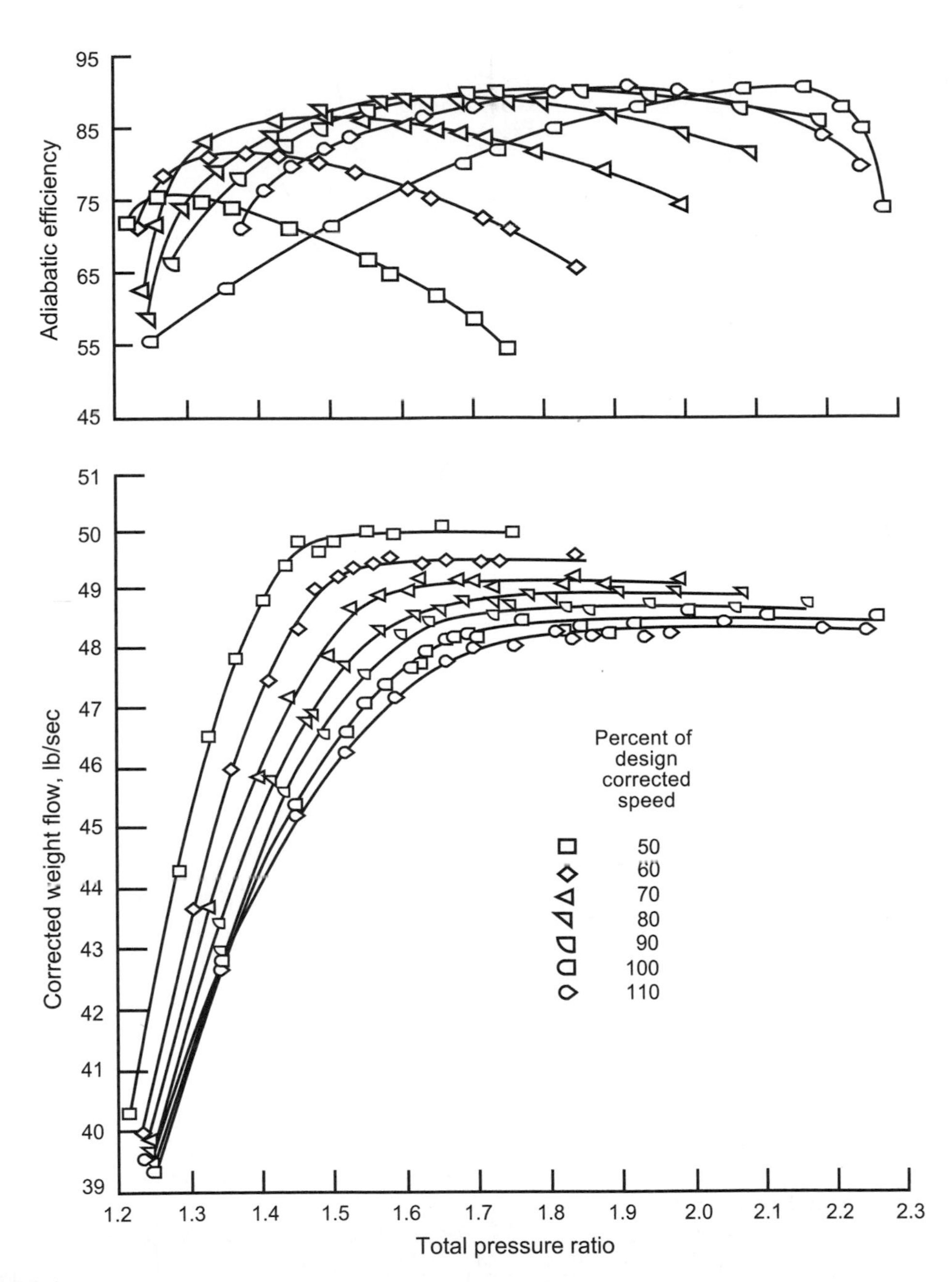


**FIGURE 9.8**

Representative turbine performance map of pressure ratio and efficiency as functions of the corrected weight flow rate through the turbine (from Plencner, 1989).

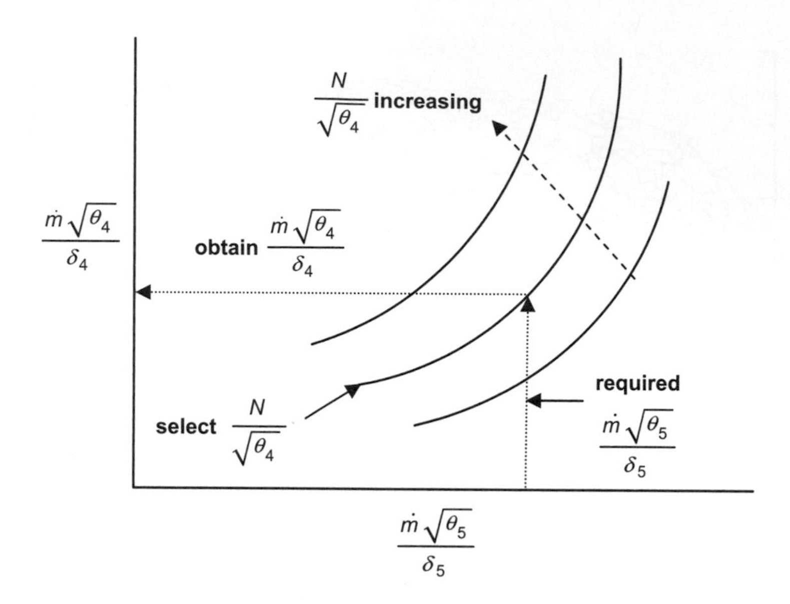


**FIGURE 9.9**

Correlation between the corrected mass flow at stations 4 and 5 for various values of corrected rpm. Shown is the required $\dot{m}\sqrt{\theta_5}/\delta_5$ from Equation (9.3) and a selected corrected rpm, which defines the corresponding $\dot{m}\sqrt{\theta_4}/\delta_4$.

pressure ratio $p_{t4}/p_{t5}$. The corrected mass flow entering the compressor may then be related to that entering the turbine by the following equation:

$$\frac{\dot{m}\sqrt{\theta_2}}{\delta_2} = \frac{\dot{m}\sqrt{\theta_4}}{\delta_4}\frac{p_{t4}}{p_{t3}}\frac{p_{t3}}{p_{t2}}\sqrt{\frac{T_{t2}}{T_{t4}}}. \tag{9.8}$$

In this equation, the corrected weight flow $\dot{m}\sqrt{\theta_4}/\delta_4$ is known and $T_{t4}$ is given. The stagnation temperature entering the compressor may be calculated, assuming that the flow through the inlet is adiabatic, so that

$$T_{t2} = T_{t0} = T_0\left(1 + \frac{\gamma - 1}{2}M_0^2\right). \tag{9.9}$$

Of course, the free stream static temperature may be determined from the altitude that was given, as was the flight Mach number. Finally, we may assume, at least as a first approximation, that the stagnation pressure across the combustor $p_{t3}/p_{t4} \sim 1$ and a generic value may be inserted since, in general, the stagnation pressure ratio across a well-designed combustor is quite close to unity. Now Equation (9.8) may be written as

$$\frac{\dot{m}\sqrt{\theta_2}}{\delta_2} = \left[\frac{\dot{m}\sqrt{\theta_4}}{\delta_4}\frac{p_{t4}}{p_{t3}}\sqrt{\frac{T_{t2}}{T_{t4}}}\right]\frac{p_{t3}}{p_{t2}} = C\frac{p_{t3}}{p_{t2}}. \tag{9.10}$$

Thus, the corrected weight flow entering the compressor is a linear function of the compressor pressure ratio with the constant of proportionality dependent on the chosen value of corrected weight flow entering the turbine. A representative compressor turbine map is illustrated in Figure 9.10.

The corrected rpm of the compressor is related to the corrected rpm of the turbine by

$$\frac{N}{\sqrt{\theta_2}} = \frac{N}{\sqrt{\theta_4}}\sqrt{\frac{T_{t4}}{T_{t2}}}. \tag{9.11}$$

We may put the linear relation of Equation (9.10) for a selected corrected mass flow entering the turbine on the compressor map of Figure 9.10, seek its intersection with the known value of $N/\theta_2^{1/2}$, and thereby determine the corresponding pressure ratio and efficiency.

Now that we have found a possible matching point on the compressor map in Figure 9.10, it is necessary to ascertain if the matching condition, $W_t = W_c$, is also satisfied at that point. The matching condition, in terms of corrected work is given by

$$\frac{W_C}{\theta_2} = \frac{W_t}{\theta_4}\frac{T_{t4}}{T_{t2}}.$$

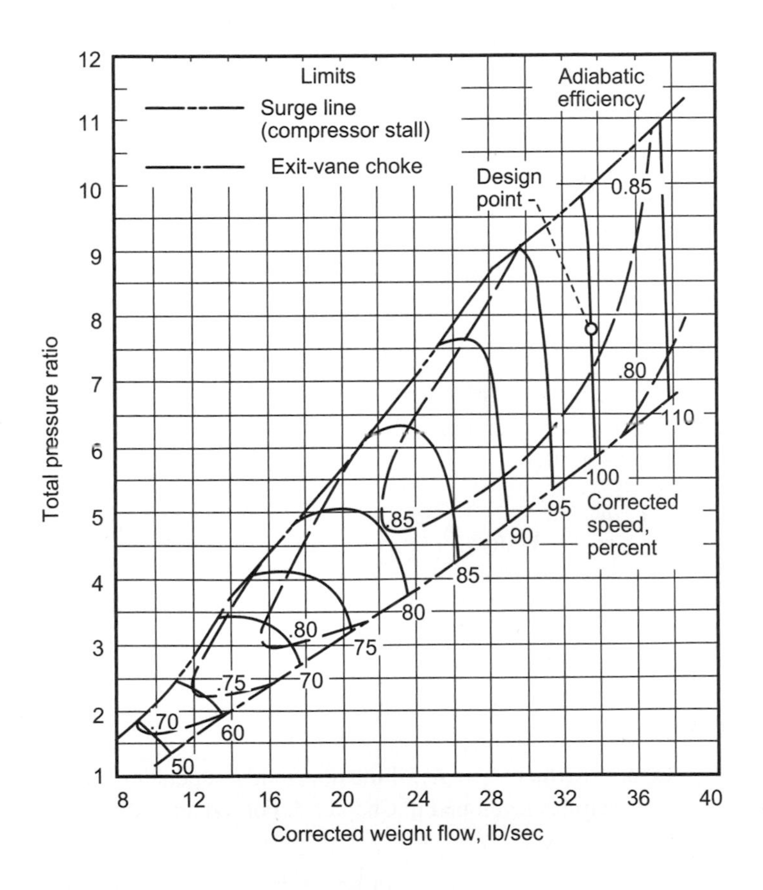


**FIGURE 9.10**

Representative compressor performance map showing pressure ratio and efficiency as a function of corrected weight flow for various corrected rotation speeds, expressed here as percent of design rpm (from Plencner, 1989).

The corrected compressor work may be written in terms of the pressure ratio as follows:

$$\frac{W_c}{\theta_2} = \frac{c_{p2}T_{ref}}{\eta_c}\left[\left(\frac{p_{t3}}{p_{t2}}\right)^{\frac{\gamma_2-1}{\gamma_2}} - 1\right]. \tag{9.12}$$

The corrected turbine work may be taken from Equation (9.6) and appears as follows:

$$\frac{W_t}{\theta_4} = \eta_t c_{p4} T_{ref}\left[1 - \left(\frac{p_{t5}}{p_{t4}}\right)^{\frac{\gamma_4-1}{\gamma_4}}\right]. \tag{9.13}$$

If the matching condition is satisfied, then the corrected mass flow $\dot{m}\sqrt{\theta_4}/\delta_4$ and corrected speed $N/\theta_2^{1/2}$ are the appropriate values, and the pressure ratios of the turbine and compressor and the associated efficiencies are determined for the case under consideration. If the compressor work is not equal to the turbine work, then the process must be repeated with a new corrected speed $N/\theta_2^{1/2}$ and corrected mass flow $\dot{m}\sqrt{\theta_4}/\delta_4$ until convergence occurs. It must be noted that if there is more than one flow path in the engine, for example, a turbofan where only part of the total airflow process goes through the core of the engine while another is by passed through the fan, then the proper matching condition is that the power generated by the turbine must equal the power required by the fan-compressor unit, as the work performed or extracted is on a per unit mass (or weight) basis.

Now that the corrected mass flow at station 2 has been determined, one may calculate the corrected mass flow in the free stream capture streamtube using the following relation:

$$\frac{\dot{m}\sqrt{\theta_0}}{\delta_0} = \frac{\dot{m}\sqrt{\theta_2}}{\delta_2}\frac{p_{t2}}{p_{t0}}\sqrt{\frac{T_{t0}}{T_{t2}}}. \tag{9.14}$$

In Equation (9.14), the terms on the right-hand side are all known because the inlet pressure recovery is given and $T_{t2} = T_{t0}$, as it is assumed that the inlet flow is adiabatic. For the corrected mass flow at station 0, we may also write

$$\frac{\dot{m}\sqrt{\theta_0}}{\delta_0} = A_0 M_0 p_{ref}\sqrt{\frac{\gamma_0}{RT_{ref}}}\left(1 + \frac{\gamma - 1}{2}M_0^2\right)^{\frac{\gamma_0+1}{2(1-\gamma_0)}}. \tag{9.15}$$

Then Equation (9.15) may be used to calculate the free stream capture area $A_0$. The actual mass flow is given by

$$\dot{m} = \frac{\dot{m}\sqrt{\theta_0}}{\delta_0}\frac{p_{t0}}{p_{ref}}\sqrt{\frac{T_{ref}}{T_{t0}}}. \tag{9.16}$$

If the nozzle is a sonic nozzle, then $A_6 = A_7$; if the nozzle is a supersonic nozzle, the exit area may be found by applying the principles developed in Chapter 5. For a matched nozzle, that is, $p_7 = p_0$, the exit area is given by

$$\frac{A_7}{A_6} = \left[\frac{\gamma_6 - 1}{2}\left(\frac{\gamma_6 + 1}{2}\right)^{\frac{\gamma_6+1}{1-\gamma_6}}\right]^{\frac{1}{2}}\left[\left(\frac{p_0}{p_{t7}}\right)^{\frac{2}{\gamma_6}} - \left(\frac{p_0}{p_{t7}}\right)^{\frac{\gamma_6+1}{\gamma_6}}\right]^{-\frac{1}{2}}.$$

## 9.4 COMPRESSOR–TURBINE MATCHING—CASE 2: MASS FLOW RATE AND ENGINE SPEED SPECIFIED

Here the flight conditions $z$ and $M_0$ are given along with the inlet pressure recovery $p_{t2}/p_{t0}$, and the mass flow rate and the engine rpm are stipulated. It is desired to determine the nozzle minimum area $A_6$ and the combustor exit stagnation temperature $T_{t4}$, as well as the nozzle exit area $A_7$ under these conditions. The nozzle flow is assumed to be isentropic so that $p_t$ and $T_t$ are both constant throughout. For this reason, the corrected mass flow is also constant throughout the nozzle. The inlet flow is assumed to be adiabatic so that the stagnation temperature is constant in the inlet and $T_{t2} = T_{t0}$. From the given information, it is possible to calculate the free streamtube capture area directly:

$$A_0 = \frac{\dot{m}}{\rho_0 a_0 M_0}. \tag{9.17}$$

It is also straightforward to calculate the corrected mass flow rates $\dot{m}\sqrt{\theta_0}/\delta_0$ and $\dot{m}\sqrt{\theta_2}/\delta_2$, as well as the corrected rpm $N/\sqrt{\theta_2}$ from the given information. The given flight condition provides the free stream pressure $p_0$ and temperature $T_0$ for any chosen atmospheric model, as well as the flight Mach number from which the stagnation pressure $p_{t0}$ and stagnation temperature $T_{t0}$ may be calculated readily using the isentropic relations

$$p_t = p_0\left(1 + \frac{\gamma - 1}{2}M_0^2\right)^{\frac{\gamma}{\gamma-1}} \tag{9.18}$$

$$T_{t0} = T_0\left(1 + \frac{\gamma - 1}{2}M_0^2\right). \tag{9.19}$$

The corrected mass flow in the free stream is

$$\frac{\dot{m}\sqrt{\theta_0}}{\delta_0} = \dot{m}\sqrt{\frac{T_{t0}}{T_{ref}}\frac{p_{ref}}{p_{t0}}}. \tag{9.20}$$

Having the free stream conditions in hand, attention turns to the inlet, for which the stagnation pressure performance $p_{t2}/p_{t0}$ is given as a function of the free stream Mach number. The inlet flow may be assumed to be adiabatic so that $T_{t2} = T_{t0}$, and then the corrected mass flow at the compressor entrance may be determined from the corrected mass flow in the free stream as follows:

$$\frac{\dot{m}\sqrt{\theta_2}}{\delta_2} = \frac{\dot{m}\sqrt{\theta_0}}{\delta_0}\sqrt{\frac{T_{t2}}{T_{t0}}\frac{p_{t0}}{p_{t2}}}. \tag{9.21}$$

In the same fashion, the corrected rpm is found as

$$\frac{N}{\sqrt{\theta_2}} = N\sqrt{\frac{T_{ref}}{T_{t2}}}. \tag{9.22}$$

The corrected mass flow and corrected rpm define a point on the compressor map, which then defines the corresponding pressure ratio $p_{t3}/p_{t2}$ and compressor efficiency $\eta_c$, as shown in Figure 9.10. Then the corrected work done by the compressor may be found from

$$\frac{W_c}{\theta_2} = \frac{c_{p2} T_{ref}}{\eta_c}\left[\left(\frac{p_{t3}}{p_{t2}}\right)^{\frac{\gamma_2 - 1}{\gamma_2}} - 1\right]. \tag{9.23}$$

The temperature at the compressor exit may be found from the definition of the actual work and Equation (9.23):

$$W_c = c_{p2}(T_{t3} - T_{t2}) \tag{9.24}$$

$$T_{t3} = T_{t2}\left\{1 + \frac{1}{\eta_c}\left[\left(\frac{p_{t3}}{p_{t2}}\right)^{\frac{\gamma_2 - 1}{\gamma_2}} - 1\right]\right\}. \tag{9.25}$$

The corrected mass flow at the compressor exit may be found from

$$\frac{\dot{m}\sqrt{\theta_3}}{\delta_3} = \frac{\dot{m}\sqrt{\theta_2}}{\delta_2}\sqrt{\frac{T_{t3}}{T_{t2}}}\frac{p_{t2}}{p_{t3}}. \tag{9.26}$$

Then the Mach number at the compressor exit may be found from the corrected mass flow equation

$$\frac{\dot{m}\sqrt{\theta_3}}{\delta_3} = \frac{p_{ref} A_3}{\sqrt{T_{ref}}}\sqrt{\frac{\gamma}{R}} M_3 \left(1 + \frac{\gamma - 1}{2} M_3^2\right)^{-\frac{\gamma_2 + 1}{2(\gamma_2 - 1)}}. \tag{9.27}$$

With the conditions at the inlet of the combustor known, it is possible to proceed across it by assuming a value for the combustor exit temperature $T_{t4}$. Assuming a constant pressure combustor, the following equations for stagnation pressure and temperature ratios are taken from Chapter 4:

$$\frac{p_{t3}}{p_{t4}} = \frac{\left(1 + \frac{\gamma_3 - 1}{2} M_3^2\right)^{\frac{\gamma_3}{\gamma_3 - 1}}}{\left(1 + \frac{\gamma_4 - 1}{2} M_4^2\right)^{\frac{\gamma_4}{\gamma_4 - 1}}} \tag{9.28}$$

$$\frac{T_{t3}}{T_{t4}} = \frac{W_3}{W_4}\frac{\gamma_4 M_4^2}{\gamma_3 M_3^2}\frac{1 + \frac{\gamma_3 - 1}{2} M_3^2}{1 + \frac{\gamma_4 - 1}{2} M_4^2}, \tag{9.29}$$

where $W$ denotes the molecular weight of the mixture. Using these equations, one may calculate the corresponding values of $M_4$ and $p_{t4}$. For the chosen value of $T_{t4}$, the corrected mass flow entering the turbine is found from

$$\frac{\dot{m}\sqrt{\theta_4}}{\delta_4} = \frac{\dot{m}\sqrt{\theta_3}}{\delta_3}\sqrt{\frac{T_{t3}}{T_{t4}}}\frac{p_{t3}}{p_{t4}}. \tag{9.30}$$

When the turbine and compressor are matched so that $W_t = W_c$, the corrected turbine work is given by

$$\frac{W_t}{\theta_4} = \frac{W_c}{\theta_2}\frac{T_{t2}}{T_{t4}}. \tag{9.31}$$

The corrected rpm is found from the given actual rpm and the known inlet conditions using

$$\frac{N}{\sqrt{\theta_4}} = \frac{N}{\sqrt{\theta_2}}\sqrt{\frac{T_{t2}}{T_{t4}}}. \tag{9.32}$$

Then, entering the turbine map, like that shown in Figure 9.8, with the corrected turbine mass flow of Equation (9.30), which is based on the assumed value of $T_{t4}$, and the corrected rpm defines the corresponding pressure ratio $p_{t4}/p_{t5}$ and turbine efficiency $\eta_t$. The corrected turbine work is given in terms of pressure ratio by Equation (9.13) and this value must satisfy the matching condition given in Equation (9.31). If it doesn't, a new value of $T_{t4}$ must be assumed and the process repeated until a match is achieved. Once a match is obtained, the pressure ratio is known and the stagnation temperature at the exit of the turbine may be found from

$$\frac{T_{t5}}{T_{t4}} = 1 - \frac{1}{c_{p4}T_{ref}}\frac{W_t}{\theta_4}. \tag{9.33}$$

The corrected mass flow in the nozzle is then found from

$$\frac{\dot{m}\sqrt{\theta_5}}{\delta_5} = \frac{\dot{m}\sqrt{\theta_4}}{\delta_4}\sqrt{\frac{T_{t5}}{T_{t4}}}\frac{p_{t4}}{p_{t5}}. \tag{9.34}$$

Using Equation (9.3) permits calculation of the nozzle throat area $A_6$, and the Mach number at the exit of the turbine $M_5$ may be found from

$$\frac{A_5}{A_6} = \frac{1}{M_5}\left[\frac{2}{\gamma+1}\left(1+\frac{\gamma-1}{2}M_5^2\right)\right]^{\frac{\gamma+1}{2(\gamma-1)}}. \tag{9.35}$$

Finally, the exit area $A_7$ may be found from Equation (9.16).

## 9.5 INLET–ENGINE MATCHING

As is clear from the preceding sections, the corrected flow properties are used by engine manufacturers to describe engine performance. In order to install an engine on an airframe successfully, attention must be paid to the integration process, and an important aspect is inlet–engine matching. Detailed discussions of inlet design for a wide variety of aircraft are presented in Goldsmith and Seddon (1993). The pertinent inlet parameters are the free stream capture area $A_0$, the inlet pressure recovery $p_{t2}/p_{t0}$, and the free stream total temperature, $T_{t0}$. The design problem increases in complexity as the flight Mach number $M_0$ increases to higher values, particularly for $M_0 > 1.5$, where the excursions of the

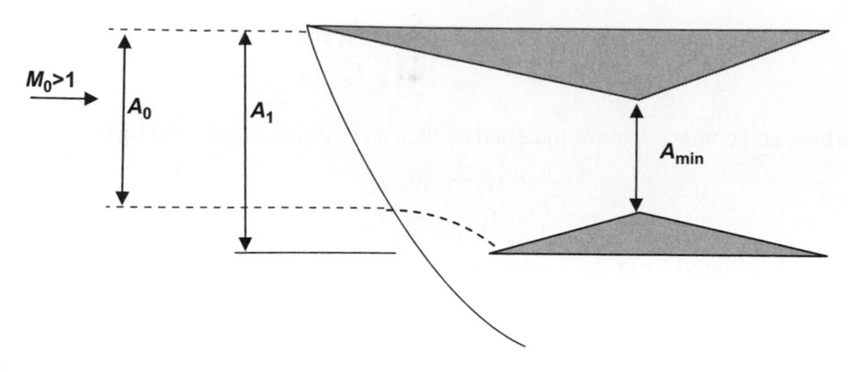


**FIGURE 9.11**

Generic inlet showing the important geometric parameters for inlet–engine integration.

required value of $A_0$ require the use of an inlet in which the ratio of inlet minimum section area to inlet area, $A_{min}/A_1$, can be varied over the flight envelope; these parameters are illustrated generically in Figure 9.11.

The corrected mass flow through any station in the engine is

$$\frac{\dot{m}\sqrt{\theta}}{\delta} = \frac{p_{ref}A}{\sqrt{T_{ref}}}\sqrt{\frac{\gamma}{R}}M\left(1+\frac{\gamma-1}{2}M^2\right)^{-\frac{\gamma+1}{2(\gamma-1)}}. \tag{9.36}$$

For the inlet, we may consider the gas to be air with $\gamma = 1.4$ so that Equation (9.36) becomes, for mass flow in kg/s,

$$\frac{\dot{m}\sqrt{\theta}}{\delta} = 416.8AM\left(1+\frac{\gamma-1}{2}M^2\right)^{-3}. \tag{9.37}$$

The corrected weight flow is a constant for isentropic flow in a duct, but in the streamtube between stations 0 and 2 the flow can, at best, be considered adiabatic because at supersonic speeds, shock waves, which change the entropy, will be present. In this case, only the stagnation temperature will be constant throughout and the corrected weight flow will be inversely proportional to the stagnation pressure as given by

$$\frac{\dot{m}\sqrt{\theta}}{\delta} \sim \frac{1}{\delta}. \tag{9.38}$$

The corrected mass flow at the entrance of the compressor is therefore

$$\frac{\dot{m}\sqrt{\theta_2}}{\delta_2} = \frac{\dot{m}\sqrt{\theta_0}}{\delta_0}\frac{p_{t0}}{p_{t2}} \tag{9.39}$$

Inlet pressure recovery is generally given as a function of flight Mach number, and the standard inlet pressure recovery ("MIL-E-5008B: Military Specifications—Engines, Aircraft, Turbojet, Model Specification For," January 1959), which is applicable for $M \leq 3$, is given by

$$\frac{p_{t2}}{p_{t0}} = 1 - 0.075(M_0 - 1)^{1.35}. \tag{9.40}$$

Combining Equations (9.37) and (9.40) and solving for the free stream capture streamtube area yield

$$A_0 = \frac{1 - 0.075(M_0 - 1)^{1.35}}{416.8 M_0 (1 + 0.2 M_0^2)^{-3}} \frac{\dot{m}' \sqrt{\theta_2}}{\delta_2}. \tag{9.41}$$

Thus the mass flow of air required by the engine is directly proportional to the free stream capture streamtube area, and the Mach number characteristics of the inlet determine the proportionality factor.

### 9.5.1 Inlet capture area

Chapter 6 presented an analysis for determining the required streamtube capture area, even for supersonic flight where the inlet pressure recovery varies with $M_0$. In order to specify the mass flow capabilities of an inlet, there are three basic flow areas we must contend with and they are related as follows:

$$A_0 = \frac{\left(\dfrac{A_0}{A_1}\right)}{\left(\dfrac{A_{min}}{A_1}\right)} A_{min},$$

where $A_0/A_1$ is a function of the flight Mach number and the inlet design, $A_{min}/A_1$ is the contraction ratio of the inlet, and $A_{min}$ is the minimum area of the inlet. The last two terms are functions of the inlet geometry. The inlet operation in supersonic flight has two basic operational conditions, as shown in Figure 9.12. In subcritical operation, the flow in the inlet is subsonic and the inlet flow is influenced readily by engine operation; in supercritical operation, the engine operation can only influence the inlet flow if the shock is ejected. Otherwise, the swallowed shock will merely change position in the region beyond the inlet minimum section.

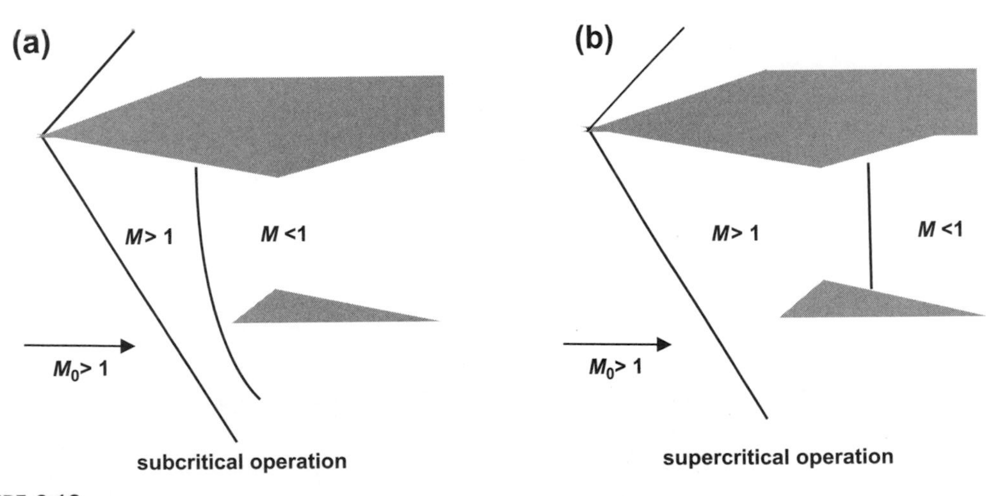


**FIGURE 9.12**

Supersonic inlet operation: (a) subcritical and (b) supercritical.

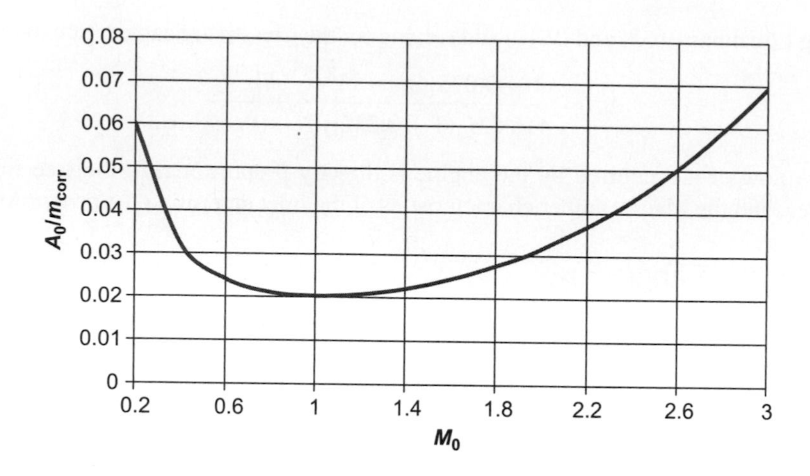


**FIGURE 9.13**

Capture streamtube area required per unit corrected weight flow according to Equation (9.51). Here $m_{corr} = \dot{m}\sqrt{\theta_2}/\delta_2$.

The required streamtube capture area per unit corrected mass flow given by Equation (9.41) is shown in Figure 9.13. This result assumes the standard inlet pressure recovery given in Equation (9.40) for supersonic flow and 100% recovery for subsonic flow.

If the inlet pressure recovery performance is less than the assumed standard, the streamtube capture area will fall below the engine requirement, as shown in Figure 9.13, and thrust will be reduced. If the inlet performance is greater than the assumed standard, the streamtube capture area will rise above the engine requirement, as shown in Figure 9.14.

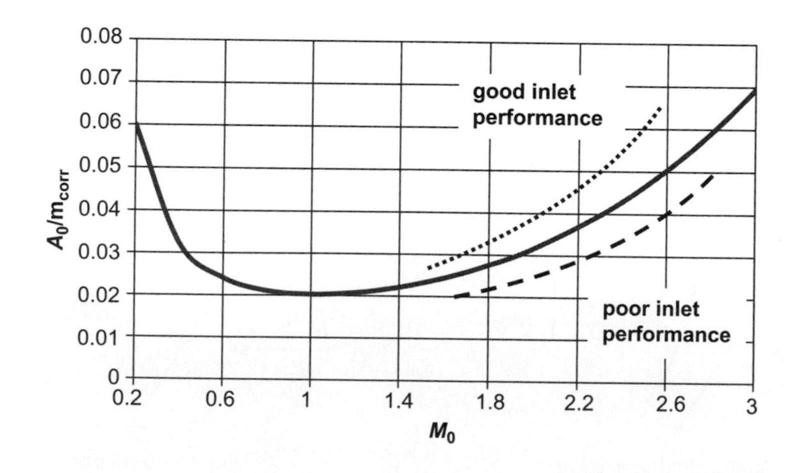


**FIGURE 9.14**

Streamtube capture area per unit corrected inlet weight flow as a function of flight Mach number showing good and poor inlet pressure recovery performance compared to the standard.

The excess mass flow taken on board in the case of better than standard performance will have to be spilled out by one or more of the following mechanisms: (a) decreasing the inlet area $A_1$ by having a variable area inlet that involves additional weight and complexity; (b) bleeding off the excess and bypassing it to the exhaust nozzle, which entails additional weight and complexity; and/or (c) spilling it overboard at the inlet lip, which is simple but results in additive drag penalties.

## 9.5.2 Internal compression shock position effects

The variation of Mach number and inlet pressure recovery as a function of the position of the shock in the inlet is shown in Figure 9.15, where $x$ denotes distance measured from the inlet minimum area. The analysis here is for an internal compression inlet, but the effects are similar in mixed compression inlets. As described previously, the stagnation pressure at the exit of the turbine, which is the stagnation pressure at the entrance of the nozzle, is

$$p_{t,5} = \frac{p_{t,5}}{p_{t,4}}\frac{p_{t,4}}{p_{t,3}}\frac{p_{t,3}}{p_{t,2}}\frac{p_{t,2}}{p_{t,1}}\frac{p_{t,1}}{p_{t,0}}p_{t,0}.$$

We know that for a fixed compressor ratio we have a fixed turbine pressure drop, while the combustion chamber pressure drop may be neglected here without a loss in generality. Assuming that the inlet itself operates with small internal loss in stagnation pressure, we find that for a fixed flight condition, that is, fixed $p_{t,0}$, the stagnation pressure at the entrance to the nozzle is proportional to the inlet pressure recovery, or $p_{t,5} \sim p_{t,2}/p_{t,0}$.

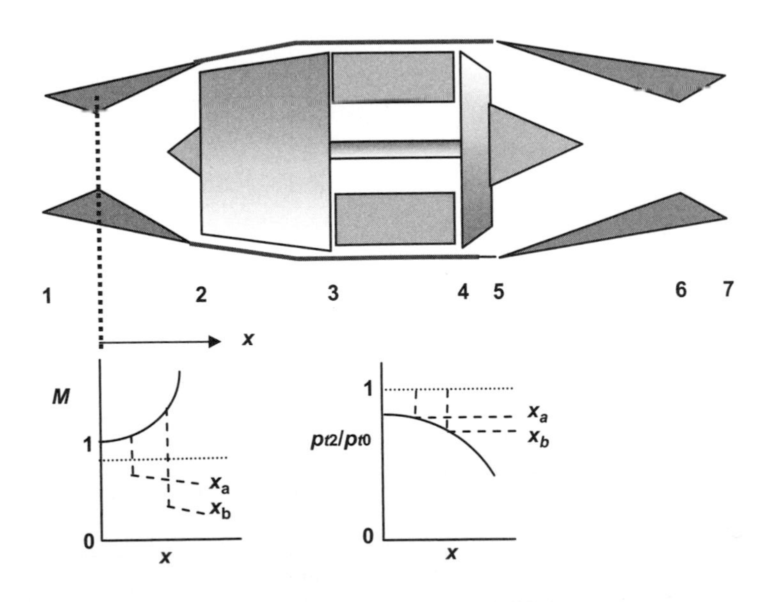


**FIGURE 9.15**

Schematic diagram of Mach number and pressure recovery as a function of shock location in the inlet.

Therefore, the stagnation pressure drops off as the shock moves down the inlet toward the compressor. The nozzle flow may be assumed to be isentropic so that $p_{t,6} = p_{t,5}$ and if we consider the nozzle minimum area to be choked, the mass flow is given by

$$\dot{m} = \frac{p_{t,6} A_6}{\sqrt{T_{t,6}}} f(\gamma, R).$$

To keep the nozzle choked, the ratio shown must remain constant so that for a fixed value of $T_{t,6} = T_{t,5}$ the minimum section area $A_6$ must increase as the shock moves further down the inlet, as shown in Figure 9.16. Because variable geometry nozzles are heavy and costly, we may inquire as to the consequence of keeping $A_6$ fixed. In that case, $T_{t,6}$ must decrease as the shock moves down the inlet, and we note that

$$T_{t,6} = T_{t,5} = \frac{T_{t,5}}{T_{t,4}} T_{t,4}.$$

Then, to decrease $T_{t,6}$, we must drop the burner outlet temperature, which is also the turbine inlet temperature, to keep the power output of the turbine appropriate to the demand of the compressor at fixed $p_{t,3}/p_{t,2}$. The drop in nozzle stagnation temperature will entail a reduction in thrust. However, controlling the turbine inlet temperature is a means of controlling the shock location in the inlet. In the cases considered, the inlet was assumed to be started. If that isn't the

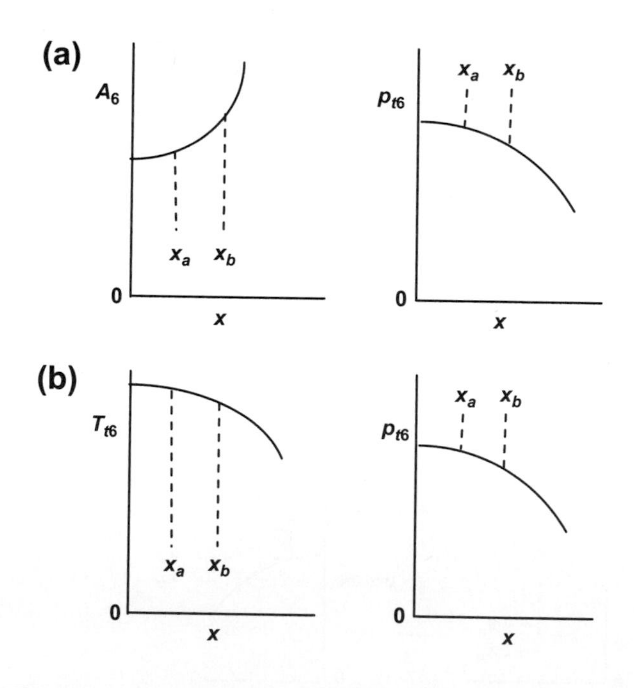


**FIGURE 9.16**

Variation of (a) nozzle minimum area and stagnation pressure and (b) nozzle stagnation temperature and pressure shown as a function of location of the shock in the inlet. The $x$ locations shown refer to shock locations in Figure 9.15.

case, the shock is ahead of the inlet lip and the pressure recovery will be the normal shock value at $M_0$, which is less than that with the inlet started, and the mass flow to the engine will drop, with mass being spilled overboard, leading to additive drag penalties as well as thrust reduction. Although the mixed compression inlet has high pressure recovery, it has sensitivity to variations in mass flow demand and requires a fairly complex control system to maintain stability of the terminal normal shock position and to avoid inlet unstarts. A detailed discussion on the inlet starting process is given by Henderson (1967).

### 9.5.3 External compression inlet installation

In the case of the simpler all external compression inlet, such as those on the Concorde $M_0 = 2$ supersonic transport, internal ramps are extended by an actuator, thereby decreasing the throat area for the reduced mass flow requirement in supersonic cruise, as shown in Figure 9.17. The ramps at the throat do not meet but instead form a gap for excess air to bleed from the throat and be bypassed to the exhaust nozzle. An additional door may be opened to bleed off additional air, as might be expected when operating at high Mach number and relatively low engine rpm. The bleed bypass to the nozzle adds weight to the engine, as well as forming a channel for coupling exhaust nozzle pressure disturbances to the inlet because the bypass flow is subsonic. The three-shock inlet provides good pressure recovery, over 90%, at the cruise flight Mach number $M_0 = 2$.

At subsonic speeds typical of cruise, the throat area is increased by retracting the ramps, as shown in Figure 9.18. This permits efficient operation at high subsonic speeds where $A = A_1$. However, at take-off and low subsonic Mach numbers, where $A_0 > A_1$, the flap door is opened inward to provide additional flow area needed during high thrust situations, such as take-off and climb.

At higher Mach numbers, $M_0 = 2.5$ or more, as required for some fighter aircraft, the pressure recovery from a three shock inlet is insufficient and a four shock inlet is necessary. In addition, the inlet for such an application must be robust, delivering high pressure recovery and relative insensitivity to

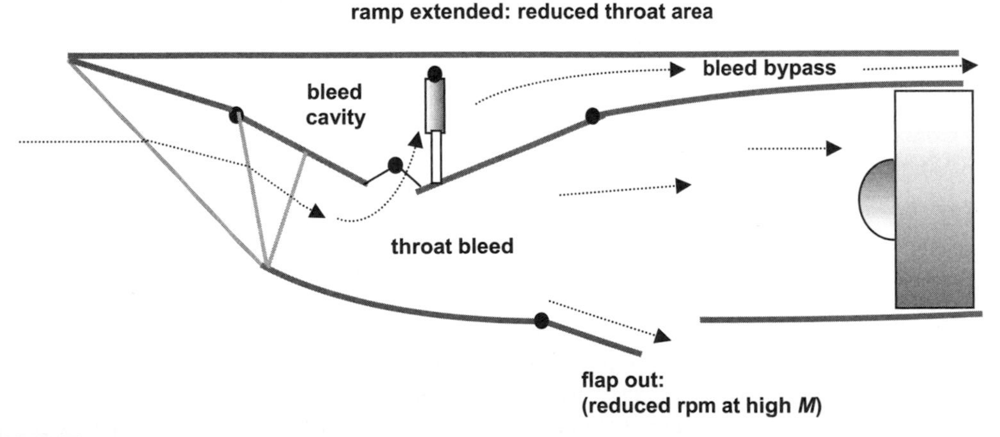


**FIGURE 9.17**

External compression inlet (e.g., Concorde) in supersonic flight showing reduced throat area, throat bleed with bypass to the exhaust nozzle, and flap door open for bleeding excess mass flow.

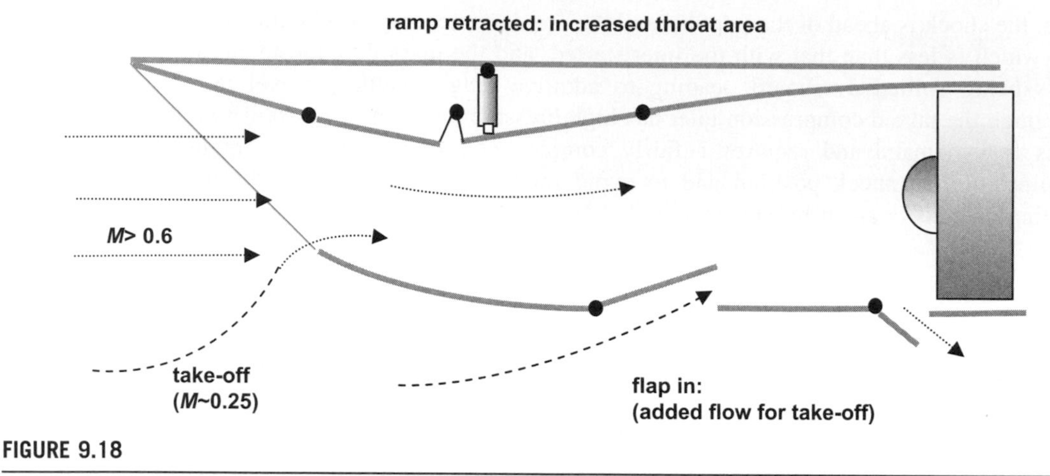


FIGURE 9.18

External compression inlet in subsonic flight showing increased throat area and reduced throat bleed gap. Also shown is the flap door opened inward to permit additional mass flow for take-off.

mass flow demands and large angle of attack excursions. A successful four shock inlet is that of Grumman's F-14 shown in Figure 9.19.

In the take-off configuration shown in Figure 9.19a, the bleed door is opened and the hinged inlet ramp is deflected upward to permit inflow of the additional air needed for high thrust situations. At high subsonic speeds, the ramp is undeflected and the bleed door is open to handle any excess air flow. At low Mach numbers close to unity the ramp and the bleed doors remain in much the same position as for high subsonic flight, as shown in Figure 9.19b. Normal shock outside the lower cowl lip of the inlet permits excess air flow to be spilled there as well as at the bleed door. In high supersonic flight, the inlet configuration is shown in Figure 9.19c.

The hinged ramp is deflected at both hinge points to permit two oblique shocks to form in addition to the cowl lip oblique shock. The final shock is a normal shock and the flow behind it is subsonic. The downward deflection of both the hinged ramp and the diffuser flap acts to reduce the throat area appropriately. Some spillage may occur around the lower cowl lip.

The Concorde, XB-70, Boeing SST, and Lockheed SR-71 inlet systems are discussed in Leynaert and colleagues (1993). The Concorde ($M_0 = 2$) is an all external compression system described previously in this chapter. The XB-70 ($M_0 = 2.7$) has a basically two-dimensional mixed compression system with deeply buried engines. The inlet benefits from precompression by the fuselage of the aircraft. Three fixed ramps provide oblique shocks to further compress the flow and variable ramps after that generate a complex system of interacting waves to complete the compression process. Much of the internal surface is porous to permit various levels of bleed for boundary layer control and mass flow adjustments. The Boeing SST design ($M_0 = 2.7$) maintains forebody precompression but instead of burying the engines in a long duct it uses engines in separate pods with axisymmetric mixed compression inlets with a translating centerbody to vary the area distribution of the inlet. A complex throat bleed system is necessary to preserve stability of the inlet flow over the flight envelope. The Lockheed SR-71, which has a substantial flight history, used an inlet similar to the Boeing SST, an axisymmetric mixed compression inlet with a translating centerbody, as shown in Figure 9.20.

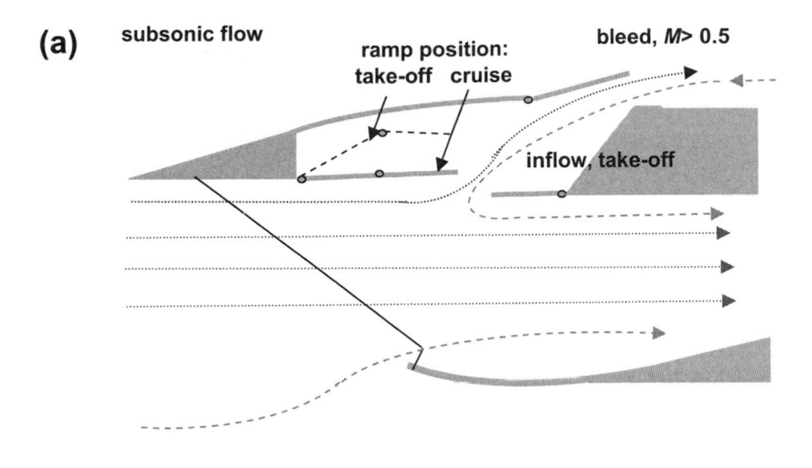


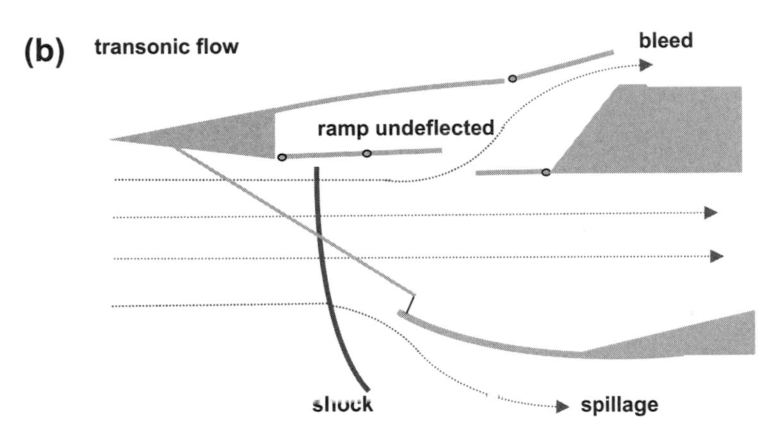


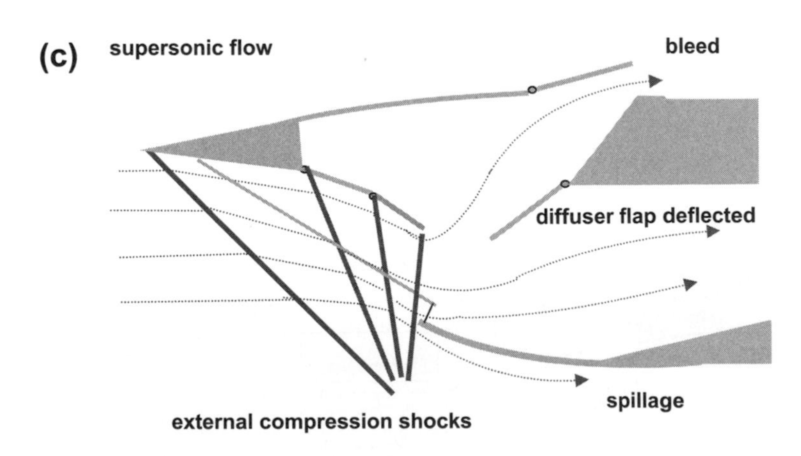


**FIGURE 9.19**

(a) Operation of an external compression inlet like the Grumman F-14 in take-off (dashed lines with arrows) and high subsonic (dotted line with arrows) flight. (b) Operation of an external compression inlet like the Grumman F-14 in transonic flight with $M_0 > 1$. (c) Operation of an external compression inlet like the Grumman F-14 in high supersonic flight.

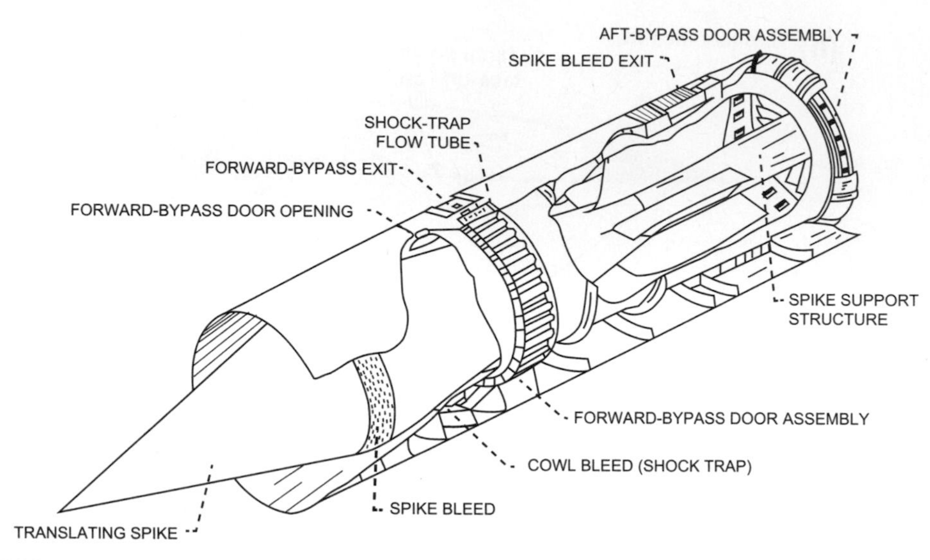


**FIGURE 9.20**

Sketch of the inlet system for the Lockheed SR-71 (courtesy of NASA Dryden Flight Research Center).

A particular feature is the ability to bypass part of the flow, mixed with throat bleed flow, to the exhaust nozzle where it is the source for excess air for the afterburner at high Mach numbers. In this sense, the engine operates mainly as a ramjet at the highest flight speeds.

## 9.6 EXAMPLE: BASIC COMPRESSOR–TURBINE MATCHING

Consider the engine shown in Figure 9.21 operating at $M_0 = 2.0$ and an altitude $z = 50$ kft (15.24 km) where $p_0 = 244$ psf (11.66 kPa) and $T_0 = 390$R (217K). The engine inlet has a pressure recovery $p_{t,2}/p_{t,0} = 0.9$, and the area of the entrance to the compressor is $A_2 = 1$ ft$^2$(0.093 m$^2$). The Mach number at the compressor face $M_2 = 0.3$, the pressure ratio of the compressor $p_{t,3}/p_{t,2} = 15$, and the compressor

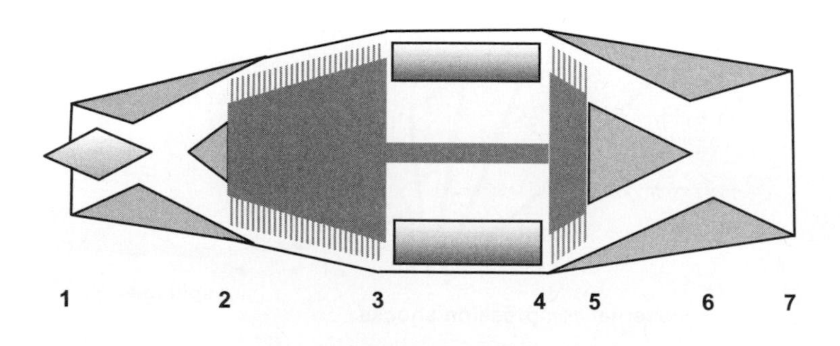


**FIGURE 9.21**

Schematic diagram of turbojet engine of the example problem.

efficiency $\eta_c = 0.85$. Assume that the burner and turbine efficiencies are $\eta_b = \eta_t = 0.95$ and that the heating value of the fuel is $HV = 18{,}900$ Btu/lb (44 MJ/kg). For the case of matched nozzle operation, find the net thrust $F_n$ and the specific fuel consumption $c_j$ as a function of the turbine inlet temperature $T_{t4}$.

The free stream stagnation pressure and temperature are given by

$$\frac{p_{t,0}}{p_0} = \left(1 + \frac{M_0^2}{5}\right)^{3.5}$$

$$\frac{T_{t,0}}{T_0} = 1 + \frac{M_0^2}{5}.$$

Therefore, in this case, $p_{t,0} = 1909$ lb/ft$^2$ (91.2 kPa) and $T_{t,0} = 702$R (391K). In the "cold" compressor section of the engine we assume constant values for the specific heat and the isentropic exponent as follows: $c_{p,c} = 0.24$ Btu/lb-R (1 kJ/kg-K) and $\gamma_c = 1.4$. In this section, the adiabatic efficiency of compression is

$$\eta_c = \frac{\left(\frac{p_{t,3}}{p_{t,2}}\right)^{\frac{\gamma_c - 1}{\gamma_c}} - 1}{\frac{T_{t,3}}{T_{t,2}} - 1}.$$

In the "hot" turbine section of the engine we assume constant values for the specific heat and the isentropic exponent as follows: $c_{p,t} = 0.276$ Btu/lb-R (1.2 kJ/kg-K) and $\gamma_t = 1.33$. In this section, the adiabatic efficiency of expansion is

$$\eta_t = \frac{1 - \frac{T_{t,5}}{T_{t,4}}}{1 - \left(\frac{p_{t,5}}{p_{t,4}}\right)^{\frac{\gamma_t - 1}{\gamma_t}}}.$$

The basic matching condition is that the compressor work done is equal to the turbine work extracted so that $W_c = c_{p,c}(T_{t,3} - T_{t,2}) = W_t = c_{p,t}(T_{t,4} - T_{t,5})$

For the given compressor pressure ratio and efficiency of 15% and 85%, respectively, we may calculate the compressor temperature ratio to be $T_{t,3}/T_{t,2} = 2.374$. Then, assuming a negligible stagnation pressure loss in the burner, $p_{t,4}/p_{t,3} \sim 1$, we may use the matching condition to write an equation for the turbine pressure ratio in terms of compressor parameters and the unknown turbine inlet temperature $T_{t,4}$:

$$\frac{p_{t,5}}{p_{t,4}} = \left\{1 - \frac{c_{p,c} T_{t,2}}{c_{p,t} T_{t,4}} \frac{1}{\eta_c \eta_t} \left[\left(\frac{p_{t,3}}{p_{t,2}}\right)^{\frac{\gamma_c - 1}{\gamma_c}} - 1\right]\right\}^{\frac{\gamma_t}{\gamma_t - 1}}$$

For any value of $T_{t,4}$, we may find the corresponding turbine temperature ratio

$$\frac{T_{t,5}}{T_{t,4}} = 1 - \eta_t \left[ 1 - \left( \frac{p_{t,5}}{p_{t,4}} \right)^{\frac{\gamma_t - 1}{\gamma_t}} \right].$$

For any turbine inlet temperature the stagnation conditions in the nozzle are known, and if we assume isentropic flow in the nozzle we may find the exit velocity from

$$V_7 = \sqrt{2c_{p,7}T_{t,7} \left[ 1 - \left( \frac{p_7}{p_{t,7}} \right)^{\frac{\gamma_7 - 1}{\gamma_7}} \right]}.$$

In this problem, we have $c_{p,7} = c_{p,t}$, $\gamma_7 = \gamma_t$, and $p_7 = p_0$, all of which are known so that $V_7$ may be found exactly. The net thrust for each value of $T_{t,4}$, assuming matched nozzle operation, may be found from

$$F_n = \dot{m}(V_7 - V_0) + A_7(p_7 - p_0) = \dot{m}(V_7 - V_0).$$

The flight speed may be found from

$$V_0 = a_0 M_0 = \sqrt{\gamma_0 R T_0} M_0.$$

For this case, $\gamma_0 = 1.4$, $R = 1716$ ft$^2$/s$^2$-R (287 m$^2$/s$^2$-K) so $V_0 = 1936$ ft/s (591 m/s). The mass flow rate may be found at station 2 from the equation for mass flow in a duct

$$\dot{m}_2 = \frac{p_{t,2}A_2}{\sqrt{T_{t,2}}} \sqrt{\frac{\gamma_2}{R}} M_2 \left( 1 + \frac{\gamma_2 - 1}{2} M_2^2 \right)^{-\frac{\gamma+1}{2(\gamma_2 - 1)}}.$$

Here we may use $\gamma_2 = \gamma_c$ and $M_2$ and $A_2$ are given. The pressure recovery is 0.9 so that $p_{t,2} = 0.9 p_{t,0}$ and $T_{t,2} = T_{t,0}$, assuming adiabatic flow through the inlet. Finally, the fuel weight flow rate is given by

$$\dot{w}_f = \dot{w}_0 \left\{ \left[ \frac{\eta_b HV}{c_{p,b}(T_{t,4} - T_{t,3})} \right] - 1 \right\}.$$

Here we may use $\dot{w}_0 = \dot{w}_2 = \dot{m}_2 g$ and $c_{p,b} = c_{p,t}$. Then the specific fuel consumption is $c_j = \dot{w}_f / F_n$.

## 9.7 THRUST MONITORING AND CONTROL IN FLIGHT

Because the thrust delivered by the engine is of primary concern in jet aircraft operation, it is important to understand how it is monitored and controlled.

$$F_g = \dot{m}_0 V_7 + A_7(p_7 - p_0).$$

The exhaust velocity is given by Equation (1.57) as

$$V_7 = \sqrt{2c_p T_{t,7}\left[1 - \left(\frac{p_7}{p_{t,7}}\right)^{\frac{\gamma_7 - 1}{\gamma_7}}\right]}.$$

The exit static to stagnation pressure ratio was given by Equation (1.59) as

$$\frac{p_7}{p_{t,7}} = \frac{p_7}{p_0}\frac{p_0}{p_{t,0}}\frac{1}{\frac{p_{t,2}}{p_{t,0}}}\frac{1}{\frac{p_{t,3}}{p_{t,2}}\frac{p_{t,4}}{p_{t,3}}\frac{p_{t,5}}{p_{t,4}}}\frac{p_{t,5}}{p_{t,7}}.$$

An engine pressure ratio ($EPR$) may be defined as follows:

$$EPR = \frac{p_{t,5}}{p_{t,2}} = \frac{p_{t,5}}{p_{t,4}}\frac{p_{t,4}}{p_{t,3}}\frac{p_{t,3}}{p_{t,2}}.$$

Then the pressure ratio

$$\frac{p_7}{p_{t7}} = \frac{p_7}{p_0}\left(\frac{p_0}{p_{t,0}}\frac{1}{\frac{p_{t2}}{p_{t,0}}}\right)\frac{1}{EPR}\frac{p_{t,5}}{p_{t,7}}.$$

The term in parentheses is a function only of flight Mach number $M_0$ for a fixed geometry aircraft engine installation, and the EPR is a function only of corrected rpm so that

$$\frac{p_7}{p_{t,7}} = \frac{p_7}{p_0}f(M_0)g\left(\frac{N}{\sqrt{\theta_2}}\right)\frac{p_{t,5}}{p_{t,7}}.$$

In keeping with our original approach of ideal operation, we expect that the stagnation pressure loss through the nozzle will be negligible, or at best a constant correction value close to unity to account for nozzle efficiency. Note that for adiabatic flow in the inlet

$$\theta_2 = \theta_0.$$

Thus the exit static to stagnation pressure ratio becomes

$$\frac{p_7}{p_{t,7}} = \frac{p_7}{p_0}f\left(M_0, \frac{N}{\sqrt{\theta_0}}\right).$$

The exit pressure $p_7$ depends on $p_{t,5}$ and nozzle geometry so for a fixed geometry the exit pressure depends only on the corrected rpm while the free stream static pressure depends on altitude. Then

$$\frac{p_7}{p_{t,7}} = f\left(M_0, \frac{N}{\sqrt{\theta_0}}, z\right).$$

By the same reasoning, for a constant geometry engine the nozzle exit pressure imbalance term

$$A_7(p_7 - p_0) = f\left(\frac{N}{\sqrt{\theta_0}}, z\right).$$

The exit velocity $V_7$ as given by Equation (1.57) also depends on the exhaust gas stagnation temperature $T_{t,7}$. In the discussion in Chapter 7 of compressor maps, an operating line for a compressor was shown in Figure 7.52. Along that design line, work done by the compressor, and therefore the compressor pressure ratio, rises with increases in weight flow and rotational speed $N$. In the compressor–turbine matching analysis discussed previously in this chapter, we showed that the corrected rpm of the compressor is related to that of the turbine by

$$\frac{N}{\sqrt{\theta_2}} = \frac{N}{\sqrt{\theta_4}}\sqrt{\frac{T_{t,4}}{T_{t,2}}}. \tag{9.42}$$

Similarly, the matching condition for the compressor–turbine pair is

$$\frac{W_c}{\theta_2} = \frac{W_t}{\theta_4}\frac{T_{t,4}}{T_{t,2}}.$$

In terms of pressure ratios, this relation becomes, using Equation (9.12), the following:

$$\frac{c_{p,2}}{\eta_c}\left[\left(\frac{p_{t,3}}{p_{t,2}}\right)^{\frac{\gamma_2-1}{\gamma_2}}-1\right] = \eta_t c_{p,4}\left[1-\left(\frac{p_{t,5}}{p_{t,4}}\right)^{\frac{\gamma_4-1}{\gamma_4}}\right]\frac{T_{t,4}}{T_{t,2}}. \tag{9.43}$$

We also know from Equation (7.12) that the turbine efficiency relates the turbine exhaust temperature $T_{t,5}$ to the turbine entry temperature $T_{t,4}$ so that Equations (9.42) and (9.43) relate the rotational speed $N$ to $EPR = (p_{t,5}/p_{t,2})$ and $T_{t,5} = T_{t,7}$. Thus the exhaust gas velocity

$$V_7 = V_7\left(M_0, \frac{N}{\sqrt{\theta_0}}, z\right).$$

Without loss in generality, we assume that the fuel flow rate is equal to the compressor air bleed rate so that the mass flow of air taken on board is equal to the mass flow of exhaust gases or

$$\dot{m}_0 = \dot{m}_7.$$

Then net thrust is given by

$$F_n = \dot{m}_0(V_7 - V_0) + A_7(p_7 - p_0).$$

The flight velocity

$$V_0 = \sqrt{\gamma R T_0}M_0.$$

The mass flow entering the engine may be written as

$$\dot{m}_0 = p_0 A_0\sqrt{\frac{\gamma}{RT_0}}M_0 = \frac{\delta_0 A_0 M_0}{\sqrt{\theta_0}}\sqrt{\frac{\gamma p_{ref}^2}{RT_{ref}}}.$$

The capture streamtube area is only a function of inlet geometry and flight Mach number:

$$A_0 = \frac{A_0}{A_1}A_1 = A_0(M_0, A_1).$$

Then, for a given aircraft with inlet area $A_1$, the ram drag is

$$\dot{m}_0 V_0 = \delta_0 \gamma p_{ref} A_1 \left( M_0^2 \frac{A_0}{A_1} \right) = \delta_0 f(M_0).$$

As mentioned previously, the net thrust finally assumes the following functional relationship:

$$\frac{F_n}{\delta_0} = f\left( M_0, \frac{N}{\sqrt{\theta_2}}, z \right).$$

Thus $N$ can be used as an indicator of engine thrust, and engine performance may be displayed as shown in Figure 9.22.

We may write, for a given flight condition $M_0$ and $z$, that

$$\frac{F_n}{\delta_0} = f\left( \frac{N}{\sqrt{\theta_0}} \right).$$

In addition to the engine speed $N$, the parameter *EPR* and exhaust gas temperature $T_{t,5}$, along with the fuel flow rate $\dot{m}_f$, are the major ones for monitoring and controlling engine performance. The air data system discussed in Appendix F provides measurements of the flight conditions while engine monitoring equipment yields $N$, stagnation pressures, and stagnation temperatures inside the engine, and the fuel delivery system described in the next section deals with the fuel flow rate.

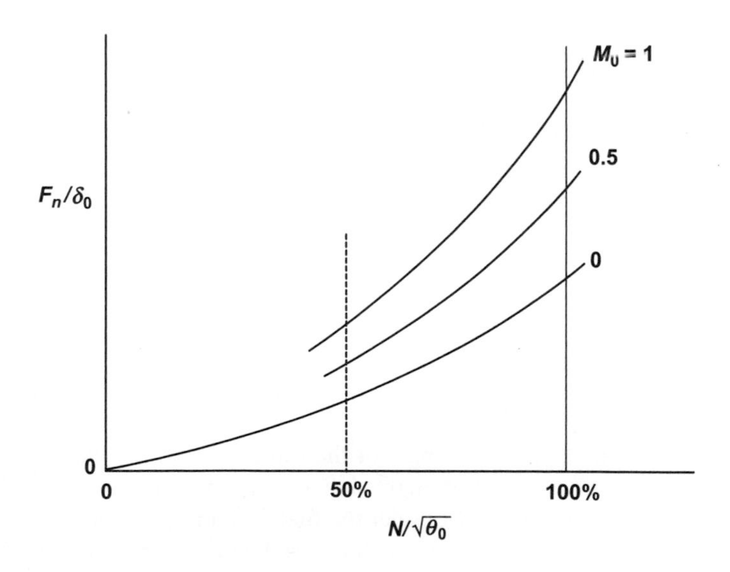


**FIGURE 9.22**

Plot of corrected thrust as a function of corrected rpm.

## 9.8 FUEL DELIVERY SYSTEMS

Computers are currently used to control thrust of jet engines as part of Full Authority Digital Engine Control (FADEC) systems. These systems can accommodate a large variety of sensor inputs to optimize engine performance and to provide information for engine health monitoring. In addition, FADEC systems incorporate substantially greater redundancy and back-up features compared to older hydromechanical control systems. Some manufacturers use $N$ as the primary control parameter. In two- or three-shaft engines, the speed of the fan driving shaft is used. Other manufacturers use the *EPR* (engine pressure ratio, $p_{t,5}/p_{t,2}$) as the primary control parameter. The control system will sense the throttle lever setting and the control parameter, $N$ or *EPR*, and will provide the appropriate fuel delivery to achieve and hold the desired setting.

The thrust ratings for commercial engines are, in decreasing order,

- maximum take-off thrust (limited by regulation to a duration of 5 min)
- maximum continuous thrust
- maximum climb thrust
- maximum cruise thrust
- flight idle thrust
- minimum idle thrust

There are also ratings for reverse thrust, namely

- maximum reverse thrust
- reverse idle thrust

In order to maintain appropriate maximum take-off thrust over a wide range of likely outside air temperatures at different airports, engines are flat-rated. That is, they will deliver a constant maximum take-off thrust up to some maximum temperature $T_0$ called the flat rate temperature. Beyond that value of outside ambient temperature the thrust will fall off as the temperature increases.

The idle thrust settings are rotational speed conditions, and $N$ is the control parameter monitored. The minimum idle thrust defines the lowest $N$ at which thrust can be produced and that value is somewhat higher than the rotational speed at which the engine speed can be accelerated. Regulations specify the time required for accelerating from minimum idle speed to 95% of take-off thrust. Because this is difficult to achieve in practice, a different idle speed is used in flight, the flight idle thrust speed. This is to permit rapid thrust delivery for recovery from a missed landing. The increased thrust permits safe acceleration to the appropriate approach speed for another landing attempt. Because landing entails speed reduction, the thrust in landing should be as low as possible; the flight idle speed is only used in the approach configuration and is often called the approach idle thrust setting. The minimum idle speed is used only in ground operations such as taxiing so as to minimize fuel burn and braking system wear.

The thrust settings are marked by the angle of the throttle lever, as illustrated schematically in Figure 9.23. The thrust control system uses engine inputs to produce the thrust level indicated by the pilot's selection of throttle lever setting. Consider the fuel system operation, as illustrated schematically in Figure 9.24. The fuel pump delivery is shown as being capable of supplying fuel at a rate proportional to the engine speed $N$. The rate of fuel flow exceeds that which would result in flame-out at the rich fuel–air ratio limit in the combustor. Similarly, there is a minimum level of fuel flow delivery

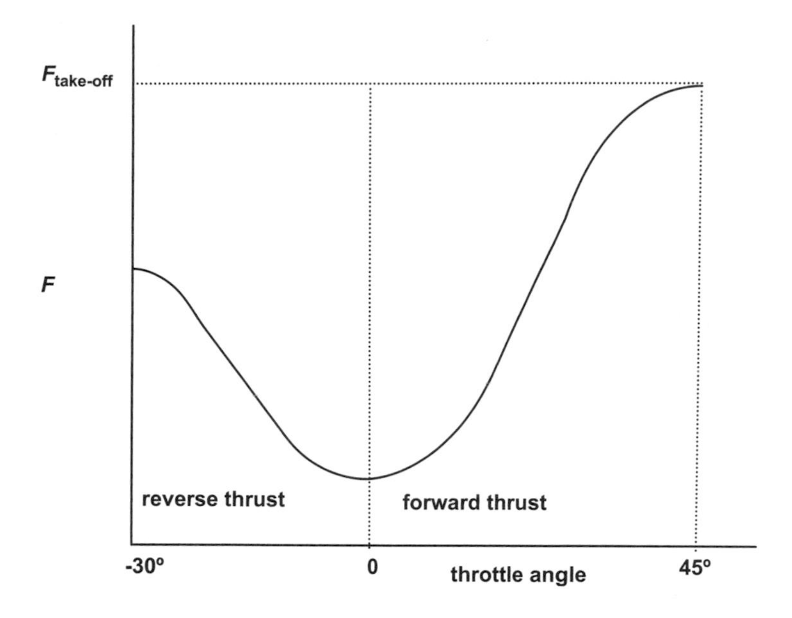


**FIGURE 9.23**

Thrust settings as a function of throttle lever angle.

below which the fuel–air ratio in the combustor is lower that the lean mixture limit and again the flame will be extinguished. The fuel flow schedule for the designed steady-state operation of the engine is shown as a solid black line; this line defines the fuel flow needed at a given speed $N$ for a specific flight condition. Assume that the engine is operating at $N = 50\%$ full speed, as indicated by point 1 on Figure 9.24.

The pilot advances the throttle lever to the $N = 100\%$ position to raise the thrust and accelerate the aircraft. The engine cannot respond instantaneously because the inertia of the rotating components of the engine must be overcome. The change in throttle setting is sensed and the control system commands the fuel pump to immediately increase the flow rate of fuel from point 1 to point 2 on the acceleration schedule shown as the solid light grey line in Figure 9.24. The pump will continue to increase the fuel flow rate according to the acceleration schedule as the engine rpm increases until point 3 is reached. The vertical distance between the steady-state operation schedule and the acceleration schedule represents the excess fuel that can be burned, raising the energy available to the turbine for increasing the rotational speed. It is desirable to have the acceleration schedule reasonably close to the rich mixture limit schedule so that rotor acceleration can be as rapid as is safely possible. When $N = 100\%$ is reached at point 3, the control system commands the pump to reduce the fuel flow rate to the steady-state value at $N = 100\%$, as indicated by point 4. In closing the throttle to reduce speed back to $N = 50\%$, the fuel flow rate is commanded to drop down to the deceleration schedule, point 5. After this drop, the pump is commanded to reduce the fuel flow according to the deceleration schedule as the rpm decreases. When the fuel flow rate reaches that appropriate for steady-state operation, point 6, the fuel flow rate is kept fixed and the rpm decreases to the steady state $N = 50\%$ value, point 1. It must be remembered that the fuel delivery schedules will change with flight

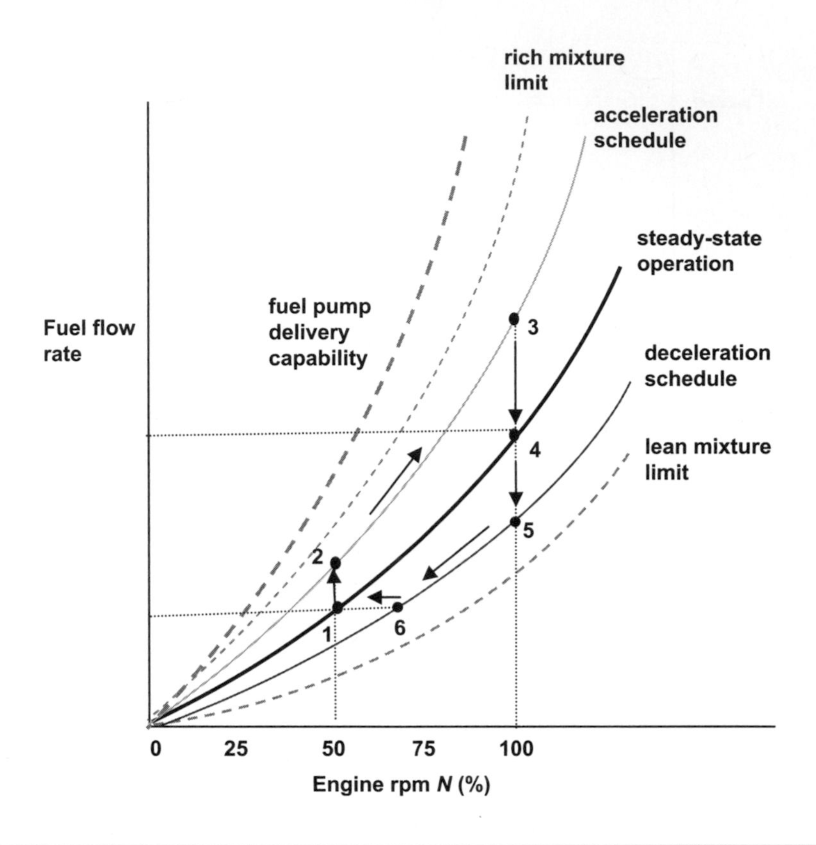


**FIGURE 9.24**

Fuel flow schedules for an engine shown as a function of engine rotational speed for one flight altitude and speed.

condition, that is, Mach number $M_0$ and altitude $z$, but the fuel pump delivery capability remains constant because the pump is a constant volume delivery device.

## 9.9 THRUST REVERSERS

To ensure good braking performance and reduce landing field length, modern turbojet and turbofan engines are equipped with thrust-reversing systems. These are mechanical devices that merely redirect the fan and/or nozzle flow away from the axial direction (call that 0°) to a nearly upstream direction by an angle on the order of 120°, thereby providing a strong negative thrust force. The thrust reversers are actuated after touchdown at which time lift dumpers, or spoilers, are deployed automatically to cut the lift and settle the aircraft firmly on the runway. The thrust reversers are used with the wheel brakes to bring the aircraft to a safe and controlled deceleration and stop. The use of negative thrust provides deceleration independent of ground conditions and is therefore helpful in wet runway situations. In addition, the use of engine thrust helps reduce wear and tear on the braking

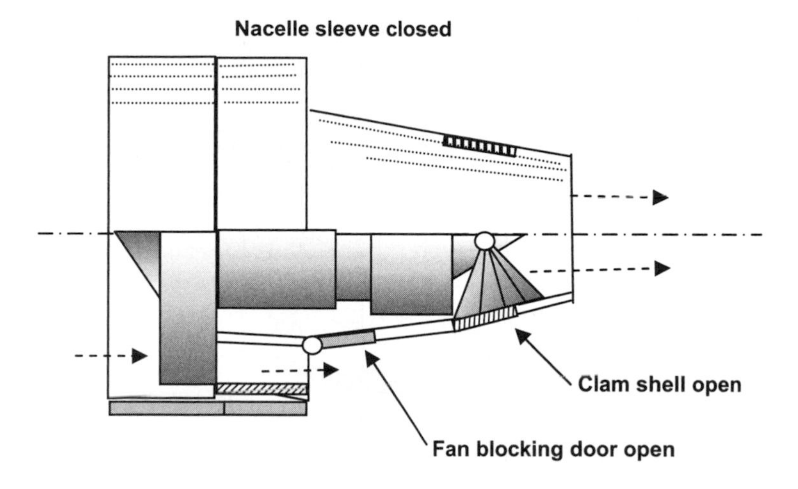


FIGURE 9.25

Split view schematic diagram of a turbofan engine in a nacelle pod under normal flight conditions. The upper half shows the outer shell, while the lower half provides an internal view. Fan flow and nozzle flow proceed in the axial direction.

system. A split view of a turbofan engine in a nacelle pod under normal flight operation is shown in Figure 9.25.

The aerodynamically clean configuration provides for full thrust generation with low nacelle drag. The post-touchdown configuration shown in Figure 9.26 shows that a movable portion of the nacelle slides back, exposing a porous cover over a cascade of turning vanes that redirect the fan flow forward as the fan blocking door is closed.

Similarly, in the nozzle, a pair of clam shell, or bucket-shaped, doors rotate to the closed position and redirect the nozzle flow upstream through another cascade of turning vanes. In modern high bypass turbofans, most of the thrust is provided by the fan so that only the fan reversing system need be used. This reduces the complexity of the thrust reversing system, while also eliminating the need for high temperature materials necessary for reversing the hot nozzle flow. Of course, turbojets must use the nozzle flow diverter system, whereas turboprop aircraft can use reverse pitch on the propeller blades to develop negative thrust. A discussion of thrust reversing systems used on commercial turbofan aircraft is presented by Linke-Diesinger (2008).

## 9.10 ESTIMATING THRUST AND SPECIFIC FUEL CONSUMPTION IN CRUISE

High bypass turbofan engines may be considered to be those with bypass ratio $\lambda > 4$ or 5; current engines have $\lambda$ values up to 10. The important engine properties to consider in the selection process for preliminary design are take-off thrust, cruise thrust, and cruise-specific fuel consumption. Take-off thrust is commonly considered to be the static thrust quoted by the manufacturer. Static thrust is the thrust measured with the engine stationary, as would be the case when the aircraft is initiating the take-off roll. The actual thrust produced drops off with forward speed, with the drop-off more pronounced

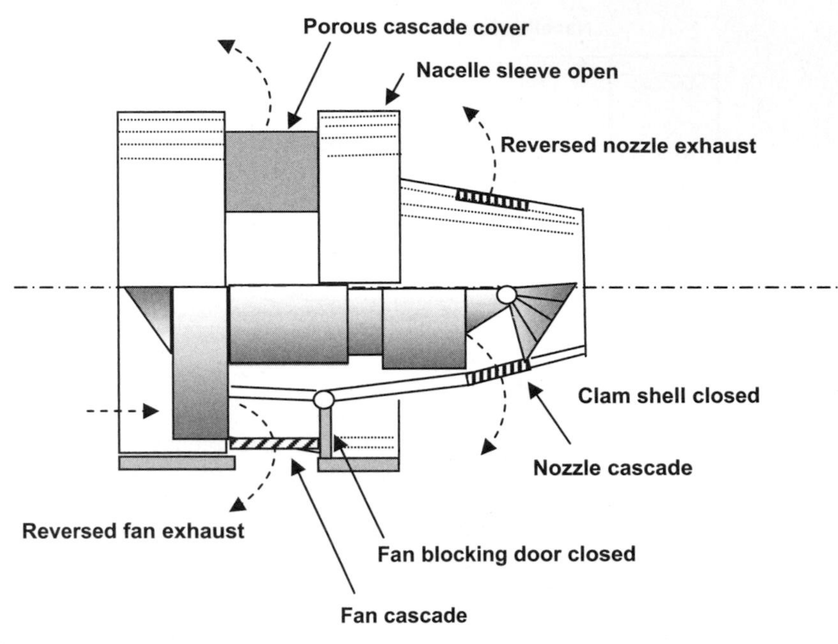


**FIGURE 9.26**

Split view schematic diagram of a turbofan engine in a nacelle pod under braking during a ground run. The upper half shows the nacelle sleeve opened with reversed flow from the fan and nozzle. The lower half shows the fan blocking door redirecting flow through a cascade and the closed clamshell doors doing the same for the nozzle flow.

for turbofans than for turbojets. As a consequence, the thrust level $\overline{F}$ used in take-off distance correlations is an average value over the take-off distance, and Torenbeek (1982) suggests it may be approximated as

$$\left(\frac{\overline{F}}{W}\right)_{to} = 0.75\left(\frac{F}{W}\right)_{to}\frac{5+\lambda}{4+\lambda}. \tag{9.44}$$

The take-off thrust to weight ratio is defined as follows:

$$\left(\frac{F}{W}\right)_{to} = \frac{F_{static}}{W_{to}}. \tag{9.45}$$

Typically, commercial aircraft use two, three, or four identical engines. The number of engines selected is often a result of engine availability, although it is generally considered preferable to have only two engines. This is the trend, as turbine engines are now so reliable that there is no significant safety advantage to having more. In addition, fewer engines mean reduced maintenance and repair costs. For the same reasons, although aircraft may be offered by the airplane manufacturer with

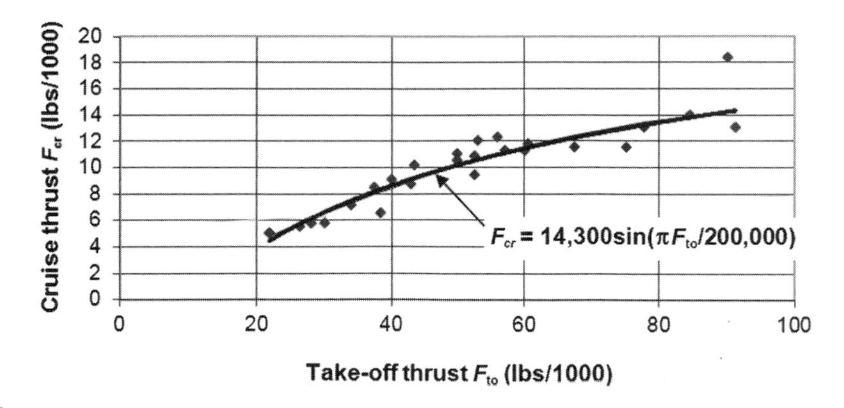


**FIGURE 9.27**

Correlation of cruise thrust with take-off thrust for 26 engines (data reported in Svoboda, 2000).

a choice of engines, on any one aircraft all the engines are the same. Thus, the number of engines used is part of the design selection process.

The cruise thrust available from the engines helps fix another boundary of the design region, and it is necessary to estimate the value for the selected engines when carrying out the performance estimates. Sometimes the engine manufacturers will quote nominal cruise thrust performance, typically giving the net thrust at a specific speed and altitude, commonly $M_0 = 0.8$ and $z = 35{,}000$ ft. If such values are given they should be used. If not, a reasonable approximation is that the cruise thrust (in lbs) is given by

$$F_{cruise} \approx 14,300 \sin\left(\frac{F_{static}}{10^5}\frac{\pi}{2}\right). \tag{9.46}$$

This result is a correlation of the performance of 26 engines as reported in Svoboda (2000) and is shown in Figure 9.27. The correlation is reasonable, mostly within $\pm 12\%$, but there are a few outlying points, with a major underprediction for the GE90, the engine with the greatest cruise thrust in Figure 9.28.

The specific fuel consumption, denoted as $c_j$, measures the weight flow rate of fuel (lb/hr) used for each unit of thrust (lb) produced and is a major figure of merit for engines. This variable is dependent on the fuel flow rate for the actual thrust level produced and is often quoted as evaluated at the maximum take-off thrust, $F_{static}$. This value of $c_{j,to}$ is around 0.3 to 0.4 for engines currently in use. However, in order to estimate fuel usage during cruise, when the majority of the fuel is consumed, it is necessary to have the value of the cruise $c_j$. This quantity, like the cruise thrust itself, is often very difficult to find in the open literature, as it is usually considered proprietary information. The database developed in Svoboda (2000) also has some information on cruise-specific fuel consumption; these data are shown in Figure 9.28. A correlation for that data is given by

$$c_{j,cruise} = 0.7 - \left(\frac{F_{cruise}}{10^5}\right). \tag{9.47}$$

In this equation, the units for thrust and for specific fuel consumption are lbs and lb/hr-lb, respectively.

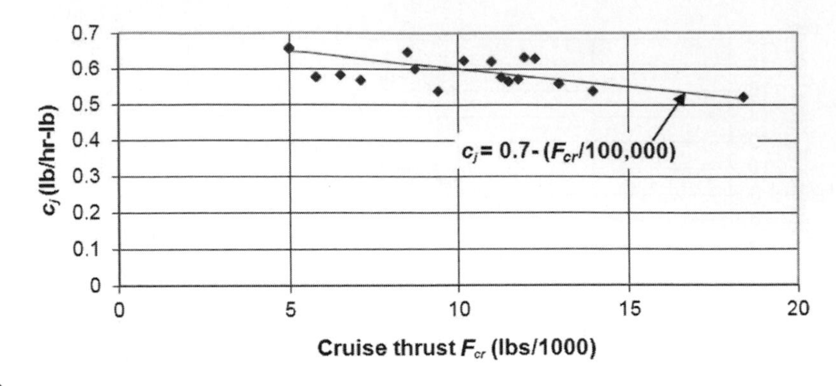


**FIGURE 9.28**

Correlation of cruise-specific fuel consumption with take-off thrust for 26 engines (data reported in Svoboda, 2000).

## 9.11 ENGINE COST

Engine maintenance costs are generally proportional to the take-off thrust, whereas the cost of the required maintenance materials tends to be a function of engine cost, $C_e$. Accurate values for price information must be sought from the manufacturer because it is not often widely publicized. However, for preliminary evaluation purposes, one may employ a correlation suggested by Jenkinson and colleagues (1999), which, in \$2008, is of the form

$$C_e = 1 + 0.956 \frac{F_{cruise}^{0.088}}{C_j^{2.58}}. \tag{9.48}$$

Applying this approximation to data for 26 turbofan engines in the database provided by Svoboda (2000) yields the engine costs shown in Figure 9.29. The circled data points enclose the actual price and the estimated price for the engines indicated. Wherever available, the database figures for cost and specific fuel consumption are used; otherwise the correlations in Equations (9.47) and (9.48) are

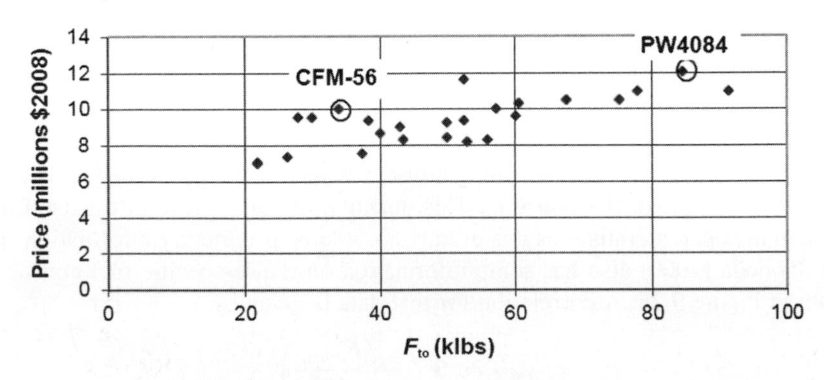


**FIGURE 9.29**

Estimated turbofan engine cost in millions of \$2008 based on Equation (9.48) is shown as a function of take-off thrust in thousands of pounds. Circled data points enclose the actual price and the estimated price for the engines indicated.

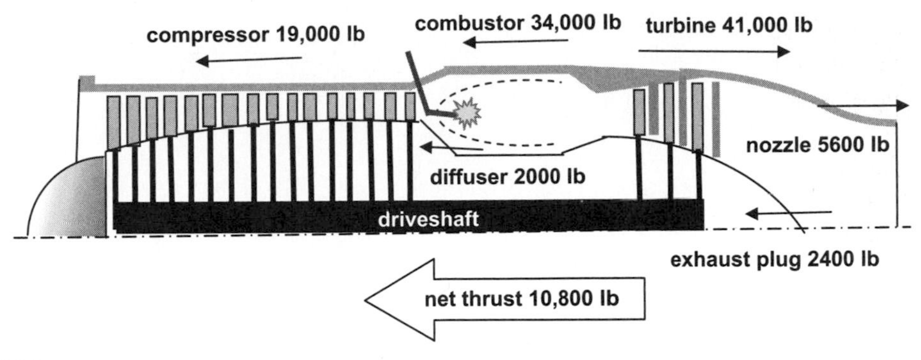


**FIGURE 9.30**

Typical pressure loads developed on jet engine components in producing 10,800 lbs of net static thrust ("The Jet Engine," 3rd ed. 1969, Rolls-Royce Ltd., Derby, UK).

employed. The engine cost so derived is shown as a function of take-off thrust because that information is available more readily. There is appreciable scatter, but the trend is evident. Actual price data should be used in the analysis wherever possible.

## 9.12 LOADS ON TURBOMACHINERY COMPONENTS

The change in momentum of the fluid passing through the various components of the engine, rotating or not, gives rise to substantial axial loads, ultimately producing a net thrust. This thrust is generally substantially smaller than many of the loads on the individual components and can be seen in Figure 9.30 ("The Jet Engine," 3rd ed. 1969, Rolls-Royce Ltd., Derby, UK). The structure to support the loads within the engine and on the interface between the engine and the airframe requires significant analysis to maintain a combination of strength and light weight.

## 9.13 NOMENCLATURE

| | |
|---|---|
| $A$ | cross-sectional area |
| $a$ | sound speed |
| C | constant |
| $c_p$ | specific heat at constant pressure |
| $F_n$ | net thrust |
| $g$ | acceleration of gravity |
| $h$ | enthalpy |
| $M$ | Mach number |
| $\dot{m}$ | mass flow |
| $m_{corr}$ | corrected mass flow |
| $N$ | engine rotational speed in rpm |

| | |
|---|---|
| $h$ | enthalpy |
| $p$ | pressure |
| $P$ | power |
| $R$ | gas constant |
| $T$ | temperature |
| $V$ | velocity |
| $W_c$ | work required by the compressor |
| $W_t$ | work delivered by the turbine |
| $\dot{w}$ | weight flow |
| $\gamma$ | ratio of specific heats |
| $\delta$ | $p_t/p_{ref}$ |
| $\rho$ | density |
| $\theta$ | $T_t/T_{ref}$ |
| $\eta$ | efficiency |
| $\lambda$ | bypass ratio |

### 9.13.1 Subscripts

| | |
|---|---|
| $ad$ | adiabatic |
| $c$ | compressor |
| $f$ | fuel |
| max | maximum condition |
| $r$ | radial direction |
| $ref$ | reference condition |
| $t$ | stagnation conditions |
| 0 | conditions in the free stream |
| 1 | conditions at the inlet entrance |
| 2 | conditions at the compressor entrance |
| 3 | conditions at the combustor entrance |
| 4 | conditions at the turbine entrance |
| 5 | conditions at the nozzle entrance |
| 6 | conditions at the nozzle minimum section |
| 7 | conditions at the nozzle exit |

## 9.14 EXERCISES

**9.1** Consider an engine with the compressor and turbine performance maps given in Figures E9.1 and E9.2. The exhaust nozzle minimum section has a diameter $d_6 = 16.25$ in. and the flight

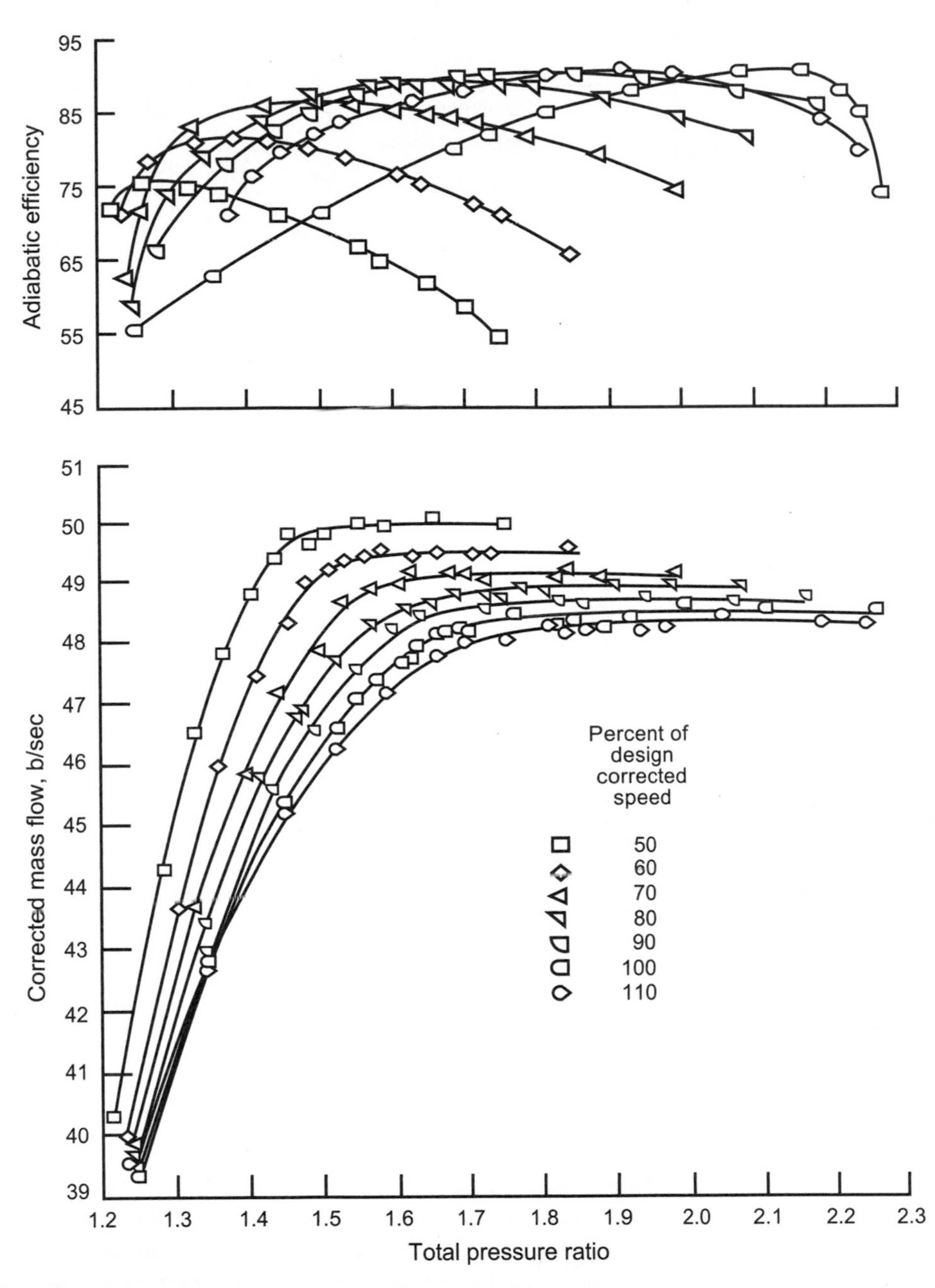


**FIGURE E9.1**

Turbine performance map for Problem 9.1.

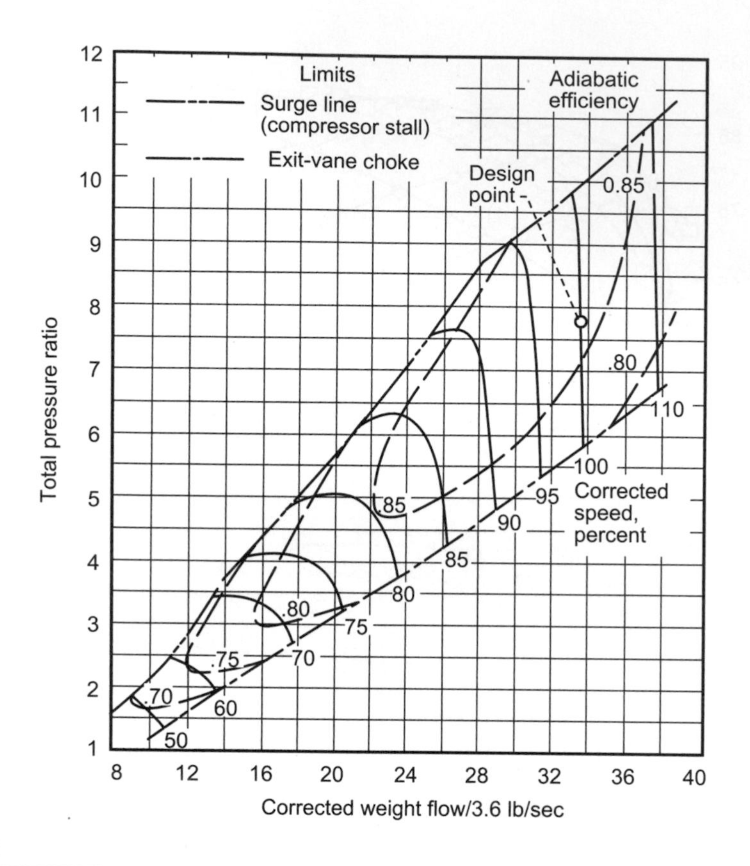


**FIGURE E9.2**

Compressor performance map for Problem 9.1. Note that for this problem the corrected weight flow scale for the compressor map has been reduced by a factor of 3.6.

condition is take-off where $M_0 = 0.25$ and the altitude is sea level at standard-day conditions. Assuming that the turbine inlet temperature is $T_{t4} = 3400\text{R}$, determine the engine operating rpm $N$ (in % of the design value), the free stream capture area $A_0$, and the net thrust produced. The nozzle may be assumed to be choked with $A_7 = A_6$. The gas properties may be assumed as follows: $\gamma_2 = 7/5$, $\gamma_4 = 4/3$, and $R = 1716\ \text{ft}^2/\text{s}^2\text{-R}$.

**9.2** Consider an engine with the compressor and turbine performance maps given in Figures E9.3 and E9.4. The exhaust nozzle minimum section has a diameter $d_6 = 41.4$ cm, and the flight condition is cruise at $M_0 = 0.8$ at an altitude of 5 km. Assuming that the turbine inlet temperature $T_{t4} = 1000°\text{C}$, determine the corrected rpm $N$ (in % of the design value) and the free stream capture area $A_0$. The nozzle may be assumed to be choked with $A_7 = A_6$. The gas properties may be assumed as follow: $\gamma_2 = 1.40$, $c_{p,2} = 1.0$ kJ/kg-K, $\gamma_4 = 1.33$, $c_{p,4} = 1.16$ kJ/kg-K, and $R = 287\ \text{m}^2/\text{s}^2\text{-K}$. The reference pressure and temperature are 101.3 kPa and 288K, respectively. The pressure and temperature at 10 km altitude are 54.0 kPa and 255K, respectively.

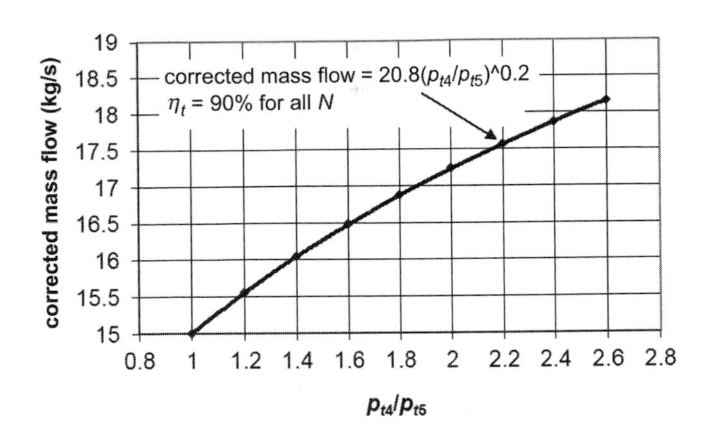


**FIGURE E9.3**

Turbine performance map for Problem 9.2. The lines for corrected speed in the range 80% $< N <$ 100% fall closely upon one another so that the curve shown is for all those values of corrected $N$. In addition, the turbine efficiency may be taken as $\eta_t = 90\%$ for all corrected speeds $N$.

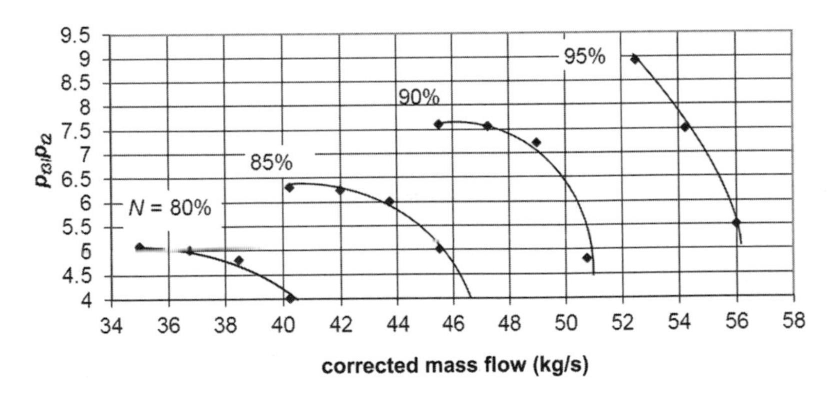


**FIGURE E9.4**

Compressor performance map for Problem 9.2. The compressor efficiency may be taken to be constant at $\eta_c = 85\%$ for all corrected speeds $N$ shown.

**9.3** Assume that a turbine performance map may be characterized by the following equation for the corrected mass flow as a function of the turbine pressure ratio:

$$\frac{\dot{m}\sqrt{\theta_4}}{\delta_4} = 15\left(\frac{p_{t4}}{p_{t5}}\right)^{0.2}.$$

This relation may be considered to apply with constant turbine efficiency $\eta_t = 90\%$ for all % corrected rpm $N$. The corresponding compressor performance map is shown in Figure E9.5.

The compressor efficiency may also be considered constant at $\eta_c = 85\%$. Assume that, for a certain nozzle minimum area, the corrected mass flow at station 5 is given (in kg/s) by

$$\frac{\dot{m}\sqrt{\theta_5}}{\delta_5} = 31.9.$$

For the case where $T_{t4} = 1273$K, find the matching point ($p_{t3}/p_{t2}$, $\%N$, and $W_c$) for the turbine–compressor unit if the aircraft is traveling at $M_0 = 0.8$ at an altitude of 5 km ($p_0 = 54$ kPa and $T_0 = 255$K). It may be assumed that the stagnation pressure across the combustor is constant ($p_{t3}/p_{t4} = 1$).

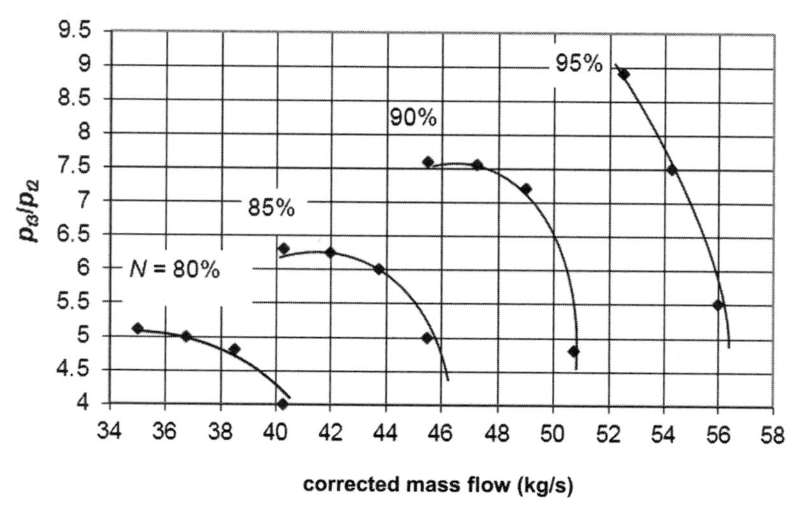


**FIGURE E9.5**

Compressor map for Exercise 9.3.

**9.4** A miniturbojet (11 in. diameter by 19 in. long; weight under 30 lb) operates at standard sea level conditions passing 2.2 lb/s of air. The axial turbine has a pressure ratio of 4.2 at a rotational speed of 60,000 rpm and the turbine inlet temperature $T_{t4} = 1750$F. Assume that the efficiency of the centrifugal compressor is $\eta_c = 85\%$, the efficiency of the axial turbine $\eta_t = 95\%$, the burner efficiency $\eta_b = 90\%$, and the heating value of the fuel $HV = 18{,}900$ Btu/lb. Determine

**(a)** The compressor pressure ratio $p_{t3}/p_{t2}$

**(b)** The static thrust $F_n$

**(c)** The specific fuel consumption $c_j$

**9.5** The large (29 m long and a take-off weight around 29,000 kg) Soviet DBR 123 unmanned reconnaissance vehicle sketched in Figure E9.6 was powered by the KR-15 jet engine and was in service between the mid-1960s and the late 1980s. The vehicle was ground launched using two solid propellant rockets, which were jettisoned after burnout. The DBR 123 could

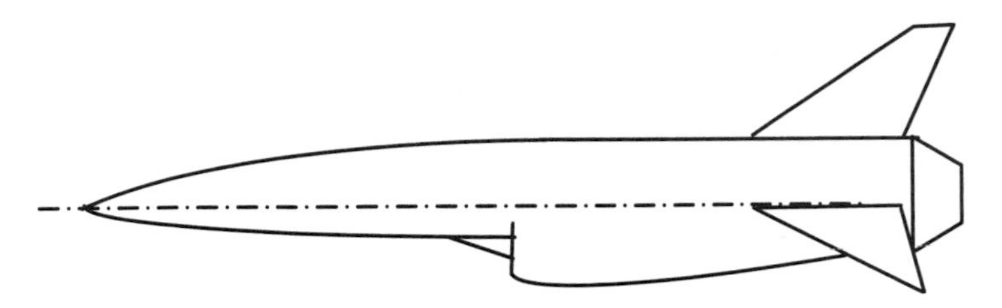

**FIGURE E9.6**

Sketch of Soviet DBR 123 unmanned reconnaissance vehicle.

achieve a range of 3000 km flying at around 22 km altitude at speed in excess of Mach 2. The maximum speed attainable was around $M_0 = 2.55$ and it carried a fuel load of 19,000 liters. Assuming that the lift to drag ratio could be represented reasonably by the equation $L/D = 4(M_0 + 3)/M_0$

**(a)** Estimate the specific fuel consumption at a constant flight speed of $M_0 = 2.25$ at the given altitude.

**(b)** Estimate the thrust of the KR-15 engine.

**(c)** Estimate the nozzle exhaust velocity $V_7$ assuming matched operation.

**9.6** In 1965, GE won the contract to develop engines for the USAF's Lockheed C-5A "Galaxy" transport, the first of the jumbo jets. The GE TF39 engines introduced the high bypass turbofan. Before the TF39, bypass ratios of turbofan engines were less than two to one. The TF39 showed that bypass ratios of 8 were possible, far more than the bypass ratio of 2 that was common at the time. This innovation resulted in specific fuel consumption as much as 25% less than previous other available engines. In the late 1960s, Lockheed and Douglas entered the commercial wide-body airliner market with their trijet offerings. GE introduced a new engine design, the CF6-50, modeled after the TF39 for use on the Douglas DC-10 and Lockheed L-1011. Consider a generic layout of the CF6-50 as shown in Figure E9.7, and assuming that the engine is operating in a ground-based test cell, determine the (a) power developed by the low-pressure and high-pressure turbines, (b) thrust developed by the engine for the case of ideal expansion of the fan flow and the core flow, and (c) specific fuel consumption.

**9.7** The MQM-74A "Chukar" target drone had a neatly tapered cigar-shaped fuselage, straight midmounted wings, an underslung jet engine with the intake under the wings, and a conventional tail configuration with the tailplanes set in an inverted vee. It was powered by a Williams International WR24-6 turbojet engine and was launched by a RATO booster from the ground or a ship. Engine characteristics: $N = 61{,}000$ rpm; compressor type: centrifugal; turbine type: axial flow with pressure ratio of 4.2; application: drones; length: 0.57 m; maximum diameter: 0.27 m; dry weight of engine: 13.6 kg; air flow rate: 1 kg/s; and turbine inlet temperature: 955°C. For sea level static conditions and assuming $p_{t4}/p_{t3} = 0.9$, $\eta_C = 75\%$, $\eta_T = 90\%$, $\eta_B = 90\%$, and

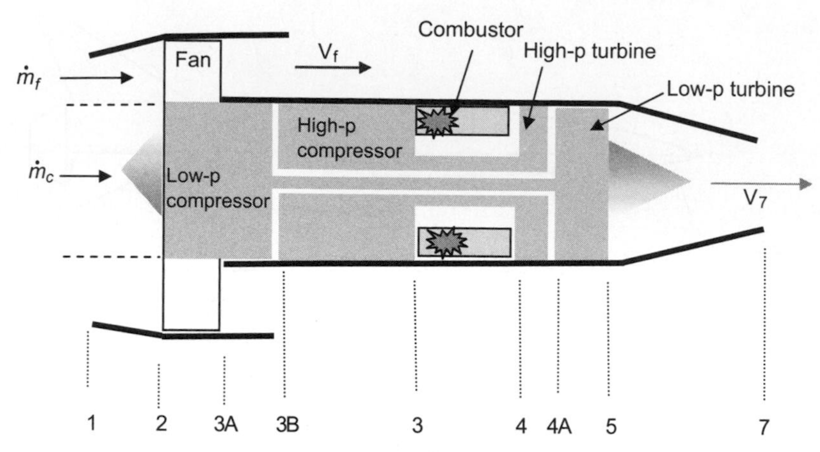


**FIGURE E9.7**

Fan pressure ratio $p_{t,3A}/p_{t,2} = 1.72$, low pressure compressor pressure ratio $p_{t3B,}/p_{t2} = 1.69$, and overall pressure ratio $p_{t,3}/p_{t,2} = 29.4$. Compressor and turbine efficiencies $\eta_c = \eta_t = 0.9$ and burner efficiency $\eta_b = 0.95$. In the compressor $\gamma = 1.4$ and in the turbine and beyond $\gamma = 1.33$. The temperature $T_{t,4} = 1360°C$ and the heating value $HV = 44$ MJ/kg. The fan flow rate $m_f = 540$ kg/s and the core flow rate $m_c = 124$ kg/s.

$HV = 44$ MJ/kg, find the (a) compressor pressure ratio $p_{t3}/p_{t2}$, (b) thrust, and (c) specific fuel consumption.

**9.8** Consider the matching of the various components of the turbojet engine with a compressor ratio $p_{t3}/p_{t2} = 15$ in two cases: (i) where the compressor and turbine are ideal, $\eta_c = \eta_t = 1$, and (ii) where $\eta_c = 0.85$ and $\eta_t = 0.95$. (a) Determine the pressure and temperature ratios across the compressor–turbine combination, $p_{t5}/p_{t2}$ and $T_{t5}/T_{t2}$, respectively. (b) Determine the fuel–air ratio $f/a$ assuming flight at $M_0 = 2$ at an altitude $z = 50{,}00$ ft where the free stream temperature $T_0 = 390$K, with $HV = 18{,}900$ Btu/lb, the stagnation temperature entering the combustor $T_{t4} = 6T_{t2}$, and a combustor efficiency $\eta_b = 0.95$. (c) Determine the net thrust $F_n$ and the specific fuel consumption $c_j$ for inlet pressure recoveries $p_{t2}/p_{t0} = 1.0$ and 0.9, respectively, if $M_3 = 0.3$ and $A_2 = 1.0$ ft$^2$ with the nozzle operating in the matched mode. (d) Briefly compare results for cases of different compressor and turbine efficiencies.

**9.9** A ramjet-powered missile flies at a Mach number $M_0 = 3$ at an altitude of 26 km, where the ambient pressure and temperature are $p_0 = 0.028$ atmospheres and $T_0 = -50°C$. Subsonic combustion takes place in the constant area region between stations 3 and 4 and results in increasing the stagnation temperature of the hot gas. The stagnation pressure recovery of the inlet, $p_{t,3}/p_{t,0}$, is that corresponding to a normal shock standing at the lip of the inlet, station 0, as shown in Figure E9.7; this corresponds to maximum mass flow entering the engine. The increase in stagnation temperature due to combustion is $T_{t,4} - T_{t,3} = 600°C$. The combustor diameter $d = 25$ cm and the inlet area ratio $A_1/A_3 = 0.68$. Assuming $R = 0.286$ kJ/kg-K and

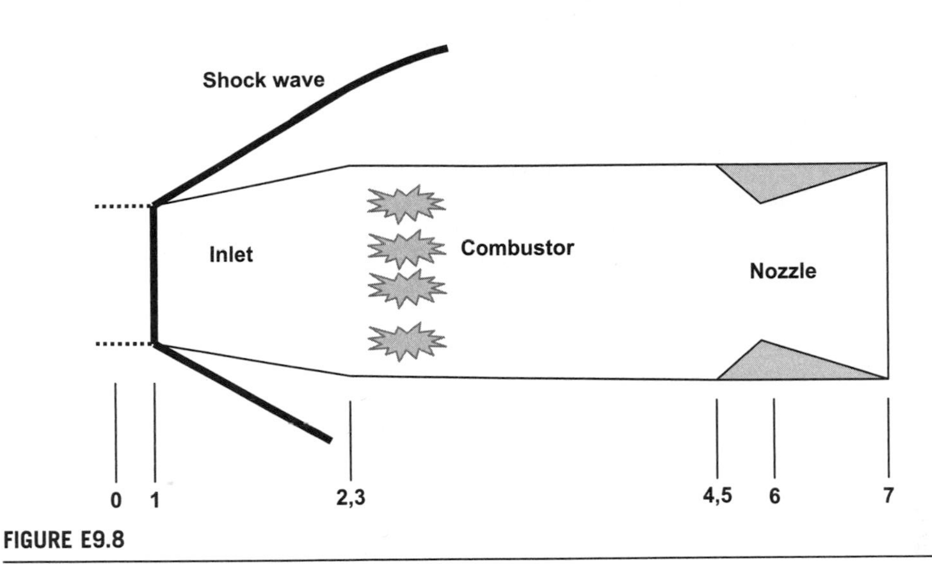


**FIGURE E9.8**

Schematic diagram of a ramjet engine.

$\gamma = 1.4$ throughout, and neglecting the (small) weight of the added fuel, provide the following information:

**(a)** For the maximum mass flow condition shown, what is the area of the nozzle minimum section, $A_6$?

**(b)** For maximum thrust the nozzle must be "matched", that is, $p_7 = p_6$. What is the exit area required to achieve this condition?

**(c)** What is the engine thrust?

**(d)** Carefully sketch the distribution of stagnation pressure and temperature from station 0 to station 7 and explain how engine thrust could be increased.

## References

*Flight*, March 1959, pp. 394–395.
*Flight*, March 18, 1960, p. 385.
Goldsmith, E. L. & Seddon, J. (Eds.). (1993). *Practical Intake Aerodynamic Design.* Cambridge, MA: Blackwell Scientific Publishers.
Hefley, R. K., & Jewell, W. F. (1972). *Aircraft Handling Qualities Data*. NASA CR-2144, December.
Henderson, L.F. (1967). A critique of the starting phenomenon on supersonic intakes. *Z. Flugwissenschaft* 15, Heft 2, pp. 57–67.
Jenkinson, L. R., Simpkin, P., & Rhodes, D. (1999). *Civil Jet Aircraft Design.* Reston, VA: AIAA.
Leynaert, J., Surber, L. E., & Goldsmith, E. L. (1993). Transport aircraft intake design. In E. L. Goldsmith & J. Seddon (Eds.), *Practical Intake Aerodynamic Design* (pp. 214–252). Cambridge MA: Blackwell Scientific Publishers.

Linke-Diesinger, A. (2008). *Systems of Commercial Turbofan Engines.* New York: Springer.

Mattingly, J. D. (1996). *Elements of Gas Turbine Propulsion.* New York: McGraw-Hill.

MIL-E-5008B: Military Specifications—Engines, Aircraft, Turbojet, Model Specification For, January 1959.

Plencner, R.M. (1989). Plotting Component Maps in the Navy/NASA Engine Program (NNEP)—A Method and Its Usage. NASA Technical Memo 101433, January.

Svoboda, C. (2000). Turbofan Engine Database as a Preliminary Design Tool. *Aircraft Design, Vol. 3*, 17–31.

The Gas Turbine Engine and Its Operation. PWA OI 200 Pratt & Whitney Aircraft, May 1974, E. Hartford, CT.

*The Jet Engine*. (1969) (3rd ed.). Derby. UK: Rolls-Royce Ltd.

Torenbeek, E. (1982). *Synthesis of Subsonic Airplane Design.* Dordrecht, The Netherlands: Kluwer Academic Publishers.

CHAPTER

# Propellers

10

## CHAPTER OUTLINE



## 10.1 CLASSICAL MOMENTUM THEORY

Consider a streamtube enclosing the steady flow processed by a thrusting propeller as first carried out in Chapter 1 for idealized flow machines. The additional assumptions we apply are that the flow machine itself is an infinitesimally thin disc, generally called the actuator disc, and that

- the thrust is distributed uniformly over the disc
- no rotation is imparted to the flow by the actuator disc
- the streamtube entering and leaving the disc defines the flow distinctly
- the pressure far ahead and far behind the disc matches the ambient value

The flow picture is shown in Figure 10.1. Constructing a cylindrical control volume of large radius $R$ around the streamtube leads to the diagram in Figure 10.2. Applying the conservation of mass principle in the control volume, which has a boundary area of $S$, and assuming that the fluid density is constant along the boundaries of the control volume, leads to

**Theory of Aerospace Propulsion.**

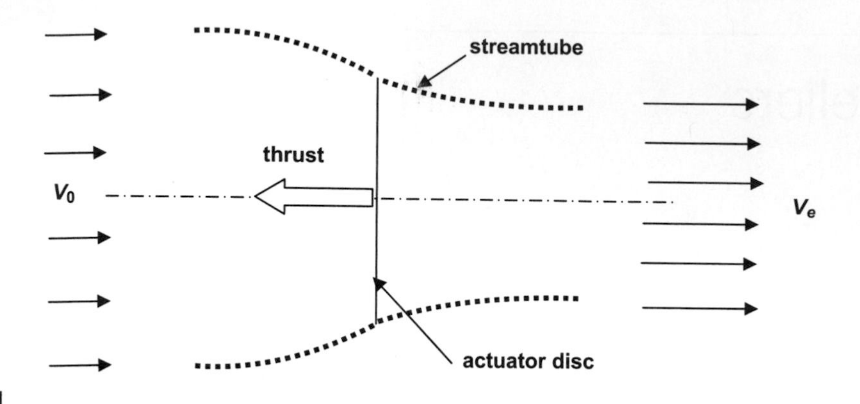


**FIGURE 10.1**

Idealized representation of flow through a propeller actuator disc.

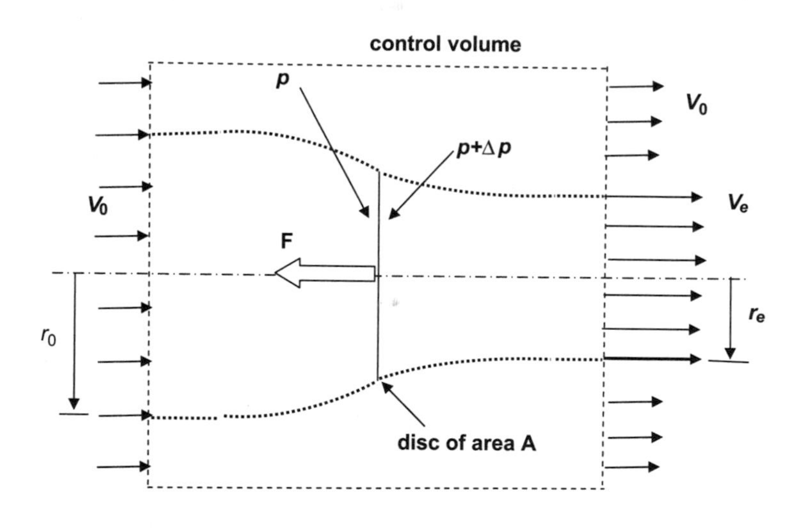


**FIGURE 10.2**

Control volume around a streamtube containing the actuator disc.

$$
\begin{aligned}
&\iint_S \rho \vec{V} \cdot \hat{n} dS = 0 \\
&-\rho V_0 \pi r_0^2 + \rho V_0 \pi (r_0^2 - r_e^2) + \rho V_e \pi r_e^2 - \rho Q = 0. \\
&Q = \pi r_e^2 (V_e - V_0).
\end{aligned}
\tag{10.1}
$$

The quantity $\rho Q$ is the inflow of mass across the horizontal boundaries of the control volume. Applying the conservation of mass in the control volume leads to

$$\begin{aligned}
&\iint_S (\rho\vec{V}\cdot\hat{n})\vec{V}\,dS = F_{fluid} \\
&-\rho V_0^2\pi r_0^2 - \rho V_0 Q + \rho V_0^2\pi\left(r_0^2 - r_e^2\right) + \rho V_e^2\pi r_e^2 = F_{fluid}. \\
&F_{fluid} = -\rho\pi r_e^2\left(V_0 V_2 - V_0^2 - V_e^2 + V_0^2\right) = -F_{disc} \\
&F_{disc} = \rho\pi r_e^2 V_e(V_e - V_0)
\end{aligned} \tag{10.2}$$

The thrust on the disc may also be written as

$$F = A\Delta p.$$

We may now drop the subscript for the force on the disc in Equation (10.2) and consider the thrust developed to be denoted simply as $F$. It is assumed here that the static and stagnation pressure jump discontinuously across the disc while the velocity remains continuous through it. The velocity at the disc is denoted by $V_1$. However, we may consider the flow to be isentropic up to the disc and then isentropic from the disc on; of course the stagnation pressures upstream and downstream, although constant, will not be the same constant. Further assuming that the flow is incompressible outside the disc, we may apply Bernoulli's equation to the upstream and downstream streamlines to obtain the following:

$$\begin{aligned}
p_0 + \frac{1}{2}\rho V_0^2 &= p + \frac{1}{2}\rho V_1^2 \\
p_e + \frac{1}{2}\rho V_e^2 &= p + \Delta p + \frac{1}{2}\rho V_1^2.
\end{aligned} \tag{10.3}$$

Then we may find the pressure jump across the disc to be

$$\Delta p = \frac{1}{2}\rho\left(V_e^2 - V_0^2\right) = \frac{1}{2}\rho(V_e - V_0)(V_e + V_0). \tag{10.4}$$

Conservation of mass through the disc requires that $\pi r^2 V_e = AV_1$ and therefore

$$V_1 = \frac{1}{2}(V_0 + V_e). \tag{10.5}$$

This shows that the velocity at the disc is the average of the upstream and downstream velocities. The actuator disc representing the propeller may be thought of as imparting an increment in velocity $V'$ to the undisturbed velocity $V_0$, that is, $V' = V_1 - V_0$. The thrust is then given by

$$F = \rho\pi r_e^2 V_e(V_e - V_0) = \rho A V_1[2(V_1 - V_0)] = 2\rho A V'(V_0 + V'). \tag{10.6}$$

We may now apply the energy equation, keeping in mind that the flow is frictionless everywhere (except at the disc, and any effects there may be considered to be incorporated into the thrust developed). Thus we have

$$\begin{aligned}
&\iint_S (\rho\vec{V}\cdot\hat{n})\frac{1}{2}\vec{V}\cdot\vec{V}\,dS = \vec{F}\cdot\vec{V} = P_{fluid} \\
&\frac{1}{2}\rho\pi r_0^2\left[-V_0^3 + \left(1 - \frac{r_e^2}{r_0^2}\right)V_0^3 + \frac{r_e^2}{r_0^2}V_e^3 - \frac{Q}{\pi r_0^2}V_0^2\right] = P_{fluid}.
\end{aligned}$$

The power absorbed by the fluid may be further reduced to

$$P_{fluid} = \frac{1}{2}\rho AV_1\left(V_2^2 - V_0^2\right) = \left[\rho AV_1(V_e - V_0)\right]\left[\frac{1}{2}(V_2 + V_0)\right].$$
$$P_{fluid} = FV_{avg} \tag{10.7}$$

Because the power is a scalar, note that only the thrust magnitude need be considered and we will drop the subscript, understanding that $P$ represents the power required. However, we do consider the power added to the fluid to be a positive quantity in the current analysis, as we do generally. We may also rewrite the power in terms of the induced velocity $V'$ as follows:

$$P = FV_0\left(1 + \frac{V'}{V_0}\right). \tag{10.8}$$

Because the ideal required power, that is, the power needed to make the disc proceed through the air at a constant speed $V_0$, is $P_i = FV_0$, we may write the propulsive efficiency as

$$\eta_p = \frac{P_i}{P} = \frac{1}{1 + \dfrac{V'}{V_0}}. \tag{10.9}$$

The thrust as given by the last term in Equation (10.6) may be written as

$$F = 2\rho AV'(V_0 + V') = \frac{1}{2}\rho AV_0^2\left[4\frac{V'}{V_0}\left(1 + \frac{V'}{V_0}\right)\right]. \tag{10.10}$$

Then we may form the nondimensional thrust coefficient

$$C_T = \frac{\dfrac{F}{A}}{\dfrac{1}{2}\rho V_0^2} = 4\frac{V'}{V_0}\left(1 + \frac{V'}{V_0}\right). \tag{10.11}$$

The quantity $F/A$ in Equation (10.11) is called disc loading and has the units of pressure; the denominator is, of course, the free stream dynamic pressure. By rearranging this equation, we may show that

$$\frac{V'}{V_0} = \frac{1}{2}\left(\sqrt{1 + C_T} - 1\right). \tag{10.12}$$

We see that the nondimensional induced velocity is a function of the thrust coefficient alone. In the same fashion, we may introduce an ideal power coefficient

$$C_{P,i} = \frac{P}{\dfrac{1}{2}\rho AV_0^3} = C_T\left(1 + \frac{V'}{V_0}\right) = \frac{1}{2}C_T\left(1 + \sqrt{1 + C_T}\right). \tag{10.13}$$

Note that for a statically thrusting propeller $V_0 = 0$ and the nondimensional coefficients don't apply. Instead, using $V_0 = 0$ in Equations (10.8) and (10.10), and using the subscript $s$ to denote static thrust conditions, yields

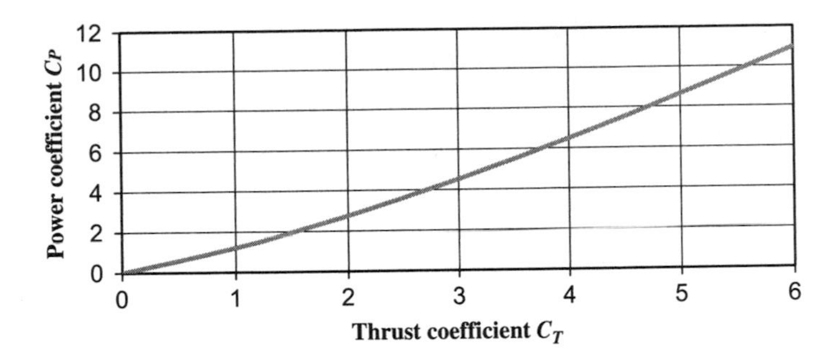


**FIGURE 10.3**

Variation of the power coefficient with the thrust coefficient.

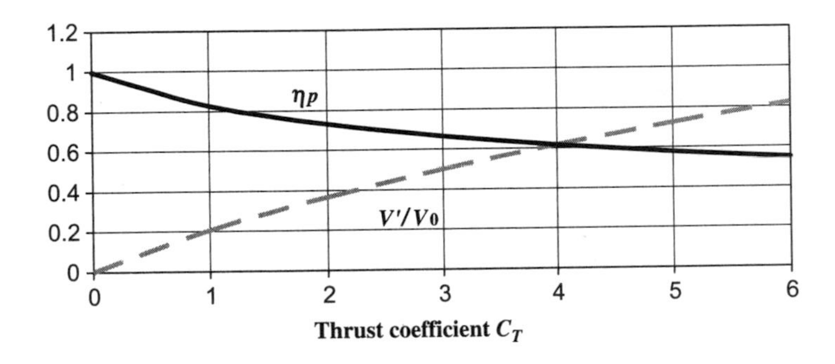


**FIGURE 10.4**

Variation of propulsive efficiency and nondimensional-induced velocity.

$$V_s' = \sqrt{\frac{F_s}{2\rho A}}$$
$$P_s = \sqrt{\frac{F_s^3}{2\rho A}}. \tag{10.14}$$

The variation of the power coefficient $C_{p,i}$ with the thrust coefficient $C_T$ is illustrated in Figure 10.3 and that of the propulsive efficiency $\eta_p$ and the induced velocity $V'$ with $C_T$ is shown in Figure 10.4.

Note that as the thrust coefficient increases, the induced velocity increases and propulsive efficiency decreases. Thus for effective operation, the disc loading $F/A$ should be small, which means that for a given thrust level the disc area should be large. This is the efficiency edge for propellers: providing a small induced velocity while processing a large mass flow of air.

We may determine the effect of flight speed on the thrust produced by the propeller, or, more accurately, the actuator disc that is emulating the propeller, by assuming that the input power to the propeller shaft is independent of speed, which is in general a good assumption. Equating the power for the moving and static cases in Equations (10.13) and (10.14) leads to

$$\sqrt{\frac{F_s^3}{2\rho A}} = \frac{1}{2}FV_0\left(1+\sqrt{\frac{2F}{\rho A}}\right), \tag{10.15}$$

which can be transformed into

$$1 = \frac{1}{2}\frac{F}{F_s}\left[\frac{V_0}{V_s'} + \sqrt{\frac{V_0^2}{V_s'^2} + 4\frac{F}{F_s}}\right]. \tag{10.16}$$

The dynamic pressure in the slipstream of the propeller is

$$q_e = \frac{1}{2}\rho(V_0 + 2V')^2 = q_0 + \frac{F}{A}.$$

This shows that the dynamic pressure in the far field behind the disc is increased over the free stream value by the disc loading.

## 10.2 BLADE ELEMENT THEORY

In order to be able to design the propeller blades, the analysis of performance must be carried to a more detailed level. The blade will be considered locally as a two-dimensional airfoil located at some distance $r$ from the axis of rotation and acting in an annular segment of thickness $dr$, as shown in Figure 10.5. The velocities and differential forces acting on the element of the propeller blade are shown in Figure 10.6.

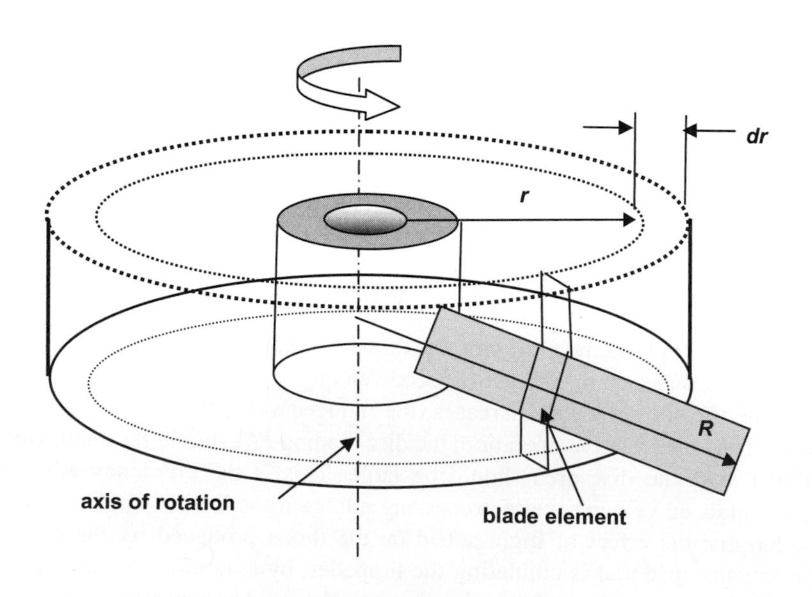


**FIGURE 10.5**

Schematic diagram of blade element in an annular region of flow.

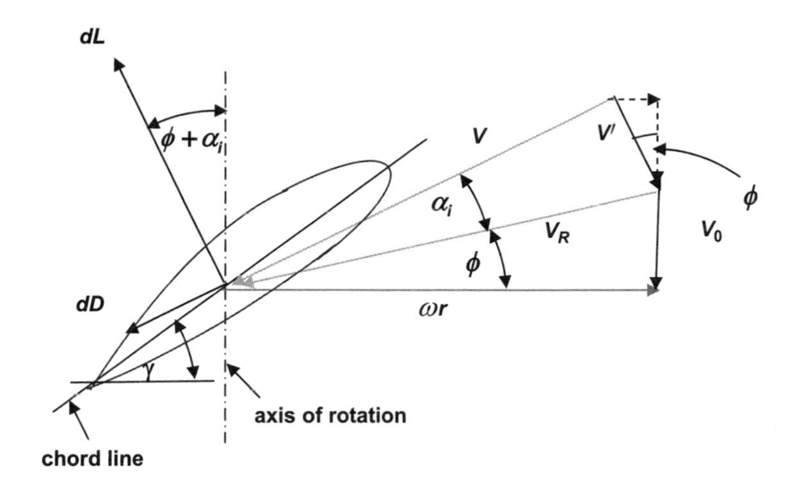


**FIGURE 10.6**

Schematic diagram of a blade element showing force and velocity components.

The velocity component $V'$ represents the generalized induced velocity in this two-dimensional approximation to the flow and illustrates that there are components of velocity induced in both axial and tangential directions. The differential lift and drag forces depend on the relative wind velocity $V$, which in turn depends on the induced velocity $V'$, which is unknown. We may determine the elemental thrust, that is, the differential axial force, produced by applying the conservation of momentum principle to the annular region of thickness $dr$ as follows:

$$dF = [\rho(2\pi r dr)(V_0 + V_a')][2V_a']. \tag{10.17}$$

Here we have neglected squares of differentials and denoted the axial component of the induced velocity $V'$ illustrated in Figure 10.6 as $V_a'$. We further note that

$$V_a' = V' \cos\phi = V_R \sin\alpha_i \cos\phi.$$

For small angles $\alpha_i$ such that $\sin\alpha_i \sim \alpha_i$ we have

$$V_a' \approx \alpha_i V_R \cos\phi. \tag{10.18}$$

The thrust then becomes

$$dF = 2\pi\rho r dr(V_0 + \alpha_i V_R \cos\phi)(2\alpha_i V_R \cos\phi). \tag{10.19}$$

However, the thrust may also be written in terms of the differential lift and drag forces, as follows:

$$dF = dL\cos(\phi + \alpha_i) - dD\sin(\phi + \alpha_i). \tag{10.20}$$

Then, substituting the airfoil coefficients for lift and drag, $c_L$ and $c_D$, and considering that there are B blades comprising the propeller, we obtain

$$dF = Bc_L(cdr)\frac{1}{2}\rho V^2[\cos(\phi + \alpha_i) - \varepsilon\sin(\phi + \alpha_i)], \tag{10.21}$$

where $c$ is the chord of the airfoil and $\varepsilon = c_D / c_L$ is the inverse of the lift to drag ratio of the airfoil. Equating the differential thrust for a single blade as given in Equations (10.19) and (10.21) we obtain

$$2a(1+a) = \frac{Bcc_L}{4\pi r}\left[(\lambda+a')^2+(1+a)^2\right][\cos(\phi+\alpha_i) - \varepsilon\sin(\phi+\alpha_i)]. \tag{10.22}$$

Equation (10.22) introduced the following nondimensional quantities:

$$\begin{aligned} \lambda &= \frac{\omega r}{V_0} \\ a' &= \frac{V'\sin\phi}{V_0} \\ a &= \frac{V'\cos\phi}{V_0}. \end{aligned} \tag{10.23}$$

The differential power imparted by the propeller is

$$dP = \omega T_q = \omega r dF_u = \omega r B[dL\sin(\phi+\alpha_i) + dD\cos(\phi+\alpha_i)]. \tag{10.24}$$

The power needed to keep the propeller flying through the air at constant speed $V$ is

$$dP = V_0 dF = 2\pi\rho r dr V_0^3(1+a)2a. \tag{10.25}$$

However, from energy principles and Equation (10.17), the power transferred to the fluid is

$$dP = (V_0 + V_a')dF = 2\pi\rho r dr V_0^3(1+a)^2\,2a. \tag{10.26}$$

Setting Equations (10.24) and (10.26) equal yields

$$(1+a)^2\,2a = \frac{Bcc_L}{4\pi r}\lambda\left(\frac{V}{V_0}\right)^2[\sin(\phi+\alpha_i) + \varepsilon\cos(\phi+\alpha_i)]. \tag{10.27}$$

Dividing Equation (10.27) by Equation (10.22) and noting, from Equation (10.23) and Figure 10.6, that

$$\left(\frac{V}{V_0}\right)^2 = (\lambda+a')^2+(1-a)^2, \tag{10.28}$$

leads to the following result:

$$\frac{\sin(\phi+\alpha_i)+\varepsilon\cos(\phi+\alpha_i)}{\cos(\phi+\alpha_i)-\varepsilon\sin(\phi+\alpha_i)} = \frac{1+a}{\lambda}. \tag{10.29}$$

For high airfoil lift to drag ratios, that is, $\varepsilon \ll 1$ small, and small values of the angles $\alpha_\iota$ and $\phi$ , this equation shows that $\alpha_\iota \sim (1+a)/\lambda$.

The thrust is then given by

$$F = \frac{1}{2}\rho B\int_0^R cc_L V^2\cos(\phi+\alpha_i)[1-\varepsilon\tan(\phi+\alpha_i)]dr. \tag{10.30}$$

Likewise, the power is given by

$$P = \frac{1}{2}\rho B\int_0^R cc_L\omega r V^2\sin(\phi+\alpha_i)[1+\varepsilon\cot(\phi+\alpha_i)]dr. \tag{10.31}$$

A detailed study of propeller optimization is presented by Adkins and Liebeck (1994). Advanced computational methods for propeller analysis are given by Bosquet and Garadarein (2000).

## 10.3 PROPELLER CHARTS AND EMPIRICAL METHODS

Propeller blades are typically characterized by their geometrical features, such as

- number of blades
- integrated design lift coefficient
- blade activity factor
- pitch to diameter ratio

The number of blades is self-evident while the integrated lift coefficient is given by

$$C_{L,d} = 3\int_0^1 c_L\left(\frac{r}{R}\right)^2 d\left(\frac{r}{R}\right). \tag{10.32}$$

This lift coefficient weights the outer portion of the blade more heavily than the inner portions, which contribute relatively little to the thrust. The activity factor is related to the power-absorbing capability of the propeller. We may examine the concept of the activity factor by considering static operation, where $V_0 = 0$ and the velocity relative to the blade is $\omega r$ so that the differential torque required by a blade element is determined by the drag of the blade and is given by

$$dT_q \approx c_D\frac{1}{2}\rho(\omega r)^2 crdr. \tag{10.33}$$

Assuming that the drag coefficient $c_D$ is constant along the blade (not completely accurate but sufficient here as a first approximation), we may calculate the torque required for a propeller with $B$ blades as follows:

$$T_q = 2\pi^2 n^2 B c_D \rho \int_{r_i}^{R} cr^3 dr, \tag{10.34}$$

where the quantity $n = \omega/2\pi$ is the rotational speed in revolutions per second. Similarly, the power required is given by

$$P = 2\pi n T_q = 4\pi^3 \rho c_D B n^3 \int_{r_i}^{R} cr^3 dr. \tag{10.35}$$

As mentioned previously, because the inner 15% of the propeller blade radius contributes little to the power, the power may be approximated by

$$P = (\rho n^3 d^5)\frac{\pi^3 c_D B}{4}\int_{0.15}^{1}\left(\frac{c}{d}\right)\left(\frac{r}{R}\right)^3 d\left(\frac{r}{R}\right). \tag{10.36}$$

Note that the power coefficient would then be given by

$$C_p = \frac{P}{\rho n^3 d^5} = \frac{\pi^3 c_D B}{4}\int_{0.15}^{1}\left(\frac{c}{d}\right)\left(\frac{r}{R}\right)^3 d\left(\frac{r}{R}\right). \tag{10.37}$$

The activity factor $AF$ is defined as follows:

$$AF = 10^5 \int_{0.15}^{1} \left(\frac{c}{d}\right)\left(\frac{r}{R}\right)^3 d\left(\frac{r}{R}\right). \tag{10.38}$$

The correspondence between this and the simple analysis set forth here illustrates the relationship between the activity factor and the power coefficient. Obviously a wider chord blade, particularly at the outboard portions of the blade, will have a higher activity factor and therefore a higher power coefficient. This explains the success of the "paddle" blades appearing on high-performance fighter aircraft in World War II. Although engine power increased continually through the war, it was impractical to increase propeller diameter on existing aircraft and difficult to employ additional blades due to increasing interference effects. As a result, recourse was made to wide chord, high activity factor propellers. Note that increasing the chord also acts to increase solidity, as would increasing the number of blades. For a constant chord blade, the solidity is

$$\sigma = \frac{Bc}{\pi R} = 1.28 \times 10^{-3} \frac{B}{\pi} AF. \tag{10.39}$$

The typical range for activity factor is about $100 < AF < 150$.

The pitch of the propeller is the distance the propeller would advance in the direction of flight assuming no slippage, that is, if we consider the propeller to advance as does a screw. For a blade set at a given pitch angle $\gamma$, as shown in Figure 10.7, the geometric pitch distance is given by

$$\delta = 2\pi R \tan \gamma. \tag{10.40}$$

The geometric pitch to diameter ratio is then simply

$$\frac{\delta}{d} = \pi \tan \gamma.$$

Thus a fine pitch refers to small pitch angles and a coarse pitch to large pitch angles. A fine pitch setting of the propeller refers to a low blade pitch angle, which is suitable for low-speed operations

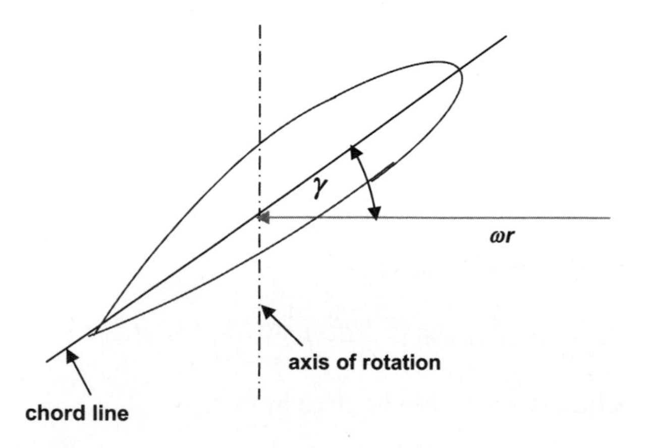


**FIGURE 10.7**

Blade section showing the pitch angle $\gamma$.

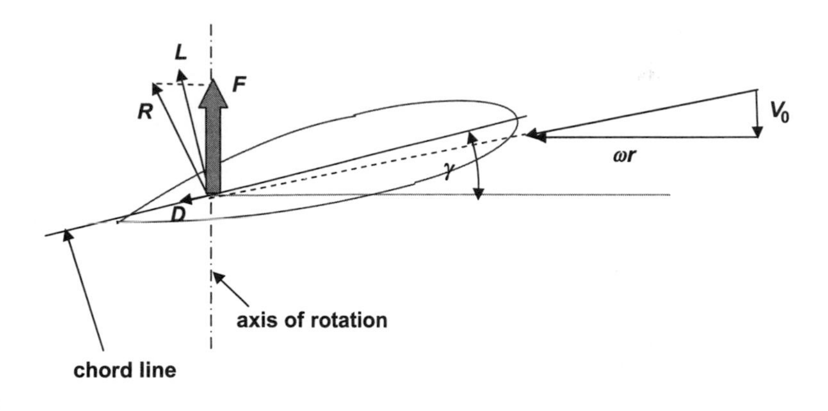


FIGURE 10.8

Fine pitch setting for high thrust at low forward speed.

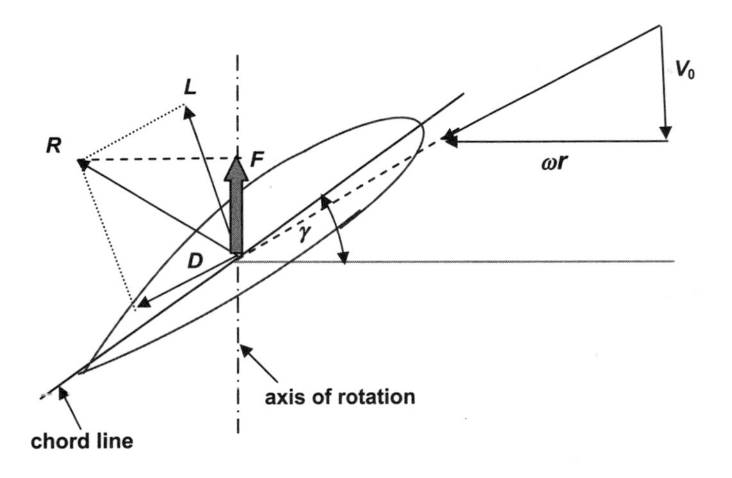


FIGURE 10.9

Coarse pitch setting for high thrust at high forward speed.

such as take-off and climb, as shown in Figure 10.8. Note that as the forward speed increases, the angle of attack of the blade would get very small and the thrust developed would drop off. Therefore, for high-speed flight, as in cruise, a coarse pitch setting would be desirable, as illustrated in Figure 10.9.

## 10.4 THE VARIABLE SPEED PROPELLER

The propeller characteristic important here is the advance ratio

$$J = \frac{V_0}{nd}. \qquad (10.41)$$

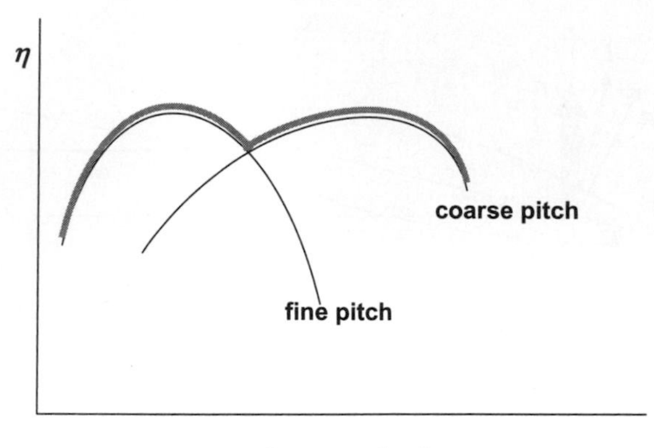


FIGURE 10.10

Propeller efficiency as a function of advance ratio for a two-speed propeller which can be set to two pitch angles.

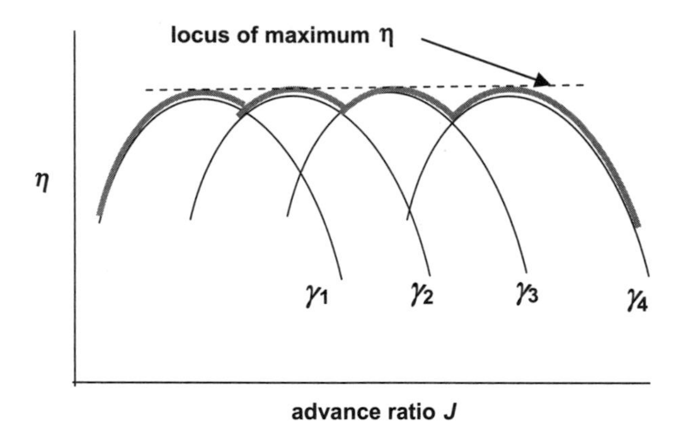


FIGURE 10.11

Propeller efficiency as a function of advance ratio for a constant speed propeller which can continuously change the pitch angle.

This quantity measures the number of diameters the propeller advances during one revolution. It provides an indication of the relative magnitude of forward flight speed compared to propeller rotational speed. Thus, low $J$ corresponds to take-off and high $J$ to cruising flight. It is advantageous to change the pitch of the propeller as required by flight conditions. The first such improvement was the two-speed propeller in which the pitch could be changed between two settings, low and high. The operation of such a propeller is shown in Figure 10.10. This type of propeller was followed by the constant speed propeller, which would change pitch during constant rotational speed of the propeller. This would permit the engine to operate at the most efficient rpm while also delivering optimum propulsive efficiency. The operation of such a propeller is illustrated in Figure 10.11.

Other nondimensional parameters used in designing and analyzing propeller performance are thrust, torque, and power coefficients, which are given, respectively, as follows:

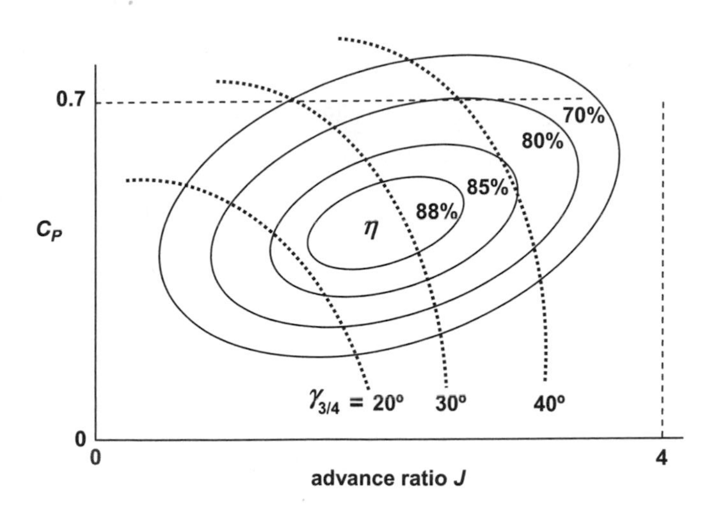


FIGURE 10.12

Typical propeller performance chart showing power coefficient as a function of advance ratio for various pitch angle at three-quarters blade radius.

$$C_T = \frac{F}{\rho n^2 d^4}$$
$$C_{Tq} = \frac{T_q}{\rho n^2 d^5}. \tag{10.42}$$
$$C_P = \frac{P}{\rho n^3 d^5}$$

The power coefficient is usually represented on a graph as a function of advance ratio with overlays of propulsive efficiency and blade pitch angle at the three-quarters radius location, $\gamma_{3R/4}$, as shown in Figure 10.12.

## 10.5 PROPELLER PERFORMANCE

### 10.5.1 Calculation of the performance of a specified propeller

The use of charts provided by the propeller manufacturer permits computation of the performance of a given propeller for various flight conditions and selected integrated design lift coefficients (Hamilton Standard, 1963). The information needed includes number of blades $B$, swept diameter $d$, desired activity factor $AF$ [Equation (10.38)], desired level(s) of the integrated design lift coefficient $C_{L,d}$ [Equation (10.32)], power available from the engine $P$, engine rpm $N$, altitude $z$, and flight speed $V_0$.

The first step is to determine the maximum integrated design lift coefficient $C_{L,d,\max}$ from Figure 10.13. Because this static performance approach is based on negligible compressibility effects, the actual design-integrated lift coefficient may not exceed the maximum value just determined from Figure 10.13. The product of the propeller speed $N$, propeller diameter $d$, and ratio of sea level standard

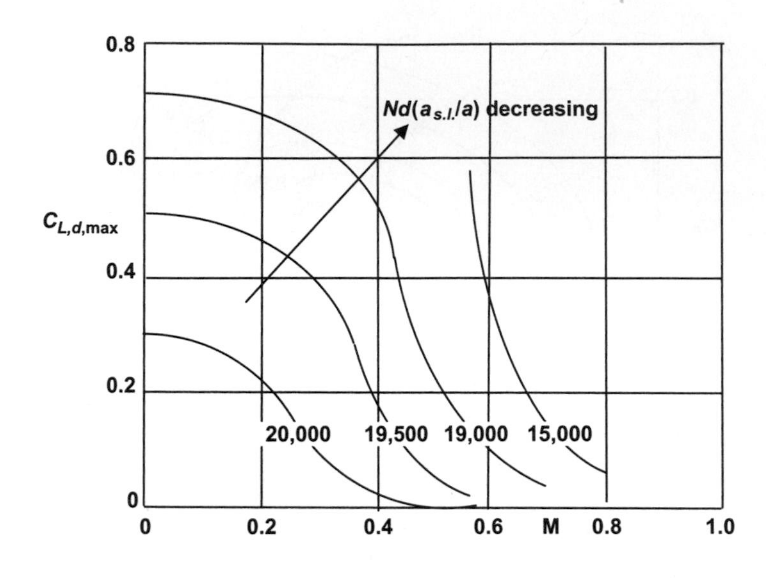


**FIGURE 10.13**

Maximum integrated design lift coefficient as a function of flight Mach number and altitude (here *d* and altitude are measured in feet and *N* in rpm).

sound speed to local sound speed $a_{s.l.}/a$ are used with the flight Mach number to find the allowable $C_{L,d,\max}$.

Using the power level available to the propeller from the engine and the rotational speed and diameter of the propeller, one may calculate the power coefficient $C_P$ from Equation (10.37). With the chosen flight speed $V_0$, one may also determine the advance ratio $J$ from Equation (10.41).Then one may enter the appropriate $C_P$-$J$ chart, like that of Figure 10.12, and obtain the propeller efficiency $\eta$. The thrust produced is simply

$$F = \eta \frac{P}{V_0}. \tag{10.43}$$

Note that if the performance under static conditions or hovering flight ($V_0 = 0$) is sought, one must enter the static thrust charts with $C_P$ and the selected integrated design lift coefficient in order to read off the ratio of thrust coefficient to power coefficient $C_T/C_P$. Then the static thrust may be developed using Equation (10.42) to obtain

$$F = \frac{C_T}{C_P} \frac{P}{nd}. \tag{10.44}$$

It must be remembered that the units used have to be consistent with those used in the curves provided by the manufacturer and that the appropriate conversions must be applied in any determination of a dimensional quantity. For example, if power is given in kW and speed in m/s, then to obtain $F$ in $N$ the right-hand side of Equation (10.43) must be multiplied by a factor of $10^3$. The same is true of Equation (10.44) as long as $N$ is in $s^{-1}$ as otherwise the units must be corrected accordingly. An example of the radial distribution of thickness to chord ratio $t/c$, blade angle $\beta$, and section lift coefficient $c_L$ of a typical propeller is given in Figure 10.14.

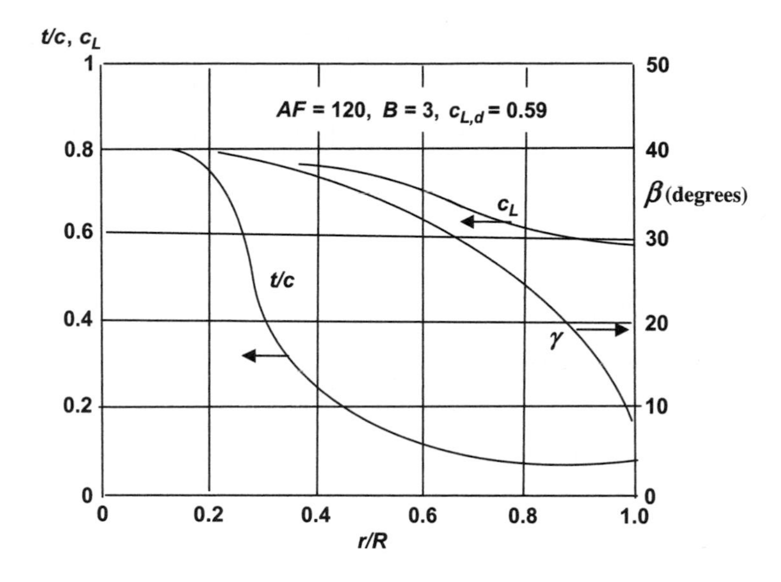


**FIGURE 10.14**

Distribution of propeller blade properties as a function of radial distance.

### 10.5.2 Selecting a propeller

One may make a reasonable estimate of the propeller required by considering the quantity

$$\frac{J}{C_P^{\frac{1}{3}}} = V_0\left(\rho\frac{d^2}{P}\right)^{\frac{1}{3}}. \tag{10.45}$$

The only parameter involving the propeller here is the diameter $d$. Analysis indicates that the best efficiency ($\eta$ around 86%) occurs for $2.4 < J/C_P^{1/3} < 3.2$, which is true for values of the parameter $C_{P,X} = C_P/X$ in the range $0.2 < C_{P,X} < 0.4$. The quantity $X$, which "adjusts" the $C_p$ value, is a function of the total activity factor $AF_{tot} = B(AF)$ where the activity factor for a single blade is given by Equation (10.38). This adjustment factor can be estimated by the following expression:

$$X = 1.49\left(\frac{AF_{tot}}{1000}\right)^{\frac{5}{4}}. \tag{10.46}$$

Choosing a value for $J/C_P^{1/3}$ from the high efficiency range mentioned previously permits one to determine the associated value of the propeller diameter $d$ from Equation (10.45) because the flight speed, altitude, and available power are known. It is obvious that the choice of $J/C_P^{1/3}$ should be somewhat on the lower side, as that leads to a smaller diameter propeller. With $d$ known for the trial propeller, one may calculate the power coefficient $C_P$ from Equation (10.37). Then the parameter $X = C_P/C_{P,X}$ may be found by selecting a value of $C_{P,X}$ from the high efficiency band mentioned previously, and the total activity factor may be estimated from Equation (10.46). Typically, for $AF_{tot} < 400$, a three-bladed propeller may be preferable, while a four-bladed propeller would be the better choice for higher $AF$. The only aspect of the trial propeller that is unknown is the integrated design lift coefficient. It is at this point, with the trial propeller characteristics settled, that one may go to the propeller performance procedures described in the previous section and refine the propeller design. Because propeller charts are presented as $C_P$ versus $J$ for specific values of $B$, $AF$, and $C_{L,d}$ and all but the last parameter is defined, the designer can plot $C_{L,d}$ versus $z$ to select the optimum propeller. The propeller should be optimized for the cruise condition because that is where the most fuel is consumed.

## 10.6 EXAMPLE: PROPELLER SELECTION

A propeller must be selected for an aircraft that will cruise at $V = 200$ kts at an altitude of 3 km. The engine provides 1175 hp (877 kW) at a shaft speed of 1105 rpm. The best propeller efficiency falls in the range $2.4 < J/C_P^{1/3} < 3.2$, and to keep the propeller diameter small, a value of $J/C_P^{1/3} = 2.6$ is assumed. Then the associated propeller diameter $d$ may be obtained by solving for it from the power coefficient equation in Equation (10.42):

$$C_P = \frac{P}{\rho n^3 d^5} = \frac{P}{\rho d^2}\left(\frac{J}{V_0}\right)^3.$$

Using this equation and using consistent units with $V_0 = 200$ kts $= 103$ m/s, $P = 8.77 \times 10^5$ Nm/s, and $\sigma = \rho/\rho_{s.l.} = 0.742$, with sea level density $\rho_{s.l.} = 1.225$ kg/m$^3$, we find

$$d = \left[\frac{1}{V_0}\left(\frac{P}{\rho_{s.l.}\sigma}\right)^{1/3}\frac{J}{C_P^{1/3}}\right]^{3/2} = \left[\frac{1}{103m/s}\left(\frac{8.77 \times 10^5 Nm/s}{0.742 \times 1.225kg/m^3}\right)^{1/3}(2.6)\right]^{3/2} = 3.94m.$$

We may then find the power coefficient from its definition as

$$C_P = \frac{P}{\rho n^3 d^5} = \frac{8.77 \times 10^5 Nm/s}{(0.742 \times 1.225kg/m^3)(115.7s^{-1})^3(3.94m)^5} = 6.56 \times 10^{-4}.$$

Propeller charts from different manufacturers may not be available in the units of choice. For example, the Hamilton Standard (1963) uses English units and normalizes the coefficients in order to deal with numbers of about order unity so that their power coefficient is defined as

$$C_{P,c} = \frac{\frac{BHP}{2000}}{\sigma\left(\frac{N}{1000}\right)^3\left(\frac{d}{10}\right)^5},$$

where $BHP$ is in shaft horsepower, $N$ is in rpm, and $d$ is in feet and the sea level density has been absorbed into the numerical coefficients. This approach has all the components being of order unity. As a result, to use the power coefficient in the charts in that reference, the result obtained with SI units must be multiplied by 246.8. In the case treated here, $C_{P,c} = 246.8(C_P) = 0.1619$. Then the advance ratio is

$$J = \frac{V_0}{nd} = \frac{103m/s}{(115.7s^{-1})(3.94m)} = 0.226.$$

Once again the manufacturer's charts use less fundamental units and set the advance ratio to

$$J_c = \frac{101.4V_{0,kts}}{Nd} = 1.417,$$

where $V_{0,kts}$ is the flight velocity in knots and $N$ is the propeller speed in rpm. The information now permits evaluation of the propeller. We can follow the analysis approach outlined in Section 10.5. The maximum design lift coefficient allowed may be estimated from Figure 10.13 (with $d$ in feet) using $Nd(a/a_{s.l.}) = (1105\text{ rpm})(12.95\text{ ft})(329/340) = 13850$ rpm-ft. As might be expected, because this value is lower than any of the curves shown, any reasonable value for the maximum design lift coefficient may be used without concern about compressibility effects. Entering the manufacturer's propeller efficiency plot, a portion of which is represented schematically in Figure 10.15, for a three-bladed propeller with $AF = 140$ and a design lift coefficient of 0.5, with the selected value $J_c = 1.417$ $C_{P,c} = 0.1619$ corresponds to an efficiency $\eta = 0.89$ and a pitch angle of 32° at the three-quarters radius.

The thrust may be found from Equation (10.43):

$$F = \frac{\eta P}{V_0} = \frac{0.89(8.77 \times 10^5 N - m/s)}{103 m/s} = 7.58 kN.$$

This process may be repeated for different propeller selections until a design considered reasonable is reached. For example, the thrust developed may be too far from that required based on the aircraft's drag or it may be desirable to attempt to reduce the blade diameter. To calculate the static thrust, one enters the manufacturer's $(C_T/C_P)_c - C_{P,c}$ with the values of $C_{P,c} = 0.1619$ and $C_{L,d,\max} = 0.5$; from

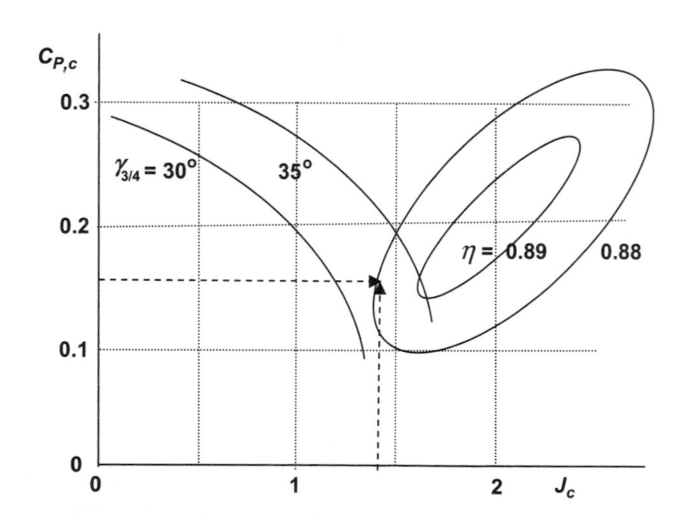


FIGURE 10.15

Propeller efficiency plot for a three-bladed propeller with $AF = 140$ and a design lift coefficient of 0.5, as presented by Hamilton Standard (1963). Arrows delineate calculated data entered to find efficiency $\eta$ and blade pitch angle at three-quarters blade radius $\gamma_{3/4}$.

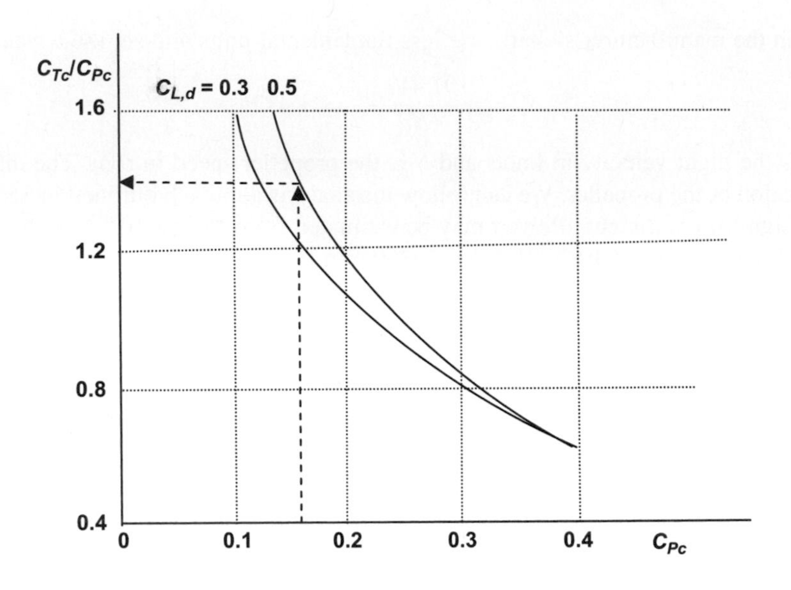


**FIGURE 10.16**

Static thrust chart for a three-bladed propeller with $AF = 140$ and various design lift coefficients, as presented by Hamilton Standard (1963). Arrows delineate calculated data entered to find $C_{Tc}/C_{Pc}$.

Figure (10.16) one would find a value of $C_{Tc}/C_{Pc} = 1.42$. This would be used in calculating the static thrust using Equation (10.44). However, continuing with the definitions in the Hamilton Standard (1963), Equation (10.44) would take the form and the static thrust would be given by

$$F = 33,000\frac{C_T}{C_P}\frac{BHP}{Nd} = 33,000(1.42)\frac{1175hp}{(1105rpm)(12.95\,ft)} = 3850lb = 17.1kN.$$

## 10.7 DUCTED PROPELLERS

Tip losses in a propeller might be reduced if a properly designed duct were placed around it so that the blade tips would just clear it. As a result, it might be possible to increase the thrust of the duct–propeller assembly over that possible with the propeller alone. With that idea in mind, one may consider a ducted propeller and an open propeller designed for the same captured free stream tube $A_0$ and the same velocity in the far wake $V_e$ as shown in Figure 10.17; keep in mind that the discs are of different diameter. We again use the concept of an actuator disc working on a steady quasi-one-dimensional incompressible inviscid flow. The jump in stagnation pressure across the actuator disc representing the rotors leads to the following pressure jump across the disc:

$$\Delta p = \frac{1}{2}\rho\left(V_e^2 - V_0^2\right). \tag{10.47}$$

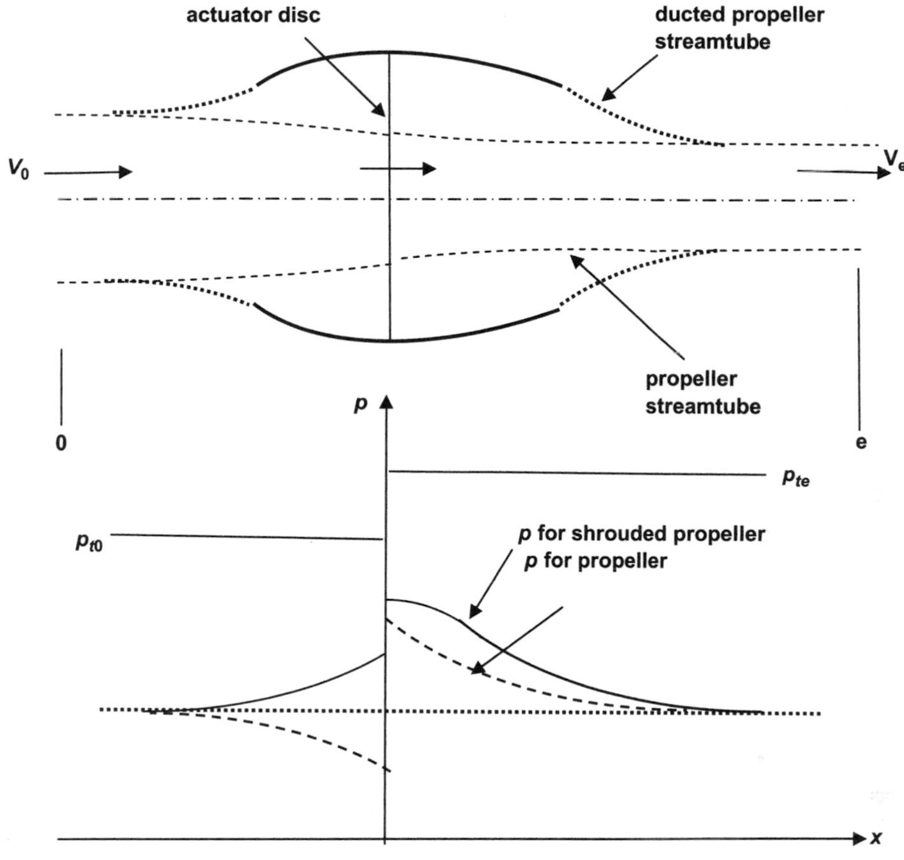


**FIGURE 10.17**

Pressure field for an open propeller and a ducted propeller processing the same streamtube.

The thrust of the open propeller is

$$F_{open} = A\Delta p. \tag{10.48}$$

The ducted propeller has an actuator disc that may be larger or smaller than that of the open propeller and we denote the thrust of its propeller alone as

$$F_{dp} = (A + \Delta A)\Delta p. \tag{10.49}$$

However, the thrust of the two *systems*, that is, the open propeller system and the propeller plus duct system, must be equal because they each process the same mass flow $m$ and from momentum conservation we have

$$F = \dot{m}(V_e - V_0). \tag{10.50}$$

Then the duct itself must somehow be generating a positive or negative thrust because the thrust of both systems are equal as given by

$$F_{duct} + (A + \Delta A)\Delta p = A\Delta p.$$

We find that the thrust contribution of the duct alone is

$$F_{duct} = -\Delta A \Delta p = -F_{open}\frac{\Delta A}{A}. \tag{10.51}$$

Combining Equations (10.47) and (10.50) and solving for the pressure rise yields

$$\Delta p = \frac{F}{A_0}\left(1 + \frac{F}{2mV_0}\right). \tag{10.52}$$

Then, using Equations (10.49) and (10.52), the ratio of the thrust of the propeller operating in the duct to the open propeller is

$$\frac{F_{dp}}{F_{open}} = \frac{\Delta p(A + \Delta A)}{F} = \frac{A + \Delta A}{A_0}\left(1 + \frac{F}{2\dot{m}V_0}\right). \tag{10.53}$$

Applying the conservation of mass from the free stream up to the actuator disc for the ducted propeller we find that

$$\rho A_0 V_0 = \rho(A + dA)V_1. \tag{10.54}$$

The quantity $V_1$ is the velocity at the disc, and we may write Equation (10.55) as

$$\frac{A + \Delta A}{A_0} = \frac{V_0}{V_1}. \tag{10.55}$$

The stagnation pressure is constant up to the disc so

$$p_0 + \frac{1}{2}\rho V_0^2 = p_1 + \frac{1}{2}\rho V_1^2. \tag{10.56}$$

We may use Equation (10.56) to define a pressure coefficient for the ducted propeller as follows:

$$C_{p,dp} = \frac{p_1 - p_0}{\frac{1}{2}\rho V_0^2} = 1 - \left(\frac{V_1}{V_0}\right)^2. \tag{10.57}$$

Using Equations (10.55) and (10.57) in Equation (10.49) yields

$$\frac{F_{dp}}{F_{open}} = \frac{1 + \sqrt{1 + C_{T,dp}}}{2\sqrt{1 - C_{p,dp}}}. \tag{10.58}$$

Here we have introduced the thrust coefficient for the ducted propeller itself

$$C_{T,dp} = \frac{F_{dp}}{\frac{1}{2}\rho A_{dp} V_0^2}. \tag{10.59}$$

The total mechanical power supplied to the fluid at the actuator disc is given by

$$P = FV_1 = \Delta p(A_{dp}V_1) = \dot{m}\frac{\Delta p}{\rho}. \tag{10.60}$$

The ideal propulsive efficiency is the ratio of useful thrust power to the total mechanical power supplied to the fluid or

$$\eta_p = \frac{FV_0}{\dot{m}\frac{\Delta p}{\rho}}. \tag{10.61}$$

Using Equations (10.52) and (10.60), we find the efficiency to be

$$\eta_p = \frac{1}{1 + \frac{F}{2\dot{m}V_0}}. \tag{10.62}$$

Note that for an open propeller the thrust is given by Equation (10.6) so that the efficiency as given by Equation (10.62) reduces to the open propeller efficiency shown in Equation (10.9).

The ratio of the ducted propeller thrust to the open propeller thrust now may be written as

$$\frac{F_{dp}}{F_{open}} = \frac{1}{\eta_p\sqrt{1 - C_{p,dp}}}. \tag{10.63}$$

Note that because the efficiency $\eta_p < 1$, the ducted propeller contribution is

$$\frac{F_{dp}}{F_{open}} \geq \frac{1}{\sqrt{1 - C_{p,dp}}}.$$

The ducted propeller pressure coefficient can be positive or negative. If $C_{p,dp} > 0$, the ducted propeller will be producing more than the total thrust of the duct–propeller combination, which means that the duct is producing drag in an amount to counteract the increased thrust of the propeller within the duct. However, if $C_{p,dp} < 1$, the ducted propeller will be producing less thrust than the open propeller, which means that the duct is producing thrust sufficient to bring the total thrust of the duct–propeller combination up to the appropriate open propeller thrust. The latter case is of most interest because one can envisage developing a given amount of thrust with a duct enclosing a propeller of diameter smaller than that required by an open propeller. Such a ducted propeller is shown in Figure 10.18.

The duct thrust produced arises from the suction force on the curved duct surface. This effect is more pronounced as the flight velocity is decreased. For this reason, ducted propellers are particularly good at low speed or in hovering situations for vertical flight.

## 10.8 TURBOPROPS

The shaft power required to drive the propeller can come from different sources: the conventional reciprocating engine, which has been the power source since the early days of flight; the electric motor, which has been receiving substantial attention in recent years; and the gas turbine engine, which we have been studying as a jet propulsion device. The advantages of the propeller in terms of low speed thrust and high propulsive, and therefore fuel, efficiency could be coupled to the smooth running, high reliability of the gas turbine while also reaping the benefit of some jet thrust. The only major technical hurdle was mating the two because the rpm of a gas turbine is

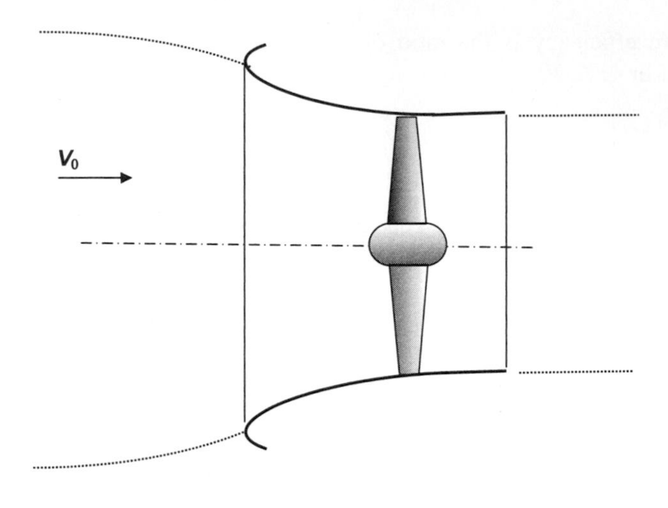


**FIGURE 10.18**

Ducted propeller with $V_1 > V_0$ and $C_{p,dp} < 0$.

about 10 times greater than that appropriate to a propeller. Thus a gear reduction unit is required between the propeller and the gas turbine. The weight and reliability of such units reduced the attractiveness somewhat, but sufficient advantages still accrued to this propeller–jet combination that their use is widespread today in applications for larger aircraft that need not fly much faster than about 300 kts.

The turboprop started out by adding the propeller and gearbox directly to a gas turbine engine and extracting the work needed for the compressor and the propeller, leaving a smaller percentage of the enthalpy rise for expansion through a nozzle for jet thrust. It became apparent that this was less effective than employing a separate shaft for the propeller assembly that would be driven by a so-called free turbine whose output power and speed could be adapted more efficiently to different shaft power requirements. The compressor–burner–turbine unit would be essentially a hot gas generator for the free turbine. A schematic diagram of a typical free turbine turboprop engine is shown in Figure 10.19.

The cycle diagram for an engine of this type is sketched in Figure 10.20. The air brought in through the inlet at station 1 is compressed from station 2 to station 3 and brought into a constant pressure combustor. Because the stagnation pressure losses are small, for clarity, stations 3 and 4 are shown on the same line of constant stagnation pressure. Fuel is added and combusted and the products exit at station 4. The compressor drive turbine extracts work sufficient to drive the compressor by expansion from station 4 to station 5. Further expansion through the free turbine from station 5 to station $5_{FT}$ extracts sufficient work to drive the propeller–gearbox combination. Then expansion continues as the flow proceeds through the nozzle from station 5 to station 6.

On the cycle diagram, the turbine provides the work necessary to drive the compressor and the free turbine provides the work to drive the propeller–gearbox combination. The division of the work sent to the propeller and that sent to the nozzle is a question of some importance. Intuitively, one would expect that as the flight speed approached the sound of speed, a greater portion of the work should be sent to the nozzle because the propeller effectiveness drops off as the flight Mach

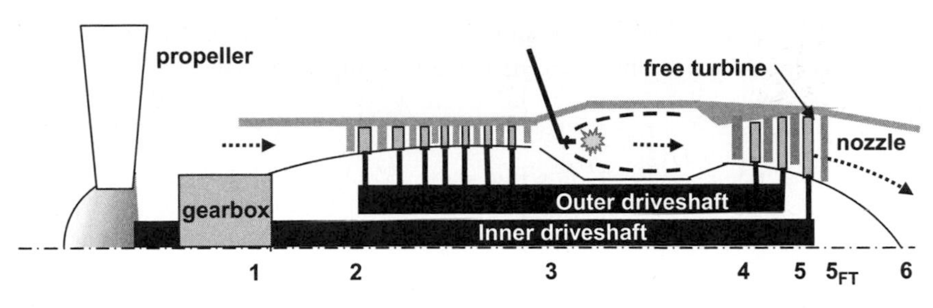


FIGURE 10.19

Turboprop engine showing free turbine driving propeller through a gearbox. Compressor–burner–turbine gas generator unit feeds free turbine and nozzle.

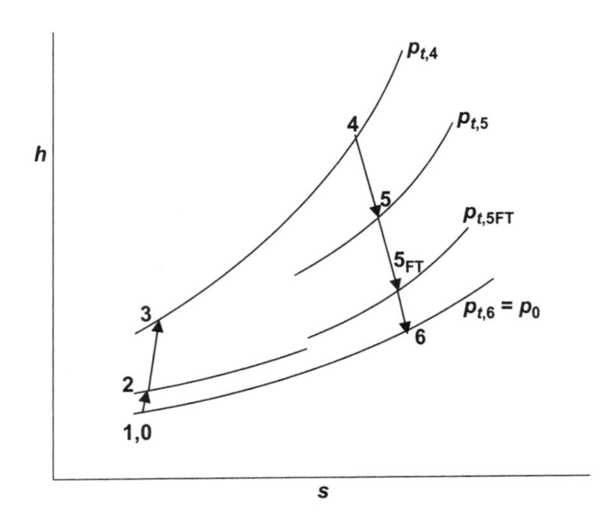


FIGURE 10.20

Cycle diagram for free turbine turboprop engine.

number increases. The tip of the propeller would reach sonic speed well before the flight Mach number does, which results in losses due to shock waves and in increased aerodynamic noise. As a consequence, the turboprop is best suited to speeds below about 350 kts or to Mach numbers of about 0.5.

The thrust power developed by the turboprop may be written as

$$TP = \eta_p P_{FT} + \dot{m} V_0 (V_e - V_0).$$

The first assumption implicit in this equation is that the fuel to air ratio (*f/a*) is about equal to the fraction of air bled from the compressor for various purposes, such as turbine blade cooling, so that the mass flow of gas passing through the engine may be considered constant. The second assumption is that the exit velocity $V_6 = V_e$, the effective exhaust velocity, because it is expected that the nozzle exit

Mach number will be subsonic. The third assumption is that the propulsive efficiency $\eta_p$ includes the propeller–gearbox combination so that $\eta_p P_{FT}$ is the power imparted to the air. Then the thrust power per unit mass is given by

$$\frac{TP}{\dot{m}} = \eta_p(h_{t,5} - h_{t,5TF}) + V_0(V_e - V_0). \tag{10.64}$$

This can be put in terms of the stagnation temperature at the entrance to the free turbine and the pressure ratio across the free turbine as follows:

$$\frac{TP}{\dot{m}} = \eta_p \eta_{ad,e} c_{p,4} T_{t,5} \left[1 - \left(\frac{p_{t,5FT}}{p_{t,5}}\right)^{\frac{\gamma_5 - 1}{\gamma_5}}\right] + V_0(V_e - V_0). \tag{10.65}$$

In this equation, we presume to know the flight conditions, the efficiencies, and the thermodynamic properties of the gas. We will consider $c_{p,2}$ and $\gamma_2$ to be constant up to station 3 (the "cold" value) and $c_{p,4}$ and $\gamma_4$ to be constant at station 4 and beyond (the "hot" value). What is not known are the following: the gas generator turbine exit stagnation temperature $T_{t,5}$, the free turbine pressure drop, and the nozzle exit velocity. We know that work done by the gas generator turbine must equal that required by the compressor. The turbine may also have to provide work for some accessories, such as pumps or generators, but for clarity and compactness that incremental work is neglected here because it is small in magnitude. Thus, equating the work of the compressor and the driving turbine leads to

$$\frac{c_{p,2} T_{t,2}}{\eta_{ad,c}} \left[\left(\frac{p_{t,3}}{p_{t,2}}\right)^{\frac{\gamma_2 - 1}{\gamma_2}} - 1\right] = c_{p,4}(T_{t,5} - T_{t,4}). \tag{10.66}$$

In this equation, we assume we have specified a compressor pressure ratio so that the only unknowns are the turbine entry temperature $T_{t,4}$ and the gas generator turbine exit temperature. The latter can be determined if we specify the former, which can be done by considering the energy balance across the combustor, which leads to

$$T_{t,4} = \frac{c_{p,2}}{c_{p,4}} T_{t3} + \frac{f}{a} \frac{\eta_b HV}{c_{p,4}}. \tag{10.67}$$

Here we need to know the compressor exit stagnation temperature, which is obtained from the compressor work equation as follows:

$$T_{t,3} = T_{t,2} \left\{1 + \frac{1}{\eta_{ad,c}} \left[\left(\frac{p_{t,3}}{p_{,2}}\right)^{\frac{\gamma_2 - 1}{\gamma_2}} - 1\right]\right\}. \tag{10.68}$$

Now knowing $T_{t,3}$ from Equation (10.68), we may find $T_{t,4}$ from Equation (10.67) and then $T_{t,5}$ from Equation (10.66). This leaves Equation (10.65) with two unknowns: free turbine pressure drop and jet exit velocity. The question would then be how should the thrust power be apportioned between free turbine shaft power and jet exhaust power so as to maximize the thrust power? A simple and

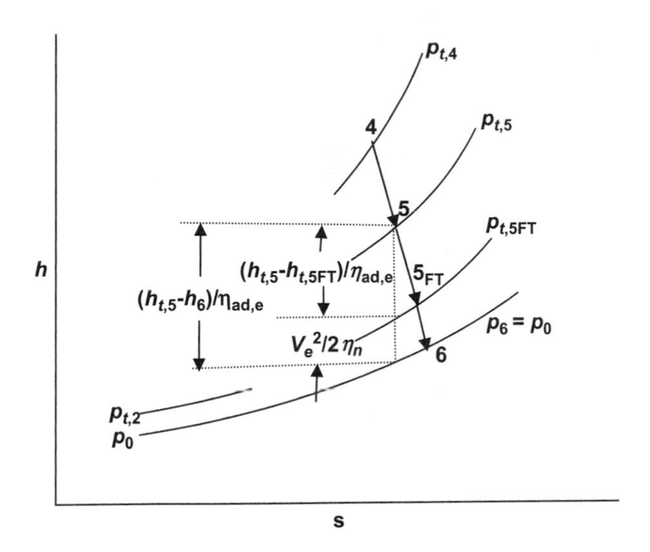


**FIGURE 10.21**

Isentropic processes across the engine from the free turbine to the nozzle exit.

reasonable approximation may be had by examining Figure 10.21, which focuses on the turbine end of the cycle diagram.

Because the stagnation enthalpy entering the free turbine $h_{t,5}$ is fixed by the other parameters, as set out earlier, we may approximate the stagnation enthalpy drop in Equation (10.64) as follows:

$$\frac{TP}{\dot{m}} \approx \eta_p \left[ (h_{t,5} - h_6) - \frac{\eta_{ad,e}}{\eta_n} \frac{1}{2} V_e^2 \right] + V_0(V_e - V_0). \tag{10.69}$$

Because only the second and third terms are variable, we may take the derivative of Equation (10.69) with respect to $V_e$ and set it to zero to find the value of the extreme point, which is

$$V_e = \frac{\eta_n}{\eta_p \eta_{ad,c}} V_0. \tag{10.70}$$

Because the second derivative of Equation (10.69) is negative, the value for $V_e$ given by Equation (10.70) is a maximum. The ratio of the nozzle expansion to the turbine expansion efficiency $\eta_n/\eta_{ad,e} \sim 1$ so that within the approximation made the optimum value of the jet exit velocity is

$$V_{e,opt} \approx \frac{V_0}{\eta_p}.$$

Thus the optimum jet velocity is about 30% to 50% greater than the flight velocity so it is clear that the jet thrust is not a major factor in a turboprop engine. More detailed analysis of the turboprop engine may be found in Mattingly (1996).

## 10.9 NOMENCLATURE

| | |
|---|---|
| $AF$ | activity factor |
| $a$ | nondimensional-induced tangential velocity component, Equation (10.23) |
| $a'$ | nondimensional-induced normal velocity component, Equation (10.23) |
| $B$ | number of blades |
| $C_p$ | propeller power coefficient |
| $C_T$ | propeller thrust coefficient |
| $C_{Tq}$ | torque coefficient |
| $c$ | blade chord length |
| $c_L$ | blade lift coefficient |
| $c_D$ | blade drag coefficient |
| $c_p$ | specific heat |
| $D$ | drag |
| $d$ | propeller diameter |
| $F$ | thrust |
| $f/a$ | fuel to air ratio |
| $g$ | acceleration of gravity |
| $HV$ | fuel heating value |
| $h$ | enthalpy |
| $J$ | advance ratio, $V_0/nD$ |
| $k$ | ratio of burning area to throat area |
| $L$ | lift |
| $M$ | Mach number |
| $\dot{m}$ | mass flow |
| $n$ | rotational speed in revolutions per second |
| $P$ | power |
| $p$ | pressure or propeller pitch |
| $Q$ | volume flux |
| $R$ | propeller radius |
| $r$ | radius |
| $S$ | surface area |
| $T$ | temperature |
| $TP$ | thrust power |
| $T_q$ | torque |
| $u$ | tangential velocity component $= \omega r$ |
| $V$ | velocity |

$V_R$ relative velocity
$z$ altitude
$\alpha_i$ angle of incidence between $V_R$ and $V_e$
$\delta$ propeller pitch distance, Equation (10.41)
$\varepsilon$ blade drag to lift ratio
$\phi$ angle between $V_e$ and $\omega r$
$\gamma$ blade pitch angle
$\lambda$ local speed ratio, Equation (10.23)
$\rho$ fluid density
$\sigma$ solidity factor, $Bc/\pi R$
$\eta$ efficiency

## 10.9.1 Subscripts

*ad,c* adiabatic compression
*ad,e* adiabatic expansion
*avg* average
*b* burner
*c* manufacturer's chart
*e* downstream conditions where $p = p_0$
*fluid* conditions in the fluid
*i* ideal
max maximum conditions
*opt* optimum
*p* propulsive
*s* static conditions
*s.l.* sea level conditions
*t* stagnation conditions
0 free stream conditions
1 conditions at actuator disc or engine inlet
2 conditions at compressor entrance
3 conditions at compressor exit and burner entrance
4 conditions at burner exit and gas generator turbine entrance
5 conditions at gas generator exit and free turbine entrance
5FT conditions at free turbine exit and nozzle entrance
6 conditions at nozzle exit

### 10.9.2 Superscripts

| | |
|---|---|
| ' | perturbation quantity |
| ^ | unit vector |
| _ | vector |

## 10.10 EXERCISES

**10.1** A pair of propellers is flying in tandem at a separation distance sufficient to eliminate mutual interference, as shown schematically in Figure E10.1. The rear propeller is of a diameter such that it just contains the slipstream of the front propeller. If the front propeller has an efficiency of 90% and both propellers provide equal thrust, calculate (a) the efficiency of the combined unit and (b) the efficiency of the rear propeller.

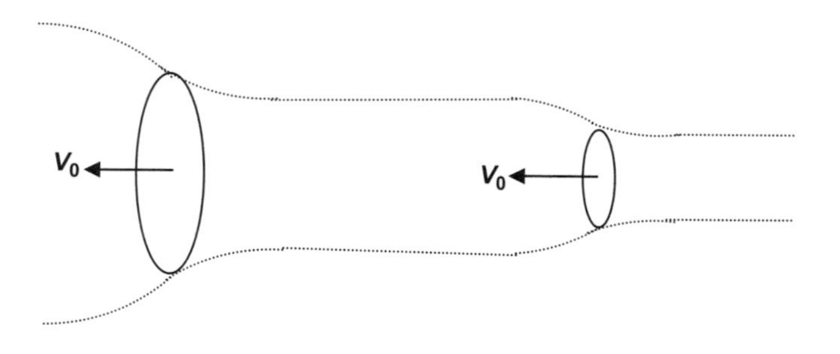


**FIGURE E10.1**

Two propellers flying at constant speed $V_0$ and therefore fixed separation.

**10.2** A three-bladed propeller with $AF = 140$, diameter of 15 ft, and design lift coefficient of 0.5 is driven by a free turbine engine at standard sea level conditions. The engine delivers 2000 hp to the propeller at a free turbine rotational speed of 15,000 rpm through a transmission with a gear reduction of 12:1. The power delivered decreases linearly to 1800 hp as the rotational speed is decreased to 10,000 rpm. Using Figure 10.16, plot the variation of the static thrust delivered by the propeller as a function of propeller rpm. For $C_{Pc} < 0.12$, you may use the relation $C_{Tc}/C_{Pc} = 0.62C_{Pc}^{-0.5}$.

**10.3** Consider the propeller chart for a three-bladed propeller shown in Figure E10.2. The speed power coefficient is $C_s = J/C_p^{1/5}$, which is a coefficient independent of the propeller diameter. A propeller must be selected for an aircraft that will cruise at $V = 200$ kts at an altitude of 3 km. The engine provides 1175 hp (877 kW) at a shaft speed of 1105 rpm. Determine the diameter, the efficiency, the blade pitch angle at three-quarters radius, and the thrust.

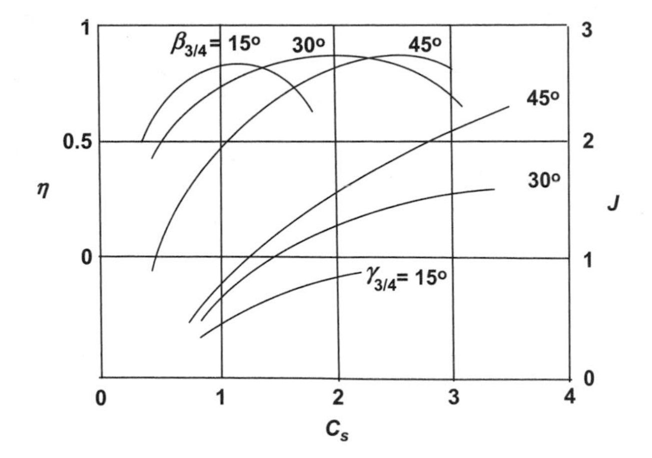


**FIGURE E10.2**

Combined chart of speed coefficient, advance ratio, and efficiency for a three-bladed propeller with various 3/4 radius pitch angles.

**10.4** A figure of merit for propeller static thrust is $\Phi = P_i/P$, the ratio of the ideal power to the actual power used in producing a given thrust $F$.

**(a)** Derive the following expression for the figure of merit:

$$\Phi = 0.798 \frac{C_T^{3/2}}{C_P}.$$

**(b)** Use the result of part (a) to develop an expression that would be useful in a preliminary estimate of rotor diameter of a helicopter of known weight and engine power.

**(c)** The Boeing V-22 Osprey has a maximum mass of 27,400 kg and is powered by two Rolls-Royce Allison engines, each providing 4590 kW of power. Estimate the required rotor diameter at sea level and compare it to the actual value. Discuss the effect of altitude on this result.

## References

Adkins, C. A., & Liebeck, R. H. (1994). Design of optimum propellers. *JournalPropulsion Power 10* (5), Sept.-Oct., 676–682.

Bosquet, J.-M., & Garadarein, P. (2000). Recent Improvements in Propeller Aerodynamic Computations. AIAA-2000-4124, 18th AIAA Applied Aerodynamics Conference, Denver, CO, August 14–17.

Hamilton Standard (1963). *Generalized Method of Propeller Performance Estimation.* Hamilton Standard Division of United Technologies. PDB 6101, Revision A, interim reissue Sept. 1974.

Mattingly, J. D. (1996). *Elements of Gas Turbine Propulsion.* New York: McGraw-Hill.

CHAPTER

# Liquid Rockets

11

## CHAPTER OUTLINE

## 11.1 LIQUID ROCKET MOTORS

The rocket is the simplest jet propulsion device in that it carries all its propellant and liberates the chemical energy in that propellant under virtually quiescent conditions in the combustion chamber. The rocket motor is basically a nozzle used to transform the heat energy in the combustion chamber gas into kinetic energy by guiding the nearly isentropic expansion of that gas through the increasing cross-sectional area of the nozzle. A schematic diagram of the conventional liquid rocket motor is shown in Figure 11.1.

Liquid fuel and oxidizer combinations have high energy density on a volumetric basis, and therefore the packaging of such propellants is efficient in terms of vehicle frontal area. The liquid propellants are brought to an injector that is designed to atomize the liquid by injection through small nozzles at high pressure while bringing them into close contact to facilitate mixing. Ignition is generally initiated by an electric arc, and the combustion reaction is then self-sustaining. The flow rate of liquid propellant brought through the feed lines is equivalent to the gas mass flow through the nozzle, and the pressure level in the combustion chamber is maintained at a high and constant level by the constraining influencer of the choked throat, where $M = 1$. The conservation of mass equation written for the combustion chamber is

$$\frac{\partial}{\partial t}\iiint \rho dv - \iint \rho \vec{V} \cdot \hat{n} dA = 0.$$

The mass in the volume of the combustion chamber will remain constant if the rate of flow into and out of the combustion chamber balances. When the fixed area throat chokes, the mass flow through it cannot increase unless there is a change in stagnation temperature or pressure. Thus, under the choked condition, the net mass flow into the chamber is zero and the mass in the chamber at any instant is fixed and given by

$$\iiint \rho dv = \rho_c v_c = \frac{p_c v_c}{R_c T_c}.$$

Inside the combustion chamber the Mach number is subsonic because the flow is accelerating to $M = 1$ at the throat. For a combustion chamber whose characteristic radius $r_c$ is considerably larger than the throat radius $r_t$, the Mach number will have a low subsonic value, except in the vicinity of the throat where $r_c$ is contracting to $r_t$. Therefore the stagnation and static pressure and temperature

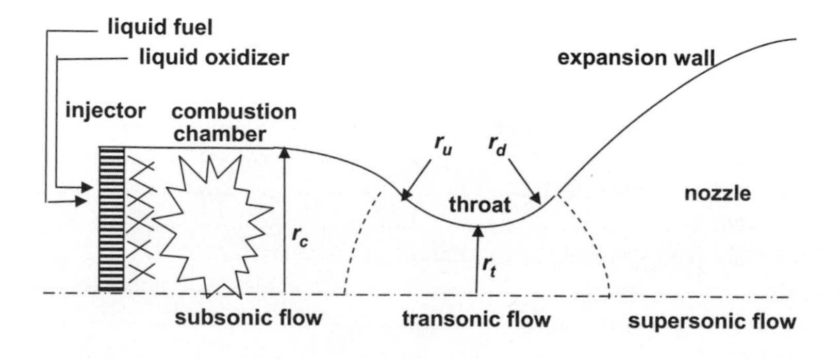


**FIGURE 11.1**

Schematic diagram of a conventional liquid rocket motor.

conditions are essentially equal and will be referred to as the combustion chamber pressure and temperature. The propellant combination basically fixes the chamber temperature, while the throat area and propellant flow rate fix the chamber pressure. The combustion chamber with its associated equipment and structure is often called the powerhead and its design is influenced by the high pressures and temperatures it must withstand. The throat area has the highest surface heating levels, which influence its design. The nozzle has relatively small loads on it because the pressure within it is dropping continually, although heating levels are still substantial. Because the nozzle is the largest component of the rocket motor, its design is constrained mainly by size and weight considerations.

### 11.1.1 Liquid rocket nozzles

The objective of the nozzle is to produce a steady flow of high-velocity gas parallel to the centerline of the nozzle in order to produce thrust along that line of action. Because all aerospace vehicle operations are constrained by weight considerations, it is desired that the nozzle be as light as possible. For given combustion chamber conditions, acceleration is determined by the rate of area change of the nozzle. Recall from the influence coefficient table in Chapter 2 that the velocity change is given by

$$\frac{dV}{V} = -\frac{1}{1-\gamma M^2}\frac{dA}{A}.$$

In addition, from Chapter 5 it is noted that the ideal mass flow through the nozzle is

$$\dot{m} = \frac{p_t A^*}{\sqrt{T_t}}\sqrt{\frac{\gamma}{R}}\left(\frac{\gamma+1}{2}\right)^{-\frac{\gamma+1}{2(\gamma-1)}}.$$

Then for a given mass flow the throat area is fixed, and for a given final velocity the exit area is fixed. Because the nozzle is basically a shell-like structure, its weight depends on the total surface area. This area $S$ of a generic nozzle may be estimated by assuming it is a frustum of a cone of length $L$ with minimum area $A^*$ and maximum area $A_e$ according to the following equation:

$$S = L\sqrt{\pi A^*}\left(1+\sqrt{\frac{A_e}{A^*}}\right) = L\sqrt{\pi A^*}(1+\sqrt{\varepsilon}),$$

where $\varepsilon = A_e/A^*$ is called the expansion ratio of the nozzle. It is clear then that as far as the nozzle weight is concerned, the only free parameter is the length of the nozzle, as typified by the complete cone length $L$. Furthermore, reducing the nozzle surface area also reduces the total heating load and skin friction drag, making the nozzle more efficient in producing the required thrust. Thus substantial effort is aimed at designing nozzles that are as short in length as possible.

### 11.1.2 Conical nozzle

Surely the simplest nozzle is a conical nozzle of the sort described earlier and shown in Figure 11.2. The expansion angle of the cone is set by the difference between the exit radius and the minimum radius of the cone, which is typically near the throat, and is defined as

$$\alpha = \arctan\left(\frac{r_e - r_{min}}{L}\right).$$

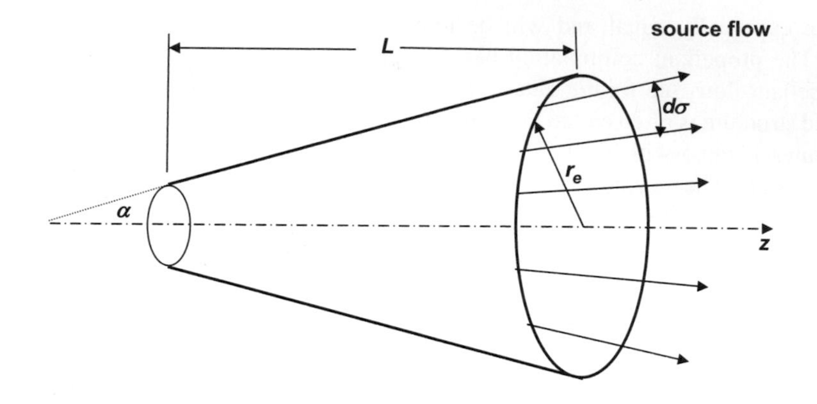


**FIGURE 11.2**

Conical nozzle of length $L$ and expansion angle $\alpha$.

The maximum thrust will occur when the exit pressure matches the ambient pressure and we may simply deal with the exit momentum. The flow in a cone may be considered to be described as a spherical source flow with the streamlines following radial lines emanating from the (virtual) tip of the cone. The elemental mass flow is given by

$$d\dot{m} = \rho_e V_e \left( 2\pi \frac{L^2}{\cos^2 \alpha} \sin \sigma d\sigma \right),$$

where the exit density and velocity are constant, $\alpha$ is the expansion angle of the cone, and $\sigma$ is the angle between the axis and any streamline. The thrust is measured parallel to the axis of the cone so that the elemental thrust is

$$dF = (V_e \cos \sigma) d\dot{m} = 2\pi \rho_e V_e^2 \frac{L^2}{\cos^2 \alpha} (\sin \sigma \cos \sigma d\sigma).$$

In order to calculate the total thrust, an integration from the axis ($\sigma = 0$) to the nozzle surface ($\sigma = \alpha$) must be performed, which leads to the result that

$$F = \pi \rho_e V_e^2 \frac{L^2}{\cos^2 \alpha} \sin^2 \alpha.$$

Note that the elemental mass flow may be integrated over the same limits to yield

$$\dot{m} = 2\pi \rho_e V_e \frac{L^2}{\cos^2 \alpha} (1 - \cos \alpha).$$

Then the thrust may be written as

$$F = \lambda \dot{m} V_e.$$

In this equation,

$$\lambda = \frac{1}{2}\frac{\sin^2 \alpha}{(1 - \cos \alpha)} = \frac{1}{2}\frac{(1 - \cos^2 \alpha)}{1 - \cos \alpha} = \frac{1 + \cos \alpha}{2}.$$

Thus $\lambda$ describes the degree of thrust loss due to angularity of the exhaust flow. For example, an expansion angle of 10° reduces thrust by 1% and an expansion angle of 20° reduces thrust by 3%; an expansion angle of 15° is common. Although this sounds like a minor loss, it must be remembered that many large rocket engines produce several hundred thousands of pounds of thrust, so actual thrust losses are thousands of pounds, which affects the launch weight capability, thereby requiring reductions in payload.

### 11.1.3 Bell nozzle

Because the nozzle length and the flow angularity are important factors in performance, it is worth assessing how well these constraints may be accommodated. To achieve short nozzle length and parallel flow at the exit for a given area ratio means that the nozzle should have a reflexive shape, as shown in Figure 11.3. This contoured nozzle is often called a bell nozzle in the rocket literature. It is characterized by a rapid expansion of area just beyond the throat, followed by a more gradual expansion thereafter and culminating in a nearly cylindrical exit section providing parallel exit flow, that is, $\alpha = 0$.

The design of smooth nozzles of this sort can be carried out by the method of characteristics, as described by Liepmann and Roshko (1957). Typically, an area ratio is chosen that will produce the desired exit conditions of Mach number and pressure, then a centerline Mach number or pressure distribution is specified, and finally the method of characteristics is used to generate the wall shape. It is generally the case that emphasis on a smoothly varying shape leads to nozzles that are much too long for practical applications. The approach of Rao (1958) for maximizing thrust for a specified exit area and length has been used widely and results in nozzles that are about 75% as long as conical

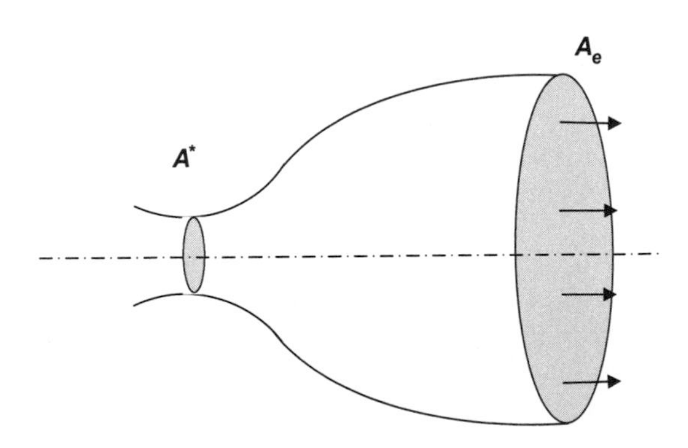


**FIGURE 11.3**

Contoured or "bell" nozzle.

nozzles. Efforts to optimize the design of these contoured nozzles concentrated on having rapid expansion in the throat region, which reduces the overall length of the nozzle. The shortest nozzle would have a sharp corner at the throat, which produces a large fraction of the total expansion required. The remainder of the expansion would take place along the contoured wall. However, a sharp throat leads to several undesirable effects, such as the possibility of separation of the wall boundary layer at the sharp corner and difficulty in providing adequate cooling to the sharp corner. The radius of curvature of the wall upstream of the throat shown in Figure 11.1 is often in the range of $0.6 < r_u/r_t < 1.5$, and the larger values can be advantageous if film cooling of the throat is desired. Reducing the radius of curvature of the throat not only reduces overall nozzle length, but also reduces the surface area exposed to high heat loads. The nature of the actual high temperature flow in the transonic region around the throat can depart significantly from the simple one-dimensional approximation so that more detailed studies are needed to finalize designs. In particular, attention to the possibility of shock waves forming in the nozzle is important.

### 11.1.4 Plug nozzle

The interest in reduced nozzle length has led to a variety of different designs. One of the most promising of these is the so-called plug nozzle, which is shown in Figure 11.4. The formation of the

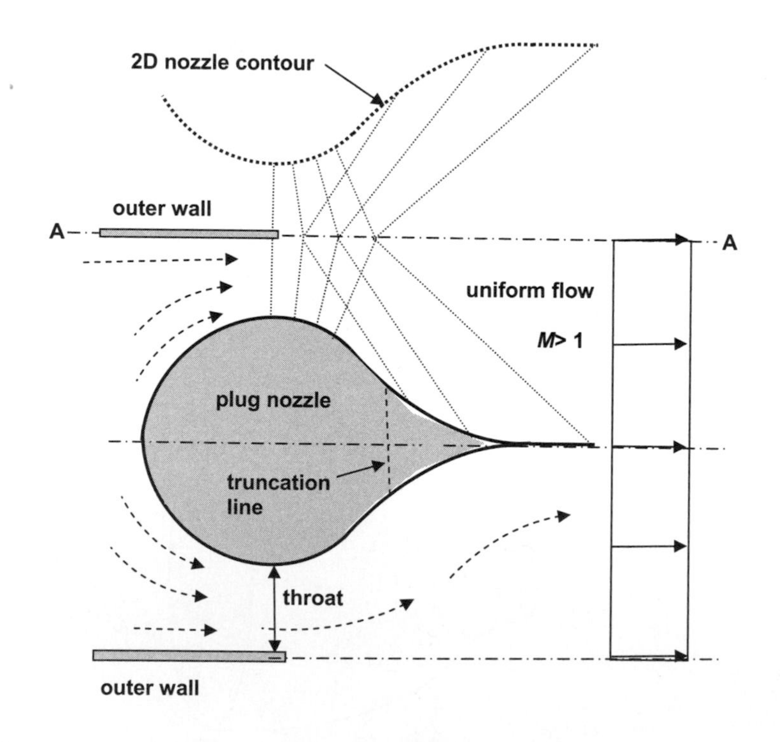


**FIGURE 11.4**

Axisymmetric plug nozzle formed by rotating a two-dimensional (2D) nozzle contour about the axis A-A. Throat size is exaggerated. Truncation reduces nozzle length.

plug nozzle contour may be visualized by considering first a two-dimensional nozzle that produces the required uniform exit Mach number. If the contour of one of the walls is simply rotated about an axis parallel to its centerline, an axisymmetric turnip-shaped body is formed. The addition of an outer cylindrical wall completes the nozzle configuration. Along the cylinder defined by the line A-A the flow is a straight streamline and uniform flow exists in the region marked in Figure 11.4. The operation of the plug nozzle is not affected much by truncating the plug some distance upstream of its tip, which permits the nozzle to be substantially shorter, by about one-third, than a bell nozzle.

The design of plug nozzles is well established, and a widely used approach is that presented by Angelino (1964). Plug nozzles have been proven in experiment but haven't yet seen common use in practical flight vehicles. Their major advantage is their reduced sensitivity to overexpansion effects, which in conventional bell nozzles can lead to flow separation in the nozzle with subsequent loss of performance. Rockets rise through the atmosphere, and because the pressure in the atmosphere drops off exponentially, nozzles will always face overexpansion at some altitude. However, plug nozzles have limitations because of the necessarily small dimensions of their annular throats, which lead to structural and thermal difficulties. Similarly, the plug surface is in contact with the hot gases and requires suitable thermal protection. A useful summary of flow and performance features of plug nozzles is given by Onofri (2002).

The aerospike nozzle is a modification of the plug nozzle, which has been tested both in an axisymmetric version and in a two-dimensional, or linear, version. Basically the plug in Figure 11.4 is truncated at the line shown, and a secondary flow is exhausted through the base to reduce the drag. This type of nozzle may be considered to be an essentially two-dimensional device as shown rather than an axisymmetric one. Such a two-dimensional version was considered for testing in flight aboard an SR-71 for possible application on an aerospace plane, such as the X-33, as described by Moes and colleagues (1996).

### 11.1.5 Extendable nozzle

The desire to keep nozzles as short as possible, yet still be able to accommodate the additional expansion necessary as altitude increases, led to the concept of an extendable, or two-position, nozzle. A schematic diagram of a two-position nozzle is shown in Figure 11.5. At low altitudes the nozzle skirt

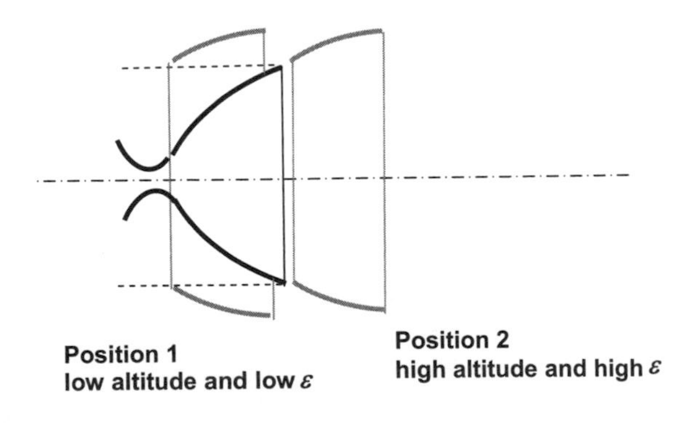


**FIGURE 11.5**

Schematic diagram of a two-position bell nozzle.

is drawn up around the powerhead and a low expansion ratio nozzle is employed. At high altitude the nozzle skirt is extended, raising the expansion ratio of the nozzle substantially to improve performance. Although performance enhancements are obtained with such an approach, it is rarely employed because of the added weight and complexity associated with such a system; see, for example, Martin (1989).

### 11.1.6 Discharge coefficient

Mass flow through the rocket engine is a quantity that can be measured quite accurately from flowmeters on the fuel and propellant lines. Likewise, chamber pressure $p_e$ and throat area may be measured, and we may introduce a mass flow coefficient

$$c_m = \frac{\dot{m}}{p_c A_t}. \tag{11.1}$$

We normally calculate the mass flow on the basis of the simplifying assumptions described previously, which results in the usual equation for the ideal nozzle mass flow

$$\dot{m}_i = \frac{p_c A_t}{\sqrt{T_c}} \sqrt{\frac{W_c}{R_u}} \Gamma. \tag{11.2}$$

In this equation, $A_t$ is the throat area, which is assumed to be choked, and

$$\Gamma = \sqrt{\gamma} \left( \frac{2}{\gamma + 1} \right)^{\frac{\gamma+1}{2(\gamma-1)}}. \tag{11.3}$$

Recall from Chapter 5 that we may approximate $\Gamma$ by the following linear relation

$$\Gamma \approx 0.192\gamma + 0.417.$$

Then we may define a discharge coefficient

$$c_d = \frac{\dot{m}}{\dot{m}_i}. \tag{11.4}$$

It is usual to find the discharge coefficient in the range $0.97 < c_d < 1.15$. The actual mass flow may be larger than that calculated for the idealized case due to friction, heat transfer, three-dimensionality, and so on, all of which tend to lead to higher gas density than in the idealized case.

### 11.1.7 Nozzle coefficient

In the same fashion, we define a nozzle coefficient

$$c_V = \frac{V_e}{V_{e,i}} \tag{11.5}$$

because it is difficult to develop an accurate method for predicting the exit velocity. We address this problem by considering the flow through the real nozzle and the flow through the idealized frictionless isentropic nozzle on an *h-s* plane as shown in Figure 11.6.

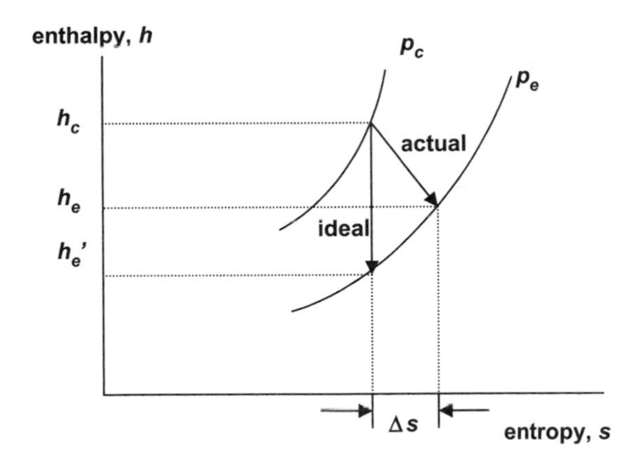


**FIGURE 11.6**

Nozzle expansion process on an *h-s* diagram.

### 11.1.8 Nozzle efficiency

The nozzle efficiency is defined as follows:

$$\eta_n = \frac{h_c - h_e}{h_c - h'_e}. \tag{11.6}$$

This is the usual definition for the efficiency of expansion. However, from the adiabatic energy equation we see that

$$\eta_n = \frac{V_e^2}{V_{e,i}^2} = c_V^2. \tag{11.7}$$

So $c_V = \sqrt{\eta_n}$ and is generally around 0.95–0.96 in a well-designed rocket nozzle. Now the equation for rocket thrust derived previously is

$$F = \lambda \frac{w}{g} V_e + A_e(p_e - p_0). \tag{11.8}$$

### 11.1.9 Nozzle thrust coefficient

In Equation (11.8) the quantity $\lambda = (1 + \cos\alpha)/2$ represents the effect of the conical-like flow discharging from the nozzle. This behavior occurs because it is generally impractical to use a fully contoured nozzle with purely axial outflow at the exit station. Instead, the nozzle has a wall angle $\alpha$ at the exit station that permits the flow to expand away from the axis. Smaller nozzles, such as those used for attitude control, are often fashioned in conical form. Naturally, angularity of this sort results in some loss of thrust, which is accounted for by the correction factor $\lambda$. Introducing our new coefficients, we may put the equation for the thrust in the form

$$F = \lambda c_d c_V \Gamma \left\{ \frac{2\gamma}{\gamma - 1} \left[ 1 - \left( \frac{p_e}{p_c} \right)^{\frac{\gamma-1}{\gamma}} \right] \right\}^{1/2} p_c A_t + A_e (p_e - p_o). \quad (11.9)$$

This equation may be normalized by the product $p_c A_t$ to yield a nondimensional thrust coefficient:

$$c_F = \frac{F}{p_c A_t} = \lambda c_d c_V \Gamma \left\{ \frac{2\gamma}{\gamma - 1} \left[ 1 - \left( \frac{p_e}{p_c} \right)^{\frac{\gamma-1}{\gamma}} \right] \right\}^{1/2} + \frac{A_e}{A_t} \left( \frac{p_e}{p_c} - \frac{p_0}{p_c} \right). \quad (11.10)$$

If the nozzle exhausts into a vacuum, $p_o = 0$, as it would in space, the thrust coefficient $c_F$ is called the vacuum thrust coefficient and then we may write

$$c_F = c_{F,vac} - \varepsilon \frac{p_0}{p_c}. \quad (11.11)$$

The quantity $\varepsilon = A_e/A_t$ is the nozzle expansion ratio. In Figure 11.7, the vacuum thrust coefficient appears as a function of $p_c/p_e$ for various values of the specific heat ratio $\gamma$.

We may use Equation (11.2) to express the product of the chamber pressure and the choked throat area in terms of the product of the ideal mass flow and a velocity parameter c* as follows:

$$F = c_F p_c A_t = c_F \frac{\dot{m}_i \sqrt{R_u T_c}}{\Gamma \sqrt{W_c}} = c_F \dot{m}_i c^* = c_F \frac{\dot{m}}{c_m} c^*. \quad (11.12)$$

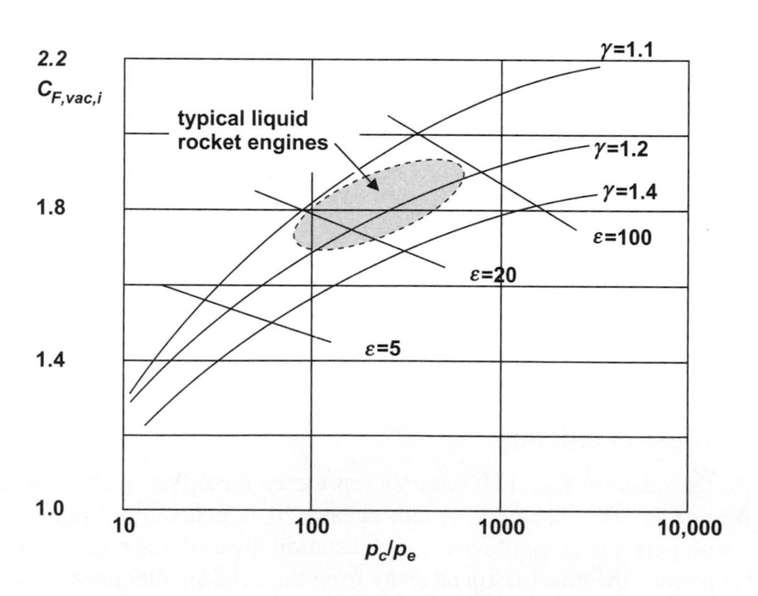


**FIGURE 11.7**

Ideal vacuum thrust coefficient as a function of nozzle pressure ratio for various values of the nozzle expansion ratio and $\gamma$.

**Table 11.1** Performance Characteristics of Various Fuel–Oxidizer Combinations

| Oxidizer | Fuel | O/F | $T_c$ (R) | $\gamma$ | c* (fps) | $I_{sp}$ (Equilibrium) | $I_{sp}$ (Frozen) |
|---|---|---|---|---|---|---|---|
| LOX | $LH_2$ | 4.0 | 5375 | 1.26 | 7985 | 391 | 388 |
| LOX | RP-1 | 2.45 | 6605 | 1.24 | 5915 | 301 | 286 |
| $LF_2$ | $LH_2$ | 8.0 | 7130 | 1.33 | 8355 | 410 | 398 |
| $LF_2$ | RP-1 | 2.8 | 7940 | 1.22 | 6390 | 326 | 304 |

*These values are for $p_c = 1000$ psi and $p_o = 14.7$ psi.*

Here we have defined the characteristic velocity

$$c^* = \frac{\sqrt{\gamma_c (R_u/W_c) T_c}}{\sqrt{\gamma}\Gamma} = \frac{a_c}{\sqrt{\gamma}\Gamma}. \tag{11.13}$$

The quantity $a_c$ is the sound speed in the combustion chamber:

$$a_c = \sqrt{\gamma_c R_c T_c}. \tag{11.14}$$

Now the thrust is made up of two factors multiplying the mass flow rate: $c^*$, which is a measure of propellant performance, and $c_F$, which is a measure of nozzle performance. Some of these characteristics are listed in Table 11.1.

The effect of pressure ratio and of the ratio of specific heats on the ratio of exit velocity to combustion chamber sound speed $V_e/a_c$ is shown in Figure 11.8. Note the substantial improvement as $\gamma$ decreases.

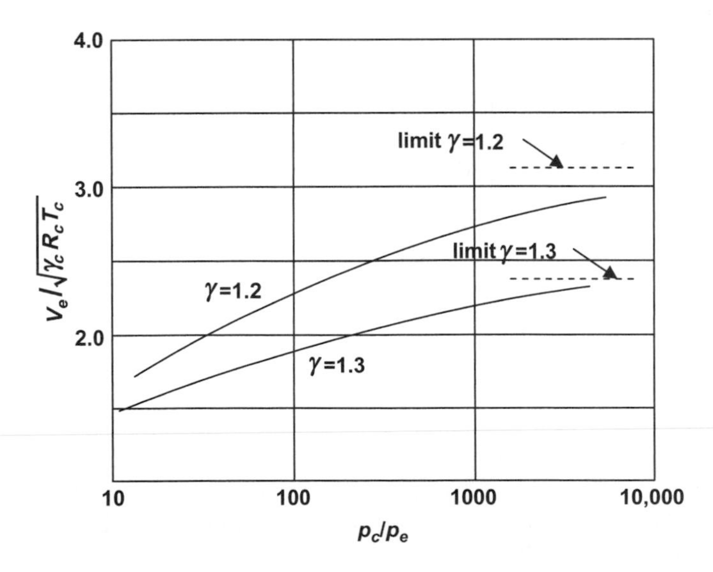


**FIGURE 11.8**

Variation of the ratio of exit velocity to combustion chamber sound speed $V_e/a_c$ as a function of pressure ratio and of the ratio of specific heats.

## 11.2 SPECIFIC IMPULSE

This is a measure of the efficiency with which the rocket engine converts *on-board* propellant into thrust. Because rocket engines carry both fuel *and* oxidizer, the combination of the weight flow rate of these two propellants must be used in the definition given by

$$I_{sp} = \frac{F}{\dot{m}_e g} = \frac{V_e}{g} + \frac{A_e}{\dot{m}_e g}(p_e - p_o).$$

Because $\dot{m}_e = \rho_e V_e A_e$, we get

$$I_{sp} = \frac{V_e}{g}\left[1 + \frac{p_e}{\rho_e V_e^2}\left(1 - \frac{p_e}{p_o}\right)\right].$$

Using the relation $\rho_e V_e^2 = \gamma_e p_e M_e^2$, the final result is

$$I_{sp} = \frac{V_e}{g}\left[1 + \left(1 - \frac{p_o}{p_e}\right)\frac{1}{\gamma_e M_e^2}\right]. \tag{11.15}$$

For an isentropic nozzle with constant $\gamma$ , the area ratio

$$\varepsilon = \frac{A_e}{A_t} = \frac{1}{M_e}\left[\frac{2}{\gamma+1}\left(1 + \frac{\gamma-1}{2}M_e^2\right)\right]^{\frac{\gamma+1}{2(\gamma-1)}}. \tag{11.16}$$

For $M_e^2 \gg 1$ (say, 5 or greater) we find the following result:

$$M_e^2 \approx \left(\frac{\gamma+1}{\gamma-1}\right)^{\frac{\gamma+1}{2}} \varepsilon^{\gamma-1}.$$

The relationship between the exit Mach number and the nozzle expansion ratio is shown in Figure 11.9 (note that the condition $M_e^2 \gg 1$ depends on $\gamma$ and to satisfy this means $M_e > 5$). We may also express the area ratio in another fashion, noting that

$$\frac{A_e}{A_t} = \frac{\dot{m}}{\rho_e V_e}\frac{p_c \Gamma}{\dot{m}\sqrt{RT_c}} = \frac{\rho_c}{\rho_e}\sqrt{RT_c}\frac{\Gamma}{V_e}.$$

Assuming isentropic flow and recalling Equations (11.12) through (11.13), we find that

$$\frac{A_e}{A_t} = \left(\frac{p_c}{p_e}\right)^{\frac{1}{\gamma}}\frac{\Gamma^2}{c_{F,i}}. \tag{11.17}$$

Here the ideal value $c_{F,i}$ is that value of $c_F$ in Equation (11.10) with matched exit pressure and all the nozzle coefficients equal to unity. In addition, because $V_e$ depends only on chamber conditions and $\varepsilon$, the specific impulse for a given engine will vary only with altitude, that is, $p_0/p_e$. Then the vacuum specific impulse is given by the following:

$$I_{sp,vac} = \frac{V_e}{g}\left(1 + \frac{1}{\gamma M_e^2}\right). \tag{11.18}$$

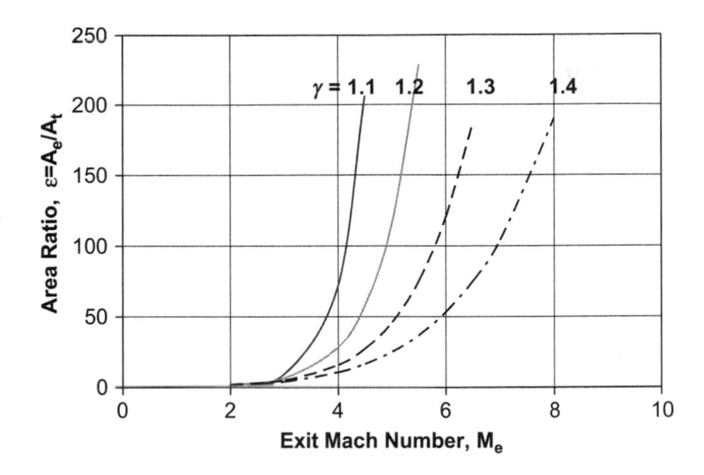


**FIGURE 11.9**

Nozzle area ratio as a function of exit Mach number for various values of the ratio of specific heats $\gamma$.

Then the ratio of actual to vacuum specific impulse is given by

$$\frac{I_{sp}}{I_{sp,vac}} = 1 - \frac{1}{1+\gamma M_e^2}\frac{p_o}{p_e} \approx 1 - \frac{1}{\gamma M_e^2}\frac{p_o}{p_e} \leq 1. \tag{11.19}$$

Therefore, the vacuum-specific impulse is always the largest specific impulse for a given engine and is the one typically quoted in the manufacturer's literature. Of course, it is a definite, unambiguous parameter that can be used to determine $I_{sp}$ at any other ambient pressure condition. If we once again assume that the exit Mach number is high, such that $M_e^2 \gg 1$, we obtain the following approximation:

$$\frac{I_{sp}}{I_{sp,vac}} \approx 1 - \varepsilon\Gamma' \frac{p_{o,s.l.}}{p_c}\frac{p_o}{p_{o,s.l.}}. \tag{11.20}$$

A good fit for the pressure variation with altitude in Earth's atmosphere is given by the exponential fit

$$\frac{p_o}{p_{o,s.l.}} = \delta = \exp(-z/z_s). \tag{11.21}$$

The usual constants employed are $p_{o,s.l.} = 95.7$ kPa and $z_s = 7300$ m, while

$$\Gamma' = \frac{1}{\gamma}\left(\frac{\gamma-1}{2}\right)^{\frac{\gamma}{\gamma-1}}\left(\frac{\gamma+1}{\gamma-1}\right)^{\frac{\gamma+1}{2(\gamma-1)}}. \tag{11.22}$$

Then, the ratio

$$\frac{I_{sp}}{I_{sp,vac}} = 1 - \frac{\varepsilon\Gamma'}{p_c}\exp\left(-\frac{z}{z_s}\right). \tag{11.23}$$

The chamber pressure here must be measured in atmospheres. The expression just given shows that for a given chamber pressure, the ratio $I_{sp}/I_{sp,vac}$ drops off with altitude more rapidly as the nozzle expansion ratio increases. However, the actual value of the specific impulse $I_{sp}$ is a function of altitude,

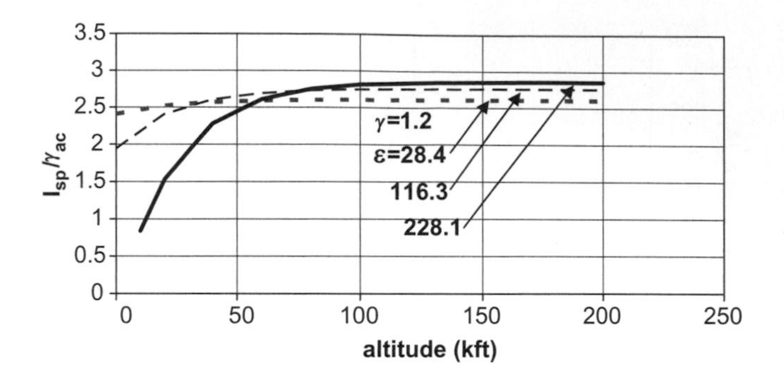


**FIGURE 11.10**

Variation of specific impulse $I_{sp}$ with altitude for fixed $\gamma = 1.2$ and chamber pressure $p_c = 200$ atm for different expansion ratios, $\varepsilon$. The specific impulse is normalized to the product of $\gamma$ and chamber sound speed $a_c$.

and this effect is not readily apparent from the equations. Therefore, a plot of these variations is shown in Figure 11.10.

It is clear that low expansion ratio nozzles are better suited to low altitudes and high expansion ratio nozzles to higher altitudes. Thus it would be desirable to have a variable expansion ratio nozzle. Consider a hypothetical rocket motor with chamber temperature $T_c = 5000$R (2778K), chamber pressure $p_c = 1000$ psia (68 atm), and $\gamma = 1.4$ (constant). Consider two contoured nozzles with different expansion ratios being used with the same combustion chamber:

- Nozzle 1, $e = 10$, $M_e = 3.92$, $p_e = 7.33$ psia, $T_e = 1227$R: $I_{sp,1} = 209.25[1 + 0.0465\,(1 - p_0/p_e)]$
- Nozzle 2, $e = 25$, $M_e = 5.00$, $p_e = 1.89$ psia, $T_e = 833$R: $I_{sp,2} = 219.91[1 + 0.0286\,(1 - p_0/p_e)]$

The performance of the two nozzles is shown in Figure 11.11, and it is seen that the lower expansion ratio outperforms the high expansion ratio nozzle for altitudes less than about 40,000 ft, while the reverse is true above that altitude. A schematic diagram of a two-position nozzle is shown in Figure 11.7. Although there are performance enhancements with such an approach, it is rarely employed because of the added weight and complexity associated with such a system; see, for example, Martin (1989).

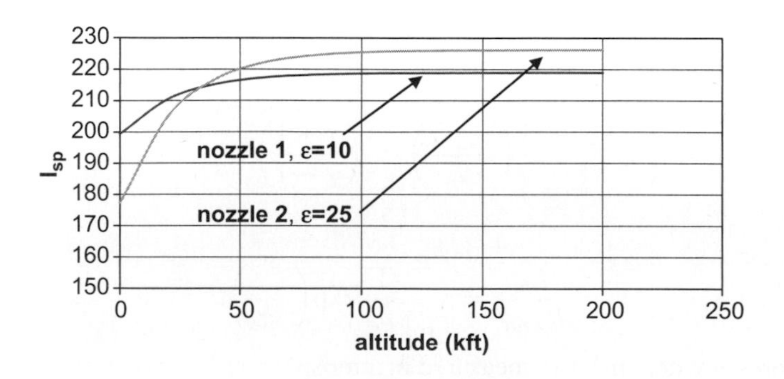


**FIGURE 11.11**

Performance of two nozzles with different expansion ratios as a function of altitude.

## 11.3 EXAMPLE: ROCKET PERFORMANCE

Typical measurements during rocket motor test firings are mass flow, thrust, and chamber pressure. In addition, the chemistry may be calculated beforehand from knowledge of the propellants and the mixture ratio used. As an example, consider an LH2-LOX rocket motor that operates at a pressure $p_c = 2.34$ MPa. Using a mixture ratio O/F = 5.33 results in the following chamber quantities: $T_c = 3270$K, $W_c = 11.7$, and $\gamma_c = 1.25$. The measured mass flow rate $m = 29$ kg/s, and the nozzle used has $c_m = 0.95$, $\eta_n = 0.9$, and an exit pressure $p_e = 96.5$ kPa. The nozzle has a wall divergence angle $\alpha = 10°$ and exhausts into an ambient whose pressure is $p_0 = 68.9$ kPa. Determine the thrust, the thrust coefficient, the exit Mach number, the effective exhaust velocity, and the specific impulse of the rocket motor under the given conditions.

First the chamber conditions may be determined. The gas constant is determined as follows:

$$R = R_u/W_c = (8.3145 \text{ kJ/kmol-K})/(11.7 \text{ kg/kmol}) = 0.711 \text{ kJ/kg} = 711 \text{ m}^2/\text{s}^2 - \text{K}.$$

The chamber sound speed is given by

$$a_c = \sqrt{\gamma_c R_c T_c} = \sqrt{1.25(711 m^2/s^2 - K)3270K} = 1705 m/s.$$

The characteristic velocity may be computed using Equation (11.13):

$$c^* = \frac{a_c}{\sqrt{\gamma}\Gamma} = \frac{a_c}{\gamma}\left(\frac{\gamma+1}{2}\right)^{\frac{\gamma+1}{2(\gamma-1)}} = \frac{a_c}{1.25}(1.125)^{4.5} = 2317 m/s.$$

The nozzle throat area may be found from Equation (11.12), noting that the measured mass flow is not equal to the ideal mass flow because the mass flow coefficient $c_m$ is not identically one. Keeping this in mind, the throat area is given by

$$A_t = \frac{\dot{m}c^*}{c_m p_c} = \frac{(29kg/s)(2317m/s)}{(0.95)2.34 \times 10^6 N/m^2} = 0.030m^2.$$

Because the nozzle is not matched, we will need to know the exit area to determine the contribution to the thrust made by the pressure imbalance at the exit plane. We first gather the following pertinent information: $p_e/p_c = 96.5/2340 = 0.0412$, $\lambda = 0.5(1 + \cos\alpha) = 0.992$, $c_V = \eta_n^{1/2} = 0.949$, $c_d = 0.95$, $\Gamma = 0.658$, and $\gamma = 1.25$.

The exit area may be found from the conservation of mass equation if we know $T_e$ and $V_e$. The exit velocity may be found from Equation (1.58), where we use $c_p = \gamma R/(\gamma - 1) = 3.555$ kJ/kg-K = 3555 m$^2$/s$^2$-K, and we note that the velocity given there is an ideal value and we must correct for angularity of the flow and the nozzle efficiency. Thus the actual velocity is $c_v \lambda V_{ideal}$, and the expression for exit velocity Equation (1.58) becomes

$$V_e = c_V \lambda \sqrt{2c_{p,c}T_c\left[1 - \left(\frac{p_e}{p_c}\right)^{\frac{\gamma-1}{\gamma}}\right]} = (0.949)(0.992)\sqrt{2(3555)(3270)(1 - 0.0412^{\cdot 2})^{0.2}}$$

$$V_e = 3117 m/s.$$

Nozzle efficiency defined by Equation (11.6) takes into account any heat transfer that has occurred so that we may write

$$\frac{T_e}{T_c} = (1-\eta_n) + \eta_n \frac{T_e'}{T_c} = 0.1 + 0.9\left(\frac{p_e}{p_c}\right)^{\frac{\gamma-1}{\gamma}} = 0.576.$$

Therefore, the actual exit temperature $T_e = 1882\text{K}$ and the exit area is

$$A_e = \frac{\dot{m}RT}{p_e V_e} = \frac{(29kg/s)(711m^2/s^2 - K)(1882K)}{(96500N/m^2)(3117m/s)} = 0.129m^2.$$

The thrust is then given by

$$F = \dot{m}V_e + A_e(p_e - p_0) = (29kg/s)(3117m/s) + 0.129(96500N/m^2 - 68900N/m^2) = 93.95kN.$$

The thrust coefficient is found from

$$c_F = \frac{F}{p_c A_t} = \frac{93950N}{2.34 \times 10^6 N/m^2(0.03m^2)} = 1.34.$$

The exit Mach number is $M_e = V_e/a_e$ and

$$a_e = \sqrt{\gamma R T_e} = \sqrt{1.25(711m^2/s^2 - K)(1882K)} = 1293m/s.$$

Therefore, $M_e = 2.41$ and the specific impulse is

$$I_{sp} = \frac{F}{\dot{m}g} = \frac{93950N}{(29kg/s)(9.807m/s^2)} = 330s.$$

## 11.4 COMBUSTION CHAMBERS

There are three important parameters in combustion: time, turbulence, and temperature. To ensure the first two, the combustion chamber must be of sufficient size and configuration to permit a reasonable residence time during which adequate mixing may occur, while the third requires that there be sufficient energy supplied to ensure that the desired reaction takes place. The residence time may be thought of as being given by $t_r = L/U$, where $L$ is a characteristic length of the chamber and $U$ is the average velocity of the reactants in the direction of $L$. One may generalize this description somewhat by considering the average velocity to be given by $U = \dot{m}/\rho_c A_c$ so that

$$t_r = \frac{L}{\dot{m}}\rho_c A = \frac{\rho_c v_c}{\dot{m}} = \frac{p_c v_c}{R_c T_c}\frac{1}{\dot{m}} = \frac{m}{\dot{m}}. \tag{11.24}$$

This equation states that the residence time is described adequately by the ratio of the instantaneous mass in the chamber at any time to the global mass flow rate through the chamber. The quantity

$$\frac{p_c}{RT_c} = \frac{p_c W_c}{R_u T_c} \approx \frac{(1.013 \times 10^7 N/m^2)\left(20\dfrac{kg}{mol}\right)}{\left(8315\dfrac{kJ}{mol - K}\right)(2778K)} \approx 8.77kg/m^3. \tag{11.25}$$

Thus the residence time $t_r \approx 8.77 v_c/\dot{m}$ and for typical flow rates of 10 to 100 kg/s, and combustion chamber volumes on the order of 0.03 $m^3$, we find residence times on the order of 25 to 2.5 ms. Chemical reaction times are orders of magnitude smaller, so combustion chambers are generally operating under conditions appropriate to equilibrium chemistry.

### 11.4.1 Propellant injectors

Injectors must provide means for rapidly gasifying and mixing the propellants in the proportions necessary to ensure good combustion. The combustion process is generally initiated by electric ignition devices and thereafter is self-sustaining. Liquid rocket motors use propellants that are introduced to the injector in liquid form. The injector may rely on exploiting the physical properties of the propellants, such as viscosity and surface tension, among others. The most important aspect to gasifying the propellants is to greatly increase the surface area they present to the surroundings so as to enhance evaporative effects. This may be accomplished by mechanical means by inducing the breakdown of a liquid stream into droplets of small diameter. The ratio of surface area to volume is inversely proportional to the diameter of any droplets formed; therefore, the smaller the droplets, the more rapidly they will evaporate. Shearing forces may be generated by forcing the liquid propellant out of a small diameter hole under high pressure. The chamber pressure for rocket motors is in the range of 20 to 200 atmospheres so the injector must be able to exert at least that much pressure on the propellants as it forces them into the combustion chamber. Small-diameter, high-velocity jets of liquid exhibit instabilities that cause the stream to break up into small droplets. An important similarity parameter is the Weber number:

$$We = \frac{\frac{1}{2}\rho V^2 r}{\tau} = \frac{q}{\left(\frac{\tau}{r}\right)},$$

where the quantity $q$ is the dynamic pressure of a gas stream impinging on a liquid stream or droplet of radius $r$ and $\tau$ is the surface tension of the liquid. Obviously the Weber number may be considered to be a ratio of aerodynamic forces to surface tension forces. The larger the Weber number, the greater the likelihood of atomization into small droplets. Typical injector schemes are shown in Figure 11.12.

If propellants are gasified readily, as are, for example, liquid oxygen and hydrogen, then the injector can rely primarily on the instability waves to encourage atomization and gaseous diffusion to accomplish the required mixing and do away with splash plates or impinging jets. The showerhead injector shown in Figure 11.12a provides fine parallel jets of liquid propellant in alternating fuel and oxidizer streams that atomize and vaporize readily, diffusing into one another to produce the requisite fuel–oxidizer mixture. Other propellants, particularly hypergolic propellants such as red fuming nitric acid (RFNA) and unsymmetrical dimethyl hydrazine (UDMH), which react upon mixing in the liquid state, are handled more effectively with an injector that brings the propellants into direct contact. The impinging jet injector shown in Figure 11.12b relies on direct collision of the streams to instigate instabilities and to force atomization mechanically, after which vaporization may proceed more rapidly. The splash plate injector provides a means for liquid–liquid mixing during the mechanical breakup process, followed by atomization and vaporization. Each injector bore is on the order of 10 mm in diameter, and the jet orifice is on the order of 1 to 3 mm. The number of jets can vary from tens to hundreds, depending on the size of the combustion chamber and the propellant flow rate

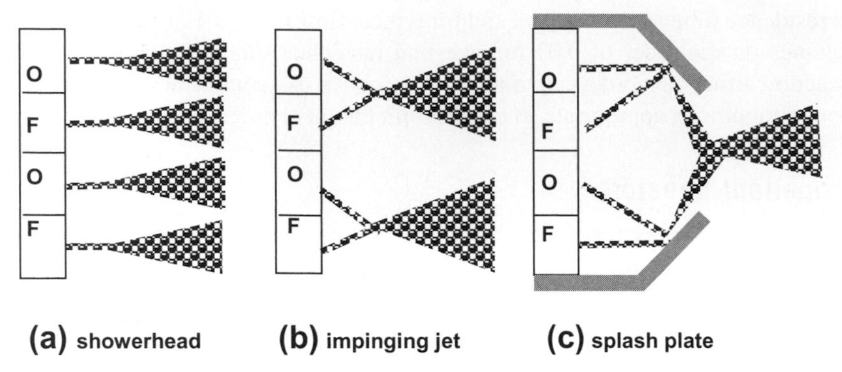


FIGURE 11.12

Schematic diagram of three types of injectors for liquid propellants. The letters O and F denote oxidizer and fuel, respectively.

required. Further details on injection systems may be found in Sutton (1992) and Barrere and colleagues (1960).

## 11.5 LIQUID ROCKET MOTOR OPERATIONAL CONSIDERATIONS

### 11.5.1 Rocket nozzle heat transfer

Several important similarity parameters in boundary layer heat transfer were discussed in Section 8.14. The first was the Stanton number, which normalizes the convective wall heat transfer by the external flow enthalpy flux based on conditions at the wall and was given in Equation (8.135) as

$$St = \frac{\dot{q}_{c,w}}{\rho_e u_e (h_w - h_{aw})}.$$

If one considers the applicability of an average specific heat, this may be rewritten as

$$\dot{q}_{c,w} = St(\rho_e u_e)\bar{c}_p(T_w - T_{aw}) = c_h(T_w - T_{aw}).$$

Then the convective heat transfer at the wall is proportional to a temperature difference and the coefficient of proportionality $c_h$ includes the local mass flux $\rho_e u_e$. The flow in rocket nozzles is characterized by rapidly varying values of pressure, temperature, and velocity, and calculating the heat transfer under such conditions is difficult. For design and comparison purposes, a simple but reliable method for determining the heat transfer in the throat and through the nozzle is desirable. A long-standing empirical correlation for the convective heat transfer coefficient $c_h$ that is useful in this regard is that due to Bartz (1957). Using the definitions of the similarity parameters in Section 8.14, we may write

$$c_h = St(\rho_e u_e)\bar{c}_p = \frac{\Pr Nu}{\mathrm{Re}}(\rho_e u_e)\bar{c}_p. \tag{11.26}$$

Using the flat plate reference enthalpy method, with Equation (8.143) and the coefficients for turbulent flow given in Table 8.1, the convective heat transfer coefficient of Equation (11.26) becomes

$$c_h = 0.296\,\mathrm{Pr}^{4/3}\mu_e^{1/5}\overline{c}_p\left(\frac{\rho}{\rho_e}\right)^{4/5}\left(\frac{\mu}{\mu_e}\right)^{1/5}\frac{(\rho_e u_e)^{4/5}}{l^{1/5}}. \tag{11.27}$$

Bartz (1957) developed a correlation that has a form very similar to that of Equation (11.27) but casts it in a form more convenient for rocket nozzle applications, which is as follows:

$$c_h = \left[\frac{0.026}{d_t^{1/5}}\left(\frac{\mu^{1/5}c_p}{\mathrm{Pr}^{3/5}}\right)_c\left(\frac{p_c}{c*}\right)^{4/5}\left(\frac{d_t}{r_u}\right)^{1/10}\right]C\left(\frac{A_t}{A}\right)^{9/10}. \tag{11.28}$$

The quantity in the square brackets of Equation (11.28) is constant for a given rocket motor geometry and chamber conditions. The quantity $r_u$ in the square brackets is the radius of curvature of the throat as shown in Figure 11.1. Here consistent use of SI units will yield $c_h$ in W/m$^2$-K. It should be noted that this equation, as presented originally by Bartz (1957), used rather mixed units where the throat diameter $d_t$ is in inches, the chamber pressure $p_c$ is in psia [and in Equation (11.28) it must appear multiplied by $g$, the gravitational acceleration in ft/s$^2$], the characteristic velocity $c^*$ is in ft/s, the viscosity $\mu$ is in lb/in-s, and the specific heat $c_p$ is in Btu/lb-R. As a result, the convective heat transfer coefficient $c_h$ has the units Btu/in$^2$-s-R.

The area ratio $A/A_t$ is a specified function of axial distance down the nozzle, and quantity $C$ is a form of $\rho\mu$ ratio and is defined as follows:

$$C = \left(\frac{\overline{\rho}}{\rho_e}\right)^{4/5}\left(\frac{\overline{\mu}}{\mu_c}\right)^{1/5}.$$

Factor $C$ has properties based on average values of temperature and a viscosity power law where $\mu \sim T^{\omega}$; the final form given by Bartz (1957) is

$$C = \left\{\left[\frac{1}{2}\frac{T_w}{T_c}\left(1+\frac{\gamma-1}{2}M^2\right)+\frac{1}{2}\right]^{\frac{2(2-\omega)}{5}}\left[1+\frac{\gamma-1}{2}M^2\right]^{\frac{2\omega}{5}}\right\}^{-1}.$$

It is clear that the heat transfer coefficient drops rapidly from a maximum at the throat as the area ratio and the Mach number in the nozzle grow with distance downstream. This correlation for the heat transfer coefficient has seen long service and does provide a reasonable estimate for conventional rocket nozzles. Typical maximum or throat values for $c_h$ are on the order of about 5 to 10 kW/m$^2$-K or about 2 to $4 \times 10^{-3}$ Btu/in$^2$-s-R. Having determined the convective heat transfer coefficient as a function of axial location, one may use the information on adiabatic wall temperature to determine the heat flux at the wall throughout the nozzle.

### 11.5.2 Nozzle and chamber cooling

Determination of the convective heat transfer coefficient of the nozzle was described in the previous subsection, and it was noted that the maximum heat transfer is essentially inversely proportional to the local cross-sectional flow area, with all other factors being held constant. As the rocket motor runs, temperatures throughout the system increase toward some equilibrium distribution, but such temperatures may locally exceed the values for which structural integrity can be maintained. Various techniques for cooling have been developed and many find application primarily because of practicality for the specific case considered. Research in this area has been intense because of the need to provide thermal protection of rocket motors of launch vehicles, as well as for the surfaces of reentry bodies. The major parameters to consider in thermal management are the time and temperature of exposure and the scale of the component to be cooled. The simplest modes of heat rejection are as follow.

- Radiation to the surroundings. As systems are reduced in scale, the surface area to volume ratio grows and makes radiation attractive. This is suitable for cases where refractory materials are suitable for the loads involved so that high temperatures can be maintained and where the surrounding ambient is relatively transparent to thermal radiation. For example, small rocket motors and the aft portions of nozzles in the space environment can employ radiative cooling.
- Conduction to the surroundings. This is the case of heat sink protection. If the exposure time is not long, a mass of material with suitable heat capacity can be put in thermal contact with the heated component to maintain its temperature at suitable levels. Such a technique is rarely applicable because of the additional weight penalty posed by the heat sink material. This was found to be equally true in the case of thermal protection of reentry vehicles.
- Change of phase cooling. An extension of heat sink cooling is to amplify the heat capacity of the heat sink material by exploiting the latent heats of melting and evaporation. In this case, the heat sink material is sacrificial and is consumed in the act of ablating to provide thermal protection. This is a technique that finds application in nozzle throats where heating, pressure, and flow velocities are high and the scale is so small that cooling access to the heated region is difficult. The mass of ablative material gives it strength, melting and evaporating give it high heat capacity, and the presence of high velocity helps remove melted material and expose virgin surface to the heat load.
- Convective cooling. Here a coolant moves over or through the heated component in order to carry away the heat and then reject it elsewhere. The most common application is regenerative cooling where fuel is passed from the tank, where it is relatively cool, through cooling tubes that are made part of the heated component, thereby absorbing the heat being conducted from the heated interior surface to the cooling tube. The fuel continues flowing through tubes distributed around the nozzle and combustion chamber, absorbing heat continually until it reaches the powerhead where it is injected back into the combustion chamber to contribute to thrust generation. There are other manifestations of this mode of cooling, namely film cooling, where the coolant is injected near the heated surface to affect cooling, and transpiration cooling, where the coolant "transpires" or passes through a porous portion of the heated surface and acts to absorb heat and flow on through the system.

There are many more variants of these basic methods of cooling, but final criteria in a practical situation are reliability, safety, and manufacturability. High performance alone is insufficient in practice, and the rule is that simplicity trumps performance.

### 11.5.3 Combustion instabilities

Determining the range of stable operation can be a major problem for new combustion chamber designs. Low-frequency (<100 Hz) "chugging" results from a coupling between the combustion process and the fuel feed system. It is often relieved by increasing the pressure drop across the injector head. High-frequency (>1 kHz) "screeching" is associated with one or more of the acoustic vibration modes within the combustion chamber: longitudinal, radial, transverse, or tangential. Ordinarily, such vibrational motion of the propellant gases would be damped by viscous effects. Intermediate frequencies may be excited, but such pressure oscillations within the chamber may interfere with the injector spray pattern but not with the propellant delivery rate or system operation.

The appearance of pressure fluctuations in the propellant feed system and in the combustion chamber itself will lead to temperature fluctuations as well. The low-frequency instability has fluctuations that have characteristic amplitudes and wave shapes. The pressure in the combustion chamber may rise rapidly and then fall off more slowly. The steep pressure rise can interfere with the delivery of propellants from the injector, and in that period of interference the propellants packed into the combustion chamber can flow out, thereby reducing the pressure. This irregularity leads to an interruption of the combustion process. When the accumulated propellant ignites, a new pressure wave is formed and the process repeats itself. The characteristic period for this process is close to the residence time. For the combustion chamber discussed in Section 11.3, this suggests a frequency of 40 to 400 Hz. If chamber pressures are increased, the pressure wave assumes a more nearly sinusoidal character with an increase in frequency and a drop in amplitude. The general trend is to see the amplitude of pressure waves suppressed by increasing the chamber pressure. As might be expected, the pressure fluctuations caused by irregularities in the combustion process give rise to temperature fluctuations in the chamber and in the nozzle. The fluctuation frequency drops with increasing characteristic residence time or length, while it increases with increasing chamber pressure. The increase in pressure may be thought of as "stiffening" the fluid in the chamber, thereby raising its vibration frequency, while the increased length is analogous to increasing the length of a beam and reducing its natural frequency.

High-frequency instabilities tend to be localized in the combustion chamber and, as a consequence, the pressure level in the combustion chamber is of less direct importance. Experiments show that increases in pressure reduce the amplitude of the fluctuations but not the frequency. The combustion chamber length $L_c$ does have a direct effect on the frequency of the fluctuations with the product of chamber length and fluctuation frequency $fL_c$ remaining constant. The characteristic velocity in the combustion chamber, which is a function of the propellant characteristics, correlates the frequency by the relation $fL_c = kc^*$, where $k$ is a constant. When the rocket motors increase in size, it is usual for the diameter to grow rather than the length. In that case, the diameter becomes the characteristic length and the product $fD_c$ is found to be constant. The geometry of the converging section of the combustion chamber leading to the throat is found to have an important influence in that impinging waves are reflected by the converging walls as well as being transmitted through the throat. As expected, the shorter the converging section, the more reflective it is and wave reflections can be reinforcing. Long smooth converging sections serve to provide damping of the waves. As always, there is a trade-off: the longer the converging section, the greater the exposure of the combustion chamber walls to the high heat transfer occasioned by combustion.

There are also instabilities arising from dynamic coupling between the engine characteristics and the vehicle's structural characteristics. Missiles are long slender structures with relatively low natural frequency. In addition, their effective stiffness decreases with powered flight time as the pressure in the propellant tanks falls. Low-frequency disturbances communicated from the engine to the fluid-filled propellant feed lines can couple to the vehicle structure, resulting in longitudinal waves propagating back and forth along the vehicle. This type of instability is called pogo instability after the bouncing movement of a pogo stick. Such instabilities can be controlled by accumulators in the propellant supply lines to reduce the natural frequency to a level below the vehicle natural frequency. The Saturn S II stage used an accumulator in the LOX line to reduce the frequency of the line to 4.5 Hz, well below the lowest natural frequency of the stage, which was 6 Hz. The space shuttle main engine (SSME) has a similar accumulator assembly in the LOX feed line and calls it the pogo suppression system.

### 11.5.4 Thrust vector control

The line of action of the thrust may be altered either by mounting the nozzle in a gimbal mechanism so that it may swivel as a unit to redirect the thrust or by upsetting the axisymmetry of the flow in the nozzle, resulting in a change in the thrust direction. One means of changing the symmetry of the flow is by injecting fluid, for example, either of the propellants, at discrete locations on the nozzle wall in order to cause local separation. Alternatively, one may have vanes in the exhaust flow, which could be deflected to cause a change in the thrust direction. These flow interaction methods are not particularly precise and because thermal management problems are associated with them, they are relegated to smaller rocket motors and applications where intermittent small changes in attitude are required, such as satellite station-keeping. Large rocket motors such as the 3.34 MN (750,000 lb) thrust Pratt & Whitney Rocketdyne RS-68 engine use a gimbaled combustion chamber and nozzle unit for pitch and yaw control. Roll control for this engine is achieved by providing a gimbaled mount for the exhaust from the fuel turbopump (Conley et al., 2002). Roll control is sometimes provided by a small, independently gimbaled, auxiliary motor. When several gimbaled rocket motors are combined at the base of a vehicle stage, they can also provide roll control without any other auxiliary mechanisms. For example, the Saturn second stage, the S II, had five gimbaled J-2 rocket motors, which provided control around all three axes.

The magnitude of the thrust vector, including thrust termination, is achieved by throttling the fuel delivery or stopping it altogether. The Boeing Rocketdyne RS-68 engine is capable of throttling the engine down to 57% of full power. The SSME can be throttled from 65% to 109% of its rated vacuum thrust of 470,000 lb (2.09 MN)

### 11.5.5 Flight environment effects

As the rocket motor with fixed expansion ratio nozzle ascends through the atmosphere the ambient pressure drops exponentially. At low altitude the nozzle is overexpanded with exit pressure lower than the ambient pressure, whereas at high altitudes the nozzle is underexpanded with exit pressure higher than ambient pressure. At low altitudes then, the nozzle boundary layer is flowing against an adverse pressure gradient in the vicinity of the nozzle exit and separation may occur. Depending on local conditions, the separated region can extend relatively far upstream in the nozzle, causing a substantial reduction in thrust. The effect is shown schematically in Figure 11.13a.

Because heating of the base of the rocket by the exhaust plume at low altitudes is ameliorated by the relatively narrow plume produced there, base heating is due primarily to radiation from the hot exhaust. However, at high altitude the exhaust plume grows ever broader due to the increasingly underexpanded operation of the nozzle and induces a bow shock ahead of it. Then recirculation regions form between the nozzle and the base, resulting in high convective heating rates, as shown schematically in Figure 11.13b. This condition is exacerbated by the clustering of multiple nozzles as

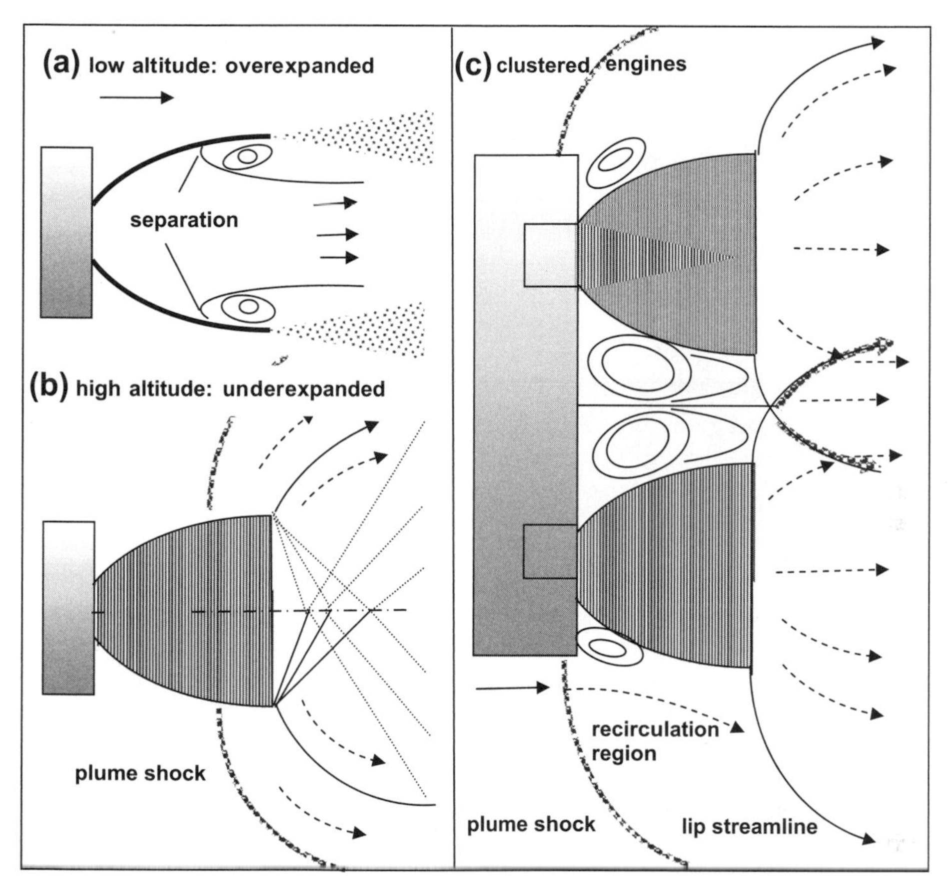


**FIGURE 11.13**

Nozzle exhaust plume characteristics as altitude increases. (a) Low-altitude plume: overexpansion leads to separation within nozzle, (b) high-altitude plume: underexpansion leads to plume enlargement, causing a bow shock, and (c) clustered engines induce recirculation regions at vehicle base.

illustrated in Figure 11.13c. For example, the Saturn V with five F-1 rocket motors clustered around the base had to operate much richer (O/F = 2.4, rather than the stoichiometric value of 3.4) to reduce the damaging convective heating encountered by the interaction between the plume and the vehicle.

## 11.6 ROCKET PROPELLANTS

A collection of general information on various rocket propellants is given in Table 11.2. Further details on the most common rocket propellants currently in use are covered in this section. Further detailed information on liquid propellants and their performance may be found in Kit and Evered (1960) and Siegel and Schieler (1964). Taking the space shuttle as an example, two major divisions for rocket

**Table 11.2** General Properties of Representative Rocket Propellants

| Type | Propellant | Energy Source | $I_{sp}$ (s) Vacuum | $F$ (lbs) Range | Specific Gravity | Advantages | Disadvantages |
|---|---|---|---|---|---|---|---|
| Solid | Organic polymers, ammonium perchlorate, and powdered Al | Chemical | 280–300 | 10 to $10^6$ | 1.8 | Simple, reliable, low cost | Limited performance, safety issues |
| Cold gas | $N_2$, $NH_3$, He, freon | High pressure | 50–75 | 0.01 to 50 | 1.46 | Simple, reliable, low cost | Low performance, high weight |
| Liquid monopropellant | $H_2O_2$<br>$N_2H_4$ | Exothermic decomposition | 150<br>200 | 0.01–0.1 | 1 | Simple, reliable, low cost | Low performance, higher weight |
| Liquid bipropellant | RP-1/$O_2$ | Chemical | 270–360 | 1 to $10^6$ | 1 | High performance | Complicated |
| | $H_2/O_2$ | Chemical | 360–450 | 1 to $10^6$ | 1.26 | Very high performance | Cryogenic, complicated |
| | UMDH/$N_2O_4$ | Chemical | 270–340 | 1 to $10^6$ | 1.14 | Storable, good performance | Complicated |
| | $N_2H_4/F_2$ | Chemical | 425 | 1 to $10^6$ | 1.1 | Very high performance | Toxic, dangerous, complicated |

propellants are represented: solid propellants used in the integrated solid rocket booster (ISRB) and liquid propellants used in the SSME, the orbital maneuvering system (OMS), and the reaction control subsystem (RCS). The discussion of solid propellants is left to the next chapter, while liquid propellants are treated here. The term liquid propellant refers to the fact that these propellants are carried in the liquid phase, and there are two minor divisions of such propellants: cryogenic liquids and conventionally storable liquids. A further subdivision is possible in that the propellants may be considered to be ignitable or hypergolic. Ignitable propellants must have a separate ignition source, whereas hypergolic propellants react spontaneously with each other. Liquid propellants are typically considered as pairs, a fuel and an oxidizer, and these are called bipropellants to distinguish them from monopropellants, liquids that can decompose chemically, accompanied by the release of heat. One other specialized type of propellant mentioned in Table 11.2 is the cold gas propellant, which is merely pressurized gas, which can be expelled through a nozzle to provide thrust without any chemical reaction.

### 11.6.1 The $H_2$–$O_2$ propellant combination

The energy per unit mass of hydrogen is about three times that of hydrocarbon fuels used in air-breathing engines, which makes it an attractive candidate for rocket engines for space launch applications. In addition, the rocket application is one where the oxidizer is carried onboard and therefore can be pure oxygen. For a stoichiometric hydrogen–oxygen propellant mixture, the chemical reaction is defined by

$$H_2 + \frac{1}{2}O_2 \rightarrow H_2O.$$

The oxidizer to fuel ratio (O/F) may be calculated from this reaction in the same manner as was done for the hydrocarbon–air reaction in Chapter 4. Thus, taking the moles of fuel and oxidizer and converting them to mass values leads to the following result:

$$\frac{m_f}{m_o} = \frac{(1kg - mol)\left(2\frac{kg}{kg - mol}\right)}{\left(\frac{1}{2}kg - mol\right)\left(32\frac{kg}{kg - mol}\right)} = \frac{1}{8}.$$

This clearly shows that O/F = 8 for the stoichiometric reaction. The adiabatic flame temperature, which was discussed in Chapter 4, for this reaction is $T_c = 5100K$ and the molecular weight of the product gas, water, is $W_c = 18$. One may calculate the limiting velocity for the product gas assuming adiabatic expansion of the gas to zero temperature and find that the maximum exhaust velocity is

$$V_{e,\lim} = \sqrt{2c_pT_c} = 5590m/s. \tag{11.29}$$

The limiting specific impulse is then

$$I_{sp,\lim} = \frac{V_e}{g} \approx 560s.$$

However, in Table 11.1, the LOX–$LH_2$ propellant combination is shown with O/F = 4. When the oxidizer to fuel ratio is reduced from the stoichiometric value, the mixture is said to be rich. In this case the equivalence ratio, discussed in Chapter 4, $\phi = 2$, which is quite rich. Consider the case where the ratio of the number of moles of oxygen to those of hydrogen is reduced from the stoichiometric value of one-half to the rich value of one-quarter, as in the following reaction:

$$4H_2 + O_2 \rightarrow b_1H_2 + b_2O_2 + b_3H_2O + b_4OH + b_5O + b_6H.$$

The fuel to oxidizer ratio may be computed as before, yielding

$$\frac{m_f}{m_o} = \frac{(4kg - mol)\left(2\frac{kg}{kg - mol}\right)}{(1kg - mol)\left(32\frac{kg}{kg - mol}\right)} = \frac{1}{4},$$

where now O/F = 4. The composition of the product gases and the corresponding adiabatic flame temperature may be calculated by the methods presented previously in Chapter 4. Results give an adiabatic flame temperature $T_c = 3075$K and a mixture molecular weight of $W_c = 9.79$. Calculating the limiting velocity for the product gas, assuming adiabatic expansion of the gas to zero temperature, shows that the maximum exhaust velocity is

$$V_{e,\lim} = \sqrt{2c_pT_c} \approx 5295m/s$$

$$I_{sp,\lim} = \frac{V_e}{g} \approx 530s.$$

It is clear that the limiting specific impulse in the rich mixture case is reduced by about 5%, while the combustion chamber temperature is reduced by almost 40%. Thus only a small performance penalty is paid for a large reduction in temperature that must be handled by the structure. The trade-off here is between stagnation temperature and molecular weight. The rich mixture combustion products have a high percentage of unburned hydrogen that reduces the molecular weight to about half the value in the stoichiometric mixture case.

### 11.6.2 Cryogenic propellants

Under standard atmospheric conditions, many good propellants would be in the gaseous form, and as such would have such low density as to make storage volume requirements that are impractical. However, in the liquid phase, the density would be high enough to overcome that problem. The difficulty with that solution is that the propellants would have to be kept at a very low temperature, that is, stored cryogenically. Characteristics of the more common cryogenic propellants are given in Table 11.3.

The SSME uses the LOX–$LH_2$ cryogenic propulsion combination, as does the RL-10 engine, first used as an upper stage in the Centaur, and the J-2, first used in the second and third stages of the Saturn V. The ability to maintain the hydrogen and oxygen in the liquid phase enabled the achievement of the high specific impulse that the hydrogen–oxygen chemistry promised. The engines just mentioned all perform in the range $425 < I_{sp} < 455$ s. The downside to the cryogenic combination is that it is difficult to store the propellants effectively over long periods of time, making it impractical for military applications where missiles must be kept launch-ready for months at a time. The oxygen–hydrogen fuel system requires an igniter to initiate the reaction, which then becomes self-sustaining. Of course, there are hydrocarbon fuels that are in the liquid phase under standard atmospheric conditions, such as RP-1, a kerosene-based fuel blended for use

**Table 11.3** Characteristics of Some Cryogenic Propellants

| Name | Type | Boiling Point (K) | Density (kg/m$^3$) | Molecular Weight | $L_V$ (kJ/kg) |
|---|---|---|---|---|---|
| Liquid oxygen (LOX) | Oxidizer | 90 | 1140 | 32.00 | 213 |
| Liquid hydrogen ($LH_2$) | Fuel | 20.3 | 69.5 | 2.01 | 450 |
| Liquid methane ($LCH_4$) | Fuel | 111.5 | 415 | 16.03 | 578 |
| Liquid fluorine ($LF_2$) | Oxidizer | 84.7 | 1514 | 38.00 | 172 |

with liquid oxygen in rocket motors. The high thrust F-1 engine used in the first stage of the Saturn V moon rocket used this propellant combination

### 11.6.3 Hypergolic propellants

The ability of a fuel–oxidizer combination to ignite spontaneously is a particular benefit, especially for space operations where many start–stop cycles may be necessary for attitude control and station-keeping. Although the $LF_2$–$LH_2$ cryogenic combination is hypergolic, the toxicity of the fuel and the products of combustion have made the use of this system impractical. In general, the commonly used hypergolic propellants are considered conventionally storable, that is, they do not need special refrigeration equipment to keep them liquid. A particular subset of so-called storable propellants are those called space storable, meaning that they can be kept liquid without special equipment in the temperatures typical of near-earth space operations, where vehicle surface temperatures are generally greater than those found in the earth's stratosphere, 216K. Commonly used hypergolic propellants are described in Table 11.4.

The space shuttle orbiter uses the monomethylhydrazine (MMH)–$N_2O_4$ combination in its OMS and RCS with $I_{sp}$ performances of 313 s and 260–280 s, respectively. More common in other rockets is the use of unsymmetrical dimethylhydrazine (UMDH)–$N_2O_4$ or UMDH/$N_2H_4$–$N_2O_4$, but the performance and density are not as attractive as the MMH–$N_2O_4$ combination. These materials are dangerous and toxic. As a result, they must be handled under strict safety protocols and are now very expensive due to stringent environmental regulations.

**Table 11.4** Characteristics of Some Common Hypergolic Propellants

| Name | Type | Formula | Boiling Point, K | Freeze Point, K | Density (kg/m$^3$) | Molecular Weight |
|---|---|---|---|---|---|---|
| Hydrazine | F | $N_2H_4$ | 386 | 274 | 1011 | 32.05 |
| Nitrogen Tetroxide | O | $N_2O_4$ | 294 | 262 | 1450 | 92.02 |
| Monomethylhydrazine (MMH) | F | $CH_3NHNH_2$ | 360 | 221 | 880 | 46.03 |
| Unsymmetrical-dimethylhydrazine (UDMH) | F | $(CH_3)_2NNH_2$ | 336 | 221 | 790 | 60.04 |
| Red fuming nitric acid (RFNA) | O | $HNO_3$ + 13% $N_2O_4$ + 3%$H_2O$ | 216 | 358 | 1550 | 63 |

## 11.7 ROCKET CHARACTERISTICS

The movement in ballistic missile and manned space access technology in the late 1950s and early 1960s resulted in the development of a number of practical rocket engines. The U.S. manned space program up through the 1960s involved the high-performance liquid bipropellant engines as shown in Table 11.5, which was taken from Williams and associates (1976).

Liquid bipropellant rocket engines built since the 1960s are included in Table 11.6.

## 11.8 PROPELLANT TANK AND FEED SYSTEM DESIGN

### 11.8.1 Propellant tank characteristics

The weight of propellant storage tanks constitutes a significant part of the total weight of a launch vehicle. Any excess weight of the fuel tanks takes away from the payload weight, and the propellant tank design must be performed carefully. The tanks are subject to both static and dynamic loads. The main static loads are as follows:

- Hydrostatic pressure—due to the state of the fluid contained
- Overpressure—sometimes used to avoid cavitation in the fuel pumps
- Payload weight—the tanks generally also serve as structural members

The dynamic loads are as follows:

- Inertia loads—due to vehicle acceleration
- Control loads—due to thrust vectoring induced bending
- Aerodynamic loads—due to vehicle geometry and wind shear

The tanks may be spherical or cylindrical with spherical end caps; other shapes are generally not used because they are usually heavier than these simpler shapes.

### 11.8.2 Tank structural analysis

We can assess the basic loading on the tanks and start by assuming that they will be basically of a thin shell configuration. Sections of a cylindrical shell and a spherical shell are shown in Figure 11.14, along with forces acting due to internal pressure. The shells are thin in the sense that the thickness is much smaller everywhere than the length and/or radius of the shell, that is, $t \ll L,R$. Also shown in Figure 11.14 is the axial load on a cylindrical shell tank with spherical shell end caps.

From the equilibrium of forces in Figure 11.14, we find that the allowable thickness $t_s$ of a spherical shell tank of radius $R_s$, based on the yield stress $\sigma_y$ of the tank material, is

$$t_s = \frac{1}{2} R_s \frac{p}{\sigma_y}. \tag{11.30}$$

Likewise, the allowable thickness $t_c$ of a circular cylindrical tank with spherical end caps all with radius $R_s$, based on the yield stress $\sigma_y$ of the tank material, is

$$t_c = R_c \frac{p}{\sigma_y} = 2t_s. \tag{11.31}$$

**Table 11.5** Liquid Bipropellant Rocket Engines of the 1960s (Williams et al., 1976)

| Engine | Manufacturer | Sea Level Thrust (klbs) | Vacuum Thrust (klbs) | Chamber Cooling Method | Oxidizer | Fuel | Vehicle | Prime Contractor | Number of Engines |
|---|---|---|---|---|---|---|---|---|---|
| LR43-NA-1 | Rocketdyne | 78 | | Regenerative | $LO_2$ | Alcohol | Redstone | Chrysler | 1 |
| LR89-NA-7 | Rocketdyne | 154 | | Regenerative | $LO_2$ | RP-1 | Atlas | GD | 2 |
| LR105-NA-7 | Rocketdyne | 57 | | Regenerative | $LO_2$ | RP-1 | Atlas | GD | 1 |
| LR87-AJ-5 | Aerojet | 215 | | Regenerative | $N_2O_4$ | $N_2H_4$-UDMH | Titan | Martin | 2 |
| LR91-AJ-5 | Aerojet | 100 | | Regenerative | $N_2O_4$ | $N_2H_4$ | Titan | Martin | 1 |
| SE-6 & -7 | Rocketdyne | | 0.025 | Ablative | $N_2O_4$ | $N_2H_4$-UDMH | Gemini | | |
| | Rocketdyne | | 0.085 | Ablative | $N_2O_4$ | $N_2H_4$-UDMH | Gemini | | |
| | Rocketdyne | | 0.1 | Ablative | $N_2O_4$ | $N_2H_4$-UDMH | Gemini | | |
| H-1 | Rocketdyne | 205 | | Regenerative | $LO_2$ | RP-1 | S-1 | Chrysler | 8 |
| | | | | | | | Uprated S-1 | Chrysler | 8 |
| F-1 | Rocketdyne | 1522 | | Regenerative | $LO_2$ | RP-1 | S-1C | Boeing | 5 |
| RL10A-3-3 | Pratt & Whitney | | 15 | Regenerative | $LO_2$ | $LH_2$ | S-IV | MDAC | 6 |
| J-2 | Rocketdyne | | 230 | Regenerative | $LO_2$ | $LH_2$ | S-IVB | MDAC | 1 |
| SE-7-1 | Rocketdyne | | 0.074 | Ablative | $N_2O_4$ | $N_2H_4$-UDMH | S-IVB | MDAC | 2 |
| S-IVB-ACE | TRW | | 0.150 | Ablative | $N_2O_4$ | $N_2H_4$-UDMH | S-IVB | MDAC | 6 |
| AJ10-137 | Aerojet | | 21.9 | Ablative | $N_2O_4$ | $N_2H_4$-UDMH | SM | RI | 1 |
| LMDE | TRW | | 10.5 | Ablative | $N_2O_4$ | $N_2H_4$-UDMH | LM descent | Grumman | 1 |
| LEM- 8258 | Textron/Bell | | 3.5 | Ablative | $N_2O_4$ | $N_2H_4$-UDMH | LM descent | Grumman | 1 |
| R-4D | CCI-Marquardt | | 0.1 | Radiation | $N_2O_4$ | $N_2H_4$-UDMH | SM | RI | 16 |
| SE-8 | Rocketdyne | | 0.93 | Ablative | $N_2O_4$ | $N_2H_4$-UDMH | CM | RI | 12 |

**Table 11.6** Some Liquid Bipropellant Rocket Engines

| Engine | Max. F vac/ s.l. (klbs) | Impulse vac/s.l. (s) | $p_c$ (psia) | A/A* | Weight (lbs) | Propellant | O/F | Exit dia. (in.) |
|---|---|---|---|---|---|---|---|---|
| RD-180 | 933/860 | 338/311 | 3722 | 36.4 | 12000 | $RP_1$-$LO_2$ | 2.72 | 124 |
| RS-68 | 751/656 | 409/357 | 1420 | 22 | 14800 | $LH_2$-$LO_2$ | 6 | 96 |
| SSME | 512/409 | 454/- | 3260 | 77.5 | 7000 | $LH_2$-$LO_2$ | 6 | 96 |
| J2-S | 265/197 | 436/320 | 1200 | 40 | 3800 | $LH_2$-$LO_2$ | 5.5 | 84 |
| RS-53 | 470/375 | 454/- | 3000 | 77.5 | 7200 | $LH_2$-$LO_2$ | 6 | |
| STME | 580/- | 424/352 | 2250 | 56 | 7100 | $LH_2$-$LO_2$ | 6 | |
| RS-27 | 232/207 | 295/263 | 700 | 8 | 2290 | $RP_1$-$LO_2$ | 2.25 | 76 |
| RS27A | 237/200 | 302/255 | 700 | 12 | 2440 | $RP_1$-$LO_2$ | 2.25 | 76 |
| F-1A | 2021/1800 | 303/270 | 1160 | 16 | 19000 | $RP_1$-$LO_2$ | 2.27 | 144 |
| F-1 | -/1522 | -/265 | 980 | 16 | 18600 | $RP_1$-$LO_2$ | 2.27 | 144 |
| J-2 | 230/- | 425/- | 763 | 27.5 | 3480 | $LH_2$-$LO_2$ | 5.5 | 81 |
| H-1 | 205 | 263 | 700 | 8 | 2000 | $LH_2$-$LO_2$ | 2.23 | 66 |
| MA-5 | -/378 | -/259 | 639 | 8 | 3140 | $RP_1$-$LO_2$ | 2.25 | 47 |
| Vulcain 2 | 305/- | 429/- | 1700 | 58.2 | 2000 | $LH_2$-$LO_2$ | 6.7 | 82 |
| Vulcain | 250/- | 439/- | 1580 | 45.1 | 1380 | $LH_2$-$LO_2$ | 5.9 | 69 |
| HM 7B | 14.6/- | 446/- | 534 | 83.1 | 364 | $LH_2$-$LO_2$ | 5 | 39 |
| Vinci | 40.5/- | 465/- | 882 | 240 | 353 | $LH_2$-$LO_2$ | 5.8 | 87 |
| AestusII/ RS72 | 12.5/- | 340/- | 871 | 300 | 305 | $MMH/N_2O_4$ | 1.9 | 51 |

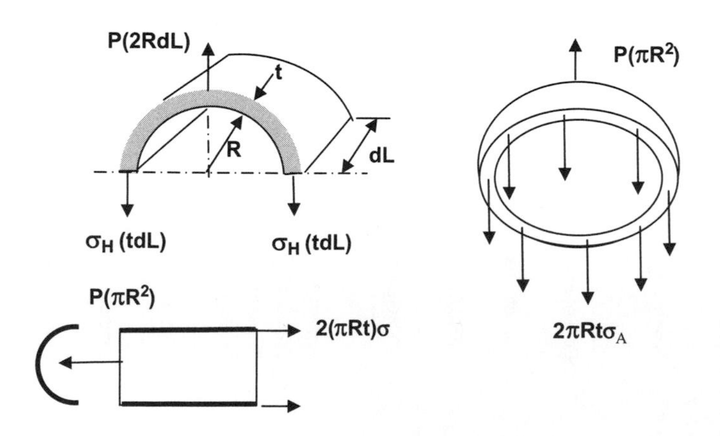


**FIGURE 11.14**

Circular cylindrical shell with forces due to internal pressure, spherical shell with forces due to internal pressure, and axial force on a circular cylindrical shell with spherical end cap under axial loading pressure.

The design yield stress for a number of common aerospace metallic materials is given in Table 11.7 (Anonymous, 1970).

Composite materials have become increasingly popular because the high tensile strength of various nonmetallic fibers such as glass, Kevlar, carbon, and other materials can be exploited with new manufacturing techniques such as filament winding. Thin fibers are wound continuously in a particular pattern around a mandrel, or form, in the shape desired and embedded in a resin to retain a structure. The fibers can be aligned to best respond to the stresses expected and thereby provide a more effective tank that can be lighter in weight than the conventional metal one.

Note that a circular cylindrical tank with spherical end caps must be twice as thick as that of a spherical tank of the same radius. However, by increasing the length of the circular cylindrical tank, the volume enclosed can be increased arbitrarily. Examining then the weight performance of the two types of tank leads to the following weights for the spherical tank and the circular cylindrical tank of overall length, respectively:

$$W_s = 2\pi R_s^3 \frac{p}{(\sigma_y/\rho)} \tag{11.32}$$

$$W_c = 2\pi R_c^3 \left(\frac{L}{R_c} - 1\right) \frac{p}{(\sigma_y/\rho)}, \tag{11.33}$$

**Table 11.7** Properties of Some Typical Aerospace Materials

| Material | Design Yield Strength (ksi) | Modulus of Elasticity (1000 ksi) | Density (lb/in$^3$) | Design Yield Strength (MPa) | Modulus of Elasticity (GPa) | Density (kg/m$^3$) |
|---|---|---|---|---|---|---|
| HY steel | | | | | | |
| HY-80 | 80 | 29.5 | 0.285 | 551 | 203 | 7880 |
| HY-130/150 | 130–150 | 29.5 | 0.285 | 896–1030 | 203 | 7880 |
| Low alloy steel | | | | | | |
| 4130 | 150–180 | 29 | 0.283 | 1030–1240 | 200 | 7820 |
| 4335V | 180–200 | 29 | 0.283 | 1240–1380 | 200 | 7820 |
| Maraging steel | | | | | | |
| Grade 250 | 240 | 27.5 | 0.289 | 1650 | 190 | 7990 |
| Grade 300 | 280 | 27.5 | 0.289 | 1930 | 190 | 7990 |
| HP steel | | | | | | |
| 9Ni-4Co-0.250 | 180–220 | 28.5 | 0.28 | 1240–1520 | 196 | 7740 |
| 9Ni-4Co-0.450 | 260–300 | 28.5 | 0.28 | 1790–2070 | 196 | 7740 |
| Titanium | | | | | | |
| Ti-6A1-4V | 150 | 16 | 0.167 | 1030 | 110 | 4620 |
| Aluminum alloys | | | | | | |
| 2000 series | 35–65 | 10.3 | 0.10 | 241–448 | 71 | 2750 |
| 6000 series | 37–47 | 10.3 | 0.10 | 255–324 | 71 | 2760 |
| 7000 series | 60–68 | 10.3 | 0.10 | 413–469 | 71 | 2760 |

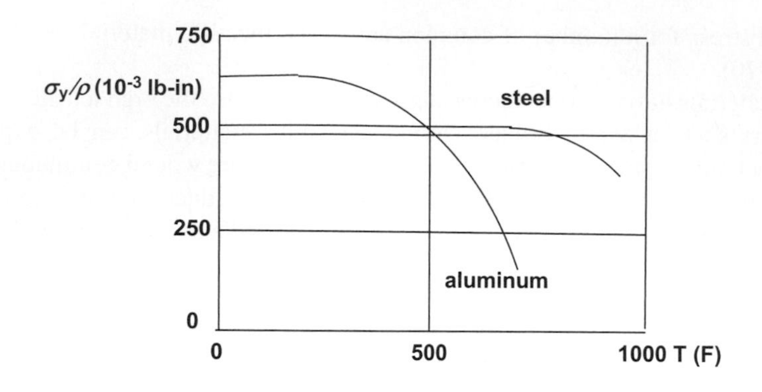


**FIGURE 11.15**

Typical behavior of the specific stress of two common tank materials is shown as a function of operating temperature.

where $\rho$ is the weight density of the tank material, and the ratio of yield stress to material density is called the specific yield stress. Typical behavior of the specific stress of two common tank materials is shown as a function of operating temperature in Figure 11.15.

For a 50-psia (345 kPa) cylindrical tank with a diameter of 8.33 ft (2.54 m), a length of 66.4 ft (20.3 m), and reasonable temperature, the minimum thickness would be 0.0167 in. (0.42 mm) for a steel tank and 0.0385 in. (0.98 mm) for an aluminum tank. Thus thin walls can support large pressures, but then thin-walled vessels are subject to buckling. The critical (buckling) axial stress for a cylinder having no stiffeners is given empirically by

$$\frac{\sigma_{crit}}{E} = 9\left(\frac{t}{R_c}\right)^2 + 0.16\left(\frac{t}{L}\right)^{1.3}. \tag{11.34}$$

Considering the example case for a steel skin with $E = 3 \times 10^7$psi, the buckling stress is 744 psi and for aluminum it would be 943 psi, with both values being much less than the yield stress for the material.

## 11.8.3 Tank weight

Because of buckling concerns, cylindrical tanks require stiffeners, both longitudinal and ring stiffeners, which again add to the weight. Of course the internal pressurization of the tanks also serves to stiffen the cylinders and add in avoiding buckling. For equal volume contained, the ratio of the weight of circular cylindrical tanks with spherical end caps to spherical tanks is

$$\frac{W_c}{W_s} = \frac{4}{3}\frac{\frac{L}{R_c} - 1}{\frac{L}{R_c} - \frac{2}{3}}. \tag{11.35}$$

Note that for $L/R_c = 2$, the tank is a spherical one and $W_c/W_s = 1$ so that the lightest tank to contain a given volume is a spherical one; however, it will also have the largest radius. Thus the

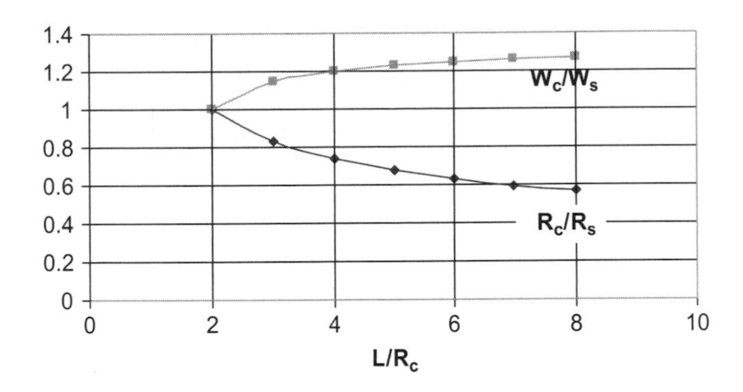


**FIGURE 11.16**

Variation of the weight ratio $W_c/W_S$ and the radius ratio $R_c/R_S$ is shown as a function of the length ratio $L/R_C$ for tanks of equal volume.

driving factor is how large a diameter tank the design can support. However, using a small-diameter tank means that the tank will be very long, with the attendant problem of buckling under bending loads. Spherical tanks are used commonly for small volume applications because of their superior weight characteristics. An illustration of the variation of the weight ratio $W_c/W_S$ and the radius ratio $R_c/R_S$ for tanks of equal volume is shown as a function of the length ratio $L/R_C$ in Figure 11.16.

Note that a 40% reduction in diameter is realized when using a length ratio of 8 at the cost of only a 30% increase in weight. The choice of tank configuration must also account for dynamic loading in addition to the hydrostatic load. Because the aerodynamic drag is proportional to the frontal area, spherical tanks will suffer a disadvantage. However, the inertial load due to acceleration will increase the pressure loading beyond it as hydrostatic level, and this increase is proportional to the length of the tank so that here the spherical tank has an edge. Mounting and securing the tanks, as shown in Figure 11.17, also provide extra weight that must be taken into account, and results are shown schematically in Figure 11.18.

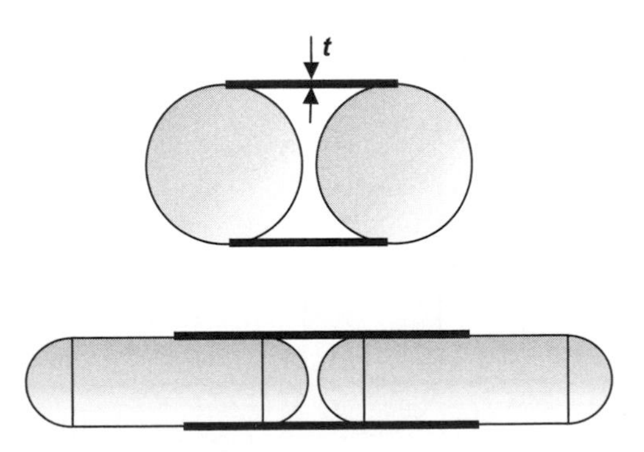


**FIGURE 11.17**

Different mounting schemes for securing tanks.

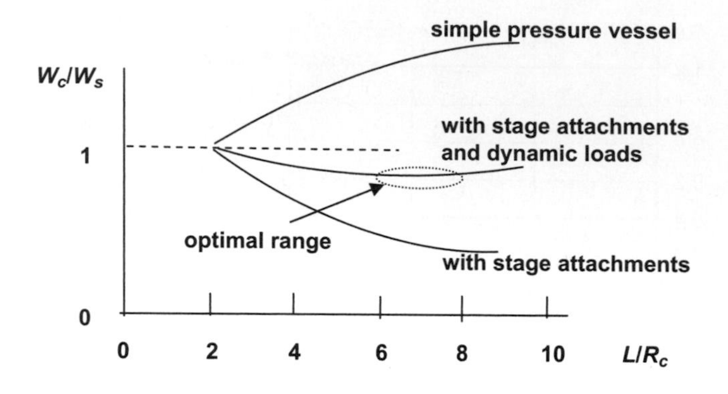


**FIGURE 11.18**

Effects of stage attachments and fittings on weight ratio $W_c/W_s$ are shown as a function of the length ratio $L/R_c$.

The weight of the cylindrical part of a tank made of material with specific weight $w_t$ is

$$W_c = \rho_t g \int_0^L \pi\, Dtdx. \tag{11.36}$$

Cylindrical tanks are generally fairly long with respect to their diameter, and their longitudinal axis is always aligned closely to the acceleration of the vehicle to which the tank is attached, as illustrated in Figure 11.19.

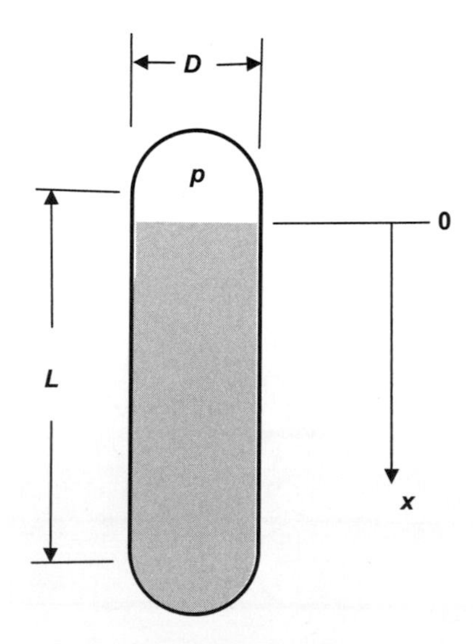


**FIGURE 11.19**

Cylindrical tank with spherical end caps containing liquid propellant of density $\rho_p$.

For completeness then, it is important to include the additional hydrostatic pressure induced by acceleration of the tank. The pressure at any point $x$ along the tank axis is $p + ng\rho_p x$, where $p$ is the ullage pressure of the gas above the free surface and $n$ is the load factor due to acceleration; $n = 1$ when the vehicle is static or moving longitudinally at constant speed in the earth's gravitational field.

The minimum thickness of the wall of the tank, which can be subjected to a longitudinal acceleration equal to $ng$, has been found to be

$$\frac{t}{D} = \frac{p + ng\rho_p x}{2\sigma_y}. \tag{11.37}$$

This shows that the thickness of the wall would have to increase as we proceed down the length of the tank. Substituting Equation (11.37) into Equation (11.36) and integrating yields

$$W_c = \pi D^2 \rho_t g \left( \frac{pL}{2\sigma_y} + \frac{\rho_p ng L^2}{4\sigma_y} \right). \tag{11.38}$$

The minimum thicknesses of the upper and lower hemispheres are given, respectively, by

$$\frac{t_{s,u}}{D} = \frac{p}{4\sigma_y}$$

$$\frac{t_{s,l}}{D} = \frac{p + \rho_p ngL}{4\sigma_y}.$$

The minimum total weight of the complete tank is then

$$W_t = \pi D^2 \rho_t g \left( L + \frac{D}{2} \right) \left( \frac{p}{2\sigma_y} + \frac{\rho_p ngL}{4\sigma_y} \right).$$

The weight of the propellant contained in the tank is

$$W_p = \rho_p g \frac{\pi D^2}{4} \left( L + \frac{D}{2} \right).$$

Then the ratio of minimum tank weight to propellant weight is

$$\frac{W_t}{W_p} = \frac{3}{\sigma_y} \frac{\rho_t}{\rho_p} \frac{2 + \frac{D}{L}}{3 + \frac{D}{L}} \left( p + \frac{1}{2} \rho_p ngL \right).$$

Obviously, a factor of safety may be introduced by replacing $\sigma_y$ by an allowable stress. Furthermore, if the acceleration load factor $n$ is very small and $D/L$ is likewise small, an estimate of the tank to propellant weight is

$$\frac{W_t}{W_p} \simeq \frac{2p}{\sigma_y} \frac{\rho_t}{\rho_p}.$$

A table showing typical ratios of tank weight to propellant weight is given in Table 11.8.

**Table 11.8** Tank to Propellant Weight Ratios for Different Systems

| Propellant Type | Internal Pressure (psia) | Specific Gravity | $W_{tank}/W_{propellant}$ |
|---|---|---|---|
| Liquid | 50 | 1 | 0.005 |
| Solid | 400 | 1 | 0.04 |
| Liquid | 50 | 0.3 | 0.0167 |
| Solid | 1000 | 1.8 | 0.056 |

### 11.8.4 Propellant feed systems

For effective operation, the pressure in the combustion chamber of the rocket motor must be high. In order to introduce propellants into the combustion chamber they must enter at an even higher pressure. The basic propellant feed systems are shown in Figure 11.20. The simplest is the pressure-fed system in Figure 11.20a, where a high-pressure tank with an inert gas such as nitrogen is fed into the fuel and oxidizer tanks through a pressure regulator. This pressure forces the liquid propellants into the combustion chamber, as shown in Figure 11.20a. This system is well suited to smaller rocket motors

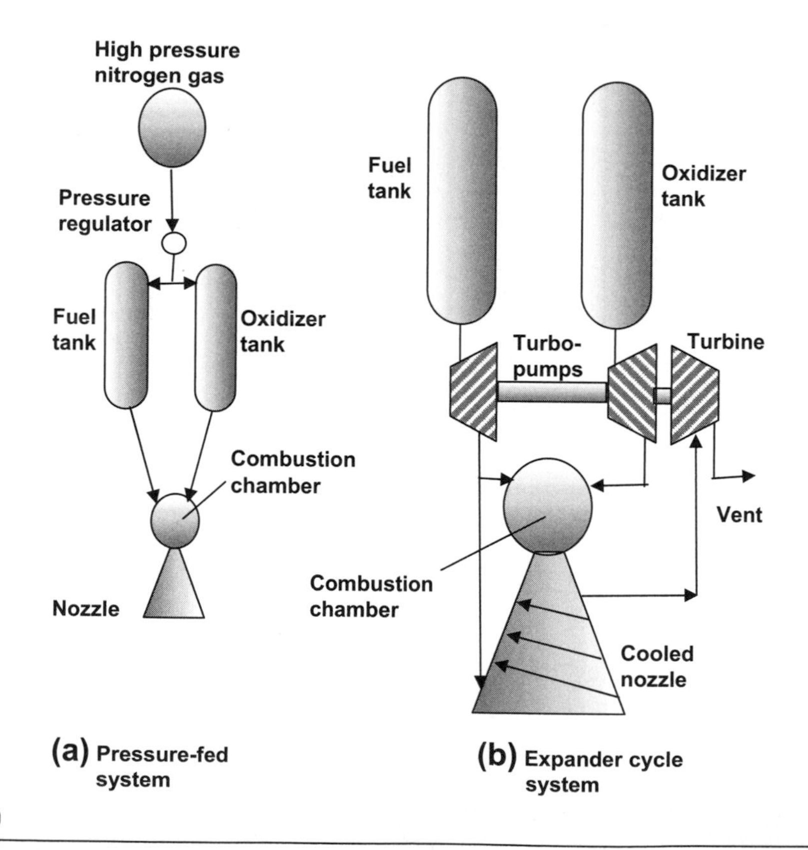


**FIGURE 11.20**

Basic propellant feed systems for liquid rocket motors.

because the amount of pressurized gas necessary to force all the propellant through the combustion chamber is relatively small.

For large rocket motors, the size and weight of a pressurized nitrogen tank become prohibitive and attention turns to a pump-fed propellant system like that shown in Figure 11.20b, which shows an expander cycle system in which the fuel is used to cool the nozzle and the heat it picks up is then expended in powering a turbine that drives the turbopumps for pumping both the fuel and the oxidizer into the combustion chamber. The expanded fuel gas enters a collector tube, which serves to inject it back into the system at some point down the nozzle where the pressure is low enough. This collector tube can often be seen running around the circumference of the nozzle at some axial location where the fuel is injected into the nozzle. There are several propellant feed systems that are variants of the expander cycle shown. The basic requirement is a hot gas source to power a turbine, which then drives the propellant pumps; this need may be filled in several ways, which are described subsequently.

The gas generator cycle uses a separate hot gas source like the hot gas from a monopropellant combustor using, for example, hydrogen peroxide. This is an improvement over the simple pressure-fed system but it requires a separate combustion and propellant system, which is not effective from a weight standpoint. An improved version of the gas generator cycle uses the main propellants to feed an auxiliary burner to generate the necessary hot gas for the drive turbine. This eliminated the need for additional tankage and handling of a separate dedicated propellant. The next improvement was the tap-off cycle, which diverts (taps off) some hot gas from the main combustor and ports it to the drive turbine products, eliminating the need for a separate combustion chamber and the associated valves and controls. However, this approach reduces thrust, as the waste gas is dumped overboard at low energy, as well as compromising the integrity of the combustion chamber with the tap-off opening. The next advancement is to produce the high temperature gas needed for the turbine by heating the fuel as it flows through a cooling jacket around the nozzle. In this fashion, the waste heat arising from thermally protecting the nozzle is used efficiently to drive the power turbine. This gas fraction that drives the turbine may be returned to the combustion chamber or ported overboard.

More complicated systems may be employed. For example, the SSME, which operates at a combustor chamber pressure over 20 MPa, uses a two-stage combustion process. In the first stage, fuel and oxidizer are fed to two preburners, the hot combustion gases of which are used to power a fuel and an oxidizer turbopump, and then are fed to the main combustor where the second, full, stage of combustion is carried out.

An empirical approximation to the weight of the pumps, turbine, and piping system is

$$W_{ps} = AQp^{\frac{2}{3}} + B,$$

where $Q_f$ is the volumetric flow rate in ft$^3$/s and $p$ is the discharge pressure in lb/ft$^2$. In English units, the quantity $A = 1$s-lb$^{1/3}$ft$^{-5/3}$ and serves to reconcile units and $B = 100$ lb. Centrifugal pumps are generally used for large flow rates and lower pressures, whereas axial pumps are used for moderate flow rates and higher pressures. The volumetric flow rate that must be handled by a pump is

$$Q_f = \frac{m_p(1+\phi)}{\rho_p}.$$

Quantity $\phi$ is the fraction of the propellant used to power the turbine, and for discharge pressures around 7 MPa is typically in the range $0.02 < \phi < 0.05$; this is small enough to make the turbopump-fed propellant feed system quite practical.

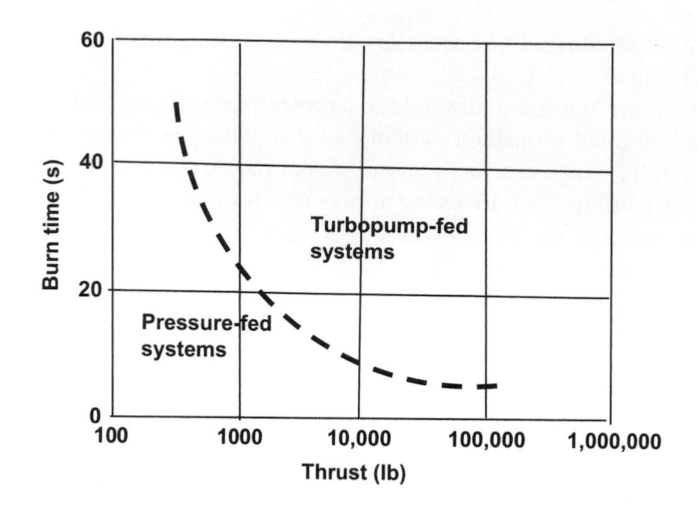


FIGURE 11.21

Typical regimes of application of different propellant feed systems.

The regimes for which pressure-fed and turbopump-fed systems are practical are depicted broadly in Figure 11.21 as a function of burn duration and thrust level.

## 11.9 NOMENCLATURE

| | |
|---|---|
| $A_e$ | nozzle exit cross-sectional area |
| $A_t$ | nozzle throat cross-sectional area |
| $a$ | sound speed |
| $c^*$ | characteristic velocity |
| $c_d$ | nozzle discharge coefficient |
| $c_F$ | nozzle thrust coefficient |
| $c_h$ | convective heat transfer coefficient |
| $c_V$ | nozzle velocity coefficient |
| $c_w$ | nozzle weight flow coefficient |
| $F$ | thrust |
| $g$ | acceleration of gravity |
| $h$ | enthalpy |
| $I_{sp}$ | specific impulse |
| $L$ | length |
| $L^*$ | characteristic length for chemical reaction |
| $L_V$ | latent heat of vaporization |
| $M$ | Mach number |
| $\dot{m}$ | mass flow |

| | |
|---|---|
| $h$ | enthalpy |
| $p$ | pressure |
| $\dot{q}_c$ | convective heat flux |
| $R$ | gas constant |
| $R_c$ | cylindrical tank radius |
| $R_s$ | spherical tank radius |
| $R_u$ | universal gas constant |
| $r$ | radius |
| $S$ | nozzle surface area |
| $T$ | temperature |
| $t_c$ | thickness of cylindrical tank skin |
| $t_s$ | thickness of spherical tank skin |
| $t_r$ | residence time |
| $V$ | velocity |
| $W_c$ | weight of cylindrical tank |
| $W_s$ | weight of spherical tank |
| $w$ | weight flow $= mg$ |
| $z$ | altitude |
| $z_s$ | atmospheric scale height |
| $\alpha$ | conical nozzle expansion angle |
| $\Gamma$ | function of $\gamma$, Equation (11.3) |
| $\Gamma'$ | function of $\gamma$, Equation (11.22) |
| $\gamma$ | ratio of specific heats |
| $\delta$ | atmospheric pressure ratio $p_0/p_{0,s.l.}$ |
| $\varepsilon$ | nozzle expansion ratio $A_e/A_t$ |
| $\phi$ | equivalence ratio |
| $\lambda$ | thrust loss factor due to flow angularity |
| $\rho$ | gas density |
| $\sigma$ | angle between nozzle axis and any streamline |
| $\sigma_A$ | axial stress |
| $\sigma_y$ | yield stress |
| $\eta$ | efficiency |

### 11.9.1 Subscripts

| | |
|---|---|
| $a$ | axial direction |
| $c$ | combustion chamber |
| $cr$ | critical |

| | |
|---|---|
| $e$ | nozzle exit |
| $f$ | fuel |
| $i$ | ideal |
| $m$ | mean |
| max | maximum conditions |
| min | minimum conditions |
| $n$ | nozzle |
| $s.l.$ | sea level conditions |
| $t$ | stagnation conditions |
| $vac$ | vacuum conditions |
| 0 | free stream conditions |

### 11.9.2 Superscript

| | |
|---|---|
| * | critical conditions, $M = 1$ |

## 11.10 EXERCISES

**11.1** An $H_2$–$O_2$ rocket engine is under development in an experimental facility and certain estimates of performance and geometry must be made.

**(a)** A combustion chamber and associated propellant feed system has been designed, built, and tested. During the test program it was found that for a combustion chamber pressure of 20 atmospheres and an injection temperature of 298.16K the $H_2$–$O_2$ reaction occurs stoichiometrically. Under these conditions, calculate the temperature in the combustion chamber.

**(b)** A rocket nozzle was fitted to the combustion chamber and the complete unit operated on a special test stand. The nozzle exit plane was monitored and the pressure there was found to be equal to that of the ambient atmosphere in the test chamber while the temperature was measured to be 1700K. Determine the specific impulse of the engine under these conditions.

**(c)** The static thrust of the rocket engine, as measured on the test stand, was found to be 100,000 lbs. Calculate the weight flow of propellant required and the diameter of the nozzle throat.

**(d)** The area ratio, exit to throat, of the nozzle on this engine was 42.3. Find the ambient pressure in the test facility.

**11.2** A rocket engine that uses $H_2$ as the fuel and $O_2$ as the oxidizer is being designed to produce 20,000 lb of thrust. The engine gas properties are assumed to be as follows: $\gamma = 1.25$ and

$W = 9.80$ lb/lb-mol; the universal gas constant $R_u = 49{,}720$ ft$^2$/s$^2$-R. The chamber pressure $p_c = 300$ psia and the nozzle exit pressure $p_e = 14.7$ psia. The engine is to consume 60 lb/s of propellant, and the nozzle is to operate in the matched mode. Under these conditions, determine the following: (a) nozzle thrust coefficient $c_F$ and specific impulse; (b) throat diameter $d_t$; (c) nozzle exit diameter $d_e$, velocity $V_e$, and temperature $T_e$; (d) chamber temperature $T_c$; and (e) fuel weight flow rate and oxidizer weight flow rate for the case where the chamber temperature (in degrees K) depends on the number of moles of $H_2$ and $O_2$ in the following fashion:

$$T_c = 4200 - 300\frac{n_{H_2}}{n_{O_2}}.$$

**11.3** A liquid propellant rocket motor using a JP-4/RFNA propellant combination must produce 19.6 kN of thrust at sea level. The characteristics of the combustion products are combustion chamber temperature $T_c = 3000$K, average specific heat ratio $\gamma = 1.2$, molecular weight $W = 27$, and chamber pressure $p_c = 19.34$ atmospheres. Calculate the following: (a) the throat and exit areas, $A_t$ and $A_e$, for matched nozzle exit flow at sea level assuming a nozzle efficiency $\eta_n = 95\%$; (b) the characteristic velocity $c^*$, the propellant mass flow rate, and the specific impulse of the engine at sea level; (c) the thrust developed at an altitude of 11.5 km where the pressure is $p_0 = 0.2$ atmospheres; and (d) for comparison purposes, calculate the thrust and exit area at this altitude for matched nozzle exit operation.

**11.4** The Rocketdyne RS-X engine was a developmental outgrowth of the RS-27 engine (Figure E11.1) used on Delta launchers. It is an RP-1/LOX engine that develops an $I_{sp} = 264$ s at sea level and an $I_{sp,vac} = 299$ s in space. The engine O/F = 2.25, and the chamber pressure of $p_c = 775$ psia feeds a nozzle with an expansion ratio $\varepsilon = 10$. The engine weighs approximately 4500 lbs and is 134 in. long and 72.6 in. wide (nozzle exit diameter). The nozzle has an angular divergence $\alpha = 12°$ at the exit plane. The combustion chamber temperature may be taken as $T_c = 6400$R, while the molecular weight and ratio of specific

**FIGURE E11.1**

The Pratt and Whitney Rocketdyne RS-27A powers the first stage of the United Launch Alliance Delta II launch vehicle.

heats are considered constants at the values 21.9 and 1.135, respectively. Experiments have shown that the nozzle discharge and velocity coefficients are $c_d = 0.05$ and $c_V = 0.903$, respectively. Under these conditions and assuming sea level static test conditions, determine the (a) exit and throat areas $A_e$ and $A_t$, (b) weight flow rate of the propellants, (c) thrust produced $F$, (d) thrust coefficient $c_F$, (e) nozzle exit Mach number $M_e$, (f) nozzle exit pressure $p_e$, (g) nozzle exit velocity $V_e$, (h) effective exhaust velocity $V_{eff}$, and (i) characteristic velocity $c^*$.

**11.5** An RP-1/LOX propellant combination (O/F = 2.28) is used in a rocket motor producing a thrust $F = 49.1$ kN at sea level. The chamber pressure $p_c = 20$ atmospheres and the combustion products have the following characteristics: the combustion chamber temperature may be taken as $T_c = 3430$K, while the molecular weight and ratio of specific heats are considered constants at the values 21.98 and 1.226, respectively, and are assumed constant throughout the expansion. Note that this last assumption is approximate and that the actual values will change somewhat as the flow proceeds down the nozzle and the pressure and temperature drop. Determine the (a) nozzle throat and exit areas for matched nozzle exit flow at sea level conditions; (b) characteristic velocity and specific impulse at sea level; and (c) evolution of $p$, $T$, $M$, and $V$ along the nozzle whose coordinates are given in the following table in terms of axial distance $x/r_t$ and corresponding radial distance $r/r_t$. The origin is at the throat, and the exit radius is not given because it needs to be calculated. Using the final dimensional results, also draw the nozzle.

| $x/r_t$ | −2.60 | −2.0 | −1.5 | −1.0 | −0.5 | 0 | 0.5 | 1.0 | 1.5 | 2.0 | 3.15 |
|---|---|---|---|---|---|---|---|---|---|---|---|
| $r/r_t$ | 2.65 | 2.5 | 2.1 | 1.65 | 1.25 | 1 | 1.15 | 1.35 | 1.5 | 1.7 | calc. |

**(d)** Find the residence time of a fluid particle in the nozzle.

**11.6** Consider the gaseous propellant system described by the reaction

$$4H_2 + O_2 \rightarrow products.$$

Note that this reaction is fuel rich and it should be possible to reduce the number of product species accordingly so as to simplify the calculation, if desired. Use $p = 20$ atm for the chamber pressure. Determine the (a) equivalence ratio $\phi$ and (b) adiabatic flame temperature $T_{ad}$. If a combustor operating under these conditions is part of a rocket engine operating on a static test stand and the nozzle is matched at an exit pressure of $p_e = p_0 = 1$ atm, calculate the following for the case of frozen flow in the nozzle and for the case of equilibrium flow in the nozzle: (c) exit temperature $T_e$, (d) mixture molecular weight at the exit $W_{e,mix}$, (e) exit velocity $V_e$, (f) specific impulse $I_{sp}$, and (g) exit area $A_e$.

**11.7** Consider a rocket motor operating with a chamber pressure and temperature of $p_c = 2.07$ MPa and $T_c = 2860$K, respectively, and a nozzle that produces the optimum thrust of 1334 kN when exhausting into the atmosphere under standard sea level conditions. Exhaust gases are characterized by a mean molecular weight and isentropic exponent $W = 21.87$ kg/kg-mol and $\gamma = 1.229$, respectively. Determine the following properties of the rocket motor: (a) thrust coefficient $c_F$, (b) characteristic exhaust velocity $c^*$, (c) nozzle minimum area $A_t$, and (d) nozzle exit area $A_e$.

**11.8** A rocket nozzle is shown (Figure E11.2) being tested in the USAF rocket test cell at the Arnold Engineering Development Center. The nozzle, which operates with a stagnation pressure $p_t = 2.76$ MPa and an area ratio $A_{exit}/A_{min} = 5.16$, was designed for optimal operation at an altitude $z = 30.5$ km, but launch takes place at sea level ($p = 101.3$ kPa). Find $F_{launch}/F_{30.5\ km}$.

**FIGURE E11.2**

A rocket nozzle operating in an under-expanded mode shows an expanding exhaust plume while firing in a rocket test cell.

## References


Angelino, G. (1964). Approximate method for plug nozzle design. *AIAA Journal, 2*(10).

Anonymous (1970). Solid Rocket Motor Metal Cases. NASA SP-8025, April.

Barrere, M., Jaumotte, A., Fraeijs De Veubeke, B., & Vandenkerckhove, J. (1960). *Rocket Propulsion*. New York: Elsevier.

Bartz, D. R. (1957). A simple equation for rapid estimation of rocket nozzle convective heat transfer coefficients. *Jet Propulsion,* January, 49–51.

Conley, D., Lee, N. Y., Portanova, P. L., & Wood, B. K. (2002). Evolved Expendable Launch Vehicle System: RS-68 Main Engine Development. 53rd International Astronautical Congress, Oct. 10-19, Houston, TX.

Kit, B., & Evered, D. S. (1960). *Rocket Propellant Handbook*. New York: Macmillan.

Liepmann, H. W., & Roshko, A. (1957). *Elements of Gasdynamics.* New York: Wiley.

Martin, J. A. (1989). Rocket nozzle expansion ratio analysis for dual fuel earth-to-orbit vehicles. *Journal Spacecraft, 26*(3), May-June, 196–198.

Moes, T. R., Cobleigh, B. R., Conners, T.R., Cox, T. H., Smith, S. C., & Shirakata, N. (1996). Wind Tunnel Development of an SR-71 Aerospike Rocket Flight Test Configuration. NASA Technical Memorandum 4749.

Onofri, M. (2002). Plug Nozzles: Summary of Flow Features and Engine Performance. AIAA 2002-0584, 40th AIAA Aerospace Sciences Meeting, Jan. 14-17, Reno, NV.

Rao, G. V. R. (1958). Exhaust nozzle contour for maximum thrust. *Jet Propulsion, 28*(6), June, 377–382.

Siegel, B., & Schieler, L. (1964). *Energetics of Propellant Chemistry.* New York: Wiley.

Sutton, G. P. (1992). *Rocket Propulsion Elements* (6th ed.). New York: Wiley.

Williams, O. S., et al. (1976). Liquid rockets in perspective: Developments in the 1960s. *Aerospace America,* March, 47–57.

CHAPTER

# Solid Propellant Rockets

# 12

## CHAPTER OUTLINE



**Theory of Aerospace Propulsion.**

## 12.1 SOLID ROCKET DESCRIPTION

The solid rocket is so called because it consists of a mixture of oxidizer and fuel in the form of a solid completely encased in a tank that serves as a pressure vessel and to which is attached a nozzle. Thus there is no need for separate tanks of oxidizer and fuel or for means to introduce them into the combustion chamber in correct ratio. The solid rocket tank is sturdy because it must contain the high-pressure gases formed by the chemical reaction of the fuel and the oxidizer and therefore it is relatively heavy. However, because of its sturdiness, it requires no special means for stabilizing it structurally. The main advantage of solid rockets is their simplicity. They have no moving parts, no injection system, and do not have to be fueled prior to launch. Therefore the storage, handling, service, and auxiliary equipment connected with solid rockets are much simpler than those for liquid rockets. Because of the low part count and few, if any, moving parts, the reliability of solid rockets is high, around 99%. For the same reasons, the payload mass ratio is high and the overall cost low compared to liquid rockets. These attributes of simplicity of structure, easy storability of the solid propellant, and virtually instantaneous availability for launch make the solid rocket motor an attractive propulsion device.

The solid propellant contains both fuel and oxidizer and they are ignited by electric or pyrotechnic means in order to permit gasification to initiate. Thereafter the propellant burns in the gas phase with the heat of the combustion being sufficient to continue the gasification process. The rocket design is very simple, as shown in Figure 12.1.

Such assets do not come without penalties. For example, because the solid propellant mixture must be cast into the required shape as one piece, there are limitations in the size of the propellant blocks achievable. Thus, application of solid rocket motors to lower thrust levels is more attractive, and clusters of smaller such motors are typically used to achieve higher thrust levels. However, there are successful large-scale solid rocket motors, such as those used to launch the space shuttle, which produce more than 1 million pounds of thrust. Another major disadvantage is that the performance, as

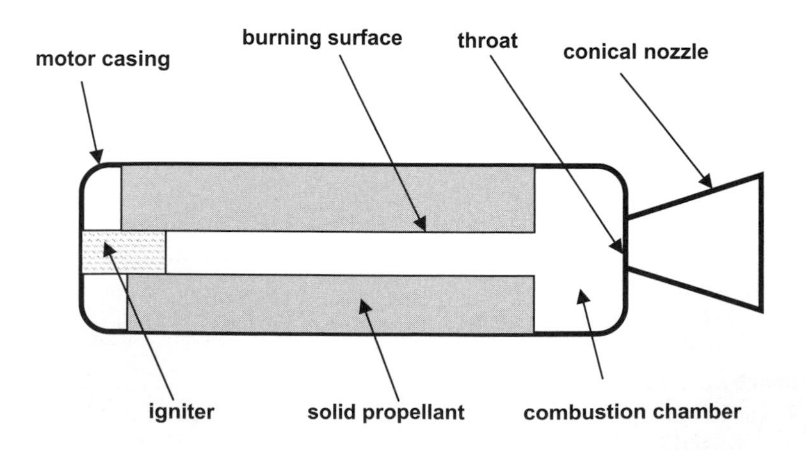


**FIGURE 12.1**

Schematic diagram of a solid propellant rocket motor.

measured by specific impulse, is lower relative to liquid rockets, and modulating and terminating the thrust are both difficult. Burn duration is relatively short compared to that in liquid rockets and therefore the total impulse delivered is smaller. Finally, there is no means for adequately cooling the nozzle of a solid rocket motor like there is in a liquid rocket motor where the liquid propellants may be circulated through cooling jackets built into the nozzle. The performance of solid propellants is quite sensitive to temperature, and the associated manufacture of the energetic materials is involved, costly, and difficult. Because the gases produced result from gasification and combustion at the burning face, rather than from feed lines from a pump or pressurized tank, the controlling factor is the burning rate $r$, which is discussed in the following section.

## 12.2 SOLID PROPELLANT GRAIN CONFIGURATIONS

The propellant grain is the premixed fuel and oxidizer bonded into an integral mass. The material is a viscous pourable material that may be cast into the appropriate shape wherein it hardens to a rubbery solid. Characteristics required of the grain are as follows:

1. It must be contained within the combustion chamber.
2. It must be retained during burning, at chamber ends and walls.
3. It must burn only at the exposed surfaces and must not explode.
4. It must ignite readily at temperatures around 250 to 350°C.
5. It must be structurally robust and resistant to cracks.

The combustion of the solid propellant generates large quantities of hot gas in very little time. The two types of grain generally used are described here.

### 12.2.1 Homogeneous propellant

The chemical molecule contains both fuel and oxidizer. These are referred to as double-base propellants; an example is JPN, the typical propellant for the rockets used in World War II. It is a mixture of nitroglycerin (43%) and nitrocellulose (51%) plus some additives (7%). Such propellants are characterized by low combustion temperature, low specific impulse ($I_{sp} < 250$ s), burning rates that are strongly temperature dependent, and almost smokeless exhaust gases.

### 12.2.2 Heterogeneous or composite propellant

Small discrete particles of oxidizer are mixed uniformly into an organic plastic-like fuel binder. The fuel is generally a thick organic liquid that may be transformed into a rubber-like substance using a catalyst. Composite propellants may be subdivided into rubber-like materials that can be poured into rocket motor casings and bond directly to the walls and rigid materials, which are mounted as subassemblies within the motor casing. An example of a composite propellant is PBAN, which is used in the solid rocket boosters for the space shuttle launch vehicle: ammonium perchlorate (70%) oxidizer and polybutadiene acrylonitrile (PBAN) (14%) and aluminum powder (16%) fuel. The inorganic oxidizer ammonium perchlorate ($NH_4ClO_4$) has 34% available oxygen by weight, but also provides hydrogen for the combustion process. Ammonium perchlorate has a molecular weight $W = 117.5$ and a density of 1950 kg/m$^3$. It is an unstable substance and poses a dangerous fire hazard. Upon heating it decomposes without melting, that is, it sublimates. During combustion in solid rockets, hydrochloric acid (HCl) is produced in the products. This material is highly corrosive and forms a tell-tale semi-opaque white cloud when exhausted into the atmosphere.

### 12.2.3 Grain cross sections

Although the outer boundary of the grain is generally circular, and that is the shape of typical casings, the inner boundaries may be quite complex because they form the burning surface and their shape is important in controlling the gas generation and pressure buildup process, which will be described in a later section. Some typical grain configurations are illustrated in Figure 12.2. The end-burning (or cigarette burning) grain is the most common and is used in a later section to analyze the chamber pressurization process. A star-shaped grain is used in the space transportation system (i.e., the space shuttle) redesigned solid rocket motor, which is discussed in more detail in a subsequent section.

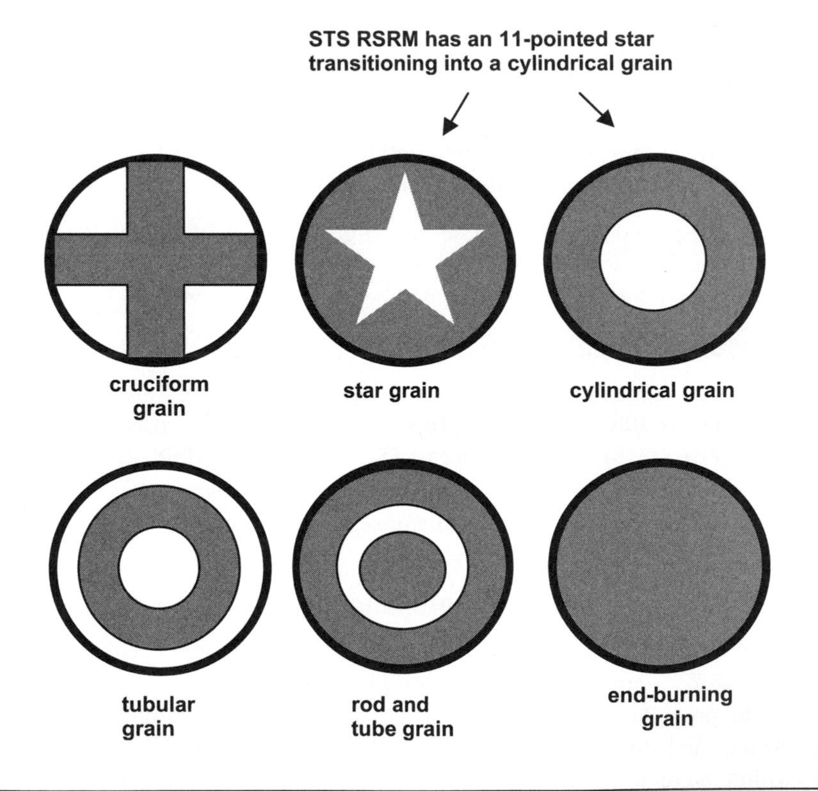


**FIGURE 12.2**

Cross-sectional view of different solid propellant grain configurations. The outer black line is the motor casing, shaded area is the propellant grain, and white areas denote the region into which the gas generated flows. The end-burning grain burns over the surface shown. The STS RSRM refers to the space transportation system (more commonly, the space shuttle) redesigned solid rocket motor.

The grain is separated from the motor casing by a layer of insulating material, which serves as a thermal barrier to limit the temperature rise of the casing. This permits the casing to retain suitable strength properties to maintain casing integrity. The grain web thickness characterizes the minimum initial thickness of the grain, measured between the burning surface and the insulating layer or between two burning surfaces. For end burning grains the web thickness is merely the length of the grain. Because the grain burning is not perfectly symmetrical, there may be some patches of unburned propellant remaining even after the web is consumed, which are referred to as slivers.

## 12.3 BURNING RATE

The burning rate ($r$) is the speed at which the burning face recedes, measured normal to that surface, and is typically measured in inches per second or centimeters per second. The burning rate is well characterized by the power law relation

$$r = ap_c^n. \tag{12.1}$$

In this equation, $p_c$ is the chamber pressure; $n$ is a propellant characteristic, which is independent of propellant temperature $T_p$; and $a$ is a temperature-dependent function, which is characteristic of the propellant. It is found that the temperature-dependent function may be represented by the following:

$$a = \frac{A}{T_1 - T_p} = a_o\left[\tau\left(T_{pi} - T_{pi}^0\right)\right]. \tag{12.2}$$

In this equation, the following notation applies:

$A$ = constant
$T_l$ = self-ignition temperature = constant
$a_0$ = $a$ at $T_p = T_{pi}^0$
$T_{pi}^0$ = reference temperature, say 15°C
$T_{pi}$ = initial temperature of propellant
$\tau$ = constant

We characterize the power-law behavior of Equation (12.1) as that shown in Figure 12.3.

By varying the nature of the propellant grain, various burn rates may be produced, as shown in Figure 12.4.

## 12.4 GRAIN DESIGN FOR THRUST-TIME TAILORING

The thrust produced by the rocket with a matched nozzle is given by

$$F = \dot{m}V_e.$$

The exit velocity is

$$V_e = \sqrt{2h_c\left(1 - \frac{h_e}{h_c}\right)}.$$

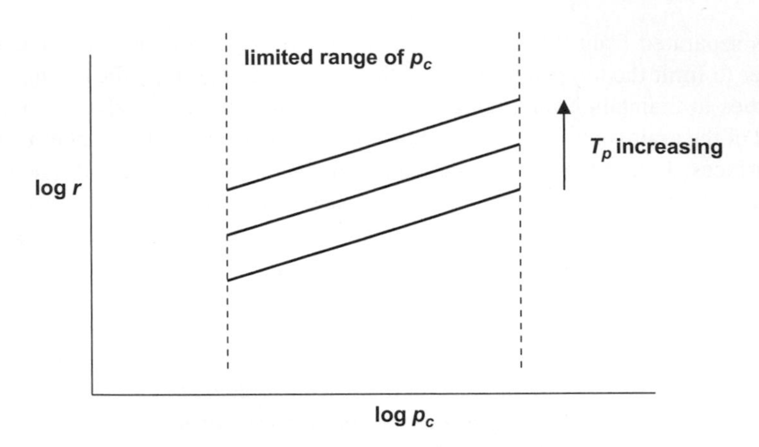


**FIGURE 12.3**

Power-law behavior of the burning rate as a function of chamber pressure.

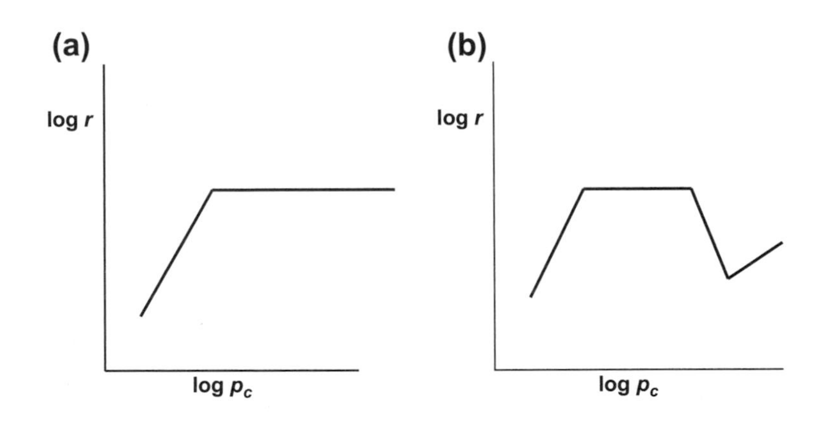


**FIGURE 12.4**

Different propellant grains and their burning characteristics: (a) plateau burning grain and (b) mesa burning grain.

Thus exit velocity depends on the heat release in the combustion chamber, which is dependent on the propellant, and the nozzle design. However, the mass flow produced is

$$\dot{m} = \rho_p(rA_b).$$

For a given propellant, the burning rate $r$ and the propellant density $\rho_p$ are essentially fixed so that the burning area $A_b$ determines the thrust signature as a function of time:

$$F(t) \sim A_b(t).$$

By designing the grain cross-section appropriately, one can tailor different thrust-time histories, as shown in Figure 12.5.

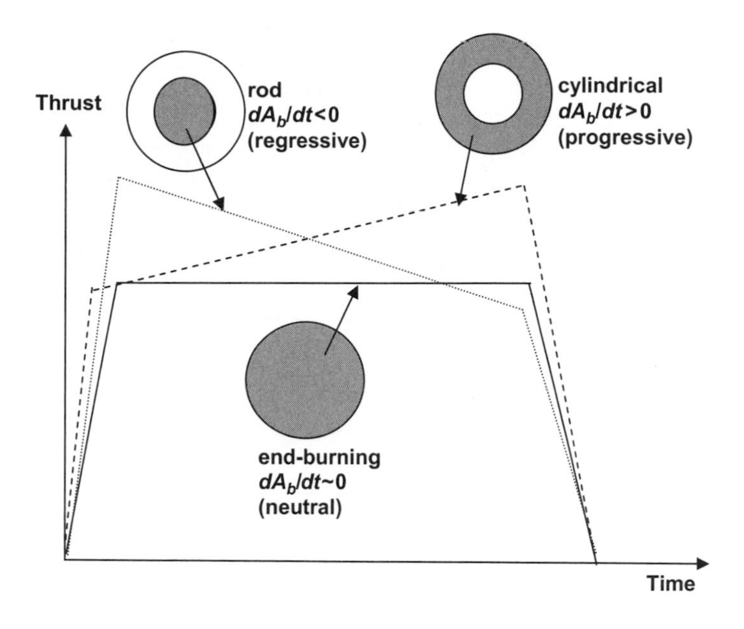


**FIGURE 12.5**

Variation of thrust-time profiles depending on grain cross section.

Similarly, the grain cross section can also be varied longitudinally by tapering the cross section or varying the grain shape. The time rate of change of the burning area so achieved can be used to influence the thrust-time history.

## 12.5 COMBUSTION CHAMBER PRESSURE

Because the combustion chamber pressure is the dominant factor in solid rockets, we must determine how this varies as the propellant is consumed. For simplicity, consider an end-burning (cigarette burning) grain where the burning surface is receding at a rate $r$, as shown in Figure 12.6.

### 12.5.1 Mass conservation analysis

If we consider the control volume to be the inside of the motor case up to the throat, we may apply the integral mass conservation equation

$$\frac{\partial}{\partial t}\int_{v} \rho dv + \int_{A} \rho \vec{V} \cdot \hat{n} dA = 0. \tag{12.3}$$

The only place where material crosses the control volume is at the throat so

$$\int_{A} \rho \vec{V} \cdot \hat{n} dA = \dot{m}_t = \frac{p_c A_t}{\sqrt{R T_c}} \Gamma. \tag{12.4}$$

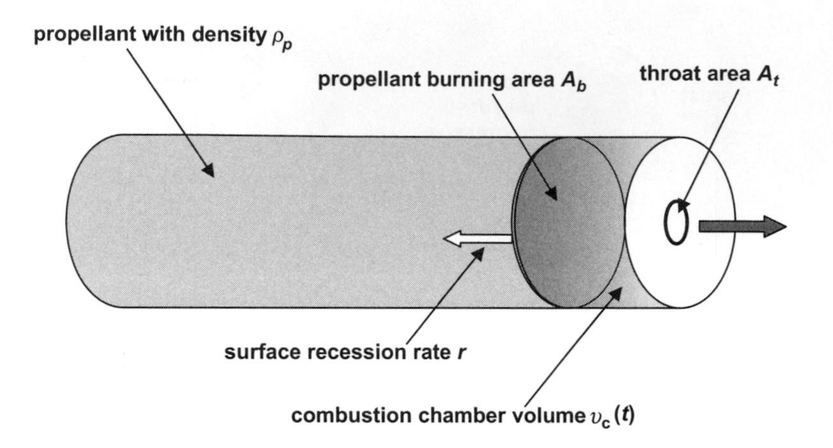


FIGURE 12.6

Schematic diagram of combustion chamber volume formed by recession of the burning surface.

We assume the throat is choked and that

$$\Gamma = \sqrt{\gamma\left(\frac{2}{\gamma+1}\right)^{\frac{\gamma+1}{\gamma-1}}}. \tag{12.5}$$

The first term of the conservation of mass equation, Equation (12.3), is made up of two parts:

$$\frac{\partial}{\partial t}\int\limits_{\upsilon_p}\rho_p d\upsilon + \frac{\partial}{\partial t}\int\limits_{\upsilon_c}\rho_c d\upsilon = \rho_p\frac{d\upsilon_p}{dt} + \frac{\partial}{\partial t}\rho_c\upsilon_c. \tag{12.6}$$

The first term on the right side of Equation (12.6) is the rate of change of propellant mass within the motor case due to transformation to gases, and noting that the propellant density doesn't change with time. The second term on the right side is true if we assume that the density in the combustion chamber is distributed uniformly throughout $\upsilon_c$.

The volume occupied by the propellant decreases as the burning face recedes. That volume may be expressed as $V_p = A_b L(t)$, where $L$ is the length of the charge. Then

$$\frac{d\upsilon_p}{dt} = A_b\frac{dL}{dt} = -rA_b \tag{12.7}$$

since the change of $L$ with respect to time is the rate at which the burning face recedes.

The mass conservation equation is then

$$-\rho_p r A_b + \frac{\partial}{\partial t}(\rho_c\upsilon_c) + \frac{p_c A_t}{\sqrt{RT_c}}\Gamma = 0. \tag{12.8}$$

The second term may be written as

$$\upsilon_c\frac{\partial\rho_c}{\partial t} + \rho_c\frac{\partial\upsilon_c}{\partial t} = \upsilon_c\frac{\partial\rho_c}{\partial t} + \rho_c r A_b. \tag{12.9}$$

This shows that the combustion chamber volume increases at exactly the same rate as the propellant volume decreases, which causes Equation (12.8) to become, assuming that the chamber temperature remains constant,

$$(\rho_c - \rho_p) r A_b + \frac{v_c}{RT_c}\frac{\partial p_c}{\partial t} + \frac{p_c A_t}{\sqrt{RT_c}}\Gamma = 0. \tag{12.10}$$

Our mass conservation equation is then

$$\frac{dp_c}{dt} + \left[(\rho_c - \rho_p) a A_b \frac{RT_c}{v_c}\right] p_c^n + \left[\frac{\sqrt{RT_c}}{v_c} A_t \Gamma\right] p_c = 0. \tag{12.11}$$

Now the combustion chamber volume is

$$v_c(t) = v_0 + \int_0^t r A_b dt = v_0 + \int_0^t A_b a p_c^n dt. \tag{12.12}$$

## 12.5.2 Equilibrium chamber pressure

Equations (12.11) and (12.12) show that an integrodifferential equation must be solved for the chamber pressure $p_c$ as a function of time. Instead of doing that, we can inquire into what happens when $p_c$ steadies out and $(dp_c/dt) = 0$. Then the equilibrium chamber pressure is

$$p_c = \left[\frac{A_b}{A_t}\frac{\sqrt{RT_c}}{\Gamma} a(\rho_p - \rho_c)\right]^{\frac{1}{1-n}}. \tag{12.13}$$

Recall that the characteristic velocity is

$$c^* = \frac{\sqrt{RT_c}}{\Gamma}. \tag{12.14}$$

Because the solid propellant density is much larger than the combustion gas density ($\rho_p \gg \rho_c$), we may neglect the latter with respect to the former and therefore Equation (12.13) becomes

$$p_c = \left[\frac{A_b}{A_t} a \rho_p c^*\right]^{\frac{1}{1-n}}. \tag{12.15}$$

Because the quantity $a\rho_p c^*$ in Equation (12.15) is a characteristic of the propellant the chamber pressure,

$$p_c \sim \left(\frac{A_b}{A_t}\right)^{\frac{1}{1-n}}.$$

In general, $0.4 < n < 0.85$, resulting in $1.667 < (1 - n)^{-1} < 6.667$, illustrating the importance of the parameter $k = A_b/A_t$, the ratio of the burning area to throat area, in establishing the equilibrium chamber pressure level in the rocket motor.

### 12.5.3 Combustion chamber stability

From Equation (12.4), we see that the maximum steady mass flow leaving through the throat, which occurs when the throat is choked, is

$$\dot{m}_{leaving} \sim p_c.$$

Similarly, Equation (12.7) shows that the steady mass rate of production of gas is

$$\dot{m}_{produced} \sim p_c^n.$$

This behavior is illustrated in Figure 12.7, and the steady-state operating point is shown as the point where the mass flow produced by burning is equal to the mass flow leaving through the choked throat. It is noted that propellants with $n < 1$ are stable and those with $n > 1$ are unstable.

To examine the reason for this conclusion regarding the stability of the combustion chamber flow, we may use the information in Figure 12.8. Consider the change in mass flow $\Delta\dot{m}$ brought about by a small change in pressure $\Delta p$ when initially at the equilibrium operating point. For stable operation $\Delta\dot{m}/\Delta p < 1$ for stability, that is, any pressure increase above the operating point leads to a gas generation rate less than that which can be expelled through the nozzle so that the pressure drops back to the operating point pressure; this occurs for $n < 1$. However, when $\Delta\dot{m}/\Delta p > 1$, that is, when any pressure increase above the operating point leads to a gas generation rate greater than that which can be expelled through the nozzle, the chamber pressure climbs above the operating point pressure and may cause explosion of the case; this occurs for $n > 1$.

Appropriate thrust and pressure levels are generally obtained by appropriate choice of $k = A_b/A_t$. The thrust may be found from the following equation:

$$F = c_F p_c A_t. \tag{12.16}$$

The thrust level is determined mainly by the combustion chamber pressure for a given propellant, as can be seen by examining Equations (12.15) and (12.16), and therefore by $k$, the ratio of burning area

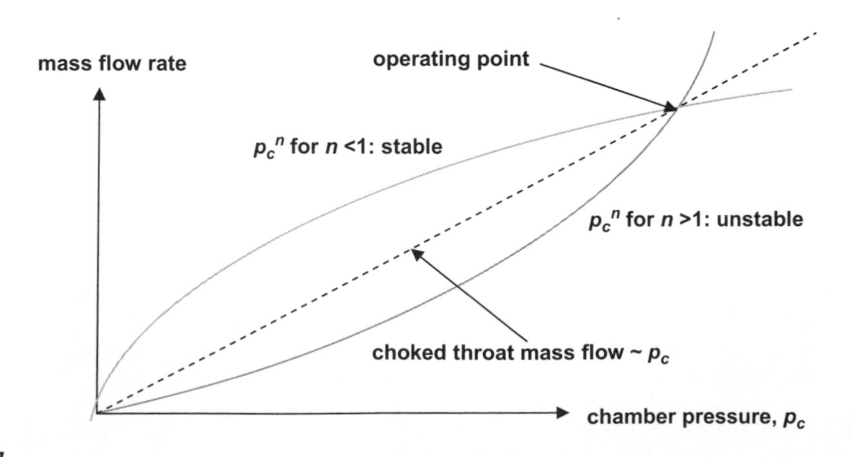


**FIGURE 12.7**

Mass flow rate behavior for burning rate exponent greater and less than unity showing stable and unstable operation.

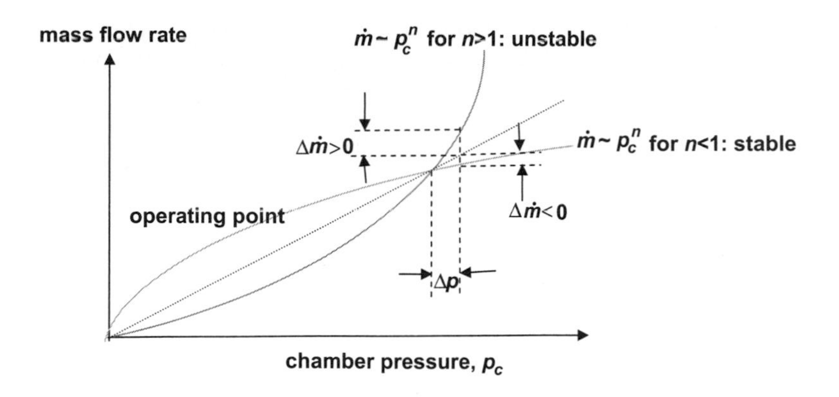


**FIGURE 12.8**

Stability of solid rocket combustion chamber pressure to pressure disturbances.

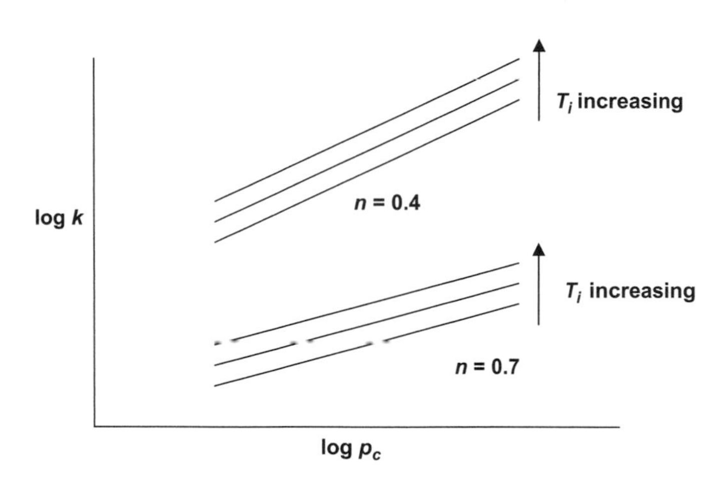


**FIGURE 12.9**

Variation of burner to throat area ratio and chamber pressure for different propellants and different initial temperatures.

to throat area. For typical stable propellants, the relationship between $k$ and $p_c$ is as shown in Figure 12.9.

### 12.5.4 Propellant performance sensitivity

One can determine the sensitivity of the propellant performance to changes in temperature by taking logarithmic derivatives of the equilibrium chamber pressure as given by Equation (12.15) under conditions that the parameter $k = A_b/A_t$ is held constant. The ratio of burning area to throat area is an important design parameter in solid rocket design and may vary between 100 and 2000. Temperature sensitivity coefficients determined in this fashion are listed in Equations (12.17) through (12.20).

$$(\pi_p)_k = \frac{1}{p_c}\left(\frac{dp_c}{dT_i}\right)_k \tag{12.17}$$

$$(\pi_r)_p = \frac{1}{r}\left(\frac{dr}{dT_i}\right)_p = \frac{1}{T_{ign} - T_i} = \tau \tag{12.18}$$

$$(\pi_r)_k = \frac{1}{r}\left(\frac{dr}{dT_i}\right)_k \tag{12.19}$$

$$(\pi_p)_k = (\pi_r)_k = \frac{1}{1-n}(\pi_r)_p. \tag{12.20}$$

## 12.6 EROSIVE BURNING

The previous treatment of burning rates ignored the motion of the hot gas over the burning area as it flows toward the nozzle. For ease of analysis, the burning rate has been evaluated as if the exposed propellant were in a large reservoir such that the gas produced moves only in a direction normal to the burning surface. However, to maximize the performance of solid rocket motors it is important to fill the casing with propellant as fully as possible. As a result, the initial cavity or port in the grain is relatively small, which leads unavoidably to high longitudinal gas velocity in the combustion cavity, particularly at the throat end of the chamber. Because the port areas of modern rockets are not much larger than the throat areas, it is likely to have near-sonic longitudinal velocities in the chamber. The typical variation of combustion chamber pressure and velocity with distance along the axis of a rocket motor with a cylindrical grain is shown in Figure 12.10. The simple one-dimensional momentum equation for frictionless flow is

$$dp + \rho V dV = 0.$$

Therefore, increases in velocity imply decreases in pressure, and because mass flow, and velocity, in the rocket motor increases with distance downstream, the pressure must decrease in that direction.

The effect of the longitudinal velocity over the burning surface is to increase the burning rate due to the enhanced convective heat transfer and viscous dissipation. This increased burning rate is called erosive burning and is usually most pronounced in slower burning propellants. The simplest means for accounting for erosive burning is to consider an increment to the usual static burning rate. For example, we may write that the total burning rate is given by the sum of the static burning rate and an erosive burning rate

$$r_{tot} = r + r_e$$

One may reasonably expect that the erosive burning rate is proportional to the heat transfer rate or $r_e \sim \dot{q}_w$. The convective heat transfer rate to a wall surface at temperature $T_w$ exposed to a gas flowing over it at temperature $T$ is taken to be proportional to the temperature difference so that

$$\dot{q}_{c,w} = c_h(T - T_w),$$

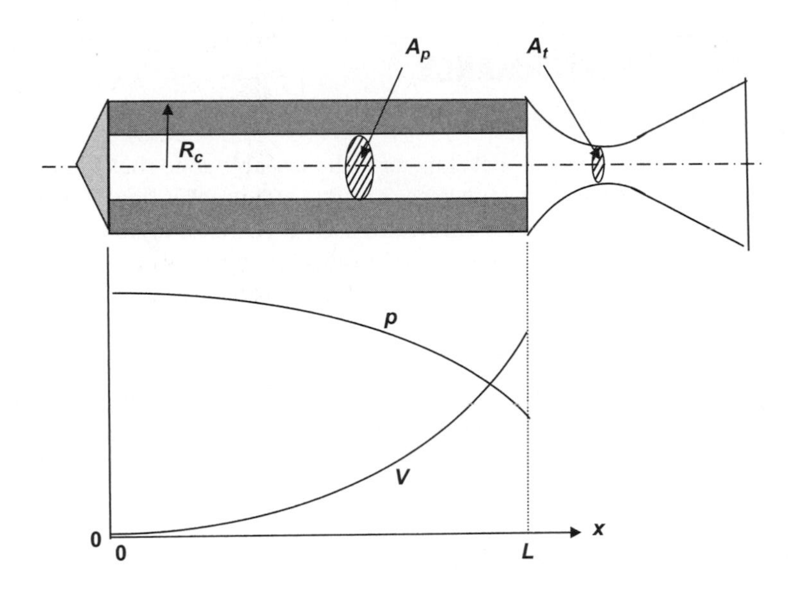


**FIGURE 12.10**

Typical longitudinal variation of pressure and velocity.

where $c_h$ is the convective heat transfer coefficient, which for flow through a circular duct may be expressed as follows (Kays and Crawford, 1993):

$$c_h = 0.026\frac{\kappa \mathrm{Pr}^{0.4}}{\mu^{0.8}}\frac{(\rho V)^{0.8}}{d^{0.2}}.$$

The quantity Pr is the Prandtl number $\mu c_p/\kappa$ where those variables denote the viscosity, specific heat, and thermal conductivity of the gas, respectively. The density and velocity of the gas are denoted by $\rho$ and $V$, respectively, and $d$ is the diameter of the duct. Then we may assume that the erosive burning rate is

$$r_e = \beta c_h(T - T_w) = \beta\left[0.026\frac{\kappa \mathrm{Pr}^{0.4}}{\mu^{0.8}}\frac{(\rho V)^{0.8}}{d^{0.2}}\right](T - T_w).$$

A similar expression that is successful in correlating experimental data is presented by Sutton (1992). The salient feature to note here is that the erosive burning increases as the flow velocity increases. Then the total burning rate may be cast functionally as follows:

$$r_{tot} \sim Ap^n + BV^{0.8}.$$

Noting, from Figure 12.10, that $p$ decreases as $V$ increases, we see that the two effects tend to counteract one another. By balancing these two effects properly, it is possible to tailor the burning history desired. In a cylindrical grain, such as the one shown in Figure 12.2, it is important to have uniform burning throughout so that the burning area reaches the motor casing simultaneously at all points thereby avoiding burn-through of the casing and failure of the motor.

## 12.7 SOLID ROCKET PERFORMANCE

### 12.7.1 Large-scale solid rocket motor

A good illustration of a large-scale solid rocket motor is the reusable solid rocket motor (RSRM) built by ATK, which is the major component of the space shuttle solid rocket booster (SRB) system. The SRB is shown schematically in Figure 12.11. The RSRM is 1513 in. (38.43 m) long and 146 in. (3.708 m) in diameter, and it is housed in the SRB, which is 1790 in. (45.47 m) long. The configuration is made up of four motor segments with an igniter system in the forward segment and an articulated nozzle attached to the aft segment.

The forward motor segment has an 11-point star grain that transitions to a cylindrical grain, the following two segments have a cylindrical grain with a double taper, and the final segment has a cylindrical grain with a triple taper. The solid propellant grain is composed of the following:

Ammonium perchlorate ($NH_4ClO_4$)—70%
Polybutadiene acrylonitrile (PBAN)—14%
Aluminum powder—16%
Iron oxidizer powder—0.07% (catalyst)

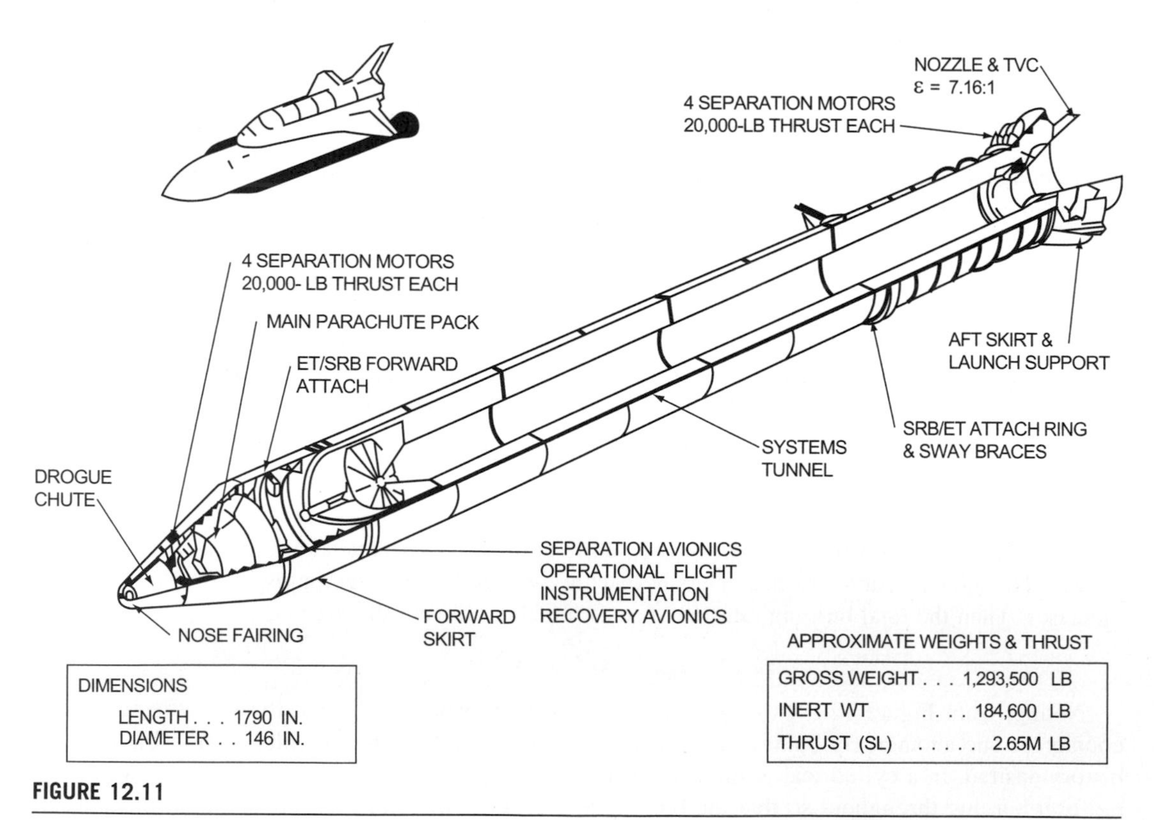


**FIGURE 12.11**

Schematic diagram of the solid rocket booster for the space shuttle (courtesy of NASA).

Ammonium perchlorate produces HCl in the combustion products and forms a white cloud when exhausted into even mildly humid air. PBAN is a polymeric rubber-based binder that also serves as fuel. Aluminum ($HV = 32$ MJ/kg) is added to the grain to enhance the thrust. The RSRM operating temperature range is from 20 to 120F (−6 to 49°C). The total weight of propellant is about $1.106 \times 10^6$ lb ($5 \times 10^5$ kg), while the total weight of the RSRM is about $1.255 \times 10^6$ lb ($5.698 \times 10^5$ kg). The burnout weight is about 141,000 lb (63,840 kg), which suggests a propellant mass fraction of roughly 89%. Other characteristics of the propellant are given in Table 12.1.

The RSRMs in the two SRBs of the space transportation system RSRM deliver a thrust-time history as shown in Figure 12.12. The pressure-time history has a similar shape with a maximum value of around 900 psia (68.9 MPa).

The average vacuum level of thrust is $2.59 \times 10^6$lb (11.52 MN), and the associated vacuum-specific impulse is 268 s. The action time, the elapsed time from a specified pressure (here 563.5 psia or 3.88 MPa) to a specified final pressure (here 22.1 psia or 152 kPa), is about 124 s.

A new large solid rocket motor manufactured by Avio in Italy, the P80, is designed for future evolution of the boosters for the ESA Ariane-5 launch vehicle. It is 3 m in diameter and 10.5 m long and has a mass of 88,000 kg. The propellant is based on hydroxyl terminated polybutadiene (HTPB) fuel and has high aluminum powder content. The maximum thrust is about 3040 kN (683 klb) and the burn time is about 107 s. The specific impulse is given as 280 s. Another large solid rocket motor is the J-1 first-stage motor developed by the National Space Development Agency of Japan. This is a 21-m-long and 1.8-m in diameter rocket motor with a weight of 71 tons (142 klb). The propellant is also based on polybutadiene fuel and delivers 159 tons of thrust at a specific impulse of 273 s with a burn time of 94 s.

### 12.7.2 Dual-thrust rocket motors

In some rocket applications, large variations in thrust may be required. Typically two levels of thrust will be needed: high thrust for accelerating the fully fueled and therefore relatively heavy rocket to the desired speed—the boost phase—and a much lower level of thrust to maintain high speed of the now much lighter rocket motor due to fuel consumption—the sustainer phase. Such a two-phase operation is

**Table 12.1** Properties Of RSRM Propellant Grain

| Propellant Property | English Units | SI Units |
|---|---|---|
| Burn rate $r$ | 0.434 in./s at 1000 psia | 1.1 cm/s at 68.9 MPa |
| Burn rate exponent $n$ | 0.35 | 0.35 |
| Density | 0.063 lbs/in$^3$ | 1742 kg/m$^3$ |
| Temperature coefficient $\pi_k$ | 0.115%/F | 0.207%/C |
| Characteristic velocity $c^*$ | 5053 ft/s | 1540 m/s |
| Adiabatic flame temperature | 6552R | 3640K |
| Effective $\gamma$ in chamber | 1.138 | 1.138 |
| Effective $\gamma$ at nozzle exit | 1.147 | 1.147 |
| Initial burning surface area to throat area ratio $k$ | 200.2 | 200.2 |
| Web thickness | 41.5 in. | 105.4 cm |

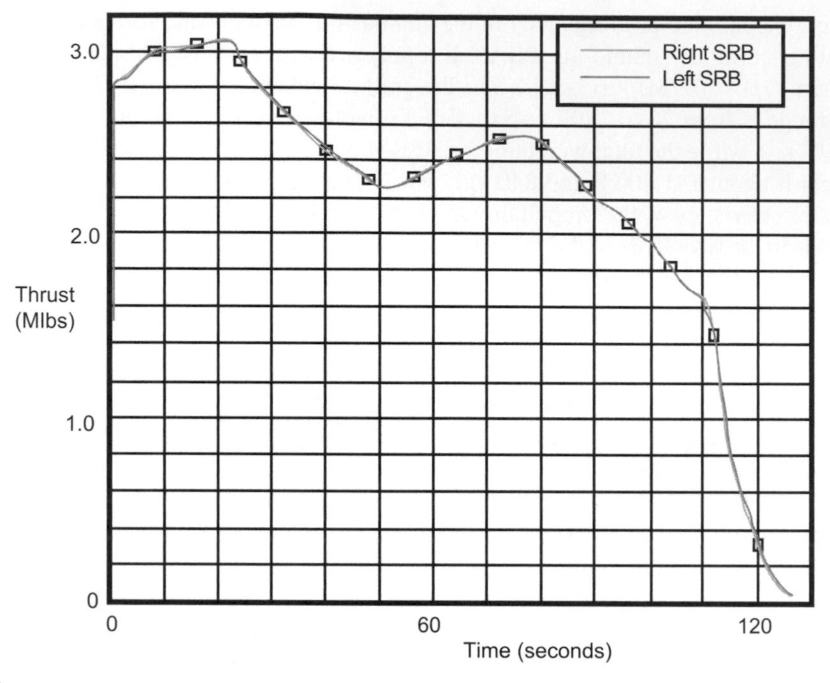


**FIGURE 12.12**

Thrust-time history for left and right RSRM in the space transportation SRBs under sea level conditions. Note that differences are imperceptible on this scale.

met relatively easily by solid rocket motors. Either grains with two different burning rates or grains with two different cross-sectional shapes can provide the modulation described. Two possible configurations are illustrated in Figure 12.13 in one of each of these cases. One has two grains with different burning rates, whereas the other has one grain with two different grain configurations. Note that the RSRM described previously has a variable grain cross section to achieve the desired performance.

## 12.7.3 Solid rocket motor casings

Materials used for the motor casing must support the grain and contain the combustion chamber pressure during operation. Materials usually used for this purpose include the steel and aluminum alloys, as well as titanium and, increasingly, composite materials. The basic mechanical properties of the metallic materials were presented in Table 11.8 of the previous chapter. Composite materials have become more popular because the high tensile strength of various nonmetallic fibers, such as glass, Kevlar, carbon, and other materials, can be exploited with new manufacturing techniques such as filament winding. Thin fibers are wound continuously in a particular pattern around a mandrel, or form, in the shape desired and embedded in a resin to retain a structure. The fibers can be aligned to best respond to the stresses expected and thereby provide a more effective casing that can be lighter in weight than the conventional metal one. This technique has been particularly successful in the production of tanks for the propellants of liquid rockets. Indeed, the motor casing of a solid rocket

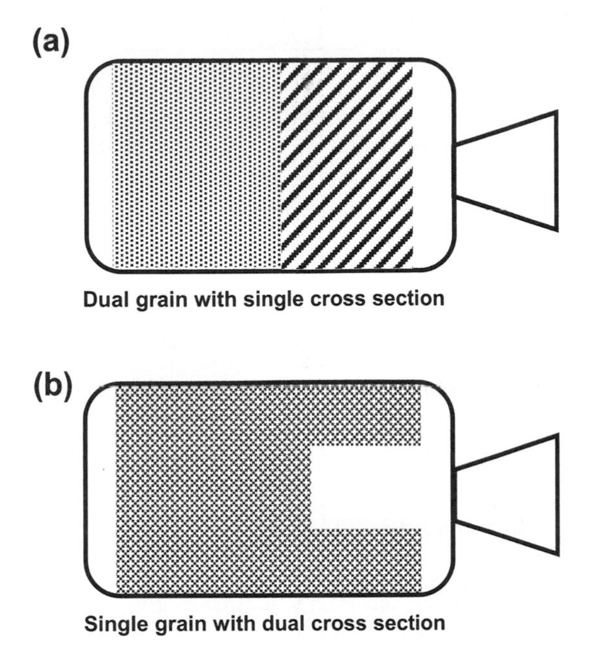


**FIGURE 12.13**

Two possible configurations for dual-thrust solid rocket motors.

motor casing is essentially a propellant tank that contains much higher pressures than liquid propellant tanks and must operate under relatively higher temperatures. Insulating layers in solid rocket motor casings are needed to keep casing material temperatures to acceptable limits.

The propellant grain is attached to the case by direct bonding to the inner wall of the casing and is used for relatively soft and resilient grains. Local structural support of the grain is employed for harder and more brittle grains. Because of the wide temperature range encountered in the grain, the temperature coefficient of expansion of the grain must be considered so that high stresses are not generated in the rocket motor casing.

The nozzle of a solid rocket motor has no means of cooling available to it so it is often constructed of some refractory material. Materials used commonly include graphite, beryllium oxide, and silicon carbide. The other alternative is to use an ablative material that is consumed during operation, absorbing some of the intense heat through change of phase. This is similar to the thermal protection method used on reentry vehicles.

## 12.8 TRANSIENT OPERATIONS

### 12.8.1 Initial pressure rise

The initial pressure rise may be obtained by rearranging the differential equation for $p_c$, Equation (12.11), and rewriting it in the following form:

$$dt = \frac{v_c}{\Gamma^2 c^* A_t} \frac{dp_c}{p_c(\rho_p c^* kap_c^{n-1} - 1)}. \tag{12.21}$$

To implement this equation, we assume the following: the initial pressure rise is so rapid that the chamber volume is the initial chamber volume, $v = v_{c,i}$, the characteristic velocity immediately achieves the equilibrium value, $c^* = c^*_{eq}$, and is constant, and that ignition brings the pressure to a level $p_{c,ign}$, which is high enough for ignition to occur. Then the implicit relation for the initial pressure rise with time is

$$t = \frac{1}{1-n} \frac{v_{c,i}}{\Gamma^2 c^* A_t} \left[ \frac{\rho_p a c^* k - p_{c,ign}^{1-n}}{\rho_p a c^* k - p_c^{1-n}} \right]. \tag{12.22}$$

This time can be reduced by diminishing $v_{c,i}$ or using a plug to prevent gases from escaping through throat until $p_c$ is reached.

### 12.8.2 Local equilibrium pressure variation

The differential pressure [Equation (12.11)] has been shown to actually be an integrodifferential equation because of the chamber volume variation with time as given by Equation (12.12). However, if the pressure variation is small, we can assume local equilibrium, that is, the pressure in the chamber at any time is equal to the equilibrium value given by Equation (12.15), as determined by the local values of all the variables in that equation. We may rewrite Equation (12.15) as

$$p_c = \left( \frac{a \rho_p c^*}{A_t} \right)^{\frac{1}{1-n}} (A_b)^{\frac{1}{1-n}}. \tag{12.23}$$

If we assume a coordinate system with $y$ denoting the distance the burning surface has receded from its original location, the burning rate is given by

$$r = \frac{dy}{dt}. \tag{12.24}$$

The burning surface area is a direct function of the coordinate $y$, that is $A_b = A_b(y)$, which means that $A_b$ is implicitly a function of time. Then for selected values of $y$, one may generate the corresponding values of $A_b$ and, using Equation (12.23), the corresponding values of $p_c = p_c(y)$. It is assumed of course that the values for $c^*$, $\rho_p$, and $a$ are known and constant for this analysis. We may then find the burning rate from Equation (12.1), that is,

$$r = a[p_c(y)]^n = r(y).$$

Then, from Equation (12.24), we may relate the burning time $t$ to the location of the burning surface $y$ as follows:

$$t = \int_0^y \frac{1}{r} d\eta = \int_0^y \frac{1}{a[p_c(\eta)]^n} d\eta. \tag{12.25}$$

Numerical integration of Equation (12.25) provides $t = t(y)$ so that we may finally find $p_c = p_c(t)$. In the most general case, it would be necessary to solve Equation (12.11) numerically.

### 12.8.3 Final pressure drop

The history of the pressure drop-off as burnout is approached may be determined by assuming that the temperature $T \sim$ constant, the chamber volume is equal to the final chamber volume $\upsilon_c = \upsilon_{final}$ and $A_b \rightarrow 0$, for which Equation (12.21) becomes

$$\frac{dp_c}{p_c} = -\frac{A_t c^* \Gamma^2}{\upsilon_c} dt. \tag{12.26}$$

Integrating this equation yields

$$\frac{p_c}{p_{c,eq}} = \exp\left[\frac{\Gamma^2 c^* A_t}{\upsilon_{final}}(t_{b.o.} - t)\right], \tag{12.27}$$

where the quantity $t_{b\text{-}o}$ = burnout time.

## 12.9 EXAMPLE: TUBULAR GRAIN ROCKET MOTOR

The solid propellant rocket motor shown in Figure 12.14 has the following characteristics:

- Sea level thrust $F = 4900$ N with a burnout time $t_{b\text{-}o} = 0.5$ s
- Chamber pressure $p_c = 9.81$ MPa with an internal chamber radius $R_i = 3.2$ cm

The propellant to be used is double-based JPN (nitrocellulose and nitroglycerin) with the following characteristics:

- Burning rate $r = (2.17 \times 10^{-9})\, p^{0.69}$ m/s at 15°C ($p$ in Pa)
- Combustion temperature $T_c = 3125$K
- Specific heat ratio $\gamma = 1.215$ and molecular weight $W = 26.4$
- Propellant density $r_p = 1620$ kg/m$^3$

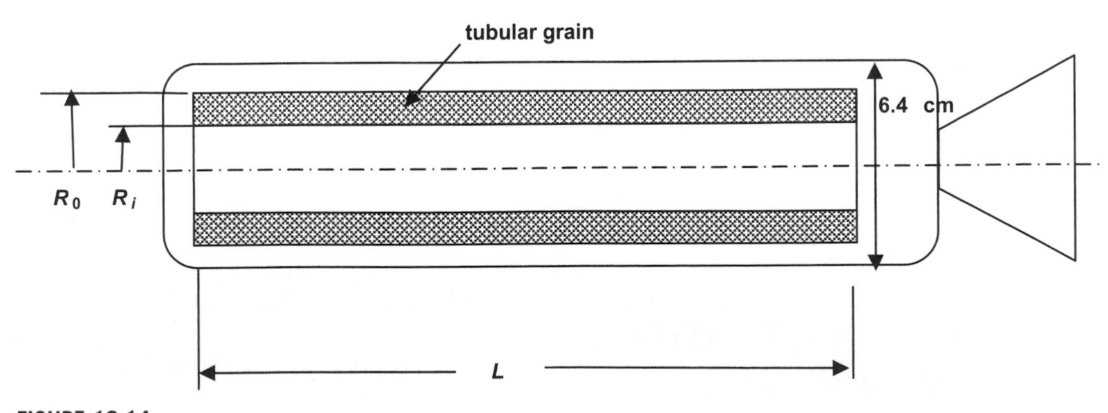


**FIGURE 12.14**

Solid propellant rocket motor with a tubular grain.

Design the dimensions of the tubular grain appropriate to the quoted performance and physical characteristics.

To find the nozzle throat area, consider a matched nozzle with $p_e = p_0 = 101$ kPa. The ideal thrust coefficient for matched operation may be found from the given values for chamber pressure and isentropic exponent to be $c_{F,I} = 1.632$. Further, we assume that the discharge coefficient $c_d$ and the exhaust flow angularity factor $l$ are both unity and select a typical value for the nozzle coefficient $c_v = 95\%$ and then find

$$A_t = F/c_v p_c c_{F,i} = 4900/(0.95 * 9.81 \times 10^6 * 1.632) = 3.22 \times 10^4 \text{ m}^2.$$

To find the burning area, we first calculate the characteristic velocity in the combustion chamber using

$$c^* = (RT_c)^{1/2}/\Gamma = (3.125 * 8315/26.4)^{1/2}/0.651$$

$$c^* = 1523 \text{ m/s}.$$

Using Equation (12.15), we see that

$$A_b/A_t = p_c^{1-n}/a\rho_p c^* = p_c/r\,\rho_p c^*$$

$$A_b/A_t = (9.81 \times 10^6 \text{ N/m}^2)/(0.0213 \text{ m/s} * 1620 \text{ kg/m}^3 * 1523 \text{ m/s}) = 186.8$$

$$A_b = 0.06 \text{ m}^2.$$

For $t_{b\text{-}o} = 0.5$ s and $r = 0.0213$ m/s, the thickness of grain is

$$\Delta R = 2(1.05) r t_{b-o} = 2.24 \text{ cm} = R_o - R_i.$$

Then, because a common rule is $R_o R_i/(R_c^2 - R_o^2) = 1$, that is, $A_p/A_b = $ constant, we have $R_o = 2.89$ cm and therefore $R_i = 0.65$ cm. The length of the grain is determined from $2\pi L(R_o + R_i) = A_b$ and therefore

$$L = (600 \text{ cm}^2)/2\pi(2.89 \text{ cm} + 0.65 \text{ cm}) = 26.98 \text{ cm}.$$

The nozzle exit area

$$A_e/A_t = (\Gamma^2/c_F)(p_e/p_c)^{-1/\gamma} = (0.651^2/1.632)(101.3/9815)^{0.823}$$

$$Ae = 11.2A_t = 3.61 \times 10^{-3} \text{ m}^2 = 36.1 \text{ cm}^2 \text{ and } d_e = 6.78 \text{ cm}.$$

## 12.10 NOZZLE HEAT TRANSFER

### 12.10.1 Heat sink nozzles

Solid rockets typically use simple conical nozzles with the 15° wall angle as described in Chapter 11. This is in part because although the exhaust stream is mostly hot gas, it also contains solid particles and even liquid droplets, depending on the propellants and additives used in the grain. A more complicated bell nozzle does not fulfill its full range of benefits with such a multiphase flow and therefore is usually not worth the additional cost. In addition, the nozzle of a solid rocket motor, unlike that of a liquid rocket motor, has no means of cooling available to it. One means to protect it thermally is to construct a material that can withstand heating without failing, that is, a material that acts as a passive heat sink by thermal conduction. The simplest case of such heat conduction is that of a semi-infinite slab with constant material properties so that the one-dimensional heat conduction equation becomes

$$\frac{\partial T}{\partial t} - \alpha \frac{\partial^2 T}{\partial y^2} = 0. \tag{12.28}$$

The quantity $\alpha = k/\rho c$ is the thermal diffusivity of the nozzle throat material. Again, in the interests of simplicity, assume that any point in the nozzle, including the throat, which has the highest heating, can be considered locally as a semi-infinite solid that is at a uniform temperature $T_{w,i}$ at the start of combustion in the solid rocket. This is tantamount to neglecting conduction in any direction other than the local normal to the surface and to having sufficient material so that material furthest from the surface remains at the initial temperature throughout the burn. Continuing with this simplified case, and further assuming that the convective heat flux at the surface $\dot{q}_{c,w}$ is invariant with time during the burn and that the nozzle material does not change phase during the heating, we may determine the temperature distribution through the nozzle wall as a function of time. The situation as described implies the following boundary conditions:

$$\begin{aligned} T(y,0) &= T_{w,i} \\ \left(\frac{\partial T}{\partial y}\right)_{y=0} &= -\frac{\dot{q}_{c,w}}{k} = const. \end{aligned} \tag{12.29}$$

Then the temperature distribution is given by

$$T(y,t) = T_{w,i}(y,0) + \frac{\dot{q}_{c,w}}{k}\left[\sqrt{\frac{4\alpha t}{\pi}}\exp\left(-\frac{y^2}{4\alpha t}\right) - y\left(1 - erf\frac{y}{\sqrt{4\alpha t}}\right)\right]. \tag{12.30}$$

Note that at the surface ($y = 0$) the wall temperature variation is

$$T(0,t) = T_{w,i}(y,0) + \dot{q}_{w,c}\sqrt{\frac{4t}{\rho c k \pi}}. \tag{12.31}$$

The rise in surface temperature at a specified axial location of the nozzle depends on the quantity $\rho ck$ and the incident heat transfer rate. Relevant material properties for several high temperature materials useful in high heat transfer applications, such as heat shields or heat sinks on reentry vehicles, are given in Table 12.2; the material properties are taken from the NASA TPSX Database. Note that the temperature of the surface rises as $T^{1/2}$ and that for a given heat transfer rate the magnitude of the temperature rise will be set by the value of $\rho ck$. To keep the surface temperature low, the value of $\rho ck$ should be large, and metals such as copper and Invar might serve as good heat sink materials. A fast burn time for a solid rocket is on the order of 10 s, and throat heat transfer rates are in

**Table 12.2** Thermal Properties of Various High Temperature Materials

| Material | $\rho$ (kg/m$^3$) | $c$ (J/kg-K) | $k$ (W/m-K) | $(\rho ck)$^0.5 |
|---|---|---|---|---|
| SLA-561V (virgin) | 264 | 1160 | 0.0592 | 134.6 |
| SLA-561V (char) | 145 | 1050 | 0.12 | 135.2 |
| SLA-561S (virgin) | 288 | 1120 | 0.0576 | 136.3 |
| SLA-561S (char) | 128 | 901 | 0.12 | 117.6 |
| Avcoat-5026H/CG (virgin) | 529 | 1610 | 0.24 | 452.1 |
| Avcoat-5026H/CG (char) | 264 | 2780 | 0.24 | 419.7 |
| RCC (reinforced CC composite) | 158 | 712 | 4 | 670.8 |
| RCC (in-plane) | 158 | 712 | 6 | 821.6 |
| AETB rigid tile | 192 | 628 | 0.064 | 87.8 |
| Cork (ground) | 150 | 2010 | 0.0431 | 114.0 |
| Cork (ground and regranulated) | 130 | 2010 | 0.0448 | 108.2 |
| Copper | 8940 | 378 | 406 | 37040.6 |
| Invar | 8000 | 515 | 10.91 | 6704.4 |

the range of 1–10 kW/cm$^2$. Under these conditions, all the materials would be well beyond a practical temperature. However, some of these materials may be suitable for thermal protection of downstream portions of the nozzle, where the heat transfer rate falls off appreciably to perhaps an order of magnitude less than at the throat.

However, materials with poor thermal conductivity and low heat capacity would be predicted to reach very high surface temperatures, as is the case for AETB-12 rigid tile and cork, and could not withstand such temperatures. These two materials are typical insulating materials that can be used for thermal protection purposes, but not as directly heated materials. Reinforced carbon–carbon composites (RCC), such as those used on the space shuttle nose and leading edge surfaces, can play a role in less-demanding heat transfer environments.

Knowing that metals would tend to be the best heat sinks it is interesting to consider the maximum (constant) heat transfer rate different materials could withstand. The maximum temperature allowable is considered to be the melting temperature $T_m$, which then sets the maximum heat transfer rate allowable as follows:

$$\dot{q}_{c,w,\max} = 0.89\sqrt{\frac{\rho ck}{t}}(T_m - T_{w,i}). \tag{12.32}$$

Some typical values for the variable portion of this equation are given in Table 12.3. Being able to sustain the maximum (constant) heat transfer rate it is necessary to have a short exposure time $t$ and a large value of $(\rho ck)^{1/2}(T_m - T_{w,i})$. The properties listed in Table 12.3 show that copper is a good heat sink candidate, as are graphite and titanium carbide. However, copper as a throat material with a nominal heat transfer rate of 1 kW/cm$^2$ would last less than 10 s. Graphite and titanium carbide would be only marginally better. However, again, for lower heat transfer rates, some of these materials may serve.

**Table 12.3** Typical Values of $\sqrt{\rho ck}(T_m - T_{w,i})$ for Several Candidate Heat Sink Materials

| Material | Approximate[a] Values for $\sqrt{\rho ck}(T_m - T_{w,i})$ [kW/cm²-s$^{0.5}$] | Approximate[a] Heat Capacity in kJ/kg |
|---|---|---|
| Copper | 3.18 | 349–465 |
| Steel | 0.378 | 209 |
| Graphite | 3.56 | 3720 |
| Titanium carbide | 3.47 | 2560 |
| Magnesium oxide | 0.468 | 3840 |
| Silicon carbide | 0.666 | 814 |
| Beryllium oxide | 0.787 | 6400 |

[a]*Taken from Dorrance (1962). The properties have been averaged over the range between* $T_m$ *and* $T_{w,i}$. *Values shown are for* $T_{w,i} = 530R$.

## 12.10.2 Melting ablator nozzles with constant heat transfer

If the heat transfer rate is too high for a heat sink thermal protection system (TPS), resulting in melting of the material, recourse may be made to an ablative TPS. The simplest such case is the melting ablator, a material whose heat of fusion may be used to expand the performance of the TPS. The governing equation for heat transfer into the material is still Equation (12.28), assuming one-dimensional heat transfer and constant material properties, but a transformation of coordinates proves helpful. We may fix our $y$ coordinate to the melting surface, which recedes as melted material is stripped away continually by boundary layer friction and new solid material is presented to the hot gas stream, as shown in Figure 12.15.

The new coordinate system is described by the relation

$$\eta = y - V_a t. \tag{12.33}$$

Applying this coordinate transformation to Equation (12.28) results in the following equation:

$$\frac{d^2T}{d\eta^2} + \frac{V_a}{\alpha}\frac{dT}{d\eta} = 0, \tag{12.34}$$

where the boundary conditions are that the receding surface is maintained at the melting temperature $T_m$, while the interior of the material far from the surface remains at the initial wall temperature $T_{w,i}$, or

$$\begin{aligned} T(0) &= T_m \\ T(\infty) &= T_{w,i} \end{aligned}. \tag{12.35}$$

The solution to Equation (12.34) is an exponential function, and when the boundary conditions are inserted, the temperature distribution becomes

$$\frac{T - T_{w,i}}{T_m - T_{w,i}} = \exp\left(-\frac{V_a}{\alpha}\eta\right). \tag{12.36}$$

In order to use this equation, we need to know the velocity of recession of the surface. Because the incident heat transfer rate in this analysis is assumed constant, the energy balance at the surface

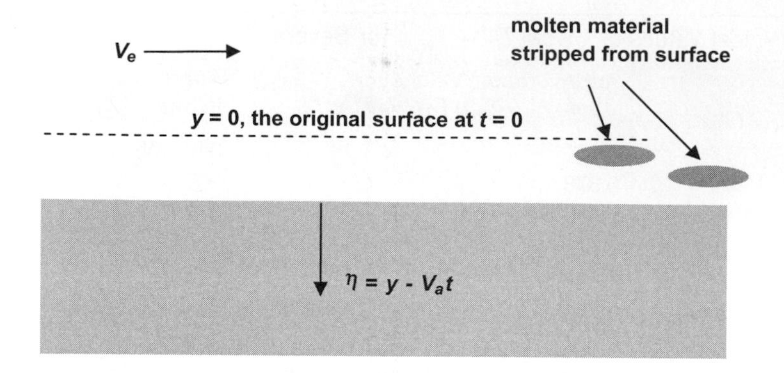


FIGURE 12.15

Moving coordinate system fixed to the receding surface of a melting ablator.

requires that the heat transfer to the surface must equal the sum of the heat conducted into the material and the heat absorbed in melting the material, which may be written as follows:

$$\dot{q}_{c,w} = \left(-k\frac{dT}{d\eta}\right)_{\eta=0} + \rho V_a L_f. \tag{12.37}$$

In Equation (12.37), $\rho$ and $L_f$ are the density and latent heat of fusion of the TPS material, respectively. The conductive heat transfer may be determined from Equation (12.36) and then the surface recession velocity may be determined from Equation (12.37), resulting in

$$V_a = \frac{q_{c,w}}{\rho L_f + \rho c(T_m - T_{w,i})}. \tag{12.38}$$

We see that for materials with high values of density, specific heat, melting point, and thermal diffusivity, the recession velocity is small and the penetration of heat into the body is small. Thus these characteristics serve to make a good ablative TPS. Recall that this analysis assumes that a constant heat flux is supplied at the surface and that the material is thick enough to permit the assumption that the heat shield acts like a semi-infinite slab, that is, the thickness $\Delta \gg V_a t$ .

## 12.10.3 Mass transfer for nozzle thermal protection

The thermal protection afforded by taking advantage of the heat of fusion of the material suggests that further advantage may accrue from exploiting the heat of vaporization of a heat shield material. It is obvious that injection of a coolant through the surface into the boundary layer reduces both skin friction and heat transfer merely by thickening the boundary layer, thereby reducing velocity and thermal gradients. In addition, the injected gas may be at lower temperature, thereby cooling the boundary layer fluid further. Aside from injecting a foreign gas into the boundary layer through the surface, one may merely let the melt material heat to vaporization and thereby carry away heat from the surface. Some mass transfer techniques for thermal protection of surfaces are as follows:

- Direct injection of a coolant through a porous surface
- Vaporization of melted surface material

- Sublimation of the solid surface material
- Pyrolysis of surface material

Illustrations of these techniques are shown in Figure 12.16.

A heat balance at the surface yields the following result:

$$(\rho V)_a Q_a^* = (\rho V)_{pa} Q_p^* = (\rho V)_i (h_i - h_w) = \dot{q}_{c,w} \tag{12.39}$$

The quantity $Q_a$* represents an effective heat of ablation for a TPS material. For example, Teflon has been characterized by an effective heat of ablation (units in Btu/lb) given by

$$Q_a^* \approx 0.48(h_{t,e} - h_w) + 564. \tag{12.40}$$

Common TPS materials are pyrolytic ablators, sometimes called charring ablators. They are primarily organic materials that decompose into charred material and gases when heated in an oxygen-poor environment. They are generally held in a supporting matrix, as shown in Figure 12.16, and are gasified by the heating, leaving behind charred material, which typically has about half the density of the virgin material. The magnitude of the effective heat of pyrolysis $Q_p$* is $5.41 \times 10^4$ kJ/kg for SLA-561V and $2.38 \times 10^4$kJ/kg for Avcoat-5026 H/CG (similar to materials used in the Apollo heat shield) and both are silicon-based ablators. A general effective heat of ablation relation is given by

$$Q^* \approx \left[L_f + c_p(T_m - T_{w,i}) + 0.65(T_w - T_m)\right] + f\left[L_v + \beta(h_{t,e} - h_w)_{0AB}\right], \tag{12.41}$$

where $L_f$ and $L_v$ denote the latent of fusion and vaporization, respectively, while $f$ denotes the fraction of material vaporized ($f = 1$ for direct coolant injection). The quantity $\beta$ is a constant in the range $1/6 < \beta < 2/3$, and the subscript $0AB$ denotes conditions with no ablation. The second term in square brackets is the dominant term for high heating rates. An example of the magnitude of the effective heat of ablation for a material composed of 30% phenolic resin and 70% fiberglass matrix is shown in Figure 12.17.

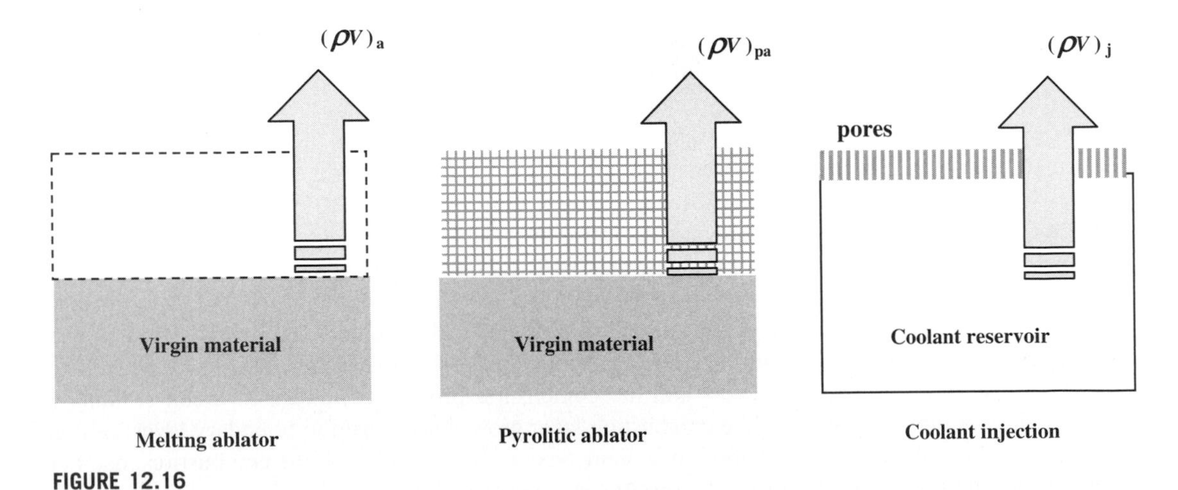


**FIGURE 12.16**

Surface thermal protection methods employing mass transfer to enhance heat removal from the surface.

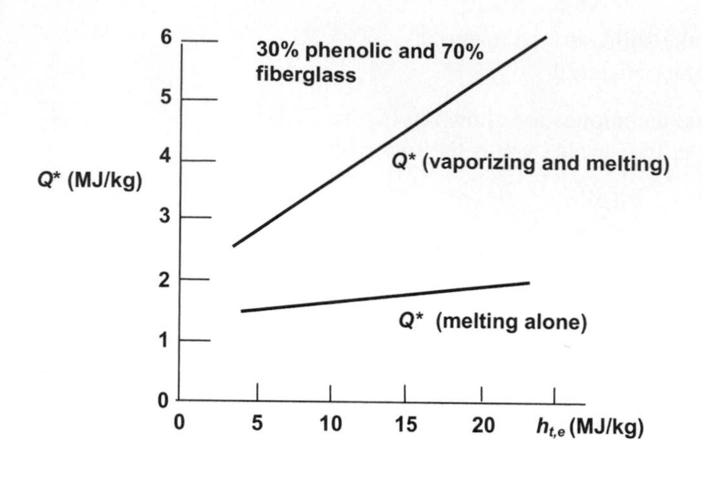


**FIGURE 12.17**

Effective heat of ablation for a material composed of 30% phenolic resin and 70% fiberglass matrix.

For solid rockets, the use of directly injected coolant is impractical because the coolant supply and injection system would not be able to exploit any onboard systems. However, liquid rocket motors can, and often do, take advantage of the existing liquids and gases stored or produced onboard. Charring ablators are the favored materials for nozzle throats in solid rockets with graphite and pyrolitic carbon composites being used commonly.

## 12.11 HYBRID ROCKETS

### 12.11.1 Hybrid rocket operation

Hybrid rockets occupy a sought-after middle ground between liquid propellant and solid propellant rockets: they use a solid fuel but a liquid oxidizer in the manner illustrated schematically in Figure 12.18. This makes them easier to handle, like solid rockets, yet more controllable, like liquid rockets. Solid propellants store safely and are simple in operation, and liquid rockets have high performance and can vary thrust by means of a throttle. On the negative side, solid propellant exhaust typically contains metallic particles and acidic exhaust gases, while liquid propellant systems, being more complicated, are expensive to operate. Although hybrid rockets are designed to combine the best attributes of pure solid or pure liquid rockets, there are still hurdles to overcome.

Conceptually, the combustion process is essentially as illustrated in Figure 12.19. Liquid oxidizer is sprayed into the combustion chamber and an igniter starts the combustion process, which liberates heat to vaporize the surface layer of solid fuel. The expanding hot fuel gas expands away from the surface and meets the spray of liquid oxidizer and the combustion process continues in a self-sustaining manner. The hot fuel gas layer and the combustion layer grow downstream in boundary layer fashion until they merge at the centerline at about five diameters (as measured across the combustion chamber between fuel surfaces) from the initial station. At some point, on the order of several tens of diameters, the oxidizer supply has been fully used and no further combustion can be maintained.

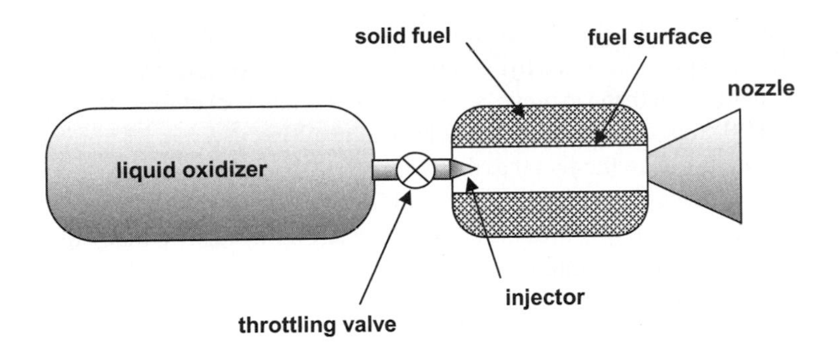


**FIGURE 12.18**

Schematic diagram of a hybrid rocket motor.

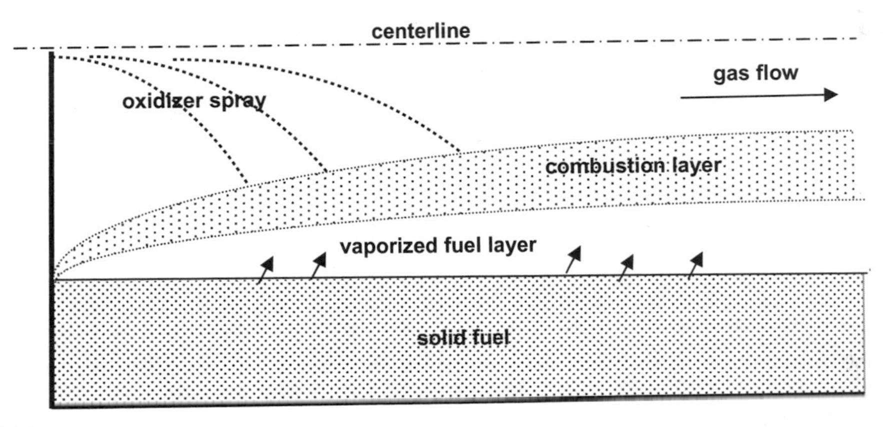


**FIGURE 12.19**

Cross section of the upstream portion of a hybrid motor combustion chamber showing hot fuel gas layer and combustion zone layer.

This combustion region is fundamentally different from that in the pure solid rocket where the oxidizer and fuel are intimately mixed and burn at the propellant surface itself. The hybrid rocket combustion zone is like the flame sheet, or zone, established by parallel streams of fuel and oxidizer, as discussed by Turns (1996). Rapid diffusion between the two is necessary to establish the appropriate mixture ratio for efficient combustion in a short distance. In this sense, the combustion process is diffusion controlled.

Currently, the typical hybrid rocket fuel is a rubberized polymer called HTPB, which stands for hydroxyl-terminated polybutadiene. It looks and feels like the eraser on a pencil. Because HTPB doesn't burn sufficiently fast, the rubbery material has a number of holes, or ports, exposing more surface area of the fuel for burning. Unfortunately, so many holes make the solid fuel structurally weak and subject to dangerous breakup as the combustion chamber builds up pressure and gas flow. A new fuel was sought;

one that burns rapidly yet without the need for many holes that weaken the solid fuel structure was needed. One new fuel is a paraffin-based hydrocarbon very similar to candle wax (Dyer et al., 2008) and chemically not much different than kerosene, a long-favored liquid rocket fuel. The new fuel burns three times faster than HTPB, reducing the need for holes and resulting in a much stronger fuel component. The increased burn rate is due to the fact that the viscosity of the hot layer of melted wax at the surface of this new solid wax fuel was low enough for the oxidizer rushing over it to make waves and whip off tiny droplets of fuel, like the spray blown off the surf in a strong wind. The many droplets provide that additional surface area needed for rapid gasification; something achieved with the old HTPB fuel only by making holes in it. The heating of the fuel does tend to weaken it and cause it to slump, but some new approaches, such as forming the wax on a graphite matrix, provide the necessary support. The effect of the higher burn rate is a smaller, lighter, and stronger rocket motor. Not only that, but the paraffin fuel is less expensive than HTPB and the exhaust is less harmful to the environment.

### 12.11.2 Hybrid rocket characteristics

Although the combustion mechanism of a hybrid rocket motor is different from that of a solid rocket, there is still the need to characterize the surface recession rate, which is about an order of magnitude smaller for hybrid rocket than solid rockets. The solid fuel must be vaporized to form the hot fuel gas layer, which requires heating of the surface. Such heating comes about from the thermal radiation from the combustion layer and convection heating from the gases flowing out toward the nozzle. Some of the fundamental developments of the contributing factors to the fuel gas generation rate may be found in Loh (1968). The surface recession rate may be described by a general empirical relation

$$r = a\left(\frac{\dot{m}}{A}\right)^{0.8}, \tag{12.42}$$

where $a$ is a constant, $A$ is the cross-sectional area of the combustion chamber, and $\dot{m}$ is combined oxidizer and fuel mass flow rate. The determination of $\dot{m}$ is quite complicated, and for illustrative purposes, Loh (1968) presents a simplified case, which is limited to certain specific geometries and propellant combinations. For a cylindrical fuel surface, the production of hot fuel gas is described by

$$\dot{m}_f = \rho_f r \pi d L.$$

Then assume that the recession rate is proportional not to the combined mass flow of fuel and oxidizer, but just that of the oxidizer alone:

$$r = a\left(\frac{\dot{m}_o}{A}\right)^{n}. \tag{12.43}$$

Then the fuel mass flow rate would be

$$\dot{m}_f = a\rho_f\left(\frac{\dot{m}_o}{A}\right)^{n}\pi d L.$$

Substituting in for the cross-sectional area, we find

$$\dot{m}_f = a\rho_f\left(\frac{4\dot{m}_o}{\pi d^2}\right)^{n}\pi d L. \tag{12.44}$$

It is clear from Equation (12.44) that if the exponent $n = 0.5$, the fuel mass flow is independent of the diameter $d$. This means that the fuel mass flow is constant even though $d$ is increasing as the fuel is consumed. Because the oxidizer mass flow rate is controlled by the throttling valve, the O/F ratio can remain constant throughout the burn. Note, however, from Equation (12.43) that the recession rate decreases in magnitude as $d$ grows due to fuel consumption. Plastics are among the fuels that have been used for hybrid rockets, and oxygen, both liquid and gaseous, as well as nitrous oxide, have been used as an oxidizer. For a fuel such as poly(methyl methacrylate)(PMM) with aluminum powder the recession rate is reasonably well represented by Equation (12.43) with $n = 0.5$ for oxidizer mass flux $\dot{m}/A > 25kg/s - m^2$.

### 12.11.3 Example: Hybrid rocket motor fuel grain design

We follow Loh (1968) in considering a cylindrical fuel element that will provide an $I_{sp} = 250$ s at a thrust of 100 lb at a chamber pressure of 300 psia using a fuel composed of 80% PMM with 20% aluminum powder. The specific weight of the fuel is 0.04 lb/in$^3$ and the oxidizer to fuel ratio O/F = 1. In Loh (1968), the weight flow is used in place of the mass flow. The thrust and $I_{sp}$ specification provides the information that the total weight flow rate of propellant is

$$\dot{w} = \frac{F}{I_{sp}} = \frac{100lb}{250s} = 0.4lbs/s.$$

With O/F = 1, the flow rates of fuel and oxidizer must be equal at 0.2 lb/s. The burn rate comes from experimental data, which show that the constant $a$ and the exponent $n$ in Equation (12.43), using $\dot{w}$ instead of $\dot{m}$, are 0.09 in$^2$/(lb-s)$^{1/2}$ (for recession rate in inches per second) and 0.5, respectively. From Equation (12.44), the required grain length $L$ is given by

$$L = \frac{\dot{w}_f}{a\rho_f\sqrt{4\pi\dot{w}_o}} = \frac{0.2lb/s}{(0.09in^2/lb^{1/2}\quad s^{1/2})(0.04lb/in^3)\sqrt{4\pi(0.2lb/s)}} = 35in.$$

The diameter necessary to accommodate an oxygen weight flow of 0.2 lb/s may be found from the recession rate [Equation (12.43)] if something is known about the likely size of the recession rate. In the case of PMM, the recession rate is around 0.01 to 0.015 in./s. Then a reasonable diameter is found to be

$$d = \frac{a}{r}\sqrt{\frac{4}{\pi}\dot{w}_o} = \frac{0.0454}{r}\frac{in^2}{s}.$$

Therefore, we would expect $d$ to be in the range 3 in. $< d <$ 4.5 in. We may determine the final diameter by integrating Equation (12.43) for a constant oxidizer flow rate to obtain

$$d^2(t_b) - d^2(0) = 8at_b\sqrt{\frac{\dot{w}_o}{\pi}}.$$

In the case where the initial diameter $d(0) = 3$ in., the final diameter after a burn time of, say, 50 s is 4.25 in. One may repeat the process for several values of recession rate, thrust levels, and burn times in order to arrive at a number of grain configurations, that is, initial and final diameters and total length, for comparative evaluation.

## 12.12 NOMENCLATURE

| | |
|---|---|
| $A$ | cross-sectional area |
| $A_b$ | propellant burning area |
| $A_e$ | nozzle exit cross-sectional area |
| $A_t$ | nozzle throat cross-sectional area |
| $a$ | constant |
| $c$ | specific heat |
| $c^*$ | characteristic velocity |
| $c_d$ | nozzle discharge coefficient |
| $c_F$ | nozzle thrust coefficient |
| $c_h$ | convective heat transfer coefficient |
| $c_V$ | nozzle velocity coefficient |
| $c_w$ | nozzle weight flow coefficient |
| $erf$ | error function |
| $F$ | thrust |
| $f$ | fraction of material vaporized |
| $g$ | acceleration of gravity |
| $h$ | enthalpy |
| $I_{sp}$ | specific impulse |
| $k$ | ratio of burning area to throat area or thermal conductivity |
| $L$ | length of the propellant charge |
| $L_f$ | latent heat of fusion |
| $M$ | Mach number |
| $\dot{m}$ | mass flow |
| $n$ | pressure exponent in burn rate |
| $p$ | pressure |
| $Q^*$ | effective heat of ablation |
| $\dot{q}_c$ | convective heat flux |
| $R$ | gas constant |
| $R_u$ | universal gas constant |
| $r$ | propellant surface recession rate |
| $T$ | temperature |
| $T_1$ | propellant self-ignition temperature |
| $T^0$ | reference temperature |
| $t$ | time |
| $t_b$ | burn time |

| | |
|---|---|
| $V$ | velocity |
| $V_a$ | recession rate of ablating surface |
| $\dot{w}$ | weight flow $= \dot{m}g$ |
| $z$ | altitude |
| $z_s$ | atmospheric scale height |
| $\alpha$ | conical nozzle expansion angle |
| $\Gamma$ | function of $\gamma$, Equation (12.5) |
| $\gamma$ | ratio of specific heats |
| $\Delta$ | slab thickness |
| $\delta$ | atmospheric pressure ratio $p_0/p_{0,s.l.}$ |
| $\varepsilon$ | nozzle expansion ratio $A_e/A_t$ |
| $\lambda$ | thrust loss factor due to flow angularity |
| $\rho$ | density |
| $\pi$ | sensitivity coefficient, Equations (12.17) through (12.20) |
| $\sigma$ | angle between nozzle axis and any streamline |
| $\upsilon$ | volume |
| $\eta$ | efficiency or transformed $y$ coordinate |

### 12.12.1 Subscripts

| | |
|---|---|
| $a$ | axial direction or ablation conditions |
| $c$ | combustion chamber |
| $cr$ | critical |
| $e$ | nozzle exit |
| $f$ | fuel |
| $i$ | ideal or initial |
| $m$ | melting |
| max | maximum conditions |
| $n$ | nozzle |
| o | oxidizer |
| $p$ | propellant |
| $s.l.$ | sea level conditions |
| $t$ | stagnation conditions |
| $w$ | wall conditions |
| $vac$ | vacuum conditions |
| 0 | free stream conditions |

### 12.12.2 Superscripts

| | |
|---|---|
| * | critical conditions, $M = 1$ |
| ^ | unit vector |
| _ | vector |

## 12.13 EXERCISES

**12.1** Calculate the dimensions of an end-burning solid propellant grain needed to pressurize the tanks of a 19.62-kN thrust liquid rocket engine. A gas generator is required to produce $33.1 \times 10^5$ N/m$^2$ for 30 s. The liquid propellants are RFNA (1520 kg/m$^3$) and a hydrocarbon fuel (1050 kg/m$^3$). The mixture ratio is 0.3, and the mass flow rate of propellant is 10.7 kg/s. The characteristics of the double base plateau burning solid propellant are burning rate $r = 0.406$ cm/s, combustion chamber temperature $T_c = 2000$K, molecular weight of the combustion gases $W = 21.2$ kg/kg-mol, ratio of specific heats $\gamma = 1.28$, and solid propellant density $\rho_p = 1560$ kg/m$^3$.

**12.2** The space shuttle's integrated solid rocket boosters (Figure E12.1) use a propellant mixture of ammonium perchlorate ($NH_4ClO_4$), PBAN, aluminum powder, and iron oxide catalyst. The characteristics of the propellant are given as follows: burn rate $r = 0.434$ in./s at 1000 psia, burn rate exponent $n = 0.35$, propellant density $\rho_p = 0.063$ lb/in$^3$, temperature coefficient of pressure $\pi_k = 0.11$(%/F), characteristic exhaust velocity $c^* = 5053$ ft/s, combustion temperature $T_c = 6092$F, and effective ratio of specific heats $\gamma_c = 1.138$ and $\gamma_e = 1.147$. The nozzle throat diameter is 53.86 in. and the exit diameter is 152.6 in.; the nozzle produces an average thrust of 2.59 million pounds at an ambient temperature of 60F. The average chamber

**FIGURE E12.1**

A launch of the Space Shuttle showing the firing of the solid rocket boosters.

pressure is 662 psia during operation. For operation at sea level take-off, determine the (a) specific impulse, (b) chamber molecular weight, (c) exit Mach number, (d) exit pressure, (e) exit velocity, (f) exit temperature, and (g) percentage change in average thrust if the ambient temperature drops to 32F.

**12.3** An end-burning grain solid propellant rocket is required to have the following characteristics at a normal ambient temperature of $T_i^o = 15°C$: sea level thrust $F = 4.91$ kN and a burn time of 15 s. The characteristics of the propellant, a potassium perchlorate–asphalt combination (76.5% $KClO_4$ and 23.5% asphalt and oil), are as follows (all pressures are expressed in atmospheres):

- Burning rate (cm/s) $r = \{0.102 \exp[0.0015(T_i - 15)]\}p^{0.745}$
- Characteristic velocity (cm/s) $c^* = \{105{,}100 \exp[0.00021(T_i - 15)]\}p^{0.015}$
- Specific heat ratio $\gamma = 1.27$
- Molecular weight $W = 30$
- Propellant density $\rho_p = 1770$ kg/m$^3$
- Chamber pressure $p_c = 135.5$ atm

Determine the (a) propellant grain dimensions $A_b$ and $L$, (b) specific impulse, (c) chamber temperature,(d) influence of a 1% variation in the burning rate of the propellant on the thrust and the chamber pressure, and (e) motor performance at $T_i = 55$ and $-25°C$.

**12.4** Consider an end-burning grain solid rocket motor with the following characteristics: $\rho_p = 1770$ kg/m$^3$, $T_p = 15°C$, $c^* = 1{,}051p^{0.015}$ in meters per second when $p$ is in atmospheres, $n = 0.745$, $a = 0.001$m/s when $p$ is in atmospheres. The sea level thrust developed is $F = 4900$N at a chamber pressure $p_c = 13.74$ MPa over a burning time $t_b = 15$ s where the combustion gases have $\gamma = 1.27$ and a molecular weight $W = 30$. Determine the (a) chamber temperature $T_c$, (b) thrust coefficient of the nozzle $c_F$, (c) throat area $A_t$, (d) burning area $A_b$, (e) length $L$ of the grain, and (e) specific impulse $I_{sp}$.

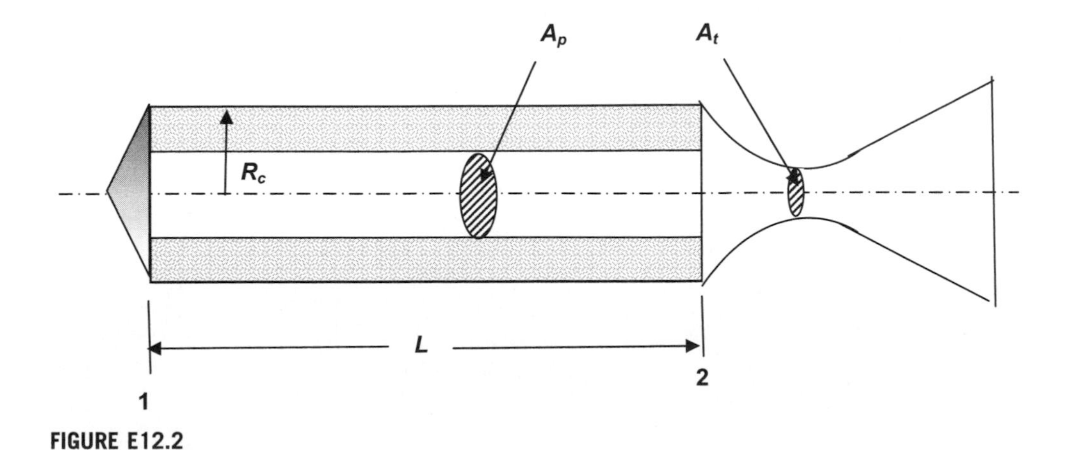


**FIGURE E12.2**

Solid rocket motor with cylindrical grain.

**12.5.** Consider the cylindrical grain solid rocket motor with constant initial port area $A_p$ as shown in Figure E12.2. For a range of chamber pressure, $p_{c,0}$ to $p_{c,1}$, the burning rate $r$ is constant, that is, $n = 0$, while the chamber temperature $T_c$ remains constant. Determine the chamber pressure variation with time within the quoted pressure range.

## References

Dorrance, W. (1962). *Viscous Hypersonic Flow.* New York: McGraw-Hill.

Dyer, J., Zilliac, G., Doran, E., Marzona, M., Lohner, K., Karlik, E., Cantwell, B., & Karabeyoglu, A. (2008). Status Update Report for the Peregrine 100km Sounding Rocket Project. AIAA2008-4829, 43rd AIAA Propulsion Meeting, Cincinnati, OH.

Kays, W. M., & Crawford, M. (1993). *Convective Heat and Mass Transfer* (3rd ed.). New York: McGraw-Hill. 1993.

Loh, W. H. T. (Ed.). (1968). *Jet, Rocket, Nuclear, Ion, and Electric Propulsion.* New York: Springer-Verlag.

NASA TPSX Material Database V4: http://tpsx.arc.nasa.gov/.

Sutton, G. P. (1992). *Rocket Propulsion Elements* (6th ed.). New York: Wiley.

Turns, S. R. (1996). *An Introduction to Combustion.* New York: McGraw-Hill.

CHAPTER

# Nuclear Rockets

13

## CHAPTER OUTLINE



## 13.1 NUCLEAR ROCKETS FOR SPACE EXPLORATION

In May 1961, after 4 months in office, President John F. Kennedy addressed Congress to outline major goals of his administration. He set four national space goals:

- manned moon landing
- development of communications
- weather satellites
- greatly increased funding for the ROVER nuclear rocket

The reason for the last item mentioned was to "give promise of some day providing a means for even more exciting and ambitious explorations of space, perhaps beyond the moon, perhaps to the very end of the solar system itself." Although this speech is often quoted to emphasize the need to take large technical risks, referring to the manned moon landing, that is, the Apollo program, little recognition is

Theory of Aerospace Propulsion.

**Table 13.1** Possible *In Situ* NTP Propellant Sources

| Propellant | Source | $I_{sp}$ Potential (s) |
|---|---|---|
| $CO_2$ | Martian atmosphere, Martian frost, Earth | 160–380 |
| $CH_4$ | Asteroids, Phobos and Deimos, Earth, outer planets | 460–670 |
| $H_2$ | Lunar polar ice, lunar silane, NEO asteroids, Earth, outer planets | 800–1200 |
| $NH_3$ | Earth, outer planets | 350–700 |
| $H_2O$ | Lunar ice, Martian ice, planetary moons | 160–240 |

given to the bold suggestion to continue the nuclear rocket program to make deep space exploration possible.

Nuclear thermal propulsion (NTP) has always been an attractive space propulsion option for a number of important reasons. First, the inherently high specific impulse capability of the nuclear rocket option reduces the in-orbit mass requirements for staging high $\Delta V$ missions. The high thrust level achievable with NTP translates into more reasonable interplanetary trip times, thus reducing one of the major dangers of deep space flight: cosmic radiation exposure. Because NTP can utilize any gaseous propellant, of course with varying levels of $I_{sp}$, it has the best potential for using *in situ* planetary propellant resources, as shown in Table 13.1.

Because of the major advantages afforded by NTP, extensive planning and research leading to ground testing of NTP systems were carried out. However, this growing development of the technology was canceled in 1972 when the signing of the Nuclear Test Ban Treaty posed insuperable constraints on such development. In addition, declining economic conditions forced NASA to cancel major lunar/planetary exploration efforts.

It is worth noting that in the course of the development of NTP, 20 reactors were designed, built, and tested in the period from 1953 to 1973 with a total expenditure of around $1.4 billion. The demonstrated performance included the following milestones; the names of the projects are given in parentheses:

- Power ($MW_t$)—1100 (NRX) to 4100 (Phoebus-2A)
- Thrust (klb)—55 (NRX) to 210 (Phoebus-2A)
- Fuel temperature (K)—2750 (PEWEE)
- Specific impulse (s)—850 vacuum (PEWEE)
- Burn endurance—1 to 2 hr
- Start/stop cycles—28 automatic cycles (XE)

The new century has seen a rise in interest in space exploration, which has rekindled some activity in nuclear propulsion, but mainly in the area of nuclear electric propulsion rather than NTP, wherein nuclear reactors would be used, but solely for the production of electric power that would be used to feed electric propulsion engines, such as those described in Chapter 14.

## 13.2 NUCLEAR ROCKET ENGINE CONFIGURATION

A typical nuclear thermal propulsion rocket engine is shown in Figure 13.1. It is evident that it is little different from a conventional liquid rocket engine, with the major difference being replacement

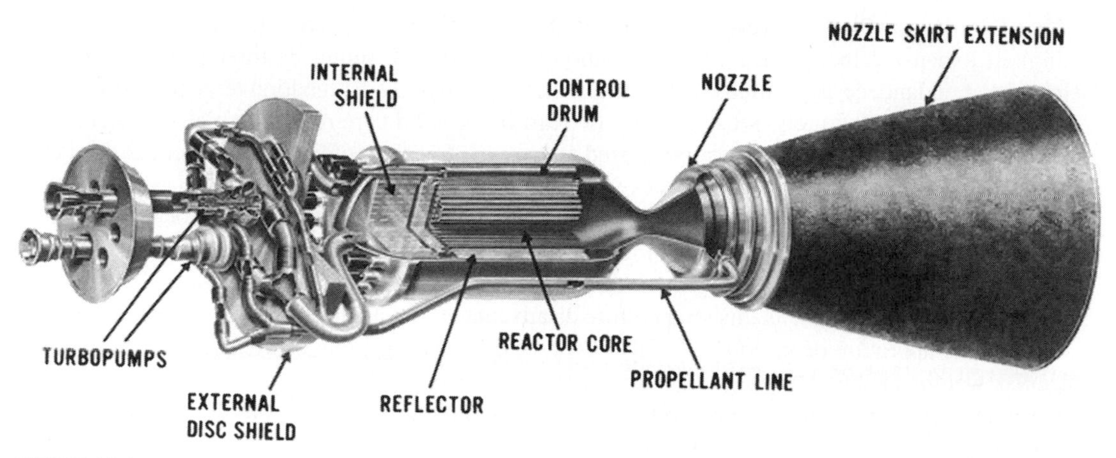


**FIGURE 13.1**

General configuration of a nuclear thermal rocket engine (courtesy of NASA).

of the liquid rocket combustion chamber with the nuclear reactor composed of the reactor core, which contains the fissionable material that provides the thermal energy to heat the propellant and the associated control features: the control drum, reflector, and radiation shields.

The nuclear rocket operates in the same fashion as a liquid rocket engine in that it adds energy to the propellant gas as it passes over the hot reactor core so that heat is transferred to the gas. The hot gas is then accelerated to high speed through an accelerating nozzle, thereby producing thrust. Although any gas can be used as a propellant, some gases are better for this purpose than others; a discussion of this issue is provided in the next section.

## 13.3 EXHAUST VELOCITY

For high specific impulse, the propellant should have low molecular weight and high stagnation temperature because the nozzle exit velocity is given by

$$V_e \sim \sqrt{\frac{T_c}{W_c}}.$$

Optimistic values for temperature and molecular weights for hydrogen–oxygen propellants in chemical rockets and for nuclear rockets using hydrogen propellant are as follows:

$$H_2 - O_2 \; T_c = 5000\text{K} \quad W_c = 18$$

$$\text{Nuclear } T_c = 2500\text{K} \quad W_c = 2$$

Therefore, the relative exhaust velocity between the two types of rocket is approximately given by

$$\frac{V_{e,chemical}}{V_{e,nuclear}} = \frac{1}{2}$$

Thus hydrogen, with the lowest molecular weight, would be a good propellant and would provide the highest exhaust velocity for a given stagnation temperature. It might be thought that in nuclear rockets the abundant energy available would mean that an even higher stagnation temperature $Tc$ could be employed. This is definitely not so because (at least in the solid core reactor, which is the only one tested experimentally to date) heat is transferred from a solid core reactor to the propellant. Thus the structural components within a nuclear rocket, unlike those in chemical rockets, must be *hotter* than the propellant, and $T_c$ cannot exceed the limiting temperature of the structure or reactor material. It is expected that reactors may be run at temperatures as high as 5500°R(3055K) for short durations. This is considerably below the temperature attained in some chemical rockets, but the use of hydrogen as the propellant more than offsets this temperature disadvantage of a nuclear rocket. Another advantage of very high temperature operation is the likelihood of dissociating some of the hydrogen so that the molecular weight of the exhaust gas is reduced to a new value in the range $1 < W_c < 2$. However, hydrogen at high temperatures is known to attack and degrade most metals and this impact must be considered if operation for sustained periods is required.

## 13.4 NUCLEAR REACTORS

A nuclear reactor is an assembly of a suitable fissionable material in a structure that contains other materials necessary to produce the desired neutron properties, to provide adequate control, and, for steady operation, to provide cooling. The attainment of the desired chain reaction depends on the proper utilization of the neutrons present. In general, a fission neutron may suffer from three fates:

**(1)** It may be captured in a process causing fission, resulting in the emission of additional neutrons.
**(2)** It may be captured in a nonfission process (such as radiative capture) either by fissionable material or by other materials.
**(3)** It may escape the reactor.

For steady operation, those excess neutrons generated by (1) must exactly balance those lost by processes (2) and (3). The first two processes depend on the nuclear properties of the fissionable and other material within the reactor and on how this material is distributed. For a given shape and material distribution pattern, it is dependent on the size of the reactor: larger reactors have lower surface-to-volume ratios and hence a lower tendency for escape of neutrons. Thus for a given shape, fuel, and so on, there is a critical size below which the neutron loss rate will not permit a steady chain reaction. The critical size does, of course, depend strongly on the constituents and design of the reactor.

To provide an adequate response rate to changes in power demand, a reactor should have some excess reactivity. That is, it should be somewhat above critical size for the particular geometry and fuel geometry distribution employed. During steady-state operation, this excess reactivity is offset by necessary neutron loss to the central system. In addition, because the amount of fissionable material decreases gradually with time, the reactor must initially have excess reactivity if it is to have a long life. In addition to burn-up of fuel, the reactor undergoes internal changes or "positioning" effects, which gradually increase the rate of nonfission neutron capture. For example, some of the fission products may have a large neutron-capture cross section. Actually, the life and total burn-up of a rocket reactor are both rather small compared to earth-bound power-generation reactors and these effects are relatively unimportant. However, a rocket reactor would be called upon to provide a very rapid power buildup. Some past nuclear propulsion programs include the following:

Project ROVER—nuclear rocket engines
Project PLUTO—nuclear ramjet
Project SNAP—nuclear auxiliary power
Project NERVA—nuclear energy rocket
Project RIFT—reactor-in-flight test

## 13.5 NUCLEAR REACTIONS

Other than the isotope carbon-12, which by definition has a mass of 12u (u denotes unified atomic mass units, equal to $1.66 \times 10^{-24}$g), atomic masses are not whole numbers. This implies that the neutrons, protons, and electrons that add up to a mass of 12u in this particular configuration do not have the same masses in other configurations. In particular, the mass of a free proton is about 1.007276u, that of a free neutron is 1.008664u, and that of an electron is 0.00054858u. Thus for the case of oxygen, consisting of eight protons, eight neutrons and eight electrons, the atomic mass is 15.9994u while the sum of the free particle masses is 16.1319u and there is a mass defect of 0.13251u when these particles exist together as the oxygen atom. According to $E = mc^2$, the mass defect is proportional to the amount of energy that would be released if an oxygen atom were formed by the assembly of individual particles. Conversely, this same amount of energy would be required to completely disassemble an oxygen atom. According to this latter interpretation, the energy increment associated with the mass defect is called binding energy.

The atomic number Z is the number of protons in the nucleus of an atom, whereas the mass number A is the sum of the number of protons and neutrons in the nucleus. From the mass of neutrons and protons described previously, it is obvious that the mass number A of an element is very close in magnitude to its atomic weight. To characterize a general element X, the following notation is typically used: ${}_{Z}X^{A}$. Thus the common form of uranium is ${}_{92}U^{238}$, signifying 92 protons and 146 neutrons in the nucleus. The atomic number Z characterizes the chemical properties of the element. Elements may have isotopes, each of which has the same atomic number Z but different mass number A. The isotopes of uranium, along with their relative occurrence in nature, are as follows:

${}_{92}U^{238}$—99.282%
${}_{92}U^{235}$—0.712%
${}_{92}U^{234}$—0.006%

The carbon isotope ${}_{6}C^{11}$ has all the same chemical properties as common carbon ${}_{6}C^{12}$. Therefore, carbon dioxide using the carbon isotope in place of common carbon can be used naturally by a plant because $Z = 6$ for either. Yet because the carbon isotope is radioactive, it can be tracked during a process such as photosynthesis.

Neutral atoms have as many orbital electrons as they have protons in the nucleus, while ions are charged and have different numbers of protons and electrons. A nuclear reaction is described as follows:

$$ {}_{92}U^{235} + {}_{0}n^{1} \rightarrow {}_{38}Sr^{90} + {}_{54}Xe^{140} + 2\,{}_{0}n^{1} + 200\,MeV, \quad (13.1) $$

where Sr denotes strontium, Xe denotes xenon, *n* denotes a neutron, and 200 MeV, or 200 million electron volts, denotes a release of energy, which will subsequently be discussed more fully. The mass number of fission fragment atoms are distributed over a wide range, with most concentrated around mass numbers 90 and 140. Most of the fission fragments are radioactive, that is, they are unstable and disintegrate spontaneously. Radioactive isotopes are those that, because of their instability, undergo spontaneous disintegration by emitting decay products. The most important of these are as follows:

- $\alpha$ particles—a helium nucleus, that is, a helium atom stripped of its two electrons that may be denoted by $[_2He^4]^{++}$. These particles are emitted at speeds of 16 to 32 km/s.
- $\beta$ particles—electrons that emanate from the nucleus where a neutron disintegrates into a proton and an electron, that is, ${}_0n^1 \rightarrow [{}_1H^1]^+ + {}_{-1}e^0$. The neutron becomes a proton plus a $\beta$ particle emitted at speeds of around 160 km/s.
- $\gamma$ rays—electromagnetic waves of extremely high frequency, that is, photons that travel at the speed of light, 299.8 km/s.

A nucleus emits a $\beta$ particle (electron) and frequently electromagnetic energy, leaving the nucleus with an additional proton. It should be recalled that the number of protons (2) determines the elemental identity of an atom.

Fast reactors use the neutrons at the high energy levels (above 1000 eV) at which they are released, whereas thermal reactors use neutrons slowed down to a low energy level (less than 1 eV) by the use of a moderator, a material that slows down neutrons passing through it.

The nuclear reaction may be described as follows:

$$ {}_0n^1 + fuel \rightarrow 2ff + (2or3)_0n^1 + energy. \quad (13.2) $$

The fuel is bombarded by neutrons in order to produce fission fragments (denoted by *ff* and having mass numbers typically between 90 and 140), more neutrons, and energy. The fuel is a heavy element such as uranium, plutonium, or thorium. For example, using uranium as fuel, we find

$$ {}_0n^1 + {}_{92}U^{235} \rightarrow {}_{42}Mo^{95} + {}_{57}La^{139} + 2_0n^1. \quad (13.3) $$

A neutron bombards the uranium, producing molybdenum, lanthanum, and two neutrons. Note that the original 92 protons have increased by 7 to 99. The $\beta$ particles (electrons) emitted are captured by the molybdenum and lanthanum so that they remain neutral, that is, uncharged. Assessing the mass (in u, unified atomic mass units) before and after the nuclear reaction is shown in Table 13.2.

**Table 13.2** Mass Balance for Nuclear Reaction

| Reactant Species | Mass (u) | Product Species | Mass (u) |
|---|---|---|---|
| ${}_{92}U^{235}$ | 235.124 | ${}_{42}Mo^{95}$ | 94.945 |
| ${}_0n^1$ | 1.009 | ${}_{57}La^{139}$ | 138.955 |
| | | $2_0n^1$ | 2.018 |
| Total mass before | 236.133 | Total mass after | 235.918 |

The mass deficit is 0.215 u, and each u = $1.660 \times 10^{-24}$g. Because $E = mc^2$, we have an energy release of $(0.3569 \times 10^{-27}$ kg$)(2.998 \times 10^8$ m/s$)^2 = 3.208 \times 10^{-11}$ J. Then, because one electron volt of energy is related to a joule of energy by 1 eV = $1.6 \times 10^{-19}$ J, the energy release may also be written as 200 MeV, as given by Equation (13.1). Thus each $U^{235}$ fission releases $3.208 \times 10^{-11}$ J and for 236.133 u, or $3.9198 \times 10^{-22}$ g, we have 8.8136 J/g participating in each fission. Thus the energy potential of each gram of $_{92}U^{235}$ is 22,730 kilowatt hours or almost one megawatt-day.

## 13.6 REACTOR OPERATION

The general nuclear reaction given by Equation (13.2) must be capable of sustaining itself, that is, there must be sufficient neutron production to provide the continued fission of the fuel. Reactors are classified as fast reactors or thermal reactors based on the energy of the neutrons causing the fission process: fast reactors use neutrons in the energy range of 100 to 20 MeV, whereas thermal reactors use slow neutrons, with energy in the range of 0.025 to 0.1 eV. Neutrons bombarding a fuel may be either scattered or absorbed, and fast neutrons are scattered predominantly by the target. As a result, it is impossible to maintain a chain reaction using the main isotope of uranium ($_{92}U^{238}$), instead it must be enriched with $_{92}U^{235}$.

To use the energy produced by fission for a rocket it must be transferred to the propellant and carried away from the reactor structure so that it will not be damaged by the high temperatures that can be achieved. Thus the propellant passes through the reactor at a rate consistent with maintaining a constant temperature in the reactor. The fission reaction is controlled primarily by controlling the net neutron production, which is apparent in the general design of a typical solid core reactor, as shown in Figure 13.2.

In Figure 13.2, the various elements depicted are described as follows:

- radiation shield—material that stops radiation from leaking out to harm structure or crew.
- reflector—reduces leakage by scattering neutrons back into the core. Materials are generally the same as those used as moderators.
- moderator—lightweight low neutron absorber that reduces the probability of neutron leakage and also reduces the energy of fast neutrons to the level of slow, or thermal, neutrons so that they may support the fission reaction. Materials include hydrogen, carbon, beryllium, and heavy water.
- fuel—enriched uranium, containing $U^{233}$, $U^{235}$, or other fissionable fuel, such as plutonium $P^{239}$. The uranium is typically dispersed throughout a graphite or metal carbide moderator. Hydrogen is the propellant of choice, and because graphite is attacked by hydrogen at temperatures above about 2200K, the operating temperature of reactors using a graphite matrix is limited.
- control rods—material with a high thermal neutron absorption cross section, which may be inserted into or withdrawn from the core to alter the progress of the fission reactions. Materials include boron, tantalum, and cadmium. Properties for these materials are described in Table 13.3.

The criticality of the reactor is characterized by the factor $k$, which is the ratio of the number of neutrons present at the end of any particular neutron-generating event to the number of neutrons at the initiation of the generation event. This factor is characterized as follows:

- $k = 1$ denotes a critical reactor where the number of neutrons are maintained at a constant level throughout the fission process.

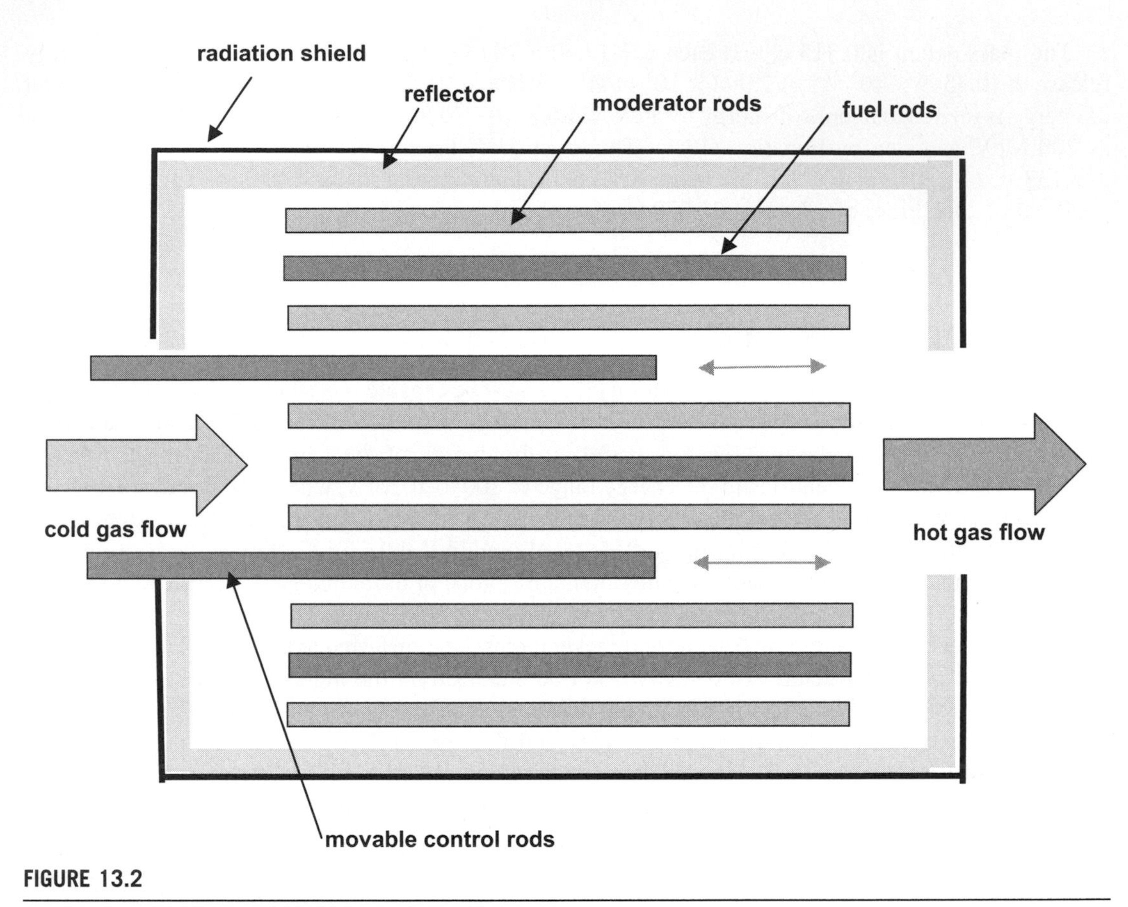


FIGURE 13.2

Schematic diagram of a thermal reactor with propellant gas flow.

**Table 13.3** Thermal Neutron Cross Sections for Various Materials

| Material | Absorption (barns)[a] | Scattering (barns)[a] | Application |
|---|---|---|---|
| Beryllium | 0.0009 | 6.9 | Moderator |
| Heavy water | 0.00092 | 15.3 | Moderator |
| Carbon | 0.0045 | 4.8 | Moderator |
| Hydrogen | 0.32 | 20–80 | Moderator |
| Tantalum | 21 | — | Control rods |
| Boron | 718 | 3.8 | Control rods |
| Cadmium | 3500 | 6.5 | Control rods |

[a] *barns* = $10^{-24}$ *cm*$^2$ *per nucleus. The macroscopic cross section is obtained by multiplying the number of molecules of the material per cm*$^3$.

- $k > 1$ denotes a supercritical reactor wherein the neutron density increases at an exponential rate.
- $k < 1$ denotes a subcritical reactor where the neutron flux and therefore the power decrease exponentially.

Let $n$ be the number of neutrons at the start of generation. The factor $k = n(t)/n_0$, where $n_0 = n(0)$, and therefore $n(k-1)$ is the number of neutrons present at the end of a generation event. We may consider that there is an interval of time $\tau$ between succeeding neutron generation events so that the rate of change of neutrons is given approximately by

$$\frac{dn}{dt} = \frac{n(k-1)}{\tau}.$$

The number of neutrons at any time is

$$n = n_0 \exp\left[\left(k-1\right)\frac{t}{\tau}\right].$$

As an example, assume $\tau = 20$ μs and the reactor criticality is 1.002, or 0.2%. Then $n/n_0 = \exp(10t)$ and after 1 s the number of neutrons has increased by a factor of 22,000 and the reactor is running away. The critical operation of a typical solid core reactor is shown schematically in Figure 13.3.

The engine may be made safe by inserting the control rods completely to reduce the number of neutrons available for initiating and maintaining a chain reaction. Withdrawing the control rods will increase the degree of reaction and the temperature in the chamber will rise. Steady-state operation can be achieved by the appropriate combination of control rod position and propellant flow rate.

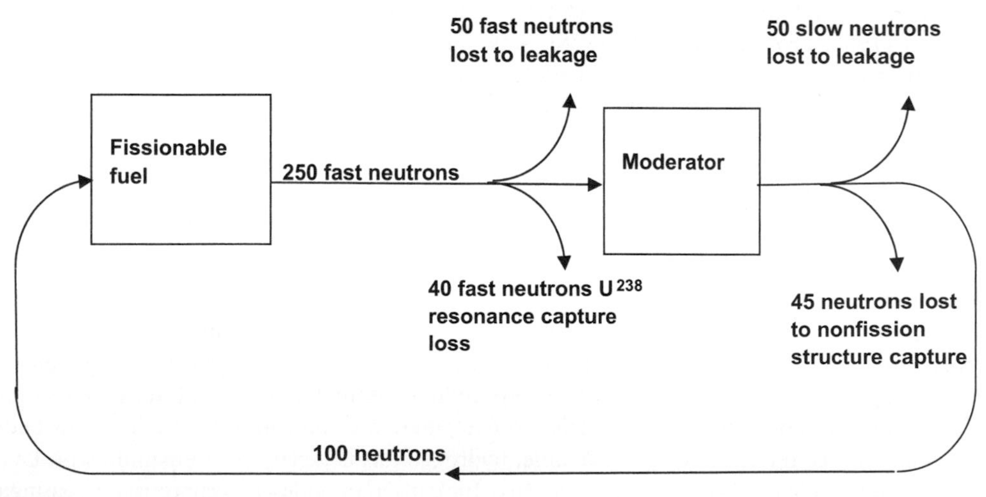


**FIGURE 13.3**

Schematic diagram of critical operation of a nuclear reactor.

## 13.7 FUELS FOR NUCLEAR PROPULSION AND POWER

Fuel material selection must be based on (a) mission requirements, including operating lifetime, number of required restarts, and the acceptability of fission product loss, as well as (b) compatibility issues, such as whether the fuel material can be formed into fuel elements with acceptable thermal stress tolerance so as to accommodate reactor transients.

Fuels with potential for catastrophic structural failure are not acceptable, which poses a serious challenge for solid graphite-based fuel elements. Smaller fuel elements with more predictable degradation properties are desirable options. Some possibilities are twisted ribbon fuel element bunches, coated particles of mixed carbides, square lattice honeycomb of cermet or mixed carbides, and open porosity foam fuel. Cermet fuels, which are ceramic fuel particles (usually uranium oxide) embedded in a metal matrix, may be the only option if complete fission products retention is needed.

Nuclear fuels are generally based on uranium oxide, nitride, or carbide. The development of the fuel microstructure was motivated by the desire to minimize mass losses of carbon by hydrogen corrosion (1500K) and vaporization loss at high temperatures (3000K). Research permitted the procession from $UC_2$ particles dispersed in a graphite matrix to SiC-coated $UC_2$ spheres in a graphite matrix to a composite fuel in the form of a solid solution of carbide fuel UC–ZrC in a graphite matrix, ending as a solid solution of carbide fuel UC–ZrC.

Fuels from the era of the ROVER and NERVA programs include $UC_2$–graphite, UC–ZrC, and $UO_2$–W–Re cermets. Fuels considered in later research include the following: UC–ZrC–NbC, UC–ZrC–TaC, (U, Zr)(C,N), tungsten-coated (U, Zr)(C,N) particles in W alloy matrix cermet, and ceramic–ceramic composites of (U, Zr)C and (U, Zr)CN in ZrC, TiN, etc.

## 13.8 NUCLEAR ROCKET PERFORMANCE

Fundamental technical considerations of nuclear rocket propulsion have been presented by Bussard and DeLauer (1958, 1965) and by Loh (1968). A broad-based technical assessment of nuclear rocket motor design is presented by Pelaccio and colleagues (1990). Some of the pertinent details of nuclear rocket performance are discussed in this section.

### 13.8.1 Solid core reactors

The ROVER or NERVA type systems used solid core reactor technology with fuel elements in rods as depicted schematically in Figure 13.2. A closer look at a typical fuel rod is given in Figure 13.4. The hexagonal fuel rods were composed of a graphite substrate with uranium carbide and zirconium carbide dispersed within it. The graphite acted as the moderator for the nuclear fission reactions. A number of coolant channels were distributed in the fuel rod through which the propellant would pass, picking up heat from the nuclear reactions proceeding within the fuel rods. It must be recognized that although any gas could be used in a nuclear rocket, there are compatibility requirements between the propellant and the nuclear fuel. For example, hydrogen gas is generally compatible with UC fuel but not $UO_2$ fuel, while just the opposite is the case for carbon dioxide gas. Appropriate coatings can alleviate this problem. The hexagonal fuel rods, which are around a meter or two in length, are closely packed to form the solid core of the reactor and require the inclusion of support elements to support the

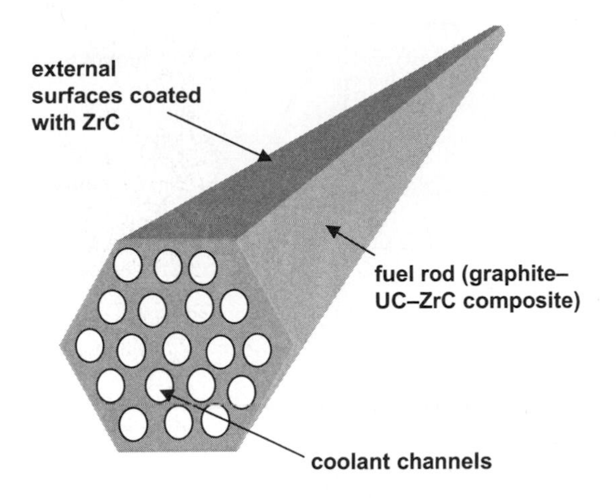


**FIGURE 13.4**

Typical fuel rod for a solid core reactor.

reactor structure, as shown in Figure 13.5. The support elements require thermal and nuclear radiation protection, which led to the complexity of their design.

Control of the criticality of the reactor in the solid core reactor for propulsion applications involves a set of rotatable control drums distributed in sectors around the reactor core, as shown in Figure 13.6. The sectors are composed of reflector material such as beryllium and each contain a drum that has a sector composed of a segment of a neutron-absorbing (or "poison") material such as boron. To start the reactor, the so-called poison plate is rotated such that only reflector material faces the core. Thus neutrons are reflected back into the core to continue the nuclear reaction. As the temperature builds, the drums may be rotated so that the poison plate can capture neutrons instead of reflecting them back into the core. The positioning must be controlled so that the reactor remains critical but does not become unstable.

The thrust to weight ratios for such reactors fall in the range of 3 to 12 with $I_{sp}$ performance in the range of 700 to 1000 s. Prototype reactors have been tested, and the design of a flight weight engine has been accomplished. Thus the solid core nuclear thermal propulsion system is a near-term option for interplanetary flight. Obviously, the strength of the structural components exposed to high temperatures for extended time periods is an issue of concern in the design process. A typical solid core nuclear rocket can operate with a gas temperature of about 2800K at around 3.5 MPa, issuing through a nozzle of expansion ratio $\varepsilon = 40$, thereby producing a specific impulse $I_{sp} = 800$ s. This is almost twice the value for the most advanced chemical rocket motor. Radiation-shielding considerations include weight and shape, as well as radiation and thermal environment compatibility. Shields for crew and payloads may be of different type than those used for engine components and propellant feed systems. The effects of various types of radiation on nuclear propulsion systems are shown in Table 13.4.

## 13.8.2 Particle bed reactors

A refinement of the solid core reactor is the particle bed reactor concept described by Powell and associates (1985), which uses coated fuel particles in a packed bed through which the propellant flows,

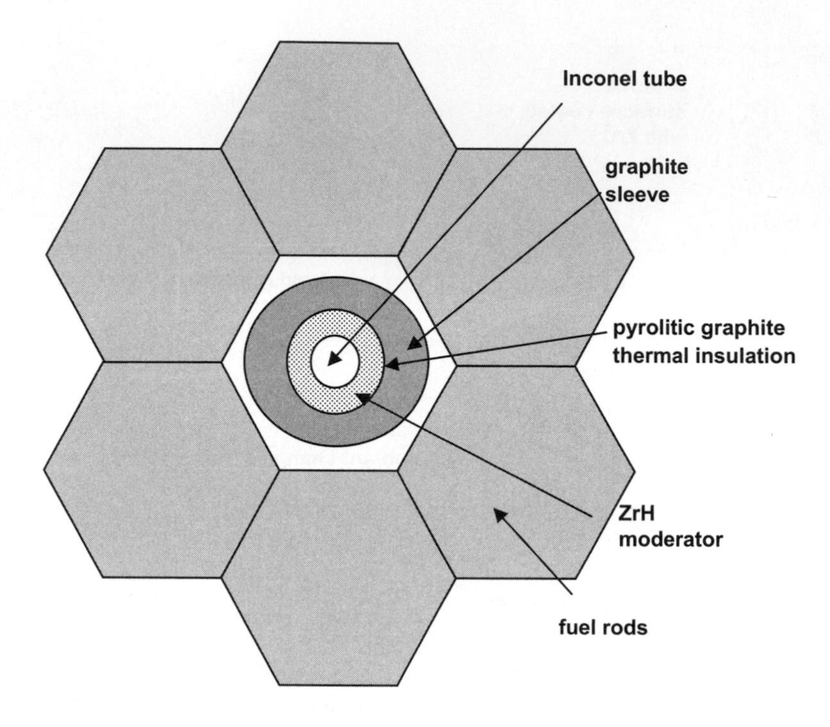


**FIGURE 13.5**

Support elements, shown here as the central feature, are surrounded by fuel elements (coolant passages are omitted from fuel rods here for clarity).

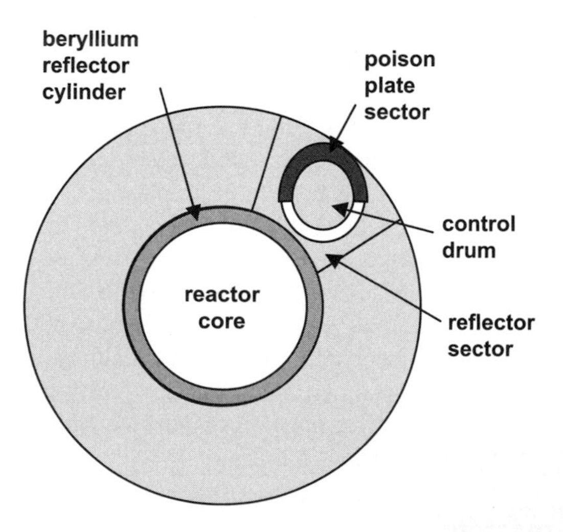


**FIGURE 13.6**

Typical reflector control drum system for solid core reactor. Only one of the control drums is depicted for clarity. Details such as core coolant channels are likewise omitted.

**Table 13.4** Typical Nuclear Rocket Reactor Radiation Effects

| Radiation Type | Effects on Metals | Effects on Organics |
|---|---|---|
| $\gamma$ rays | None | Property degradation |
| Neutrons | Embrittlement | Property degradation |
| $\beta$ particles (electrons) | None | Property degradation |

cooling the reactor and heating the propellant for acceleration through a nozzle. The application of this concept for space propulsion is described by Ludewig and colleagues (1996). These references show that the major limitations of the proven solid core reactor in space applications is the low thrust to weight ratio, much less than that of the latest chemical rockets. This is true because the solid core reactor is handicapped by the long narrow coolant passages in the fuel rods. First, the relatively low surface area to volume ratio of the passages limits the heat transfer to the propellant passing through them. Second, the flow rates of propellant are relatively low because of the large pressure drop associated with the small-diameter passages.

To address these shortcomings of the solid core reactor, several important design changes were introduced (Ludewig et al., 1996; Powell et al., 1985). First, to improve the surface to volume ratio and thereby improve the heat transfer characteristics of the reactor, the fuel element was formed by packing small-diameter (0.4 mm) particles into the annular space between two porous tubes. This provides a ratio of fuel surface area to particle bed volume of about 100 $cm^2/cm^3$. The fuel particles are spheres of $UC_2$ encapsulated by graphite and finally coated with zirconium carbide. Encapsulation serves to avoid the loss of fission products. The hydrogen gas propellant flows into the annulus of the fuel element, passing over the particles and out of the central core and to the nozzle. Second, instead of having the propellant flow axially down the channel in a fuel rod it flows radially through the packed bed of fuel spheres and into the central channel, which directs it to the nozzle. This permits less pressure drop and therefore higher propellant flow rates. Both improvements permit construction of a reactor with a much higher power density, in the range of megawatts per liter (the usual units used in reactor technology). The fuel elements are encased in a hexagonal block of moderator material, as shown in Figure 13.7. A number of these structures are closely packed within the rocket motor casing, as shown in Figure 13.8, and the hot gas flowing out of each is directed to the nozzle.

A third improvement is the use of lithium hydride as the moderator material. The LiH has a low density (approximately 800 $kg/m^3$) and can withstand relatively high temperatures (around 1000K). The major technical features of the particle bed reactor described here have been resolved (Ludewig et al., 1996), and the particle bed reactor theoretically makes possible a nuclear rocket with a thrust to weight ratio of 40 and a specific impulse of 1000 s with a start-up time of several seconds.

### 13.8.3 Propellant feed systems

Candidate propellant feed system cycles typically include the following: (a) a gas generator system, which uses a separate gas generator to drive the feed turbopump; (b) a staged combustion system, derived from the space shuttle main engine system, which uses combustion products from a separate preburner to drive the feed turbopump; and (c) an expander cycle, which uses the reactor heated

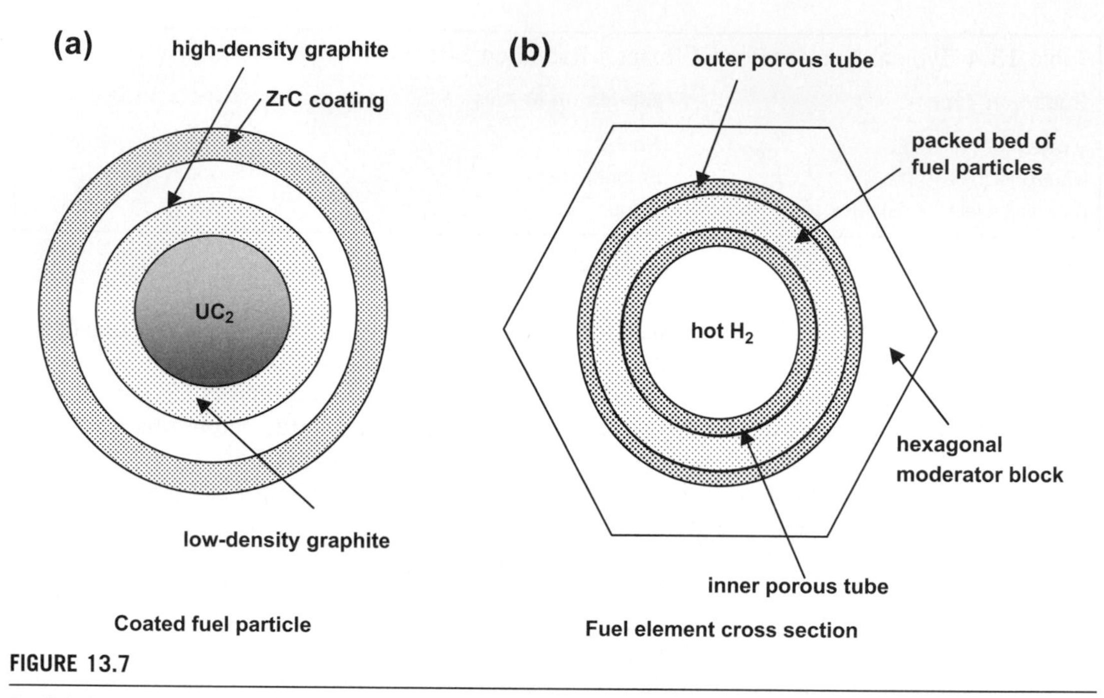


**FIGURE 13.7**

Particle bed reactor fuel element design (Ludewig et al., 1996; Powell et al., 1985).

propellant to drive the feed turbopump directly. The expander cycle shown in Figure 13.9 is considered the best alternative for a solid core nuclear thermal rocket motor (Pelaccio et al., 1990). This cycle employs a turbopump to feed propellant to the nozzle and reactor structure as a coolant. The heated propellant gas then provides the power to drive the turbopump.

## 13.8.4 Comparison of nuclear and chemical rockets

A comparison of the basic performance characteristics of solid or particle bed nuclear thermal rockets and chemical rockets is given in Table 13.5. The major disadvantage of the nuclear rocket is weight, which is largely due to the thermal and radiation-shielding requirements of the nuclear rocket. Means for reducing the weight of nuclear rockets is a major driver for research into such systems.

Despite the weight factor, the specific impulse performance is still a major attraction of nuclear rockets. Consider the payload performance of launch systems using nuclear rockets compared to those using chemical rockets. The simple rocket performance equation yields the ratio of payload to initial launch weight, $W_{pay}/W_0$, for a specified velocity increase $\Delta V = V_f - V_0$ as

$$\frac{W_{pay}}{W_0} = 1 - \exp\left(-\frac{\Delta V}{gI_{sp}}\right).$$

Payload fractions for different launch applications and rocket motors are shown in Table 13.6.

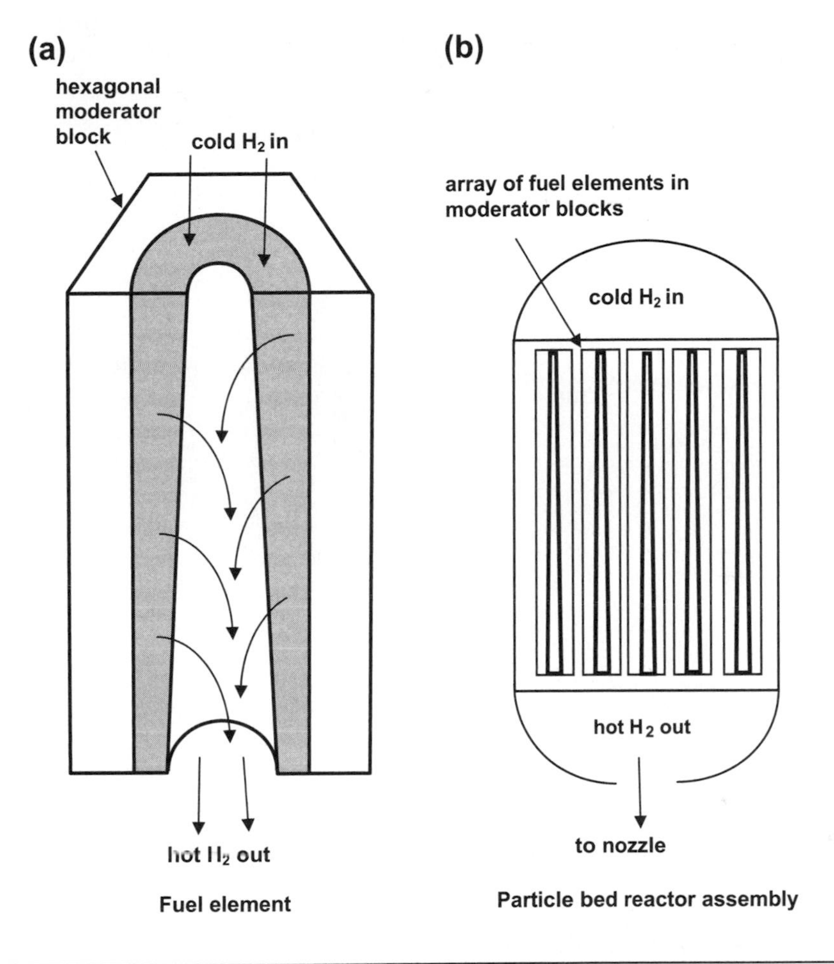


**FIGURE 13.8**

Fuel element assembly: (a) individual fuel element and (b) array of fuel elements to form a particle bed reactor.

## 13.9 GAS CORE NUCLEAR ROCKETS

Time in transit is a major issue for interplanetary space missions:

- In manned exploration missions, the duration of exposure to space radiation and microgravity conditions constitutes a health hazard for the crew (Sforza et al., 1993).
- Unmanned missions to asteroids, for example, to mitigate the hazard due to Earth-crossing asteroids, require fast response to ensure success (Sforza and Remo, 1996).

To minimize flight time, high speed and therefore high thrust are required. At the same time, efficiency considerations demand low fuel consumption.

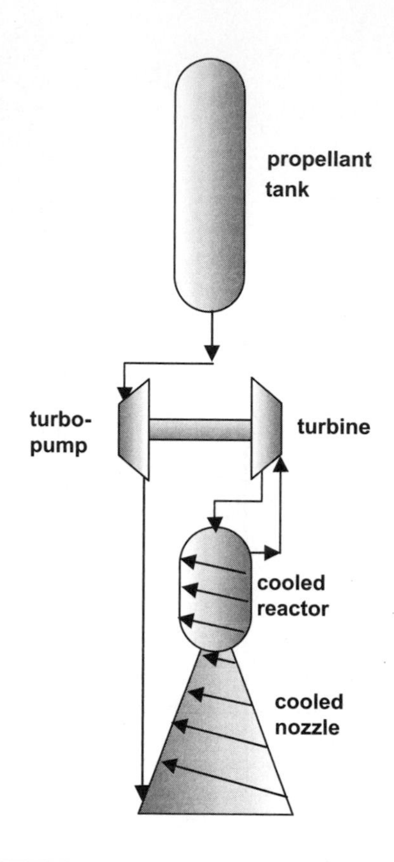


FIGURE 13.9

An expander cycle propellant feed system for a solid core nuclear rocket motor.

Table 13.5 Comparison of Nuclear Thermal Rockets and Chemical Rockets

| Parameter | Nuclear Thermal Rocket | Chemical Rocket |
|---|---|---|
| $I_{sp}$ performance | Solid core > 800 s<br>Gas core > 1500 s | 250–450 s |
| Thrust to weight ratio | $2 < F/W < 5$ | $F/W > 50$ |
| Propellant | Monopropellant: $H_2$, other gases | Bipropellant: LOX–$LH_2$,<br>LOX–hydrocarbon, etc. |
| Chamber pressure | 2 to 7 MPa | 3.5 to 245 MPa |
| Chamber temperature | 2500 to 3200K | 2700 to 3800K |

The ultimate performance of a solid core nuclear reactor is limited by the loss of structural strength as temperature is increased. A gas core reactor uses the nuclear fuel in the gas phase, e.g., uranium hexafluoride. Although there is still the need for a moderating and reflecting encapsulation of the fissionable gas, heat transfer takes place between the hot fuel and the cool propellant while both are in the gaseous state. The heated mixture of propellant and fuel is exhausted through a nozzle at much

**Table 13.6** Payload Fractions for Dfferent Launch Applications and Rocket Motors

| Application | ΔV (m/s) | Solid Rocket $I_{sp} = 250$ s | Liquid Rocket $I_{sp} = 450$ s | Nuclear Rocket $I_{sp} = 750$ s |
|---|---|---|---|---|
| Intercontinental ballistic missile | 4,500 | 0.840 | 0.639 | 0.458 |
| Low Earth orbit | 9,000 | 0.975 | 0.870 | 0.706 |
| Lunar round trip | 18,000 | 0.999 | 0.983 | 0.913 |

higher temperatures than that feasible if there were a solid heat exchanger involved, as in a solid core reactor. Thus the rocket motor case is the only structure exposed to high temperature, and it is possible to have several cooling schemes to provide the necessary thermal protection. Typically somc sort of inertial confinement of the gas is provided, usually in the form of a vortex flow, as shown in Figure 13.10. The gas core rocket is an open cycle system and therefore radioactive fuel is exhausted into the ambient. The application is for interplanetary missions in the deep space environment in which such emissions are acceptable. Such operations are described by Ragsdale (1990).

A typical gas core nuclear rocket can operate with a gas temperature of about 8600K at around 3.5 MPa issuing through a nozzle of expansion ratio $\varepsilon = 40$, thereby producing a specific impulse $I_{sp} = 2200$ s. Part of the increase can be attributed to the hydrogen being completely dissociated so that the effective molecular weight of the nozzle exhaust gas is approximately unity. This is almost five times the value for the most advanced chemical rocket motor.

As pointed out in the example, gas core fission reactors are capable of specific impulse levels on the order of 1500 to more than 2500 s (Schwenk and Franklin, 1970). Such performance is necessary for missions foreseen as part of interplanetary space exploration for the reasons given at the start of this section. Both open and closed cycle (light bulb) reactor designs have been considered and each has its own advantages. Attention here is focused on the open cycle reactor, a system that is limited in

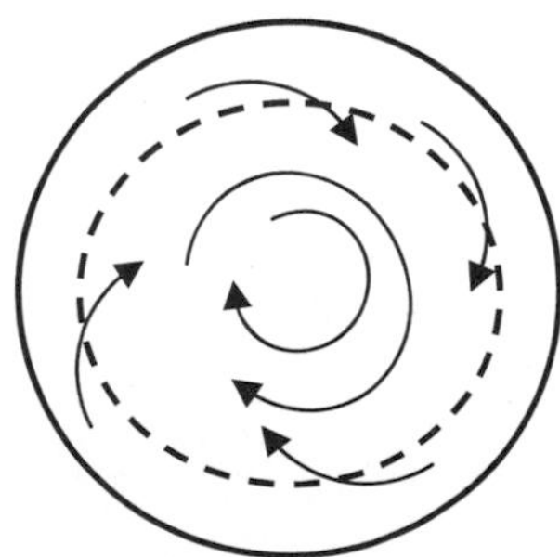


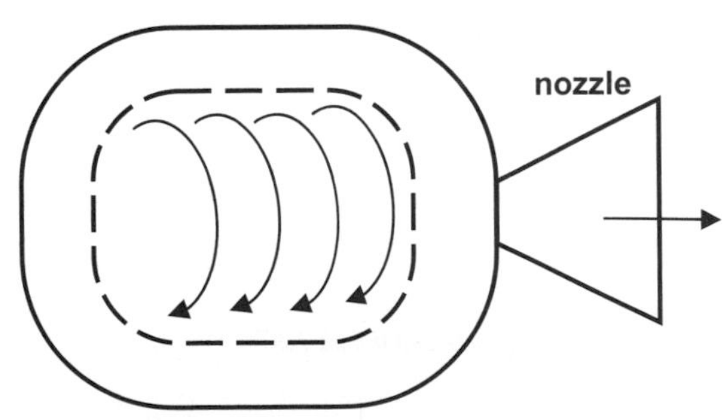


**FIGURE 13.10**

Schematic diagram of gas core nuclear rocket motor.

practical realization by one fundamental problem: effective containment of the uranium plasma. Containment must be one that minimizes the loss of the uranium fuel while reducing the thermal impact of the fission fuel on the structural components of the reactor chamber.

### 13.9.1 Base bleed fuel confinement

An innovative approach to dealing with this problem involves the transfer of technology from the field of projectile ballistics. A method of performance enhancement recently employed for ballistic projectiles is that of "base bleed." This is a technique whereby the base pressure is altered by injecting a small amount of gas through the base of the projectile into the recirculating flow region. The bleed parameter, which is the ratio of mass flow injected into the base recirculation region to the mass flow in the surrounding free stream passing through an area equivalent to the total base area, is found to be on the order of $10^{-2}$ or less for a stable recirculation zone (Sahu and Chow, 1991). Uniform injection over the entire base surface causes the recirculation region to detach from the base as a bubble. In this state, the injected flow stabilizes the recirculation bubble by flowing around it, thereby buffering it from the outer shear layer, and proceeding downstream (Strahle, 1991).

This is the basic mechanism suggested (Sforza et al., 1993) for the hydrodynamic containment of the fuel cloud in an open cycle gas core NTP system. In the application of this concept to NTP the detached flow bubble contains the uranium plasma in a fashion that minimizes fuel losses and maintains its position remote from structural surfaces, as depicted schematically in Figure 13.11. The size of the bubble depends solely on the base dimension so that scaling for criticality is assured and effective radiant heating of the surrounding propellant is readily achieved. In the application of base bleed to projectiles, it is sufficient to maintain a constant injection rate to meet the performance

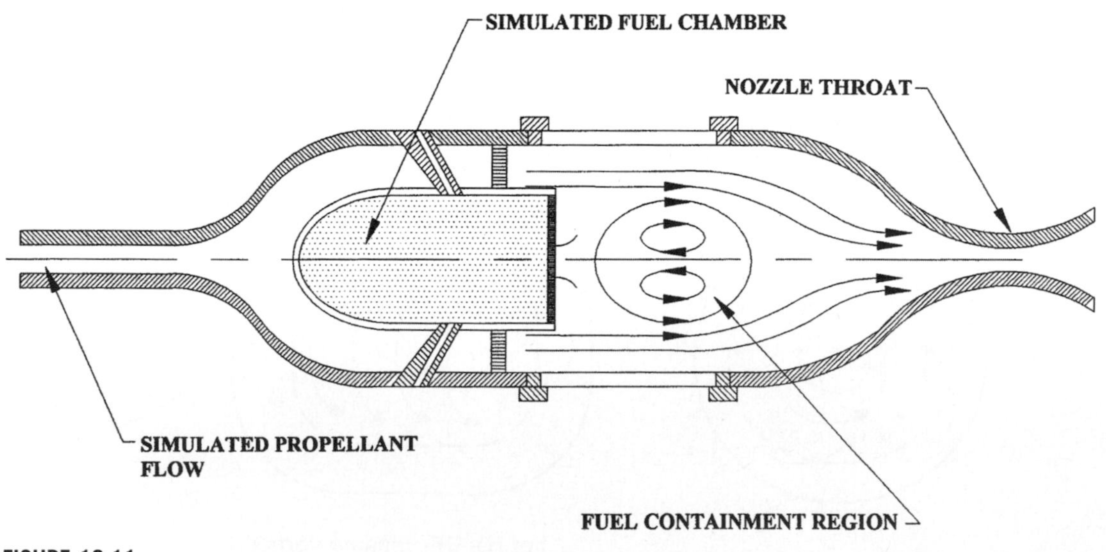


**FIGURE 13.11**

Base bleed stabilized gas core nuclear rocket motor. Uranium hexafluoride fuel is injected through the porous base of the fuel chamber into a surrounding flow of propellant.

objectives desired. However, in the present NTP application, such an approach would waste the fuel gas that stabilizes the fuel-laden bubble and passes on downstream. To optimize this process, a periodic injection scheme may be more appropriate. A laboratory experimental facility of a gas core fission reactor incorporating this new approach for hydrodynamic confinement of the fuel cloud by means of a base injection stabilized recirculation bubble has been developed, fabricated, and demonstrated under cold flow conditions (Sforza et al., 1994).

## 13.10 NUCLEAR RAMJETS

As has been shown elsewhere in the text, flight at high enough supersonic speeds obviates the need for a compressor to provide high-pressure air to a combustor, which is the principle of the ramjet. If the source of heat were a nuclear reactor rather than a combustor, one would have a nuclear ramjet. In 1957, Project PLUTO was initiated to explore the use of such a ramjet for a supersonic low-altitude missile. Developments in the project led to the TORY IIC nuclear ramjet shown in Figure 13.12. It was operated successfully at full power for 5 min at a reactor power of 513 MW, which would result in the production of 35,000 lbs of thrust. Although much progress was made in terms of technical design, concerns about fission products being exhausted into the atmosphere caused the program to be terminated in 1964.

### 13.10.1 A nuclear ramjet for planetary exploration

Since the early 2000s, interest in the long-term study of planetary atmospheres revealed a need for a probe that had the following characteristics:

- three-dimensional mobility for atmospheric mapping
- capability to explore for many months

**FIGURE 13.12**

TORY IIC nuclear ramjet at the Nevada test site.

**Table 13.7** Properties of the Atmospheres of Several Planets

| Planet | Radius (km) | Gas Composition | Molecular Weight | R ($m^2/s^2$-K) | $\gamma$ | $T$ (K) | $a$ (km/s) |
|---|---|---|---|---|---|---|---|
| Venus | 6,052.8 | $N_2$, $CO_2$ | 30 | 286 | 1.4 | 735 | 0.542 |
| Earth | 6,378.1 | $N_2$, $O_2$ | 29 | 287 | 1.4 | 185 | 0.273 |
| Mars | 3,397 | $CO_2$, $N_2$, A | 44 | 190 | 1.3 | 186 | 0.214 |
| Jupiter | 71,492 | $H_2$, He | 2 | 4170 | 1.4 | 137 | 0.894 |
| Titan | 2,575 | $N_2$, $CH_4$ | 27 | 309 | 1.4 | 92 | 0.199 |
| Saturn | 60,268 | $H_2$, He | 2 | 4170 | 1.4 | — | — |

- simple, reliable, small mass and size
- scalable to different planets

One proposed solution to this need is an autonomous supersonic aircraft powered by a nuclear ramjet engine. Such an aircraft would use the planet's atmospheric gas captured by the inlet as the propellant to which a nuclear reactor would add heat. Because essentially any gas could be used as a propellant in a nuclear reactor, the nuclear ramjet could operate in any known planetary atmosphere. The composition of the planetary atmosphere of several of the closer planets in the solar system is described in Table 13.7. Jupiter and Saturn are of particular interest because of the high concentration of hydrogen in their atmospheres. The large moons of these planets, such as Titan, may also be candidates for exploration. A nuclear ramjet for just such missions is described by Maise and colleagues (2003).

In Maise and colleagues (2003), the so-called Jupiter Flyer, shown in Figure 13.13, was designed to be capable of being transported in a spacecraft the size of that which performed the Galileo mission to Jupiter. The payload of 900 kg was composed of

- Ramjet Flyer 300 kg
- Thermal Shield 300 kg
- Companion Orbiter 200 kg

Upon arrival at Jupiter, the companion orbiter and the entry capsule containing the Jupiter Flyer separate with the former entering an orbit around Jupiter and the latter entering Jupiter's atmosphere. When the entry capsule slows to Mach 3, the Jupiter Flyer separates from thermal shield, the nuclear engine starts, and the Jupiter Flyer commences atmospheric mapping, sending data to the companion orbiter for relaying to earth stations.

The compact reactor was provided by a modification of the MITEE (MIniature ReacTor EnginE) nuclear powerplant described in Powell and associates (1999). This lightweight reactor is a derivative of the ultralight particle bed reactor described previously. The propellant is the ambient atmosphere, and the outlet temperature is reduced from 3000 to 1500K. Similarly, the power density is derated from 10 to 2 MW/liter to increase the operating life from hours to months. The general cross section of the reactor is shown in Figure 13.14, and the typical packing of fuel elements is clear.

However, the fuel element in the MITEE reactor is unique in that it consists of a thin porous annular tube of cermet fuel through which the propellant passes and is heated, as shown in Figures 13.15 and 13.16.

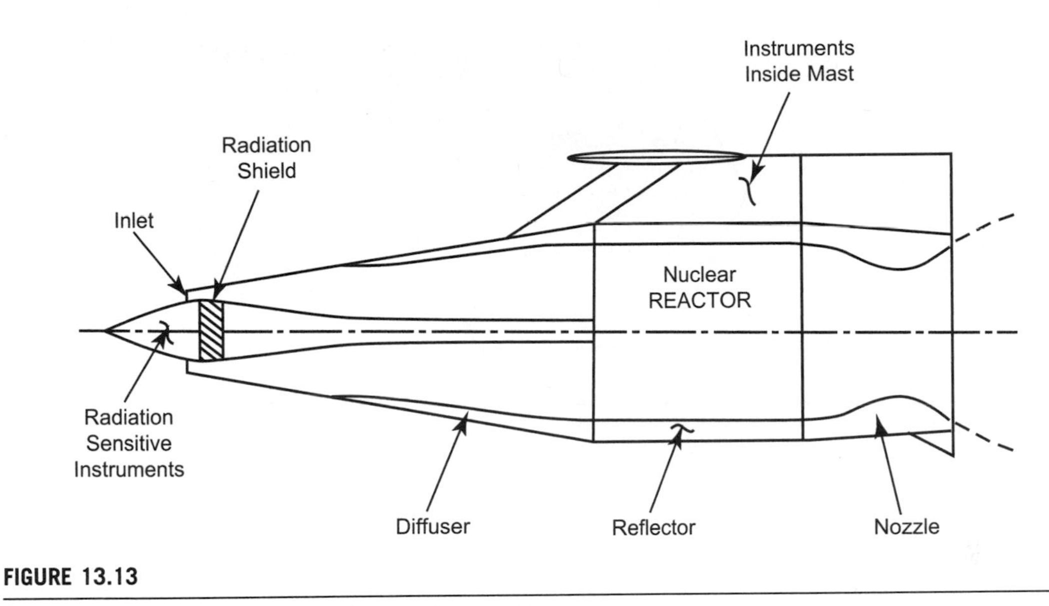


**FIGURE 13.13**

Schematic diagram of the Jupiter Flyer nuclear ramjet.

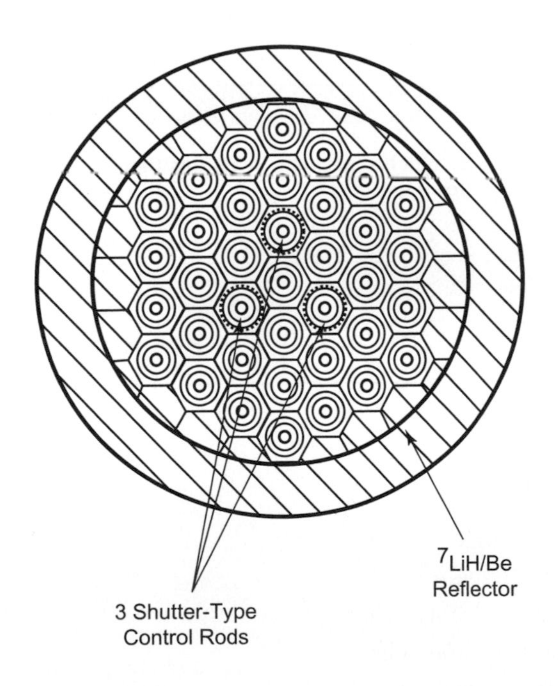


**FIGURE 13.14**

Cross section of the lightweight MITEE reactor.

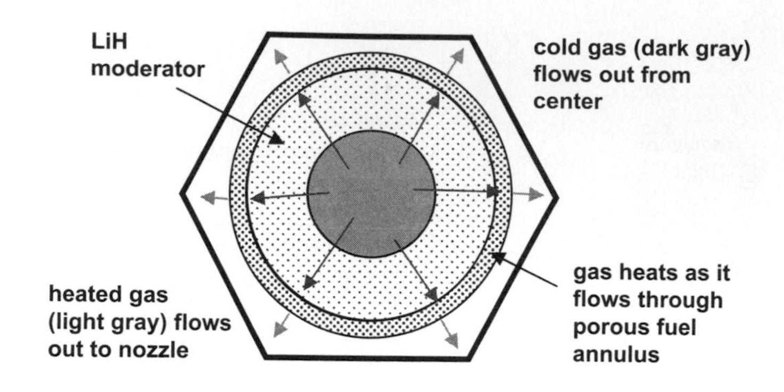


**FIGURE 13.15**

Cross section of the MITEE reactor fuel element.

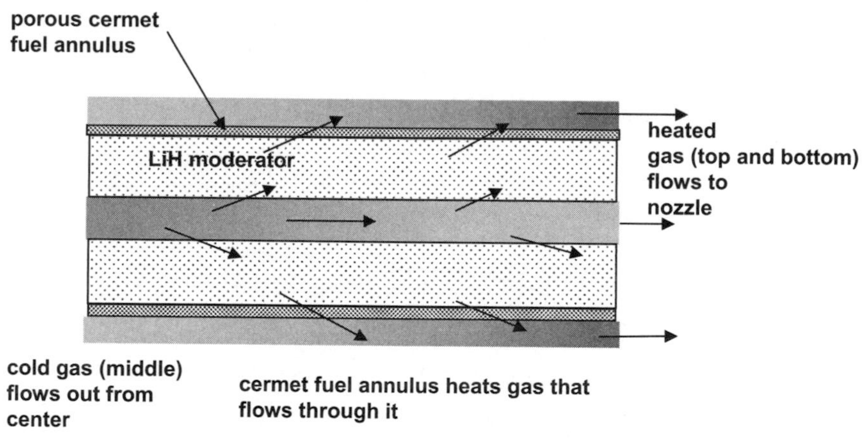


**FIGURE 13.16**

Axial cross section of MITEE fuel element.

## 13.11 NOMENCLATURE

| | |
|---|---|
| $A$ | mass number |
| amu | atomic mass units |
| $c$ | speed of light |
| $I_{sp}$ | specific impulse |
| $k$ | reactor criticality factor |
| MeV | million electron volts |
| $n$ | number of neutrons |
| $T$ | temperature |

| | |
|---|---|
| $t$ | time |
| $V$ | velocity |
| W | molecular weight |
| Z | atomic number |
| $\varepsilon$ | nozzle expansion ratio |
| $\tau$ | characteristic time between fission events |

## 13.12 EXERCISES

**13.1** A nuclear thermal rocket using hydrogen as the propellant has a specific impulse $I_{sp,NUKE} = 900$s and a liquid rocket using LH2-LOX as the propellant has a specific impulse $I_{sp,LOX} = 450$s when both use the same nozzle and operate under the same ambient pressure conditions with the nozzle matched. The liquid rocket engine operates with a chamber temperature $T_c = 3600$K. Calculate a reasonable estimate of the chamber temperature at which the nuclear rocket is operating.

**13.2** A fast trip to Mars, including landing and return, in a spacecraft with initial and final mass $M_i$ and $M_f$, respectively, requires a spacecraft velocity increment

$$\Delta V = gI_{sp} \ln\left(\frac{M_i}{M_f}\right) = 40,000 \frac{m}{s}$$

The ideal mission would require 22,000 m/s, about half that of the fast mission.(a) For both missions compare the required propellant mass fraction $M_p/M_i$ required for chemical, solid-core nuclear, gas-core nuclear, and electric propulsion devices having Isp = 450s, 1000s, 2000s, and 10,000s, respectively.

**13.3** A nuclear ramjet like that shown in Figure 13.13 is considered for flight in the stratosphere of the earth's atmosphere. The reactor directly heats the air passing through it and the reactor is kept at a constant temperature $T_r$ in the range $1500\text{K} < T_r < 1900\text{K}$. Show that the optimum specific net thrust $F_n/\dot{m}_0$ occurs for flight Mach numbers near $M_0 = 3$. It may be assumed that the exhaust nozzle is always matched so that $p_7 = p_0$ and that the stagnation pressure ratio across the reactor $p_{t,ri}/p_{t,re} = 2$ and remains essentially constant.

## References

Bussard, R. W., & DeLauer, R. D. (1958). *Nuclear Rocket Propulsion*. New York: McGraw-Hill.
Bussard, R. W., & DeLauer, R. D. (1965). *Fundamentals of Nuclear Flight*. New York: McGraw-Hill.
Loh, W. H. T. (1968). *Jet, Rocket, Nuclear, Ion, and Electric Propulsion*. New York: Springer-Verlag.
Ludewig, H., Powell, J. R., Todosow, M., Maise, G., Barletta, R., & Schweitzer, D. G. (1996). Design of particle bed reactors for the space nuclear thermal propulsion program. *Progress Nuclear Eng, 30*(1), 1–65.

Maise, G., Powell, J., Paniagua, J., Kush, E., Sforza, P., Ludewig, H., & Dowling, T. (2003). Application of the MITEE Nuclear Ramjet for Ultra Long Range Flyer Missions in the Atmospheres of Jupiter and Other Giant Planets. 54th International Astronautical Congress, Bremen, Germany, September-October.

Pelaccio, D., Perry, F., Oeding, B., & Scheil, C. (1990). Nuclear Engine Design Technical Assessment Considerations. AIAA 90-1950, AIAA-SAE-ASME-ASEE 26th Joint Propulsion Conference, Orlando, FL.

Powell, J., Paniagua, J., Maise, G., Ludewig, H., & Todosow, M. (1999). High performance nuclear thermal propulsion system for near term exploration missions to 100AU and beyond. *Acta Astronautica, 44*(2-4). Jan.-Feb, 159–66.

Powell, J.R., & Horn, F.L. (1985). High Power Density Reactors Based on Direct Cooled Particle Beds. Proceedings of the 2nd Symposium on Space Nuclear Power Systems, Albuquerque, NM.

Ragsdale, R. (1990). Open Cycle Gas Core Nuclear Rockets. Proceedings of the Nuclear Thermal Propulsion Workshop, NASA Lewis Research Center, 343–357.

Sahu, J., & Chow, W. L. (1991). A review of the fluid dynamic aspects of the effect of base bleed. In K. K. Kuo & J. N. Fleming (Eds.), *Base Bleed* (pp. 81–92). New York: Hemisphere Publishing Corp.

Schwenk, F.C., & Franklin, C.E. (1979). Comparison of Closed and Open Cycle Systems. NASA SP-236, Research on Uranium Plasmas and Their Technological Applications, 3–12.

Sforza, P. M., Cresci, R. J., Artz, J., & Castrogiovanni, A. (1994). Recirculation Containment for Gas Core Fission Rockets. AIAA Paper No. 94-2899, 30th AIAA/ASME/SAE/ASEE Joint Propulsion Conference. Indianapolis, IN June, 1–9.

Sforza, P.M., Cresci, R.J., & Girlea, F. (1993). Fuel Efficient Hydrodynamic Containment for Gas Core Fission Reactor Rocket Propulsion. IAF93-R.1.427, 44th International Astronautical Federation Congress, Graz, Austria, October, 1–11.

Sforza, P. M., & Remo, J. L. (1996). Propulsion options for missions to near earth objects. *Acta Astronautica, 39*(7), 517–528.

Sforza, P. M., Shooman, M., & Pelaccio, D. (1993). A safety and reliability analysis for space nuclear thermal propulsion systems. *Acta Astronautica, Vol. 30*, 67–83, Also in *Space Safety and Rescue* 1992, Vol. 84, Science Technology Series, 1994, 29–50.

Strahle, W. C. (1991). Base burning performance at Mach 3. In K. K. Kuo & J. N. Fleming (Eds.), *Base Bleed* (pp. 217–225). New York: Hemisphere Publishing Corp.

CHAPTER

# Space Propulsion

# 14

## CHAPTER OUTLINE



## 14.1 SPACE PROPULSION SYSTEMS

The thrust force for typical space propulsion reaction engines with constant exhaust velocity $V_e$ is

$$F = \dot{m}_{pro} V_e, \tag{14.1}$$

where the quantity $m_{pro}$ is the constant propellant flow rate passing through the engine. The figure of merit for space propulsion systems is the specific impulse, which measures the total impulse delivered by the engine per unit weight of propellant consumed, as shown here:

$$I_{sp} = \frac{\int F dt}{g_E \int \dot{m} dt} = \frac{V_e}{g_E}, \tag{14.2}$$

where $g_E$ is the gravitational acceleration at the surface of the Earth; the units of $I_{sp}$ are therefore given in seconds. The power in the engine exhaust is given by

$$P_e = \frac{1}{2}\dot{m}_{pro}V_e^2. \tag{14.3}$$

This power may also be written, in the units of kW, as follows:

$$P_e = \frac{g_E F I_{sp}}{2000}. \tag{14.4}$$

The mass of the spacecraft powered by the engine may be written as

$$M_0 = M_{pay} + M_{pro} + M_{eng} + M_{str}. \tag{14.5}$$

The subscripts *pay*, *pro*, *eng*, and *str* refer to the payload, propellant, engine system, and structure, respectively. In terms of mass fractions, Equation (14.5) may be written as

$$1 = \mu_{pay} + \mu_{pro} + \mu_{eng} + \mu_{str}. \tag{14.6}$$

Typically it is assumed that the engine system mass is proportional to the power delivered to the engine unit so that

$$M_{eng} = \alpha(\eta P), \tag{14.7}$$

where the primary power of the engine system is $P$ and $\eta$ is the efficiency with which that power is transformed and delivered to the engine itself. Quantity $\alpha$ is the constant of proportionality and is called the specific mass of the power source; it has the units of kg/kW. It is the second figure of merit for the space propulsion system. The power in the exhaust stream is then given by

$$P_e = \eta'(\eta P),$$

where $\eta'$ is the efficiency with which the power in the engine is converted to power in the exhaust jet. The thrust to weight ratio of the engine is the third figure of merit and is related to $I_{sp}$ and $\alpha$ through Equations (14.2), (14.4), and (14.7) and is given by

$$\frac{F}{M_{eng} g_E} = \frac{2000\eta'}{\alpha I_{sp} g_E^2}. \tag{14.8}$$

The important parameters in Equation (14.8) are related graphically in Figure 14.1. Also shown is the corresponding parameter space for a number of different operational and conceptual propulsion systems. A natural classification of these systems arises from their characteristic thrust to weight ratios. Type I systems are those that have high thrust to weight ratios and low specific mass but are limited by relatively low specific impulse achievable, whereas type II systems have low thrust to weight ratios and high specific impulse but are limited by relatively high specific mass. From Equation (14.2) it is clear that to deliver the same total impulse, type I systems will consume far larger quantities of propellant than type II systems. This has important consequences for overall mission performance and vehicle initial mass. A roadmap for future development of in-space propulsion systems is given by Meyer and colleagues (2010).

The mass fraction distribution of the vehicle may now be written as follows:

$$1 = \mu_{pay} + \frac{\alpha g_E I_{sp} F}{2000\eta' M_0} + \frac{F t_b}{g_E I_{sp} M_0} + \mu_{str}, \tag{14.9}$$

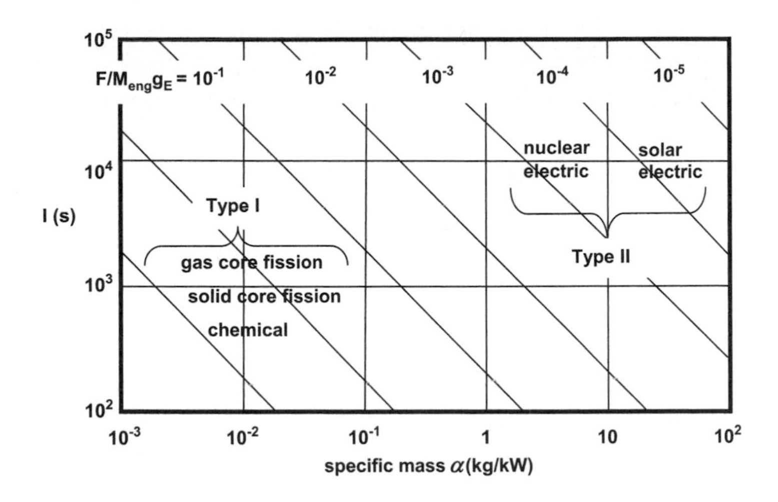


**FIGURE 14.1**

Classification of propulsion systems.

where the quantity $t_b$ represents the burn time of the engine, that is, the duration of operation during which the propellant is consumed. The mass fractions $\mu_{pay}$ (on the order of 0.01) and $\mu_{str}$ (on the order of 0.1) are generally small compared to the other two mass fractions. As a first-order approximation, we may neglect these two mass fractions with respect to unity and rewrite Equation (14.9) as follows:

$$\frac{\alpha g_E I_{sp} F}{2000\eta' M_0} + \frac{F t_p}{g_E I_{sp} M_0} \approx 1. \tag{14.10}$$

Using Equation (14.8) and Equation (14.10) and taking $g_E \sim 10 \text{ m/s}^2$, we may approximate the propellant burn time as

$$t_p = \frac{\alpha I_{sp}^2}{20\eta'}\left(\frac{1-\mu_{eng}}{\mu_{eng}}\right). \tag{14.11}$$

Considering typical values of the variables in Equation (14.11) for different propulsion systems, it may be shown that type I systems have $t_p$ measured in minutes or hours, whereas type II systems have $t_p$ measured in months or years. From Equation (14.9) it is clear that for fixed values of the payload mass and structural mass, the initial mass will be minimized at some optimum value of $I_{sp}$ because the engine system mass grows linearly with $I_{sp}$, whereas propellant mass is inversely proportional to $I_{sp}$.

## 14.2 ELECTRIC PROPULSION SYSTEMS

In theory, electric propulsion has been considered for many years. It was only in the middle of the 20th century that research sufficient to develop a practical device was carried out. The problem with electric propulsion was the usual one, lack of flight weight components, particularly electric power supplies. Stuhlinger (1991) provides a concise history of the development of electric propulsion concepts. We may get an introductory idea of what sort of performance is required of an electric propulsion system by interpreting Equation (14.11) in terms of the specific impulse, and therefore the exhaust velocity, that would be required for good performance of an electric propulsion system. This approach yields

$$V_e = \sqrt{\frac{t_p}{\alpha}\left[\frac{20g_E^2\eta'\mu_{eng}}{1-\mu_{eng}}\right]}. \tag{14.12}$$

For typical values of the parameters in square brackets in Equation (14.12), the optimal exhaust velocity is given approximately by

$$V_e \approx \sqrt{1000\frac{t_p}{\alpha}}. \tag{14.13}$$

An equivalent result is given by Stuhlinger (1991) as a useful estimate of the optimal exhaust velocity of an electric propulsion system for bringing a rocket powered by electric propulsion to its maximum velocity. For example, at $\alpha = 10$ kg/kW, the best exhaust velocity for a 1-year burn time would be about 56,000 m/s. We will examine the electric propulsion systems that can provide that kind of performance. As far as thrust is concerned, we may solve Equation (14.8) to obtain the general result

$$F = 2000\frac{\eta'\eta P}{V_e}.$$

Here we see that the thrust level is inversely proportional to the exhaust velocity; note that the primary power $P$ delivered to the propulsion system is measured in kW. Because efficiencies are $O(10^{-1})$, it is clear that thrust levels will be small. For the example mission considered, a 1-kW power supply would provide a thrust measured in Newtons.

Jahn (1968) defines electric propulsion as "the acceleration of gases for propulsion by electric heating and/or by electric and magnetic body forces." Thus we speak of electrothermal, electrostatic, or electromagnetic propulsion devices. A useful overview of electric propulsion for spacecraft was presented by Martinez-Sanchez and Pollard (1998).

Electrothermal rocket engines are very similar in principle to chemical and nuclear thermal rockets, differing only in using electrical heating to raise the temperature of the propellant prior to accelerating it in a nozzle. The electric heating may be achieved by simple resistance heating, as in the "resistojet," by the breakdown of the propellant caused by a high energy arc passing through it, as in the arcjet, or by passing radio-frequency (RF) electromagnetic waves through the propellant to heat it. These devices can produce high thrust levels at reasonably high specific impulse, but limitations posed by the high heating makes them suffer the same shortcomings as chemical and nuclear thermal rocket engines.

Electrostatic propulsion depends on the attraction or repulsion of electrically charged particles and uses this effect to accelerate a stream of positively charged ions in order to produce thrust. The so-called ion space propulsion engines are based on this effect and such engines first flew in space in the mid-1960s. The electrostatic engines accelerate ions to very high exhaust speeds so that the specific impulse generated is very high, meaning that the engine is very fuel efficient. However, because the mass flow of particles accelerated is small, only low thrust values are achievable. Recall that though the thrust is small, the burn time is long and therefore the total impulse delivered to the spacecraft can

**Table 14.1** Typical Performance of Electric Propulsion Systems

| Type | Propellant | Energy Source | $I_{sp,vac}$ (s) | Thrust (*N*) | Density (kg/m$^3$) |
|---|---|---|---|---|---|
| Resistojet | $N_2$, $NH_3$, $N_2O_4$, $H_2$ | Resistive $\eta = 0.9$ | 150–700 | 0.005–0.5 | 280, 600, 1000, 19 |
| Arcjet | $NH_3$, $H_2$, $N_2H_4$ | Arc heat $\eta = 0.3$ | 450–1500 | 0.05–5 | 600, 19, 1000 |
| Ion | Hg, A,Xe, Cs | Electrostatic $\eta = 0.75$ | 2000–6000 | $5 \times 10^{-6} - 0.5$ | 13,500, 440, 273, 187 |
| MHD | A | Magnetic | 2000 | 25–200 | 440 |
| Pulsed plasma | Teflon | Magnetic | 1500 | $5 \times 10^{-6} - 5 \times 10^{-3}$ | 220 |

**Table 14.2** Applications for Different Electric Propulsion Systems

| Application | Electrothermal | Electrostatic | Electromagnetic |
|---|---|---|---|
| Orbital operations: attitude control, drag compensation, trajectory modification, orbit transfer | Yes | Yes | Yes |
| Interplanetary operations: planetary missions, deep space missions, long-term science missions | No | Yes | No |

be large, resulting in good performance for long-term missions. Goebel and Katz (2008) provide a comprehensive study of electrostatic propulsion.

Electromagnetic propulsion devices depend on the creation of a magnetic body force on a conducting gas, or plasma, by passing it through a field with crossed electric and magnetic fields. This is an attractive method but suffers from the need to produce plasma with high conductivity; this generally requires high temperatures as well as seeding with an additional substance.

Typical characteristics of the various electric propulsion devices are described subsequently, and typical performance capabilities are summarized in Table 14.1.

In general, all the electric propulsion systems can be used successfully for near-earth satellite operations, but for long-term missions beyond earth only electrostatic systems can be used. A description of the applicability of these systems is given in Table 14.2.

## 14.3 ELECTROTHERMAL PROPULSION DEVICES

Rather than relying on chemical energy for the production of high chamber temperatures, a reliance that is limited by the available propellant combinations, one may turn to the deposition of energy into a passive propellant by electrical means. The electrically heated propellant is then expanded through a supersonic nozzle to produce thrust. Three popular classes of electrothermal propulsion devices are as follows:

- Resistojets: propellant gas is passed over an electrically heated surface and heat is transferred from the surface to the gas
- Arcjets: propellant gas is passed through an electric arc discharge, which releases energy that the gas absorbs, raising its temperature
- Radio-frequency excited jets: propellant gas is heated by RF radiation

These devices are limited by the temperature that can be tolerated by the pressure chamber and nozzle material. Because these devices operate with hot high-pressure gas being exhausted through a supersonic nozzle, they are subject to the same losses that afflict chemical and nuclear rockets, which operate on the same physical principle. Performance is limited by heat transfer losses, which may occur in several ways. First, the heater element may conduct or radiate heat to the casing and other elements of the thruster, which in turn radiates that energy into space. Second, the hot gas produced by the electric heating will also transfer heat to the thruster walls by both convection and radiation, and this heat will ultimately by directed out to space. Finally, the hot propellant gas will also radiate heat, which escapes axially out the exhaust nozzle.

Other losses are a result of the nature of rapidly expanded flows. Energy in the translational, rotational, vibrational, dissociational, and electronic degrees of freedom of the gas molecules increases as the gas is heated. Normally, when sufficient interparticle collisions are in evidence, as is the case under high-pressure conditions, equilibration among the various modes occurs almost instantaneously.

With a propellant heated to a given enthalpy $h_c$ in a settling chamber where the velocity is negligible, the stagnation enthalpy in the chamber may be considered to be virtually equal to it, that is, $h_t = h_c$. The conservation of energy equation shows that the velocity achievable after adiabatic expansion to the nozzle exit enthalpy $h_e$ is given by

$$V_e = \sqrt{2(h_t - h_e)}.$$

If reaction rates are sufficiently rapid, equilibrium chemical composition will be in evidence throughout the flow and the expansion will be an isentropic one. However, if equilibration of the vibrational and electronic degrees of freedom lags that of the translational degrees of freedom, energy will be "frozen" in these lagging degrees of freedom and not available to the expansion process, resulting in a lower exhaust velocity than that in the equilibrium flow. Therefore, the thrust and specific impulse will be reduced. For the case where hydrogen is the propellant the equilibrium mole fraction of atomic hydrogen is shown for various temperatures and pressures of interest in Figure 14.2.

At low pressures, the degree of dissociation is high and substantial amounts of H are present in the flow. As the temperature in the flow drops during expansion through the nozzle, atomic hydrogen must recombine to molecular hydrogen to release the chemical energy and transfer it ultimately to kinetic energy of the exhaust flow. The exit velocity achievable with expansion to, say, $T_e = 298$K is shown in Figure 14.3. Note that there is a clear increase in exit velocity, and therefore specific impulse, at the lower chamber pressures where substantially more atomic hydrogen is produced than is the case at higher pressures. The practical problem with achieving these increases in exit velocity is that the expansion process drops the pressure as well as the temperature and the resulting low density reduces the interparticle collision frequency and recombination is delayed. As discussed in Chapter 4, the chemical composition of the flow may "freeze" at some intermediate point within the nozzle and deny

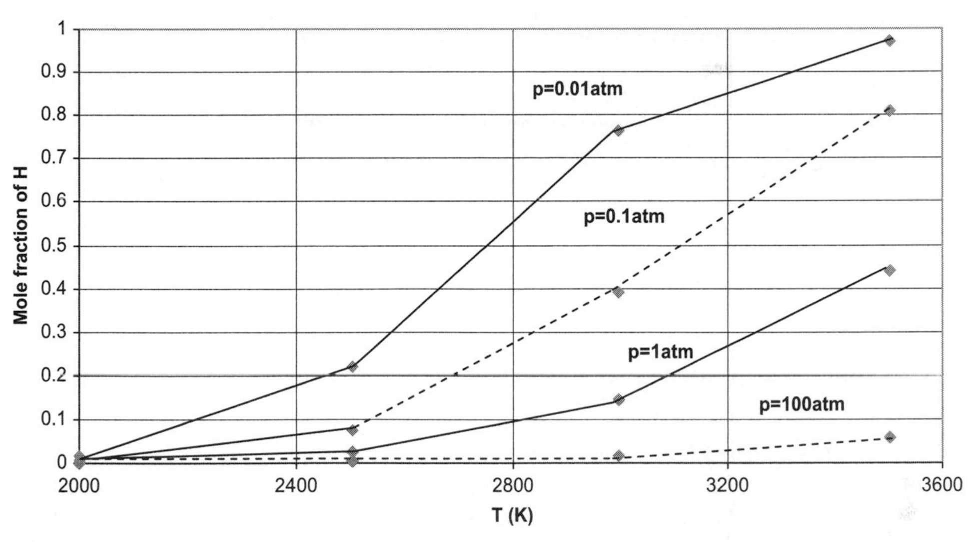


**FIGURE 14.2**

Equilibrium mole fraction of atomic hydrogen at four temperatures and pressures. Points connected by straight lines are for illustrative purposes.

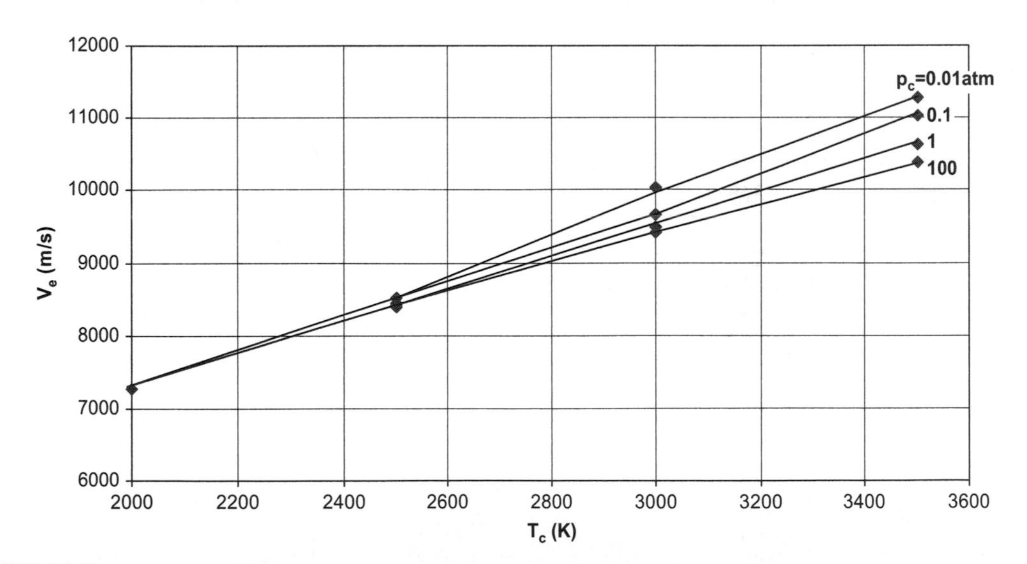


**FIGURE 14.3**

Exit velocity achievable for equilibrium expansion from $T_C$ to 298K for several chamber pressures.

the flow the additional energy of recombination. It is equally important to note that this case is concerned with hydrogen as the fuel and it is heated by some external source. Chemical rockets using the energy of combustion of hydrogen and oxygen are limited by a relatively large molecular weight of the exhaust gases. For example, stoichiometric hydrogen and oxygen will yield water with $W = 18$, while

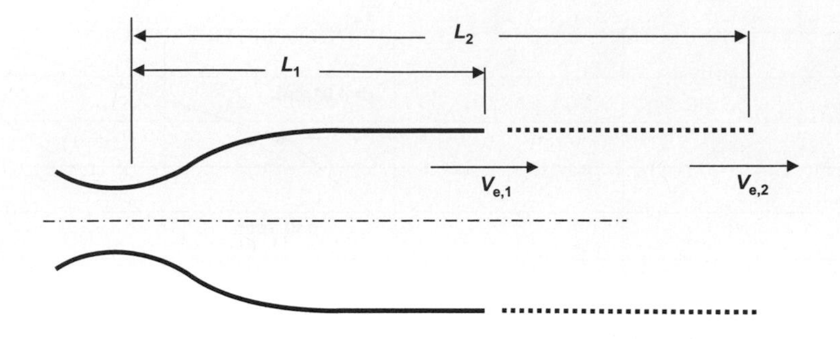


FIGURE 14.4

Lengthened exhaust nozzle provides additional residence time so that $V_e$ reaches full potential.

pure molecular hydrogen has $W = 2$. Thus the square root of the ratio results in a factor of 3 so that the specific impulse for hydrogen will be three times that for water developed by combustion, that is, around 900 s versus 300 s. The hydrogen–oxygen system can get up to 450 s because it runs fuel rich and the products of combustion have a lower molecular weight than pure water.

The first Damkohler parameter $D_1$ expresses the ratio of residence time of a particle flowing through the nozzle to the chemical reaction time for the particles as given by

$$D_1 = \frac{\tau_f}{\tau_c} = \frac{(L/V)}{\tau_c}.$$

One way to increase the flow residence time so that equilibration of all the internal degrees of freedom occurs is to lengthen the rocket nozzle, as shown in Figure 14.4. In practice, losses incurred by increased friction and heat transfer are found to be more than offset by the ability to ensure more complete recombination in the nozzle.

### 14.3.1 Resistojets

The simplest electrothermal device is the resistojet, shown schematically in Figure 14.5. It consists of an electric heater that is fed propellant in a pressure casing that includes a nozzle. The propellant flow

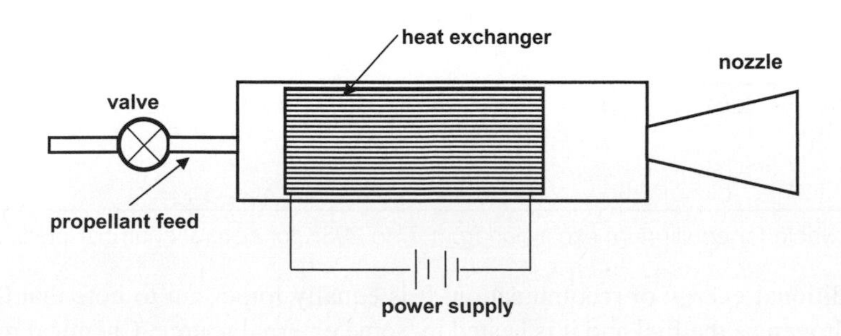


FIGURE 14.5

Schematic diagram of a resistojet.

is controlled by a valve so that the thrust level can be altered during operation. In a sense, the resistojet is like the solid core nuclear thermal propulsion system in that a propellant is simply heated by conduction and convection; only in this case, the thermal power is electric rather than that arising from radioactive decay. The propellant may be passed around the nozzle to provide some cooling for the nozzle walls. These devices are often used because of their inherent simplicity, despite the fact that they are limited in thrust capability.

### 14.3.2 Arcjets

Another common electrothermal propulsion device for space propulsion is the arcjet, a schematic diagram of which is shown in Figure 14.6. The high voltage arc breaks down the gas into plasma, an electrically conductive gas, which is accelerated by the nozzle as in other thermal propulsion systems, such as chemical and nuclear rockets. Propellants for arcjets include hydrazine ($N_2H_4$), ammonia ($NH_3$), and hydrogen ($H_2$). An arcjet using ammonia as the propellant produced a thrust of 2N at a specific impulse of 800 s. The power to the thruster was 26 kW and the ammonia mass flow was 0.25 mg/s. The engine was operated in space as part of a high power in-space flight experiment in 1999. The design of the arcjet is described by Vaugh and colleagues (1993).

The thermal constraint on electrothermal propulsion devices arises from the high temperatures caused by the arc. Typical temperature profiles in the throat region of the nozzle are shown in

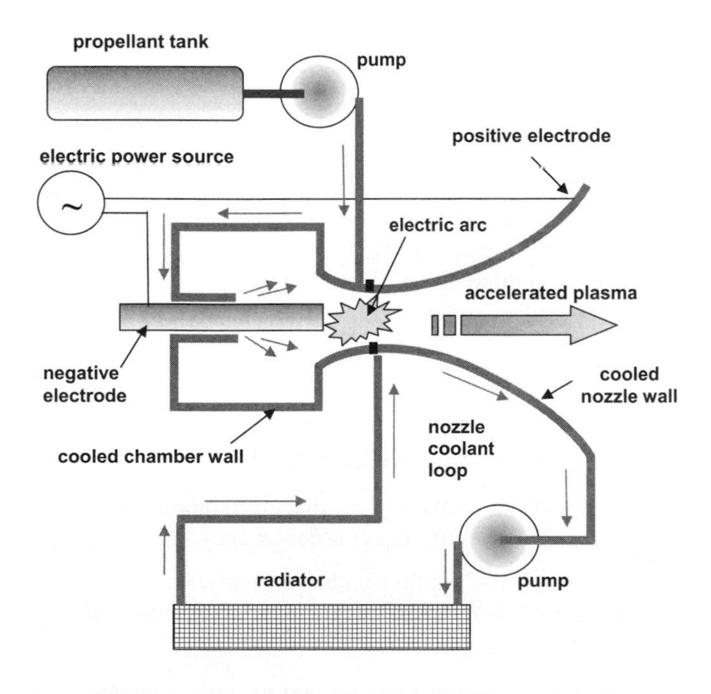


**FIGURE 14.6**

Schematic diagram of an electrothermal propulsion device, the arcjet.

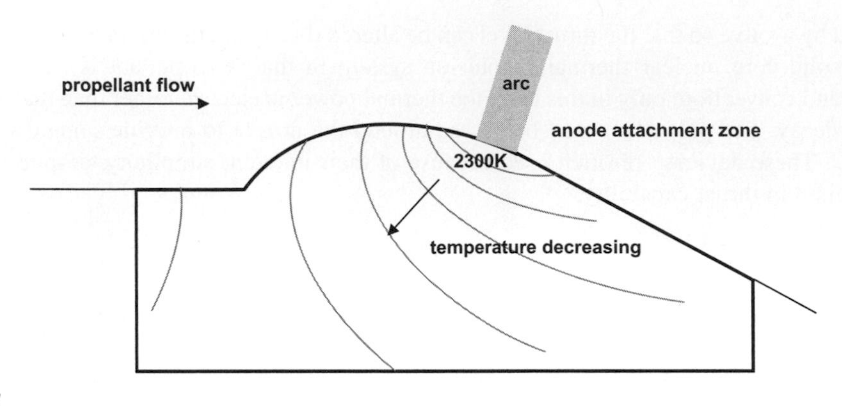


FIGURE 14.7

Temperature variation in the nozzle throat insert of an arcjet.

Figure 14.7. The arc attachment point is not truly steady but moves about with very high temperatures in the region of attachment. Active nozzle throat cooling is not practical, and there is reliance only on the heat sink capacity of the throat material and the cooling effect of the erosion (ablation) process itself so that the life of the arcjet is limited by anode erosion. An overview of arcjets has been given by Birkan and Myers (1996).

### 14.3.3 Radio-frequency and microwave excited jets

If instead of a direct current supplied to a gap in order to produce an arc and heat the propellant gas, one may use alternating current to generate high-frequency waves in the RF regime of around 1 MHz and even higher into the microwave band in order to produce heating in the propellant. This approach, which requires no electrodes, offers the possibility of a longer rocket motor lifetime, an important issue for deep space exploration where missions are measured in years. Satellite station keeping and orbit maintenance over long periods of time are also areas for application of more reliable propulsion systems. An evaluation of RF and microwave sources for ion rockets may be found in Foster and Patterson (2002).

## 14.4 ELECTROSTATIC PROPULSION DEVICES

The temperature limitations of electrothermal rockets may be avoided if acceleration of the propellant is achieved by electric body forces. The ion rocket accomplishes this by using

- an ion source to produce a stream of positively charged particles
- a negatively charged grid electrode to accelerate the ions electrostatically
- an electron source to neutralize the accelerated ions

Thus there is no physical nozzle or pressure chamber, and the only temperature limitations are on the ion source device.

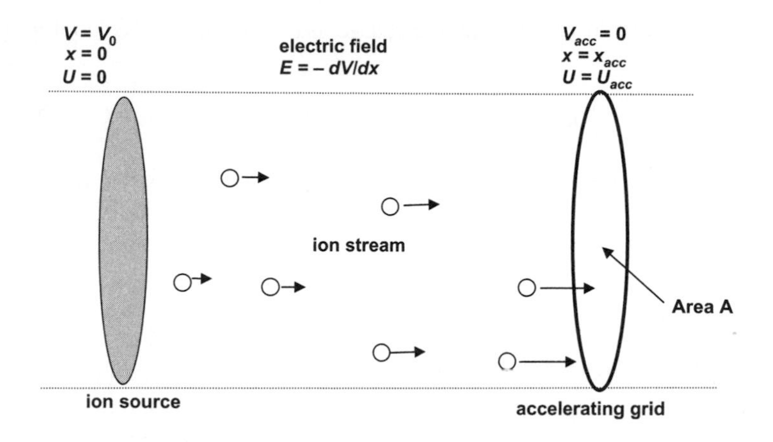


**FIGURE 14.8**

Schematic diagram of an ion propulsion device.

### 14.4.1 One-dimensional electrostatic thruster

Consider a one-dimensional geometry in which ions of negligible kinetic energy (ion velocity $U_0 = 0$) are produced at $x = 0$, where the electric potential is $V_0$ volts. An accelerating grid with potential $V = 0$ is located at $x = x_{acc}$, as shown in Figure 14.8. The mass flow of ions of mass $M$ through an area $A$ is given by

$$\dot{m} = NMUA, \tag{14.14}$$

where $N$ is the ion number density and $U$ is the ion velocity. At the accelerating grid location, $x = x_{acc}$, the velocity of the ions is $U_{acc}$, while the velocity $U_0 = 0$ at the ion source. The momentum conservation principle may be used to define the thrust as follows:

$$F = \dot{m}(U_{acc} - U_0) = \dot{m}U_{acc}. \tag{14.15}$$

Using mass flow Equation (14.14) permits the thrust per unit area to be written as

$$\frac{F}{A} = NMU_{acc} = (qNU_{acc})\frac{M}{q}U_{acc} = j\frac{M}{q}U_{acc}. \tag{14.16}$$

### 14.4.2 Ion stream speed

The ion current is given by the rate at which charge passes through a unit area or

$$j = qNU.$$

Thus $j$ is the ion current in coulombs/s/m$^2$ or amperes/m$^2$. As far as units are concerned, recall that coulombs = (farad)(volt), while a flow of charge in coulomb/s = amperes. Energy conservation requires that the kinetic energy of each ion is equal to the work done in moving the charge across a potential drop as given by the following relation:

$$\frac{1}{2}MU_{acc}^2 = q(V_0 - V_{acc}). \tag{14.17}$$

The electron charge, which is equal to the ion charge, is $q = 1.602 \times 10^{-19}$coulombs. Then energy of the moving ion, in kJ, is

$$\frac{1}{2}MU_{acc}^2 = 1.602 \times 10^{-16}(V_0 - V_{acc}). \tag{14.18}$$

For the maximum potential drop across the grids, $V_{acc} = 0$ and the ion speed at the accelerating grid is

$$U_{acc} = \sqrt{3.204 \times 10^{-16}\frac{V_0}{M}}. \tag{14.19}$$

In Equation (14.19), the ion mass must be measured in kilograms and we know that is given by

$$M = \frac{W \times 10^{-3}}{A} - zM_e, \tag{14.20}$$

where $W$ is the molecular weight of the ion species, $A$ is Avogadro's number ($6.0248 \times 10^{23}$ molecules per mole), $z$ is the degree of ionization, that is, the number of electrons stripped from the species to form the ion, and $M_e$ is the electron mass, $9.1084 \times 10^{-31}$ kg. Obviously, the electron mass may be neglected and $M = 1.660 \times 10^{-27}W$ measured in kilograms. The ion velocity, in meters per second, is then

$$U_{acc} = 1.389 \times 10^4\sqrt{\frac{V_0}{W}}. \tag{14.21}$$

Thus the exhaust speed of the ions, and therefore the specific impulse, is set primarily by the ratio of the potential drop across the grid to the molecular weight of the ion species.

Recall that in the discussion on spacecraft performance in Section 14.2 it was pointed out that solving Equation (14.8) for thrust shows it to be inversely proportional to the exhaust velocity. This shows there is a need to balance high fuel efficiency ($I_{sp}$) with the thrust level required to assure appropriate trip times. For reasonable grid voltages on the order of 1000 V or more, it turns out to be beneficial to have ions formed from relatively heavy species with a molecular weight $W$ around 100. This would yield exhaust velocities around 40,000 m/s, which is on the order desired for a 1-year mission. Cesium ($W = 132.9$) had been the propellant of choice but it is toxic and difficult to handle. Currently xenon ($W = 131.3$) is favored for ion generation in practical ion engines. Although it is rare, it is a noble gas, which is chemically inert, safe to handle, and easy to store.

### 14.4.3 Electric field and ion current

Because the electron mass is so small compared to the ion mass, their contribution to the thrust is negligible even though they are accelerated to much higher speeds by the grid. The electric field is given by

$$\vec{E} = -\nabla V. \tag{14.22}$$

In the one-dimensional case considered here.

$$\vec{E} = E\hat{i} = -\frac{dV}{dx}\hat{i}. \tag{14.23}$$

Then, taking the divergence of the electric field, we get

$$\nabla \cdot \vec{E} = -\frac{d^2V}{dx^2} = \frac{Nq}{\varepsilon_0}. \tag{14.24}$$

The quantity $\varepsilon_0$ is the electrical permittivity of free space ($8.85 \times 10^{-12}$ farad/m). First, consider the case of a constant electric field strength $E$ for which the potential drop across the grids may be found by integrating the electric field [Equation (14.23)] to obtain

$$E = \frac{V_0 - V_{acc}}{x_{acc}} = \frac{V_0}{x_{acc}} = const. \tag{14.25}$$

Note that for a constant electric field, Equation (14.24) shows that there will be no ions ($N = 0$) moving across the space between the grids and thus no current flow. This is the case of a pure electrostatic field. The maximum field strength occurs for $V_{acc} = 0$ as given in Equation (14.25) and depicted in Figure 14.8 and $E = V_0/x_{acc}$ is the electric field strength in volts per meter.

Now consider a less restrictive case where the electric field may vary with distance. The charge density in Equation (14.24) may be written in terms of the current density

$$Nq = \frac{j}{U_{acc}}. \tag{14.26}$$

When there is current flow, the electric field is not constant. Then, using Equations (14.26) and (14.17), written for a generic point $x$ in the flow rather than the end point $x_{acc}$, Equation (14.24) yields the following differential equation:

$$\frac{d^2V}{dx^2} = -\frac{j}{\varepsilon_0}\sqrt{\frac{M}{2q(V_0 - V)}}, \tag{14.27}$$

where $V(0) = V_0$ and $V(x_{acc}) = 0$ are the boundary conditions on the potential. Multiplying Equation (14.27) through by $2(dV/dx)$ and integrating to a general point $x$ where the potential is $V$ yields

$$\left(\frac{dV}{dx}\right)^2 - \left(\frac{dV}{dx}\right)_0^2 = \frac{4j}{\varepsilon_0}\sqrt{\frac{M}{2q}}(V_0 - V). \tag{14.28}$$

The left-hand side of Equation (14.28) is maximum for $E_0 = 0$ because the term

$$\left(\frac{dV}{dx}\right)_0^2 = (-E_0)^2 \geq 0.$$

Then, using $E_0 = 0$, we may integrate Equation (14.28) between $x = 0$ and $x = x_{acc}$ to obtain the following:

$$\frac{4}{3}V_0^{\frac{3}{4}} = 2\sqrt{\frac{j}{\varepsilon_0}}\left(\frac{M}{2q}\right)^{\frac{1}{4}} x_{acc}.$$

Then the current is given by

$$j = \frac{4}{9}\varepsilon_0\sqrt{\frac{2q}{M}}\frac{V_0^{\frac{3}{2}}}{x_{acc}^2}. \tag{14.29}$$

The result for current is called the charge-limited current, and the expression in Equation (14.29) is called Child's law. This is the maximum current that can be drawn for a given potential drop across a specified gap.

### 14.4.4 Performance implications

Now we may return to the thrust per unit area as given by Equation (14.26) to find that it may also be written as

$$\frac{F}{A} = j\frac{M}{q}U_{acc} = \frac{8}{9}\varepsilon_0\left(\frac{V_0}{x_{acc}}\right)^2. \tag{14.30}$$

Thus the maximum thrust per unit area depends on the maximum ion current per unit area. This in turn indicates that the thrust per unit area depends on the electric field between the ion source and the accelerating grid and the gap width between the two grids.

Although the thrust does not depend on the charge to mass ratio $M/q$, Equation (14.17) shows that the ion jet velocity, and therefore the specific impulse, does:

$$U_{acc} = \sqrt{2\frac{q}{M}V_0}. \tag{14.31}$$

The exhaust stream power required [Equation (14.3)] per unit area also depends on the ion charge to mass ratio

$$\frac{P_e}{A} = \frac{1}{2}\frac{FU_{acc}}{A} = \frac{4}{9}\varepsilon_0\sqrt{2\frac{q}{M}\frac{V_0^{\frac{5}{2}}}{x_{acc}^2}}. \tag{14.32}$$

Note that this is not the same power as the prime power $P$ [Equation (14.7)] that must be delivered to the ion engine from the onboard powerplant.

Equations (14.30)–(14.32) may be used to provide an estimate of the range of propulsion capabilities of the electrostatic ion rocket. It is considered impractical to achieve an electric field of more than about $10^7$ volts/m, even in a hard vacuum. The maximum electric field will occur at the accelerating electrode where

$$E_{acc} = -\left(\frac{dV}{dx}\right)_{acc} = \frac{4}{3}\frac{V_0}{x_{acc}}.$$

Thus the limiting thrust per unit area practically achievable is

$$\frac{F}{A} = \frac{1}{2}\varepsilon_0 E_a^2 \approx 440N/m^2. \tag{14.33}$$

As pointed out by Jahn (1968), even this optimistic limit will likely be reduced, as the voltage needed for practical spacing between the ion source and the accelerating electrode may be too high. Typically, gaps of 2 to 5 mm are generally found to be robust enough for reliable thrusters. The accelerating voltages corresponding to breakdown for these cases are 15,000 and 37,500 V, respectively.

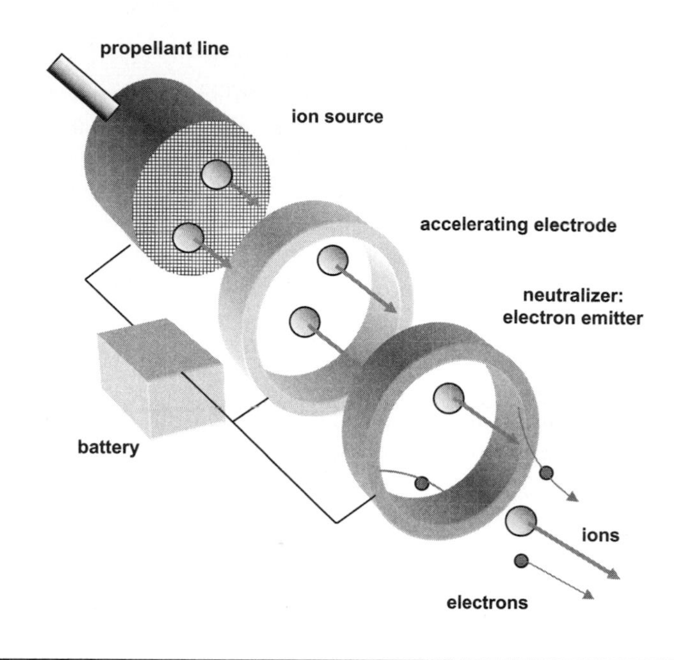


FIGURE 14.9

Schematic diagram of an electrostatic propulsion system: an ion rocket.

### 14.4.5 Surface contact source of ions

The electrostatic propulsion systems that have been discussed must have the capability to produce a sufficiently large flow of ions with reasonable power demands in order to be practical. Because the missions for such propulsion systems may have durations on the order of $10^3$ to $10^4$ hr, the electrostatic ion rocket must have high reliability. A particularly simple ion source is a heated metal that is in contact with a suitable vapor; this is called a surface contact source. One of the most useful such sources is cesium vapor in contact with a heated tungsten surface. In particular, the cesium vapor is forced through a porous tungsten plug. This source is used in a type of ion engine currently used in space propulsion applications, as shown schematically in Figure 14.9.

Ion rocket engines eliminate the problem of freezing the energy in the internal degrees of freedom of the propellant gas because there is no nozzle and therefore no mechanical expansion of the gas to cause the rapidly decreasing density that reduces the collision frequency. However, the high temperatures at which the ion sources must operate can cause thermal radiation losses. These engines also suffer from the need to inject electrons into the exiting beam in order to neutralize it electrically. Without this neutralization, the spacecraft would build up an unacceptable negative charge. In addition, the space-charge limitation on current density given in Equation (14.29) constrains the allowable diameter of the ion beam, making it necessary to use a number of such engines in parallel to provide a useful level of thrust.

### 14.4.6 Example: Surface contact source dimensions

These heated element surface contact ion generators are so efficient in producing ions that propellant losses are negligible. Another positive factor is that the velocity dispersion in the exhaust stream may readily be kept low enough to be equally negligible. The main performance limiter is the thermal radiation loss arising from the need to heat the porous tungsten plug. That means that the energy expenditure per ion produced is relatively large. This energy per ion produced drops as the ion current density is increased. However, electrode life decreases due to erosion as the current density increases. Thus there is a region for trade-off between energy efficiency and electrode lifetime that must be considered within the context of particular mission requirements. One approach for extending electrode life is to increase their mass, but that brings in the limitations posed by space-charge limited ion flow, as given by Equation (14.29).

For a typical cesium ion engine with a potential of 2000 V, an ion current $j = 200$ A/m$^2$, and a charge to mass ratio $q/M = 7.24 \times 10^5$ C/kg, the electrode gap would be charge limited according to Equation (14.29) as follows:

$$x_{acc}^2 = \frac{4}{9}\varepsilon_0\sqrt{2\frac{q}{M}\frac{V^2}{j^2}}.$$

The electrical permittivity of free space $\varepsilon_0 = 8.85 \times 10^{-12}$ farad/m $= 8.85 \times 10^{-12}$ C/m-V so that the gap size equation becomes

$$x_{acc}^2 = \frac{4}{9}(8.85 \times 10^{-12} C/m - V)\sqrt{2(7.24 \times 10^5 C/kg)}\frac{(2000V)^{3/2}}{(200A/m^2)} = 2.117 \times 10^{-6}\frac{C^{3/2}V^{1/2}m}{kg^{1/2}A}.$$

Using square brackets to denote units, they may be collected as follows:

$$\left[\frac{C^{3/2}V^{1/2}m}{kg^{1/2}A}\right] = \left[\frac{C}{A}\right]\left[\left(\frac{CV}{kg}\right)^{1/2}\right][m] = \left[\frac{A-s}{A}\right]\left[\left(\frac{A-V-s}{kg}\right)^{1/2}\right][m] = [m-s]\left[\left(\frac{J}{kg}\right)^{1/2}\right] = m^2.$$

Then the accelerating gap $x_{acc} = 1.455 \times 10^{-3}$ m $= 0.1455$ cm. Practical gap sizes are somewhat larger than this, as pointed out in the previous section. The general scale of the electrode is set by the gap size so there is not going to be much mass available to tolerate loss by erosion. If one wishes to increase the electrode size, the gap must also be increased and with it higher voltages must be provided. The result usually is not advantageous, as higher voltages lead to greater erosion.

### 14.4.7 Electron bombardment source of ions

Ionization may be produced by an arc that is produced between an annular anode and a central cathode in an arc chamber through which cesium or xenon gas passes at low pressure, around 1 kPa. An axial magnetic field generated by a solenoid situated around the anode serves to constrain the arc magnetically. Electrons emitted by the cathode initially move radially out toward the anode, but the magnetic field causes them to also execute a spiraling motion, which extends their residence time in the chamber. This extended duration permits many more collisions, with the low-pressure gas producing dissociation and ionization. An electric field produced between two porous screens at one

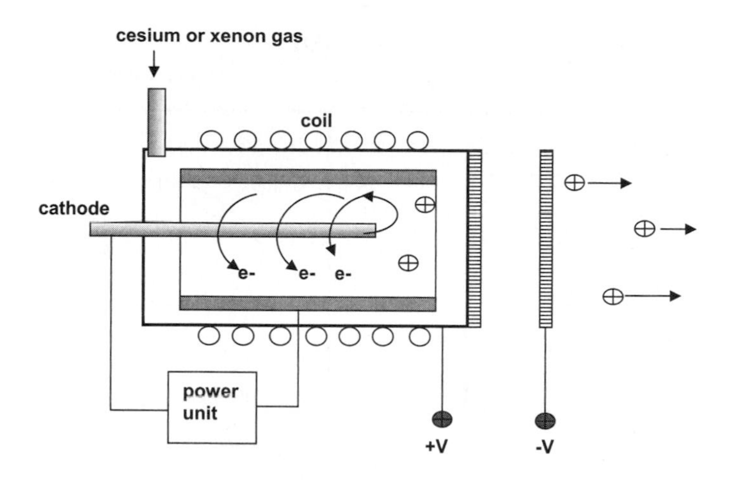


**FIGURE 14.10**

Schematic diagram of an electron bombardment ion source thruster.

end of the arc chamber draws the ions toward them and accelerates them out of the system to produce thrust as in the surface contact source ion engine. A schematic diagram of the system is shown in Figure 14.10.

## 14.4.8 Hall thruster

The Hall thruster was invented in the late 1950s. Until the mid-1990s, it had been developed primarily in Russia where it was called the stationary plasma thruster (SPT). Since the early 1980s, Russia has placed more than 100 Hall thrusters in orbit. As pointed out in Section 14.4.3, in the electrostatic ion thruster a positive charge builds up in the space between the grids, limiting the ion current and, therefore, the magnitude of the thrust that can be attained. In a Hall thruster, electrons injected into and trapped within a radial magnetic field act to neutralize the space charge. The magnitude of the applied magnetic field is approximately 100–200 gauss, strong enough to trap the electrons by causing them to spiral around the field lines in the coaxial channel. The magnetic field and a trapped electron cloud together serve as a virtual cathode. The ions, too heavy to be affected by the field, continue their journey through the virtual cathode. The movement of the positive and negative electrical charges through the system results in a net force (thrust) on the thruster in a direction opposite that of the ion flow.

In the stationary plasma thruster, as shown in Figure 14.11, the cathode provides a source of free electrons, while the feed tube supplies xenon propellant to an anode that also acts to distribute the propellant throughout the annular discharge chamber. The inner and outer magnets provide an essentially radial magnetic field that traps the free electrons migrating from the cathode. The electrons collide with the xenon atoms ionizing them, permitting them to be accelerated by the electric field and exhausted from the device providing thrust. The thrust can be throttled by varying the flow rate of the xenon propellant. Note that the cathode electron supply also serves to neutralize the exhaust flow.

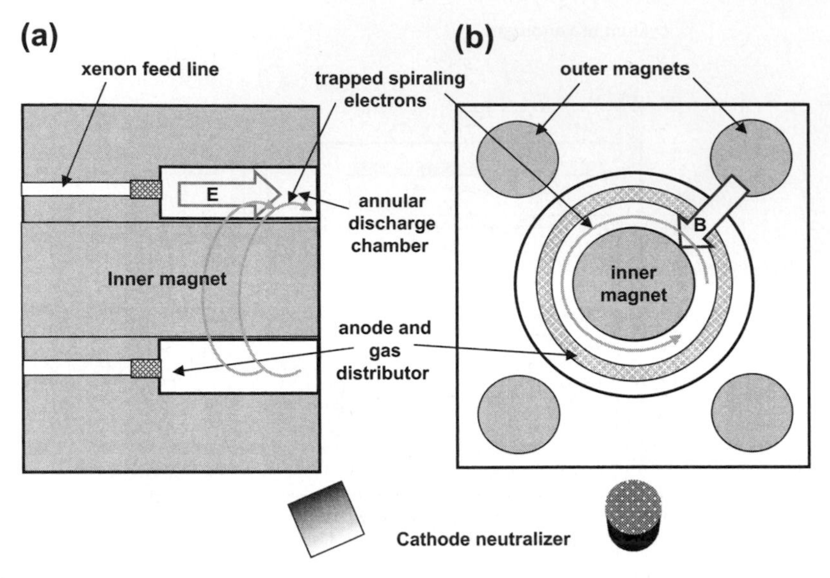


**FIGURE 14.11**

Schematic diagram of the stationary plasma thruster showing (a) cross section and (b) front view. Inner and outer magnets provide the radial magnetic field *B*. The cathode supplies electrons that stream toward the anode providing the electric field *E* and are trapped in helical paths (light gray lines) by the magnetic field.

Stationary plasma, or Hall, thrusters have found application in satellite station-keeping. The typical performance of these thrusters includes high reliability (nominally 4000 on–off cycles) and thrust levels $F$ (in $N$) for power input $P$ (in $W$) given approximately by

$$F = 6.67 \times 10^{-5} P.$$

Typical power levels are on the order of kilowatts, but higher levels are possible. The xenon propellant mass flow is typically on the order of several milligrams per second ($10^{-6}$ kg/s), and the specific impulse is around 1500 s. The mass of such a thruster is around 3.5 kg, and the characteristic dimension is about 15 cm, roughly half that of an ion rocket of the same thrust.

### 14.4.9 An ion rocket for a deep space mission

The Deep Space 1 (DS1) spacecraft was the first project launched in 1998 as part of NASA's New Millennium program. Its mission was to evaluate 12 new technologies onboard the spacecraft and to use these technologies to perform a flyby of the asteroid Braille (1992 KD). The extended mission called for a flyby of the comet Borelly. DS1 used the NSTAR ion engine as primary propulsion for its mission. The electron bombardment ion engine propelled DS1 263,179,600 km at speeds up to 4.5 km/s. The NSTAR engine, developed by the ion propulsion program at the NASA Glenn Research Center, has a 30-cm beam diameter and weighs only 8.3 kg and is shown firing in Figure 14.12.

**FIGURE 14.12**

The DS1 ion rocket firing in the laboratory at the NASA Glenn Research Center (courtesy of NASA).

The engine was powered by two solar arrays, which output 2500 W. The ion engine used 2100 W of this power. The xenon ions travel up to 35 km/s, which means the specific impulse can reach a value of up to $I_{sp} = 3570$ s, and the engine produced 92 mN of thrust. Due to the engine's high efficiency, it needed only 81 kg (178 lbs) of xenon propellant to achieve its mission. If there were more xenon on board the engine could have produced a top speed of 30 km/s (67100 mph). It provided thrust for 678 days, far longer than any propulsion system had ever been operated

## 14.5 ELECTROMAGNETIC PROPULSION DEVICES

One may go beyond simple electrostatic forces to accelerate charged particles and include the body force arising from crossed electric and magnetic fields. One such device is a magnetoplasmadynamic (MPD) accelerator. One embodiment of such a device is shown in Figure 14.13.

Here an electric arc again transforms the propellant gas into plasma, which is then accelerated by the *jxB* force. Usually the electrical conductivity of the plasma must be enhanced by the addition of an easily ionized species, such as potassium chloride, a process known as "seeding." These devices have the same limitation of composition "freezing" common to all mechanically accelerated flows and, in addition, the problem of excessive wall temperatures. In these systems the applied electric and magnetic fields are steady, although pulsed versions are also being studied.

Different thruster configurations have been studied along with various propellants. Lithium vapor gave high performance in studies performed in Russia at power levels around 100 kW with jet velocities up to 50 km/s and propulsive jet efficiencies of up to 45%, yielding specific impulse up to

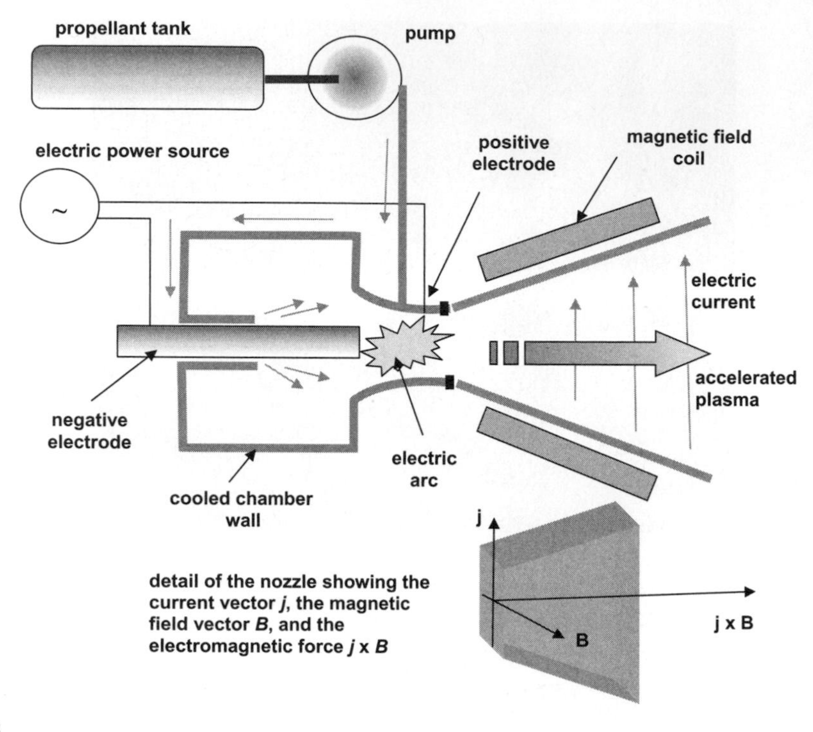


**FIGURE 14.13**

Schematic diagram of an electromagnetic propulsion device, the magnetoplasmadynamic accelerator.

5000 s. A difficulty with lithium as a propellant is that it may condense on external spacecraft components such as power arrays. Hydrogen will not condense in this fashion and is an attractive candidate for high-power MPD thrusters. Tests at NASA Glenn Research Center have shown exhaust velocities of 100 km/s at thrust levels of 100 N. The power required is on the megawatt scale to achieve the demonstrated specific impulse of about 10,000 s. A 200-kW MPD test article is shown in Figure 14.14.

### 14.5.1 Pulsed plasma thrusters

Pulsed electromagnetic devices use a switching circuit to generate a sharply defined discharge pulse with duration on the order of milliseconds. These devices can use a "seeded" propellant gas or the products of ablation of a solid dielectric such as Teflon. Because of the short duration pulses, the unsteady electromagnetic thruster doesn't suffer from the temperature limitations of the steady counterpart. Repetitive pulsing, at a sufficiently high rate, can provide an approximately steady thrust force. In addition, the thrust produced can be modulated very accurately by virtue of the ease with which pulse trains can be controlled.

A schematic diagram of a pulsed plasma thruster (PPT) is shown in Figure 14.15. A power supply provides a high voltage drop across two electrodes positioned close to the propellant source, for

FIGURE 14.14

A 200-kW MPD thruster for testing at NASA Glenn Research Center (courtesy of NASA).

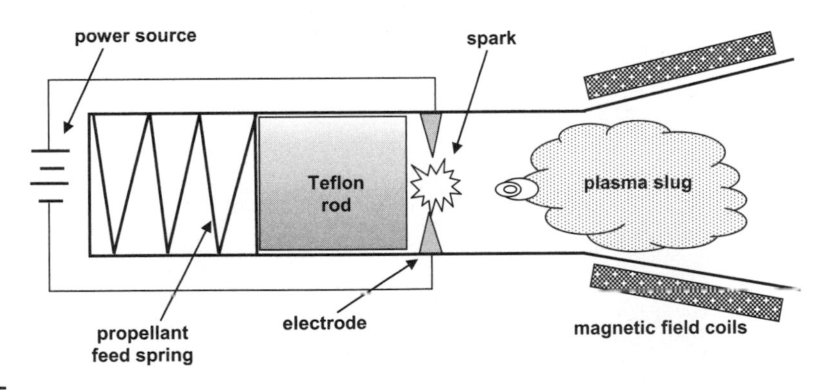


FIGURE 14.15

Schematic diagram of a pulsed plasma thruster.

example, as shown here, a solid Teflon rod. The electrodes, mounted close to the propellant, produce a spark that acts briefly to ablate and ionize some of the surface of the propellant, thereby creating a slug of neutral plasma. This finite volume of plasma is then accelerated out of the thruster by the ***j***x***B*** force. This force is described more accurately in the diagram of Figure 14.16. As the propellant is consumed, a spring forces the remaining solid propellant forward, providing a constant fuel source without the need for any moving parts.

The Earth Observing 1 (EO-1) spacecraft, shown in Figure 14.17, was launched in 2000. It uses one dual-axis PPT for pitch axis control and momentum management. The thruster was developed at the NASA Glenn Research Center and produced by Primex Aerospace Company and is capable of producing a thrust of 0.86 mN with an exhaust velocity over 13,700 m/s, while consuming only 70 W

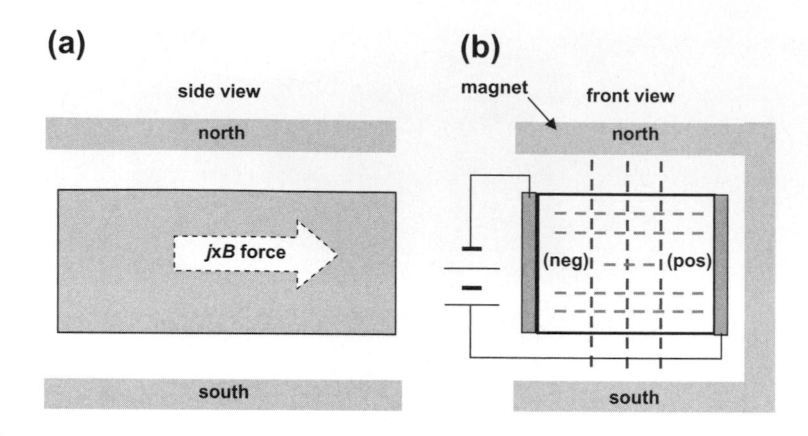


**FIGURE 14.16**

Schematic diagram of the manner in which crossed electric and magnetic fields may be produced to provide a *jxB* electromagnetic accelerating force.

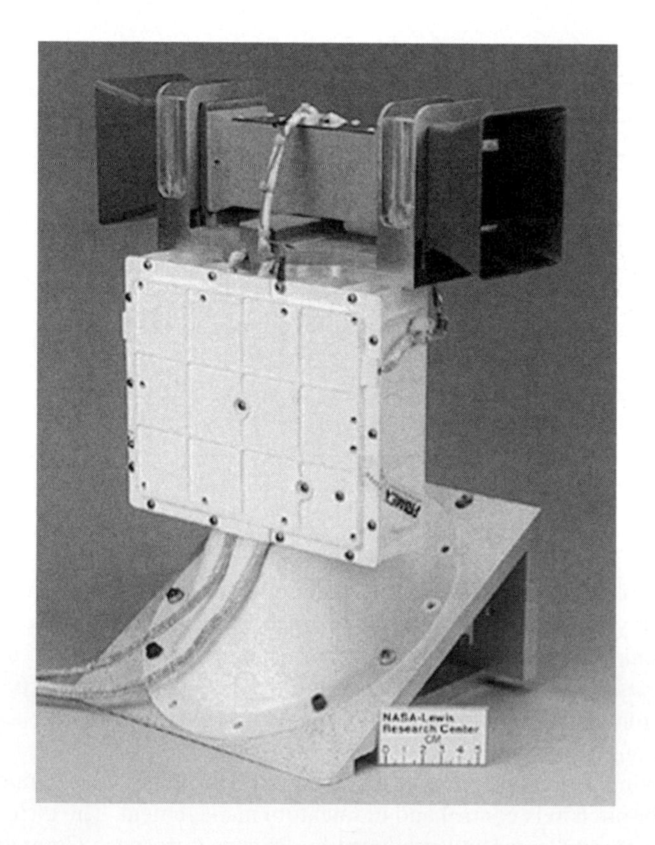


**FIGURE 14.17**

EO-1 Earth orbiter flight-pulsed plasma thruster (courtesy of NASA).

of power. The fuel used is Teflon, and the thrust developed, $F$(N), is related to the input power, $P$(W), by

$$F = 1.23 \times 10^{-5} P.$$

This is a substantially lower thrust to power ratio than the Hall thruster. The unit weight is 4.9 kg, and the propellant load is 0.07 kg per thruster. The production of precisely controlled impulse bits provides accurate throttling for attitude control. The specific impulse can be varied from 660 to 1350s.

## 14.6 NOMENCLATURE

| | |
|---|---|
| $A$ | area or the unit of current, amperes |
| $B$ | magnetic field intensity |
| $E$ | electric field strength |
| $F$ | thrust |
| $g_E$ | gravitational acceleration at Earth's surface |
| $h$ | enthalpy |
| $I_{sp}$ | specific impulse |
| $j$ | ion current |
| $L$ | length |
| $M$ | mass |
| $m$ | mass flow |
| $N$ | ion number density |
| $q$ | electron charge |
| $P$ | prime power |
| $P_e$ | power in the exhaust jet |
| $T$ | temperature |
| $t$ | time |
| $t_p$ | propellant burn time |
| $U$ | ion velocity |
| $V$ | electric potential |
| $V_e$ | exhaust velocity |
| $z$ | degree of ionization |
| $\alpha$ | proportionality constant in Equation (14.7) |
| $\eta$ | efficiency of power transformation by prime powerplant |
| $\eta'$ | efficiency of power transformation by engine |
| $\varepsilon_0$ | electrical permittivity of free space |
| $\tau_f$ | characteristic flow time |
| $\tau_c$ | characteristic chemical reaction time |

### 14.6.1 Subscripts

| | |
|---|---|
| *acc* | conditions at the accelerating grid |
| *e* | exhaust conditions |
| *eng* | engine |
| *pay* | payload |
| *pro* | propellant |
| *str* | structure |
| *t* | stagnation conditions |
| 0 | initial value |

## 14.7 EXERCISES

**14.1** An electrostatic thruster produces 0.5 N thrust for 3 hr and consumed a total of 0.5 kg of propellant using a power of 4 kW. (a) Find the specific impulse $I_{sp}$ of the thruster and (b) find $\eta'$, the efficiency of the propulsive jet

**14.2** A spacecraft with a solar electric powerplant with a specific mass $\alpha = 15$ kg/kW is considered for a space mission that requires 4.8 months. (a) Estimate the optimum specific impulse to carry out this mission. (b) Assuming that the efficiency of the propulsive jet $\eta' = 50\%$, determine the maximum thrust that can be developed if the solar array power delivered to the engine is 25 kW. (c) Find the total impulse delivered over the entire mission. (d) Calculate the mass of the powerplant and of the propellant required for the mission.

**14.3** A resistojet is to be employed on a satellite for attitude control in low earth orbit. It will use gaseous nitrogen from the orbit modification propellant pressurization system. The thrust level required is 100 mN, and the propellant feed is available at 1 MPa. If the heater is limited to a maximum temperature of 500°C, determine (a) the specific impulse, (b) the mass flow of nitrogen, (c) the mass of nitrogen used if the design calls for a total delivered impulse of $10^4$ N-s, and (d) the nozzle dimensions for a practical device.

**14.4** Determine the operating time required for an ion engine with a specific impulse $I_{sp} = 10{,}000$s to accelerate a vehicle through a velocity increment $\Delta V = 5$ km/s if the thrust to weight ratio of the vehicle is $10^{-4}$. If an MPD engine with $I_{sp} = 1000$ s is substituted with the vehicle thrust to weight ratio now equal to $10^{-3}$, how long would acceleration take? Estimate the propellant mass fraction needed in each case.

## References

Birkan, M. A., & Myers, R. M. (1996). An introduction to arcjets and arc heaters. *Journal Propulsion Power, 12*(6), 1010.

Foster, J.E., & Patterson, M.J. (2002). Microwave ECR Ion Thruster Development Activities at NASA GRC. AIAA-2002-3837.

Goebel, D. M., & Katz, I. (2008). *Fundamentals of Electric Propulsion: Ion and Hall Thrusters*. New York: Wiley.

Jahn, R. G. (1968). *Physics of Electric Propulsion*. New York: McGraw-Hill.

Martinez-Sanchez, M., & Pollard, J. E. (1998). Spacecraft electric propulsion: An overview. *Journal Propulsion Power, 14*(5), Sept.-Oct, 688–699.

Meyer, M., Johnson, L., Palaszewski, B., Goebel, D., White, H., & Coote, D. (2010). Draft In-Space Propulsion Systems Roadmap, Technology Area 02. NASA, November.

Stuhlinger, E. (1991). Origin and Early Development of Electric Propulsion Concepts. AIAA 91–3442, AIAA/NASA/OAI Conference on Advanced SEI Technologies, Cleveland, Ohio.

Vaughn, C.E., Cassady, R.J., & Fisher, J.R. (1993). The Design Fabrication and Test of a 26kw Arcjet and Power Conditioning Unit. IEPC 93-048, Proceedings of the International Electric Propulsion Conference, 448–459.

CHAPTER

# 15 Propulsion Aspects of High-Speed Flight

## CHAPTER OUTLINE



## 15.1 FLIGHT TIME

The basic advantage of high-speed flight is reduced travel time. To illustrate this, let us consider the cruise time (in hours) for long-range flights over a range $R$ (in statute miles) as a function of cruise Mach number:

$$t \approx \frac{R}{667}\frac{1}{M}. \tag{15.1}$$

This relationship is illustrated in Figure 15.1 with representative performance points for the Boeing 747, a supersonic commercial transport (SST), and a hypothetical Mach 7 hypersonic transport (HST).

It is clear that for long-range flights (6000–9000 miles) there are enormous time advantages for high-speed transport. Also seen clearly in Figure 15.1 are points where $dt/dM = -1$, that is, the "knee" of the curve. This represents the point of diminishing returns and is a typical "rule-of-thumb" design point for any design process in which no extremum appears. The "knee" points shown suggest that reasonable design Mach numbers for 6000- and 9000-mile ranges are around 2.2 and 3, respectively; increases beyond these values buy increasingly smaller time savings. Note that $dt/dM = -1$ denotes that each unit increase in Mach number saves 1 hr in time. It is interesting to see that the SST is fairly close to the suggested Mach number for the range of 6000 miles, while the HST is far away, not only for 6000 miles, but even for 9000 miles. At the larger ranges, a Mach 3.5-4.0 aircraft would seem to be more suitable. Because there are not many routes calling for 9000-mile range capability, it seems unlikely that a hypersonic transport aircraft would ever be attractive enough to pursue for any commercial market. An important simplification used here is that only the cruise portion of flight is considered. Obviously, the higher the Mach number, the greater the



Theory of Aerospace Propulsion.

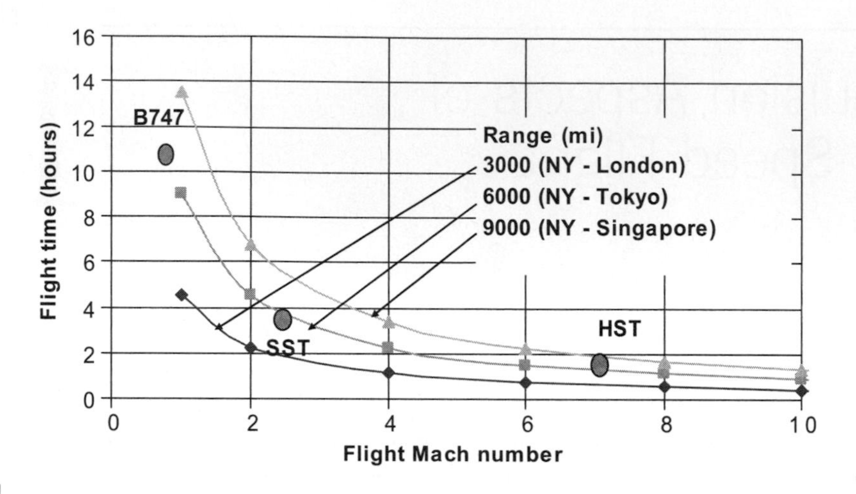


**FIGURE 15.1**

Cruise time for various routes and flight Mach numbers. Shown for reference are representative points for a Boeing 747 (B747), a supersonic transport (SST), and a hypersonic transport (HST).

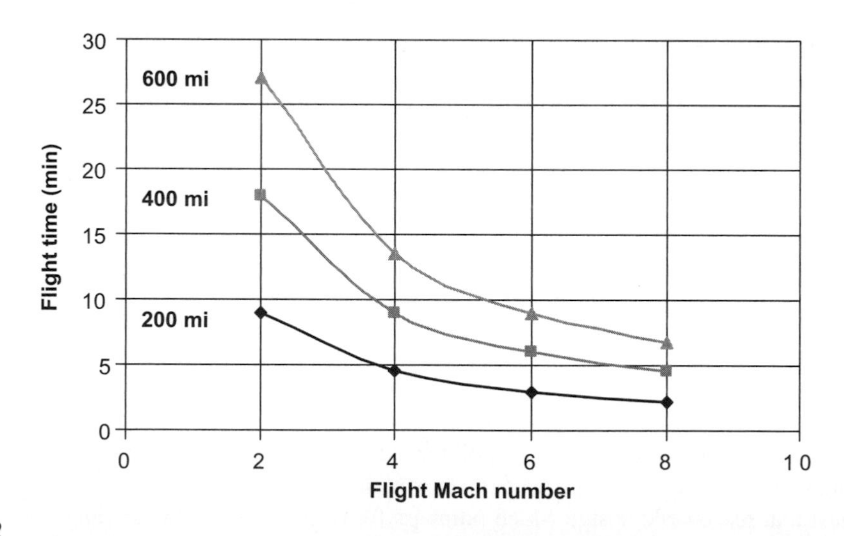


**FIGURE 15.2**

Flight time as a function of Mach number for intermediate range missiles.

proportion of flight time spent accelerating and decelerating, and therefore the time-saving attributes of the aircraft will diminish accordingly.

In the case of a missile that is much more limited in range, the picture changes somewhat, as shown in Figure 15.2. Now the knee in the curve occurs at higher Mach numbers, say, 4 to 4.5.

## 15.2 FLIGHT PRODUCTIVITY

The productivity of a commercial transport may be measured in seat miles per hour of flight, or

$$P = \frac{RN}{t} = 667\,MN, \tag{15.2}$$

where $N$ is the number of passengers. Therefore, for two different aircraft to provide the same productivity, $MN$ must be a constant. The typical proposed SST is an $M = 2.4$, $N = 300$ passenger aircraft so that $MN = 720$. Therefore, a conventional subsonic aircraft with $M = 0.85$ must carry $N = 847$ passengers to generate the same productivity. This is the reason that there is continuing interest in very large aircraft. Of course, there are also reasons connected with costs: capital costs, which are dependent on weight and speed, and operating costs, which are dependent on the amount of fuel burned, which, in turn, depends on speed. An appreciation of the capital costs is provided by the results described in Large and colleagues (1976), where the total cost for 100 aircraft (in 1975 dollars) is suggested to be correlated by

$$TC_{100} = 5.49W^{0.73}V^{0.74}. \tag{15.3}$$

This equation was developed on the basis of an analysis of 24 military aircraft of all types, including high-performance aircraft such as the F-4, F-14, F-111, and the B-58. The quantity $W$ is the aircraft empty weight in pounds, and $V$ is the speed of the aircraft in knots. Because only ratios of costs will be considered here, the coefficient in the equation is not important, as it sets the cost in a given currency year. This result indicates that the ratio of the cost of an aircraft, which operates in one speed range to that of one that operates in another, is

$$\frac{TC_1}{TC_2} = \left(\frac{W_1}{W_2}\right)^{0.73}\left(\frac{M_1}{M_2}\right)^{0.74}. \tag{15.4}$$

Note that use is made of the fact that because the speed of sound in the stratosphere is roughly constant, the ratio of the flight speeds is equal to the ratio of flight Mach numbers. For two airplanes of the same weight,

$$\frac{TC_1}{TC_2} = \left(\frac{M_1}{M_2}\right)^{0.74}. \tag{15.5}$$

Assuming a supersonic cruise airliner with $M_1 = 2.4$ and a conventional airliner with $M_2 = 0.85$, the ratio of costs is

$$\frac{TC_1}{TC_2} = 2.16.$$

This suggests that the capital cost of the supersonic airplane is twice that of the subsonic airplane of the same empty weight. However, if we compare airplanes of the same productivity, the cost ratio is

$$\frac{TC_1}{TC_2} = \left(\frac{W_1}{W_2}\right)^{0.73}\left(\frac{N_2}{N_1}\right)^{0.74}. \tag{15.6}$$

Because $W \sim N$, this indicates that

$$\frac{TC_1}{TC_2} \approx 1.$$

## 15.3 FUEL BURN

Therefore, on the basis of productivity, the capital costs are quite similar for both a supersonic and a subsonic airliner. However, the operating costs depend strongly on fuel cost, which, in turn, depends on the weight of fuel burned, which, in cruise, is approximated by

$$W_{fuel} = c_j F t = c_j D t = c_j \frac{1.5RW}{1000M\frac{L}{D}}. \tag{15.7}$$

The ratio of fuel burned is, for the same productivity and range, given by

$$\frac{W_{fuel,1}}{W_{fuel,2}} = \frac{c_{j,1}}{c_{j,2}}\left(\frac{M_2}{M_1}\right)^2 \frac{\left(\frac{L}{D}\right)_2}{\left(\frac{L}{D}\right)_1}. \tag{15.8}$$

The lift to drag ratio for high subsonic mach numbers and beyond is often assumed to be of the form

$$\frac{L}{D} = k_1 \frac{M+3}{M}. \tag{15.9}$$

The factor $k_1$ lies somewhere between 3 and 4, but on a relative basis the magnitude of this constant is irrelevant, and the fuel burn ratio is

$$\frac{W_{fuel,1}}{W_{fuel,2}} = \frac{c_{j,1}}{c_{j,2}} \frac{M_2}{M_1} \frac{M_2+3}{M_1+3}. \tag{15.10}$$

For the Mach numbers in the two cases chosen, that is, $M_1 = 2.4$ and $M_2 = 0.85$,

$$\frac{W_{fuel,1}}{W_{fuel,2}} = 0.25 \frac{c_{j,1}}{c_{j,2}}. \tag{15.11}$$

The ratio of $c_j$ for the supersonic case (which involves afterburning) to that for the subsonic case is around 3 to 5 so that

$$\frac{3}{4} < \frac{W_{fuel,1}}{W_{fuel,2}} < \frac{5}{4}. \tag{15.12}$$

Thus there is likely to be a cruise penalty for supersonic flight but it isn't as great as might be imagined. However, the fuel burn difference in the acceleration phase definitely favors the subsonic airplane.

## 15.4 FLIGHT RANGE

The maximum range $R_m$ (in miles) of an aircraft may be estimated by the following approximate form of the Breguet equation:

$$R_{\max} = 0.19 I_{sp} M \left(\frac{L}{D}\right) \log_e \left(\frac{1}{1 - \frac{W_{fuel}}{W}}\right), \tag{15.13}$$

where it is assumed that $M$ and $L/D$ are essentially constant and that $W_{fuel}$ denotes total fuel weight for cruise and $W$ the total weight of the aircraft at the start of cruise. Note that the importance of propulsion, aerodynamics, and structures are represented by the terms $I_{sp}$, $ML/D$, and $W_{fuel}/W$, respectively. The logarithmic term varies rapidly with the fuel fraction, as shown in Figure 15.3.

It is possible to estimate the utility of air-breathing engines versus rockets by assuming a mission of equal cruise range, Mach number, and $L/D$, which leads to the result that

$$\left(1 - \frac{W_{propellant,\ rocket}}{W}\right) = \left(1 - \frac{W_{fuel,\ jet}}{W}\right)^{\frac{I_{sp,\ jet}}{I_{sp,\ rocket}}}, \tag{15.14}$$

where $W_{propellant,\ rocket}$ is the propellant (fuel plus oxidizer) weight needed for the rocket engine and $W_{fuel,\ jet}$ is the propellant weight (fuel only) for the air-breathing jet engine. Because we have already shown that $I_{sp,jet} > I_{sp,rocket}$ until very high flight velocity is reached, we see that air-breathing propulsion yields a weight advantage for aircraft until these very high flight speeds are attained. This is the reason for continuing effort on ramjet and scramjet engines for the lower stages of launch vehicles. The

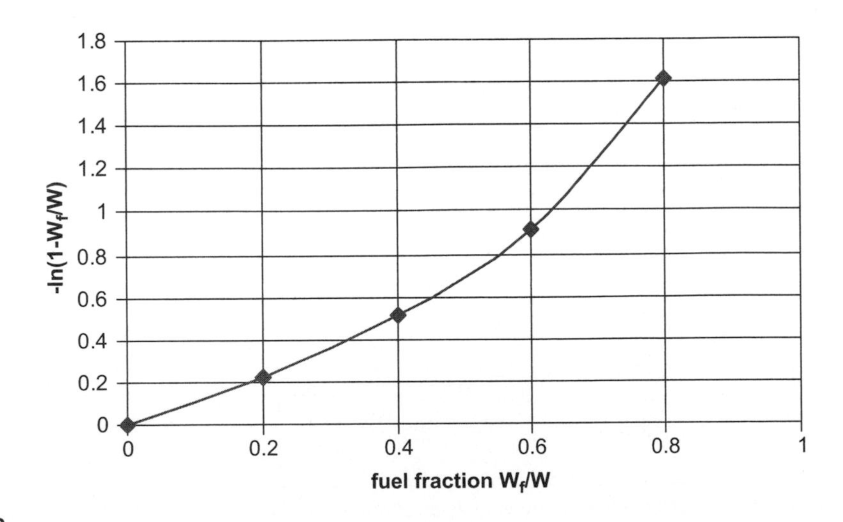


**FIGURE 15.3**

Variation of logarithmic function of the fuel fraction with the fuel fraction itself.

relative values of specific impulse achievable with different air-breathing jet propulsion engines using hydrocarbon fuel and hydrogen fuel, as a function of flight Mach number, may be approximated by the relation

$$I_{sp,ab} \approx \frac{k}{M}. \tag{15.15}$$

The constant $k$ is about 7000 s for hydrocarbon fuels and about 21,000 s for hydrogen fuel. The range of specific impulse for rockets is 250 s $< I_{sp,r} <$ 450 s and is independent of Mach number. The relationship between the propellant weight ratio for a rocket and the fuel weight ratio for an air breather is shown in Figure 15.4.

The importance of the fuel used can be illustrated by noting that the overall efficiency of an engine was given earlier by

$$\eta_o = \eta_p \eta_{th} = \frac{DV}{\dot{m}\Delta Q}. \tag{15.16}$$

This is a measure of thrust power obtained for a given rate of heat energy addition accomplished by the fuel. An analysis of supersonic aircraft empty weight (e.g., that due to Roskam, 1986) as a function of gross take-off weight leads to a simple approximation for the propellant fraction, which results in the following rough estimate for air-breathing engines:

$$R = \eta_p \eta_{th} \eta_b H \left(\frac{L}{D}\right)\left(0.65 - 1.44\frac{W_{structure}}{W}\right), \tag{15.17}$$

where $H = 778HV/5280$ is the heating value of the fuel expressed in miles and has the typical values $H_{HC} = 2800$ miles for hydrocarbon fuels and $H_H = 7300$ miles for hydrogen fuel so that the absolute maximum range is

$$R_{\max} = 0.65\eta_o \eta_b H \left(\frac{L}{D}\right). \tag{15.18}$$

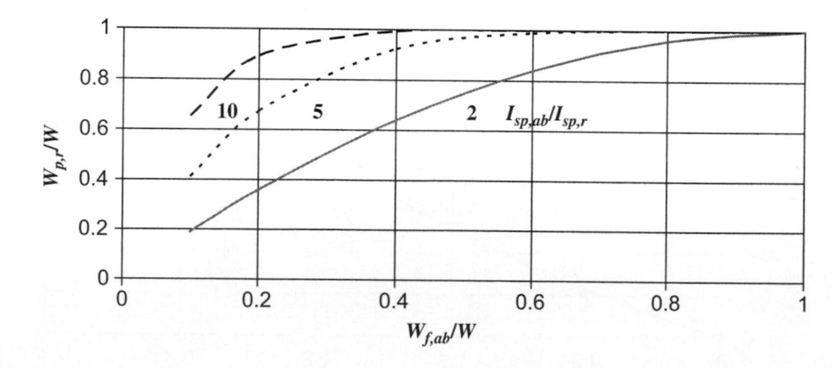


**FIGURE 15.4**

Propellant weight fraction for a rocket-powered vehicle as a function of the fuel weight fraction for an air-breathing powered vehicle for different specific impulse ratios for the two propulsion systems. The cruise range, Mach number, and $L/D$ are identical for both vehicles.

Approximate analyses (e.g., Kuchemann, 1978) of high-speed vehicles suggest that

$$\eta_o \approx \frac{M}{M+3}. \tag{15.19}$$

This can be written as

$$\eta_o = \frac{F_{net}V}{\dot{w}_{fuel}\eta_b HV} = \frac{F_{net}}{\dot{w}_{propellant}} \frac{V}{\frac{\dot{w}_{fuel}}{\dot{w}_{propellant}}\eta_b HV}. \tag{15.20}$$

or as follows:

$$\eta_o = \frac{1.285MI_{sp}}{\frac{\dot{w}_{fuel}}{\dot{w}_{propellant}}\eta_b HV}. \tag{15.21}$$

The subscript *propellant* denotes the fuel plus the oxidizer for rockets and just the fuel for air-breathing jets. Thus the maximum range is

$$R_{max} = \eta_o \eta_b \frac{\dot{w}_{fuel}}{\dot{w}_{propellant}} \frac{L}{D} H \log_e \left( \frac{1}{1 - \frac{W_{propellant}}{W}} \right). \tag{15.22}$$

Note that for air-breathing engines the propellant flow rate is the same as the fuel flow rate and, as mentioned previously, that the lift to drag ratio is assumed to be given by

$$\frac{L}{D} \approx 4\frac{M+3}{M}. \tag{15.23}$$

Then, incorporating all the previous approximations, and further assuming 100% burner efficiency, results in $R_{max} \sim 2.6\ H$, or $R_{max,HC} = 7300$ miles for hydrocarbon fuel and $R_{max,H} = 19{,}000$ miles for hydrogen fuel. Thus we see that high-speed flight is an attractive possibility for future aircraft systems.

## References

Kuchemann, D. (1978). The Aerodynamic Design of Aircraft, Pergamon Press, New York.

Large, J. P., Campbell, H. G., & Gates, D. (1976). Parametric Equations for Estimating Aircraft Airframe Costs. RAND Corporation Report R-1693-1-PA&E, February.

Roskam, J. (1986). Rapid sizing method for airplanes. *J. Aircraft, 23*(7), 554–560.

APPENDIX

# Shock Waves, Expansions, Tables and Charts

## A.1 NORMAL SHOCK WAVE RELATIONS

Regions of isentropic flow may be separated by discontinuities within which the entropy jumps. These abrupt transitions from one value of entropy to a higher one are called shock waves. Flow properties such as velocity, pressure, and temperature also change abruptly across shock waves. One-dimensional flow equations describing the conservation of mass, momentum, and energy may be written, respectively, as follows:

$$d(\rho u) = 0$$

$$dp + \dot{m}\, du = 0$$

$$dh + u du = 0.$$

Integrals of the equations are given by

$$\rho u = \dot{m} = \text{constant}$$

$$p + \dot{m}u = p + \rho u^2 = \text{constant}$$

$$h + \frac{u^2}{2} = \text{constant}.$$

Consider a discontinuity, as shown in Figure A.1, where subscript 1 denotes conditions upstream of the discontinuity and subscript 2 denotes conditions downstream.

The integrals of the motion yield the following relations:

$$\rho_1 u_1 = \rho_2 u_2$$

$$p_1 + \rho_1 u_1^2 = p_2 + \rho_2 u_2^2$$

$$h_1 + \frac{u_1^2}{2} = h_2 + \frac{u_2^2}{2}.$$

The density ratio across the shock is given by the continuity equation as

$$\frac{\rho_1}{\rho_2} = \varepsilon = \frac{u_2}{u_1}.$$

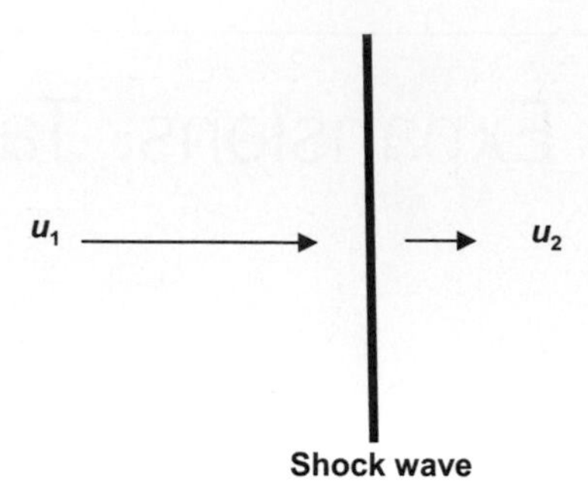


**FIGURE A.1**

One-dimensional flow through a shock wave.

Then the change in flow variables across the shock wave is given by

$$p_2 - p_1 = \rho_1 u_1^2 - \rho_2 u_2^2 = m u_1 (1 - \varepsilon)$$

$$h_2 - h_1 = \frac{1}{2}(u_1^2 - u_2^2) = \frac{1}{2}u_1^2(1 - \varepsilon^2)$$

$$u_2 - u_1 = -u_1(1 - \varepsilon).$$

Concentrating on the enthalpy jump, we find that

$$h_2 - h_1 = \frac{1}{2}u_1 \cdot u_1(1 - \varepsilon^2) = \frac{1}{2}\left[\frac{p_2 - p_1}{m(1 - \varepsilon)}\right]u_1(1 - \varepsilon)(1 + \varepsilon)$$

$$h_2 - h_1 = \frac{p_2 - p_1}{2\rho_1} \cdot (1 + \varepsilon).$$

This last equation is called the Hugoniot equation and it relates thermodynamic state properties *alone* across a normal shock; no velocities are involved. Note that in the limit

$$\lim_{\varepsilon \to 1} h_2 - h_1 \to \frac{p_2 - p_1}{\rho_1} = dh = \frac{dp}{\rho}.$$

This is the case of a wave of infinitesimal strength, that is, an isentropic Mach wave. The sound speed $a$ is given by

$$a^2 = \left(\frac{\partial p}{\partial \rho}\right)_s = \gamma \frac{p}{\rho},$$

where $\gamma$ is the isentropic exponent in the relation $p\rho^{-\gamma}$ = constant, and we may set $\gamma = c_p/c_v$ only if the gas is calorically perfect, that is, $c_p$ and $c_v$ are constant. Conservation equations, along with the perfect gas law, may be used to find the following thermodynamic state properties across a normal shock:

$$\frac{T_2}{T_1} = \left(\frac{a_2}{a_1}\right)^2 = \frac{[2\gamma M_1^2 - (\gamma - 1)][(\gamma - 1)M_1^2 + 2]}{(\gamma + 1)^2 M_1^2} \tag{A.1}$$

$$\frac{p_2}{p_1} = 1 + \frac{2\gamma}{\gamma + 1}(M_1^2 - 1) \tag{A.2}$$

$$\frac{\rho_2}{\rho_1} = \frac{u_1}{u_2} = \frac{q_1}{q_2} = \frac{\gamma + 1}{(\gamma - 1) + 2/M_1^2}. \tag{A.3}$$

The Mach number behind the shock is determined from the Mach number ahead of the shock according to the following relation:

$$M_2 = u_2/a_2 = \left\{ \frac{(\gamma - 1)M_1^2 + 2}{2\gamma M_1^2 - (\gamma - 1)} \right\}^{1/2}. \tag{A.4}$$

Because $M_1 > 1$, Equation (A.4) shows that $M_2 < 1$. The pressure coefficient across the shock is given by

$$C_p = \frac{p_2 - p_1}{q_1} = \frac{4}{\gamma + 1}\left( \frac{M_1^2 - 1}{M_1^2} \right). \tag{A.5}$$

The normal shock process is illustrated on the $h,s$ plane in Figure A.2 where the flow is shown proceeding from state 1 ahead of the shock to state 2 behind the shock. The adiabatic compression produced by the shock raises internal energy at the expense of the kinetic energy, although the total energy is conserved. Because the energy is conserved, the stagnation enthalpy is constant; therefore, if the gas is calorically perfect, the stagnation temperature is constant and

$$\frac{T_{t,2}}{T_{t,1}} = 1. \tag{A.6}$$

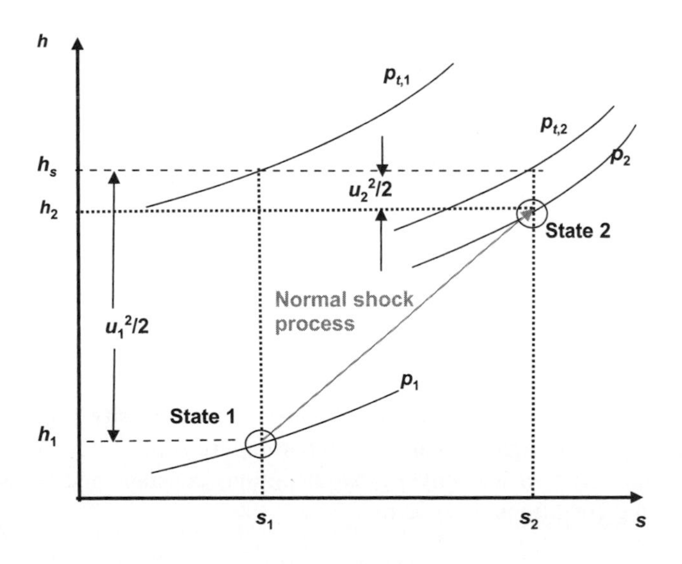


**FIGURE A.2**

Adiabatic normal shock process for flow along a streamline passing through a shock is shown as the flow proceeds from state 1 in the free stream to state 2 behind the shock wave.

The ratio of stagnation pressure to stagnation pressure in the regions on either side of the shock is given by the isentropic relation

$$\frac{p_t}{p} = \left(1 + \frac{\gamma - 1}{2} M^2\right)^{\frac{\gamma}{\gamma - 1}}. \tag{A.7}$$

Then the ratio of stagnation pressure behind the shock to that ahead of the shock is

$$\frac{p_{t,2}}{p_{t,1}} = \frac{p_{t,2}}{p_2}\frac{p_1}{p_{t,1}}\frac{p_2}{p_1} = \left(\frac{1 + \frac{\gamma - 1}{2} M_2^2}{1 + \frac{\gamma - 1}{2} M_1^2}\right)^{\frac{\gamma}{\gamma - 1}} \left(1 + \frac{2\gamma}{\gamma + 1}(M_1^2 - 1)\right). \tag{A.8}$$

From the equations for the Mach number across the shock, Eq. (A-7) may be written as

$$\frac{p_{t,2}}{p_2} = \left(\frac{(\gamma + 1)^2 M_1^2}{4\gamma M_1^2 - 2(\gamma - 1)}\right)^{\frac{\gamma}{\gamma - 1}}. \tag{A.9}$$

Thus the stagnation pressure ratio may be found from the upstream Mach number as follows:

$$\frac{p_{t,2}}{p_{t,1}} = \left\{\frac{(\gamma + 1)^2 M_1^2}{[2\gamma M_1^2 - (\gamma - 1)][2 + (\gamma - 1)M_1^2]}\right\}^{\frac{\gamma}{\gamma - 1}} \left\{\frac{2\gamma M_1^2 - (\gamma - 1)}{\gamma + 1}\right\}. \tag{A.10}$$

For inlet flows in air-breathing engines where it is reasonable to consider $\gamma = 7/5$, the stagnation pressure ratio becomes

$$\frac{p_{t,2}}{p_{t,1}} = \left[\frac{36 M_1^2}{(7M_1^2 - 1)(M_1^2 + 5)}\right]^{\frac{7}{2}} \left[\frac{7M_1^2 - 1}{6}\right]. \tag{A.11}$$

For hot nozzle flows where it is reasonable to consider $\gamma = 4/3$, the stagnation pressure ratio becomes

$$\frac{p_{t,2}}{p_{t,1}} = \left[\frac{49 M_1^2}{(8M_1^2 - 1)(M_1^2 + 6)}\right]^{4} \left[\frac{8M_1^2 - 1}{7}\right]. \tag{A.12}$$

## A.2 OBLIQUE SHOCK WAVE RELATIONS

Consider the normal shock moving parallel to itself with constant velocity $V_{1,t}$. To an observer of this shock wave, the velocity field might appear as shown in Figure A.3.

The tangential component of velocity is, by definition, constant and because mass must be conserved, the following conditions must be met:

$$\rho_1 V_{1n} = \rho_2 V_{2n}.$$

$$V_{1t} = V_{2t}.$$

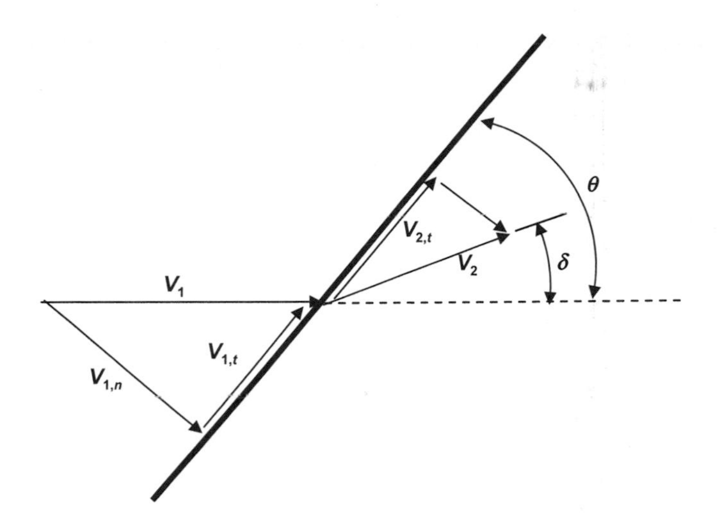


**FIGURE A.3**

Schematic diagram of an oblique shock wave.

Using these conditions and applying the tangential momentum conservation equation to the flow across the shock yields

$$\begin{aligned} \rho_1 V_{1n} &= \rho_2 V_{2n} \\ p_1 + \rho_1 V_{1n}^2 &= p_2 + \rho_2 V_{2n}^2 \\ \rho_1 V_{1n} V_{1t} &= \rho_2 V_{2n} V_{2t} \\ h_1 + \frac{1}{2}(V_{1n}^2 + V_{1t}^2) &= h_2 + \frac{1}{2}(V_{2n}^2 + V_{2t}^2). \end{aligned}$$

The Hugoniot relation remains the same as before because it includes neither constants of the flow nor velocities. All relations for normal shocks may be used for oblique shocks with $M_1$ replaced by $M_{1n} = M_1 \sin\theta$ except for the ratios of static to stagnation pressure, which are given by

$$p_2/p_{t,1} = \frac{2\gamma M_1^2 \sin^2\theta - (\gamma-1)}{\gamma+1}\left[\frac{2}{(\gamma-1)M_1^2+2}\right]^{\frac{\gamma}{\gamma-1}}$$

$$p_2/p_{t,2} = \left\{\frac{2[2\gamma M_1^2\sin^2\theta - (\gamma-1)][(\gamma-1)M_1^2\sin^2\theta+2]}{(\gamma+1)^2 M_1^2\sin^2\theta[(\gamma-1)M_1^2+2]}\right\}^{\frac{\gamma}{\gamma-1}}$$

$$p_{t,2}/p_1 = \left[\frac{\gamma+1}{2\gamma M_1^2\sin^2\theta-(\gamma-1)}\right]^{\frac{1}{\gamma-1}}\left\{\frac{(\gamma+1)M_1^2\sin^2\theta[(\gamma-1)M_1^2+2]}{2[(\gamma-1)M_1^2\sin^2\theta+2]}\right\}^{\frac{\gamma}{\gamma-1}}.$$

Along a streamline behind the shock the compressible Bernoulli equation (for $\gamma$ = constant) is

$$\frac{\gamma}{\gamma-1}\frac{p_2}{\rho_2}+\frac{1}{2}(V_{2n}^2+V_{2t}^2) = \frac{\gamma}{\gamma-1}\frac{p_{t,2}}{\rho_{t,2}}.$$

Now conditions behind the shock are known for all properties (static, not total) and, along a streamline with constant entropy, we have

$$\frac{p_{t,2}}{\rho_{t,2}} = \frac{p_{t,2}}{p_2}\cdot\frac{p_2}{\rho_2}\cdot\frac{\rho_2}{\rho_{t,2}} = \frac{p_{t,2}}{p_2}\cdot\frac{p_2}{\rho_2}\cdot\left(\frac{p_2}{p_{t,2}}\right)^{\frac{1}{\gamma}}$$

$$\frac{p_{t,2}}{\rho_{t,2}} = \left(\frac{p_{t,2}}{p_2}\right)^{\frac{\gamma-1}{\gamma}}\frac{p_2}{\rho_2}.$$

Then we may continue with the development to find

$$\frac{\gamma}{\gamma-1}\frac{p_2}{\rho_2}+\frac{1}{2}V_2^2 = \frac{\gamma}{\gamma-1}\left(\frac{p_{t,2}}{p_2}\right)^{\frac{\gamma-1}{\gamma}}\frac{p_2}{\rho_2}$$

$$\frac{\gamma}{\gamma-1}\frac{p_2}{\rho_2}\left[\left(\frac{p_{t,2}}{p_2}\right)^{\frac{\gamma-1}{\gamma}}-1\right] = \frac{1}{2}V_2^2.$$

Therefore, the ratio of static to stagnation pressure becomes

$$\frac{p_{t,2}}{p_2} = \left(1+\frac{\gamma-1}{2\gamma}V_2^2\frac{\rho_2}{p_2}\right)^{\frac{\gamma}{\gamma-1}}.$$

Finally, this relation may be put in terms of the Mach number downstream of the shock:

$$\frac{p_{t,2}}{p_2} = \left(1+\frac{\gamma-1}{2}M_2^2\right)^{\frac{\gamma}{\gamma-1}}.$$

Thus the stagnation pressure behind the shock can be found, assuming $\gamma$ = constant, where $\gamma$ is the isentropic exponent. Note that this equation is the same as the one-dimensional isentropic flow relations developed previously.

From Figure A.3 we may form the following equation for the difference between the shock angle $\theta$ and the flow deflection angle $\delta$ as follows:

$$\tan(\theta-\delta) = \frac{V_{2n}}{V_{2t}} = \varepsilon\frac{V_{1n}}{V_{1t}} = \varepsilon\tan\theta.$$

Using the trigonometric relation for the tangent of the difference between $\theta$ and $\delta$ yields

$$\frac{\tan\theta-\tan\delta}{1+\tan\theta\tan\delta} = \varepsilon\tan\theta.$$

Solving this equation for the flow deflection angle results in the following:

$$\tan\delta = \frac{(1-\varepsilon)\tan\theta}{1+\varepsilon\tan^2\theta}.$$

For given values of shock angle $\theta$ and upstream Mach number $M_1$, we can find $\varepsilon$ and therefore $\delta$. Then the pressure coefficient

$$C_p = \frac{p_2 - p_1}{q_1} = \frac{\rho_1 V_{1n}^2 - \rho_2 V_{2n}^2}{\frac{1}{2}\rho_1 V_1^2} = 2(1-\varepsilon)\sin^2\theta.$$

This is an important relation in that it sets the stage for the Newtonian theory of hypersonic flow where the basic assumption is that $\varepsilon \rightarrow 0$ in the limit as $M \rightarrow \infty$ so that the pressure coefficient across the shock is solely dependent on the shock angle. Now we may also relate the shock angle and the flow deflection angle by the following equation:

$$\sin^2\theta = \frac{\sin^2\delta}{(1-\varepsilon)^2\cos^2(\theta-\delta)}.$$

Then the pressure coefficient across the shock may be written as

$$C_p = \frac{2\sin^2\delta}{(1-\varepsilon)\cos^2(\theta-\delta)}.$$

The general behavior of the shock wave angle $\theta$ as a function of the flow deflection angle $\delta$ is illustrated for various Mach numbers in a gas with $\gamma = 7/5$ in Figure A.4. Solutions to the equation relating the angles have two branches, yielding two values for the shock wave angle for a given flow deflection angle and Mach number. The locus of the smaller deflections is called the weak branch and only this solution branch is observed in steady supersonic flows.

## A.3 PRANDTL–MEYER EXPANSION

When the oblique shock wave shown in Figure A.3 is generated by a very small deflection angle $\delta$, the pressure and density increase is very small. In the limit of a vanishingly small deflection angle $d\delta$, it is an isentropic Mach wave and the inclination angle of the wave is the Mach angle $\mu = \arcsin(1/M)$. The energy equation requires that the stagnation enthalpy is constant or

$$dh_t = 0 = c_p dT + V dV. \quad \text{(A.13)}$$

The isentropic relation between temperature and pressure is

$$\frac{dT}{T} = \frac{\gamma - 1}{\gamma}\frac{dp}{p}. \quad \text{(A.14)}$$

Using the definition of sound speed $a^2 = \gamma RT$ and substituting Equation (A.13) into Equation (A.14) yields the following relationship between pressure and velocity:

$$\frac{dp}{p} = -\gamma M^2 \frac{dV}{V}. \quad \text{(A.15)}$$

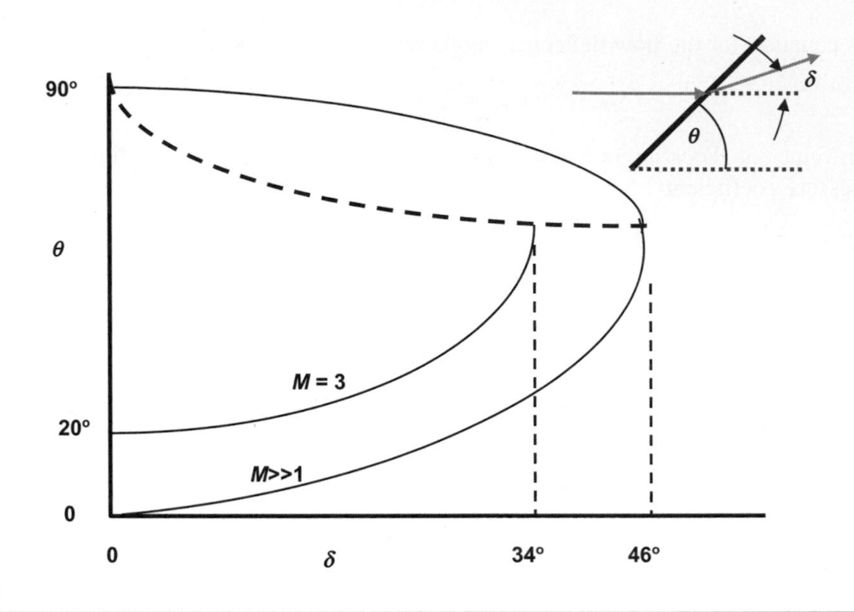


**FIGURE A.4**

The shock wave angle $\theta$ as a function of flow deflection $\delta$ for various Mach numbers and $\gamma = 7/5$. Vertical tangents denote separation of solutions into weak solutions (lower branch) and strong solutions (upper branch).

The Mach number relation $M = V/a$ may be written as

$$\frac{dM}{M} = \frac{dV}{V} - \frac{1}{2}\frac{dp}{p} + \frac{1}{2}\frac{d\rho}{\rho}. \tag{A.16}$$

The isentropic relation may be put in the form

$$\frac{d\rho}{\rho} = \frac{1}{\gamma}\frac{dp}{p}. \tag{A.17}$$

Using Equations (A.15) and (A.17) in Equation (A.16) yields

$$\frac{dM}{M} = \left(1 + \frac{\gamma - 1}{2}M^2\right)\frac{dV}{V}. \tag{A.18}$$

The tangential components of velocity along the wave are equal, $V_{1t} = V_{2t}$, and after setting $V_1 = V$, $V_2 = V + dV$, $\theta = \mu$, and $\delta = d\delta$ in Figure A.4, this may be written as

$$V\cos\mu = (V + dV)\cos(\mu - d\delta). \tag{A.19}$$

Expanding Equation (A.19) and neglecting second-order terms yields

$$\frac{dV}{V} = -d\delta\tan\mu = -\frac{1}{\sqrt{1 - M^2}}d\delta. \tag{A.20}$$

Substituting Equation (A.20) into Equation (A.18) provides the following relationship between the infinitesimal change in flow direction and the corresponding change in Mach number:

$$d\delta = -\frac{\sqrt{M^2-1}}{1+\frac{\gamma-1}{2}M^2}\frac{dM}{M}. \tag{A.21}$$

Taking the Mach number $M = 1$ for zero flow deflection ($\delta = 0$) and integrating over an angle $\nu$ where the Mach number has increased to $M$ yields the Prandtl–Meyer relation

$$\nu = \sqrt{\frac{\gamma+1}{\gamma-1}}\tan^{-1}\sqrt{\frac{\gamma-1}{\gamma+1}(M^2-1)} - \tan^{-1}\sqrt{M^2-1}. \tag{A.22}$$

This process is shown in Figure A.5, where a sonic flow, $M = 1$, along a horizontal wall encounters a flow deflection $\delta = \nu$ and accelerates to a new Mach number $M$. For a given value of $\gamma$ one may generate values of $\nu$ for each $M$ and its corresponding $\mu$. Such a table of values appears later for standard air with $\gamma = 1.4$. The relationship between $M$ and $\mu$ and the unit circle are illustrated in Figure A.6.

It may be pointed out here that according to Equation (A.22) a given perfect gas may be turned through some limiting angle before $M \to \infty$. That limiting deflection for standard air with $\gamma = 1.4$ is given by

$$\nu_{max} = \left(\sqrt{\frac{\gamma-1}{\gamma+1}} - 1\right) \times 90^\circ = 130.45^\circ(\gamma = 1.4). \tag{A.23}$$

An expansion process is isentropic under conditions of either equilibrium chemistry or frozen chemistry, that is, in the absence of finite rate processes, friction, conduction, and so on. Therefore,

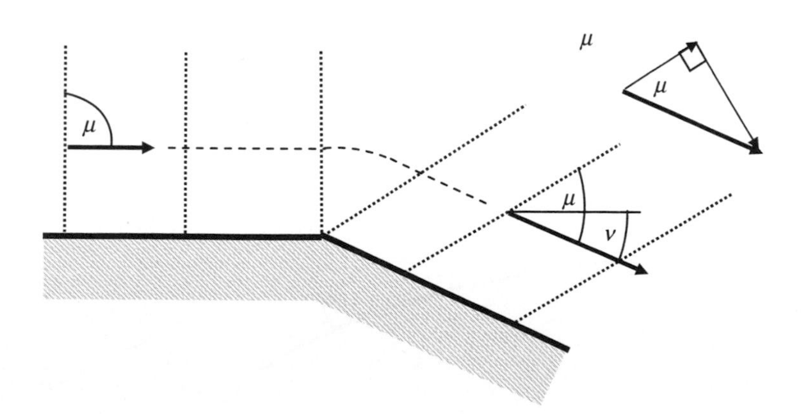


**FIGURE A.5**

Prandtl–Meyer expansion over a sharp corner.

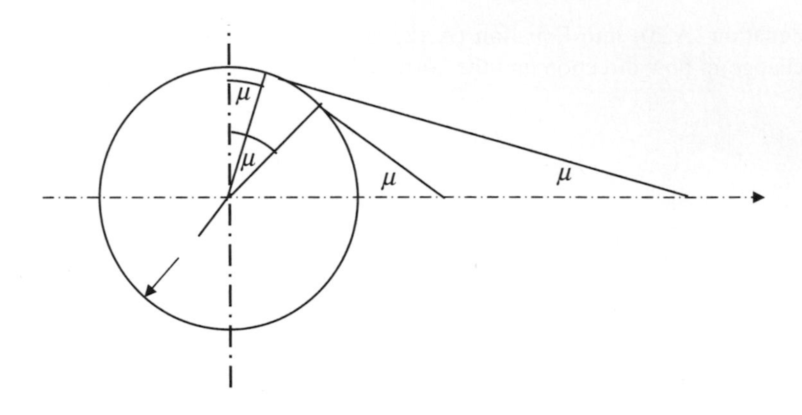


**FIGURE A.6**

Mach angles and unit circle.

the stagnation pressure is constant, and the ratio of static to stagnation temperature may be calculated from

$$\left(\frac{p}{p_t}\right)^{\frac{\gamma-1}{\gamma}} = \frac{1}{\gamma+1}\left\{1+\cos\left[2\sqrt{\frac{\gamma-1}{\gamma+1}}\left(\nu+\arctan\sqrt{M^2-1}\right)\right]\right\}. \tag{A.24}$$

We may consider the supersonic flow over the smoothly curving wall shown in Figure A.7, where the flow deflection is $\Delta\nu$.

The Prandtl–Meyer angle at station 2 is given by

$$\nu_2 = \Delta\nu + \nu_1. \tag{A.25}$$

Thus, for convex surfaces, the value of $\nu_2$ is known if $\Delta\nu = \Delta\delta$ is known and $M_2$ may be found, not directly, but generally by an iterative or interpolative process, as it is not possible to rearrange Equation (A.22). Then $p_2$ may be found from

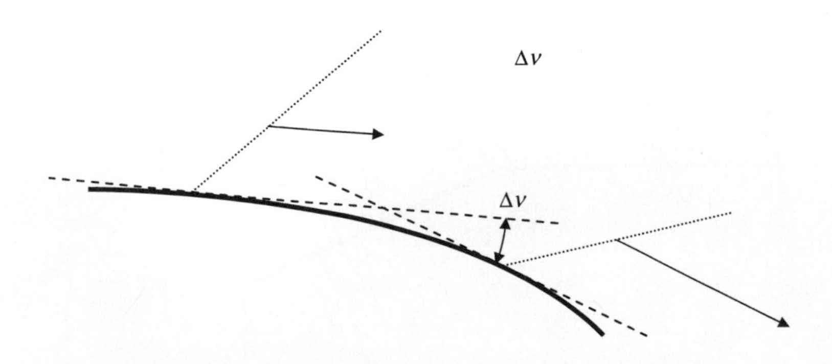


**FIGURE A.7**

Prandtl–Meyer flow over a smooth convex surface.

$$p_2 = \left(\frac{p_2}{p_t}\right)\left(\frac{p_1}{p_t}\right)^{-1} p_1.$$

Alternatively, knowing the Mach number at points 1 and 2 permits one to determine the pressure ratio directly in terms of $M$ and $\gamma$:

$$\frac{p_2}{p_1} = \left(\frac{1+\frac{\gamma-1}{2}M_2^2}{1+\frac{\gamma-1}{2}M_1^2}\right)^{\frac{-\gamma}{\gamma-1}}. \tag{A.26}$$

An expansion of the pressure expression for small deflections $\Delta\nu = \delta \ll 1$ leads to

$$\frac{p_2 - p_1}{p_1} \approx \frac{-\gamma M_1^2}{\sqrt{M_1^2-1}}\Delta\nu = \frac{-\gamma M_1^2}{\sqrt{M_1^2-1}}\delta. \tag{A.27}$$

The surface pressure coefficient for small flow deflections is then given by

$$C_p = \frac{p_2 - p_1}{p_1}\frac{p_1}{q_1} = \frac{-2\delta}{\sqrt{M_1^2-1}}. \tag{A.28}$$

This is the classical linearized supersonic result.

## A.4 TABLES AND CHARTS FOR ISENTROPIC COMPRESSIBLE GAS FLOWS AND SHOCK WAVES IN A GAS WITH $\gamma = 1.4$

The following tables and charts are excerpted from NACA TR 1135 for the Mach number range $0 \leq M \leq 3$

REPORT 1135—NATIONAL ADVISORY COMMITTEE FOR AERONAUTICS

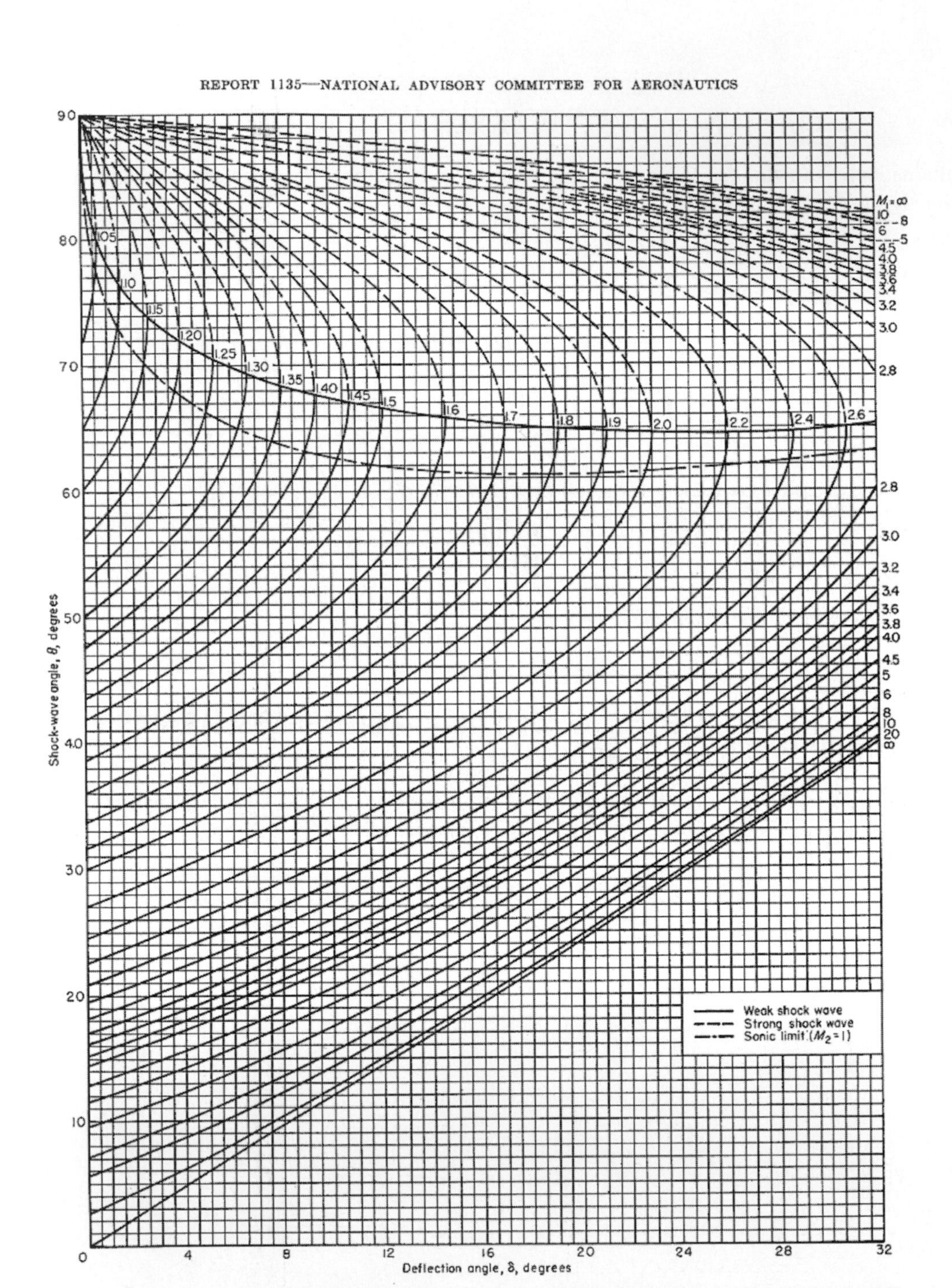


CHART 2.—Variation of shock-wave angle with flow-deflection angle for various upstream Mach numbers. Perfect gas, $\gamma = 7/5$.

EQUATIONS, TABLES, AND CHARTS FOR COMPRESSIBLE FLOW

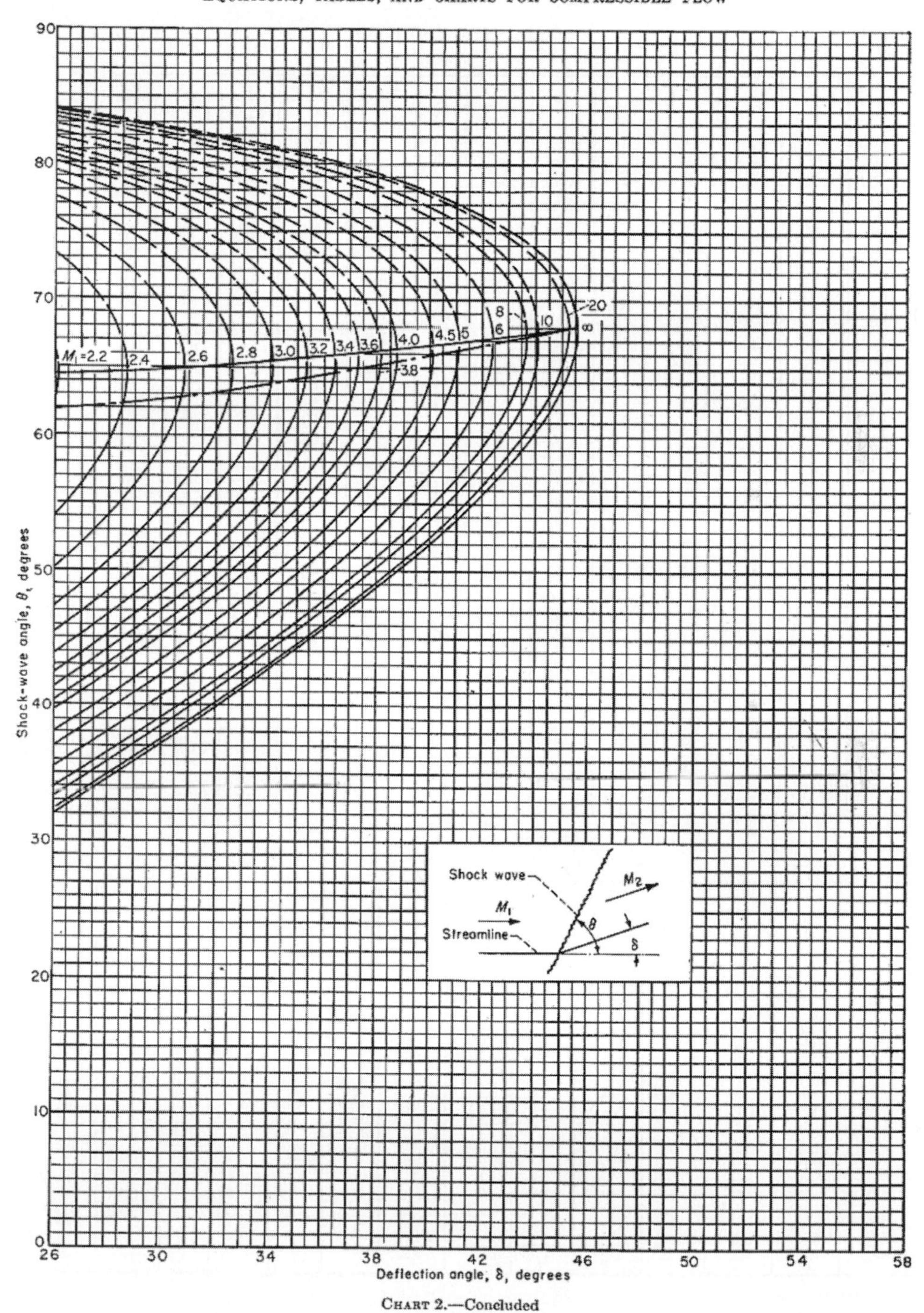


CHART 2.—Concluded

## EQUATIONS, TABLES, AND CHARTS FOR COMPRESSIBLE FLOW

### TABLE I.—SUBSONIC FLOW

$\gamma=7/5$

| $M$ | $\frac{p}{p_t}$ | $\frac{\rho}{\rho_t}$ | $\frac{T}{T_t}$ | $\beta$ | $\frac{q}{p_t}$ | $\frac{A}{A_*}$ | $\frac{V}{a_*}$ |
|---|---|---|---|---|---|---|---|
| 0 | 1.0000 | 1.0000 | 1.0000 | 1.0000 | 0 | ∞ | 0 |
| .01 | .9999 | 1.0000 | 1.0000 | 1.0000 | .7000 $^{-4}$ | 57.8738 | .01095 |
| .02 | .9997 | .9998 | .9999 | .9998 | .2799 $^{-3}$ | 28.9421 | .02191 |
| .03 | .9994 | .9996 | .9998 | .9995 | .6296 $^{-3}$ | 19.3005 | .03286 |
| .04 | .9989 | .9992 | .9997 | .9992 | .1119 $^{-2}$ | 14.4815 | .04381 |
| .05 | .9983 | .9988 | .9995 | .9987 | .1747 $^{-2}$ | 11.5914 | .05476 |
| .06 | .9975 | .9982 | .9993 | .9982 | .2514 $^{-2}$ | 9.6659 | .06570 |
| .07 | .9966 | .9976 | .9990 | .9975 | .3418 $^{-2}$ | 8.2915 | .07664 |
| .08 | .9955 | .9968 | .9987 | .9968 | .4460 $^{-2}$ | 7.2616 | .08758 |
| .09 | .9944 | .9960 | .9984 | .9959 | .5638 $^{-2}$ | 6.4613 | .09851 |
| .10 | .9930 | .9950 | .9980 | .9950 | .6951 $^{-2}$ | 5.8218 | .10944 |
| .11 | .9916 | .9940 | .9976 | .9939 | .8399 $^{-2}$ | 5.2992 | .12035 |
| .12 | .9900 | .9928 | .9971 | .9928 | .9979 $^{-2}$ | 4.8643 | .13126 |
| .13 | .9883 | .9916 | .9966 | .9915 | .1169 $^{-1}$ | 4.4969 | .14217 |
| .14 | .9864 | .9903 | .9961 | .9902 | .1353 $^{-1}$ | 4.1824 | .15306 |
| .15 | .9844 | .9888 | .9955 | .9887 | .1550 $^{-1}$ | 3.9103 | .16395 |
| .16 | .9823 | .9873 | .9949 | .9871 | .1760 $^{-1}$ | 3.6727 | .17482 |
| .17 | .9800 | .9857 | .9943 | .9854 | .1983 $^{-1}$ | 3.4635 | .18569 |
| .18 | .9776 | .9840 | .9936 | .9837 | .2217 $^{-1}$ | 3.2779 | .19654 |
| .19 | .9751 | .9822 | .9928 | .9818 | .2464 $^{-1}$ | 3.1123 | .20739 |
| .20 | .9725 | .9803 | .9921 | .9798 | .2723 $^{-1}$ | 2.9635 | .21822 |
| .21 | .9697 | .9783 | .9913 | .9777 | .2994 $^{-1}$ | 2.8293 | .22904 |
| .22 | .9668 | .9762 | .9904 | .9755 | .3276 $^{-1}$ | 2.7076 | .23984 |
| .23 | .9638 | .9740 | .9895 | .9732 | .3569 $^{-1}$ | 2.5968 | .25063 |
| .24 | .9607 | .9718 | .9886 | .9708 | .3874 $^{-1}$ | 2.4956 | .26141 |
| .25 | .9575 | .9694 | .9877 | .9682 | .4189 $^{-1}$ | 2.4027 | .27217 |
| .26 | .9541 | .9670 | .9867 | .9656 | .4515 $^{-1}$ | 2.3173 | .28291 |
| .27 | .9506 | .9645 | .9856 | .9629 | .4851 $^{-1}$ | 2.2385 | .29364 |
| .28 | .9470 | .9619 | .9846 | .9600 | .5197 $^{-1}$ | 2.1656 | .30435 |
| .29 | .9433 | .9592 | .9835 | .9570 | .5553 $^{-1}$ | 2.0979 | .31504 |
| .30 | .9395 | .9564 | .9823 | .9539 | .5919 $^{-1}$ | 2.0351 | .32572 |
| .31 | .9355 | .9535 | .9811 | .9507 | .6293 $^{-1}$ | 1.9765 | .33637 |
| .32 | .9315 | .9506 | .9799 | .9474 | .6677 $^{-1}$ | 1.9219 | .34701 |
| .33 | .9274 | .9476 | .9787 | .9440 | .7069 $^{-1}$ | 1.8707 | .35762 |
| .34 | .9231 | .9445 | .9774 | .9404 | .7470 $^{-1}$ | 1.8229 | .36822 |
| .35 | .9188 | .9413 | .9761 | .9367 | .7879 $^{-1}$ | 1.7780 | .37879 |
| .36 | .9143 | .9380 | .9747 | .9330 | .8295 $^{-1}$ | 1.7358 | .38935 |
| .37 | .9098 | .9347 | .9733 | .9290 | .8719 $^{-1}$ | 1.6961 | .39988 |
| .38 | .9052 | .9313 | .9719 | .9250 | .9149 $^{-1}$ | 1.6587 | .41039 |
| .39 | .9004 | .9278 | .9705 | .9208 | .9587 $^{-1}$ | 1.6234 | .42087 |
| .40 | .8956 | .9243 | .9690 | .9165 | .1003 | 1.5901 | .43133 |
| .41 | .8907 | .9207 | .9675 | .9121 | .1048 | 1.5587 | .44177 |
| .42 | .8857 | .9170 | .9659 | .9075 | .1094 | 1.5289 | .45218 |
| .43 | .8807 | .9132 | .9643 | .9028 | .1140 | 1.5007 | .46257 |
| .44 | .8755 | .9094 | .9627 | .8980 | .1187 | 1.4740 | .47293 |
| .45 | .8703 | .9055 | .9611 | .8930 | .1234 | 1.4487 | .48326 |
| .46 | .8650 | .9016 | .9594 | .8879 | .1281 | 1.4246 | .49357 |
| .47 | .8596 | .8976 | .9577 | .8827 | .1329 | 1.4018 | .50385 |
| .48 | .8541 | .8935 | .9560 | .8773 | .1378 | 1.3801 | .51410 |
| .49 | .8486 | .8894 | .9542 | .8717 | .1426 | 1.3595 | .52433 |
| 0.50 | 0.8430 | 0.8852 | 0.9524 | 0.8660 | 0.1475 | 1.3398 | 0.53452 |
| .51 | .8374 | .8809 | .9506 | .8602 | .1525 | 1.3212 | .54469 |
| .52 | .8317 | .8766 | .9487 | .8542 | .1574 | 1.3034 | .55483 |
| .53 | .8259 | .8723 | .9468 | .8480 | .1624 | 1.2865 | .56493 |
| .54 | .8201 | .8679 | .9449 | .8417 | .1674 | 1.2703 | .57501 |
| .55 | .8142 | .8634 | .9430 | .8352 | .1724 | 1.2550 | .58506 |
| .56 | .8082 | .8589 | .9410 | .8285 | .1774 | 1.2403 | .59507 |
| .57 | .8022 | .8544 | .9390 | .8216 | .1825 | 1.2263 | .60505 |
| .58 | .7962 | .8498 | .9370 | .8146 | .1875 | 1.2130 | .61501 |
| .59 | .7901 | .8451 | .9349 | .8074 | .1925 | 1.2003 | .62492 |
| .60 | .7840 | .8405 | .9328 | .8000 | .1976 | 1.1882 | .63481 |
| .61 | .7778 | .8357 | .9307 | .7924 | .2026 | 1.1767 | .64466 |
| .62 | .7716 | .8310 | .9286 | .7846 | .2076 | 1.1657 | .65448 |
| .63 | .7654 | .8262 | .9265 | .7766 | .2127 | 1.1552 | .66427 |
| .64 | .7591 | .8213 | .9243 | .7684 | .2177 | 1.1452 | .67402 |
| .65 | .7528 | .8164 | .9221 | .7599 | .2227 | 1.1356 | .68374 |
| .66 | .7465 | .8115 | .9199 | .7513 | .2276 | 1.1265 | .69342 |
| .67 | .7401 | .8066 | .9176 | .7424 | .2326 | 1.1179 | .70307 |
| .68 | .7338 | .8016 | .9153 | .7332 | .2375 | 1.1097 | .71263 |
| .69 | .7274 | .7966 | .9131 | .7238 | .2424 | 1.1018 | .72225 |
| .70 | .7209 | .7916 | .9107 | .7141 | .2473 | 1.0944 | .73179 |
| .71 | .7145 | .7865 | .9084 | .7042 | .2521 | 1.0873 | .74129 |
| .72 | .7080 | .7814 | .9061 | .6940 | .2569 | 1.0806 | .75076 |
| .73 | .7016 | .7763 | .9037 | .6834 | .2617 | 1.0742 | .76019 |
| .74 | .6951 | .7712 | .9013 | .6726 | .2664 | 1.0681 | .76958 |
| .75 | .6886 | .7660 | .8989 | .6614 | .2711 | 1.0624 | .77894 |
| .76 | .6821 | .7609 | .8964 | .6499 | .2758 | 1.0570 | .78825 |
| .77 | .6756 | .7557 | .8940 | .6380 | .2804 | 1.0519 | .79753 |
| .78 | .6691 | .7505 | .8915 | .6258 | .2849 | 1.0471 | .80677 |
| .79 | .6625 | .7452 | .8890 | .6131 | .2894 | 1.0425 | .81597 |
| .80 | .6560 | .7400 | .8865 | .6000 | .2939 | 1.0382 | .82514 |
| .81 | .6495 | .7347 | .8840 | .5864 | .2983 | 1.0342 | .83426 |
| .82 | .6430 | .7295 | .8815 | .5724 | .3027 | 1.0305 | .84335 |
| .83 | .6365 | .7242 | .8789 | .5578 | .3069 | 1.0270 | .85239 |
| .84 | .6300 | .7189 | .8763 | .5426 | .3112 | 1.0237 | .86140 |
| .85 | .6235 | .7136 | .8737 | .5268 | .3153 | 1.0207 | .87037 |
| .86 | .6170 | .7083 | .8711 | .5103 | .3195 | 1.0179 | .87929 |
| .87 | .6106 | .7030 | .8685 | .4931 | .3235 | 1.0153 | .88818 |
| .88 | .6041 | .6977 | .8659 | .4750 | .3275 | 1.0129 | .89703 |
| .89 | .5977 | .6924 | .8632 | .4560 | .3314 | 1.0108 | .90583 |
| .90 | .5913 | .6870 | .8606 | .4359 | .3352 | 1.0089 | .91460 |
| .91 | .5849 | .6817 | .8579 | .4146 | .3390 | 1.0071 | .92332 |
| .92 | .5785 | .6764 | .8552 | .3919 | .3427 | 1.0056 | .93201 |
| .93 | .5721 | .6711 | .8525 | .3676 | .3464 | 1.0043 | .94065 |
| .94 | .5658 | .6658 | .8498 | .3412 | .3500 | 1.0031 | .94925 |
| .95 | .5595 | .6604 | .8471 | .3122 | .3534 | 1.0022 | .95781 |
| .96 | .5532 | .6551 | .8444 | .2800 | .3569 | 1.0014 | .96633 |
| .97 | .5469 | .6498 | .8416 | .2431 | .3602 | 1.0008 | .97481 |
| .98 | .5407 | .6445 | .8389 | .1990 | .3635 | 1.0003 | .98325 |
| .99 | .5345 | .6392 | .8361 | .1411 | .3667 | 1.0001 | .99165 |
| 1.00 | .5283 | .6339 | .8333 | .0000 | .3698 | 1.0000 | 1.00000 |

### TABLE II.—SUPERSONIC FLOW

$\gamma=7/5$

| $M$ or $M_1$ | $\frac{p}{p_t}$ | $\frac{\rho}{\rho_t}$ | $\frac{T}{T_t}$ | $\beta$ | $\frac{q}{p_t}$ | $\frac{A}{A_*}$ | $\frac{V}{a_*}$ | $\nu$ | $\mu$ | $M_2$ | $\frac{p_2}{p_1}$ | $\frac{\rho_2}{\rho_1}$ | $\frac{T_2}{T_1}$ | $\frac{p_{t_2}}{p_{t_1}}$ | $\frac{p_1}{p_{t_2}}$ |
|---|---|---|---|---|---|---|---|---|---|---|---|---|---|---|---|
| 1.00 | 0.5283 | 0.6339 | 0.8333 | 0 | 0.3698 | 1.000 | 1.00000 | 0 | 90.00 | 1.000 | 1.000 | 1.000 | 1.000 | 1.000 | 0.5283 |
| 1.01 | .5221 | .6287 | .8306 | .1418 | .3728 | 1.000 | 1.00831 | .04473 | 81.93 | .9901 | 1.023 | 1.017 | 1.007 | 1.000 | .5221 |
| 1.02 | .5160 | .6234 | .8278 | .2010 | .3758 | 1.000 | 1.01658 | .1257 | 78.64 | .9805 | 1.047 | 1.033 | 1.013 | 1.000 | .5160 |
| 1.03 | .5099 | .6181 | .8250 | .2468 | .3787 | 1.001 | 1.02481 | .2294 | 76.14 | .9712 | 1.071 | 1.050 | 1.020 | 1.000 | .5100 |
| 1.04 | .5039 | .6129 | .8222 | .2857 | .3815 | 1.001 | 1.03300 | .3510 | 74.06 | .9620 | 1.095 | 1.067 | 1.026 | .9999 | .5039 |
| 1.05 | .4979 | .6077 | .8193 | .3202 | .3842 | 1.002 | 1.04114 | .4874 | 72.25 | .9531 | 1.120 | 1.084 | 1.033 | .9999 | .4980 |
| 1.06 | .4919 | .6024 | .8165 | .3516 | .3869 | 1.003 | 1.04925 | .6367 | 70.63 | .9444 | 1.144 | 1.101 | 1.039 | .9997 | .4920 |
| 1.07 | .4860 | .5972 | .8137 | .3807 | .3895 | 1.004 | 1.05731 | .7973 | 69.16 | .9360 | 1.169 | 1.118 | 1.046 | .9996 | .4861 |
| 1.08 | .4800 | .5920 | .8108 | .4079 | .3919 | 1.005 | 1.06533 | .9680 | 67.81 | .9277 | 1.194 | 1.135 | 1.052 | .9994 | .4803 |
| 1.09 | .4742 | .5869 | .8080 | .4337 | .3944 | 1.006 | 1.07331 | 1.148 | 66.55 | .9196 | 1.219 | 1.152 | 1.059 | .9992 | .4746 |
| 1.10 | .4684 | .5817 | .8052 | .4583 | .3967 | 1.008 | 1.08124 | 1.336 | 65.38 | .9118 | 1.245 | 1.169 | 1.065 | .9989 | .4689 |
| 1.11 | .4626 | .5766 | .8023 | .4818 | .3990 | 1.010 | 1.08913 | 1.532 | 64.28 | .9041 | 1.271 | 1.186 | 1.071 | .9986 | .4632 |
| 1.12 | .4568 | .5714 | .7994 | .5044 | .4011 | 1.011 | 1.09699 | 1.735 | 63.23 | .8966 | 1.297 | 1.203 | 1.078 | .9982 | .4576 |
| 1.13 | .4511 | .5663 | .7966 | .5262 | .4032 | 1.013 | 1.10479 | 1.944 | 62.25 | .8892 | 1.323 | 1.221 | 1.084 | .9978 | .4521 |
| 1.14 | .4455 | .5612 | .7937 | .5474 | .4052 | 1.015 | 1.11256 | 2.160 | 61.31 | .8820 | 1.350 | 1.238 | 1.090 | .9973 | .4467 |
| 1.15 | .4398 | .5562 | .7908 | .5679 | .4072 | 1.017 | 1.12029 | 2.381 | 60.4[illegible] | .8750 | 1.376 | 1.255 | 1.097 | .9967 | .4413 |
| 1.16 | .4343 | .5511 | .7879 | .5879 | .4090 | 1.020 | 1.12797 | 2.607 | 59.55 | .8682 | 1.403 | 1.272 | 1.103 | .9961 | .4360 |
| 1.17 | .4287 | .5461 | .7851 | .6074 | .4108 | 1.022 | 1.13561 | 2.839 | 58.73 | .8615 | 1.430 | 1.290 | 1.109 | .9953 | .4307 |
| 1.18 | .4232 | .5411 | .7822 | .6264 | .4125 | 1.025 | 1.14321 | 3.074 | 57.94 | .8549 | 1.458 | 1.307 | 1.115 | .9946 | .4255 |
| 1.19 | .4178 | .5361 | .7793 | .6451 | .4141 | 1.026 | 1.15077 | 3.314 | 57.18 | .8485 | 1.485 | 1.324 | 1.122 | .9937 | .4204 |
| 1.20 | .4124 | .5311 | .7764 | .6633 | .4157 | 1.030 | 1.15828 | 3.558 | 56.44 | .8422 | 1.513 | 1.342 | 1.128 | .9928 | .4154 |
| 1.21 | .4070 | .5262 | .7735 | .6812 | .4171 | 1.033 | 1.16575 | 3.806 | 55.74 | .8360 | 1.541 | 1.359 | 1.134 | .9918 | .4104 |
| 1.22 | .4017 | .5213 | .7706 | .6989 | .4185 | 1.037 | 1.17319 | 4.057 | 55.05 | .8300 | 1.570 | 1.376 | 1.141 | .9907 | .4055 |
| 1.23 | .3964 | .5164 | .7677 | .7162 | .4198 | 1.040 | 1.18057 | 4.312 | 54.39 | .8241 | 1.598 | 1.394 | 1.147 | .9896 | .4006 |
| 1.24 | .3912 | .5115 | .7648 | .7332 | .4211 | 1.043 | 1.18792 | 4.569 | 53.75 | .8183 | 1.627 | 1.411 | 1.153 | .9884 | .3958 |

268292—54——4

TABLE II.—SUPERSONIC FLOW—Continued

$\gamma=7/5$

| $M$ or $M_1$ | $\frac{p}{p_t}$ | $\frac{\rho}{\rho_t}$ | $\frac{T}{T_t}$ | $\beta$ | $\frac{q}{p_t}$ | $\frac{A}{A_*}$ | $\frac{V}{a_*}$ | $\nu$ | $\mu$ | $M_2$ | $\frac{p_2}{p_1}$ | $\frac{\rho_2}{\rho_1}$ | $\frac{T_2}{T_1}$ | $\frac{p_{t_2}}{p_{t_1}}$ | $\frac{p_1}{p_{t_2}}$ |
|---|---|---|---|---|---|---|---|---|---|---|---|---|---|---|---|
| 1.25 | .3861 | .5067 | .7619 | .7500 | .4223 | 1.047 | 1.19523 | 4.830 | 53.13 | .8126 | 1.656 | 1.429 | 1.159 | .9871 | .3911 |
| 1.26 | .3809 | .5019 | .7590 | .7666 | .4233 | 1.050 | 1.20249 | 5.093 | 52.53 | .8071 | 1.686 | 1.446 | 1.166 | .9857 | .3865 |
| 1.27 | .3759 | .4971 | .7561 | .7829 | .4244 | 1.054 | 1.20972 | 5.359 | 51.94 | .8016 | 1.715 | 1.463 | 1.172 | .9842 | .3819 |
| 1.28 | .3708 | .4923 | .7532 | .7990 | .4253 | 1.058 | 1.21690 | 5.627 | 51.38 | .7963 | 1.745 | 1.481 | 1.178 | .9827 | .3774 |
| 1.29 | .3658 | .4876 | .7503 | .8149 | .4262 | 1.062 | 1.22404 | 5.898 | 50.82 | .7911 | 1.775 | 1.498 | 1.185 | .9811 | .3729 |
| 1.30 | .3609 | .4829 | .7474 | .8307 | .4270 | 1.066 | 1.23114 | 6.170 | 50.28 | .7860 | 1.805 | 1.516 | 1.191 | .9794 | .3685 |
| 1.31 | .3560 | .4782 | .7445 | .8462 | .4277 | 1.071 | 1.23819 | 6.445 | 49.76 | .7809 | 1.835 | 1.533 | 1.197 | .9776 | .3642 |
| 1.32 | .3512 | .4736 | .7416 | .8616 | .4283 | 1.075 | 1.24521 | 6.721 | 49.25 | .7760 | 1.866 | 1.551 | 1.204 | .9758 | .3599 |
| 1.33 | .3464 | .4690 | .7387 | .8769 | .4289 | 1.080 | 1.25218 | 7.000 | 48.75 | .7712 | 1.897 | 1.568 | 1.210 | .9738 | .3557 |
| 1.34 | .3417 | .4644 | .7358 | .8920 | .4294 | 1.084 | 1.25912 | 7.280 | 48.27 | .7664 | 1.928 | 1.585 | 1.216 | .9718 | .3516 |
| 1.35 | .3370 | .4598 | .7329 | .9069 | .4299 | 1.089 | 1.26601 | 7.561 | 47.79 | .7618 | 1.960 | 1.603 | 1.223 | .9697 | .3475 |
| 1.36 | .3323 | .4553 | .7300 | .9217 | .4303 | 1.094 | 1.27286 | 7.844 | 47.33 | .7572 | 1.991 | 1.620 | 1.229 | .9676 | .3435 |
| 1.37 | .3277 | .4508 | .7271 | .9364 | .4306 | 1.099 | 1.27968 | 8.128 | 46.88 | .7527 | 2.023 | 1.638 | 1.235 | .9653 | .3395 |
| 1.38 | .3232 | .4463 | .7242 | .9510 | .4308 | 1.104 | 1.28645 | 8.413 | 46.44 | .7483 | 2.055 | 1.655 | 1.242 | .9630 | .3356 |
| 1.39 | .3187 | .4418 | .7213 | .9655 | .4310 | 1.109 | 1.29318 | 8.699 | 46.01 | .7440 | 2.087 | 1.672 | 1.248 | .9607 | .3317 |
| 1.40 | .3142 | .4374 | .7184 | .9798 | .4311 | 1.115 | 1.29987 | 8.987 | 45.58 | .7397 | 2.120 | 1.690 | 1.255 | .9582 | .3280 |
| 1.41 | .3098 | .4330 | .7155 | .9940 | .4312 | 1.120 | 1.30652 | 9.276 | 45.17 | .7355 | 2.153 | 1.707 | 1.261 | .9557 | .3242 |
| 1.42 | .3055 | .4287 | .7126 | 1.008 | .4312 | 1.126 | 1.31313 | 9.565 | 44.77 | .7314 | 2.186 | 1.724 | 1.268 | .9531 | .3205 |
| 1.43 | .3012 | .4244 | .7097 | 1.022 | .4311 | 1.132 | 1.31970 | 9.855 | 44.37 | .7274 | 2.219 | 1.742 | 1.274 | .9504 | .3169 |
| 1.44 | .2969 | .4201 | .7069 | 1.036 | .4310 | 1.138 | 1.32623 | 10.146 | 43.98 | .7235 | 2.253 | 1.759 | 1.281 | .9476 | .3133 |
| 1.45 | .2927 | .4158 | .7040 | 1.050 | .4308 | 1.144 | 1.33272 | 10.438 | 43.60 | .7196 | 2.286 | 1.776 | 1.287 | .9448 | .3098 |
| 1.46 | .2886 | .4116 | .7011 | 1.064 | .4306 | 1.150 | 1.33917 | 10.731 | 43.23 | .7157 | 2.320 | 1.793 | 1.294 | .9420 | .3063 |
| 1.47 | .2845 | .4074 | .6982 | 1.077 | .4303 | 1.156 | 1.34558 | 11.023 | 42.86 | .7120 | 2.354 | 1.811 | 1.300 | .9390 | .3029 |
| 1.48 | .2804 | .4032 | .6954 | 1.091 | .4299 | 1.163 | 1.35195 | 11.317 | 42.51 | .7083 | 2.389 | 1.828 | 1.307 | .9360 | .2996 |
| 1.49 | .2764 | .3991 | .6925 | 1.105 | .4295 | 1.169 | 1.35828 | 11.611 | 42.16 | .7047 | 2.423 | 1.845 | 1.314 | .9329 | .2962 |
| 1.50 | .2724 | .3950 | .6897 | 1.118 | .4290 | 1.176 | 1.36458 | 11.905 | 41.81 | .7011 | 2.458 | 1.862 | 1.320 | .9298 | .2930 |
| 1.51 | .2685 | .3909 | .6868 | 1.131 | .4285 | 1.183 | 1.37083 | 12.200 | 41.47 | .6976 | 2.493 | 1.879 | 1.327 | .9266 | .2898 |
| 1.52 | .2646 | .3869 | .6840 | 1.145 | .4279 | 1.190 | 1.37705 | 12.495 | 41.14 | .6941 | 2.529 | 1.896 | 1.334 | .9233 | .2866 |
| 1.53 | .2608 | .3829 | .6811 | 1.158 | .4273 | 1.197 | 1.38322 | 12.790 | 40.81 | .6907 | 2.564 | 1.913 | 1.340 | .9200 | .2835 |
| 1.54 | .2570 | .3789 | .6783 | 1.171 | .4266 | 1.204 | 1.38936 | 13.086 | 40.49 | .6874 | 2.600 | 1.930 | 1.347 | .9166 | .2804 |
| 1.55 | .2533 | .3750 | .6754 | 1.184 | .4259 | 1.212 | 1.39546 | 13.381 | 40.18 | .6841 | 2.636 | 1.947 | 1.354 | .9132 | .2773 |
| 1.56 | .2496 | .3710 | .6726 | 1.107 | .4252 | 1.219 | 1.40152 | 13.677 | 39.87 | .6809 | 2.673 | 1.964 | 1.361 | .9097 | .2744 |
| 1.57 | .2459 | .3672 | .6698 | 1.210 | .4243 | 1.227 | 1.40755 | 13.973 | 39.56 | .6777 | 2.709 | 1.981 | 1.367 | .9061 | .2714 |
| 1.58 | .2423 | .3633 | .6670 | 1.223 | .4235 | 1.234 | 1.41353 | 14.269 | 39.27 | .6746 | 2.746 | 1.998 | 1.374 | .9026 | .2685 |
| 1.59 | .2388 | .3595 | .6642 | 1.236 | .4226 | 1.242 | 1.41948 | 14.564 | 38.97 | .6715 | 2.783 | 2.015 | 1.381 | .8989 | .2656 |
| 1.60 | .2353 | .3557 | .6614 | 1.249 | .4216 | 1.250 | 1.42539 | 14.861 | 38.68 | .6684 | 2.820 | 2.032 | 1.388 | .8952 | .2628 |
| 1.61 | .2318 | .3520 | .6586 | 1.262 | .4206 | 1.258 | 1.43127 | 15.156 | 38.40 | .6655 | 2.857 | 2.049 | 1.395 | .8915 | .2600 |
| 1.62 | .2284 | .3483 | .6558 | 1.275 | .4196 | 1.267 | 1.43710 | 15.452 | 38.12 | .6625 | 2.895 | 2.065 | 1.402 | .8877 | .2573 |
| 1.63 | .2250 | .3446 | .6530 | 1.287 | .4185 | 1.275 | 1.44290 | 15.747 | 37.84 | .6596 | 2.933 | 2.082 | 1.409 | .8838 | .2546 |
| 1.64 | .2217 | .3409 | .6502 | 1.300 | .4174 | 1.284 | 1.44866 | 16.043 | 37.57 | .6568 | 2.971 | 2.099 | 1.416 | .8799 | .2519 |
| 1.65 | .2184 | .3373 | .6475 | 1.312 | .4162 | 1.292 | 1.45439 | 16.338 | 37.31 | .6540 | 3.010 | 2.115 | 1.423 | .8760 | .2493 |
| 1.66 | .2151 | .3337 | .6447 | 1.325 | .4150 | 1.301 | 1.46008 | 16.633 | 37.04 | .6512 | 3.048 | 2.132 | 1.430 | .8720 | .2467 |
| 1.67 | .2119 | .3302 | .6419 | 1.337 | .4138 | 1.310 | 1.46573 | 16.928 | 36.78 | .6485 | 3.087 | 2.148 | 1.437 | .8680 | .2442 |
| 1.68 | .2088 | .3266 | .6392 | 1.350 | .4125 | 1.319 | 1.47135 | 17.222 | 36.53 | .6458 | 3.126 | 2.165 | 1.444 | .8640 | .2417 |
| 1.69 | .2057 | .3232 | .6364 | 1.362 | .4112 | 1.328 | 1.47693 | 17.516 | 36.28 | .6431 | 3.165 | 2.181 | 1.451 | .8598 | .2392 |
| 1.70 | .2026 | .3197 | .6337 | 1.375 | .4098 | 1.338 | 1.48247 | 17.810 | 36.03 | .6405 | 3.205 | 2.198 | 1.458 | .8557 | .2368 |
| 1.71 | .1996 | .3163 | .6310 | 1.387 | .4085 | 1.347 | 1.48798 | 18.103 | 35.79 | .6380 | 3.245 | 2.214 | 1.466 | .8516 | .2344 |
| 1.72 | .1966 | .3129 | .6283 | 1.399 | .4071 | 1.357 | 1.49345 | 18.397 | 35.55 | .6355 | 3.285 | 2.230 | 1.473 | .8474 | .2320 |
| 1.73 | .1936 | .3095 | .6256 | 1.412 | .4056 | 1.367 | 1.49889 | 18.689 | 35.31 | .6330 | 3.325 | 2.247 | 1.480 | .8431 | .2296 |
| 1.74 | .1907 | .3062 | .6229 | 1.424 | .4041 | 1.376 | 1.50429 | 18.981 | 35.08 | .6305 | 3.366 | 2.263 | 1.487 | .8389 | .2273 |
| 1.75 | .1878 | .3029 | .6202 | 1.436 | .4026 | 1.386 | 1.50966 | 19.273 | 34.85 | .6281 | 3.406 | 2.279 | 1.495 | .8346 | .2251 |
| 1.76 | .1850 | .2996 | .6175 | 1.448 | .4011 | 1.397 | 1.51499 | 19.565 | 34.62 | .6257 | 3.447 | 2.295 | 1.502 | .8302 | .2228 |
| 1.77 | .1822 | .2964 | .6148 | 1.460 | .3996 | 1.407 | 1.52029 | 19.855 | 34.40 | .6234 | 3.488 | 2.311 | 1.509 | .8259 | .2206 |
| 1.78 | .1794 | .2931 | .6121 | 1.473 | .3980 | 1.418 | 1.52555 | 20.146 | 34.18 | .6210 | 3.530 | 2.327 | 1.517 | .8215 | .2184 |
| 1.79 | .1767 | .2900 | .6095 | 1.485 | .3964 | 1.428 | 1.53078 | 20.436 | 33.96 | .6188 | 3.571 | 2.343 | 1.524 | .8171 | .2163 |
| 1.80 | .1740 | .2868 | .6068 | 1.497 | .3947 | 1.439 | 1.53598 | 20.725 | 33.75 | .6165 | 3.613 | 2.359 | 1.532 | .8127 | .2142 |
| 1.81 | .1714 | .2837 | .6041 | 1.509 | .3931 | 1.450 | 1.54114 | 21.014 | 33.54 | .6143 | 3.655 | 2.375 | 1.539 | .8082 | .2121 |
| 1.82 | .1688 | .2806 | .6015 | 1.521 | .3914 | 1.461 | 1.54626 | 21.302 | 33.33 | .6121 | 3.698 | 2.391 | 1.547 | .8038 | .2100 |
| 1.83 | .1662 | .2776 | .5989 | 1.533 | .3897 | 1.472 | 1.55136 | 21.590 | 33.12 | .6099 | 3.740 | 2.407 | 1.554 | .7993 | .2080 |
| 1.84 | .1637 | .2745 | .5963 | 1.545 | .3879 | 1.484 | 1.55642 | 21.877 | 32.92 | .6078 | 3.783 | 2.422 | 1.562 | .7948 | .2060 |
| 1.85 | .1612 | .2715 | .5936 | 1.556 | .3862 | 1.495 | 1.56145 | 22.163 | 32.72 | .6057 | 3.826 | 2.438 | 1.569 | .7902 | .2040 |
| 1.86 | .1587 | .2686 | .5910 | 1.568 | .3844 | 1.507 | 1.56644 | 22.449 | 32.52 | .6036 | 3.870 | 2.454 | 1.577 | .7857 | .2020 |
| 1.87 | .1563 | .2656 | .5884 | 1.580 | .3826 | 1.519 | 1.57140 | 22.735 | 32.33 | .6016 | 3.913 | 2.469 | 1.585 | .7811 | .2001 |
| 1.88 | .1539 | .2627 | .5859 | 1.592 | .3808 | 1.531 | 1.57633 | 23.019 | 32.13 | .5996 | 3.957 | 2.485 | 1.592 | .7765 | .1982 |
| 1.89 | .1516 | .2598 | .5833 | 1.604 | .3790 | 1.543 | 1.58123 | 23.303 | 31.94 | .5976 | 4.001 | 2.500 | 1.600 | .7720 | .1963 |
| 1.90 | .1492 | .2570 | .5807 | 1.616 | .3771 | 1.555 | 1.58609 | 23.586 | 31.76 | .5956 | 4.045 | 2.516 | 1.608 | .7674 | .1945 |
| 1.91 | .1470 | .2542 | .5782 | 1.627 | .3753 | 1.568 | 1.59092 | 23.869 | 31.57 | .5937 | 4.089 | 2.531 | 1.616 | .7627 | .1927 |
| 1.92 | .1447 | .2514 | .5756 | 1.639 | .3734 | 1.580 | 1.59572 | 24.151 | 31.39 | .5918 | 4.134 | 2.546 | 1.624 | .7581 | .1909 |
| 1.93 | .1425 | .2486 | .5731 | 1.651 | .3715 | 1.593 | 1.60049 | 24.432 | 31.21 | .5899 | 4.179 | 2.562 | 1.631 | .7535 | .1891 |
| 1.94 | .1403 | .2459 | .5705 | 1.662 | .3696 | 1.606 | 1.60523 | 24.712 | 31.03 | .5880 | 4.224 | 2.577 | 1.639 | .7488 | .1873 |
| 1.95 | .1381 | .2432 | .5680 | 1.674 | .3677 | 1.619 | 1.60993 | 24.992 | 30.85 | .5862 | 4.270 | 2.592 | 1.647 | .7442 | .1856 |
| 1.96 | .1360 | .2405 | .5655 | 1.686 | .3657 | 1.633 | 1.61460 | 25.271 | 30.68 | .5844 | 4.315 | 2.607 | 1.655 | .7395 | .1839 |
| 1.97 | .1339 | .2378 | .5630 | 1.697 | .3638 | 1.646 | 1.61925 | 25.549 | 30.51 | .5826 | 4.361 | 2.622 | 1.663 | .7349 | .1822 |
| 1.98 | .1318 | .2352 | .5605 | 1.709 | .3618 | 1.660 | 1.62386 | 25.827 | 30.33 | .5808 | 4.407 | 2.637 | 1.671 | .7302 | .1806 |
| 1.99 | .1298 | .2326 | .5580 | 1.720 | .3598 | 1.674 | 1.62844 | 26.104 | 30.17 | .5791 | 4.453 | 2.652 | 1.679 | .7255 | .1789 |
| 2.00 | .1278 | .2300 | .5556 | 1.732 | .3579 | 1.688 | 1.63299 | 26.380 | 30.00 | .5774 | 4.500 | 2.667 | 1.688 | .7209 | .1773 |
| 2.01 | .1258 | .2275 | .5531 | 1.744 | .3559 | 1.702 | 1.63751 | 26.655 | 29.84 | .5757 | 4.547 | 2.681 | 1.696 | .7162 | .1757 |
| 2.02 | .1239 | .2250 | .5506 | 1.755 | .3539 | 1.716 | 1.64201 | 26.929 | 29.67 | .5740 | 4.594 | 2.696 | 1.704 | .7115 | .1741 |
| 2.03 | .1220 | .2225 | .5482 | 1.767 | .3518 | 1.730 | 1.64647 | 27.203 | 29.51 | .5723 | 4.641 | 2.711 | 1.712 | .7069 | .1726 |
| 2.04 | .1201 | .2200 | .5458 | 1.778 | .3498 | 1.745 | 1.65090 | 27.476 | 29.35 | .5707 | 4.689 | 2.725 | 1.720 | .7022 | .1710 |
| 2.05 | .1182 | .2176 | .5433 | 1.790 | .3478 | 1.760 | 1.65530 | 27.748 | 29.20 | .5691 | 4.736 | 2.740 | 1.729 | .6975 | .1695 |
| 2.06 | .1164 | .2152 | .5409 | 1.801 | .3458 | 1.775 | 1.65967 | 28.020 | 29.04 | .5675 | 4.784 | 2.755 | 1.737 | .6928 | .1680 |
| 2.07 | .1146 | .2128 | .5385 | 1.812 | .3437 | 1.790 | 1.66402 | 28.290 | 28.89 | .5659 | 4.832 | 2.769 | 1.745 | .6882 | .1665 |
| 2.08 | .1128 | .2104 | .5361 | 1.824 | .3417 | 1.806 | 1.66833 | 28.560 | 28.74 | .5643 | 4.881 | 2.783 | 1.754 | .6835 | .1651 |
| 2.09 | .1111 | .2081 | .5337 | 1.835 | .3396 | 1.821 | 1.67262 | 28.829 | 28.59 | .5628 | 4.929 | 2.798 | 1.762 | .6789 | .1636 |
| 2.10 | .1094 | .2058 | .5313 | 1.847 | .3376 | 1.837 | 1.67687 | 29.097 | 28.44 | .5613 | 4.978 | 2.812 | 1.770 | .6742 | .1622 |
| 2.11 | .1077 | .2035 | .5290 | 1.858 | .3355 | 1.853 | 1.68110 | 29.364 | 28.29 | .5598 | 5.027 | 2.826 | 1.779 | .6696 | .1608 |
| 2.12 | .1060 | .2013 | .5266 | 1.869 | .3334 | 1.869 | 1.68530 | 29.631 | 28.14 | .5583 | 5.077 | 2.840 | 1.787 | .6649 | .1594 |
| 2.13 | .1043 | .1990 | .5243 | 1.881 | .3314 | 1.885 | 1.68947 | 29.897 | 28.00 | .5568 | 5.126 | 2.854 | 1.796 | .6603 | .1580 |
| 2.14 | .1027 | .1968 | .5219 | 1.892 | .3293 | 1.902 | 1.69362 | 30.161 | 27.86 | .5554 | 5.176 | 2.868 | 1.805 | .6557 | .1567 |

EQUATIONS, TABLES, AND CHARTS FOR COMPRESSIBLE FLOW

TABLE II.—SUPERSONIC FLOW—Continued

$\gamma = 7/5$

| $M$ or $M_1$ | $\frac{p}{p_t}$ | $\frac{\rho}{\rho_t}$ | $\frac{T}{T_t}$ | $\beta$ | $\frac{q}{p_t}$ | $\frac{A}{A_*}$ | $\frac{V}{a_*}$ | $\nu$ | $\mu$ | $M_2$ | $\frac{p_2}{p_1}$ | $\frac{\rho_2}{\rho_1}$ | $\frac{T_2}{T_1}$ | $\frac{p_{t_2}}{p_{t_1}}$ | $\frac{p_1}{p_{t_2}}$ |
|---|---|---|---|---|---|---|---|---|---|---|---|---|---|---|---|
| 2.15 | .1011 | .1946 | .5196 | 1.903 | .3272 | 1.919 | 1.69774 | 30.425 | 27.72 | .5540 | 5.226 | 2.882 | 1.813 | .6509 | .1553 |
| 2.16 | .9956 -1 | .1925 | .5173 | 1.915 | .3252 | 1.935 | 1.70183 | 30.689 | 27.58 | .5525 | 5.277 | 2.896 | 1.822 | .6464 | .1540 |
| 2.17 | .9802 -1 | .1903 | .5150 | 1.926 | .3231 | 1.953 | 1.70589 | 30.951 | 27.44 | .5511 | 5.327 | 2.910 | 1.831 | .6419 | .1527 |
| 2.18 | .9649 -1 | .1882 | .5127 | 1.937 | .3210 | 1.970 | 1.70992 | 31.212 | 27.30 | .5498 | 5.378 | 2.924 | 1.839 | .6373 | .1514 |
| 2.19 | .9500 -1 | .1861 | .5104 | 1.948 | .3189 | 1.987 | 1.71393 | 31.473 | 27.17 | .5484 | 5.429 | 2.938 | 1.848 | .6327 | .1502 |
| 2.20 | .9352 -1 | .1841 | .5081 | 1.960 | .3169 | 2.005 | 1.71791 | 31.732 | 27.04 | .5471 | 5.480 | 2.951 | 1.857 | .6281 | .1489 |
| 2.21 | .9207 -1 | .1820 | .5059 | 1.971 | .3148 | 2.023 | 1.72187 | 31.991 | 26.90 | .5457 | 5.531 | 2.965 | 1.866 | .6236 | .1476 |
| 2.22 | .9064 -1 | .1800 | .5036 | 1.982 | .3127 | 2.041 | 1.72579 | 32.250 | 26.77 | .5444 | 5.583 | 2.978 | 1.875 | .6191 | .1464 |
| 2.23 | .8923 -1 | .1780 | .5014 | 1.993 | .3106 | 2.059 | 1.72970 | 32.507 | 26.64 | .5431 | 5.636 | 2.992 | 1.883 | .6145 | .1452 |
| 2.24 | .8785 -1 | .1760 | .4991 | 2.004 | .3085 | 2.078 | 1.73357 | 32.763 | 26.51 | .5418 | 5.687 | 3.005 | 1.892 | .6100 | .1440 |
| 2.25 | .8648 -1 | .1740 | .4969 | 2.016 | .3065 | 2.096 | 1.73742 | 33.018 | 26.39 | .5406 | 5.740 | 3.019 | 1.901 | .6055 | .1428 |
| 2.26 | .8514 -1 | .1721 | .4947 | 2.027 | .3044 | 2.115 | 1.74125 | 33.273 | 26.26 | .5393 | 5.792 | 3.032 | 1.910 | .6011 | .1417 |
| 2.27 | .8382 -1 | .1702 | .4925 | 2.038 | .3023 | 2.134 | 1.74504 | 33.527 | 26.14 | .5381 | 5.845 | 3.045 | 1.919 | .5966 | .1405 |
| 2.28 | .8251 -1 | .1683 | .4903 | 2.049 | .3003 | 2.154 | 1.74882 | 33.780 | 26.01 | .5368 | 5.898 | 3.058 | 1.929 | .5921 | .1394 |
| 2.29 | .8123 -1 | .1664 | .4881 | 2.060 | .2982 | 2.173 | 1.75257 | 34.032 | 25.89 | .5356 | 5.951 | 3.071 | 1.938 | .5877 | .1382 |
| 2.30 | .7997 -1 | .1646 | .4859 | 2.071 | .2961 | 2.193 | 1.75629 | 34.283 | 25.77 | .5344 | 6.005 | 3.085 | 1.947 | .5833 | .1371 |
| 2.31 | .7873 -1 | .1628 | .4837 | 2.082 | .2941 | 2.213 | 1.75999 | 34.533 | 25.65 | .5332 | 6.059 | 3.098 | 1.956 | .5789 | .1360 |
| 2.32 | .7751 -1 | .1609 | .4816 | 2.093 | .2920 | 2.233 | 1.76366 | 34.783 | 25.53 | .5321 | 6.113 | 3.110 | 1.965 | .5745 | .1349 |
| 2.33 | .7631 -1 | .1592 | .4794 | 2.104 | .2900 | 2.254 | 1.76731 | 35.031 | 25.42 | .5309 | 6.167 | 3.123 | 1.974 | .5702 | .1338 |
| 2.34 | .7512 -1 | .1574 | .4773 | 2.116 | .2879 | 2.274 | 1.77093 | 35.279 | 25.30 | .5297 | 6.222 | 3.136 | 1.984 | .5658 | .1328 |
| 2.35 | .7396 -1 | .1556 | .4752 | 2.127 | .2859 | 2.295 | 1.77453 | 35.526 | 25.18 | .5286 | 6.276 | 3.149 | 1.993 | .5615 | .1317 |
| 2.36 | .7281 -1 | .1539 | .4731 | 2.138 | .2839 | 2.316 | 1.77811 | 35.771 | 25.07 | .5275 | 6.331 | 3.162 | 2.002 | .5572 | .1307 |
| 2.37 | .7168 -1 | .1522 | .4709 | 2.149 | .2818 | 2.338 | 1.78166 | 36.017 | 24.96 | .5264 | 6.386 | 3.174 | 2.012 | .5529 | .1297 |
| 2.38 | .7057 -1 | .1505 | .4688 | 2.160 | .2798 | 2.359 | 1.78519 | 36.261 | 24.85 | .5253 | 6.442 | 3.187 | 2.021 | .5486 | .1286 |
| 2.39 | .6948 -1 | .1488 | .4668 | 2.171 | .2778 | 2.381 | 1.78869 | 36.504 | 24.73 | .5242 | 6.497 | 3.199 | 2.031 | .5444 | .1276 |
| 2.40 | .6840 -1 | .1472 | .4647 | 2.182 | .2758 | 2.403 | 1.79218 | 36.746 | 24.62 | .5231 | 6.553 | 3.212 | 2.040 | .5401 | .1266 |
| 2.41 | .6734 -1 | .1456 | .4626 | 2.193 | .2738 | 2.425 | 1.79563 | 36.988 | 24.52 | .5221 | 6.609 | 3.224 | 2.050 | .5359 | .1257 |
| 2.42 | .6630 -1 | .1439 | .4606 | 2.204 | .2718 | 2.448 | 1.79907 | 37.229 | 24.41 | .5210 | 6.666 | 3.237 | 2.059 | .5317 | .1247 |
| 2.43 | .6527 -1 | .1424 | .4585 | 2.215 | .2698 | 2.471 | 1.80248 | 37.469 | 24.30 | .5200 | 6.722 | 3.249 | 2.069 | .5276 | .1237 |
| 2.44 | .6426 -1 | .1408 | .4565 | 2.226 | .2678 | 2.494 | 1.80587 | 37.708 | 24.19 | .5189 | 6.779 | 3.261 | 2.079 | .5234 | .1228 |
| 2.45 | .6327 -1 | .1392 | .4544 | 2.237 | .2658 | 2.517 | 1.80924 | 37.946 | 24.09 | .5179 | 6.836 | 3.273 | 2.088 | .5193 | .1218 |
| 2.46 | .6229 -1 | .1377 | .4524 | 2.248 | .2639 | 2.540 | 1.81258 | 38.183 | 23.99 | .5169 | 6.894 | 3.285 | 2.098 | .5152 | .1209 |
| 2.47 | .6133 -1 | .1362 | .4504 | 2.259 | .2619 | 2.564 | 1.81591 | 38.420 | 23.88 | .5159 | 6.951 | 3.298 | 2.108 | .5111 | .1200 |
| 2.48 | .6038 -1 | .1346 | .4484 | 2.269 | .2599 | 2.588 | 1.81921 | 38.655 | 23.78 | .5149 | 7.009 | 3.310 | 2.118 | .5071 | .1191 |
| 2.49 | .5945 -1 | .1332 | .4464 | 2.280 | .2580 | 2.612 | 1.82249 | 38.890 | 23.68 | .5140 | 7.067 | 3.321 | 2.128 | .5030 | .1182 |
| 2.50 | .5853 -1 | .1317 | .4444 | 2.291 | .2561 | 2.637 | 1.82574 | 39.124 | 23.58 | .5130 | 7.125 | 3.333 | 2.138 | .4990 | .1173 |
| 2.51 | .5762 -1 | .1302 | .4425 | 2.302 | .2541 | 2.661 | 1.82898 | 39.357 | 23.48 | .5120 | 7.183 | 3.345 | 2.147 | .4950 | .1164 |
| 2.52 | .5674 -1 | .1288 | .4405 | 2.313 | .2522 | 2.686 | 1.83219 | 39.589 | 23.38 | .5111 | 7.242 | 3.357 | 2.157 | .4911 | .1155 |
| 2.53 | .5586 -1 | .1274 | .4386 | 2.324 | .2503 | 2.712 | 1.83538 | 39.820 | 23.28 | .5102 | 7.301 | 3.369 | 2.167 | .4871 | .1147 |
| 2.54 | .5500 -1 | .1260 | .4366 | 2.335 | .2484 | 2.737 | 1.83855 | 40.050 | 23.18 | .5092 | 7.360 | 3.380 | 2.177 | .4832 | .1138 |
| 2.55 | .5415 -1 | .1246 | .4347 | 2.346 | .2465 | 2.763 | 1.84170 | 40.280 | 23.09 | .5083 | 7.420 | 3.392 | 2.187 | .4793 | .1130 |
| 2.56 | .5332 -1 | .1232 | .4328 | 2.357 | .2446 | 2.789 | 1.84483 | 40.509 | 22.99 | .5074 | 7.479 | 3.403 | 2.198 | .4754 | .1122 |
| 2.57 | .5250 -1 | .1218 | .4309 | 2.367 | .2427 | 2.815 | 1.84794 | 40.736 | 22.91 | .5065 | 7.539 | 3.415 | 2.208 | .4715 | .1113 |
| 2.58 | .5169 -1 | .1205 | .4289 | 2.378 | .2409 | 2.842 | 1.85103 | 40.963 | 22.81 | .5056 | 7.599 | 3.426 | 2.218 | .4677 | .1105 |
| 2.59 | .5090 -1 | .1192 | .4271 | 2.389 | .2390 | 2.869 | 1.85410 | 41.189 | 22.71 | .5047 | 7.659 | 3.438 | 2.228 | .4639 | .1097 |
| 2.60 | .5012 -1 | .1179 | .4252 | 2.400 | .2371 | 2.896 | 1.85714 | 41.415 | 22.62 | .5039 | 7.720 | 3.449 | 2.238 | .4601 | .1089 |
| 2.61 | .4935 -1 | .1166 | .4233 | 2.411 | .2353 | 2.923 | 1.86017 | 41.639 | 22.53 | .5030 | 7.781 | 3.460 | 2.249 | .4564 | .1081 |
| 2.62 | .4859 -1 | .1153 | .4214 | 2.422 | .2335 | 2.951 | 1.86318 | 41.863 | 22.44 | .5022 | 7.842 | 3.471 | 2.259 | .4526 | .1074 |
| 2.63 | .4784 -1 | .1140 | .4196 | 2.432 | .2317 | 2.979 | 1.86616 | 42.086 | 22.35 | .5013 | 7.903 | 3.483 | 2.269 | .4489 | .1066 |
| 2.64 | .4711 -1 | .1128 | .4177 | 2.443 | .2298 | 3.007 | 1.86913 | 42.307 | 22.26 | .5005 | 7.965 | 3.494 | 2.280 | .4452 | .1058 |
| 2.65 | .4639 -1 | .1115 | .4159 | 2.454 | .2280 | 3.036 | 1.87208 | 42.529 | 22.17 | .4996 | 8.026 | 3.505 | 2.290 | .4416 | .1051 |
| 2.66 | .4568 -1 | .1103 | .4141 | 2.465 | .2262 | 3.065 | 1.87501 | 42.749 | 22.08 | .4988 | 8.088 | 3.516 | 2.301 | .4379 | .1043 |
| 2.67 | .4498 -1 | .1091 | .4122 | 2.476 | .2245 | 3.094 | 1.87792 | 42.968 | 22.00 | .4980 | 8.150 | 3.527 | 2.311 | .4343 | .1036 |
| 2.68 | .4429 -1 | .1079 | .4104 | 2.486 | .2227 | 3.123 | 1.88081 | 43.187 | 21.91 | .4972 | 8.213 | 3.537 | 2.322 | .4307 | .1028 |
| 2.69 | .4362 -1 | .1067 | .4086 | 2.497 | .2209 | 3.153 | 1.88368 | 43.405 | 21.82 | .4964 | 8.275 | 3.548 | 2.332 | .4271 | .1021 |
| 2.70 | .4295 -1 | .1056 | .4068 | 2.508 | .2192 | 3.183 | 1.88653 | 43.621 | 21.74 | .4956 | 8.338 | 3.559 | 2.343 | .4236 | .1014 |
| 2.71 | .4229 -1 | .1044 | .4051 | 2.519 | .2174 | 3.213 | 1.88936 | 43.838 | 21.65 | .4949 | 8.401 | 3.570 | 2.354 | .4201 | .1007 |
| 2.72 | .4165 -1 | .1033 | .4033 | 2.530 | .2157 | 3.244 | 1.89218 | 44.053 | 21.57 | .4941 | 8.465 | 3.580 | 2.364 | .4166 | .9998 -1 |
| 2.73 | .4102 -1 | .1022 | .4015 | 2.540 | .2140 | 3.275 | 1.89497 | 44.267 | 21.49 | .4933 | 8.528 | 3.591 | 2.375 | .4131 | .9929 -1 |
| 2.74 | .4039 -1 | .1010 | .3998 | 2.551 | .2123 | 3.306 | 1.89775 | 44.481 | 21.41 | .4926 | 8.592 | 3.601 | 2.386 | .4097 | .9860 -1 |
| 2.75 | .3978 -1 | .9994 -1 | .3980 | 2.562 | .2106 | 3.338 | 1.90051 | 44.694 | 21.32 | .4918 | 8.656 | 3.612 | 2.397 | .4062 | .9792 -1 |
| 2.76 | .3917 -1 | .9885 -1 | .3963 | 2.572 | .2089 | 3.370 | 1.90325 | 44.906 | 21.24 | .4911 | 8.721 | 3.622 | 2.407 | .4028 | .9724 -1 |
| 2.77 | .3858 -1 | .9778 -1 | .3945 | 2.583 | .2072 | 3.402 | 1.90598 | 45.117 | 21.16 | .4903 | 8.785 | 3.633 | 2.418 | .3994 | .9658 -1 |
| 2.78 | .3799 -1 | .9671 -1 | .3928 | 2.594 | .2055 | 3.434 | 1.90868 | 45.327 | 21.08 | .4896 | 8.850 | 3.643 | 2.429 | .3961 | .9591 -1 |
| 2.79 | .3742 -1 | .9566 -1 | .3911 | 2.605 | .2039 | 3.467 | 1.91137 | 45.537 | 21.00 | .4889 | 8.915 | 3.653 | 2.440 | .3928 | .9526 -1 |
| 2.80 | .3685 -1 | .9463 -1 | .3894 | 2.615 | .2022 | 3.500 | 1.91404 | 45.746 | 20.92 | .4882 | 8.980 | 3.664 | 2.451 | .3895 | .9461 -1 |
| 2.81 | .3629 -1 | .9360 -1 | .3877 | 2.626 | .2006 | 3.534 | 1.91669 | 45.954 | 20.85 | .4875 | 9.045 | 3.674 | 2.462 | .3862 | .9397 -1 |
| 2.82 | .3574 -1 | .9259 -1 | .3860 | 2.637 | .1990 | 3.567 | 1.91933 | 46.161 | 20.77 | .4868 | 9.111 | 3.684 | 2.473 | .3829 | .9334 -1 |
| 2.83 | .3520 -1 | .9158 -1 | .3844 | 2.647 | .1973 | 3.601 | 1.92195 | 46.368 | 20.69 | .4861 | 9.177 | 3.694 | 2.484 | .3797 | .9271 -1 |
| 2.84 | .3467 -1 | .9059 -1 | .3827 | 2.658 | .1957 | 3.636 | 1.92455 | 46.573 | 20.62 | .4854 | 9.243 | 3.704 | 2.496 | .3765 | .9209 -1 |
| 2.85 | .3415 -1 | .8962 -1 | .3810 | 2.669 | .1941 | 3.671 | 1.92714 | 46.778 | 20.54 | .4847 | 9.310 | 3.714 | 2.507 | .3733 | .9147 -1 |
| 2.86 | .3363 -1 | .8865 -1 | .3794 | 2.679 | .1926 | 3.706 | 1.92970 | 46.982 | 20.47 | .4840 | 9.376 | 3.724 | 2.518 | .3701 | .9086 -1 |
| 2.87 | .3312 -1 | .8769 -1 | .3777 | 2.690 | .1910 | 3.741 | 1.93225 | 47.185 | 20.39 | .4833 | 9.443 | 3.734 | 2.529 | .3670 | .9026 -1 |
| 2.88 | .3263 -1 | .8675 -1 | .3761 | 2.701 | .1894 | 3.777 | 1.93479 | 47.388 | 20.32 | .4827 | 9.510 | 3.743 | 2.540 | .3639 | .8966 -1 |
| 2.89 | .3213 -1 | .8581 -1 | .3745 | 2.711 | .1879 | 3.813 | 1.93731 | 47.589 | 20.24 | .4820 | 9.577 | 3.753 | 2.552 | .3608 | .8906 -1 |
| 2.90 | .3165 -1 | .8489 -1 | .3729 | 2.722 | .1863 | 3.850 | 1.93981 | 47.790 | 20.17 | .4814 | 9.645 | 3.763 | 2.563 | .3577 | .8848 -1 |
| 2.91 | .3118 -1 | .8398 -1 | .3712 | 2.733 | .1848 | 3.887 | 1.94230 | 47.990 | 20.10 | .4807 | 9.713 | 3.773 | 2.575 | .3547 | .8790 -1 |
| 2.92 | .3071 -1 | .8307 -1 | .3696 | 2.743 | .1833 | 3.924 | 1.94477 | 48.190 | 20.03 | .4801 | 9.781 | 3.782 | 2.586 | .3517 | .8732 -1 |
| 2.93 | .3025 -1 | .8218 -1 | .3681 | 2.754 | .1818 | 3.961 | 1.94722 | 48.388 | 19.96 | .4795 | 9.849 | 3.792 | 2.598 | .3487 | .8675 -1 |
| 2.94 | .2980 -1 | .8130 -1 | .3665 | 2.765 | .1803 | 3.999 | 1.94966 | 48.586 | 19.89 | .4788 | 9.918 | 3.801 | 2.609 | .3457 | .8619 -1 |
| 2.95 | .2935 -1 | .8043 -1 | .3649 | 2.775 | .1788 | 4.038 | 1.95208 | 48.783 | 19.81 | .4782 | 9.986 | 3.811 | 2.621 | .3428 | .8563 -1 |
| 2.96 | .2891 -1 | .7957 -1 | .3633 | 2.786 | .1773 | 4.076 | 1.95449 | 48.980 | 19.75 | .4776 | 10.06 | 3.820 | 2.632 | .3398 | .8507 -1 |
| 2.97 | .2848 -1 | .7872 -1 | .3618 | 2.797 | .1758 | 4.115 | 1.95688 | 49.175 | 19.68 | .4770 | 10.12 | 3.829 | 2.644 | .3369 | .8453 -1 |
| 2.98 | .2805 -1 | .7788 -1 | .3602 | 2.807 | .1744 | 4.155 | 1.95925 | 49.370 | 19.61 | .4764 | 10.19 | 3.839 | 2.656 | .3340 | .8398 -1 |
| 2.99 | .2764 -1 | .7706 -1 | .3587 | 2.818 | .1729 | 4.194 | 1.96162 | 49.564 | 19.54 | .4758 | 10.26 | 3.848 | 2.667 | .3312 | .8345 -1 |
| 3.00 | .2722 -1 | .7623 -1 | .3571 | 2.828 | .1715 | 4.235 | 1.96396 | 49.757 | 19.47 | .4752 | 10.33 | 3.857 | 2.679 | .3283 | .8291 -1 |
| 3.01 | .2682 -1 | .7541 -1 | .3556 | 2.839 | .1701 | 4.275 | 1.96629 | 49.950 | 19.40 | .4746 | 10.40 | 3.866 | 2.691 | .3255 | .8238 -1 |
| 3.02 | .2642 -1 | .7461 -1 | .3541 | 2.850 | .1687 | 4.316 | 1.96861 | 50.142 | 19.34 | .4740 | 10.47 | 3.875 | 2.703 | .3227 | .8186 -1 |
| 3.03 | .2603 -1 | .7382 -1 | .3526 | 2.860 | .1673 | 4.357 | 1.97091 | 50.333 | 19.27 | .4734 | 10.54 | 3.884 | 2.714 | .3200 | .8134 -1 |
| 3.04 | .2564 -1 | .7303 -1 | .3511 | 2.871 | .1659 | 4.399 | 1.97319 | 50.523 | 19.20 | .4729 | 10.62 | 3.893 | 2.726 | .3172 | .8083 -1 |

## A.5 NOMENCLATURE

| | |
|---|---|
| $A$ | cross-sectional area |
| $a$ | sound speed |
| $C_p$ | pressure coefficient |
| $c_p$ | specific heat at constant pressure |
| $g$ | acceleration of gravity |
| $h$ | enthalpy |
| $L$ | length |
| $M$ | Mach number |
| $\dot{m}$ | mass flux |
| $p$ | pressure |
| $q$ | dynamic pressure |
| $R$ | gas constant |
| $s$ | entropy |
| $T$ | temperature |
| $u$ | velocity in one-dimensional flow |
| $V$ | velocity in two-dimensional flow |
| $W$ | molecular weight |
| B | Prandtl–Meyer function |
| $\delta$ | flow deflection angle |
| $\gamma$ | ratio of specific heats |
| $\varepsilon$ | density ratio |
| $\rho$ | gas density |
| $\mu$ | Mach angle, $\sin^{-1}(1/M)$ or gas viscosity |
| $\nu$ | Prandtl–Meyer angle |
| $\theta$ | shock angle |

### A.5.1 Subscripts

| | |
|---|---|
| $a$ | ambient |
| max | maximum conditions |
| $n$ | normal to shock |
| $t$ | stagnation conditions or tangent to shock |
| $x$ | axial direction |

1 upstream of shock
2 downstream of shock

### A.5.2 Superscript

( )* critical conditions, $M = 1$

APPENDIX

# B Properties of Hydrocarbon Fuel Combustion

## B.1 TABLES AND CHARTS OF SOME THERMODYNAMIC PROPERTIES

Hippensteele and Colladay (1978) used chemical equilibrium computations performed using the code presented by Svehla and McBride (1973) to prepare an interpolation code to provide the thermodynamic properties of the combustion of ASTM-A-1 fuel with air at various values of fuel to air ratio (f/a) and pressure. The chemical formula for air is assumed to be $C_{0.00030}N_{1.56176}O_{0.41959}Ar_{0.00932}$ and for ASTM-A-1, $CH_{1.9185}$. Some data from Hippensteele and Colladay (1978) are collected here for the reasonably representative conditions of f/a = 0 (pure air heating), 0.02 (primary plus secondary air), and 0.06817(stoichimetric). The pressures chosen are $p$ = 1, 3, and 20 atm. The thermodynamic properties are presented as functions of the temperature over a range of 500K $< T <$ 2800K in Tables B.1 to B.3 and as plots in Figures B.1 to B.3.

**Table B.1** Thermodynamic Properties of Products of Combustion of ASTM-A-1 Fuel for Fuel to Air Ratio (f/a) = 0.06817 (Stoichiometric) at Different Pressures ($c_p$ is in kJ/kg-K)

| $T$(K) | $c_p$ 20 atm | $c_p$ 3 atm | $c_p$ 1 atm | $W$ 20 atm | $W$ 3 atm | $W$ 1 atm | $\gamma$ 20 atm | $\gamma$ 3 atm | $\gamma$ 1 atm |
|---|---|---|---|---|---|---|---|---|---|
| 500 | 1.113 | 1.113 | 1.113 | 28.97 | 28.97 | 28.97 | 1.348 | 1.348 | 1.348 |
| 600 | 1.143 | 1.143 | 1.143 | 28.97 | 28.97 | 28.97 | 1.335 | 1.335 | 1.335 |
| 800 | 1.206 | 1.206 | 1.206 | 28.97 | 28.97 | 28.97 | 1.313 | 1.313 | 1.313 |
| 1000 | 1.262 | 1.262 | 1.262 | 28.97 | 28.97 | 28.97 | 1.294 | 1.294 | 1.294 |
| 1200 | 1.307 | 1.307 | 1.307 | 28.97 | 28.97 | 28.97 | 1.282 | 1.282 | 1.282 |
| 1400 | 1.345 | 1.347 | 1.348 | 28.96 | 28.97 | 28.97 | 1.272 | 1.271 | 1.271 |
| 1600 | 1.384 | 1.393 | 1.401 | 28.96 | 28.97 | 28.97 | 1.262 | 1.261 | 1.259 |
| 1800 | 1.438 | 1.471 | 1.502 | 28.96 | 28.96 | 28.95 | 1.252 | 1.246 | 1.242 |
| 2000 | 1.529 | 1.621 | 1.706 | 28.94 | 28.92 | 28.9 | 1.237 | 1.226 | 1.216 |
| 2200 | 1.685 | 1.887 | 2.074 | 28.89 | 28.83 | 28.77 | 1.219 | 1.2 | 1.187 |
| 2400 | 1.925 | 2.297 | 2.634 | 28.79 | 28.64 | 28.51 | 1.198 | 1.176 | 1.162 |
| 2600 | 2.250 | 2.835 | 3.354 | 28.61 | 28.33 | 28.07 | 1.18 | 1.159 | 1.147 |
| 2800 | 2.635 | 3.444 | 4.160 | 28.34 | 27.86 | 27.43 | 1.168 | 1.149 | 1.139 |

**Table B.2** Thermodynamic Properties of Products of Combustion of ASTM-A-1 Fuel for Fuel to Air Ratio (f/a) = 0.02 at Different Pressures ($c_p$ is in kJ/kg-K)

| *T*(K) | $c_p$ 20 atm | $c_p$ 3 atm | $c_p$ 1 atm | *W* 20 atm | *W* 3 atm | *W* 1 atm | γ 20 atm | γ 3 atm | γ 1 atm |
|---|---|---|---|---|---|---|---|---|---|
| 500 | 1.036 | 1.058 | 1.058 | 28.97 | 28.97 | 28.97 | 1.373 | 1.373 | 1.373 |
| 600 | 1.080 | 1.080 | 1.080 | 28.97 | 28.97 | 28.97 | 1.362 | 1.362 | 1.362 |
| 800 | 1.132 | 1.132 | 1.132 | 28.97 | 28.97 | 28.97 | 1.34 | 1.34 | 1.34 |
| 1000 | 1.180 | 1.180 | 1.180 | 28.97 | 28.97 | 28.97 | 1.322 | 1.322 | 1.322 |
| 1200 | 1.218 | 1.218 | 1.218 | 28.97 | 28.97 | 28.97 | 1.308 | 1.308 | 1.308 |
| 1400 | 1.255 | 1.255 | 1.256 | 28.97 | 28.97 | 28.97 | 1.297 | 1.297 | 1.297 |
| 1600 | 1.292 | 1.294 | 1.295 | 28.97 | 28.97 | 28.97 | 1.286 | 1.286 | 1.285 |
| 1800 | 1.333 | 1.339 | 1.344 | 28.97 | 28.96 | 28.96 | 1.275 | 1.274 | 1.273 |
| 2000 | 1.380 | 1.400 | 1.421 | 28.96 | 28.96 | 28.95 | 1.264 | 1.261 | 1.257 |
| 2200 | 1.443 | 1.499 | 1.560 | 28.95 | 28.94 | 28.92 | 1.252 | 1.244 | 1.236 |
| 2400 | 1.535 | 1.672 | 1.823 | 28.93 | 28.88 | 28.84 | 1.238 | 1.223 | 1.209 |
| 2600 | 1.676 | 1.961 | 2.269 | 28.88 | 28.77 | 28.65 | 1.222 | 1.2 | 1.183 |
| 2800 | 1.883 | 2.383 | 2.891 | 28.78 | 28.55 | 28.31 | 1.206 | 1.18 | 1.164 |

**Table B.3** Thermodynamic Properties of Products of Combustion of ASTM-A-1 Fuel for Fuel to Air Ratio (f/a) = 0 at Different Pressures ($c_p$ is in kJ/kg-K)

| *T*(K) | $c_p$ 20 atm | $c_p$ 3 atm | $c_p$ 1 atm | *W* 20 atm | *W* 3 atm | *W* 1 atm | γ 20 atm | γ 3 atm | γ 1 atm |
|---|---|---|---|---|---|---|---|---|---|
| 500 | 1.030 | 1.030 | 1.030 | 28.96 | 28.96 | 28.96 | 1.387 | 1.387 | 1.387 |
| 600 | 1.049 | 1.068 | 1.053 | 28.96 | 28.96 | 28.96 | 1.376 | 1.376 | 1.376 |
| 800 | 1.087 | 1.079 | 1.099 | 28.96 | 28.96 | 28.96 | 1.354 | 1.354 | 1.354 |
| 1000 | 1.143 | 1.143 | 1.143 | 28.96 | 28.96 | 28.96 | 1.336 | 1.336 | 1.336 |
| 1200 | 1.178 | 1.178 | 1.178 | 28.96 | 28.96 | 28.96 | 1.322 | 1.322 | 1.322 |
| 1400 | 1.213 | 1.213 | 1.213 | 28.96 | 28.96 | 28.96 | 1.31 | 1.31 | 1.31 |
| 1600 | 1.248 | 1.249 | 1.249 | 28.96 | 28.96 | 28.96 | 1.299 | 1.299 | 1.299 |
| 1800 | 1.285 | 1.287 | 1.289 | 28.96 | 28.96 | 28.96 | 1.288 | 1.288 | 1.287 |
| 2000 | 1.324 | 1.340 | 1.340 | 28.96 | 28.96 | 28.96 | 1.277 | 1.276 | 1.274 |
| 2200 | 1.369 | 1.393 | 1.421 | 28.96 | 28.95 | 28.95 | 1.267 | 1.263 | 1.258 |
| 2400 | 1.425 | 1.488 | 1.563 | 28.95 | 28.93 | 28.91 | 1.256 | 1.247 | 1.238 |
| 2600 | 1.502 | 1.643 | 1.808 | 28.93 | 28.88 | 28.82 | 1.244 | 1.229 | 1.214 |
| 2800 | 1.612 | 1.883 | 2.193 | 28.89 | 28.77 | 28.63 | 1.231 | 1.209 | 1.192 |

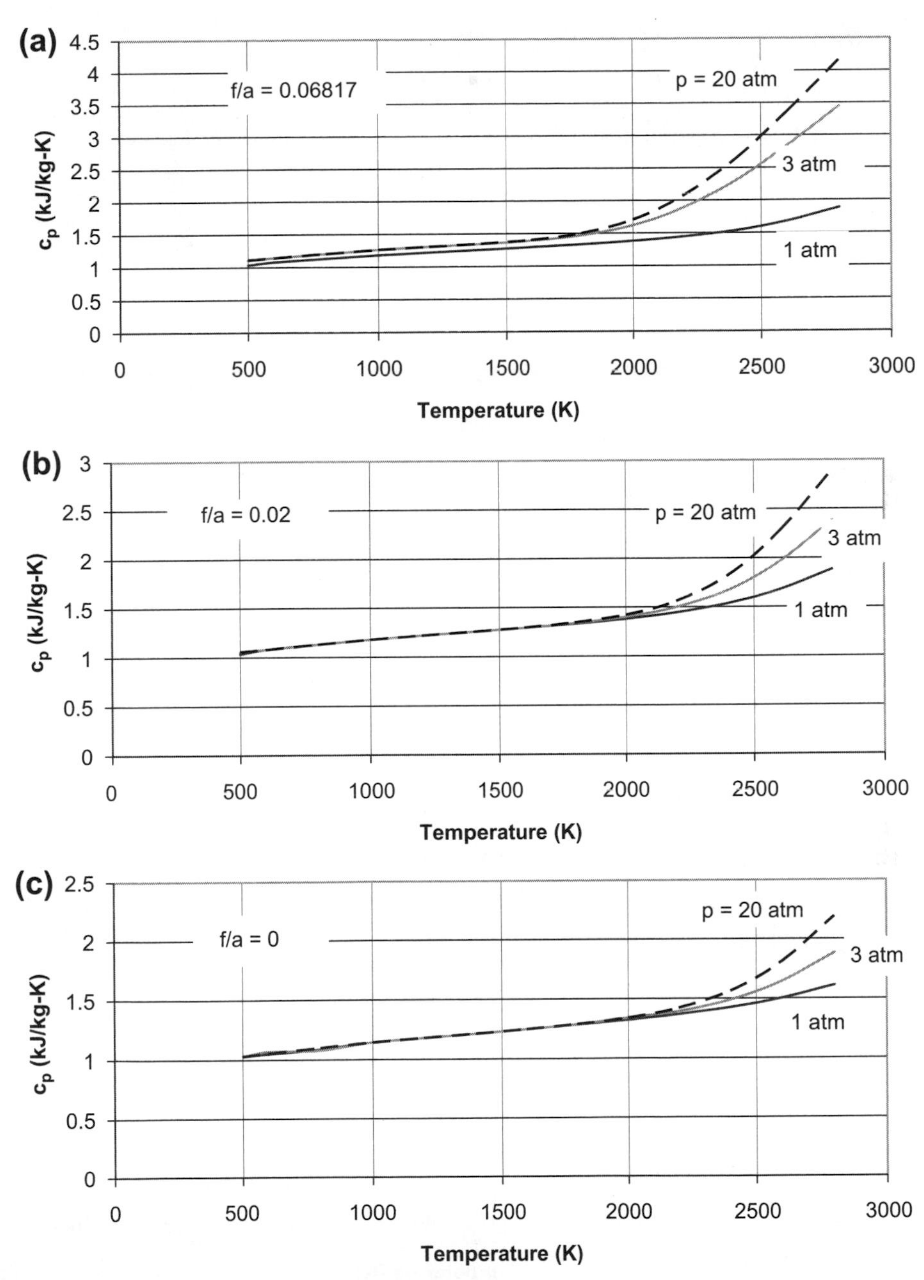


**FIGURE B.1**

Specific heat of the products of combustion of ASTM-1 fuel for fuel to air ratios (a) f/a = 0.06817 (stoichiometric), (b) f/a = 0.02, and (c) f/a = 0 at different pressures.

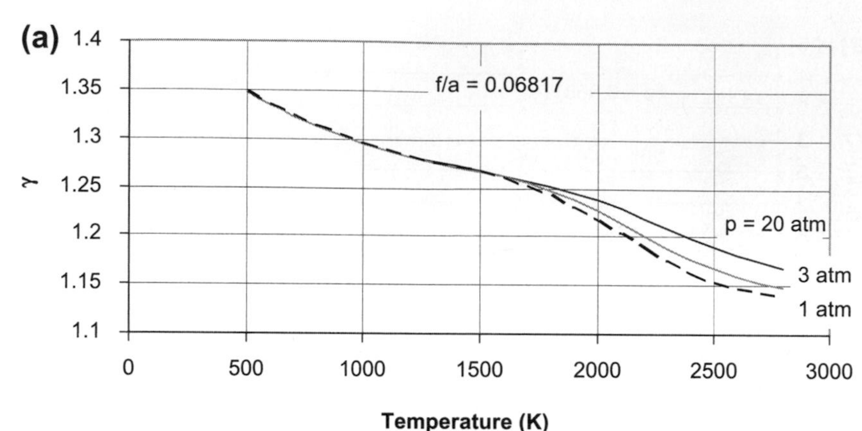


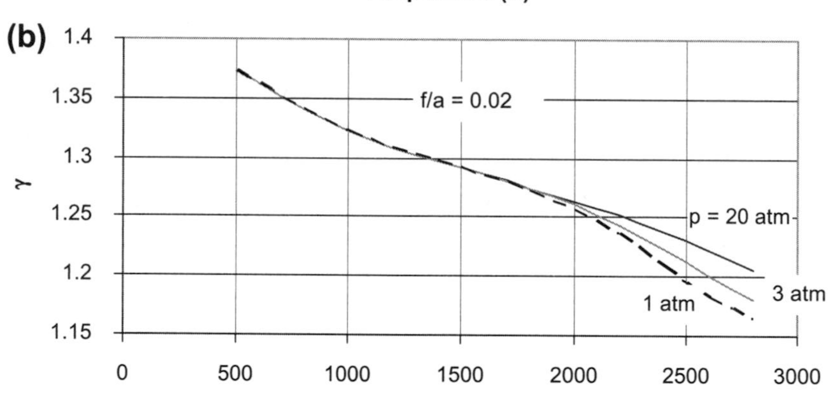


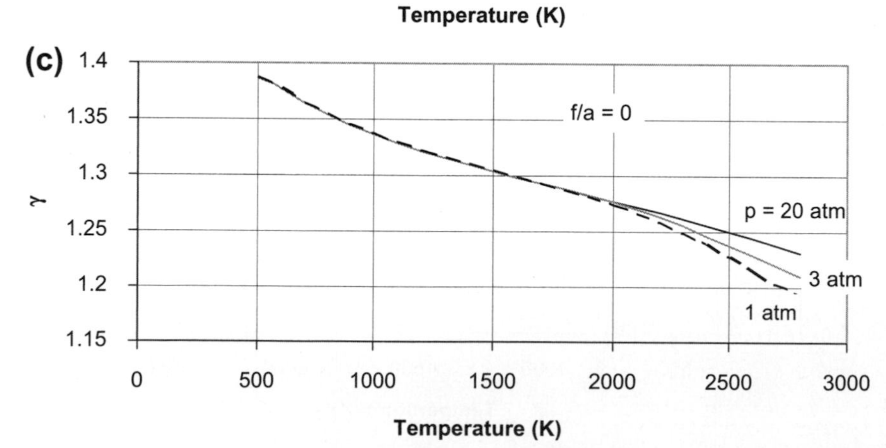


**FIGURE B.2**

Isentropic exponent of the products of combustion of ASTM-1 fuel for fuel to air ratios (a) f/a = 0.06817 (stoichiometric), (b) f/a = 0.02, and (c) f/a = 0 at different pressures.

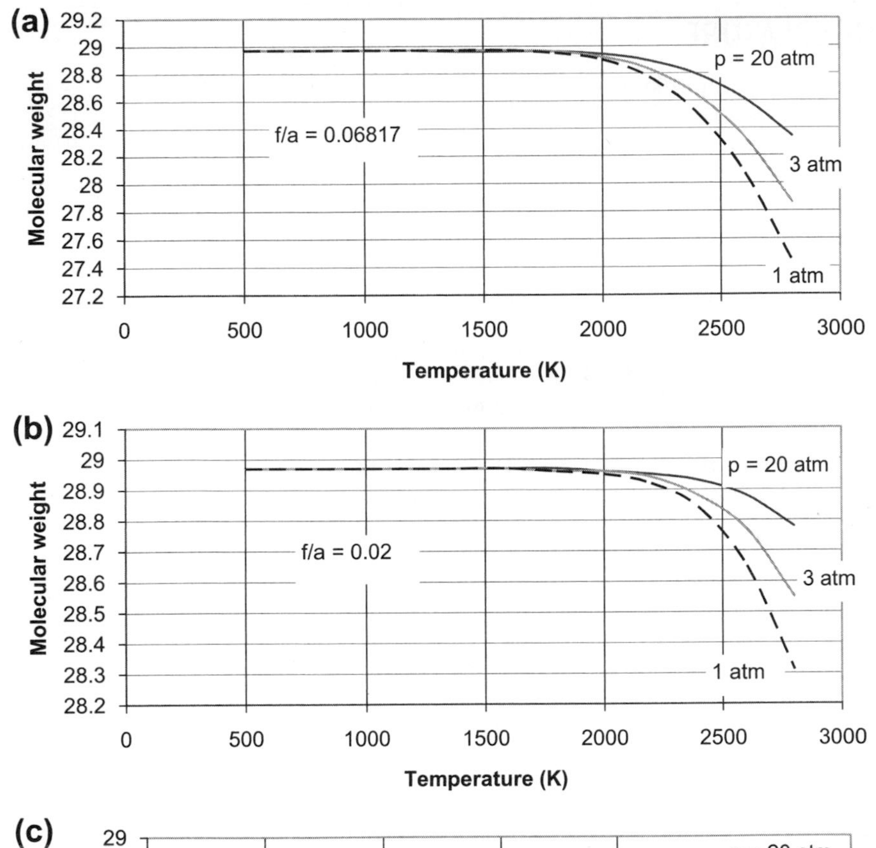


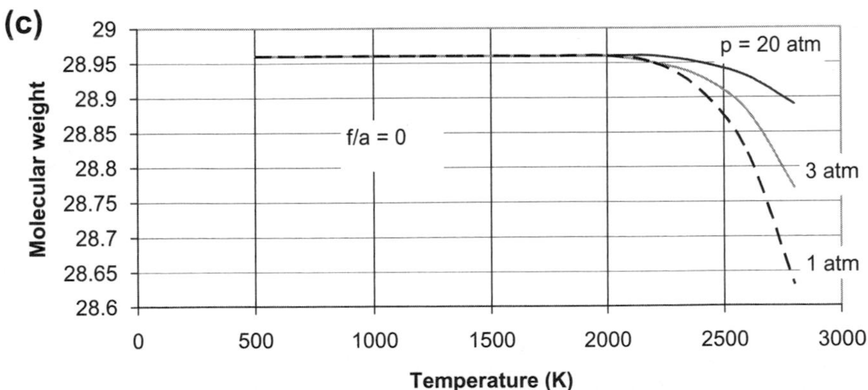


**FIGURE B.3**

Molecular weight of the products of combustion of ASTM-1 fuel for fuel to air ratios (a) f/a = 0.06817 (stoichiometric), (b) f/a = 0.02, and (c) f/a = 0 at different pressures.

## B.2 NOMENCLATURE

| | |
|---|---|
| $c_p$ | specific heat, kJ/kg-K |
| f/a | fuel to air ratio |
| $W$ | molecular weight |
| $\gamma$ | ratio of specific heats |

## References

Hippensteele, S. A., & Colladay, R. S. (1978). Computer Program for Obtaining Thermodynamic and Transport Properties of Air and Products of Combustion of ASTM-A-1 Fuel and Air. NASA Technical Paper 1160, March.

Svehla, R. A., & McBride, B. J. (1973). Fortran IV Computer Program for Calculation of Thermodynamic and Transport Properties of Complex Chemical Systems. NASA TN D-7056.

APPENDIX

# Earth's Atmosphere

# C

## C.1 ATMOSPHERIC ENVIRONMENT

A useful representation of the Earth's atmosphere appears in Figure C.1, taken from NASA. The composition of the sensible atmosphere is essentially constant and by mole fraction it is composed of 78% nitrogen, 21% oxygen, and 1% other gases (argon 0.93%, $CO_2$ 0.03%, and neon, helium, krypton, hydrogen, xenon, and ozone in increasingly smaller amounts). For most thermochemical purposes, the atmosphere is considered a binary mixture of 79% nitrogen and 21% oxygen. This fixed composition approximation makes the temperature distribution a reliable means for dividing up the various important regions in the atmosphere. The pressure and density in the regions of specified temperature behavior may be determined from the equation of state and the conditions of hydrostatic equilibrium. Note that the nominal edge of the atmosphere, as far as aerodynamic entry is concerned, is taken as 100 km, or just around the thin layer known as the mesopause.

## C.2 EQUATION OF STATE AND HYDROSTATIC EQUILIBRIUM

If one considers the atmosphere to behave as a perfect gas, then

$$p = \rho RT = \rho \frac{R_u}{W_m} T.$$

The molecular weight of the mixture of atmospheric gases is essentially constant up to 100 km and is given by $W_m = 28.96$, and the atmospheric gas constant $R = 0.287$ kJ/kg-K (or 1716 $ft^2/s^2$-R). The hydrostatic equation for the atmosphere is

$$dp = -\rho g dz.$$

Because the gravitational acceleration depends on altitude, a new altitude function may be defined. This is the *geopotential* altitude, $h$, and it is related to the geometric altitude by the relation

$$g_E dh = g dz,$$

where $g_E = 9.087$ m/s$^2$ (or 32.15 ft/s$^2$) is the gravitational acceleration at the surface of the Earth, $z = 0$. The gravitational acceleration varies with altitude according to Newton's law of gravitation and may be written as

$$g = \frac{g_E R_E^2}{(R_E + z)^2} = g_E \frac{1}{\left(1 + \frac{z}{R_E}\right)^2}.$$

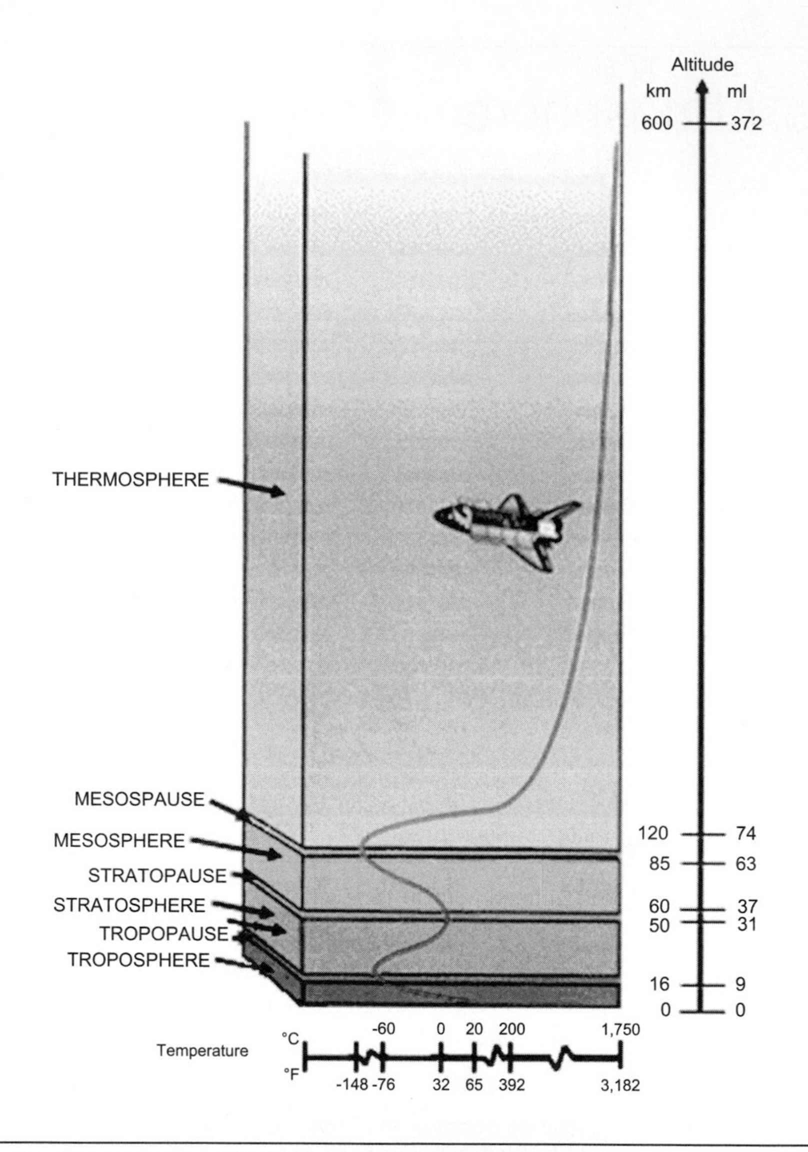


**FIGURE C.1**

Definition of the Earth's atmosphere in terms of temperature (courtesy NASA).

The radius of the Earth is taken to be $R_E = 6357$ km so that for low Earth orbits (LEO) the ratio $z/R_E \sim 0.06$, and therefore the difference between $h$ and $z$ is relatively small. Integrating the relation between $h$ and $z$ yields the geopotential altitude

$$h = z\frac{1}{1+\dfrac{z}{R_E}}.$$

Thus the difference between geometric and geopotential altitudes for a LEO of up to 400 km is less than 6.3%.

Using the equation of state in the hydrostatic equilibrium equation yields

$$\frac{dp}{p} = -g\frac{dz}{RT} = -g_E\frac{dh}{RT}.$$

With an appropriate relation for temperature in terms of geopotential altitude, $T = T(h)$, one may integrate the hydrostatic equation and find $p = p(h)$ and $\rho = \rho(h)$. The 1976 Standard Atmosphere defines atmospheric layers, each with $T = T_i + \lambda_i(h - h_i)$ , where $T_i$ is the temperature of the start of layer $i$, $h_i$ is the altitude at the start of layer $i$, and $\lambda_i$ is the lapse rate, that is, $dT/dh$, in that layer. Integration of the hydrostatic equation for nonzero $\lambda$ yields

$$p = p_i\left[\frac{T_i}{T_i + \lambda_i(h - h_i)}\right]^{\frac{g_0}{R\lambda_i}}.$$

In isothermal layers where $\lambda = 0$, the temperature $T = T_i$ = constant, and the pressure is instead given by

$$p = p_i \exp\left[-\frac{g_E}{RT_i}(h - h_i)\right].$$

The temperature at the Earth's surface is taken as $T = 15°C = 288.15K$ and $g_E/R = 34.17$ K/km.

The properties of this atmospheric model in the various layers are given in Table C.1.

The distribution of pressure, in kPa, in the various layers is then given by the following:

Layer 1 (0 to 11 km):

$$p = 101.3\left(\frac{288.15}{288.15 - 6.5h}\right)^{\frac{34.17}{-6.5}}$$

**Table C.1** Definition of the Layers in the 1976 Standard Atmosphere[a]

| Layer | Geopotential Altitude, *h* (km) | Geopotential Altitude, *h* (kft) | Lapse Rate, $\lambda_i$ (K/km) | Thermal Type |
|---|---|---|---|---|
| 1 | 0 | 0 | −6.5 | Neutral |
| 2 | 11 | 36.1 | 0 | Isothermal |
| 3 | 20 | 65.6 | +1.0 | Inversion |
| 4 | 32 | 105 | +2.8 | Inversion |
| 5 | 47 | 154 | 0 | Isothermal |
| 6 | 51 | 167 | −2.8 | Neutral |
| 7 | 71 | 233 | −2.0 | Neutral |
| 8 | 84.85 | 278 | +1.65 | Inversion |
| 9 | 100 | | | |

[a]*The eighth layer lapse rate is based on the 1962 Standard Atmosphere. The difference between 1976 and 1962 Standard Atmospheres is small for altitudes* $h < 150$ *km.*

Layer 2 (11 to 20 km):

$$p = 22.62\exp\left(\frac{-34.17[h-11]}{216.65}\right)$$

Layer 3 (20 to 32 km):

$$p = 5.47\left(\frac{216.65}{216.65+[h-20]}\right)^{\frac{34.17}{1}}$$

Layer 4 (32 to 47 km):

$$p = 0.8669\left(\frac{228.65}{228+2.8[h-32]}\right)^{\frac{34.17}{2.8}}$$

Layer 5 (47 to 51 km):

$$p = 0.1107\exp\left(\frac{-34.17[h-47]}{270.65}\right)$$

Layer 6 (51 to 71 km):

$$p = 0.06681\left(\frac{270.65}{270.65-2.8[h-51]}\right)^{\frac{34.17}{-2.8}}$$

Layer 7 (71 to 84.85 km):

$$p = 0.003946\left(\frac{214.65}{214.65-2[h-71]}\right)^{\frac{34.17}{-2}}$$

Layer 8 (84.85 to 100 km)

$$p = 0.0003724\left(\frac{186.95}{186.95+1.65[h-84.85]}\right)^{\frac{34.17}{1.65}}$$

The density may be found from the equation of state. Graphs of the variation of the important variables are presented in Figures C.2 to C.5. Some of the more important features of the atmosphere may be determined from these figures. First, from Figure C.2 it is clear that the temperature is constant at about 216K in the lower reaches of the stratosphere, in the altitude range of 10 to 20 km. This is the region of high-speed manned flight, from jet airliners to military aircraft up to the Mach 3 SR-71 Blackbird. Variations of the pressure and density, as given by Figures C.3 and C.4, respectively, appear to be exponential in nature. This behavior is best illustrated in Figure C.6 where the actual density ratio is compared to an exponential approximation $\sigma = \exp(-h/H)$. The quantity $H$ is called the scale height of the atmosphere, and a common value used is $H = 7.16$ km. This approximation is introduced in Chapter 3. Finally, because the speed of sound is proportional to the square root of the temperature, and the temperature through the Standard Atmosphere has

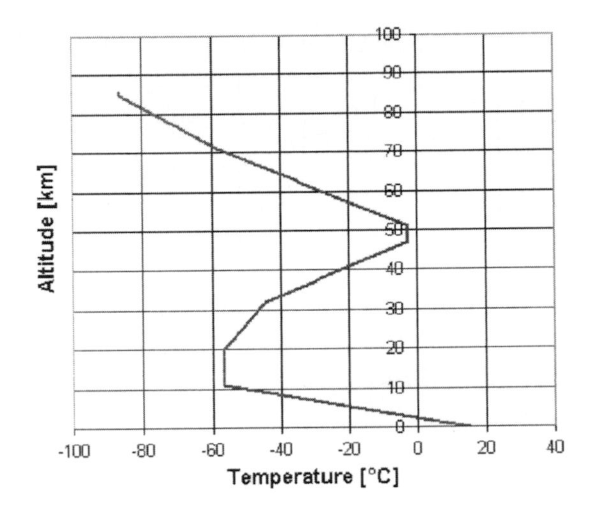


FIGURE C.2

The 1976 U.S Standard Atmosphere temperature distribution.

little variation, the speed of sound is relatively constant up to 100 km altitude. It is common to assume a constant value for preliminary design purposes, which can be used with an error on the order of $\pm 10\%$ over the whole range. A table of the pressure, density, and temperature ratios is given in Table C.2.

Note that the model atmosphere does not account for day-to-day variations in the characteristics of the atmosphere. More detailed models for narrower geographical and seasonal data are usually given

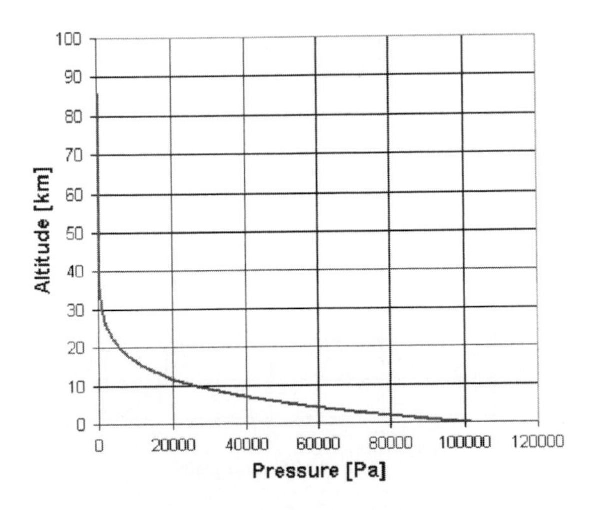


FIGURE C.3

The 1976 U.S Standard Atmosphere pressure distribution.

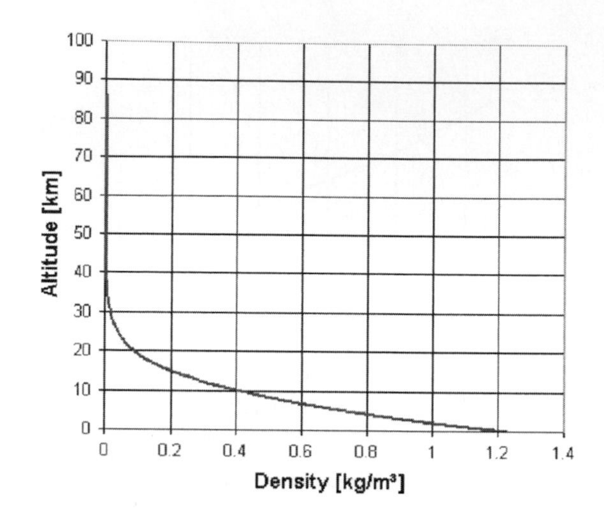


**FIGURE C.4**

The 1976 U.S Standard Atmosphere density distribution.

as midlatitude winter, midlatitude summer, subarctic winter, subarctic summer, and tropical and will show deviations from the U.S. Standard Atmosphere. Comprehensive atmospheric properties are available as part of extensive atmospheric propagation codes, such as MODTRAN and FASCOD, which are described in MODTRAN/FASCOD.

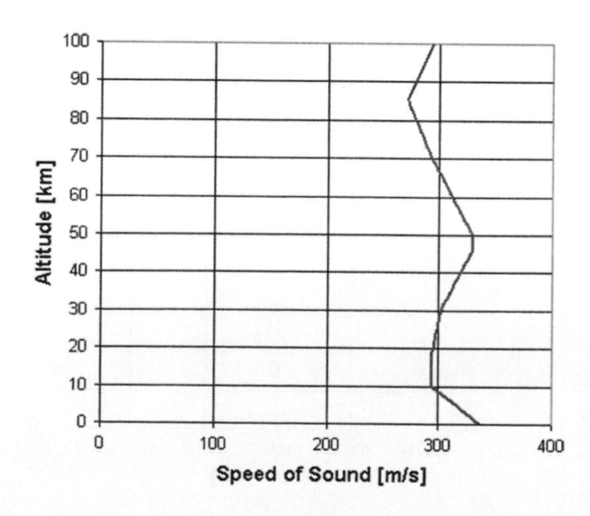


**FIGURE C.5**

The 1976 U.S Standard Atmosphere sound speed distribution.

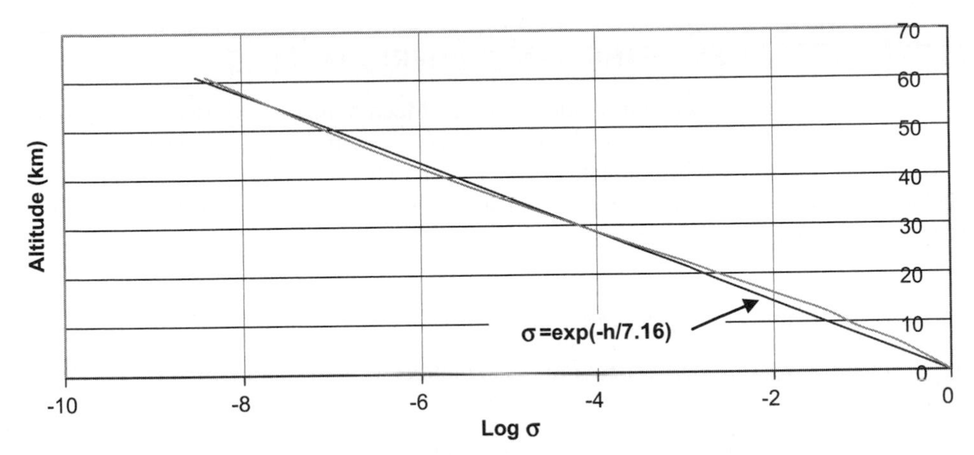


**FIGURE C.6**

Comparison of the actual density ratio to an exponential approximation.

**Table C.2** Atmospheric Properties up to 200,000 ft Altitude

| alt. (kft) | $\sigma = \rho/\rho(0)$ | $\delta = p/p(0)$ | $\theta = T/T(0)$ | *a* (fps) | alt. (km) |
|---|---|---|---|---|---|
| 0 | 1 | 1 | 1 | 1016.0 | 0 |
| 10 | 0.73859 | 0.68783 | 0.9313 | 980.5 | 3.048 |
| 20 | 0.53316 | 0.45991 | 0.8626 | 943.6 | 6.096 |
| 30 | 0.34473 | 0.29754 | 0.794 | 905.3 | 9.144 |
| 40 | 0.24708 | 0.18577 | 0.7519 | 881.0 | 12.192 |
| 50 | 0.15311 | 0.11512 | 0.7519 | 881.0 | 15.24 |
| 60 | 0.094919 | 0.071367 | 0.7519 | 881.0 | 18.288 |
| 70 | 0.058565 | 0.04429 | 0.7563 | 883.6 | 21.336 |
| 80 | 0.03606 | 0.027649 | 0.7668 | 889.7 | 24.384 |
| 90 | 0.02236 | 0.017379 | 0.7772 | 895.7 | 27.432 |
| 100 | 0.01396 | 0.010997 | 0.7877 | 901.7 | 30.48 |
| 110 | 0.008692 | 0.007011 | 0.8066 | 912.5 | 33.528 |
| 120 | 0.005428 | 0.004537 | 0.8359 | 928.9 | 36.576 |
| 130 | 0.003446 | 0.002982 | 0.8652 | 945.0 | 39.624 |
| 140 | 0.002222 | 0.001988 | 0.8944 | 960.9 | 42.672 |
| 150 | 0.001454 | 0.001343 | 0.9237 | 976.5 | 45.72 |
| 160 | 0.000977 | 0.000918 | 0.9393 | 984.7 | 48.768 |
| 170 | 0.000672 | 0.000628 | 0.9354 | 982.6 | 51.816 |
| 180 | 0.000471 | 0.000427 | 0.9063 | 967.2 | 54.864 |
| 190 | 0.000327 | 0.000287 | 0.8772 | 951.6 | 57.912 |
| 200 | 0.000224 | 0.00019 | 0.8481 | 935.7 | 60.96 |

*Standard sea level values:* $p(0) = 2116\ lb/ft^2 = 101.3\ kPa$
$\rho(0) = 0.002377\ lb\text{-}s^2/ft^4 = 1.225\ kg/m^3$
$T(0) = 518.67R = 218.15K$

## C.3 FLOW PROPERTIES USING ATMOSPHERIC MODELS

Free stream flow parameters such as Reynolds number, Mach number, and dynamic pressure involve the atmospheric properties discussed in the previous sections. It is often convenient to have representations for these properties in an analytic form, and a number of such approximations are presented subsequently. In the following material, the English system of units is used, as these units still appear quite regularly in aerospace applications.

### C.3.1 Reynolds number and Mach number

The ratio of the unit Reynolds number Re/$l$ to Mach number $M$ is a function of atmospheric properties alone and may be expressed as follows:

$$\frac{\mathrm{Re}/l}{M} = \frac{\rho a}{\mu}.$$

The general variation of this ratio with altitude is shown in Figure C.7. On the scale shown, the difference between the curve using the 1976 Standard Atmosphere and an approximate model is not apparent.
The simplest analytic approximation for Re/$Ml$ in the units of ft$^{-1}$ is given in Equation (C.1), where $h$ is measured in thousands of feet (kft):

$$\frac{\mathrm{Re}}{Ml} = 7,167,914 \exp\left(-\frac{h}{24,000}\right). \tag{C.1}$$

A more accurate model, with the same units, is given by

$$\frac{\mathrm{Re}}{Ml} = 7,167,914\left[1 + \sin\left(\frac{\pi h}{100,000}\right)\right]\exp\left(-\frac{h}{24,000}\right). \tag{C.2}$$

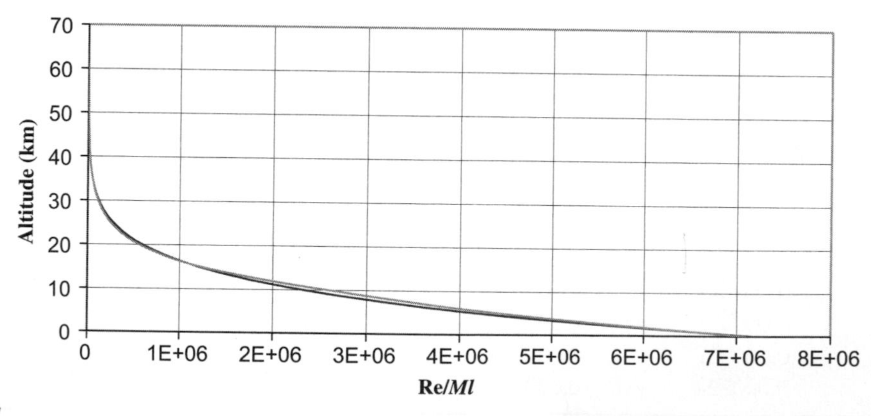


**FIGURE C.7**

Variation with altitude of the ratio of unit Reynolds number to Mach number for the 1976 Standard Atmosphere and the approximation in Equation (C.1).

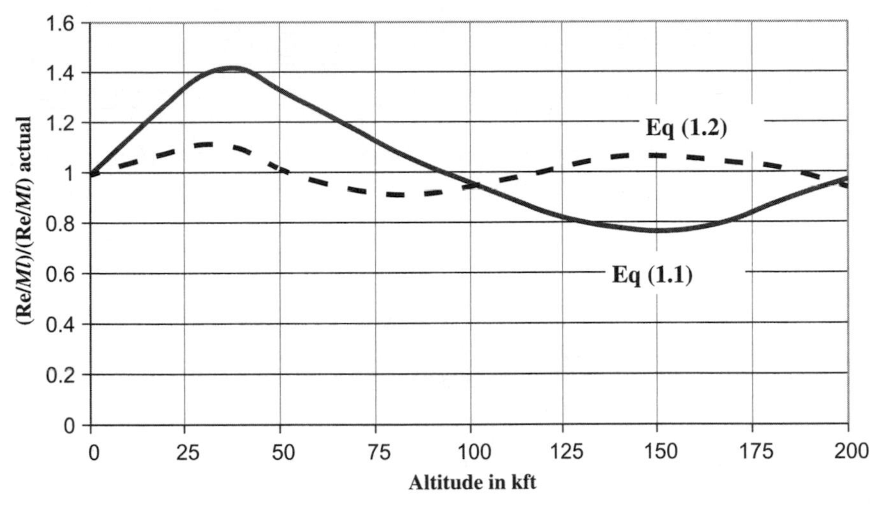


**FIGURE C.8**

Accuracy of approximate Equations (C.1)[Eq (1.1)] and (C.2)[Eq (1.2)].

The accuracy of each of these two approximations is compared to the actual result in Figure C.8. Note that Equation (C.2) keeps the error to $\pm 10\%$ over the entire altitude range shown.

## C.3.2 Dynamic pressure

The variation of the ratio of dynamic pressure to the square of the Mach number is

$$\frac{q}{M^2} = \frac{1}{2}\gamma p. \tag{C.3}$$

The general variation of this ratio with altitude is shown in Figure C.9. An approximate analytic correlation for this equation is

$$\frac{q}{M^2} = 1392.3\left[1 + 0.175\sin\left(\frac{\pi h}{80,000}\right)\right]\exp\left(-\frac{h}{23,000}\right), \tag{C.4}$$

where $q$ is measured in psf (lbs/ft$^2$) and $h$ in thousands of feet (kft). On the scale shown in Figure C.9, the difference between the curve using the 1976 Standard Atmosphere and an approximate model is not apparent. The accuracy of Equation (C.4) is illustrated in Figure C.10, where it is seen that the error is everywhere less than $\pm 8\%$.

The dynamic pressure is related to the lift of a vehicle, and during cruise the dynamic pressure may be expressed by

$$q = \frac{L}{C_L S} = \frac{1}{C_L}\left(\frac{W}{S}\right)_{cruise}.$$

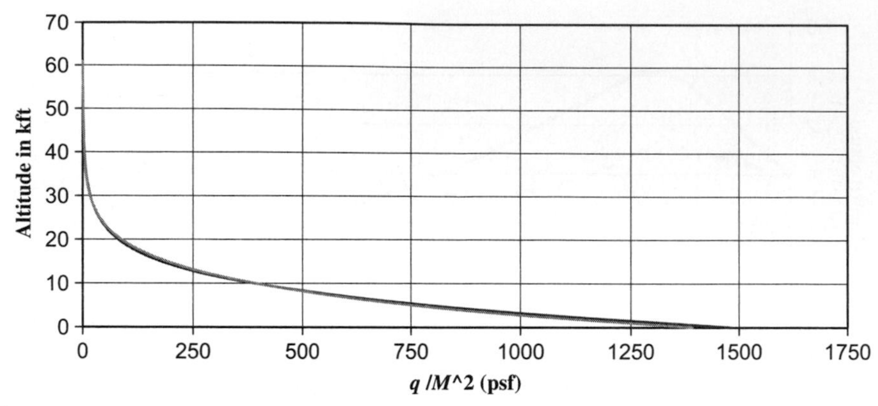


**FIGURE C.9**

Variation with altitude of the ratio of dynamic pressure $q$ (in psf) to the square of the Mach number for the 1976 Standard Atmosphere and the approximation in Equation (C.4).

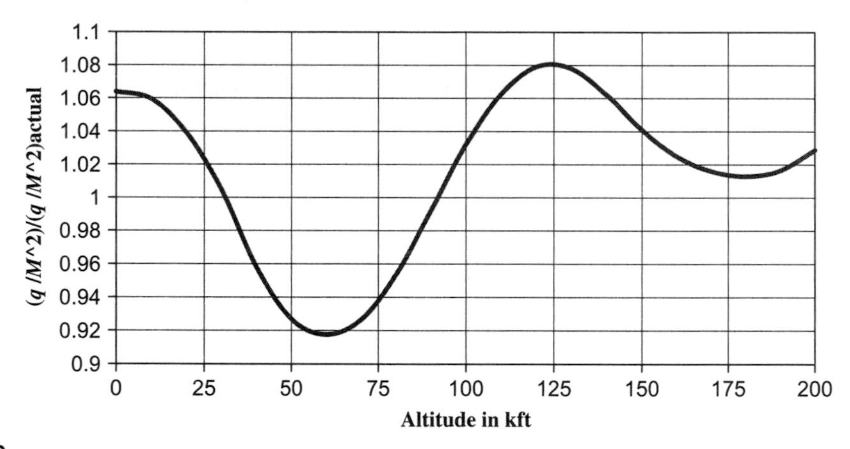


**FIGURE C.10**

Accuracy of approximate Equation (C.4).

Therefore, for a given wing (or, more generally, planform) loading the dynamic pressure encountered fixes the required lift coefficient $C_L$ for equilibrium flight. It is useful to display contours of constant dynamic pressure as a function of altitude and Mach number. This can be accomplished readily by means of the analytic approximations presented earlier, the results of which are displayed in Figures C.11 and C.12 for different flight Mach number ranges.

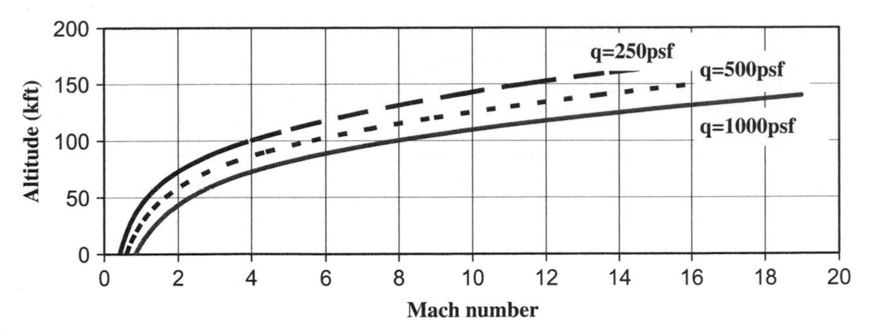


**FIGURE C.11**

Contours of constant dynamic pressure for practical flight conditions at high Mach numbers.

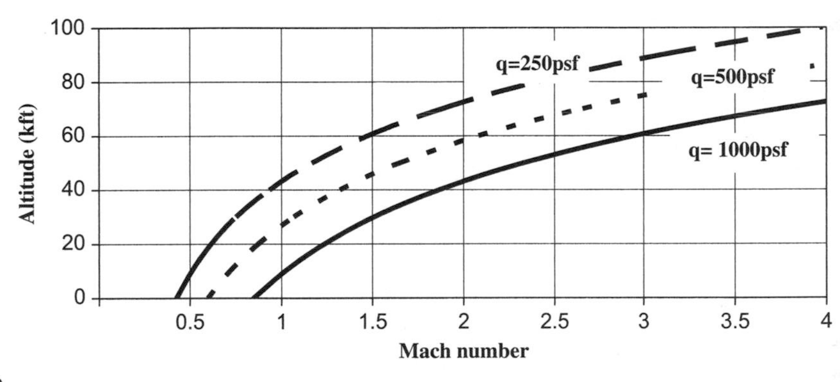


**FIGURE C.12**

Contours of constant dynamic pressure for practical flight conditions at low Mach numbers.

## C.3.3 Atmospheric pressure and density

In many flight applications, particularly spacecraft launch and reentry, the atmospheric density is approximated by a simple exponential function of altitude. However, this usually yields relatively high errors if used over too wide an altitude range. Here we suggest the following approximations to the altitude profiles (with $h$ in kft) of atmospheric pressure and density:

$$\delta = \frac{p(h)}{p(0)} = 0.94\left[1 + 0.175\sin\left(\frac{\pi h}{80}\right)\right]\exp\left(-\frac{h}{23}\right) \tag{C.5}$$

$$\sigma = \frac{\rho(h)}{\rho(0)} = 1.12\left[1 + 0.175\sin\left(\frac{\pi h}{80}\right)\right]\exp\left(-\frac{h}{23}\right). \tag{C.6}$$

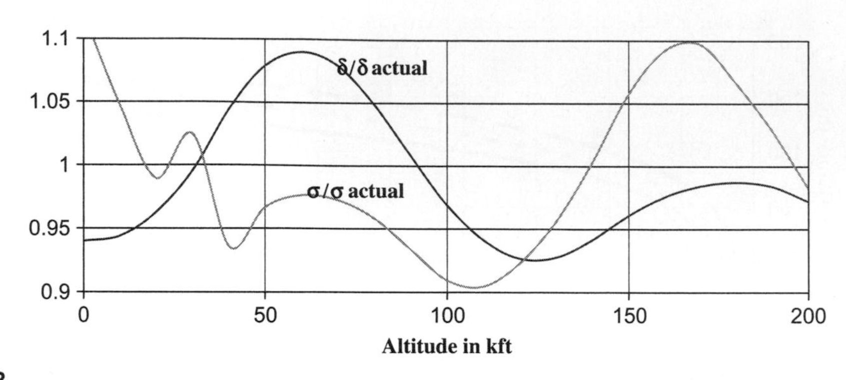


**FIGURE C.13**

Comparison of approximate and actual atmospheric pressure and density ratios.

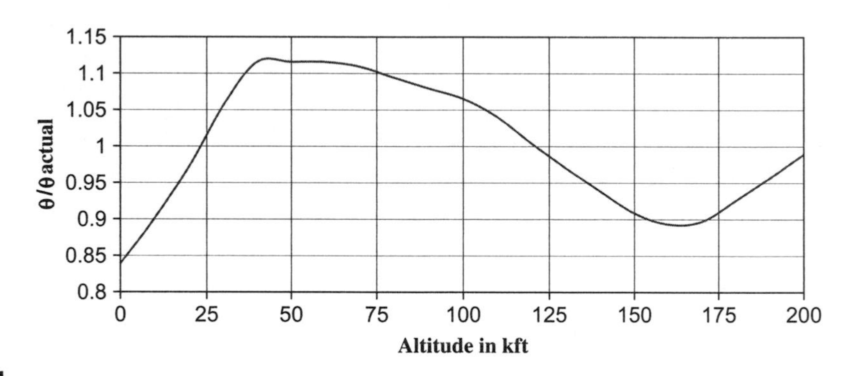


**FIGURE C.14**

Comparison of approximate and actual atmospheric temperature ratio.

The accuracy of the expressions given in Equations (C.5) and (C.6) is illustrated in Figure C.13 and it is clear that again the approximations are within $\pm 10\%$ of the actual Standard Atmosphere model.

Of course, the equation of state then requires that the temperature of the atmosphere be constant at $T = 436\text{R}$ and therefore that the sound speed also be constant with a value of $a = 1023$ ft/s. The nondimensional temperature obtained from the state equation and Equations (C.5) and (C.6) is

$$\theta = \frac{T(h)}{T(0)} = 0.839.$$

A comparison of this result with the actual temperature variation in the atmosphere is shown in Figure C.14 and reveals an accuracy of about $\pm 12\%$ over the whole altitude range, except for the lowest 10 kft, where the error goes up to 16% as sea level is approached.

## C.4 NOMENCLATURE

| | |
|---|---|
| $a$ | sound speed |
| $C_L$ | lift coefficient |
| $g$ | acceleration of gravity |
| $g_E$ | acceleration of gravity at $z = 0$ |
| $h$ | geopotential altitude |
| $l$ | length |
| $M$ | Mach number |
| $p$ | pressure |
| $q$ | dynamic pressure |
| Re | Reynolds number |
| $R_u$ | universal gas constant |
| $S$ | surface area |
| $T$ | temperature |
| $W$ | weight |
| $W_m$ | molecular weight |
| $z$ | geometric altitude |
| $\delta$ | pressure ratio $p/p(0)$ |
| $\gamma$ | ratio of specific heats |
| $\sigma$ | density ratio $\rho/\rho(0)$ |
| $\lambda$ | temperature lapse rate |
| $\rho$ | gas density |
| $\mu$ | gas viscosity |
| $\theta$ | temperature ratio $T/T(0)$ |

## References

MODTRAN/FASCOD (http://www.ontar.com/Software/ProductDetails.aspx?item=pub_ModtranFascod_Documentation).

NASA (http://www.nasa.gov/audience/forstudents/9-12/features/912_liftoff_atm.html).

APPENDIX

# Boost Phase and Staging of Rockets

D

## D.1 SIMPLIFIED BOOST ANALYSIS INCLUDING LIFT AND DRAG

The boost phase of an orbital space mission is critical to the design process, as the initial weight of the vehicle, which is generally related to its cost, is at its maximum at lift-off. The payload of the launch system is typically a small fraction of the total weight. Boost trajectory and associated nomenclature are shown in Figure D.1.

In the ideal case of a propulsion system with constant specific impulse over the entire speed and altitude range (the conventional rocket satisfies this ideal fairly well), the thrust force, $F$, and mass consumption rate remain constant and

$$m = m_0 + \int_0^t \dot{m}dt = m_0\left(1 + \frac{\dot{m}}{m_0}t\right) = m_0\left(1 - \frac{F}{m_0 g_E I_{sp}}t\right). \tag{D.1}$$

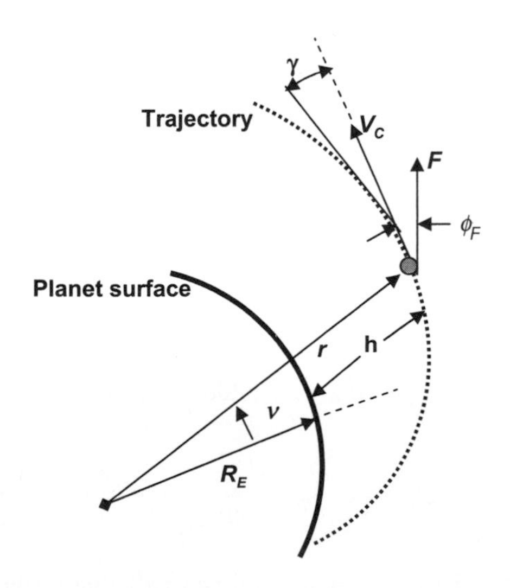


**FIGURE D.1**

Schematic diagram of a boost trajectory with zero lift.

The specific impulse here is defined as

$$I_{sp} = \frac{F}{\dot{m} g_E}. \tag{D.2}$$

The general equations of motion are as follows:

$$\begin{aligned} a_t &= \frac{dV_C}{dt} = \frac{F_t}{m} \\ a_n &= V_C\left(\frac{d\gamma}{dt} - \frac{d\nu}{dt}\right) = \frac{F_n}{m}. \end{aligned} \tag{D.3}$$

The equations of motion for flight over a nonrotating planet are given here:

$$\begin{aligned} \dot{V}_C &= \frac{F\cos\phi_F}{m} - \frac{D}{m} - \frac{k\sin\gamma}{r^2} \\ \dot{\gamma} &= \frac{V_C\cos\gamma}{r} + \frac{F\sin\phi_F}{V_C m} + \frac{L}{V_C m} - \frac{k\cos\gamma}{V_C r^2} \\ \dot{r} \equiv \dot{h} &= V_C \sin\gamma \\ \dot{\nu} &= \frac{V_C\cos\gamma}{r} \\ \dot{m} &= -\frac{F}{I_{sp}}. \end{aligned} \tag{D.4}$$

These equations may be simplified for typical boost trajectories. The boost phase of an orbital space mission is critical to the design process, as the initial weight of the vehicle, which is generally related to its cost, is at its maximum at lift-off. The payload of the launch system is typically a small fraction of the total weight.

Obviously, the burn time of the rocket engine is limited by the amount of fuel available so that

$$m_{b-o} = m_0 - m_{fuel} = m_0 - F t_{b-o}/g_E I_{sp}, \tag{D.5}$$

where the subscript *b-o* denotes burnout, that is, shutdown of the engine. The specific impulse here is defined as

$$I_{sp} = \frac{F}{\dot{m} g_E}. \tag{D.6}$$

We assume again that the boost phase occurs such that $\phi_F = 0$, that is, the thrust vector is aligned with the velocity vector.

## D.2 APPROXIMATE BOOST ANALYSIS

The equations of motion for flight utilizing both lift and drag are given here, still under the assumption that the thrust is aligned with the trajectory, $\phi_F = 0$.

$$\dot{V}_C = \frac{g_E\left(\frac{F}{W}\right)_0}{1-\left(\frac{F}{W}\right)_0\frac{t}{I_{sp}}} - g_E \sin\gamma - \frac{g_E\rho_0\sigma}{2B}V_c^2$$

$$\dot{\gamma} = \frac{V_C\cos\gamma}{R_E} - \frac{g_E\cos\gamma}{R_E^2 V_C} + \frac{g_E\rho_0\sigma}{2B}V_c \quad , \tag{D.7}$$

$$\dot{r} \equiv \dot{h} = V_C \sin\gamma$$

$$\dot{x} = \frac{R_E}{r}V_C\cos\gamma = V_C\cos\gamma$$

where $B$ is the ballistic coefficient $W/C_DA$. Vehicles with a high degree of reusability are expected to make use of air-breathing engines to reduce the required propellant load along with lifting wings to aid in ascent by countering gravitational effects. However, with lift and drag proportional to $V_C^2$ and $m$ varying with time, at least as given in Equation (D.5), the equation set is more closely coupled than before and is best computed numerically. Furthermore, for air-breathing engines, the specific impulse $I_{sp}$ is a function of flight speed $V_c$, and Equation (D.1) must be adjusted accordingly.

The velocity equation is, using the gravitational constant $k = g_E R_E^2$, given by

$$\frac{dV_c}{dt} = \frac{F}{m} - \frac{k}{r^2}\sin\gamma - \frac{D}{m}, \tag{D.8}$$

which may be written as

$$V_C - V_{C,0} = g_E\int_0^t \frac{F-D}{W_0 - \dot{W}t}dt - g_E\int_0^t \sin\gamma\, dt. \tag{D.9}$$

The subscript zero denotes the initial state, that is, at time $t = 0$. If the thrust of the rocket is essentially constant and we consider average values for the drag and the flight path angle, the equation becomes

$$V_C - V_{C,0} = g_E\left[\frac{F - D_{avg}}{W_0}\right]\int_0^t \frac{1}{1-\frac{\dot{W}}{W_0}}dt - g_E t\sin\gamma_{avg}. \tag{D.10}$$

The definition of the specific impulse is $I_{sp} = \frac{F}{\dot{W}}$, so the velocity change is then

$$V_C - V_{C,0} = g_E I_{sp}\left[1 - \frac{D_{avg}}{F}\right]\log_e\left\{1-\left(\frac{F}{W}\right)_0\frac{t}{I_{sp}}\right\}^{-1} - g_E t\sin\gamma_{avg}. \tag{D.11}$$

The drag to thrust ratio is

$$\frac{D}{F} = \frac{C_D q S}{F} = \frac{C_D S V_C^2 \rho_0 e^{-\frac{h}{H}}}{2W_0 \left(\frac{F}{W}\right)_0} = \frac{\rho_0 V_C^2 e^{-\frac{h}{H}}}{2B_0 \left(\frac{F}{W}\right)_0}. \tag{D.12}$$

The initial thrust to weight ratio is of order unity, while the maximum dynamic pressure is typically around 500 lbs/ft$^2$ so that the drag to thrust ratio $D/F < 500\ C_D/B_0$. We can further safely assume that the drag coefficient is of order unity, while the ratio of the reference area (typically the maximum cross-sectional area of the launch vehicle) to total volume is approximately $l^{-1}$, where $l$ is a characteristic overall length of the vehicle. Thus the drag to thrust ratio may be expressed as $D/F \sim q(l\rho_{str})^{-1}$, where the term $\rho_{str}$ represents the average density of the complete launch vehicle at launch. We assume that a reasonable estimate of $\rho_{str}$ is at least the density of the propellant mixture. Table D.1 shows that the propellant density is greater than 50 lbs/ft$^3$.

Thus the drag to thrust ratio $D/F < 500/(50l)$, where $l$ is generally on the order of 100 ft, so $D/F < 10^{-1}$ and we may safely neglect drag from preliminary design studies of launch performance. The equation for the velocity change achieved by the application of constant thrust is given by

$$V_C - V_{C,0} = g_E I_{sp} \log_e \left[1 - \left(\frac{F}{W}\right)_0 \frac{t}{I_{sp}}\right]^{-1} - g_E t \sin \gamma_{avg}. \tag{D.13}$$

Expanding this approximate solution for $\left(\frac{F}{W}\right)_0 \frac{t}{I_{sp}} < 1$ (which is valid for times on the order of hundreds of seconds) yields the following:

$$V_C - V_{C,0} = g_E \left[\left(\frac{F}{W}\right)_0 t + \left(\frac{F}{W}\right)_0^2 \frac{t^2}{2I_{sp}} + \ldots - \left(\sin \gamma_{avg}\right) t\right]. \tag{D.14}$$

This demonstrates that the weight term is significant only at the earliest times, as $\sin \gamma_{avg}$ is always less than unity. Therefore, for preliminary design purposes, we may choose a value for $\sin \gamma_{avg}$, say, 0.5.

The fuel burn during the time period from 0 to $t$ is given by

$$W_f = -\int_0^t \dot{W} dt = \int_0^t \frac{F}{I_{sp}} dt = \frac{F}{I_{sp}} t. \tag{D.15}$$

**Table D.1** Propellant Densities for Several Fuel and Oxidizer Combinations

| Oxidizer | Density (lbs/ft$^3$) | Fuel | Density (lbs/ft$^3$) | O/F Ratio | Mixture Density (lbs/ft$^3$) |
|---|---|---|---|---|---|
| LOX | 71.2 | $LH_2$ | 4.34 | 4 | 57.8 |
| LOX | 71.2 | RP1 | 49.9 | 2.6 | 65.3 |
| Ammonium perchlorate | (composite solid) | Polybutadiene and aluminum | (composite solid) | 9 | 1105 |

Because the fuel weight cannot equal the total initial weight,

$$\frac{W_f}{W_0} = \left(\frac{F}{W}\right)_0 \frac{t}{I_{sp}} < 1.$$

The two previous equations provide a bound for the burn time considered for the rocket and ultimately for the total weight of the rocket. Typical initial values for the thrust to weight ratio and the $I_{sp}$ are about 1.3 and 400 s, respectively, so maximum burn times are around 300 s.

## D.3 STAGING OF ROCKETS

The change in velocity of a rocket-propelled vehicle given by Equation (D.14) may be generalized to the case where the rocket motor is initiated at a given time $t = t_i$. At this point the velocity of the vehicle is $V_C = V_{C,i}$ and the thrust to weight ratio and specific impulse are constant and equal to the values at the time $t = t_i$. In addition, the average value of the flight path angle is specified for this particular segment of the flight. In this case,

$$V_C - V_{C,i} = g_E I_{sp,i} \log_e \left[ \frac{1 - \left(\frac{F}{W}\right)_i \frac{t_i}{I_{sp,i}}}{1 - \left(\frac{F}{W}\right)_i \frac{t}{I_{sp,i}}} \right] - g_E (t - t_i) \sin \gamma_{avg,i}. \quad \text{(D.16)}$$

Similarly, the amount of fuel burned is given by

$$W_f = \frac{F}{I_{sp}}(t - t_i). \quad \text{(D.17)}$$

The weight of a launch vehicle decreases as the fuel burns, but the structure and tankage weight remains fixed as the vehicle rises. Because this weight is parasitic, it would be more efficient to jettison this weight continuously during ascent. Such a procedure is not practical, but an approximation to this is a jettisoning of discrete portions of the structure and tankage along the flight path when the fuel in them is consumed completely. We may use Equations (D.16) and (D.17) to account for the contribution of these individual (and possibly different) propulsive elements to the overall performance of a launch vehicle. This concept is known as staging.

## D.4 SINGLE STAGE TO ORBIT (SSTO)

Assume that the complete trajectory from launch to orbital speed is to be achieved with a launch stack consisting of a payload and a propulsive unit, such that the weight may be described by

$$W_0 = W_{pl} + W_{pu} = W_{pay} + W_{str} + W_{eng} + W_{fuel}. \quad \text{(D.18)}$$

The weight of the payload, which here includes the entire payload package, is to be placed in orbit at the circular speed appropriate to the orbit altitude, given by

$$V_{C,cir} = \sqrt{gr} = \sqrt{\frac{g_E R_E}{1+\frac{h}{R_E}}} \approx \sqrt{g_E R_E}\left(1-\frac{h}{2R_E}\right) \approx \sqrt{g_E R_E}. \quad \text{(D.19)}$$

Note that for low Earth orbit (LEO) the final approximation for the circular velocity is within about 3% of the exact value and is suitable for preliminary design purposes. The terms $W_{eng}$ and $W_{fuel}$ denote the weight of the operational motor and the weight of the propellant combination, respectively, while $W_{str}$ denotes the weight of the structure, tankage, and all the remaining systems and parts of the launch vehicle. In the simplest case, the engines are assumed to have a constant specific impulse, and the initial thrust to weight ratio and average flight path angle are selected. Then Equation (D.16) may be used to solve for the burn time required to accelerate from $V_{C,I} = 0$ to $V_C = V_{C,cir}$ and Equation (D.17) to solve for the required fuel fraction.

Let us assume the following data for our hypothetical single stage rocket: $(F/W)_0 = 1.3$, $I_{sp} = 450$ s, so that our equations are as follows:

$$\begin{aligned} V_C &= -4.413 \log_e(1-0.002889t) - 0.009806t \sin\gamma_{avg} \\ \frac{W_{fuel}}{W_0} &= 0.002889t. \end{aligned} \quad \text{(D.20)}$$

Results for the velocity history are shown in Figure D.2 and illustrate the effect of gravity. We may assume that the middle value, $\sin\gamma_{avg} = 0.5$, is a reasonable estimate for the gravitational effect and note that the circular velocity would be reached after the elapse of about 300 s. Therefore, this should be the burnout time at which all the fuel used is expended, and the amount of fuel may now be calculated.

From the second part of Equation (D.20) we see that for a burnout time $t_{b\text{-}o} = 300$ s the fuel fraction is $W_{fuel}/W_0 = 0.8667$. Then, from Equation (D.18), the following weight balance is determined:

$$1 = \frac{W_{pay}}{W_0} + \frac{W_{str}}{W_0} + \frac{W_{eng}}{W_0} + 0.8667. \quad \text{(D.21)}$$

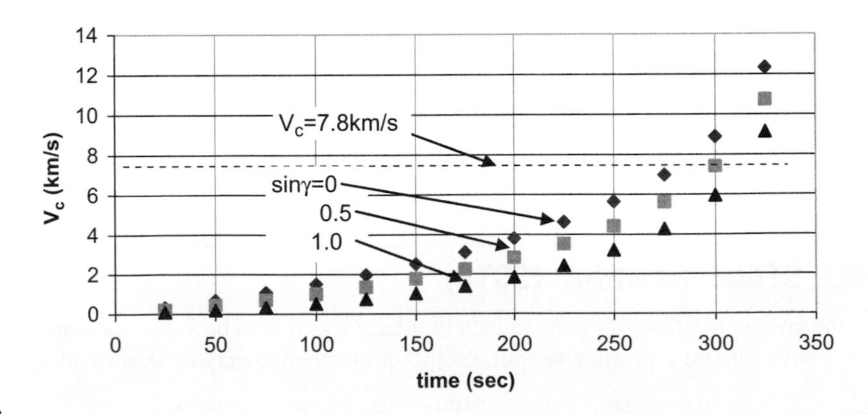


**FIGURE D.2**

Velocity history of a single stage rocket with $(F/W)_0 = 1.3$ and $I_{sp} = 450$ s for three values of $\sin\gamma_{avg}$. The circular velocity of 7.8 km/s is shown as a dashed line.

The payload weight is part of the mission specification, but the engine weight is correlated to the engine thrust as follows:

$$\frac{W_{eng}}{W_0} = \frac{KF^{0.856}}{W_0} = \frac{K}{W_0^{0.144}}\left(\frac{F}{W}\right)_0^{0.856} \tag{D.22}$$

Data from seven $LH_2$-LOX engines and eight RP1-LOX engines (plus one CH4-LOX engine) were taken from various sources, and the engine weight as a function of thrust is shown in Figures D.3 and D.4.

The value of $K$ is 0.095 for LOX–$LH_2$ propellants and 0.07 for LOX–RP1 propellants. Because we have specified $I_{sp} = 450$, we are considering a LOX–$LH_2$ propellant combination and will use $K = 0.095$. Note that all weights in this discussion are in pounds.

The structural weight of the vehicle, however, is much more difficult to estimate, but it should depend on the weight of fuel carried. The ratio of structural weight to propellant weight as a function of propellant weight was determined for 34 different vehicles and is tabulated in Table D.2, and the ratio of structural weight to propellant weight is shown as a function of propellant weight (Figure D.5). Much of these data are available in Isakowitz (1995).

Also shown in Figure D.5 is a correlation based on fuel weight (in lbs) and given by

$$\frac{W_{str}}{W_{fuel}} = \frac{28}{\sqrt{W_{fuel}}}. \tag{D.23}$$

Using data for stages from 26 operational rockets with different propellant combinations (with propellant weight varying from 4500 to 1,800,000 lbs) and including the engines, a crude estimate for the sum of these weights suggests that $(W_{str} + W_{eng})/W_{fuel} = 0.1$. This appears to be a reasonable estimate for a LOX–$LH_2$ propellant combination with propellant weights greater than 200,000 lbs. For lower propellant weights, the ratio varies between about 0.12 and 0.17, with no clearly definable trend. These trends may be seen in Figure D.6.

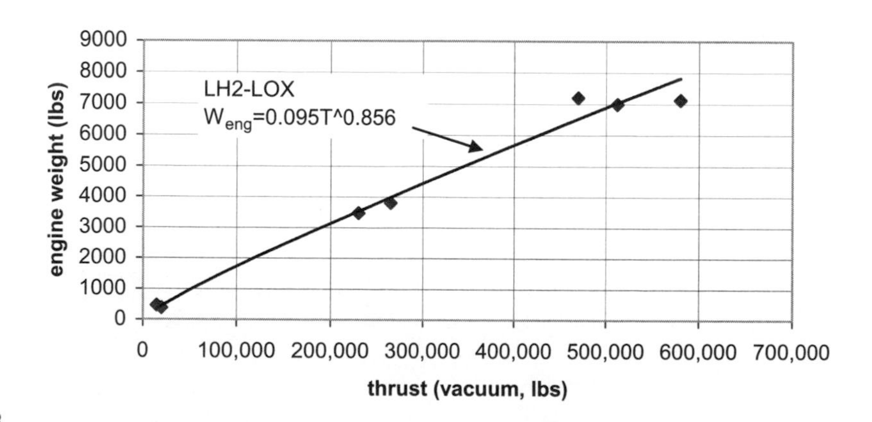


**FIGURE D.3**

Engine weight as a function of vacuum thrust for seven $LH_2$–LOX engines.

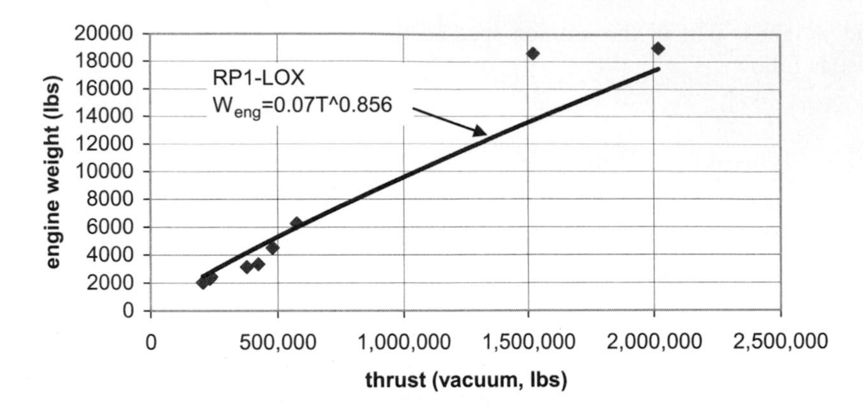


**FIGURE D.4**

Engine weight as a function of vacuum thrust for eight LOX–RP1 engines and one LOX–CH4 engine.

**Table D.2** Ascent Vehicles and Associated Propellant and Structural Weight

| Vehicle | Stage | Propellant | $W_p$ (lb x E-5) | $W_s$ (lb x E-3) | Ws/Wp |
|---|---|---|---|---|---|
| Atlas E | Booster | RP1-LOX | 2.488 | 1.79 | 0.0719 |
| Atlas IIAS | Booster | RP1-LOX | 3.445 | 2.38 | 0.0691 |
| Atlas IIAS | Centaur | $LH_2$-LOX | 0.37 | 0.48 | 0.1297 |
| Delta 7925 | First | RP1-LOX | 2.111 | 1.31 | 0.0621 |
| Delta 7925 | Second | $N_2O_4$-UDH | 0.1337 | 0.203 | 0.1518 |
| Delta 7295 | Third | HTPB | 0.0443 | 0.0291 | 0.0657 |
| Titan II | First | $N_2O_4$-UDH | 2.6 | 0.9 | 0.0346 |
| Titan II | Second | $N_2O_4$-UDH | 0.59 | 0.6 | 0.1017 |
| Titan IV | First | $N_2O_4$-UDH | 3.4 | 1.9 | 0.0559 |
| Titan IV | Second | $N_2O_4$-UDH | 0.772 | 0.98 | 0.1269 |
| Titan IV | Centaur 3 | $LH_2$-LOX | 0.4488 | 0.772 | 0.1720 |
| Saturn V | First | RP1-LOX | 45.84 | 28.8 | 0.0628 |
| Saturn V | Second | $LH_2$-LOX | 9.93 | 7.8 | 0.0785 |
| Saturn V | Third | $LH_2$-LOX | 2.38 | 2.5 | 0.1050 |
| Shuttle | Ext. tank | $LH_2$-LOX | 15.89 | 6.6 | 0.0415 |
| Ariane | First | $N_2O_4$-UDH | 5.14 | 3.9 | 0.0759 |
| Ariane | Second | $N_2O_4$-UDH | 0.776 | 0.73 | 0.0941 |
| Ariane | Third | $LH_2$-LOX | 0.238 | 0.29 | 0.1218 |
| H-2 | First | $LH_2$-LOX | 1.9 | 2.6 | 0.1368 |
| H-2 | Second | $LH_2$-LOX | 0.37 | 0.6 | 0.1622 |
| Energia | Strap-on | RP1-LOX | 7.05 | 7.8 | 0.1106 |
| Energia | Second | $LH_2$-LOX | 18.1 | 18.5 | 0.1022 |

**Table D.2** Ascent Vehicles and Associated Propellant and Structural Weight (*Continued*)

| Vehicle | Stage | Propellant | $W_p$ (lb x E-5) | $W_s$ (lb x E-3) | Ws/Wp |
|---|---|---|---|---|---|
| Proton | First | $N_2O_4$-UDH | 9.24 | 6.83 | 0.0739 |
| Proton | Second | $N_2O_4$-UDH | 3.441 | 2.58 | 0.0750 |
| Proton | Third | $N_2O_4$-UDH | 1.16 | 1.34 | 0.1155 |
| Proton | Fourth | RP1-LOX | 0.332 | 0.74 | 0.2229 |
| Atlas IIAS | Centaur | $LH_2$-LOX | 0.37 | 0.48 | 0.1297 |
| Titan IV | Centaur 3 | $LH_2$-LOX | 0.4488 | 0.772 | 0.1720 |
| Saturn V | Second | $LH_2$-LOX | 9.93 | 7.8 | 0.0785 |
| Saturn V | Third | $LH_2$-LOX | 2.38 | 2.5 | 0.1050 |
| Ariane | Third | $LH_2$-LOX | 0.238 | 0.29 | 0.1218 |
| H-2 | First | $LH_2$-LOX | 1.9 | 2.6 | 0.1368 |
| H-2 | Second | $LH_2$-LOX | 0.37 | 0.6 | 0.1622 |
| Energia | Second | $LH_2$-LOX | 18.1 | 18.5 | 0.1022 |

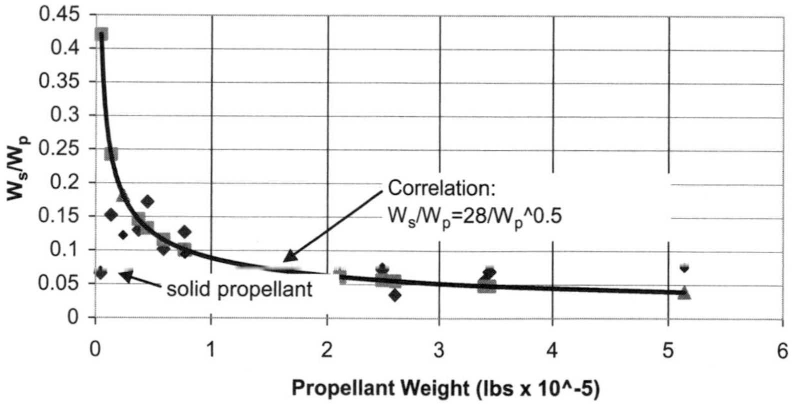

**FIGURE D.5**

The ratio of structural weight to propellant weight as a function of propellant weight for 26 vehicles (data taken from Table D.2).

Inserting the ratio $(W_{str} + W_{eng})/W_{fuel} = 0.1$ into the weight balance in Equation (D.21) yields

$$1 = \frac{W_{pay}}{W_0} + 0.1(0.8667) + 0.8667. \tag{D.24}$$

Substituting in the values that have been chosen for this example yields

$$W_0 = 23,275 + .9534W_0. \tag{D.25}$$

With the payload weight in pounds specified, Equation (D.25) is an equation for the initial weight of the entire stack. In general, for a four-person crew, one may expect a spacecraft weight of about

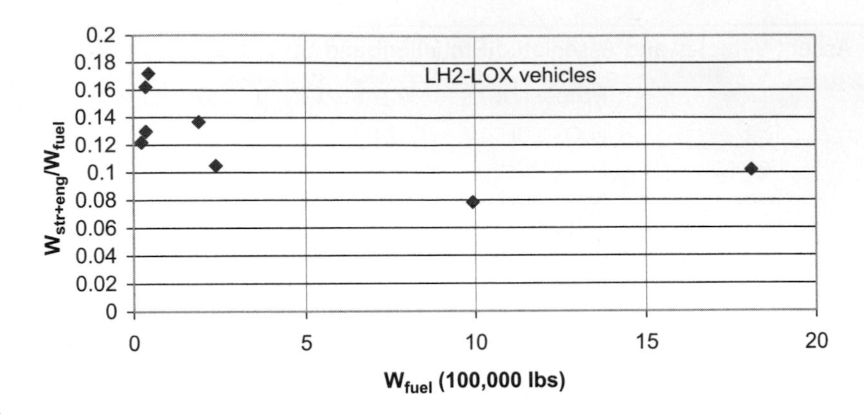


**FIGURE D.6**

Variation of structure and engine weight as a function of fuel weight for LOX-$LH_2$ rockets.

21,800 lbs. Adding about 370 lbs per crew member to account for their weight, their spacesuits, and related personal equipment results in a total spacecraft weight of about 23,280 lbs. The solution to Equation (D.25), when $W_{pay} = 23{,}280$ lbs, is $W_O = 499{,}500$ lbs, and the thrust required is therefore $F =$ 649,400 lbs. This is about the thrust level for three J-2 engines, each of which has a specific impulse of 427 s and a thrust of 230,000 lbs.

## D.5 TWO-STAGE VEHICLE TO ORBIT

Suppose that the ascent vehicle is made up of two stages, each with identical specific impulse and initial thrust to weight ratio, and further assume that we retain the same average value for the gravitational attraction, that is, sin $\gamma_{avg} = 0.5$. The first stage velocity history would be exactly the same as in the SSTO, but because we are using two stages here we would not continue the first stage burn all the way to orbit, but instead truncate it at some time, say $t_{b\text{-}o,1} = 200$ s. At that time the velocity of the complete vehicle has been calculated to be $V_C = 2.82$ km/s (9250 fps). The fuel fraction consumed during this 200-s burn may be determined from Equation (D.17) as $W_{fuel}/W_O = 0.5778$. This situation is illustrated in Figure D.7.

Let us rewrite Equation (D.18) for the weight of the vehicle as follows:

$$\begin{aligned} W_0^1 &= W_{pay} + W_{0,2} + W_{str,1} + W_{eng,1} + W_{fuel,1} \\ W_0^I &= W_0^{II} + W_{str,1} + W_{eng,1} + W_{fuel,1} \;=\; W_0^{II} + W_{0,1}, \end{aligned} \tag{D.26}$$

where $W_0{}^I$ denotes the initial weight of the entire launch stack, which includes the weight of the payload, the initial weight of the second stage, $W_{0,2}$, and the weight of the first stage, $W_{0,1}$, that is, the sum of the weight of the structure, the engines, and the fuel for the first stage. We denote the weight of the second stack as $W_0{}^{II}$; this is an attempt to differentiate between a (propulsion) stage and a (boosted) stack. Adaptation of Equation (D.21) for this shorter burn time results in the following:

$$W_0^I = W_0^{II} + 0.1(.5778W_0^I) + 0.5778W_0^I. \tag{D.27}$$

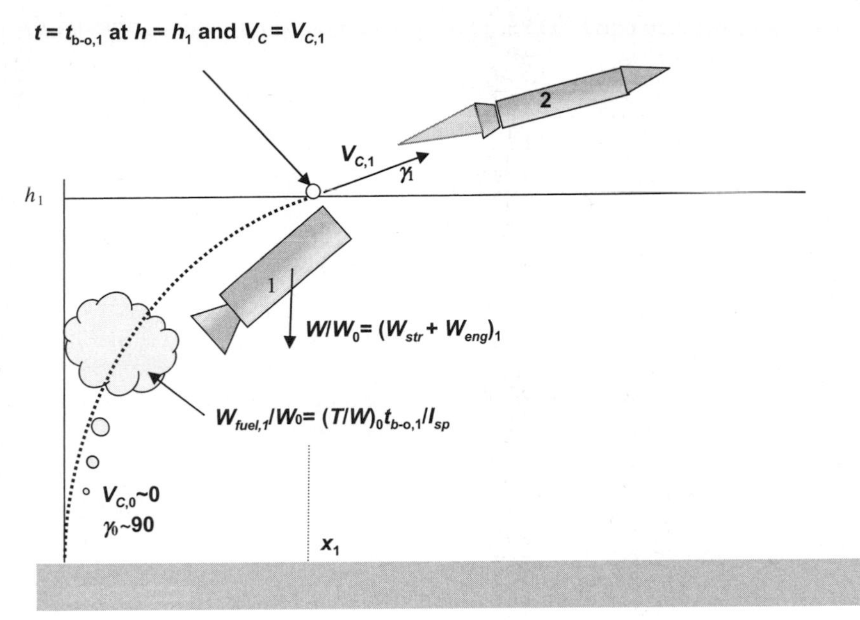


**FIGURE D.7**

Notional illustration of the first stage burnout and initiation of the second stage.

This equation can now be considered an equation for $W_0{}^{II}$ in terms of $W_0{}^{I}$. In order to find a relationship for $W_0{}^{II}$, we recall that this stack is starting at time $t = 200$ s with $V_C = 2.82$ km/s. If we use these conditions in Equation (D.16), we obtain

$$V_C^{II} - 2.82 = 4.413 \log_e \left[ 1 - \left( \frac{F}{W} \right)_0^{II} \frac{t - 200}{I_{sp}^{II}} \right]^{-1} - \frac{0.009807}{2}(t - 200). \tag{D.28}$$

Circular velocity is achieved at $t = 465$ s. The fuel fraction burned during operation of the second stage is

$$\frac{W_{fuel,2}}{W_0^{II}} = \frac{1.3}{450}(t - 200) = 0.7656. \tag{D.29}$$

Then, following Equation (D.25), the weight equation for the second stage may be written as

$$1 = \frac{W_{pay}}{W_0^{II}} + 0.1(0.7656) + 0.7656. \tag{D.30}$$

Noting that the payload has been assumed to be 23,275 lbs, we can determine that the weight of the second stage is $W_0{}^{II} = 147{,}500$ lbs. This value may be inserted into Equation (D.27) to determine the weight of the entire launch stack, $W_0{}^{I} = 404{,}800$ lbs. The thrust required for the first stage is therefore $F = 526{,}200$ lbs, which is about the thrust level of two J-2S engines ($I_{sp} = 435$ and $F = 265{,}000$ lbs). The second stage thrust required is 191,750 lbs, which can be satisfied readily by a single J2-S engine.

Thus to launch the same payload into LEO, the two-stage system weighs almost 20% less than a single stage system.

## D.6 THREE-STAGE VEHICLE TO ORBIT

The same process may be used to account for a third stage. Assuming that the first stage again burns for 200 s, the resulting velocity at first stage burnout is $V_C = 2.883$ km/s. The fuel fraction burned is also as it was in the two stage case, $W_{fuel,b\text{-}o,1} = 0.5778$ and

$$W_0^I = W_0^{II} + 0.1(0.5778)W_0^I + .5778W_0^I. \tag{D.31}$$

This equation yields $W_0{}^{II} = 0.3644W_0{}^I$. When the first stage drops off and the second stage rocket fires, we may take the burnout time for it to be, say, $t_{b\text{-}o,2} = 360$ s and

$$V_C = 4.413\log_e[1 - .002889(t-200)]^{-1} - \frac{.009807}{2}(t-200) + 2.883. \tag{D.32}$$

Therefore, $V_{C,b\text{-}o,2} = 4.981$ km/s and

$$\frac{W_{fuel,b-o,2}}{W_0^{II}} = \frac{1.3}{450}(t_{b-o,2} - 200) = 0.4622. \tag{D.33}$$

Then the weight of the second stack is

$$W_0^{II} = W_0^{III} + W_{str,2} + W_{eng,2} + W_{fuel,2}. \tag{D.34}$$

Substituting in our known or assumed values, we obtain

$$W_0^{II} = W_0^{III} + 0.1(0.4622)W_0^{II} + 0.4662W_0^{II}. \tag{D.35}$$

This equation results in $W_0{}^{III} = 0.4916W_0{}^{II} = 0.4916(0.3644W_0{}^I) = 0.1791W_0{}^I$. Now the third stage fires and accelerates the vehicle to the circular velocity, $V_C = 7.8$ km/s. One may determine the burnout time by using the velocity relation

$$V_{C,3} = 4.413\log_e[1 - .002889(t-360)]^{-1} - \frac{0.009807}{2}(t-360) + 4.778. \tag{D.36}$$

The resulting time is $t_{b\text{-}o,3} = 565$ s and $W_{fuel,3}/W_0{}^{III} = 0.002889(565\text{-}360) = 0.5922$. The weight of the third stack is then

$$W_0^{III} = 23,275 + 0.1(.5992)W_0^{III} + .5992W_0^{III}. \tag{D.37}$$

The resulting weight is $W_0{}^{III} = 66,770$ lbs, and now from the result obtained by means of Equation (D.35), we find that $W_0{}^I = 372,800$ lbs and $W_0{}^{II} = 135,700$ lbs. Therefore, $F^I = 484,700$, $F^{II} = 176,400$ lbs, and $F^{III} = 86,710$ lbs. The requirement for the first stack can be accommodated easily by two J-2 engines, but that for the second and third stacks are not met readily by existing LOX–$LH_2$ engines. The second stage might be able to use two Russian 11D-57 engines yielding 176,000 lbs of thrust. The third stage could be powered by two Pratt & Whitney RL-60 engines (currently under development), providing a total of about 100,000 lbs of thrust.

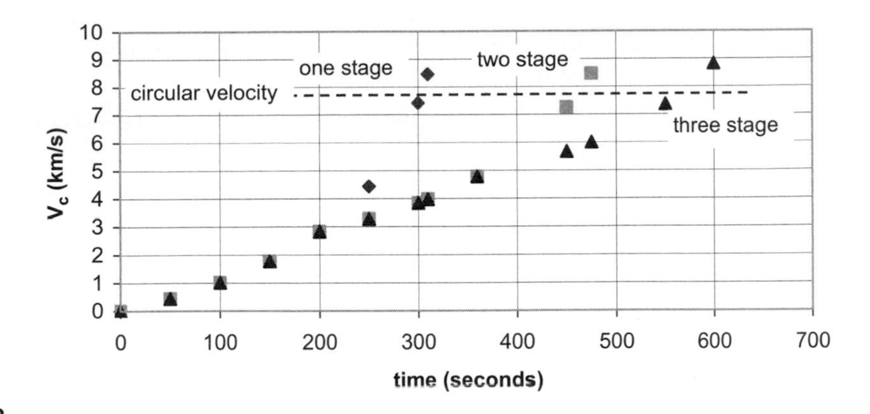


**FIGURE D.8**

Velocity histories for one-, two-, and three-stage rockets with constant specific impulse of 450 s and a payload weight of 23,275 lbs.

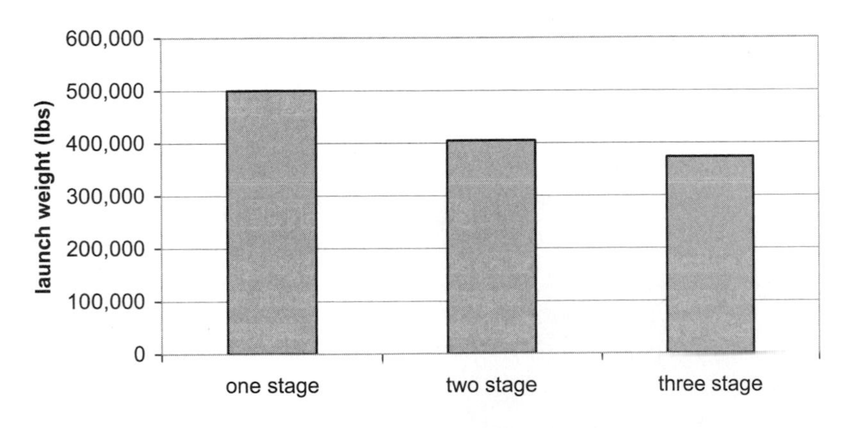


**FIGURE D.9**

Comparison of launch weights for one-, two-, and three-stage rockets with a constant specific impulse of 450 s and a payload weight of 23,275 lbs.

The reduction in launch weight achieved using three stages instead of one is 25%, while the reduction in weight compared to two stages is about 8%. The velocity histories for the three different configurations are shown in Figure D.8, and the comparison of launch weights is shown in Figure D.9. Some characteristics of the generic vehicle considered earlier and some operational vehicles are shown in Tables D.3 and D.4.

## D.7 GENERAL LAUNCH VEHICLE DESIGN CONSIDERATIONS

A view of the launch of Space Shuttle Atlantis (STS-115) on September 9, 2006, as seen from the International Space Station, is shown in Figure D.10. A useful reference for space shuttle operations and specifications may be found in the Space Shuttle Reference Manual.

**Table D.3** Breakdown of Weights for Three Different Launch Vehicles with Identical Payloads, Specific Impulses, and Initial Thrust to Weight Ratios

| | One Stage | Two Stage | Three Stage |
|---|---|---|---|
| $W_{pay}$ (lbs) | 23,275 | 23,275 | 23,275 |
| $W_{str+eng}$ (lbs) | 43,290 | 34,680 | 31,770 |
| $W_{fuel}$ (lbs) | 432,900 | 346,800 | 317,700 |
| $W_0$ | 499,500 | 404,800 | 372,800 |

**Table D.4** Breakdown of Weights for Four Operational Systems

| | Space Shuttle ($LOX-LH_2$ and Solid Rockets) | Soyuz (LOX-RP1) | Gemini-Titan II ($UDMH-N_2O_4$) | Mercury-Atlas (LOX-RP1) |
|---|---|---|---|---|
| $W_{pay}$ (lbs) | 151,000 | 13,800 | 7,000 | 2,900 |
| $W_{str+eng}$ (lbs) | 392,600 | 28,100 | 15,000 | 22,200 |
| Stage | 1.5 | 2 | 2 | 2 |

**FIGURE D.10**

View of the launch of Space Shuttle Atlantis (STS-115) from the International Space Station (NASA).

A general description of the launch vehicle design process has been presented by Blair and colleagues (2001). The movement in ballistic missile and manned space access technology in the late 1950s and early 1960s resulted in the development of a number of practical rocket engines. The U.S.-manned space program up through the 1960s involved high-performance liquid bipropellant engines, which are described in Williams and associates (1976).

## D.8 NOMENCLATURE

| | |
|---|---|
| $A$ | cross-sectional area |
| $a$ | sound speed |
| $a_n$ | normal component of acceleration |
| $a_t$ | tangential component of acceleration |
| $B$ | ballistic parameter $W/C_DA$ |
| $C_D$ | drag coefficient |
| $D$ | drag |
| $F$ | force |
| $g_E$ | acceleration of gravity at Earth's surface |
| $h$ | altitude |
| $H$ | scale height of the atmosphere |
| $I_{sp}$ | specific impulse |
| $k$ | gravitational constant |
| $L$ | lift |
| $M$ | Mach number |
| $m$ | mass |
| $\dot{m}$ | mass flow |
| $p$ | pressure |
| $R$ | gas constant |
| $R_u$ | universal gas constant |
| $r$ | radial coordinate |
| $F$ | thrust |
| $t$ | time |
| $V_C$ | velocity of center of mass |
| $W$ | weight |
| $\gamma$ | flight path angle |
| $\delta$ | atmospheric pressure ratio $p_0/p_{0,s.l.}$ |
| $\nu$ | polar angle coordinate |
| $\rho$ | gas density |
| $\sigma$ | density ratio $\rho/\rho_0$ |

### D.8.1 Subscripts

| | |
|---|---|
| $avg$ | average |
| $cir$ | circular orbit conditions |

| | |
|---|---|
| *eng* | engine |
| *fuel* | fuel |
| max | maximum conditions |
| *pay* | payload |
| *str* | structure |
| *vac* | vacuum conditions |
| 0 | initial conditions or sea level conditions |
| 1 | first stage |
| 2 | second stage |
| 3 | third stage |

### D.8.2 Superscripts

| | |
|---|---|
| * | critical conditions, $M = 1$ |
| *I* | conditions in first stack |
| *II* | conditions in second stack |
| *III* | conditions in third stack |

## References

Blair, J. C., Ryan, R. S., Schutzenhofer, L. A., & Humphries, W. R. (2001). Launch Vehicle Design Process: Characterization, Technical Integration, and Lessons Learned. NASA/TP–2001–210992, May.

Isakowitz, S.J. (1995). International Reference Guide to Space Launch Systems. AIAA, Reston, VA.

Space Shuttle Reference Manual (http://science.ksc.nasa.gov/shuttle/technology/sts-newsref/stsref-toc.html#).

Williams, O.S., et al. (1976). Liquid rockets in perspective: Developments in the 1960s. *Aerospace America*, March, 47–57.

APPENDIX

# Safety, Reliability, and Risk Assessment

# E

## E.1 SYSTEM SAFETY AND RELIABILITY

Reliability and safety of space missions have always been major goals and the design challenge this represents is substantial. This appendix provides an introductory analysis of the reliability and safety of a manned space vehicle making a round trip to the International Space Station (ISS). The material follows closely the approach of Sforza and colleagues (1993), which examines a round trip to Mars using nuclear thermal propulsion for outbound and inbound transit between low Earth orbit and Mars.

In general, we speak of reliability as the probability of mission success for a given mission time period, $t$, and use the symbol $R(t)$, or sometimes $P(t)$. In conjunction with space missions, it is common to speak of two aspects of mission success. First is the probability of crew safety, that is, the probability that no serious injury to or death of a crew member occurs during the mission. Second is the probability of mission success, that is, the successful accomplishment of a specific set of objectives for the mission. To generalize the notation to include both mission success and crew safety, we introduce the following:

$X_i$ = mission success during phase $i$
$x_i$ = mission is safe during phase $i$
$R_{ms}$= probability of mission success
$R_{cs}$= probability of crew safety

If we make the simplifying assumption that mission success and crew safety depend on serial success of all the mission phases $1 \leq i \leq n$, then the probabilities become

$$R_{ms} = Pr(X_1 X_2 \ldots X_n) \quad \text{(E.1)}$$

$$R_c = Pr(x_1 x_2 \ldots x_n). \quad \text{(E.2)}$$

A further simplifying assumption made commonly is that each of the mission phases are independent, in which case Equations (E.1) and (E.2) become

$$R_{ms} = Pr(X_1)Pr(X_2)\ldots Pr(X_n) \quad \text{(E.3)}$$

$$R_{cs} = Pr(x_1)Pr(x_2)\ldots Pr(x_n). \quad \text{(E.4)}$$

Table E.1 shows the major phases that are representative of a round trip mission to the ISS using a two-stage rocket for the launch. A similar table would be constructed for a specific space

**Table E.1** Representative Mission Phases for Two-Stage Rocket Round Trip to ISS

| Phase No. | Description | Propulsion Status | Duration |
|---|---|---|---|
| 1 | Ignition of ME to first stage separation | Main engines and boosters active | 125 s |
| 2 | Ignition of second stage to MECO | Second stage engines active | 400 s |
| 3 | MECO to ISS docking | OMS engines active | 700 s |
| 4 | Orbital operations | Inactive | 7 days |
| 5 | Deorbit burn to reentry interface | OMS engines active | 20 min |
| 6 | Reentry | Inactive | 10 min |
| 7 | Terminal operations | Inactive | 15 min |
| 8 | Final approach and landing | Inactive | 10 min |

transportation system and associated mission details. The top-level functional failures that could lead to a loss-of-vehicle (LOV) event are as follows:

- Propulsion failure—propulsive thrust must be maintained above certain levels to assure ability to abort the mission safely or complete it
- Vehicle configuration failure—control features, such as thrust vectoring, must operate within certain limits to maintain vehicle control
- Containment failure—escape of high-energy gases or solid particles from their proper paths may damage other systems
- Vehicle environment failure—life support system must operate within certain limits to assure crew safety
- Externally initiated failure—lightning strikes, wind shear, space debris, and so on are external events that may precipitate system failures and lead to LOV

One of the major risk factors for space transportation systems is that of propulsion failure. In the generic round trip to orbit mission considered here there are several phases where rocket engines are firing (phases 1, 2, 3, and 5) and several where they are inactive (phases 4, 6, 7, and 8). This appendix considers the critical thrusting phases of the mission, focusing on the reliability of the propulsion system that operates during these phases. The success of these phases depends on the successful operation of a number of systems, such as those listed in Table E.2.

**Table E.2** Major Systems Functioning during Thrusting Phases

| Symbol | System |
|---|---|
| $Y_1$ | Propulsion |
| $Y_2$ | Communications |
| $Y_3$ | Life support |
| $Y_4$ | Power and thermal management |
| $Y_5$ | Navigation, guidance, and control |

One of the important uses of a top–down analysis, as outlined earlier, is to help select overall goals and apportion the mission success among the various mission phases and the systems functions operative within them. To reiterate, we assume a simplified model of a phase in which all systems are independent and must operate successfully or safely. The following equations, which are analogous to Equations (E.3) and (E.4) are applicable for a powered phase, such as phase 1, the first stage ascent:

$$Pr(X_1) = Pr(Y_1)\, Pr(Y_2)\, Pr(Y_3)\, Pr(Y_4)\, Pr(Y_5) \tag{E.5}$$

$$Pr(Y_1) = Pr(y_1)\, Pr(y_2)\, Pr(y_3)\, Pr(y_4)\, Pr(y_5) \tag{E.6}$$

The symbols are defined as follows:

$Y_i$ = success of system $i$
$y_i$ = safe performance of system $i$
$Pr(X_1)$ = probability of mission success for the first stage ascent phase
$Pr(x_1)$ = safe performance of the first stage ascent phase

The complete system reliability and safety models involve equations such as Equations (E.5) and (E.6) for all the other phases, as well as Equations (E.3) and (E.4), or Equations (E.1) and (E.2) if it isn't possible to make the assumption that all the phases are independent.

## E.2 APPORTIONING MISSION RELIABILITY

The task of choosing appropriate goals for $R_{ms}$ and $R_{cs}$ for the entire mission and apportioning these goals among the various mission phases is difficult. In general, such a task becomes a top–down iterative process of apportionment based on evolving system details alternating with bottom–up reliability prediction of the latest models of the system. Of course, it is expected that these estimates will tend to converge as the project progresses. We begin the top–down process by considering Equation (E.3). For the present purposes, it is convenient to separate the equation into two parts, one in which propulsion is inactive and one in which it is active. Thus Equation (E.3) becomes

$$R_{ms} = [Pr(X_4)\, Pr(X_6)\, Pr(X_7)\, Pr(X_8)]\, [Pr(X_1)\, Pr(X_2)\, Pr(X_3)\, Pr(X_5)]. \tag{E.7}$$

We may designate the unpowered part by subscript 1 and the powered part by subscript 2, as follows:

$$R_{ms} = [R_1][R_2]. \tag{E.8}$$

The apportionment process should be based on historical data for similar projects, as well as on models for the variation of phase reliabilities with system costs, weights, and other factors. Because such information is seldom available during the initial system design process, carefully chosen simplifying assumptions will often be necessary.

One way to proceed is to attempt to bracket the reliability estimates for the propulsion system by considering two extremes. In one case the propulsion system has low reliability compared to the other systems and is therefore the limiting system factor. In the case where the propulsion reliability is relatively low, we may set $R_1 = 1$ and also set the probabilities for all the powered phases equal: $Pr(X_1) = Pr(X_1) = Pr(X_1) = Pr(X_1) = P_{low}$. Then Equation (E.8) yields

**Table E.3** Relationship between Mission Reliability and Phase Reliability

| $R_{ms}$ | 1-$R_{ms}$ | $P_{low}$ | 1-$P_{low}$ | $P_{same}$ | 1-$P_{same}$ |
|---|---|---|---|---|---|
| 1.000 | 0.000 | 1.000 | 0.000 | 1.00000 | 0.00000 |
| 0.999 | 0.001 | 0.9997 | 0.0003 | 0.99987 | 0.00013 |
| 0.995 | 0.005 | 0.9987 | 0.0013 | 0.99937 | 0.00063 |
| 0.990 | 0.010 | 0.9974 | 0.0026 | 0.99874 | 0.00126 |
| 0.950 | 0.050 | 0.9872 | 0.0128 | 0.99361 | 0.00639 |
| 0.900 | 0.100 | 0.9740 | 0.0260 | 0.98692 | 0.01308 |

$$R_{ms} = (P_{low})^4 \tag{E.9a}$$

$$P_{low} = (R_{ms})^{1/4}. \tag{E.9b}$$

In the other extreme, where the propulsion reliability is about the same as the other systems and all phases of the mission are about equally reliable, the following result is found:

$$R_{ms} = (P_{same})^8 \tag{E.10a}$$

$$P_{same} = (R_{ms})^{1/8}. \tag{E.10b}$$

It is helpful to compare the relationship between mission and powered phase reliabilities as defined by the apportionment rules in Equations (E.9b) and (E.10b), which is shown in Table E.3. These results show that for a mission success goal of $R_{ms} = 0.95$, the phase reliability for the propulsion system should be between 0.99361 and 0.9872. The corresponding phase unreliability ranges between 0.00639 and 0.0128, a ratio of about 2.

We must choose the goal for $R_{ms}$ as the start of the planning process. The best way to set such a goal is to study past projects, including their goals and reliability, as depicted in Table E.4 for some manned missions. Note that the Apollo and space shuttle have demonstrated a mission reliability of around 0.95.

## E.3 RELIABILITY FUNCTION

In order to complete the apportionment process and relate system reliability goals to system failure rates we must introduce and summarize a few additional aspects of reliability theory; see, for example, Shooman (1990). The reliability function $R(t)$ is defined as the probability of success for a time period, $t$. The random variable of interest, $t$, is the time to failure. Using standard random variable mathematics, we can relate $R(t)$ to the cumulative density function $F(t)$, the probability of failure in the time interval from 0 to $t$, and $F(t)$ to the probability distribution function f(t) as follows:

$$R(t) = 1 - F(t) = 1 - \int_0^t f(x)dx. \tag{E.11}$$

**Table E.4** Reliabilities of Some Past Manned Space Missions

| Project | Year | Goal | Success Ratio | Demonstrated Reliability[a] | Reference |
|---|---|---|---|---|---|
| Apollo | 1965–1975 | 0.95 | 17/18 | One failure (Apollo 13) out of 18, $R = 0.944$ | "World Almanac" |
| Space shuttle | 1981–2011 | — | 133/135 | Two failures (Challenger and Columbia) out of 135 flights, $R = 0.9852$ | www.nasa.gov/mission_pages/shuttle/shuttlemissions/list_main.html |
| Delta booster | 1960–1990 | — | — | 0.9403 | Isakowitz, 1991 |
| Atlas booster | 1957–1990 | — | — | 0.8063 | Isakowitz, 1991 |
| One to 5 J-2 engines | 1966–1975 | — | 93/96 | 0.9688 | SAIC |
| Two to six RL-10 engines | 1961–1986 | — | 198/198 | 199/200 0.9950 | SAIC |
| Three SSME engines | 1981–1989 | — | 85/86 | 0.9884 | SAIC |

[a]*Binomial distribution assumed in calculating demonstrated reliability; see Section E.4.*

Although Equation (E.11) is mathematically complete, reliability engineers find it better to deal with the hazard, or failure rate, function $z(t)$, which is the fraction of survivors at time $t$ that fail per unit time. We then define

$$z(t) = Lim_{\Delta t \to 0} \frac{failures}{survivors \times \Delta t}. \tag{E.12}$$

Then, for $N$ items placed on test we can write

$$z(t) = Lim_{\Delta t \to 0} \frac{N[f(t) - f(t + \Delta t)]}{NR(t)\Delta t} = \frac{1}{R(t)}\frac{df(t)}{dt} = -\frac{1}{R(t)}\frac{dR(t)}{dt}. \tag{E.13}$$

Solution of the differential equation in Equation (E.13), subject to the condition $R(0) = 1$, that is, the item is initially good, yields

$$R(t) = \exp\left[-\int_0^t z(x)dx\right]. \tag{E.14}$$

Another variable often used in reliability studies is the mean time to failure (MTTF) which, using the definition of the mean of a random variable, is given by

$$MTTF = \int_0^\infty tf(t)dt = \int_0^\infty R(t)dt. \tag{E.15}$$

## E.4 FAILURE RATE MODELS AND RELIABILITY ESTIMATION

Many failure rate models have been used for components and systems, but the simple constant failure rate $z(t)$ = constant is suitable for a first-order approximate analysis and is often reasonably accurate. With this assumption, we have the following relations:

$$z(t) = \lambda \tag{E.16a}$$

$$R(t) = e^{-\lambda t} \tag{E.16b}$$

$$MTTF = \lambda^{-1} \tag{E.16c}$$

Combining Equations (E.5) and (E.16) yields

$$R_{ms} = e^{-\lambda_1 t'_1} e^{-\lambda_2 t'_2} \ldots e^{-\lambda_4 t'_4} = \exp\left[\sum_{i=1}^{4} -\lambda_i t'_i\right]. \tag{E.17}$$

In this equation, $\lambda_i$ is the constant failure rate for phase $i$ and $t_i'$ is the time of operation from the start of phase $i$.

Reliability estimates are best prepared from experimental data. If it is known from prior tests and analysis that the reliability of a component or system is governed by a constant failure rate, as in Equation (E.16), statistical theory suggests estimating the parameter $\lambda$ by the ratio of failures to total operating hours. This maximum likelihood estimate (MLE) is essentially the same as that on the right-hand side of Equation (E.12). In the event the database shows no failures the MLE = 0, which will lead to results that are difficult to justify. A better procedure in this case is to assume a value of 0.33 for the number of failures; see, for example, Welker and Lipow (1974).

In some cases, the number of components tested, $n$, and the number of successes, $r$, are reported, but not the number of test hours. There one may assume that the binomial distribution is applicable and that the success parameter, $p$, is given by the MLE formula

$$p = \frac{r}{n}. \tag{E.18}$$

In the event that there are zero failures such that $p = 1$, Equation (E.18) is modified using Bayesian estimation principles to yield

$$p = \frac{r+1}{n+2}. \tag{E.19}$$

Results quoted in Table E.4 were obtained using Equation (E.18). In cases where the operating time is also specified, Equation (E.12) would be used.

## E.5 APPORTIONMENT GOALS

The apportionment carried out in Section E.2 resulted in a bracketing of the phase reliability $P$ by $P_{low}$ and $P_{same}$ given by Equations (E.9b) and (E.10b) and by Table E.3. We may now continue the apportionment process to a lower level so that we can focus on, say, propulsion system reliability.

Averaging the demonstrated reliabilities of Apollo and the space shuttle as given in Table E.4 yields a value of $R = 0.963$. For the current mission, we assume a mission goal of $R = 0.96$, which results in a range of powered propulsion phase reliability between 0.9898 and 0.9949. The major systems of such a phase, say $X_1$, are shown in Table E.2, and again we may consider two extremes.

If the propulsion system is the limiting factor, we assume that the phase reliability is equal to the reliability of the propulsion systems alone, that is, $P(X_1) = P(Y_1) = P_{low} = 0.9898$, and that the other reliabilities of all the other systems are all equal to unity, that is, $Pr(Y_2) = Pr(Y_3) = P(Y_4) = Pr(X_5) = 1$. At the other extreme, where all the systems are considered equally reliable, $P(X_1) = [P(Y_1)]^5 = 0.9898$, so that $P(Y_1) = 0.9980$. Now the reliability for the propulsion system must lie in the range between 0.9898 and 0.9980.

We now turn to estimating the difficulty in achieving the bracketing reliability goals just calculated. We may make use of the database in Table E.4 in which we see that five engine systems have an average demonstrated reliability of 0.94, and the associated failure probability is 0.06. Based on previous data, we see that it is not likely that we can meet our required reliability level. Even omitting Atlas booster data, as it encompasses the earliest efforts at high-performance lifting capability, the average of the remaining four sets of data is still only 0.9731 and lies outside the desired bracket. This suggests that a more detailed and sophisticated analysis is necessary for estimating reliability. Additional studies of this topic as it relates to missions to the space station and to Mars may be found in Shooman and Sforza (1994, 2002).

## E.6 OVERVIEW OF PROBABILISTIC RISK ASSESSMENT

Carrying out a probabilistic risk assessment requires imagining scenarios, that is, sequences of events, which lead to undesired consequences. There is an initiating, or triggering, event that sets into motion the subsequent series of events ultimately resulting in an end state. Triggering events are usually malfunctions or failures of any type that cause a deviation from the intended operation. End states are defined by the decision makers and illustrate the nature of the undesired consequence. For example, in manned space flight, a particularly important end state is LOV.

Scenarios are generally presented in diagrammatic form, the development of which is not unique and often represents, to some extent, a degree of creativity that tends to come with experience. A hierarchy of diagrams can be used effectively to form a framework for a rigorous review of the risk exposure in complicated systems. They may be described as follow.

- Master logic diagram (MLD): This is a hierarchy of initiating events illustrating the various manners in which a damage result might occur. There is a compilation of functional categories of perturbations to the system under consideration and spreading from these are characterizations of components for each of these functions. The MLD starts with a damage event of major concern and then the subsequent events that are necessary, but not sufficient to cause the major event are taken in greater and greater detail. A portion of such a master logic diagram is shown in Figures E.1, E.2, and E.3.
- Functional event sequence diagram (FESD): Initiators in the MLD are screened to determine which of them are so highly unlikely that they may be eliminated from deeper analysis. Then the remaining trigger events are evaluated in the sense of establishing the sequence of events that

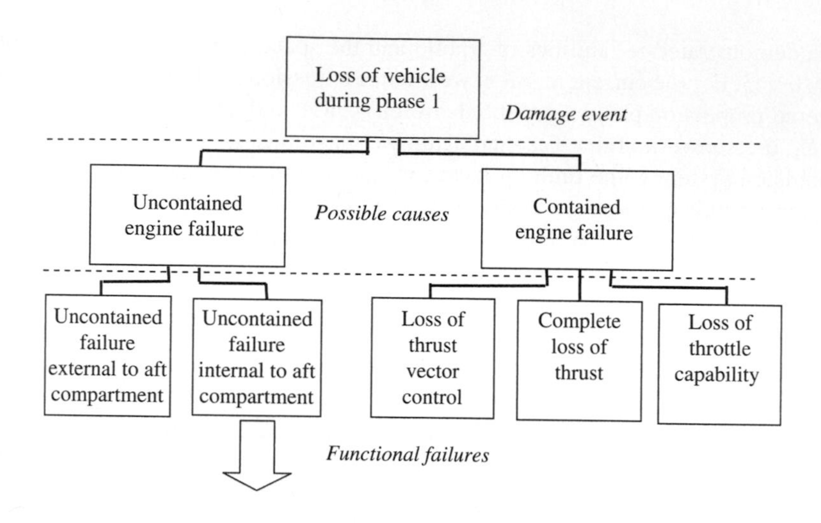


**FIGURE E.1**

A portion of a master logic diagram is shown. The top level is the damage event, and the next level is the possible cause for the event. The next level is the functional failures proceeding from the possible causes. The arrow points toward the next level of the MLD continued in Figure E.2.

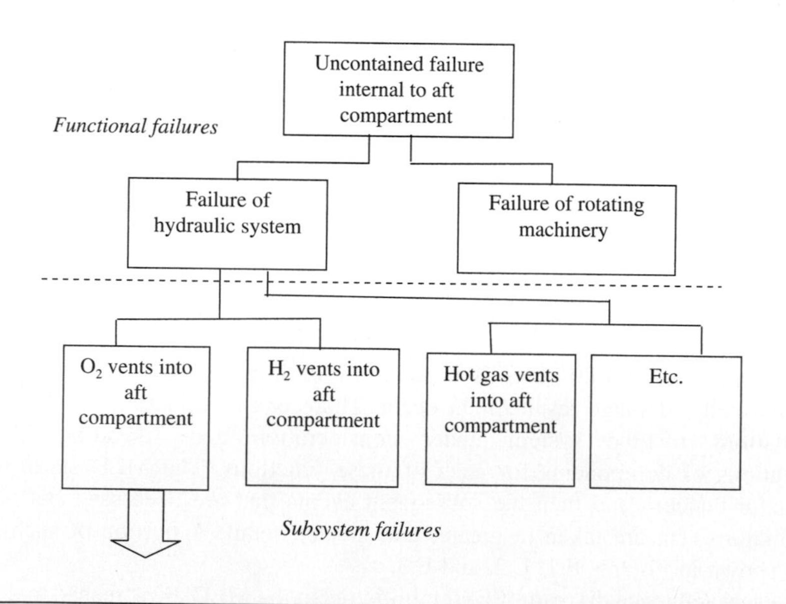


**FIGURE E.2**

Continuation of the MLD shows the flow from functional failures down to subsystem failures. The arrow points to the next level of the MLD shown in Figure E.3.

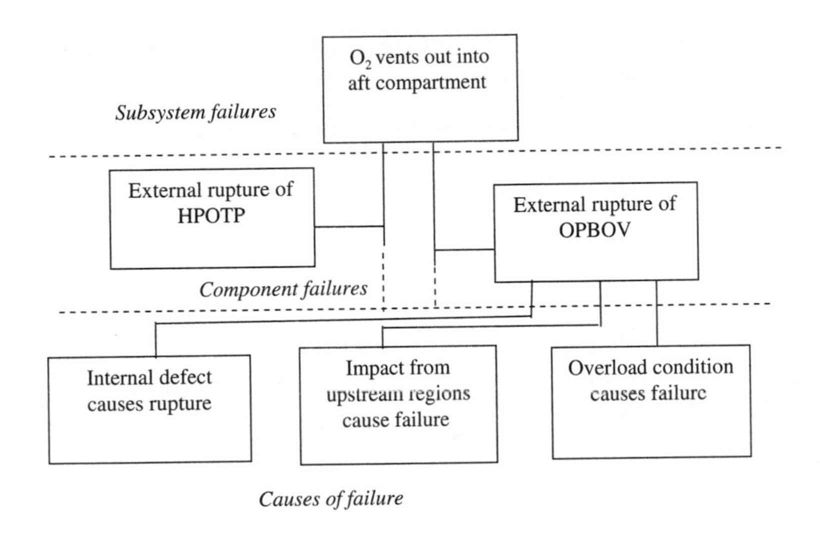


**FIGURE E.3**

Conclusion of the MLD showing flow from subsystem failures to component failures to causes of failure.

would follow. These events are formulated in a manner that provides a binary outcome, such as yes or no or failure or success. The FESD provides a useful interface for reliability and systems engineers to pool expert knowledge.

- Event tree: This is a sort of detailed flow chart of binary decisions based on the FESD that may be developed easily for computer modeling. All the different event paths may be traced down to where they lead to successes or failures.
- Fault tree: This is similar to an event tree except that here only failure sequences are considered, and the components or systems that fail are examined in detail. Once again the possible modes of failure of the component or system are treated in a binary manner and every possible source of failure is traced.

## E.7 TOP FUNCTIONAL FAILURES FOR THE SPACE SHUTTLE

Master logic diagrams are based on functional failures, and the following five are identified for the space shuttle:

- Propulsion failure—without proper levels of propulsion during powered flight phases mission or abort success cannot be achieved
- Vehicle configuration failure—without appropriate control authority throughout the flight mission failure can ensue
- Energetic gas and debris containment failure—departure of gases and debris from planned and controlled paths may destroy vehicle components and lead to mission failure
- Orbiter environmental support failure—without proper life support system functions, the crew can be lost and mission failure will be certain

- Externally caused failure—events outside the vehicle and its command and control system, such as lightning, extreme wind shear, space debris, and so on, can lead to mission failure

These functional failures are applicable to manned space missions in general and may be used in the construction of an MLD for specific cases.

A recent evaluation of safety issues in NASA's aerospace programs is discussed by the NASA Aerospace Safety Advisory Panel in its annual report for 2010.

## E.8 WEIBULL DISTRIBUTION

One of the random variables found to be valuable in the modeling of extreme events is the Weibull random variable. Its probability density function is given by

$$f(t) = \frac{g}{t_c}\left(\frac{t}{t_c}\right)^{g-1} \exp\left[-\left(\frac{t}{t_c}\right)^{g}\right], \tag{E.20}$$

where $t$ represents time, $t_c$ is a characteristic time for the event being modeled, and $g$ is a shape factor that best captures the characteristics of the event being modeled; they are all positive quantities. The general shape of the Weibull distribution is shown in Figure E.4 for several values of the parameters $t_c$ and $g$.

The probability that the system being modeled fails before some time $t$ is given by

$$P(T \leq t) = F_T(t) = \int_0^t f_T(\tau)dt, \tag{E.21}$$

where $T$ is the time to failure and $f_T(t)$ is the probability density function of the time to failure, while $F_T(t)$ is the cumulative distribution function of the time to failure, $T$. Then the probability that the system will operate after the time $t$ is given by

$$P(T > t) = R(t) = 1 - F_T(t). \tag{E.22}$$

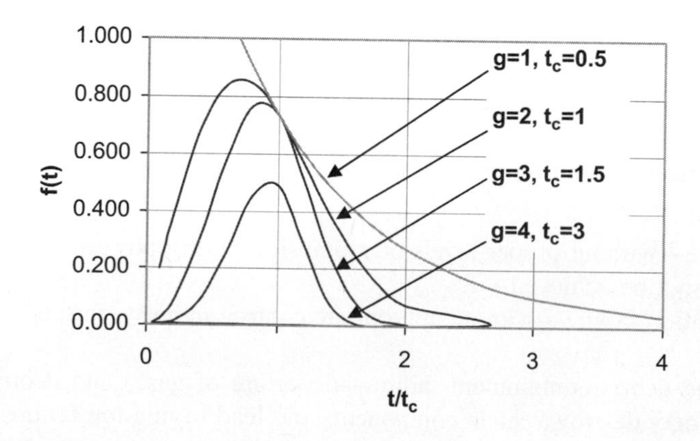


**FIGURE E.4**

Weibull distribution function as a function of normalized time for various values of the parameters $t_c$ and $g$.

The quantity $R(t)$ is called the reliability of the system being modeled. Then the mean, or expected, time to failure is given by the integral of the first moment of the failure probability density function as follows:

$$E\{t\} = \int_0^\infty t f_T(\tau)dt = \int_0^\infty t\frac{dF_T(t)}{dt}dt = -\int_0^\infty t\frac{dR(t)}{dt}dt. \tag{E.23}$$

Then, integration by parts yields

$$E\{t\} = \int_0^\infty R(t)dt. \tag{E.24}$$

For the Weibull distribution of Equation (E.20), the expected time to failure and its variance are

$$\begin{aligned} E\{T\} &= c\Gamma\left(1-\frac{1}{g}\right) \\ Var(t) &= c^2\left[\Gamma\left(1+\frac{2}{g}\right) - \Gamma^2\left(1+\frac{1}{g}\right)\right]. \end{aligned} \tag{E.25}$$

The gamma function is a standard function, is tabulated in handbooks, and is generally available in computer mathematics libraries; it is defined as follows:

$$\Gamma(x) = \int_0^\infty y^{x-1}e^{-y}dy. \tag{E.26}$$

The cumulative distribution of the Weibull probability density function is

$$F_T = 1 - \exp\left[-\left(\frac{T}{t_c}\right)^g\right]. \tag{E.27}$$

Then the reliability of a system that follows a Weibull distribution is given by

$$R(t) = \exp\left[-\left(\frac{T}{t_c}\right)^g\right]. \tag{E.28}$$

The failure rate at any time $t$ is given by the ratio of the probability density function of the time to failure of the system to the reliability of the system and is given by

$$z(t) = \frac{f_T(t)}{R(t)} = \frac{g}{t_c}\left(\frac{t}{t_c}\right)^{g-1}. \tag{E.29}$$

Failure rates for various shape factors and the same characteristic time are shown in Figure E.5. Note that the failure rate for a shape factor $g = 1$ is constant at $h(t) = 1/t_c$. A constant failure rate is typical of system behavior at times after the point of being "broken in" but before the point of being "worn out." For shape factors $g < 1$, the failure rate is inversely proportional to time so that failures are decreasing in number as time goes on. This is typical of early times, before the system is "broken in," when the faulty components of the system are weeded out. Conversely, for shape factors $g > 1$, the failure rate increases with time, as would be expected as the system "wears out."

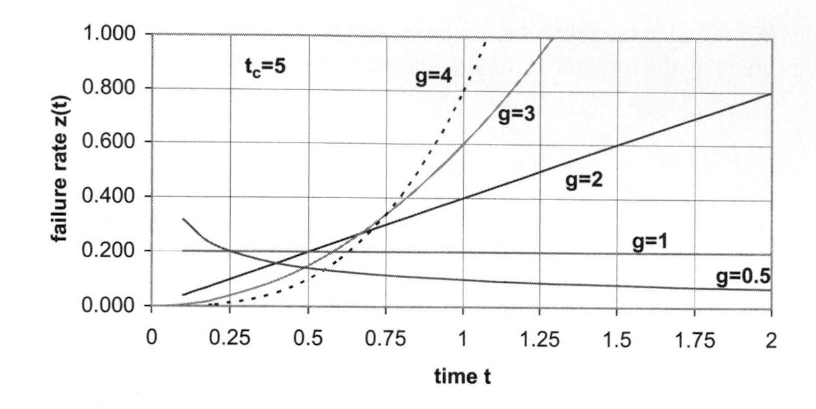


**FIGURE E.5**

Failure rate as a function for various shape factors and a constant value of characteristic time $t_c = 5$.

## References

www.nasa.gov/mission_pages/shuttle/shuttlemissions/list_main.html.

Isakowitz, S. J. (1991). *International Reference Guide to Space Launch Systems*. Reston, VA: American Institute of Aeronautics and Astronautics.

NASA Aerospace Safety Advisory Panel: Annual Report for 2010 (http://oiir.hq.nasa.gov/asap/documents/2010_ASAP_Annual_Report.pdf).

SAIC (1989). NASA Quantifiable Failure Rate Data Base for Space Flight Equipment. Science Applications International Corporation Report No. SAICNY-89-10-43A and 43B.

Sforza, P. M., Shooman, M. L., & Pelaccio, D. G. (1993). A safety and reliability analysis for space nuclear thermal propulsion systems. *Acta Astronautica, 30*, 67–83.

Shooman, M. L. (1990). *Probabilistic Reliability: An Engineering Approach*. Melbourne, FL: Krieger.

Shooman, M.L., & Sforza, P.M. (1994). Reliability-Driven Design of Nuclear Powered Spacecraft. Proceedings of the 1994 Annual Reliability and Maintainability Symposium, 94RM-164, 1–7.

Shooman, M.L., & Sforza, P.M. (2002). A Reliability Driven Mission for the Space Station. Proceedings of the 2002 Annual Reliability and Maintainability Symposium, Philadelphia, PA.

Welker, E.L., & Lipow, M. (1974). Estimating the Exponential Failure Rate from Data with No Failure Events. Proceedings of the Annual Reliability and Maintainability Symposium, 420–427.

*World Almanac*. (1990). New York: Pharos Books, 145–146.

APPENDIX

# F Aircraft Performance

## F.1 RANGE EQUATION

In the cruise configuration we can estimate airplane performance characteristics assuming steady and approximately level flight in the $x$–$z$ plane, where $x$ is the local horizontal direction and $z$ is the local vertical direction. Treating the airplane as a point mass the equilibrium equations along and normal to the flight path are given as follows:

$$F = D + W \sin\phi$$

$$L = W \cos\phi,$$

where the angle $\phi$ is the flight path angle, that is, the angle between the flight trajectory and the local horizontal, or $\tan\phi = dz/dx$.

The weight of the aircraft is described readily in terms of the fixed weight (the operating empty weight plus the payload weight), $W_{OE} + W_{PL}$, and the time-varying weight (the fuel weight), $W_F$, as follows:

$$W = W_{OE} + W_{PL} + W_F(t). \tag{F.1}$$

The rate at which propellant is used in quasi-steady operation is assumed to be linearly related to the net thrust produced so that the rate of change of the weight of the aircraft is

$$dW/dt = dW_F/dt = -c_j F \text{ (in lbs/hr)}, \tag{F.2}$$

where quantity $c_j$ is the specific fuel consumption in lbs fuel per hour per lb thrust. This equation may also be written as

$$dW_F/dt = -F/I, \tag{F.3}$$

where $I$ is the specific impulse, measured in pounds of thrust per pounds of fuel consumed per second. Thus, the relationship between it and specific fuel consumption is

$$I = (c_j/3600)^{-1}. \tag{F.4}$$

Now the rate of fuel consumption may be described as follows:

$$\frac{dW_F}{dt} = -c_j F = -F/I = -c_j(D + W\sin\phi).$$

Therefore,

$$dt = -\frac{W}{c_j(D + W\sin\phi)}\frac{dW_F}{W} = -\frac{1}{c_j\left(\frac{D}{W} + \sin\phi\right)}\frac{dW}{W}. \tag{F.5}$$

The ratio $D/W = D \cos \phi / L = \cos \phi/(L/D)$ and $L/D$ is the lift to drag ratio, or aerodynamic efficiency. We now must make some assumption as to the flight path. In this quasi-steady approximation it is taken for granted that the flight path angle $\phi \ll 1$, so that

$$\sin \phi = \phi + \phi^3/3 + \ldots \cong \phi \quad \text{and} \quad \cos \phi = 1 - \phi^2/2 + \ldots \cong 1.$$

Therefore, the lift and weight are always in balance, that is, $L = W$, to $O(\phi^2)$. Because the airplane's weight is decreasing continually as fuel is burned, the airplane will rise slowly during the course of the flight. This rise will be shown to be consistent with the assumption that $\phi \ll 1$.

Furthermore, the horizontal speed $V = dx/dt$ and the flight Mach number is $M = V/a$ so that the equation for the time differential may be rearranged as follows:

$$\frac{dW}{W} = -\frac{c_j}{aM}\left(\frac{1}{L/D} + \phi\right)dx.$$

The range achievable is

$$R = \int_0^R dx = \int_{W_1}^{W_2} -\frac{aM(L/D)}{c_j(1 + \phi L/D)}\frac{dW}{W}. \qquad \text{(F.6)}$$

The quantities $W_1$ and $W_2$ denote the aircraft weight at the start and end of the cruise segment, respectively. The simplest solution to this equation is achieved for flight where the quantity multiplying $dW/W$ in the integrand is constant. Flight at constant Mach number provides constant $ML/D$. Jet aircraft have made cruising above the weather, that is, above the troposphere, a common occurrence. Flight in this region of the tropopause and the start of the stratosphere (roughly 30,000 ft $< z <$ 40,000 ft) is characterized by nearly constant temperature (412R $> T >$ 393R) so that the speed of sound, $a$, is also essentially constant (995 fps $> a >$ 968 fps). Furthermore, the specific fuel consumption $c_j$ for jet engines in this flight region is also roughly constant, as discussed in Chapter 9. Typically $c_j$ is about 0.5 to 0.6 lbs fuel/lb thrust/hr for modern turbofan engines at cruise Mach numbers between 0.75 and 0.85, and $\phi$, the flight path angle, is, as will be shown subsequently, very small (on the order of $10^{-4}$) and essentially constant over the cruise range. Using a consistent set of units we find, for $L/D$ set to a reasonable value of approximately 15, that the term $\phi(L/D)$ is about $10^{-3}$ and therefore maybe neglected with respect to unity in the bracketed term in the denominator of the integrand of the range equation.

## F.1.1 Factors influencing range

The range in miles, with $a$, the sound speed, in mph (say 669 mph), is given by

$$R = \frac{aM(L/D)}{c_j}\left(\log_e \frac{W_1}{W_2}\right).$$

This is one form of Brequet's range equation. Because $W_2 = W_1 - W_F$, the range may also be written in the alternative form

$$R = -\frac{aM(L/D)}{c_j}\log_e\left(1 - \frac{W_F}{W_1}\right). \qquad \text{(F.7)}$$

This suggests what may appear obvious: to get long range carry more fuel! However, this is not always a good design solution. We therefore rewrite the range equation as follows:

$$R = \frac{aM(L/D)}{c_j}\log_e\left(1+\frac{W_F}{W_{OE}+W_{PL}}\right) \approx \frac{aM(L/D)}{c_j}\left(\frac{W_F}{W_{OE}+W_{PL}} - \frac{W_F^2}{2(W_{OE}+W_{PL})^2} + \ldots\right).$$

Note that for small values of $W_F/(W_{OE}+W_{PL})$, changes in that parameter yield linear changes in range while this advantage drops off for larger values. Certainly, it is not desirable in general to increase range by decreasing the payload. And, at some point, increases in $W_F$ or reductions in $W_{OE}$ are less effective in increasing range than increases in *L*/*D* or reductions in $c_j$. From another perspective, this illustrates that a point is reached where sacrifices in payload and/or structural weight are not very efficient in producing range.

As pointed out previously, the reduction in weight of the airplane as it burns fuel causes it to rise slowly. Solving the range equation for $W_2/W_1$ yields

$$\frac{W_2}{W_1} = \exp\left(-\frac{Rc_j}{aM(L/D)}\right).$$

For our assumption of flight at constant *M* and *L*/*D*, the lift coefficient $C_L$ also remains constant and the range equation becomes

$$\frac{W_2}{W_1} = \frac{C_{L,2}q_2S}{C_{L,1}q_1S} = \frac{C_{L,2}p_2M_2^2}{C_{L,1}p_1M_1^2} = \frac{p_2}{p_1} = \exp\left(-\frac{Rc_j}{aM(L/D)}\right).$$

The atmospheric pressure may be approximated by the exponential function so that

$$\frac{p_0\exp(-z_2/H)}{p_0\exp(-z_1/H)} = \exp\left(-\frac{z_2-z_1}{H}\right) = \exp\left(-\frac{Rc_j}{aM(L/D)}\right).$$

This gives then for the flight path angle

$$\tan\phi \approx \phi = \frac{z_2-z_1}{R} = \frac{c_jH}{aM(L/D)}. \tag{F.8}$$

For typical cruise values of the parameters in the previous equation, the number of feet rise in altitude is approximately equal to the number of miles covered in range, that is, $\Delta z$ measured in feet is approximately equal to *R* measured in miles.

## F.2 TAKE-OFF PERFORMANCE

The take-off may be considered in two parts: a ground run and an air run, as shown schematically in Figure F.1.

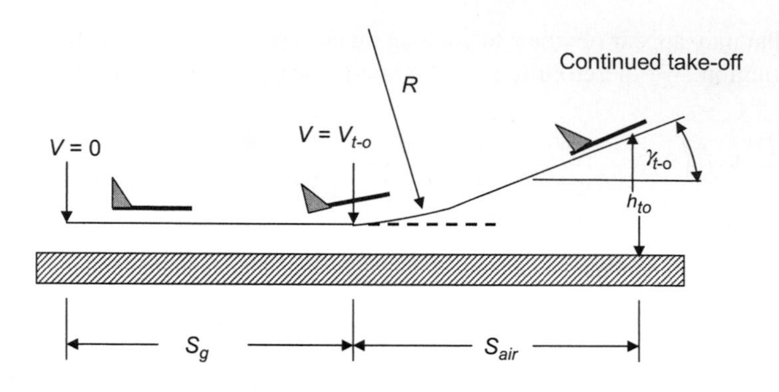


**FIGURE F.1**

Take-off procedure showing ground run and air run.

## F.2.1 Ground run

During the ground run it is assumed that all forces contributing to the acceleration of the aircraft are parallel to the ground, resulting in the following equation of motion:

$$m\frac{dV}{dt} = mV\frac{dV}{dx} = \frac{g}{W}(F - D - D_{roll}). \tag{F.9}$$

The weight in this equation is the take-off weight and is about 2.5% less than the ramp weight. The thrust may be represented by an average value as suggested by Torenbeek (1982), namely

$$F_{avg} = 0.75F_{static}\frac{(5+\lambda)}{(4+\lambda)}. \tag{F.10}$$

The drag is given by

$$D = c_D qS = \left(c_{D,0} + \frac{c_L^2}{\pi eA}\right)_{t-o} qS. \tag{F.11}$$

Here the profile drag coefficient and the lift coefficient are for the take-off condition, that is, flaps and landing gear down. The drag due to rolling resistance is given by

$$D_{roll} = \mu(L - W) = \mu(c_L qS - W)_{t-o}. \tag{F.12}$$

The equation of motion, Equation (F.9), may be used to find the ground run

$$S_g = \int_0^{S_g} dx = \int_0^{V_{t-o}^2} \frac{dV^2}{2g\left(\frac{F - D - D_{roll}}{W}\right)}. \tag{F.13}$$

A simple approach is now to assume that the denominator in the integral in Equation (F.13) may be approximated by an average value, that is, a constant:

$$\left(\frac{F - D - D_{roll}}{W}\right)_{avg} = \frac{F_{avg}}{W} - \mu'. \tag{F.14}$$

The first term on the right side of Equation (8.14) uses the average thrust value taken directly from Equation (F.10) and $\mu' = 0.02 + 0.01c_{L,\max,\ t\text{-}o}$. Then the ground run is calculated as

$$S_g = \frac{V_{t-o}^2}{2g\left[0.75\left(\frac{F_{static}}{W}\right)\frac{5+\lambda}{4+\lambda} - \left(0.02 + 0.01c_{L,\max,t-o}\right)\right]}. \tag{F.15}$$

## F.2.2 Air run

After aircraft rotation the climb is assumed to proceed along a circular arc transitioning to a constant climb angle, $\gamma_{to}$. Various approaches may be taken to determine the air run, and Torenbeek (1982, p.167) presents the following approximate result:

$$S_a = \frac{V_{to}^2}{g\sqrt{2}} + \frac{h_{to}}{\gamma_{to}}. \tag{F.16}$$

The approximation for the climb angle at the obstacle height, $h_{to} = 35$ feet, is

$$\gamma_{to} = 0.9\left(\frac{F_{avg}}{W}\right)_{to} - \frac{0.3}{\sqrt{A}} = 0.9\left(0.75F_{static}\frac{5+\lambda}{4+\lambda}\right) - \frac{0.3}{\sqrt{A}}. \tag{F.17}$$

These results for the air and ground run are combined for the total take-off distance. Shevell (1989) points out that the following correlation works well in predicting the take-off distance requirements for air transports in Federal Air Regulation (FAR) Part 25:

$$S_{to} = S_g + S_a \approx 27.1\frac{(W/S)_{to}}{\sigma c_{L,\max,to}\left(\frac{F_{avg}}{W}\right)_{to}}. \tag{F.18}$$

# F.3 CLIMB AND DESCENT

It is typical to climb as rapidly as possible to the cruising altitude for a number of reasons, including limiting the noise annoyance caused in the vicinity of the airport. The climb is carried out at constant indicated, or calibrated, airspeed in order to afford the pilot the simplest means of monitoring the climb. This airspeed is within 10% of the equivalent airspeed for the usual flight envelope of commercial airliners, the difference arising from a correction to account for compressibility effects. The onboard measurement of airspeed is discussed in greater detail in the last section of this appendix. If one climbs at constant calibrated airspeed, and it is assumed that this is approximately the equivalent airspeed, then the true airspeed is not constant, but increasing with altitude, that is, there is some acceleration. Because $V = V_E/\sigma^{1/2}$, the nondimensional magnitude of the acceleration is

$$\frac{1}{g}\frac{dV}{dt} = \left(\frac{V_E^2}{gH}\right)\sin\gamma,$$

where $\gamma$ is the angle of climb, that is, the ratio of the vertical velocity component of the aircraft velocity (the rate of climb $RC$) to the actual velocity, and $H = 24{,}000$ ft (7.32 km) is the scale height of the (assumed exponential) atmosphere. Then $g^{-1}dV/dt = 3.65 \times 10^{-8}\,(V_E RC)$ with $V_E$ measured in kts and $RC$ in ft/min. Taking large values of say 400 kts and 6000 ft/min for $V_E$ and $RC$, respectively, yields $g^{-1}dV/dt = 0.088$. Even for the extreme case considered, this value of acceleration can still be considered small enough compared to unity to be neglected in a preliminary calculation of the rate of climb.

Under these assumptions the vertical speed of the aircraft may be put in terms of the time rate of change of altitude, z. Thus the rate of climb $RC$ is given by

$$RC = \frac{dz}{dt} = V_{climb}\frac{F - D}{W} = V_{climb}\left[\delta(F/W)_{to} - \frac{1}{(L/D)}\right]. \tag{F.19}$$

The rate of climb is measured in feet per minute (fpm) so $V_{climb}$ must be corrected appropriately. Note that in order to climb there must be excess thrust. When the excess thrust is such that the altitude at which the rate of climb is down to 100 fpm is called the service ceiling for the aircraft. The climb is generally performed with the throttles at the maximum climb power setting.

A major factor in climb formulation is the variation of thrust with speed and altitude. For turbojets and turbofans the former is less important than the latter. Typically the thrust is assumed to be constant with Mach number, although this is less accurate as the bypass ratio of the engine increases. The relationship between sea level take-off thrust and thrust produced at altitude appears to be reasonably approximated by $T/T_{sea\ level} = \delta$, the atmospheric pressure ratio. A fairly good analytic expression for the pressure ratio is given by $\delta = \exp(-z/H)$, with $z$ in feet and $H = 24{,}000$ ft. This expression can be used to facilitate analytic integration of the rate of climb equation, if desired.

Similarly, the rate of descent $RD$ is determined from the rate of change of altitude as follows:

$$RD = \frac{dz}{dt} = V_{descent}\frac{-D}{W} = -V_{descent}\frac{1}{(L/D)}. \tag{F.20}$$

The rate of descent is also measured in fpm, so $V_{descent}$ must be corrected appropriately. In descent the idle thrust power setting is used so that the thrust may be neglected in the calculation.

### F.3.1 Climb profile

A typical flight profile for the ascent of a commercial airliner may be expressed as follows:

| | |
|---|---|
| Sea level $< z <$ 10,000 ft | $V_{climb} = 250$ kts |
| 10,000 $< z <$ 25,000 ft | $V_{climb} = 320$ kts |
| 25,000 $< z <$ cruise altitude | $V_{climb} = a_{climb} M_{climb} = a_{cr} M_{cr}$ |

An illustration of the different climb segments, along with the pressure, density, and temperature profiles in the atmosphere, is given in Figure F.2.

The climb is usually carried out at constant equivalent airspeed (EAS) up to about 25,000 feet so that the pilot can monitor and control the ascent more easily. This means that the true airspeed of the aircraft is rising continually during the climb to 25,000 feet. Thereafter the climb is carried out at

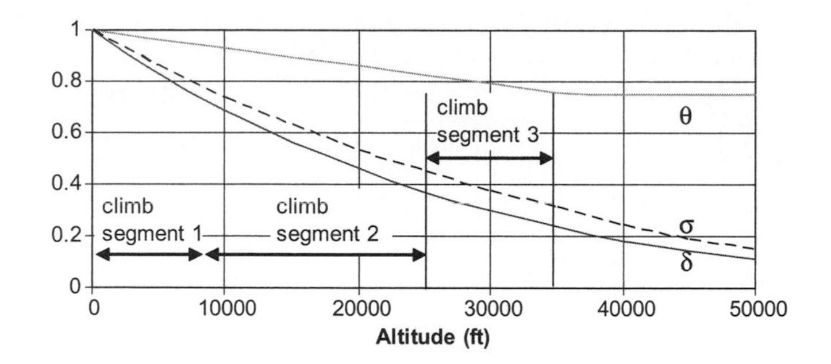


**FIGURE F.2**

Typical climb segments shown along with nondimensional thermodynamic properties of the standard atmosphere.

constant Mach number, which is close to a constant speed climb, as the sound speed only drops around 5% by the time 35,000 feet is reached.

Note that the lift coefficient needed, assuming that the climb angle is shallow and $L \sim W$, is $C_L = 2(W/S)/\rho_0 V_E^2$. The true airspeed therefore will be increasing during the climb so $C_L$ will be higher than the design value for cruise, at least for segments 1 and 2, and therefore $L/D$ in those two segments will be less than the maximum, which will lengthen the time to climb.

## F.3.2 Time to climb

The rate of climb equation for time elapsed in climb from altitude $z_1$ to altitude $z_2$ may be expressed as follows:

$$t = \int_{z_1}^{z_2} \frac{dz}{101.36\frac{V_{E,climb}}{\sqrt{\sigma}}\left[\delta\left(\frac{F}{W}\right)_{to} - \frac{1}{L/D}\right]}. \tag{F.21}$$

Here, with $z$ in feet and $V_{E,climb}$ in kts, the time to climb, in minutes, is determined. The appropriate (constant) value of $V_{E,climb}$ (in kts) must be used in the three altitude segments to obtain the time interval in minutes.

For a hypothetical airplane with $(T/W)_{to} = 0.3$, $M_{cr} = 0.8$, and $L/D$ as given here the climb performance is as follows:

| | |
|---|---|
| Sea level $< z <$ 10,000 ft: | $V_{E,climb} = 250$ kts, $L/D = 14.3$, $\Delta t = 2.07$ min |
| 10,000 $< z <$ 25,000 ft: | $V_{E,climb} = 320$kts, $L/D = 14.3$, $\Delta t = 4.65$ min |
| 25,000 $< z <$ cruise altitude: | $V_{E,climb} = 474$ kts, $L/D = 15.2$, $\Delta t = 10.23$ min |
| Total time to climb | $\Sigma\Delta t = 16.95$ min |

As mentioned previously, the $L/D$ during segments 1 and 2 is chosen to be lower than the assumed cruise value of 15.2. The climb history for this example case is shown in Figure F.3. Here it is seen that

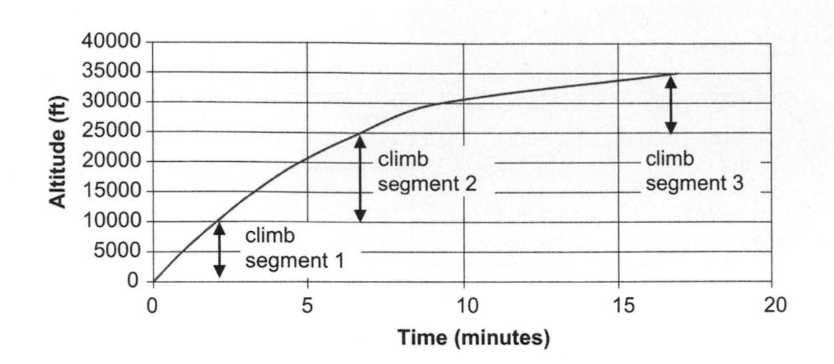


**FIGURE F.3**

Time to climb of a typical case for a turbofan airliner.

the last segment of the climb takes more time than the first two combined. This is due to the drop in engine thrust with altitude, as mentioned previously. Note that the denominator in Equation (F.21) can become very small as $\delta$ drops with altitude. This makes the integrand large and in turn increases the time required to climb higher. The integration of Equation (F.21) may be carried out analytically using the exponential atmosphere approximation but it is rather involved. Here it is simpler to carry out the integration numerically using the actual standard atmospheric properties.

## F.3.3 Distance to climb

The distance to climb is given by

$$X_{climb} = \int_{z_1}^{z_2} \frac{dx/dt}{dz/dt} dz = \int_{z_1}^{z_2} dz \frac{V_{climb}}{V_{climb}\left[\delta(F/W)_{to} - 1/(L/D)\right]}. \tag{F.22}$$

This equation may be readily integrated numerically, as information for the integrand is already available from Equation (F.21). The aircraft trajectory for the example in Figure F.3 is illustrated in Figure F.4. This assumes that the aircraft travels in a plane and that airspeed in climb is equal to the true ground speed. Such an assumption does not account for winds.

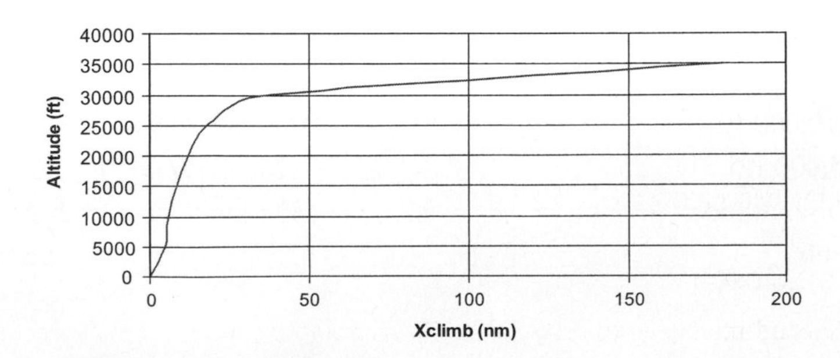


**FIGURE F.4**

Trajectory of the aircraft climb profile illustrated in Figure F.3.

## F.3.4 Fuel to climb

A reasonable approximation to the fuel used in climbing to the cruise altitude is given by $W_{F,used} = c_j F \Delta t$, using consistent units. However, because the thrust decreases with altitude, it is preferable to calculate the fuel usage directly by integration as shown in Equation (F.23), where $c_j$ is divided by 60 to keep the time units consistent as minutes:

$$\frac{W_{Fused}}{W_{to}} = \int_{z_1}^{z_2} \frac{\frac{c_j}{60}\delta\left(\frac{F}{W}\right)_{to} dz}{101.36\frac{V_{E,climb}}{\sqrt{\sigma}}\left[\delta\left(\frac{F}{W}\right)_{to} - \frac{1}{L/D}\right]}. \quad \text{(F.23)}$$

## F.3.5 Descent profile and performance

Similarly, a typical flight profile for the descent of a commercial airliner may be described as follows:

| | |
|---|---|
| Cruise altitude $> z >$ 10,000 ft: | $V_{descent} = 250 + (V_{cr} - 250)(z/z_{cr})$ with $V$ in kts |
| 10,000 ft $> z >$ sea level: | $V_{descent} = 250$ kts |

## F.3.6 Time to descend

We assume that the airspeed is reduced linearly with altitude so that $V = 250 + fz$, where $f = [(V_{cr} - 250)/z_{cr}]$ and the cruise speed is $V_{cr}$ and is given in kts. The rate of descent [Equation (F.20)] for time elapsed in descending from altitude $z_1$ to altitude $z_2$ may be integrated for the descent segment of linearly decreasing flight speed to yield:

$$\Delta t = -9.87 \times 10^{-3}\frac{(L/D)}{f}\left\{\log_e\left[\frac{1 + kz_2}{1 + kz_1}\right]\right\}. \quad \text{(F.24)}$$

where $k = [(V_{cr}/250) - 1]/z_{cr}$ and, with $V_{cr}$ in kts and $z$ in feet, the time interval is in minutes. The lift to drag ratio is assumed constant, but because flight speed is decreasing and density is increasing, the actual value during descent will be less than the cruise value. The units must be consistent so that a value of time elapsed in minutes (or hours) can be calculated correctly. For the hypothetical airplane considered previously, we find $f = 0.0064$ and $k = 2.56 \times 10^{-5}$ so that descent from $z_1 = 35{,}000$ ft to $z_2 = 10{,}000$ ft, assuming $L/D = 14$ requires $\Delta t = 0.148$ hr $= 8.9$ min. From $z_1 = 10{,}000$ ft to $z_2 = 0$ ft, the equation for $\Delta t$ is just

$$\Delta t = \frac{L}{D}\frac{(z_2 - z_1)}{101.36 V_{descent}}. \quad \text{(F.25)}$$

For this portion of the descent, $\Delta t = 5.52$ min and the total time for descent is $8.9 + 5.52 = 14.42$ min.

## F.3.7 Distance to Descend

The distance to descend may be found in the same fashion:

$$dD_{descent} = \frac{dx/dt}{dz/dt}dz = V_{descent}\frac{dz}{RD} = \frac{L}{D}\frac{z_{cr} - z_{airport}}{6080}. \quad \text{(F.26)}$$

Thus, the distance to descend is determined approximately by the glide ratio, $(L/D)^{-1}$. For the example aircraft considered previously, we have $D_{descent} = 14(35{,}000\text{-}0)/6080 = 80.6$ nautical miles.

## F.4 LANDING PERFORMANCE

The landing process begins about 8 to 10 miles away from the airport, and the sequence of events is described schematically in Figure F.5. The landing field length is denoted by $S_l = S_a + S_g$, that is, the sum of an air run and a ground run. These two portions of the landing process are treated sequentially here.

### F.4.1 Air run

The air run may be considered to start at the $h_l = 50$-ft obstacle, so that in an $x$–$y$ coordinate system with $x = 0$ at $y = h_l$, we have $V = V_a$, which, according to FAR Part 25, must be at least 1.3 $V_s$. The aircraft executes a gradual flare of large radius $R$, turning the flight path angle from $\gamma$ around 2 or 3° down to zero, at which time the aircraft has slowed to $V_l = 1.2V_s$ to 1.25 $V_s$ and settled onto the runway, ending the air run.

A simplified analysis of the air run considers the flight path angle to change so slowly that the angular acceleration may be ignored in the force balance, leading to the following appropriate equations:

$$F + W \sin\gamma - D = mdV/dt \tag{F.27}$$

$$L - W \cos\gamma = md(Rd\gamma/dt)/dt \cong 0. \tag{F.28}$$

During the flare, $dV/dt = VdV/ds$ along the trajectory. The distance traveled during the air run from $V = V_a$ at $s_a = 0$, and $V = V_l$ at $s = S_a$, may be found by integrating the equation

$$F - D + W\gamma = mVdV/ds = (W/2g)dV^2/ds \tag{F.29}$$

$$L - W = 0. \tag{F.30}$$

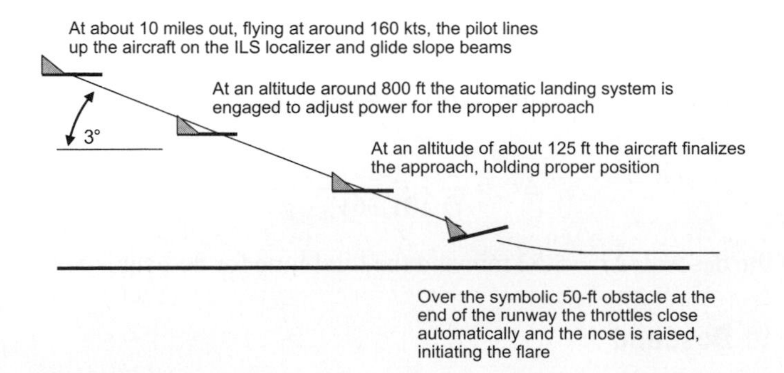


**FIGURE F.5**

Initial stages of the landing process leading up to the flare and actual landing. The angle of the flight path and all distances are exaggerated for clarity.

In these equations, we may make the assumption that the flight path angle is small enough to replace the sine and cosine functions properly with the leading terms of their expansions. This leads to the following equation:

$$dV^2/ds = 2g[(F - D)/L + \gamma]. \tag{F.31}$$

The engines are kept at the approach idle thrust setting during the flare so that the first term in the square brackets is an effective drag to lift ratio, that is, $(F - D)/L = -(D/L)_{eff}$, and may be considered constant during the flare. If it is assumed that the actual, slowly varying, small flight path angle may be replaced by an average value, say $\gamma_a$, then the equation may be integrated to yield

$$V_l^2 - V_a^2 = 2g[-(D/L)_{eff} + \gamma_a]s_a. \tag{F.32}$$

This may be rearranged to give the air run

$$S_a = (1/2g)\{[V_l^2 - V_a^2]/[-(D/L)_{eff} + \gamma_a]\}. \tag{F.33}$$

In terms of the stall characteristics of the aircraft, this equation becomes

$$S_a = [(W/S)_l/(4g\rho_0\sigma C_{L,l\max})]/[-(D/L)_{eff} + \gamma_a]. \tag{F.34}$$

### F.4.2 Ground run

On the simplest level, one may assume that the average deceleration during the ground run, $dV/dt = d(V^2/2)/dx = \text{constant} = -a$. Integrating from the $V = V_l$ (the landing speed) at $x = 0$ to $V = 0$ at the end of the landing where $x = S_g$ yields $S_g = (V_l^2)/2a$. For a representative case of $V_l = 240$ km/hr = 149 mph = 218.77 ft/s and $a = 10$ft/s/s, we find the ground run to be $S_g = 2394.6$ ft. The distance covered during the ground run in terms of stall characteristics of the aircraft becomes $S_g = 1.44(W/S)_l/[a\rho_{sl}\sigma C_{L,l\max}]$.

At the next level of approximation, the dynamics of the situation require that

$$dV/dt = (1/2g)dV^2/dx = (F/W)_l - (D/W)_l - (F_{brake}/W)_l. \tag{F.35}$$

In this equation we do know that $(D/W)_l = C_{D,l}\rho_{sl}\sigma V^2/2(W/S)_l$, where the drag coefficient will depend on the specific landing configuration (flap and slat deflection and possibly aileron droop) and spoiler deflection. The braking force may be written in terms of the normal force the wheels exert on the runway according to $F_{brake} = \mu_{brake}\,(W - L)$, where the braking coefficient of friction lies in the range of $0.4 < \mu_{brake} < 0.6$ for concrete runways. Then we may expand $L$ so that we have

$$(F_{brake}/W)_l = \mu_{brake}(1 - L/W) = \mu_{brake}[1 - C_{L,l}\rho 0\sigma V^2/2(W/S)_l]. \tag{F.36}$$

The thrust in landing is the idle thrust so we may take $(F/W)_l = \text{constant}$, and our deceleration equation is

$$a = (1/2)dV^2/dx = g[(T/W)_l - \mu_{brake}] - g[1 - \mu_{brake}(C_{L,l}/C_{D,l})]C_{D,l}\rho_0\sigma V^2/2(W/S)_l \tag{F.37}$$

Physically, this equation is of the form $a = a_1 + a_2$, where $a_1$ is the deceleration due to thrust reversal (for which $F/W < 0$) and braking, and $a_2$ is the deceleration due to aerodynamic drag less the effect of aerodynamic lift on reducing the normal force between the wheels and the runway. Mathematically, Equation (F.37) is of the form

$$Y' + AY + B = 0, \tag{F.38}$$

with $Y = V^2$, $A = [1 - \mu_{brake}(L/D)_l\,]\, C_{D,l}\, \rho_0 \sigma\, g/(W/S)_l$ and $B = -2g\,[(T/W)_l - \mu_{brake}]$. The solution in terms of ground distance is

$$V^2 = [V_l^2 + B/A]\exp(-Ax) - B/A. \tag{F.39}$$

The total ground distance is then determined by setting $V = 0$ and is given by

$$S_g = A^{-1}\log_e[1 + (A/B)V_l^2]. \tag{F.40}$$

The lift to drag ratio in landing depends on the combination of high lift devices acting in consort with spoilers, and the thrust to weight ratio depends on the capability of thrust reversers, if used.

The landing distance $S_l$ is the sum of the air distance $S_{air}$ from the 50-ft obstacle $h_l$ and the ground distance $S_{run}$. Following the approach of Shevell (1989), we may write

$$\begin{aligned} S_l &= S_{air} + S_{run} \\ S_{air} &= \frac{1}{\dfrac{1}{(L/D)_l} - \left(\dfrac{F}{W}\right)_l}\left[50 + \frac{1}{2g}\left(V_{50}^2 - V_l^2\right)\right] \\ S_{run} &= \frac{V_l^2}{2\bar{a}} \\ \bar{a} &= \mu g\left(1 - \frac{L}{W}\right)_{eff} + g\left(\frac{D}{W}\right)_{eff}, \end{aligned} \tag{F.41}$$

where $\bar{a}$ is the average deceleration caused by the effective rolling friction drag and the effective aerodynamic drag during the rollout over the ground run. As the aircraft touches down, spoilers or lift dumpers are deployed automatically, and the lift decreases to zero so that the normal force is equal to the landing weight of the aircraft alone. Values for the coefficient of friction, $\mu$, range from $0.4 < \mu < 0.6$ for dry concrete, $0.2 < \mu < 0.3$ for wet concrete, and as little as $\mu = 0.1$ for ice. The average deceleration generally lies in the range $0.3 < \bar{a} < 0.4$. The quantity $V_{50}$ is the airspeed over the 50-ft obstacle and is required by FAR Part 25 to be at least $1.3V_l$.

A further requirement in FAR Part 25 is that the landing field length be a factor of 8/5 of the demonstrated landing distance $S_l$. Shevell (1989) offers an approximation for the FAR landing distance in the form FAR $S_l = A + BV_s^2$. Values for $A$ and $B$ are as follows: for two wheel trucks, $A = 4000$, $B = 0.333$, and for four-wheel trucks, $A = 1400$, $B = 0.4$.

## F.5 AIR DATA SYSTEM

The altitude, airspeed, climb speed, Mach number, temperature, and related flight data are provided through the processing of measured raw data by a computer-controlled instrumentation system called the air data system (Collinson, 2003). The pilot monitors the airspeed by means of an instrument called an airspeed indicator and its display is called the indicated airspeed. The true airspeed, which is the speed of the aircraft relative to the undisturbed atmosphere through which it is passing, may be found by applying the principles of compressible flow. For isentropic flow along a streamline, the relationship between the stagnation pressure $p_t$ and the static pressure $p$ depends on the Mach number $M$ and the ratio of specific heats of the gas. The pressure coefficient at the stagnation point is given here:

$$C_{p,t} = \frac{p_t - p}{q} = \frac{2}{\gamma M^2}\left[\left(1+\frac{\gamma-1}{2}M^2\right)^{\frac{\gamma}{\gamma-1}} - 1\right]. \tag{F.42}$$

The quantity $q$ is the dynamic pressure

$$q = \frac{1}{2}\rho V^2 = \frac{1}{2}\gamma p M^2. \tag{F.43}$$

Because the atmosphere of interest is air, we may set $\gamma = 1.4$ so that the pressure coefficient equation becomes

$$C_{p,t} = \frac{p_t - p}{q} = \frac{10}{7M^2}\left[\left(1+\frac{M^2}{5}\right)^{\frac{7}{2}} - 1\right] = \left(1+\frac{M^2}{4}+\frac{M^4}{40}+\frac{M^6}{1600}+\ldots\right). \tag{F.44}$$

This equation shows the effect of compressibility on the measurement of velocity by considering only the difference in stagnation and static pressures. Of course, when the Mach number is small, the difference between stagnation and static pressures is approximately equal to the dynamic pressure and one may compute the true airspeed directly. Flying at a speed of 200 mph at sea level is equivalent to a Mach number of about 0.3, which yields an error of 2.3% over the incompressible value for the pressure coefficient and this is considered negligible. However, for commercial jetliners cruising at $M = 0.8$, the error jumps to about 17% and cannot be ignored.

If one measures $p_t$ with a pitot tube and pressure sensor and then $p$ with an appropriate static pressure port and pressure sensor, one can determine the Mach number from Equation (F.42). Because a static pressure port must be located somewhere on the aircraft, it is unlikely to measure the exact static pressure in the undisturbed air, and therefore the output of the static pressure sensor would have to be calibrated over the expected speed range in order to correct for this problem. The instrument using just these pressure measurements is called a Machmeter. Measuring the static temperature $T$ then permits the true air speed (TAS) to be determined using the definition of Mach number: $M = V/a$. The relationship between static and stagnation temperature in isentropic flow is

$$\frac{T}{T_t} = 1+\frac{\gamma-1}{2}M^2 = 1+\frac{M^2}{5}. \tag{F.45}$$

The stagnation temperature $T_t$ is measured relatively easily, for example, with a thermocouple. Then the true airspeed is $V = M\sqrt{\gamma RT} = 29M\sqrt{T}$ in knots when $T$ is measured in degrees Rankine.

The calibrated airspeed (CAS) arises from calculating the velocity from the isentropic relation

$$V^2 = \left(\frac{2\gamma}{\gamma-1}\right)\frac{p}{\rho}\left[\left(1+\frac{p_t-p}{p}\right)^{\frac{\gamma-1}{\gamma}} - 1\right]. \tag{F.46}$$

If one uses the sea level pressure and density rather than the actual values in all but the pressure difference term in Equation (F.46) produces the calibrated airspeed $V_{CAL}$, which is described by the following equation:

$$V_{CAL}^2 = \left(\frac{2\gamma}{\gamma-1}\right)\frac{p_0}{\rho_0}\left[\left(1+\frac{p_t-p}{p_0}\right)^{\frac{\gamma-1}{\gamma}}-1\right]. \tag{F.47}$$

The equivalent airspeed (EAS), which is a measure of the dynamic pressure of the airstream, is defined in terms of the true airspeed as follows:

$$V_E^2 = \sigma V^2 = \frac{\rho}{\rho_0}\left(\frac{2\gamma}{\gamma-1}\right)\frac{p}{\rho}\left[\left(1+\frac{p_t-p}{p}\right)^{\frac{\gamma-1}{\gamma}}-1\right] \tag{F.48}$$

Then the EAS is related directly to the CAS by $V_E = FV_{CAL}$, and the correction factor $F$ is given by

$$F^2 = \frac{\frac{p}{\rho_0}\left(1+\frac{p_t-p}{p}\right)^{\frac{\gamma-1}{\gamma}}}{\frac{p_0}{\rho_0}\left(1+\frac{p_t-p}{p_0}\right)^{\frac{\gamma-1}{\gamma}}}. \tag{F.49}$$

Using $\gamma = 1.4$ and introducing the atmospheric pressure ratio $\delta = p/p_{s.l.}$ we can simplify the correction function as follows:

$$F^2 = \delta\frac{\left(1-\frac{p_t-p}{p}\right)^{\frac{2}{7}}-1}{\left(1-\delta\frac{p_t-p}{p}\right)^{\frac{2}{7}}-1}. \tag{F.50}$$

Thus the compressibility correction is also corrected for altitude and one can compute $V_{CAL}$ and find the correction factors for incorporation into the data-processing algorithms of the air data system. This approach doesn't require measurement of temperature and uses the pressure altitude, as given by $\delta$, which is used to provide the pilot with an altitude reading.

A sample calculation was carried out for the case of flight at an altitude of 35,000 ft and results are shown in Figure F.6. The calibrated airspeed and therefore the indicated airspeed read by the pilot are approximately equal to the equivalent airspeed and thus provide a dynamically appropriate indication of the aircraft performance, as aircraft loads are proportional to the square of the EAS.

Note that in Figure F.6 the TAS is 70% greater than the CAS. The value of the TAS is in the navigation of the aircraft because it, with information about the prevailing winds, provides the ground speed of the aircraft. The advent of improved sensors and navigational aids such as the global positioning system has served to alter the avionics suite common to airliners; for more information, see Collinson (2003).

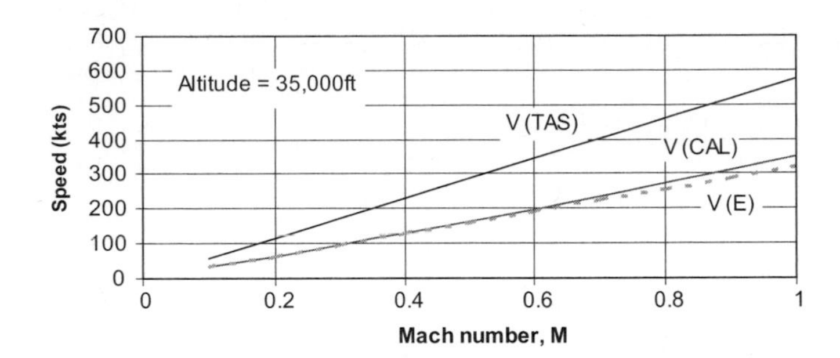


**FIGURE F.6**

Illustration of the relationship among true airspeed (TAS), calibrated airspeed (CAL), and equivalent airspeed (E) as a function of Mach number at an altitude of 35,000 ft (10.67 km).

## F.6 NOMENCLATURE

| | |
|---|---|
| $A$ | aspect ratio |
| $a$ | deceleration |
| $C_D$ | drag coefficient |
| $C_{D0}$ | zero-lift drag coefficient |
| $C_L$ | lift coefficient |
| $C_p$ | pressure coefficient |
| $c_j$ | specific fuel consumption |
| $D$ | drag |
| $e$ | span efficiency factor |
| $F$ | thrust |
| $g$ | acceleration of gravity |
| $h_{to}$ | take-off obstacle height |
| $H$ | scale height of atmosphere |
| $I$ | specific impulse |
| $k$ | constant |
| $L$ | lift |
| $M$ | Mach number |
| $p$ | pressure |
| $q$ | dynamic pressure |
| $R$ | range |
| $S$ | surface area |
| $S_a$ | air run |
| $S_g$ | ground run |
| $T$ | temperature |

| | |
|---|---|
| $t$ | time |
| $V$ | velocity |
| $V_E$ | equivalent velocity, $V\sigma^{1/2}$ |
| $W$ | weight |
| $z$ | altitude |
| $\gamma$ | flight path angle |
| $\delta$ | atmospheric pressure ratio $p_0/p_0.$ |
| $\lambda$ | engine bypass ratio |
| $\mu$ | rolling resistance |
| $\rho$ | gas density |
| $\sigma$ | density ratio $\rho/\rho_0$ |
| $\theta$ | temperature ratio $T/T_0$ |

## F.6.1 Subscripts

| | |
|---|---|
| *avg* | average |
| *cr* | cruise conditions |
| *eff* | effective |
| *F* | fuel |
| *g* | ground run |
| *l* | landing conditions |
| max | maximum conditions |
| *OE* | operating empty |
| *PL* | payload |
| *t* | stagnation conditions |
| *to* | take-off |
| 0 | sea level conditions |
| 1 | start of cruise |
| 2 | end of cruise |

## References


Collinson, R. P. G. (2003). *Introduction to Avionics Systems* (2nd ed.). Dordrecht, The Netherlands: Kluwer Academic Publishers.

Shevell, R. S. (1989). *Fundamentals of Flight*. Englewood Cliffs, NJ: Prentice-Hall.

Torenbeek, E. (1982). *Synthesis of Subsonic Airplane Design*. Dordrecht, The Netherlands: Kluwer Academic Publishers.

APPENDIX

# Thermodynamic Properties of Selected Species

G

## G.1 TABLES OF THERMODYNAMIC PROPERTIES

The NASA Glenn Research Center provides an online calculator called NASA CEA (Chemical Equilibrium with Applications), as well as a download of the computer codes for general use. The Web site also includes a thermodynamic property database called THERMOBUILD. For convenience, thermodynamic data for selected species appearing in the example problems and exercises of the text have been taken from that database and provided in the following tables, including enthalpy [H(T)-H(298K)], heat of formation [$\Delta H_f$], and equilibrium constant [$K_p$].

## G.2 PROPERTIES OF SELECTED SPECIES

Tables follow for bromine (Br, $Br_2$), hydrogen bromide (HBr), carbon monoxide (CO), carbon dioxide ($CO_2$), fluorine (F, $F_2$), hydrogen fluoride (HF), hydrogen (H, $H_2$), nitrogen (N, $N_2$), nitrous oxide (NO), nitrogen dioxide ($NO_2$), oxygen, (O, $O_2$), water ($H_2O$), and hydroxyl (OH).

Thermodynamic Functions Calculated from Coefficients for Br

| $T$ | $C_p$ | H-H298 | S | −(G-H298)/$T$ | $H$ | $\Delta H_f$ | log $K_p$ |
|---|---|---|---|---|---|---|---|
| deg-K | J/mol-K | kJ/mol | J/mol-K | J/mol-K | kJ/mol | kJ/mol | |
| 0 | 0 | -6.197 | 0 | Infinite | 105.673 | 117.933 | Infinite |
| 300* | 20.786 | 0.038 | 175.148 | 175.019 | 111.908 | 111.838 | −14.3112 |
| 500 | 20.798 | 4.196 | 185.767 | 177.374 | 116.066 | 108.465 | −6.6199 |
| 700 | 20.934 | 8.366 | 192.780 | 180.829 | 120.236 | 105.104 | −3.4271 |
| 900 | 21.382 | 12.591 | 198.087 | 184.097 | 124.461 | 101.799 | −1.7095 |
| 1100 | 22.260 | 16.948 | 202.456 | 187.049 | 128.818 | 98.626 | −0.6511 |
| 1300 | 23.535 | 21.521 | 206.273 | 189.718 | 133.391 | 95.668 | 0.0591 |
| 1500 | 25.163 | 26.386 | 209.751 | 192.160 | 138.256 | 93.004 | 0.5646 |
| 1700 | 26.961 | 31.597 | 213.010 | 194.423 | 143.467 | 90.684 | 0.9410 |
| 1900 | 28.757 | 37.170 | 216.107 | 196.544 | 149.040 | 88.727 | 1.2311 |

(Continued)

Thermodynamic Functions Calculated from Coefficients for Br (*Continued*)

| *T* | $C_p$ | H-H298 | *S* | −(G-H298)/*T* | *H* | $\Delta H_f$ | log $K_p$ |
|---|---|---|---|---|---|---|---|
| deg-K | J/mol-K | kJ/mol | J/mol-K | J/mol-K | kJ/mol | kJ/mol | |
| 2100 | 30.428 | 43.091 | 219.069 | 198.549 | 154.961 | 87.118 | 1.4613 |
| 2300 | 31.903 | 49.328 | 221.905 | 200.458 | 161.198 | 85.825 | 1.6483 |
| 2500 | 33.144 | 55.837 | 224.618 | 202.283 | 167.707 | 84.804 | 1.8033 |
| 2700 | 34.140 | 62.569 | 227.208 | 204.034 | 174.439 | 84.006 | 1.9340 |
| 2900 | 34.898 | 69.477 | 229.676 | 205.718 | 181.347 | 83.383 | 2.0456 |
| 3100 | 35.438 | 76.514 | 232.022 | 207.340 | 188.384 | 82.890 | 2.1422 |
| 3300 | 35.783 | 83.639 | 234.249 | 208.904 | 195.509 | 82.485 | 2.2267 |
| 3500 | 35.963 | 90.816 | 236.361 | 210.413 | 202.686 | 82.132 | 2.3011 |
| 3700 | 36.008 | 98.015 | 238.361 | 211.870 | 209.885 | 81.801 | 2.3672 |
| 3900 | 35.947 | 105.212 | 240.255 | 213.278 | 217.082 | 81.468 | 2.4263 |
| 4100 | 35.808 | 112.389 | 242.050 | 214.638 | 224.259 | 81.114 | 2.4794 |
| 4300 | 35.616 | 119.532 | 243.751 | 215.953 | 231.402 | 80.727 | 2.5274 |
| 4500 | 35.391 | 126.633 | 245.365 | 217.224 | 238.503 | 80.298 | 2.5709 |
| 4700 | 35.150 | 133.687 | 246.899 | 218.455 | 245.557 | 79.822 | 2.6104 |
| 4900 | 34.905 | 140.693 | 248.359 | 219.646 | 252.563 | 79.297 | 2.6465 |
| 5000 | 34.784 | 144.177 | 249.062 | 220.227 | 256.047 | 79.016 | 2.6634 |

*Assigned reference phase change at 265.90K.*

Thermodynamic Functions Calculated from Coefficients for $Br_2$

| *T* | $C_p$ | H-H298 | *S* | −(G-H298)/T | *H* | $\Delta H_f$ | log $K_p$ |
|---|---|---|---|---|---|---|---|
| deg-K | J/mol-K | kJ/mol | J/mol-K | J/mol-K | kJ/mol | kJ/mol | |
| 0 | 0 | −9.725 | 0 | Infinite | 21.185 | 45.705 | Infinite |
| 300* | 36.075 | 0.067 | 245.692 | 245.470 | 30.977 | 30.837 | −0.5106 |
| 500 | 37.082 | 7.403 | 264.409 | 249.602 | 38.313 | 23.110 | 1.4123 |
| 700 | 37.465 | 14.862 | 276.953 | 255.722 | 45.772 | 15.509 | 2.0012 |
| 900 | 37.697 | 22.379 | 286.399 | 261.533 | 53.289 | 7.966 | 2.2010 |
| 1100 | 38.011 | 29.946 | 293.990 | 266.766 | 60.856 | 0.472 | 2.2481 |
| 1300 | 38.079 | 37.558 | 300.348 | 271.457 | 68.468 | −6.976 | 2.2258 |
| 1500 | 38.163 | 45.180 | 305.801 | 275.681 | 76.090 | −14.415 | 2.1695 |
| 1700 | 38.423 | 52.836 | 310.592 | 279.512 | 83.746 | −21.820 | 2.0958 |
| 1900 | 38.846 | 60.560 | 314.887 | 283.013 | 91.470 | −29.155 | 2.0138 |
| 2100 | 39.373 | 68.381 | 318.800 | 286.238 | 99.291 | −36.395 | 1.9283 |
| 2300 | 39.942 | 76.312 | 322.407 | 289.228 | 107.222 | −43.524 | 1.8421 |
| 2500 | 40.500 | 84.357 | 325.761 | 292.018 | 115.267 | −50.540 | 1.7568 |

Thermodynamic Functions Calculated from Coefficients for $Br_2$ (*Continued*)

| $T$ | $C_p$ | H-H298 | $S$ | −(G-H298)/T | $H$ | $\Delta H_f$ | log $K_p$ |
|---|---|---|---|---|---|---|---|
| deg-K | J/mol-K | kJ/mol | J/mol-K | J/mol-K | kJ/mol | kJ/mol | |
| 2700 | 41.002 | 92.508 | 328.897 | 294.635 | 123.418 | −57.449 | 1.6734 |
| 2900 | 41.415 | 100.752 | 331.842 | 297.101 | 131.662 | −64.266 | 1.5923 |
| 3100 | 41.714 | 109.067 | 334.615 | 299.432 | 139.977 | −71.012 | 1.5138 |
| 3300 | 41.884 | 117.429 | 337.229 | 301.645 | 148.339 | −77.710 | 1.4379 |
| 3500 | 41.916 | 125.811 | 339.695 | 303.749 | 156.721 | −84.388 | 1.3646 |
| 3700 | 41.806 | 134.185 | 342.022 | 305.756 | 165.095 | −91.074 | 1.2939 |
| 3900 | 41.557 | 142.524 | 344.217 | 307.672 | 173.434 | −97.796 | 1.2256 |
| 4100 | 41.177 | 150.800 | 346.286 | 309.506 | 181.710 | −104.580 | 1.1595 |
| 4300 | 40.677 | 158.987 | 348.236 | 311.262 | 189.897 | −111.454 | 1.0956 |
| 4500 | 40.072 | 167.063 | 350.072 | 312.947 | 197.973 | −118.437 | 1.0335 |
| 4700 | 39.383 | 175.010 | 351.800 | 314.564 | 205.920 | −125.551 | 0.9733 |
| 4900 | 38.631 | 182.812 | 353.426 | 316.117 | 213.722 | −132.809 | 0.9147 |
| 5000 | 38.240 | 186.656 | 354.202 | 316.871 | 217.566 | −136.496 | 0.8860 |

*Assigned reference phase change at 265.90K.

Thermodynamic Functions Calculated from Coefficients for HBr

| $T$ | $C_p$ | H-H298 | $S$ | −(G-H298)/$T$ | $H$ | $\Delta H_f$ | log $K_p$ |
|---|---|---|---|---|---|---|---|
| deg-K | J/mol-K | kJ/mol | J/mol-K | J/mol-K | kJ/mol | kJ/mol | |
| 0 | 0 | −8.648 | 0 | Infinite | −44.938 | −28.444 | Infinite |
| 300* | 29.142 | 0.054 | 198.881 | 198.701 | −36.236 | −36.333 | 9.3091 |
| 500 | 29.455 | 5.903 | 213.814 | 202.008 | −30.387 | −40.930 | 6.6456 |
| 700 | 30.431 | 11.882 | 223.864 | 206.889 | −24.408 | −45.413 | 5.3640 |
| 900 | 31.710 | 18.095 | 231.665 | 211.558 | −18.195 | −49.694 | 4.5782 |
| 1100 | 32.910 | 24.561 | 238.147 | 215.819 | −11.729 | −53.780 | 4.0335 |
| 1300 | 33.937 | 31.248 | 243.731 | 219.694 | −5.042 | −57.721 | 3.6270 |
| 1500 | 34.773 | 38.123 | 248.648 | 223.233 | 1.833 | −61.564 | 3.3079 |
| 1700 | 35.445 | 45.147 | 253.043 | 226.486 | 8.857 | −65.343 | 3.0483 |
| 1900 | 35.992 | 52.292 | 257.016 | 229.494 | 16.002 | −69.081 | 2.8311 |
| 2100 | 36.449 | 59.537 | 260.641 | 232.290 | 23.247 | −72.793 | 2.6455 |
| 2300 | 36.842 | 66.867 | 263.975 | 234.903 | 30.577 | −76.487 | 2.4842 |
| 2500 | 37.191 | 74.271 | 267.062 | 237.353 | 37.981 | −80.169 | 2.3420 |
| 2700 | 37.507 | 81.742 | 269.936 | 239.662 | 45.452 | −83.839 | 2.2152 |
| 2900 | 37.799 | 89.272 | 272.627 | 241.843 | 52.982 | −87.499 | 2.1009 |
| 3100 | 38.073 | 96.860 | 275.157 | 243.912 | 60.570 | −91.149 | 1.9972 |

(*Continued*)

Thermodynamic Functions Calculated from Coefficients for HBr (*Continued*)

| $T$ | $C_p$ | H-H298 | $S$ | −(G-H298)/$T$ | $H$ | $\Delta H_f$ | log $K_p$ |
|---|---|---|---|---|---|---|---|
| deg-K | J/mol-K | kJ/mol | J/mol-K | J/mol-K | kJ/mol | kJ/mol | |
| 3300 | 38.330 | 104.500 | 277.545 | 245.878 | 68.210 | −94.790 | 1.9023 |
| 3500 | 38.572 | 112.191 | 279.808 | 247.753 | 75.901 | −98.423 | 1.8149 |
| 3700 | 38.799 | 119.928 | 281.958 | 249.544 | 83.638 | −102.049 | 1.7341 |
| 3900 | 39.009 | 127.709 | 284.006 | 251.260 | 91.419 | −105.670 | 1.6589 |
| 4100 | 39.200 | 135.531 | 285.961 | 252.905 | 99.241 | −109.287 | 1.5887 |
| 4300 | 39.370 | 143.388 | 287.832 | 254.486 | 107.098 | −112.905 | 1.5229 |
| 4500 | 39.516 | 151.277 | 289.626 | 256.009 | 114.987 | −116.526 | 1.4610 |
| 4700 | 39.635 | 159.193 | 291.347 | 257.476 | 122.903 | −120.155 | 1.4026 |
| 4900 | 39.724 | 167.129 | 293.000 | 258.892 | 130.839 | −123.797 | 1.3472 |
| 5000 | 39.756 | 171.103 | 293.803 | 259.583 | 134.813 | −125.623 | 1.3206 |

*Assigned reference phase change at 265.90K.*

Thermodynamic Functions Calculated from Coefficients for CO

| $T$ | $C_p$ | H-H298 | $S$ | −(G-H298)/$T$ | $H$ | $\Delta H_f$ | log $K_p$ |
|---|---|---|---|---|---|---|---|
| deg-K | J/mol-K | kJ/mol | J/mol-K | J/mol-K | kJ/mol | kJ/mol | |
| 0 | 0 | −8.671 | 0 | Infinite | −119.206 | −113.813 | Infinite |
| 300 | 29.143 | 0.054 | 197.840 | 197.660 | −110.481 | −110.524 | 23.9125 |
| 500 | 29.794 | 5.931 | 212.837 | 200.975 | −104.604 | −110.013 | 16.2370 |
| 700 | 31.171 | 12.021 | 223.070 | 205.897 | −98.514 | −110.480 | 12.9489 |
| 900 | 32.572 | 18.399 | 231.077 | 210.633 | −92.136 | −111.430 | 11.1102 |
| 1100 | 33.710 | 25.032 | 237.729 | 214.972 | −85.503 | −112.601 | 9.9287 |
| 1300 | 34.569 | 31.864 | 243.433 | 218.922 | −78.671 | −113.888 | 9.1017 |
| 1500 | 35.213 | 38.846 | 248.427 | 222.530 | −71.690 | −115.248 | 8.4882 |
| 1700 | 35.704 | 45.940 | 252.866 | 225.843 | −64.596 | −116.670 | 8.0132 |
| 1900 | 36.083 | 53.120 | 256.859 | 228.901 | −57.415 | −118.155 | 7.6336 |
| 2100 | 36.384 | 60.368 | 260.485 | 231.739 | −50.168 | −119.703 | 7.3223 |
| 2300 | 36.626 | 67.669 | 263.807 | 234.385 | −42.866 | −121.316 | 7.0617 |
| 2500 | 36.826 | 75.015 | 266.869 | 236.863 | −35.520 | −122.995 | 6.8398 |
| 2700 | 36.995 | 82.398 | 269.710 | 239.192 | −28.137 | −124.742 | 6.6482 |
| 2900 | 37.140 | 89.812 | 272.359 | 241.389 | −20.723 | −126.555 | 6.4806 |
| 3100 | 37.266 | 97.253 | 274.840 | 243.468 | −13.283 | −128.433 | 6.3325 |
| 3300 | 37.379 | 104.717 | 277.173 | 245.441 | −5.818 | −130.377 | 6.2003 |
| 3500 | 37.480 | 112.203 | 279.376 | 247.318 | 1.668 | −132.384 | 6.0815 |
| 3700 | 37.572 | 119.709 | 281.461 | 249.107 | 9.173 | −134.453 | 5.9739 |
| 3900 | 37.657 | 127.232 | 283.441 | 250.818 | 16.696 | −136.584 | 5.8758 |
| 4100 | 37.737 | 134.771 | 285.326 | 252.455 | 24.236 | −138.774 | 5.7859 |

Thermodynamic Functions Calculated from Coefficients for CO (*Continued*)

| $T$ | $C_p$ | H-H298 | $S$ | −(G-H298)/$T$ | $H$ | $\Delta H_f$ | log $K_p$ |
|---|---|---|---|---|---|---|---|
| deg-K | J/mol-K | kJ/mol | J/mol-K | J/mol-K | kJ/mol | kJ/mol | |
| 4300 | 37.813 | 142.326 | 287.126 | 254.026 | 31.791 | −141.022 | 5.7030 |
| 4500 | 37.885 | 149.896 | 288.846 | 255.536 | 39.361 | −143.326 | 5.6263 |
| 4700 | 37.956 | 157.480 | 290.495 | 256.989 | 46.945 | −145.686 | 5.5549 |
| 4900 | 38.027 | 165.079 | 292.078 | 258.389 | 54.543 | −148.099 | 5.4883 |
| 5000 | 38.063 | 168.883 | 292.847 | 259.070 | 58.348 | −149.325 | 5.4566 |

Thermodynamic Functions Calculated from Coefficients for $CO_2$

| $T$ | $C_p$ | H-H298 | $S$ | −(G-H298)/$T$ | $H$ | $\Delta H_f$ | log $K_p$ |
|---|---|---|---|---|---|---|---|
| deg-K | J/mol-K | kJ/mol | J/mol-K | J/mol-K | kJ/mol | kJ/mol | |
| 0 | 0 | −9.365 | 0 | Infinite | −402.875 | −393.142 | Infinite |
| 300 | 37.220 | 0.069 | 214.017 | 213.788 | −393.441 | −393.511 | 68.6661 |
| 500 | 44.624 | 8.307 | 234.898 | 218.284 | −385.203 | −393.655 | 41.2565 |
| 700 | 49.561 | 17.756 | 250.748 | 225.382 | −375.754 | −393.970 | 29.5031 |
| 900 | 52.999 | 28.032 | 263.643 | 232.496 | −365.478 | −394.394 | 22.9671 |
| 1100 | 55.420 | 38.888 | 274.527 | 239.174 | −354.622 | −394.830 | 18.8033 |
| 1300 | 57.130 | 50.153 | 283.932 | 245.353 | −343.357 | −395.252 | 15.9174 |
| 1500 | 58.374 | 61.709 | 292.199 | 251.059 | −331.801 | −395.665 | 13.7989 |
| 1700 | 59.310 | 73.482 | 299.565 | 256.340 | −320.028 | −396.090 | 12.1772 |
| 1900 | 60.033 | 85.419 | 306.203 | 261.245 | −308.091 | −396.548 | 10.8954 |
| 2100 | 60.604 | 97.485 | 312.240 | 265.819 | −296.025 | −397.056 | 9.8565 |
| 2300 | 61.065 | 109.653 | 317.775 | 270.099 | −283.857 | −397.628 | 8.9971 |
| 2500 | 61.443 | 121.905 | 322.883 | 274.120 | −271.605 | −398.272 | 8.2742 |
| 2700 | 61.760 | 134.226 | 327.624 | 277.910 | −259.284 | −398.995 | 7.6573 |
| 2900 | 62.033 | 146.606 | 332.047 | 281.493 | −246.904 | −399.800 | 7.1244 |
| 3100 | 62.273 | 159.037 | 336.192 | 284.890 | −234.473 | −400.686 | 6.6593 |
| 3300 | 62.492 | 171.514 | 340.092 | 288.118 | −221.996 | −401.654 | 6.2497 |
| 3500 | 62.69 | 184.033 | 343.775 | 291.194 | −209.477 | −402.701 | 5.8859 |
| 3700 | 62.898 | 196.593 | 347.265 | 294.132 | −196.917 | −403.823 | 5.5606 |
| 3900 | 63.100 | 209.192 | 350.581 | 296.942 | −184.318 | −405.017 | 5.2679 |
| 4100 | 63.310 | 221.833 | 353.742 | 299.636 | −171.677 | −406.276 | 5.0029 |
| 4300 | 63.534 | 234.517 | 356.763 | 302.224 | −158.993 | −407.595 | 4.7617 |
| 4500 | 63.777 | 247.248 | 359.656 | 304.712 | −146.262 | −408.965 | 4.5413 |
| 4700 | 64.045 | 260.030 | 362.435 | 307.110 | −133.480 | −410.380 | 4.3390 |
| 4900 | 64.343 | 272.868 | 365.110 | 309.423 | −120.642 | −411.830 | 4.1525 |
| 5000 | 64.505 | 279.310 | 366.412 | 310.550 | −114.200 | −412.565 | 4.0646 |

Thermodynamic Functions Calculated from Coefficients for F

| $T$ | $C_p$ | H-H298 | $S$ | −(G-H298)/$T$ | $H$ | $\Delta H_f$ | log $K_p$ |
|---|---|---|---|---|---|---|---|
| deg-K | J/mol-K | kJ/mol | J/mol-K | J/mol-K | kJ/mol | kJ/mol | |
| 0 | 0 | −6.518 | 0 | Infinite | 72.862 | 77.274 | Infinite |
| 300 | 22.743 | 0.042 | 158.893 | 158.752 | 79.422 | 79.393 | −10.8251 |
| 500 | 22.100 | 4.528 | 170.365 | 161.309 | 83.908 | 80.587 | −5.2588 |
| 700 | 21.629 | 8.897 | 177.719 | 165.009 | 88.277 | 81.442 | −2.8415 |
| 900 | 21.356 | 13.193 | 183.119 | 168.460 | 92.573 | 82.106 | −1.4858 |
| 1100 | 21.195 | 17.447 | 187.388 | 171.527 | 96.827 | 82.653 | −0.6166 |
| 1300 | 21.092 | 21.675 | 190.919 | 174.246 | 101.055 | 83.123 | −0.0112 |
| 1500 | 21.023 | 25.886 | 193.932 | 176.675 | 105.266 | 83.533 | 0.4352 |
| 1700 | 20.974 | 30.085 | 196.560 | 178.863 | 109.465 | 83.893 | 0.7782 |
| 1900 | 20.939 | 34.276 | 198.891 | 180.851 | 113.656 | 84.212 | 1.0500 |
| 2100 | 20.913 | 38.461 | 200.986 | 182.671 | 117.841 | 84.503 | 1.2708 |
| 2300 | 20.893 | 42.642 | 202.887 | 184.347 | 122.022 | 84.780 | 1.4539 |
| 2500 | 20.878 | 46.819 | 204.629 | 185.901 | 126.199 | 85.060 | 1.6082 |
| 2700 | 20.865 | 50.993 | 206.235 | 187.349 | 130.373 | 85.357 | 1.7401 |
| 2900 | 20.855 | 55.165 | 207.726 | 188.703 | 134.545 | 85.684 | 1.8541 |
| 3100 | 20.847 | 59.335 | 209.116 | 189.976 | 138.715 | 86.055 | 1.9539 |
| 3300 | 20.840 | 63.504 | 210.419 | 191.176 | 142.884 | 86.478 | 2.0420 |
| 3500 | 20.834 | 67.672 | 211.645 | 192.311 | 147.052 | 86.962 | 2.1204 |
| 3700 | 20.829 | 71.838 | 212.803 | 193.387 | 151.218 | 87.513 | 2.1908 |
| 3900 | 20.825 | 76.003 | 213.899 | 194.411 | 155.383 | 88.132 | 2.2544 |
| 4100 | 20.822 | 80.168 | 214.941 | 195.388 | 159.548 | 88.823 | 2.3122 |
| 4300 | 20.818 | 84.332 | 215.932 | 196.320 | 163.712 | 89.586 | 2.3650 |
| 4500 | 20.816 | 88.495 | 216.879 | 197.213 | 167.875 | 90.418 | 2.4136 |
| 4700 | 20.813 | 92.658 | 217.784 | 198.069 | 172.038 | 91.317 | 2.4585 |
| 4900 | 20.811 | 96.821 | 218.651 | 198.892 | 176.201 | 92.280 | 2.5001 |
| 5000 | 20.810 | 98.902 | 219.072 | 199.291 | 178.282 | 92.784 | 2.5198 |

Thermodynamic Functions Calculated from Coefficients for $F_2$

| $T$ | $C_p$ | H-H298 | $S$ | −(G-H298)/$T$ | $H$ | $\Delta H_f$ | log $K_p$ |
|---|---|---|---|---|---|---|---|
| deg-K | J/mol-K | kJ/mol | J/mol-K | J/mol-K | kJ/mol | kJ/mol | |
| 0 | 0 | −8.825 | 0 | Infinite | −8.825 | 0 | Infinite |
| 300 | 31.338 | 0.058 | 202.986 | 202.793 | 0.058 | 0 | 0 |
| 500 | 34.259 | 6.643 | 219.742 | 206.455 | 6.643 | 0 | 0 |
| 700 | 35.838 | 13.670 | 231.548 | 212.020 | 13.670 | 0 | 0 |
| 900 | 36.743 | 20.935 | 240.672 | 217.411 | 20.935 | 0 | 0 |

Thermodynamic Functions Calculated from Coefficients for $F_2$ (*Continued*)

| *T* | $C_p$ | H-H298 | *S* | −(G-H298)/*T* | *H* | $\Delta H_f$ | log $K_p$ |
|---|---|---|---|---|---|---|---|
| deg-K | J/mol-K | kJ/mol | J/mol-K | J/mol-K | kJ/mol | kJ/mol | |
| 1100 | 37.352 | 28.347 | 248.107 | 222.337 | 28.347 | 0 | 0 |
| 1300 | 37.798 | 35.863 | 254.384 | 226.797 | 35.863 | 0 | 0 |
| 1500 | 38.208 | 43.464 | 259.822 | 230.846 | 43.464 | 0 | 0 |
| 1700 | 38.574 | 51.144 | 264.628 | 234.543 | 51.144 | 0 | 0 |
| 1900 | 38.854 | 58.888 | 268.934 | 237.941 | 58.888 | 0 | 0 |
| 2100 | 39.011 | 66.677 | 272.832 | 241.081 | 66.677 | 0 | 0 |
| 2300 | 39.027 | 74.483 | 276.383 | 243.999 | 74.483 | 0 | 0 |
| 2500 | 38.898 | 82.278 | 279.632 | 246.721 | 82.278 | 0 | 0 |
| 2700 | 38.630 | 90.033 | 282.617 | 249.271 | 90.033 | 0 | 0 |
| 2900 | 38.238 | 97.722 | 285.364 | 251.667 | 97.722 | 0 | 0 |
| 3100 | 37.739 | 105.321 | 287.898 | 253.924 | 105.321 | 0 | 0 |
| 3300 | 37.154 | 112.812 | 290.240 | 256.055 | 112.812 | 0 | 0 |
| 3500 | 36.505 | 120.178 | 292.407 | 258.071 | 120.178 | 0 | 0 |
| 3700 | 35.813 | 127.411 | 294.417 | 259.982 | 127.411 | 0 | 0 |
| 3900 | 35.097 | 134.502 | 296.284 | 261.796 | 134.502 | 0 | 0 |
| 4100 | 34.375 | 141.449 | 298.021 | 263.521 | 141.449 | 0 | 0 |
| 4300 | 33.663 | 148.253 | 299.642 | 265.164 | 148.253 | 0 | 0 |
| 4500 | 32.971 | 154.916 | 301.156 | 266.731 | 154.916 | 0 | 0 |
| 4700 | 32.311 | 161.443 | 302.576 | 268.226 | 161.443 | 0 | 0 |
| 4900 | 31.686 | 167.842 | 303.909 | 269.655 | 167.842 | 0 | 0 |
| 5000 | 31.388 | 170.996 | 304.546 | 270.347 | 170.996 | 0 | 0 |

Thermodynamic Functions Calculated from Coefficients for H

| *T* | $C_p$ | H-H298 | *S* | −(G-H298)/*T* | *H* | $\Delta H_f$ | log $K_p$ |
|---|---|---|---|---|---|---|---|
| deg-K | J/mol-K | kJ/mol | J/mol-K | J/mol-K | kJ/mol | kJ/mol | |
| 0 | 0 | −6.197 | 0 | Infinite | 211.801 | 216.035 | Infinite |
| 300 | 20.786 | 0.038 | 114.846 | 114.718 | 218.037 | 218.011 | −35.3769 |
| 500 | 20.786 | 4.196 | 125.465 | 117.073 | 222.195 | 219.253 | −20.1575 |
| 700 | 20.786 | 8.353 | 132.459 | 120.526 | 226.352 | 220.478 | −13.5970 |
| 900 | 20.786 | 12.510 | 137.682 | 123.782 | 230.509 | 221.671 | −9.9319 |
| 1100 | 20.786 | 16.667 | 141.854 | 126.701 | 234.666 | 222.808 | −7.5871 |
| 1300 | 20.786 | 20.825 | 145.326 | 129.307 | 238.824 | 223.866 | −5.9557 |
| 1500 | 20.786 | 24.982 | 148.301 | 131.646 | 242.981 | 224.837 | −4.7539 |
| 1700 | 20.786 | 29.139 | 150.902 | 133.762 | 247.138 | 225.721 | −3.8311 |
| 1900 | 20.786 | 33.296 | 153.214 | 135.690 | 251.295 | 226.525 | −3.0997 |
| 2100 | 20.786 | 37.454 | 155.295 | 137.460 | 255.453 | 227.255 | −2.5057 |

(*Continued*)

Thermodynamic Functions Calculated from Coefficients for H (*Continued*)

| $T$ | $C_p$ | H-H298 | $S$ | −(G-H298)/$T$ | $H$ | $\Delta H_f$ | log $K_p$ |
|---|---|---|---|---|---|---|---|
| deg-K | J/mol-K | kJ/mol | J/mol-K | J/mol-K | kJ/mol | kJ/mol | |
| 2300 | 20.786 | 41.611 | 157.186 | 139.094 | 259.610 | 227.918 | −2.0135 |
| 2500 | 20.786 | 45.768 | 158.919 | 140.612 | 263.767 | 228.521 | −1.5989 |
| 2700 | 20.786 | 49.926 | 160.519 | 142.028 | 267.924 | 229.068 | −1.2448 |
| 2900 | 20.786 | 54.083 | 162.004 | 143.355 | 272.082 | 229.564 | −0.9388 |
| 3100 | 20.786 | 58.240 | 163.390 | 144.603 | 276.239 | 230.014 | −0.6718 |
| 3300 | 20.786 | 62.397 | 164.690 | 145.781 | 280.396 | 230.420 | −0.4367 |
| 3500 | 20.786 | 66.555 | 165.913 | 146.897 | 284.553 | 230.784 | −0.2281 |
| 3700 | 20.786 | 70.712 | 167.068 | 147.957 | 288.711 | 231.108 | −0.0418 |
| 3900 | 20.786 | 74.869 | 168.162 | 148.965 | 292.868 | 231.394 | 0.1256 |
| 4100 | 20.786 | 79.026 | 169.202 | 149.927 | 297.025 | 231.642 | 0.2768 |
| 4300 | 20.786 | 83.184 | 170.192 | 150.847 | 301.182 | 231.855 | 0.4142 |
| 4500 | 20.786 | 87.341 | 171.137 | 151.728 | 305.340 | 232.032 | 0.5394 |
| 4700 | 20.786 | 91.498 | 172.041 | 152.573 | 309.497 | 232.174 | 0.6540 |
| 4900 | 20.786 | 95.655 | 172.907 | 153.385 | 313.654 | 232.284 | 0.7594 |
| 5000 | 20.786 | 97.734 | 173.327 | 153.780 | 315.733 | 232.327 | 0.8089 |

Thermodynamic Functions Calculated from Coefficients for HF

| $T$ | $C_p$ | H-H298 | $S$ | −(G-H298)/$T$ | $H$ | $\Delta H_f$ | log $K_p$ |
|---|---|---|---|---|---|---|---|
| deg-K | J/mol-K | kJ/mol | J/mol-K | J/mol-K | kJ/mol | kJ/mol | |
| 0 | 0 | −8.599 | 0 | Infinite | −281.899 | −273.252 | Infinite |
| 300 | 29.138 | 0.054 | 173.957 | 173.778 | −273.246 | −273.302 | 47.9523 |
| 500 | 29.172 | 5.884 | 188.848 | 177.080 | −267.416 | −273.679 | 28.9093 |
| 700 | 29.350 | 11.732 | 198.685 | 181.924 | −261.568 | −274.277 | 20.7330 |
| 900 | 29.827 | 17.645 | 206.112 | 186.507 | −255.655 | −274.961 | 16.1798 |
| 1100 | 30.548 | 23.679 | 212.164 | 190.638 | −249.621 | −275.653 | 13.2750 |
| 1300 | 31.405 | 29.874 | 217.336 | 194.356 | −243.426 | −276.315 | 11.2589 |
| 1500 | 32.258 | 36.241 | 221.890 | 197.730 | −237.059 | −276.935 | 9.7770 |
| 1700 | 33.043 | 42.772 | 225.977 | 200.817 | −230.528 | −277.516 | 8.6414 |
| 1900 | 33.747 | 49.453 | 229.691 | 203.664 | −223.847 | −278.062 | 7.7429 |
| 2100 | 34.370 | 56.266 | 233.100 | 206.307 | −217.034 | −278.570 | 7.0143 |
| 2300 | 34.920 | 63.196 | 236.252 | 208.776 | −210.104 | −279.038 | 6.4113 |
| 2500 | 35.406 | 70.229 | 239.184 | 211.092 | −203.071 | −279.456 | 5.9039 |
| 2700 | 35.836 | 77.354 | 241.926 | 213.276 | −195.946 | −279.819 | 5.4711 |
| 2900 | 36.220 | 84.561 | 244.500 | 215.341 | −188.739 | −280.117 | 5.0976 |

Thermodynamic Functions Calculated from Coefficients for HF (*Continued*)

| $T$ | $C_p$ | H-H298 | $S$ | $-$(G-H298)/$T$ | $H$ | $\Delta H_f$ | log $K_p$ |
|---|---|---|---|---|---|---|---|
| deg-K | J/mol-K | kJ/mol | J/mol-K | J/mol-K | kJ/mol | kJ/mol | |
| 3100 | 36.566 | 91.840 | 246.927 | 217.302 | −181.460 | −280.346 | 4.7720 |
| 3300 | 36.879 | 99.185 | 249.223 | 219.167 | −174.115 | −280.498 | 4.4856 |
| 3500 | 37.167 | 106.590 | 251.402 | 220.948 | −166.710 | −280.569 | 4.2319 |
| 3700 | 37.434 | 114.050 | 253.475 | 222.650 | −159.250 | −280.558 | 4.0055 |
| 3900 | 37.684 | 121.562 | 255.452 | 224.282 | −151.738 | −280.463 | 3.8024 |
| 4100 | 37.922 | 129.123 | 257.342 | 225.849 | −144.177 | −280.285 | 3.6193 |
| 4300 | 38.150 | 136.730 | 259.154 | 227.356 | −136.570 | −280.024 | 3.4533 |
| 4500 | 38.370 | 144.382 | 260.893 | 228.808 | −128.918 | −279.683 | 3.3022 |
| 4700 | 38.585 | 152.078 | 262.567 | 230.209 | −121.222 | −279.266 | 3.1641 |
| 4900 | 38.794 | 159.816 | 264.179 | 231.563 | −113.484 | −278.775 | 3.0375 |
| 5000 | 38.896 | 163.701 | 264.964 | 232.223 | −109.599 | −278.503 | 2.9781 |

Thermodynamic Functions Calculated from Coefficients for $H_2$

| $T$ | $C_p$ | H-H298 | $S$ | $-$(G-H298)/$T$ | $H$ | $\Delta H_f$ | log $K_p$ |
|---|---|---|---|---|---|---|---|
| deg-K | J/mol-K | kJ/mol | J/mol-K | J/mol-K | kJ/mol | kJ/mol | |
| 0 | 0 | −8.468 | 0 | Infinite | −8.468 | 0 | Infinite |
| 300 | 28.849 | 0.053 | 130.859 | 130.682 | 0.053 | 0 | 0 |
| 500 | 29.254 | 5.883 | 145.740 | 133.975 | 5.883 | 0 | 0 |
| 700 | 29.444 | 11.748 | 155.607 | 138.823 | 11.748 | 0 | 0 |
| 900 | 29.873 | 17.676 | 163.053 | 143.413 | 17.676 | 0 | 0 |
| 1100 | 30.567 | 23.717 | 169.112 | 147.551 | 23.717 | 0 | 0 |
| 1300 | 31.421 | 29.914 | 174.286 | 151.275 | 29.914 | 0 | 0 |
| 1500 | 32.305 | 36.287 | 178.845 | 154.653 | 36.287 | 0 | 0 |
| 1700 | 33.144 | 42.833 | 182.940 | 157.744 | 42.833 | 0 | 0 |
| 1900 | 33.916 | 49.540 | 186.669 | 160.596 | 49.540 | 0 | 0 |
| 2100 | 34.618 | 56.395 | 190.099 | 163.244 | 56.395 | 0 | 0 |
| 2300 | 35.254 | 63.383 | 193.277 | 165.719 | 63.383 | 0 | 0 |
| 2500 | 35.832 | 70.493 | 196.241 | 168.044 | 70.493 | 0 | 0 |
| 2700 | 36.361 | 77.713 | 199.019 | 170.236 | 77.713 | 0 | 0 |
| 2900 | 36.848 | 85.034 | 201.635 | 172.313 | 85.034 | 0 | 0 |
| 3100 | 37.301 | 92.450 | 204.107 | 174.285 | 92.450 | 0 | 0 |
| 3300 | 37.728 | 99.953 | 206.453 | 176.164 | 99.953 | 0 | 0 |
| 3500 | 38.135 | 107.540 | 208.685 | 177.959 | 107.540 | 0 | 0 |
| 3700 | 38.525 | 115.206 | 210.815 | 179.678 | 115.206 | 0 | 0 |
| 3900 | 38.902 | 122.949 | 212.853 | 181.327 | 122.949 | 0 | 0 |
| 4100 | 39.269 | 130.766 | 214.807 | 182.913 | 130.766 | 0 | 0 |

(*Continued*)

| Thermodynamic Functions Calculated from Coefficients for $H_2$ (*Continued*) | | | | | | | |
|---|---|---|---|---|---|---|---|
| ***T*** | ***$C_p$*** | **H-H298** | ***S*** | **−(G-H298)/*T*** | ***H*** | **$\Delta H_f$** | **log $K_p$** |
| **deg-K** | **J/mol-K** | **kJ/mol** | **J/mol-K** | **J/mol-K** | **kJ/mol** | **kJ/mol** | |
| 4300 | 39.627 | 138.656 | 216.686 | 184.440 | 138.656 | 0 | 0 |
| 4500 | 39.975 | 146.616 | 218.495 | 185.914 | 146.616 | 0 | 0 |
| 4700 | 40.312 | 154.645 | 220.241 | 187.338 | 154.645 | 0 | 0 |
| 4900 | 40.637 | 162.740 | 221.928 | 188.715 | 162.740 | 0 | 0 |
| 5000 | 40.793 | 166.812 | 222.750 | 189.388 | 166.812 | 0 | 0 |

| Thermodynamic Functions Calculated from Coefficients for N | | | | | | | |
|---|---|---|---|---|---|---|---|
| ***T*** | ***$C_p$*** | **H-H298** | ***S*** | **−(G-H298)/*T*** | ***H*** | **$\Delta H_f$** | **log $K_p$** |
| **deg-K** | **J/mol-K** | **kJ/mol** | **J/mol-K** | **J/mol-K** | **kJ/mol** | **kJ/mol** | |
| 0 | 0 | −6.197 | 0 | Infinite | 466.483 | 470.818 | Infinite |
| 300 | 20.786 | 0.038 | 153.431 | 153.302 | 472.718 | 472.692 | −79.2955 |
| 500 | 20.786 | 4.196 | 164.049 | 155.657 | 476.876 | 473.920 | −46.3394 |
| 700 | 20.786 | 8.353 | 171.043 | 159.110 | 481.033 | 475.065 | −32.1786 |
| 900 | 20.786 | 12.510 | 176.267 | 162.366 | 485.190 | 476.079 | −24.2931 |
| 1100 | 20.780 | 16.667 | 180.438 | 165.286 | 489.347 | 476.967 | −19.2650 |
| 1300 | 20.784 | 20.823 | 183.909 | 167.891 | 493.503 | 477.752 | −15.7778 |
| 1500 | 20.790 | 24.981 | 186.884 | 170.230 | 497.661 | 478.458 | −13.2165 |
| 1700 | 20.792 | 29.139 | 189.486 | 172.345 | 501.819 | 479.105 | −11.2551 |
| 1900 | 20.791 | 33.297 | 191.799 | 174.274 | 505.977 | 479.703 | −9.7046 |
| 2100 | 20.792 | 37.455 | 193.879 | 176.043 | 510.135 | 480.265 | −8.4480 |
| 2300 | 20.800 | 41.614 | 195.771 | 177.678 | 514.294 | 480.797 | −7.4086 |
| 2500 | 20.821 | 45.776 | 197.506 | 179.196 | 518.456 | 481.308 | −6.5347 |
| 2700 | 20.860 | 49.944 | 199.110 | 180.612 | 522.624 | 481.805 | −5.7894 |
| 2900 | 20.923 | 54.122 | 200.603 | 181.940 | 526.802 | 482.295 | −5.1463 |
| 3100 | 21.012 | 58.315 | 202.001 | 183.190 | 530.995 | 482.785 | −4.5855 |
| 3300 | 21.131 | 62.529 | 203.318 | 184.370 | 535.209 | 483.284 | −4.0923 |
| 3500 | 21.283 | 66.770 | 204.566 | 185.489 | 539.450 | 483.798 | −3.6549 |
| 3700 | 21.468 | 71.044 | 205.753 | 186.552 | 543.724 | 484.335 | −3.2644 |
| 3900 | 21.688 | 75.359 | 206.889 | 187.566 | 548.039 | 484.904 | −2.9136 |
| 4100 | 21.941 | 79.721 | 207.980 | 188.536 | 552.401 | 485.512 | −2.5966 |
| 4300 | 22.228 | 84.138 | 209.031 | 189.465 | 556.818 | 486.165 | −2.3087 |
| 4500 | 22.547 | 88.615 | 210.049 | 190.357 | 561.295 | 486.872 | −2.0461 |
| 4700 | 22.893 | 93.158 | 211.037 | 191.216 | 565.838 | 487.637 | −1.8054 |
| 4900 | 23.265 | 97.774 | 211.999 | 192.045 | 570.454 | 488.467 | −1.5840 |
| 5000 | 23.459 | 100.110 | 212.471 | 192.449 | 572.790 | 488.908 | −1.4799 |

| Thermodynamic Functions Calculated from Coefficients for NO | | | | | | | |
|---|---|---|---|---|---|---|---|
| ***T*** | $C_p$ | **H-H298** | ***S*** | **–(G-H298)/*T*** | ***H*** | $\Delta H_f$ | **log $K_p$** |
| **deg-K** | **J/mol-K** | **kJ/mol** | **J/mol-K** | **J/mol-K** | **kJ/mol** | **kJ/mol** | |
| 0 | 0 | –9.179 | 0 | Infinite | 82.092 | 90.767 | Infinite |
| 300 | 29.858 | 0.055 | 210.933 | 210.749 | 91.327 | 91.272 | –15.2453 |
| 500 | 30.494 | 6.061 | 226.259 | 214.136 | 97.332 | 91.334 | –8.8864 |
| 700 | 32.032 | 12.311 | 236.758 | 219.171 | 103.582 | 91.364 | –6.1599 |
| 900 | 33.423 | 18.862 | 244.983 | 224.025 | 110.133 | 91.400 | –4.6446 |
| 1100 | 34.476 | 25.657 | 251.798 | 228.473 | 116.929 | 91.439 | –3.6800 |
| 1300 | 35.234 | 32.632 | 257.622 | 232.520 | 123.904 | 91.474 | –3.0119 |
| 1500 | 35.790 | 39.738 | 262.705 | 236.213 | 131.009 | 91.500 | –2.5217 |
| 1700 | 36.210 | 46.940 | 267.211 | 239.600 | 138.211 | 91.509 | –2.1469 |
| 1900 | 36.537 | 54.216 | 271.257 | 242.723 | 145.487 | 91.496 | –1.8509 |
| 2100 | 36.797 | 61.550 | 274.927 | 245.618 | 152.821 | 91.455 | –1.6114 |
| 2300 | 37.010 | 68.931 | 278.285 | 248.314 | 160.203 | 91.384 | –1.4137 |
| 2500 | 37.189 | 76.352 | 281.378 | 250.837 | 167.623 | 91.283 | –1.2477 |
| 2700 | 37.343 | 83.805 | 284.246 | 253.207 | 175.077 | 91.151 | –1.1065 |
| 2900 | 37.478 | 91.288 | 286.920 | 255.441 | 182.559 | 90.988 | –0.9850 |
| 3100 | 37.600 | 98.796 | 289.423 | 257.554 | 190.067 | 90.795 | –0.8794 |
| 3300 | 37.711 | 106.327 | 291.777 | 259.557 | 197.598 | 90.574 | –0.7868 |
| 3500 | 37.816 | 113.880 | 293.999 | 261.462 | 205.151 | 90.328 | –0.7050 |
| 3700 | 37.916 | 121.453 | 296.104 | 263.278 | 212.724 | 90.057 | –0.6322 |
| 3900 | 38.014 | 129.046 | 298.102 | 265.013 | 220.317 | 89.764 | –0.5671 |
| 4100 | 38.112 | 136.659 | 300.006 | 266.674 | 227.930 | 89.451 | –0.5086 |
| 4300 | 38.212 | 144.291 | 301.823 | 268.267 | 235.562 | 89.121 | 0.4557 |
| 4500 | 38.316 | 151.944 | 303.563 | 269.798 | 243.215 | 88.776 | –0.4077 |
| 4700 | 38.426 | 159.618 | 305.231 | 271.270 | 250.889 | 88.419 | –0.3639 |
| 4900 | 38.544 | 167.315 | 306.835 | 272.689 | 258.586 | 88.054 | –0.3239 |
| 5000 | 38.607 | 171.172 | 307.614 | 273.380 | 262.444 | 87.869 | –0.3051 |

| Thermodynamic Functions Calculated from Coefficients for $NO_2$ | | | | | | | |
|---|---|---|---|---|---|---|---|
| ***T*** | $C_p$ | **H-H298** | ***S*** | **–(G-H298)/*T*** | ***H*** | $\Delta H_f$ | **log $K_p$** |
| **deg-K** | **J/mol-K** | **kJ/mol** | **J/mol-K** | **J/mol-K** | **kJ/mol** | **kJ/mol** | |
| 0 | 0 | –10.208 | 0 | Infinite | 23.985 | 37.000 | Infinite |
| 300 | 37.235 | 0.069 | 240.401 | 240.172 | 34.262 | 34.181 | –9.1283 |
| 500 | 43.659 | 8.167 | 260.943 | 244.610 | 42.360 | 33.319 | –6.7779 |
| 700 | 48.603 | 17.425 | 276.473 | 251.580 | 51.618 | 33.150 | –5.7869 |

*(Continued)*

Thermodynamic Functions Calculated from Coefficients for $NO_2$ (*Continued*)

| *T* | $C_p$ | H-H298 | *S* | −(G-H298)/*T* | *H* | $\Delta H_f$ | log $K_p$ |
|---|---|---|---|---|---|---|---|
| deg-K | J/mol-K | kJ/mol | J/mol-K | J/mol-K | kJ/mol | kJ/mol | |
| 900 | 51.823 | 27.492 | 289.105 | 258.559 | 61.685 | 33.329 | −5.2362 |
| 1100 | 53.924 | 38.081 | 299.723 | 265.104 | 72.274 | 33.675 | −4.8828 |
| 1300 | 55.355 | 49.017 | 308.854 | 271.149 | 83.210 | 34.102 | −4.6354 |
| 1500 | 56.383 | 60.196 | 316.851 | 276.720 | 94.389 | 34.573 | −4.4515 |
| 1700 | 57.172 | 71.554 | 323.958 | 281.867 | 105.747 | 35.058 | −4.3089 |
| 1900 | 57.827 | 83.056 | 330.354 | 286.640 | 117.249 | 35.540 | −4.1948 |
| 2100 | 58.414 | 94.681 | 336.171 | 291.085 | 128.874 | 36.012 | −4.1011 |
| 2300 | 58.974 | 106.420 | 341.510 | 295.240 | 140.613 | 36.473 | −4.0228 |
| 2500 | 59.531 | 118.270 | 346.450 | 299.142 | 152.463 | 36.931 | −3.9561 |
| 2700 | 60.100 | 130.233 | 351.053 | 302.819 | 164.426 | 37.393 | −3.8986 |
| 2900 | 60.689 | 142.312 | 355.369 | 306.296 | 176.505 | 37.868 | −3.8484 |
| 3100 | 61.300 | 154.510 | 359.436 | 309.594 | 188.703 | 38.368 | −3.8041 |
| 3300 | 61.933 | 166.833 | 363.288 | 312.732 | 201.026 | 38.903 | −3.7647 |
| 3500 | 62.586 | 179.285 | 366.951 | 315.727 | 213.478 | 39.482 | −3.7292 |
| 3700 | 63.256 | 191.869 | 370.447 | 318.591 | 226.062 | 40.115 | −3.6971 |
| 3900 | 63.939 | 204.588 | 373.795 | 321.337 | 238.781 | 40.808 | −3.6678 |
| 4100 | 64.630 | 217.445 | 377.010 | 323.975 | 251.638 | 41.569 | −3.6409 |
| 4300 | 65.325 | 230.440 | 380.104 | 326.514 | 264.633 | 42.402 | −3.6161 |
| 4500 | 66.021 | 243.575 | 383.090 | 328.962 | 277.768 | 43.312 | −3.5929 |
| 4700 | 66.713 | 256.848 | 385.976 | 331.327 | 291.041 | 44.303 | −3.5713 |
| 4900 | 67.399 | 270.259 | 388.770 | 333.615 | 304.452 | 45.375 | −3.5510 |
| 5000 | 67.738 | 277.016 | 390.135 | 334.732 | 311.209 | 45.942 | −3.5412 |

Thermodynamic Functions Calculated from Coefficients for $N_2$

| *T* | $C_p$ | H-H298 | *S* | −(G-H298)/*T* | *H* | $\Delta H_f$ | log $K_p$ |
|---|---|---|---|---|---|---|---|
| deg-K | J/mol-K | kJ/mol | J/mol-K | J/mol-K | kJ/mol | kJ/mol | |
| 0 | 0 | −8.670 | 0 | Infinite | −8.670 | 0 | Infinite |
| 300 | 29.125 | 0.054 | 191.790 | 191.610 | 0.054 | 0 | 0 |
| 500 | 29.582 | 5.911 | 206.740 | 194.918 | 5.911 | 0 | 0 |
| 700 | 30.754 | 11.937 | 216.866 | 199.814 | 11.937 | 0 | 0 |
| 900 | 32.090 | 18.223 | 224.758 | 204.511 | 18.223 | 0 | 0 |
| 1100 | 33.242 | 24.760 | 231.313 | 208.804 | 24.760 | 0 | 0 |
| 1300 | 34.147 | 31.503 | 236.943 | 212.710 | 31.503 | 0 | 0 |
| 1500 | 34.842 | 38.405 | 241.880 | 216.277 | 38.405 | 0 | 0 |
| 1700 | 35.377 | 45.429 | 246.275 | 219.553 | 45.429 | 0 | 0 |

Thermodynamic Functions Calculated from Coefficients for $N_2$ (*Continued*)

| $T$ | $C_p$ | H-H298 | $S$ | −(G-H298)/$T$ | $H$ | $\Delta H_f$ | log $K_p$ |
|---|---|---|---|---|---|---|---|
| deg-K | J/mol-K | kJ/mol | J/mol-K | J/mol-K | kJ/mol | kJ/mol | |
| 1900 | 35.795 | 52.548 | 250.234 | 222.577 | 52.548 | 0 | 0 |
| 2100 | 36.127 | 59.741 | 253.833 | 225.385 | 59.741 | 0 | 0 |
| 2300 | 36.395 | 66.994 | 257.132 | 228.005 | 66.994 | 0 | 0 |
| 2500 | 36.615 | 74.296 | 260.176 | 230.458 | 74.296 | 0 | 0 |
| 2700 | 36.800 | 81.638 | 263.002 | 232.765 | 81.638 | 0 | 0 |
| 2900 | 36.957 | 89.014 | 265.637 | 234.942 | 89.014 | 0 | 0 |
| 3100 | 37.093 | 96.419 | 268.106 | 237.003 | 96.419 | 0 | 0 |
| 3300 | 37.213 | 103.850 | 270.429 | 238.959 | 103.850 | 0 | 0 |
| 3500 | 37.320 | 111.303 | 272.622 | 240.821 | 111.303 | 0 | 0 |
| 3700 | 37.417 | 118.777 | 274.698 | 242.596 | 118.777 | 0 | 0 |
| 3900 | 37.506 | 126.270 | 276.671 | 244.294 | 126.270 | 0 | 0 |
| 4100 | 37.589 | 133.779 | 278.548 | 245.919 | 133.779 | 0 | 0 |
| 4300 | 37.667 | 141.305 | 280.340 | 247.479 | 141.305 | 0 | 0 |
| 4500 | 37.743 | 148.846 | 282.055 | 248.978 | 148.846 | 0 | 0 |
| 4700 | 37.818 | 156.402 | 283.698 | 250.420 | 156.402 | 0 | 0 |
| 4900 | 37.893 | 163.973 | 285.275 | 251.811 | 163.973 | 0 | 0 |
| 5000 | 37.932 | 167.764 | 286.041 | 252.488 | 167.764 | 0 | 0 |

Thermodynamic Functions Calculated from Coefficients for O

| $T$ | $C_p$ | H-H298 | $S$ | −(G-H298)/$T$ | $H$ | $\Delta H_f$ | log $K_p$ |
|---|---|---|---|---|---|---|---|
| deg-K | J/mol-K | kJ/mol | J/mol-K | J/mol-K | kJ/mol | kJ/mol | |
| 0 | 0 | −6.725 | 0 | Infinite | 242.450 | 246.790 | Infinite |
| 300 | 21.901 | 0.041 | 161.196 | 161.061 | 249.216 | 249.188 | −40.3292 |
| 500 | 21.257 | 4.343 | 172.200 | 163.513 | 253.518 | 250.476 | −22.9357 |
| 700 | 21.040 | 8.570 | 179.312 | 167.070 | 257.745 | 251.495 | −15.4455 |
| 900 | 20.944 | 12.767 | 184.587 | 170.401 | 261.942 | 252.320 | −11.2687 |
| 1100 | 20.889 | 16.950 | 188.784 | 173.375 | 266.125 | 253.016 | −8.6027 |
| 1300 | 20.863 | 21.125 | 192.271 | 176.021 | 270.300 | 253.622 | −6.7522 |
| 1500 | 20.848 | 25.296 | 195.256 | 178.392 | 274.471 | 254.165 | −5.3920 |
| 1700 | 20.836 | 29.465 | 197.865 | 180.533 | 278.640 | 254.652 | −4.3498 |
| 1900 | 20.828 | 33.631 | 200.182 | 182.481 | 282.806 | 255.088 | −3.5255 |
| 2100 | 20.826 | 37.796 | 202.266 | 184.268 | 286.971 | 255.475 | −2.8571 |
| 2300 | 20.832 | 41.962 | 204.161 | 185.916 | 291.137 | 255.815 | −2.3042 |
| 2500 | 20.849 | 46.130 | 205.898 | 187.446 | 295.305 | 256.112 | −1.8392 |
| 2700 | 20.876 | 50.302 | 207.504 | 188.874 | 299.477 | 256.370 | −1.4426 |

(*Continued*)

Thermodynamic Functions Calculated from Coefficients for O (*Continued*)

| $T$ | $C_p$ | H-H298 | $S$ | −(G-H298)/$T$ | $H$ | $\Delta H_f$ | log $K_p$ |
|---|---|---|---|---|---|---|---|
| deg-K | J/mol-K | kJ/mol | J/mol-K | J/mol-K | kJ/mol | kJ/mol | |
| 2900 | 20.915 | 54.481 | 208.997 | 190.210 | 303.656 | 256.591 | −1.1004 |
| 3100 | 20.965 | 58.669 | 210.393 | 191.468 | 307.844 | 256.781 | −0.8021 |
| 3300 | 21.025 | 62.868 | 211.706 | 192.655 | 312.043 | 256.944 | −0.5398 |
| 3500 | 21.094 | 67.079 | 212.945 | 193.780 | 316.254 | 257.082 | −0.3074 |
| 3700 | 21.172 | 71.306 | 214.119 | 194.848 | 320.481 | 257.202 | −0.0999 |
| 3900 | 21.257 | 75.548 | 215.236 | 195.865 | 324.723 | 257.304 | 0.0863 |
| 4100 | 21.348 | 79.809 | 216.301 | 196.836 | 328.984 | 257.394 | 0.2544 |
| 4300 | 21.444 | 84.088 | 217.320 | 197.765 | 333.263 | 257.474 | 0.4070 |
| 4500 | 21.544 | 88.387 | 218.298 | 198.656 | 337.562 | 257.546 | 0.5460 |
| 4700 | 21.645 | 92.706 | 219.237 | 199.512 | 341.881 | 257.612 | 0.6732 |
| 4900 | 21.748 | 97.045 | 220.141 | 200.336 | 346.220 | 257.675 | 0.7901 |
| 5000 | 21.799 | 99.222 | 220.581 | 200.736 | 348.397 | 257.705 | 0.8450 |

Thermodynamic Functions Calculated from Coefficients for OH

| $T$ | $C_p$ | H-H298 | $S$ | −(G-H298)/$T$ | $H$ | $\Delta H_f$ | log $K_p$ |
|---|---|---|---|---|---|---|---|
| deg-K | J/mol-K | kJ/mol | J/mol-K | J/mol-K | kJ/mol | kJ/mol | |
| 0 | 0 | −8.813 | 0 | Infinite | 28.465 | 37.039 | Infinite |
| 300 | 29.879 | 0.055 | 183.924 | 183.740 | 37.333 | 37.280 | −5.6640 |
| 500 | 29.495 | 5.981 | 199.069 | 187.106 | 43.260 | 37.276 | −3.0662 |
| 700 | 29.656 | 11.888 | 209.005 | 192.022 | 49.167 | 37.042 | −1.9562 |
| 900 | 30.267 | 17.874 | 216.523 | 196.663 | 55.152 | 36.692 | −1.3446 |
| 1100 | 31.124 | 24.011 | 222.678 | 200.850 | 61.289 | 36.321 | −0.9592 |
| 1300 | 32.070 | 30.330 | 227.954 | 204.623 | 67.609 | 35.973 | −0.6951 |
| 1500 | 32.964 | 36.835 | 232.607 | 208.050 | 74.114 | 35.663 | −0.5032 |
| 1700 | 33.759 | 43.509 | 236.783 | 211.189 | 80.788 | 35.384 | −0.3577 |
| 1900 | 34.452 | 50.332 | 240.576 | 214.086 | 87.610 | 35.123 | −0.2436 |
| 2100 | 35.057 | 57.284 | 244.055 | 216.777 | 94.563 | 34.869 | −0.1520 |
| 2300 | 35.587 | 64.350 | 247.268 | 219.290 | 101.628 | 34.615 | −0.0768 |
| 2500 | 36.057 | 71.515 | 250.255 | 221.649 | 108.793 | 34.355 | −0.0142 |
| 2700 | 36.477 | 78.769 | 253.046 | 223.873 | 116.047 | 34.084 | 0.0388 |
| 2900 | 36.859 | 86.103 | 255.667 | 225.976 | 123.382 | 33.800 | 0.0841 |
| 3100 | 37.209 | 93.511 | 258.137 | 227.972 | 130.789 | 33.501 | 0.1232 |
| 3300 | 37.535 | 100.986 | 260.473 | 229.871 | 138.264 | 33.188 | 0.1572 |
| 3500 | 37.840 | 108.523 | 262.691 | 231.684 | 145.802 | 32.860 | 0.1871 |
| 3700 | 38.129 | 116.121 | 264.802 | 233.418 | 153.399 | 32.517 | 0.2135 |

Thermodynamic Functions Calculated from Coefficients for OH (*Continued*)

| $T$ | $C_p$ | H-H298 | $S$ | −(G-H298)/$T$ | $H$ | $\Delta H_f$ | log $K_p$ |
|---|---|---|---|---|---|---|---|
| deg-K | J/mol-K | kJ/mol | J/mol-K | J/mol-K | kJ/mol | kJ/mol | |
| 3900 | 38.404 | 123.774 | 266.816 | 235.079 | 161.052 | 32.159 | 0.2369 |
| 4100 | 38.666 | 131.481 | 268.743 | 236.675 | 168.760 | 31.787 | 0.2578 |
| 4300 | 38.915 | 139.240 | 270.591 | 238.209 | 176.518 | 31.401 | 0.2765 |
| 4500 | 39.151 | 147.046 | 272.365 | 239.688 | 184.325 | 31.000 | 0.2934 |
| 4700 | 39.373 | 154.899 | 274.073 | 241.115 | 192.177 | 30.586 | 0.3086 |
| 4900 | 39.579 | 162.795 | 275.718 | 242.494 | 200.073 | 30.157 | 0.3224 |
| 5000 | 39.675 | 166.757 | 276.518 | 243.167 | 204.036 | 29.937 | 0.3288 |

Thermodynamic Functions Calculated from Coefficients for $O_2$

| $T$ | $C_p$ | H-H298 | $S$ | −(G-H298)/$T$ | $H$ | $\Delta H_f$ | log $K_p$ |
|---|---|---|---|---|---|---|---|
| deg-K | J/mol-K | kJ/mol | J/mol-K | J/mol-K | kJ/mol | kJ/mol | |
| 0 | 0 | −8.680 | 0 | Infinite | −8.680 | 0 | Infinite |
| 300 | 29.388 | 0.054 | 205.331 | 205.150 | 0.054 | 0 | 0 |
| 500 | 31.092 | 6.086 | 220.698 | 208.527 | 6.086 | 0 | 0 |
| 700 | 32.990 | 12.500 | 231.472 | 213.615 | 12.500 | 0 | 0 |
| 900 | 34.361 | 19.244 | 239.939 | 218.557 | 19.244 | 0 | 0 |
| 1100 | 35.333 | 26.219 | 246.934 | 223.099 | 26.219 | 0 | 0 |
| 1300 | 36.006 | 33.356 | 252.894 | 227.235 | 33.356 | 0 | 0 |
| 1500 | 36.553 | 40.613 | 258.086 | 231.010 | 40.613 | 0 | 0 |
| 1700 | 37.057 | 47.975 | 262.692 | 234.471 | 47.975 | 0 | 0 |
| 1900 | 37.545 | 55.435 | 266.840 | 237.664 | 55.435 | 0 | 0 |
| 2100 | 38.020 | 62.992 | 270.621 | 240.625 | 62.992 | 0 | 0 |
| 2300 | 38.484 | 70.642 | 274.101 | 243.387 | 70.642 | 0 | 0 |
| 2500 | 38.933 | 78.384 | 277.328 | 245.975 | 78.384 | 0 | 0 |
| 2700 | 39.366 | 86.215 | 280.341 | 248.410 | 86.215 | 0 | 0 |
| 2900 | 39.780 | 94.129 | 283.169 | 250.711 | 94.129 | 0 | 0 |
| 3100 | 40.175 | 102.125 | 285.835 | 252.892 | 102.125 | 0 | 0 |
| 3300 | 40.549 | 110.198 | 288.359 | 254.965 | 110.198 | 0 | 0 |
| 3500 | 40.904 | 118.344 | 290.755 | 256.942 | 118.344 | 0 | 0 |
| 3700 | 41.239 | 126.558 | 293.037 | 258.832 | 126.558 | 0 | 0 |
| 3900 | 41.556 | 134.838 | 295.217 | 260.643 | 134.838 | 0 | 0 |
| 4100 | 41.854 | 143.179 | 297.302 | 262.381 | 143.179 | 0 | 0 |
| 4300 | 42.135 | 151.579 | 299.302 | 264.052 | 151.579 | 0 | 0 |
| 4500 | 42.400 | 160.032 | 301.224 | 265.661 | 160.032 | 0 | 0 |
| 4700 | 42.649 | 168.537 | 303.073 | 267.214 | 168.537 | 0 | 0 |
| 4900 | 42.884 | 177.091 | 304.855 | 268.714 | 177.091 | 0 | 0 |
| 5000 | 42.997 | 181.385 | 305.723 | 269.446 | 181.385 | 0 | 0 |

Thermodynamic Functions Calculated from Coefficients for $H_2O$

| $T$ | $C_p$ | H-H298 | $S$ | −(G-H298)/$T$ | $H$ | $\Delta H_f$ | log $K_p$ |
|---|---|---|---|---|---|---|---|
| deg-K | J/mol-K | kJ/mol | J/mol-K | J/mol-K | kJ/mol | kJ/mol | |
| 0 | 0 | −9.904 | 0 | Infinite | −251.730 | −238.922 | Infinite |
| 300 | 33.596 | 0.062 | 189.037 | 188.830 | −241.764 | −241.844 | 39.7840 |
| 500 | 35.225 | 6.925 | 206.529 | 192.680 | −234.901 | −243.827 | 22.8831 |
| 700 | 37.499 | 14.192 | 218.734 | 198.460 | −227.634 | −245.632 | 15.5809 |
| 900 | 39.998 | 21.939 | 228.456 | 204.079 | −219.887 | −247.185 | 11.4957 |
| 1100 | 42.574 | 30.197 | 236.732 | 209.281 | −211.629 | −248.456 | 8.8808 |
| 1300 | 45.065 | 38.964 | 244.049 | 214.077 | −202.862 | −249.455 | 7.0622 |
| 1500 | 47.318 | 48.206 | 250.659 | 218.521 | −193.620 | −250.213 | 5.7237 |
| 1700 | 49.292 | 57.872 | 256.705 | 222.663 | −183.954 | −250.774 | 4.6975 |
| 1900 | 50.996 | 67.905 | 262.283 | 226.543 | −173.921 | −251.179 | 3.8858 |
| 2100 | 52.458 | 78.254 | 267.461 | 230.197 | −163.572 | −251.462 | 3.2278 |
| 2300 | 53.709 | 88.874 | 272.291 | 233.649 | −152.952 | −251.656 | 2.6837 |
| 2500 | 54.777 | 99.726 | 276.814 | 236.924 | −142.100 | −251.785 | 2.2263 |
| 2700 | 55.690 | 110.775 | 281.065 | 240.037 | −131.051 | −251.871 | 1.8366 |
| 2900 | 56.474 | 121.993 | 285.073 | 243.007 | −119.833 | −251.932 | 1.5005 |
| 3100 | 57.149 | 133.357 | 288.862 | 245.844 | −108.469 | −251.981 | 1.2077 |
| 3300 | 57.736 | 144.847 | 292.454 | 248.561 | −96.979 | −252.031 | 0.9504 |
| 3500 | 58.252 | 156.447 | 295.866 | 251.167 | −85.379 | −252.091 | 0.7224 |
| 3700 | 58.712 | 168.144 | 299.116 | 253.672 | −73.682 | −252.167 | 0.5190 |
| 3900 | 59.129 | 179.929 | 302.218 | 256.083 | −61.897 | −252.265 | 0.3364 |
| 4100 | 59.514 | 191.794 | 305.185 | 258.406 | −50.032 | −252.388 | 0.1716 |
| 4300 | 59.875 | 203.733 | 308.028 | 260.648 | −38.093 | −252.538 | 0.0220 |
| 4500 | 60.221 | 215.743 | 310.758 | 262.815 | −26.083 | −252.716 | −0.1144 |
| 4700 | 60.556 | 227.820 | 313.384 | 264.912 | −14.006 | −252.919 | −0.2393 |
| 4900 | 60.883 | 239.964 | 315.914 | 266.942 | −1.862 | −253.147 | −0.3541 |
| 5000 | 61.045 | 246.061 | 317.146 | 267.934 | 4.235 | −253.269 | −0.4080 |

## Reference

NASA CEA: http://www.grc.nasa.gov/WWW/CEAWeb/ceaHome.htm.

# Index

Note: Page numbers followed by f indicate figures and t indicate tables

## D



## E



## F

## J



## K



## L



## M



## N

## O



## P

## Q



## R



## S

## T

## V



## W